G. Ludwig

Foundations of Quantum Mechanics I

Translated by Carl A. Hein

Springer-Verlag
New York Heidelberg Berlin

G. Ludwig
Institut für Theoretische Physik
Universität Marburg
Renthof 7
Federal Republic of Germany

Carl A. Hein (*Translator*)
Formerly with
Massachusetts Institute of Technology
Lincoln Laboratory
Lexington, MA
U.S.A.

Editors

Wolf Beiglböck
Institut für Angewandte Mathematik
Universität Heidelberg
Im Neuenheimer Feld 5
D-6900 Heidelberg 1
Federal Republic of Germany

Elliott H. Lieb
Department of Physics
Joseph Henry Laboratories
Princeton University
Princeton, NJ 08540
U.S.A.

Tullio Regge
Istituto de Fisica Teorica
Universita di Torino
C. so M. d'Azeglio, 46
10125 Torino
Italy

Walter Thirring
Institut für Theoretische Physik
der Universität Wien
Boltzmanngasse 5
A-1090 Wien
Austria

Library of Congress Cataloging in Publication Data
Ludwig, Gunther, 1918–
Foundations of quantum mechanics.
(Texts and monographs in physics)
Translation of: Die Grundlagen der
Quantenmechanik.
Bibliography: p.
Includes index.
1. Quantum theory. I. Title. II. Series.
QC174.12.L8313 1982 530.1′2 82-10437

ISBN 978-3-642-86753-8 ISBN 978-3-642-86751-4 (eBook)
DOI 10.1007/978-3-642-86751-4

Original German edition: *Die Grundlagen der Quantenmechanik*. Berlin–Heidelberg–New York: Springer-Verlag, 1954.

Typeset by Composition House Ltd., Salisbury, England.

9 8 7 6 5 4 3 2 1

Texts and Monographs in Physics

Dedicated to my wife

Preface

This book is the first volume of a two-volume work on the Foundations of Quantum Mechanics, and is intended as a new edition of the author's book *Die Grundlagen der Quantenmechanik* [37] which was published in 1954. In this two-volume work we will seek to obtain an improved formulation of the interpretation of quantum mechanics based on experiments. The second volume will appear shortly.

Since the publication of [37] there have been several attempts to develop a basis for quantum mechanics which is, in the large part, based upon the work of J. von Neumann [38]. In particular, we mention the books of G. W. Mackey [39], J. Jauch [40], C. Piron [41], M. Drieschner [9], and the original work of S. P. Gudder [42], D. J. Foulis and C. H. Randall [43], and N. Zierler [44]. Here we do not seek to compare these different formulations of the foundations of quantum mechanics. We refer interested readers to [45] for such comparisons.

In this book we shall seek only to develop a well-defined formulation for the foundations of quantum mechanics and to examine the implications of such a formulation towards the most important applications of quantum mechanics in a consistent manner. This formulation will be based only on the objective, that is, the so-called classical mode of description of the apparatuses. In this respect this book represents a systematic mathematical, as well as conceptual, formulation of the original viewpoint of N. Bohr in which it is assumed that it is necessary to use the classical mode of description in order to describe the measurement process in quantum mechanics (see, for example, the extensive discussion in M. Jammer and E. Scheibe [14]).

In our approach to developing a formulation of the foundations of quantum mechanics we shall not present a precise mathematical description of the macroscopic measurement apparatus. Instead, we shall only assume that there exists an objective characterization of the mode of operation of the apparatus. In [13] we have shown that it is possible to derive quantum mechanics without making reference to microsystems, by using only the description of macrosystems in terms of state spaces. There the reader will also find a derivation of the Hilbert space structure from general laws concerning the interactions of macrosystems. In this book we will make use of these results in III, §3 without proof.

At several places in this book the reader will find references to the book *Grundstrukturen einer Physikalischen Theorie* [1]. Previous knowledge of [1] is not necessary for an understanding of this book. Readers who are familiar with [1] can easily recognize how the general structure of a physical theory is realized for the case of quantum mechanics. Readers who wish to study [1] later will have the advantage that they will have an example to illustrate the general description in [1].

The formulation of quantum mechanics presented here is the last step of developments since 1964 which the interested reader can find in [48]. (In [48] there is also some previous work which has led to the results described here.) In a certain sense, this presentation together with [1] and [13] represents a greatly improved "second edition" of [17].

In Appendices I–V we have provided a summary of important mathematical results which may be unfamiliar to some readers. Appendix V will appear in Volume II. For readers who are unfamiliar with the mathematical results in Appendix V, we suggest that they "take them on faith" until Volume II appears.

References in the text are made as follows: For references to other sections of the same chapter, we shall only list the section number of the reference, for example, §5.3. For references to other chapters, the chapter is also given; for example, IV, §7.2 refers to Chapter IV, Section 7.2. The formulas are numbered as follows: (5.7.10) refers to the 10th formula in Section 5.7 of the current chapter. References to formulas in other chapters are given, for example, by IV, (5.7.10). References to the Appendix are given by AIV, §2, where AIV denotes Appendix IV.

I would like to express my deep gratitude to Mr. Carl A. Hein for the difficult job of translating the manuscript from German to English. He had the difficult assignment of finding suitable English language expressions for the somewhat new and sometimes difficult concepts and ideas used in a new conceptual framework for quantum mechanics. This was possible only because of his deep understanding of the text. I would also like to thank him for his patience in accommodating my wishes and a substantial number of revisions in the German text while the translation was under way.

I hope that the present book together with [13] will lead to further interest and research into the foundations of quantum mechanics, especially in the direction of a relativistic theory for quantum mechanics, that is, relativistic quantum field theory (see [46]).

Marburg, January 1982 G. LUDWIG

Contents

CHAPTER I

The Problem: An Axiomatic Basis for Quantum Mechanics

The historical path of discovery for a new physical theory is, for the most part, a complicated one. At first new concepts are tentatively introduced. By a lengthy process, involving trial, error, insight, and revision, these concepts are modified and become more clearly defined and familiar. As an understanding of the postulated structure of the theory develops, it is possible by careful application of the new concepts to learn how to avoid error and to develop an interpretation of the new theory. Such has been the case for quantum mechanics. In this book we shall not present a heuristic path to quantum mechanics as a means of developing a theory of electrons, atoms, ... (in general: microsystems). Instead, we shall assume that the reader has already had extensive contact with quantum mechanics, and has studied one or more of the elementary texts. If this is not the case, we recommend that the reader either study such a text before reading this book, or use one of the elementary texts in conjunction with this book. In this way, the reader will discover the vagueness inherent in the usual fundamental concepts which are used to formulate quantum mechanics. For the reader who seeks an elementary text which considers some of the problems to be discussed in this book, we recommend Volume 3 of [2].

In the following chapters we shall attempt to clarify the fundamental concepts of quantum mechanics and present a thorough and systematic axiomatic formulation of quantum mechanics.

1 The Axiomatic Formulation of a Physical Theory

We cannot consider the general structure of a physical theory in detail in this book. We refer readers who are interested in such questions to [1]. Here we shall only describe in broad terms what we mean by an *axiomatic basis* of a physical theory, and what we seek to accomplish when we formulate an axiomatic basis for a physical theory.

A *physical theory* (abbreviated $\mathscr{PT}$) consists of three essential parts, a *mathematical theory* ($\mathscr{MT}$), a *domain of reality* ($\mathscr{W}$), which we seek to describe by $\mathscr{MT}$, and a set of *mapping principles* ($\mathscr{MAP}$) where the latter is needed in order to describe the relationship between $\mathscr{W}$ and $\mathscr{MT}$. The mapping principles also describe what is frequently called the *interpretation* of $\mathscr{MT}$. It is here that we encounter the difficult problems in quantum mechanics. Concepts such as *observable*, *state*, *property*, and *object* are ill-suited to serve as a conceptual basis for an interpretation of quantum mechanics.

(1) We expect that the difficulties of interpretation of a $\mathscr{PT}$ should be minimized when an axiomatic basis for $\mathscr{PT}$ is found.

A $\mathscr{MT}$ as part of a $\mathscr{PT}$ can, in principle, be arbitrarily defined. In the historical evolution of physics, the mathematical theories seldom appear in the form of an axiomatic basis. This situation frequently leads to conflicting views concerning the interpretation of the theory. In particular there may be disputes about a given structure in $\mathscr{MT}$. Is it accidental, thereby having nothing to do with the real structure of the world as described by $\mathscr{W}$? In [1] it is shown how it is possible to thoroughly treat such problems if $\mathscr{MT}$ is an axiomatic basis. Here we shall only provide a brief exposition of the principle of an axiomatic basis in order to apply it to quantum mechanics in the following chapters. By doing so we shall give an explicit formulation of the mapping principles (see II). In this manner the application of the principle of an axiomatic basis can be understood without the more general and detailed analysis of [1].

(2) The introduction of new physical concepts in the domain of a $\mathscr{PT}$ may be carried out in a simple and transparent manner when the $\mathscr{MT}$ is an axiomatic basis.

If $\mathscr{MT}$ is an axiomatic basis, then we can introduce new physical concepts for which the physical meaning will naturally follow from the mapping principles $\mathscr{MAP}$ and mathematical constructions in $\mathscr{MT}$. Again, it is not necessary to study the general principles presented in [1] because it is not difficult to understand the concrete derivation of the new physical concepts (such as ensemble, effect, observable, position observable, etc.) which are presented in the following chapters.

The reader who is familiar with the "stories" which are usually told in order to explain (in a rough manner) the physical meaning of these concepts will appreciate the conceptual clarity which results from the construction of a $\mathscr{PT}$ with $\mathscr{MT}$ as an axiomatic basis.

What do we mean by the expression "axiomatic basis"? First it is essential that this expression describe certain aspects of the form of $\mathscr{M}\mathscr{T}$ itself. Thus it is essential that the $\mathscr{M}\mathscr{T}$ be studied in its own right, that is, detached from its relationship to physics. Thus we postulate that $\mathscr{M}\mathscr{T}$ should take the form of what Bourbaki [4] calls a "theory of species of structure" Σ and denotes by $\mathscr{T}_\Sigma$. Since we are studying both physical and mathematical theories, we prefer to write $\mathscr{M}\mathscr{T}_\Sigma$ instead of $\mathscr{T}_\Sigma$. Thus an axiomatic basis should be of the form $\mathscr{M}\mathscr{T}_\Sigma$ where Σ denotes the appropriate species of structure. The form of $\mathscr{M}\mathscr{T}_\Sigma$ within a $\mathscr{P}\mathscr{T}$ is somewhat more specialized than that introduced by Bourbaki in [4]. The specialization consists in the fact that, for the auxiliary base sets of Σ, we shall only use the set of real numbers **R** and for $\mathscr{M}\mathscr{T}$ we require only the use of set theory.

It is not necessary for the reader to study either [4] or [1] because it is not difficult to exhibit the nature of a $\mathscr{M}\mathscr{T}_\Sigma$ and because in II (supplemented by VII, §1 and axioms introduced later) we will explicitly construct the mathematical description $\mathscr{M}\mathscr{T}_\Sigma$ corresponding to quantum mechanics.

In $\mathscr{M}\mathscr{T}_\Sigma$ we shall assume the usual formulations of mathematical logic and set theory, and the usual formulation of the real number system **R**. Then we shall introduce what we shall call base sets, for which no internal structure is specified in advance. We obtain an internal structure for the base sets by introducing relations on the sets, and by postulating axioms for them. This is how we shall introduce the structure for the $\mathscr{M}\mathscr{T}_\Sigma$ corresponding to quantum mechanics beginning in II and continuing in VII, §1. A simple example for a $\mathscr{M}\mathscr{T}_\Sigma$ is a group, that is a set with a relation (called multiplication) which satisfies the axioms for a group. In addition to the requirement that $\mathscr{M}\mathscr{T}$ be of the form $\mathscr{M}\mathscr{T}_\Sigma$ we must also make requirements concerning the relation between the mapping principles $\mathscr{M}\mathscr{A}\mathscr{P}$ and $\mathscr{M}\mathscr{T}_\Sigma$.

We shall use the concept of mapping principles instead of the concept of interpretation because we require a neutral term, one which does not evoke differing preconceived notions among various readers. We note that inherent in the concept of mapping principles is that something is being mapped. We shall only require that what is being mapped must be expressed in terms of experiment and experience without the need for the application of the new theory. We may find that we do need older theories, that is, theories which are already known to express what is being mapped. We may illustrate this requirement with the aid of a familiar theory. For example, it is possible to specify the position of the planets without requiring the use of Newton's mechanics and Newton's law of gravitation. The position of the planets at different times provides the experimental material which can be compared to Newton's theory, that is, which is to be mapped into the mathematical framework $\mathscr{M}\mathscr{T}$ of Newton's theory. We shall use the expression *fundamental domain* $\mathscr{G}$ to denote those facts that can be specified in advance of a particular $\mathscr{P}\mathscr{T}$ and which are mapped into the mathematical framework $\mathscr{M}\mathscr{T}_\Sigma$ of $\mathscr{P}\mathscr{T}$. In §2 we shall seek to describe the fundamental domain for the case of quantum mechanics. The fundamental domain consists of all that can be determined before the application of the theory. The fundamental domain will later be expanded *with the help of the theory* to the domain of reality $\mathscr{W}$ where the latter includes all real aspects of the world which can be described by the theory.

The mapping principles are rules by which the elements of $\mathscr{G}$, that is, the stated facts, may be translated into the language of $\mathscr{M}\mathscr{T}_\Sigma$. We obtain an axiomatic basis $\mathscr{M}\mathscr{T}_\Sigma$ if and only if the translation of the determined facts (as elements of $\mathscr{G}$) into the language of $\mathscr{M}\mathscr{T}_\Sigma$ makes use only of the undefined basis sets and the undefined relations of Σ. The axioms required for $\mathscr{M}\mathscr{T}_\Sigma$ are not deduced from experiment, but are guessed from experiment by a process involving trial and error, intuition, and insight. It is important to keep this in mind in the following chapters (see also [1], §5). The axioms used in $\mathscr{M}\mathscr{T}_\Sigma$ should not be derived from philosophical *a priori* principles (for example, from forms of pure sensible intuition) believing that these principles and forms are necessary to develop physics. Here we note that some authors [9] take an *a priori* viewpoint concerning the fundamental structure of quantum mechanics.

By an axiomatic basis we mean a realization $\mathscr{M}\mathscr{T}_\Sigma$ of $\mathscr{M}\mathscr{T}$ together with a set of mapping principles $\mathscr{M}\mathscr{A}\mathscr{P}$ which have the form described above. It is obvious that many problems concerning the physical meaning associated with mathematical structures are easier to solve within the framework of an axiomatic basis because we permit only sets and relations in Σ (except for certain idealizations) which are interpreted physically by use of the mapping principles. Again, we refer interested readers to [1].

At this stage we expect that many readers will not find the above formulation of the general structure of a physical theory clear. On the basis of the formulation of quantum mechanics which will be presented in the following chapters, we expect that such readers will appreciate the conceptual clarity of our axiomatic formulation of quantum mechanics as compared to the usual one. As we progress we also expect that such readers will obtain a better understanding of the nature of an axiomatic basis. In this respect, this book can also be used in order to obtain a better understanding of [1].

2 The Fundamental Domain for Quantum Mechanics

The simplicity we find in the case of classical mechanics results from the following fact: It is possible to describe the measurement of the position of a particle (point mass) at various times without making use of the laws of mechanics. In other words, the formulation of a theory of measurement for a position–time measurement does not require the use of the laws of mechanics (see [2], II). Thus, the measured positions of a particle at different times are appropriate for inclusion in the fundamental domain of classical mechanics.

In quantum mechanics the situation is completely different. Nevertheless, in the historical development of quantum mechanics we find that there is a strong tendency to imitate the procedures of classical mechanics, where it is assumed that we know how it is possible to measure position, momentum, and many other quantities. Here the term "observable" is used to describe the quantities to be measured, and the measured values of the observables are assumed to be the experimental material which is to be compared with the theory. Unfortunately, it is not possible to state precisely what is meant by the concept of an observable.

The origin of this difficulty is easily ascertained. As we have seen, in classical mechanics it is possible to develop a theory of measurement of the position of the particles at various times without the application of mechanics. In this way it is possible to give meaning to the concept of the observable position at various times and their measurements before developing the theory of classical mechanics. In quantum mechanics this is not possible, because there is no measurement apparatus for microsystems such as electrons, atoms, molecules, etc., whose function can be explained without the use of quantum theory. Thus we find that the concept of observables and their corresponding measurement values are ill-suited for inclusion into the fundamental domain for quantum mechanics. The claim that quantum mechanics is only concerned with what can be measured (that is, the measurement values obtained from a scale on a measurement apparatus) is false because we cannot explain what the measurement values represent without the use of quantum theory.

Here we shall not review the vast body of literature devoted to the theory of measurement in quantum mechanics. Instead, we shall seek to obtain a suitable substitute for the concept of an observable (and their associated measurement values) for inclusion into the fundamental domain of quantum mechanics.

A second concept—that of a "state"—is often used as an aid in the interpretation of quantum mechanics. A microsystem is said to be in one of its possible states. But what do we mean by the notion of a state? In classical mechanics it is possible to characterize the state of a system by the positions and velocities of the individual particles of the system at a given time, that is, by a point in phase space. We shall not describe the various attempts to develop a quantum mechanical notion of a state because it is clear that such a notion will make explicit use of the structure of quantum mechanics. Thus we find that the notion of a state is also ill-suited for inclusion in the fundamental domain of quantum mechanics.

Thus, if we seek to formulate quantum mechanics in terms of an axiomatic basis, we have little to begin with other than what an experimental physicist would call experiments with a single microsystem. The term "single" is a qualitative designation which is used only to differentiate these experiments from those that treat a large number of interacting systems as a whole, that is, a macrosystem composed of a collection of microsystems. The term "microsystem" is also a qualitative designation which is used to emphasize the fact that we do not assume that quantum mechanics is a suitable theory for the description of macrosystems (for example, the earth). Indeed, quantum mechanics is inadequate for a theoretical description of macrosystems. We cannot discuss the problem of the relationship between quantum mechanics and a more comprehensive theory of macrosystems in this book. The reader will find an introduction to this problem in [2], XV, [5], [7], [13], [27] and some comments in XVIII.

By experiments with "single" microsystems we do not mean that we consider only a "single" experiment with a "single" microsystem. Since statistics plays a central role in quantum mechanics (see II), we must consider experiments with "large numbers" of microsystems. It is important to understand that experiments with "large numbers" of microsystems can be frequently understood in terms of repeated measurements with a single microsystem. This situation is familiar to every experimental physicist. An electron beam can be considered to be the result of a multiple process in which a single electron is "produced" provided that the mutual interaction between individual electrons can be neglected (see XVI). Even the so-called ideal gas can be approximately treated as a collection of many single atoms (see XV, §2) since the mutual interactions of the atoms in an ideal gas are negligible.

For the fundamental domain of quantum mechanics we shall choose the class of experiments with individual microsystems, and the relative frequencies of the phenomena associated with multiple repetitions of these experiments.

In this book we shall not describe the vast variety of such experiments. Several examples are briefly described in XI–XVII. In II we shall develop the general structure of such experiments as the basis for the formulation of quantum mechanics. In preparation for this task we find it necessary to make the meaning of the expression "experiments with microsystems" more precise. We shall begin by describing the structure of such experiments in more detail.

In order to carry out experiments with individual microsystems it is necessary to have such systems at hand. Often such microsystems can be found in nature, for example, in interstellar space. There they are sufficiently separated so that their mutual interactions can be neglected. In fact, their mutual interactions will be smaller than what can be produced in many experiments in the laboratory. On Earth such microsystems must be produced in the laboratory. Often they may be obtained naturally, as for example, from the decay of radioactive substances. Sometimes it is necessary to produce them using a complicated apparatus which is very expensive to build. Often a rarefied (ideal) gas will be suitable for many purposes. Here we shall use the generic term *preparation procedure* to denote the various methods of obtaining microsystems.

Thus some preparation procedures will require the use of a special apparatus (a giant accelerator), while others will only require the use of the sun, which emits such microsystems as light quanta, charged particles (solar wind) and neutrinos.

We shall now state an important requirement for the development of an axiomatic basis for quantum mechanics: It must be possible to describe the structure of the preparation apparatus, and the time-dependent physical process by which the preparation apparatus operates without the use of quantum mechanics. In brief, we require that the so-called pre-theories for quantum mechanics permit the description of the structure and the operation of the apparatus and the characterization of the preparation procedure (for a description of the concept of a pre-theory see [1]). In other words we require that the preparation procedures belong to the fundamental domain of quantum mechanics.

In order to prevent misconceptions concerning the characterization and description of a preparation procedure, we find it necessary to give an example of what is *not* part of the characterization of a preparation procedure. If, for the purpose of illustration, electrons are to be produced, then the specification of the spin of the prepared electron or the description of the physical process of emission of the electron from, for example, a heated cathode is not permitted as part of the description of the preparation procedure. However, all macroscopic processes which take place in the operation of the preparation apparatus, including those instructions which can be stored on magnetic tapes and in other memory devices and executed in sequence by a computer belong to the description of the preparation procedures.

The concept of a preparation procedure permits the description of complicated experimental arrangements such as one composed of an accelerator, a target and a special selection apparatus which selects the desired microsystem. In addition, it is possible to combine two or more preparation procedures into a new preparation procedure. Such combinations of preparation procedures are commonly found in scattering experiments (see XVI).

At present the formulation of a preparation procedure may appear to be too general. We find it necessary to impose an additional restriction—that of reproducibility. The latter notion is related to the relative frequencies of the various phenomena associated with repetitions of an individual experiment (see II).

By making the assumption that a microsystem is produced in a preparation procedure we do not mean that we know, in a particular case, which preparation procedure was used to produce a particular microsystem. In a test of a physical theory we require that only *known* facts are to be mapped by means of the mapping principles into the mathematical language of the theory $\mathcal{M}\mathcal{T}_\Sigma$.

We shall now consider the following question: How do we use the prepared microsystems to investigate the structure of microsystems? In the second and crucial part of such experiments we require the use of an apparatus which measures the microsystems and their structure. If we wish to interpret the macroscopic physical processes associated with the second apparatus as a measurement of the structure of a microsystem, we need to make use of quantum mechanics. Therefore we permit only the inclusion of the macroscopic processes associated with such an apparatus as part of the fundamental domain of quantum mechanics. Would it then be correct to say that the measured value obtained by a measuring apparatus may be compared with the predictions of quantum theory? It is correct in that the measurement values (scale values) are the result of macroscopic processes associated with the apparatus, and are therefore parts of the fundamental domain. What is *not* part of the fundamental domain is the interpretation of these scale values as a measurement of a property of a microsystem (that is, the "result" of the measurement of an "observable").

Since, in our description of the fundamental domain, we cannot say what is (or was) measured by the apparatus, we find it necessary to introduce the expression *registration apparatus* (instead of measuring apparatus) to describe the second part of the experimental arrangement.

It is in this sense that every experimental arrangement of an experiment with a single microsystem consists of a preparation apparatus and a registration apparatus. It is not necessary that the experimental arrangement be man-made. Indeed, the preparation apparatus can be a star or a galaxy. In order to prevent misunderstanding, it is necessary to note that the expression "single microsystem" also applies to what is called a "composite microsystem" in the sense of VIII. If, for example, we study electron–proton scattering, the "single" microsystems are electron–proton pairs.

In this book we shall not describe the construction of a typical registration apparatus. Instead, we shall give a few familiar examples: a scintillation counter, or array of such counters, a cloud chamber, a bubble chamber, a spectroscope, a photographic plate, etc. In addition, we give an example of a registration apparatus which makes use of complicated electromagnetic fields—such as a mass spectrograph.

It can be argued whether every experimental arrangement for an experiment with a single microsystem consists of a preparation apparatus and a registration apparatus. This is indeed the case. However, this does not mean that it is possible to uniquely divide a complicated experiment into preparation and registration parts. In Figure 1 we have an experimental arrangement which consists of three parts. In part (1) we produce the microsystem *a*. In part (2) we produce microsystem *b* as the result of the interaction of the macroscopic apparatus (2) with *a* (where *b* can be the same as *a*). The microsystem *b* is then registered by (3). We may consider (1) as the preparation apparatus for the microsystem *a*, and (2) plus (3) as the registration apparatus for *a*. We may also consider (1) plus (2) as the preparation apparatus for system *b* and (3) as the registration apparatus for system *b*.

Thus we find that the preparation–registration structure provides a conceptual basis for experiments with microsystems. It is possible to invent experiments in which it is not possible to speak of preparation and registration of microsystems. For example, consider an apparatus having two parts (1) and (2), and suppose that they interact by exchanging microsystems, and that the emission of microsystems by (1) is influenced by the microsystems produced by (2) and vice versa. For such a system it is not possible to specify which microsystem is the subject of the experiment. Is it the one which goes from (1) to (2) or the one which goes from (2) to (1)? Here we do

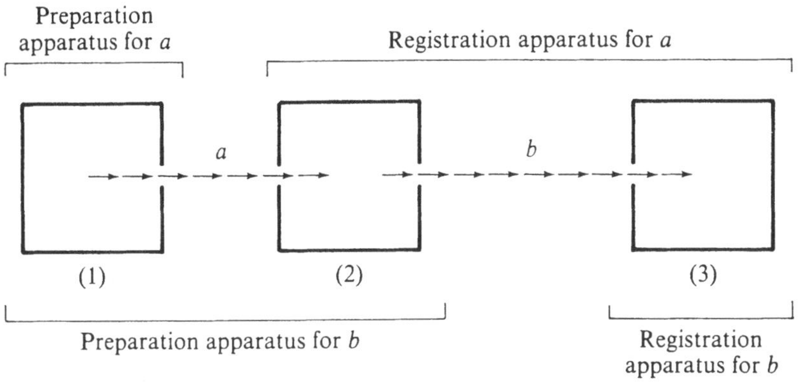

Figure 1

not mean to suggest that a more comprehensive theory (which includes quantum mechanics as a special case) will be unable to treat such complicated interaction problems. We only suggest that such experiments are ill-suited for the immediate goal of formulating an axiomatic basis for quantum mechanics.

By our reference to the possibility that the interaction between the apparatus (1) and (2) need not be directed, we may be led to question the assumption of the existence of microsystems, or at least to seek a better understanding of the concept of a microsystem.

The inclusion of the preparation apparatus and the registration apparatus (and the associated physical processes which can be described without the use of quantum mechanics) into the fundamental domain is not affected by the question of the existence of microsystems. The directedness of the interaction can be described in terms of the pre-theory of quantum mechanics, that is, without the need of quantum mechanical theory. Indeed, it can be described without the need of the concept of a microsystem. We shall not discuss these matters here; they are discussed in [2], XVI, [3], [6], [7], and briefly in XVIII. In [13] the axiomatic basis of quantum mechanics is formulated without the need for the assumption of the existence of microsystems. There we find that no special assumption is needed in order to describe the directed interaction between the preparation apparatus and the registration apparatus by means of a "carrier of interaction." In II we shall introduce the fundamental set M of "interaction carriers." The introduction of M does not violate our intention to develop an axiomatic basis provided we do not make additional assertions about the elements of M other that they depend on the preparation apparatus, the registration apparatus and their associated macroscopic processes. The introduction of M will also permit us to make the formulation of an axiomatic basis of quantum mechanics given here easier to understand than the one given in [13]. Thus the reader who has understood the foundations of the theory of microsystems given here will obtain a better understanding of the formulation presented in

[2], XVI and the more detailed presentation given in [13]. In II we will call the carriers of interactions (elements of M) "microsystems" even though this word will denote special carriers of interaction which are characterized by the axioms given in III, §3 and §5. The introduction of the set M of microsystems should not be understood as implying that such microsystems exist in every individual experiment because the "vacuum" can also be considered to be a "type" of microsystem. Here we do not exclude the possibility that the preparation apparatus does not always interact with a registration apparatus.

In summary, we will present the axiomatic basis for quantum mechanics in this book. The axiomatic basis will be constructed using the fundamental domain which consists of all those aspects of the preparation and registration procedures for microsystems which can be described without the use of quantum mechanics.

3 The Measurement Problem

Have we, by our restriction to the fundamental domain described above, eliminated the measurement problem which was described in the beginning of §2? Have we eliminated the problem in such a way that the interaction between the microsystem and the measurement apparatus can be analyzed without the need for quantum theory? Of course not. What we have done is to place the problem where it belongs—namely, with the developing theory. The mapping principles are no longer burdened with the problem of providing a theoretical description of either the effect of the microsystem on the registration apparatus or the dependence of the microsystem on the structure of the preparation apparatus.

What is the status of such a theoretical description in the arena of quantum mechanics?

In II and III we shall begin our axiomatic formulation of quantum mechanics by introducing structural rules governing the preparation and registration processes. These structural rules will be very general, and will be analogous to those found in thermostatics. These rules will not specify how individual cases of preparation and registration are structured, just as the fundamental structure of thermostatics does not specify the equation of state for a given substance.

The theory presented in II–III and extended in VII, §1 describes the preparation and registration of microsystems but is not as complete as we might wish. In opposition to such a wish of completeness, in IV we shall proceed in the opposite direction. We shall attempt to eliminate, as far as possible, the preparation and registration process in order to obtain a theory of the structure of microsystems which is independent of accidental aspects of the structure of the apparatuses. The extent to which this is possible will be discussed in IV.

In thermodynamics the equation of state is obtained from experimental data. By analogy with thermodynamics we will, in the present state of the theory, take the mode of operation of the special preparation and registration procedures from experiment, and make use of additional assumptions (see, for example, XI, §1 and §2 and XVI, §1 and §2). This "taking" of special structure from experiment is not without considerable cost. As a result, the current status of the theory is not satisfactory. Thus we shall attempt to describe the problems of preparation and registration more precisely. We shall make such a detailed investigation of the problems of preparation and registration in order to obtain a more comprehensive theory than that which was developed in II–XVI.

We shall begin these investigations in XVII. There we shall find that quantum mechanics cannot present a closed theory (more precisely—is not a g.G.-closed theory in the sense of [1], §8 and §10) of the preparation and registration process. This is perhaps disappointing. In fact, it merely demonstrates the fact that quantum mechanics is not a theory which can describe everything from a microsystem to a macrosystem.

In XVIII we shall analyze the situation in quantum mechanics in its relationship to other physical theories. Thus, at the end of this book we return to the problems posed at the beginning.

CHAPTER II

Microsystems, Preparation, and Registration Procedures

We shall now present a "theoretical" description of experiments with individual microsystems, a description which is expressed in terms of a mathematical framework $\mathcal{M}\mathcal{T}_\Sigma$. We shall introduce mathematical entities to which the individual microsystems and the experimental procedures which are to be applied to them are to be mapped. We shall use the methodology which was briefly outlined in I, and is developed in greater detail in [2], II and in [1]. Here a familiarity with I will suffice in order to understand (at least in an intuitive way) the relationship between the mathematical theory and the physical reality it describes.

We shall develop the mathematical theory as systematically as possible. At the beginning we shall introduce a number of axioms which we shall need in order to formulate the foundations of quantum mechanics in a transparent manner. Later in the development of the theory we shall only motivate the selection of new axioms. Nevertheless the reader who is mathematically inclined will find it easy to verify that the mathematical theory is of the form $\mathcal{M}\mathcal{T}_\Sigma$ as described in I, §1 and described in greater detail in [1]. The mapping principles will be presented in a more intuitive manner. The reader who has read the definite presentation in [1], and is familiar with notion of a "concise formulation" presented there will be able to formulate the mapping principles in a precise way. Even those readers who are satisfied with obtaining a more intuitive understanding of the relationship between physics and mathematics will not find it difficult to understand this "physical interpretation" of quantum mechanics given in the following chapters because this presentation is easier to understand that the usual one.

At various places we shall use expressions such as "physical reality" and deduce new physical concepts while making explicit reference to the precise formulations and methods of [1], §10 without making explicit applications of them. We do so in order to keep the size of this book within reasonable bounds, and not to stray from the intended scope of the book. In the title of the present chapter we have introduced the term "microsystem" somewhat prematurely, because we shall not define this concept until the following chapter. The expression "physical system" or "carrier of interaction" would be more appropriate in this chapter. However, since we are concerned only with the applications of the general methods and concepts described here to this special case, we shall use the expression "microsystems."

This presentation is closely related to that found in [2], XIII. In fact, an understanding of the subject matter presented in [2], XI–XIII will greatly facilitate the understanding of the formulation of quantum mechanics presented in this book. In this respect, this book is a continuation of [2], XIII.

1 The Concept of a Physical Object

When we introduce a general concept such as a "physical object" we do not intend to present an analysis of the meaning of such concepts as they are used in physics. Instead, we intend to formulate the concept anew, independent of the fact that the new formulation may not agree with the usual one in all particulars.

In [1], §10.5 we set forth the requirement that the "new" physical concepts are to be introduced into the mathematical theory $\mathcal{M}\mathcal{T}_\Sigma$ by means of a set together with a "structure." Here the term "new" refers only to the definition of the concept (that is, of the set and the structure). We have to assume that we already know how to assign physical meaning to the structure terms and to the elements of the set. How this may be done is illustrated by the concept of a "physical system" which is defined in §4. For a detailed description of the method we refer readers to [1], §10.5, §11, and §12.

We shall now consider a set M, the elements of which we wish to call "physical objects."

Here, in order to prevent misunderstandings, we warn the reader of the opinion (see [1], §10.5) that the expression "physical object" is used to describe all aspects of "physical reality" which are to be mapped to an element of a set. In mathematics, a set and the elements of a set are often loosely called "mathematical objects." Thus, it should not be misleading to use the expression "physical object" to refer to the physical reality associated with these mathematical objects. Thus we may call an element of a set M (or preferably the "physical reality" which is mapped to the element) a physical object (see [1], §5 and §10.5) if and only if the "physical reality" associated with the elements of M have (intuitively speaking) *objective properties*. In the mathematical framework we shall express the notion of a *property* in the following terms:

Let a structure $\mathscr{E}$ be defined on the set M as follows: Let $\mathscr{E} \subset \mathscr{P}(M)$, that is, $\mathscr{E}$ is a collection of subsets of M. Let $a, b \in \mathscr{E}$; let $M \backslash a$ denote the complement of a in M. Let $\mathscr{E}$ satisfy the following axioms:

AE 1. If $a \in \mathscr{E}$ then $M \backslash a \in \mathscr{E}$.
AE 2. If $a, b \in \mathscr{E}$ then $a \cap b \in \mathscr{E}$.

Physically, the elements of $\mathscr{E}$ represent definite properties. By this we mean that the mapping principles must specify what aspects of "physical reality" are to be mapped to the elements of $\mathscr{E}$ and M. The mapping principles must also specify what real relationships between an element $x \in M$ and an element $a \in \mathscr{E}$ are to be identified with the statement "x has the property a." The latter statement is mapped to the mathematical relation $x \in a$ where $x \in M$ and $a \in \mathscr{E}$. In a more general context M and $\mathscr{E}$ may represent (in the sense of [1], §10.5) sets of real but only indirectly determinable aspects of "physical reality." In other words, the statement "x has the property a" may be only indirectly verified (see [1], §10.5 and §10.6).

The axioms AE 1 and AE 2 have the following intuitive meaning:

AE 1 states that all objects which do not have the property a share a common property—which we denote by "not a."

AE 2 states that all objects which have both properties a and b have a common property—which we denote by $a \cap b$.

It is important to note that these statements do not constitute a proof of these axioms. A careless reading of these axioms may lead the reader to conclude that they are merely consequences of logic. Such is not the case; AE 1 and AE 2 cannot be derived from the logical axioms of mathematics (see [1], §4.3) and therefore must be asserted as axioms.

The concept of a "physical object" which is defined only by the elements of a set M together with a structure $\mathscr{E}$ characterized by axioms AE 1 and AE 2 is too general. The above concept of a property is also too general. In addition to describing the object itself, it may also be used to describe a physical system with respect to its environment. But objective properties should exhibit an independence of the environment. Therefore we shall find it necessary to formulate the notion of "independence of the environment" in terms of the mathematical theory $\mathscr{M}\mathscr{T}_\Sigma$.

It is customary to express this independence of the environment in the following terms: The properties of a given system are "objective" and they can be determined by suitable measurements. It is, however, not clear what we mean by a "suitable measurement."

We have not yet formulated the concept of "objectivity" (that is, independence of the environment) in a mathematical way. It is clear that "M together with $\mathscr{E}$" by itself is not sufficient for this formulation because the "interaction with the environment" must first be described if we wish to define the notion of "objective," (that is, independent of the environment). In §4 we shall define the notion of a "physical system" and describe the interaction of the system with the environment. In III, §4.1 we shall continue this discussion in order to obtain a suitable definition of a physical *object*.

We shall now proceed as if the term "objective property" has already been defined in the theory. If $\mathscr{E}$ is a set of "objective" properties, we shall call the elements of M "physical objects."

After we have introduced the above definition of the concept of a "physical object" it is important to put aside all intuition and preconceptions about physical objects (despite the fact that they were used in order to formulate the new concept—see [1], §5 and [2], III, §4) in order to obtain a correct understanding of the new concept. Thus our notion of a "physical object" is defined in terms of M, $\mathscr{E}$, axioms AE 1 and AE 2, and a definition of the notion of "objective" which will be introduced later. It is important to emphasize the fact that this concept depends not only on the elements of the set M but also on $\mathscr{E}$. In more precise terms we must speak of "physical objects with respect to the property structure $\mathscr{E}$." Such a distinction is not necessary when the choice of $\mathscr{E}$ is clear and unambiguous.

We note the fact that axioms AE 1 and AE 2 are equivalent to the statement that $\mathscr{E}$ is a Boolean ring of sets (see AI, §4).

For those readers who have not yet achieved an understanding of the remarkable features of quantum mechanics it might appear that (intuitively, on the basis of preconceived notions associated with macroscopic physical objects) it is possible to construct quantum mechanics on the basis of "micro-objects," that is, upon $\mathscr{E}$, M, and AE 1 and AE 2 in such a way that the behavior of micro-objects is *completely* determined by the properties as defined by $\mathscr{E}$. What is meant by this complete determination is delineated in III, §4.1. Such a procedure will lead to contradiction with experience, as we shall find in IV, §8.1.

2 Selection Procedures

We shall now consider the discussion of experiments with microsystems which we began in I, §2. We shall not begin by introducing a number of intuitive concepts which establish a connection between measurements of microsystems and "measurement of properties" (see III, §4.1). Instead we shall proceed cautiously and seek only to obtain a mathematical representation of the preparation and registration of microsystems. According to the description of the experimental procedures presented in I, §2, the preparation and registration procedures have a common attribute—they result in the selection of microsystems. We shall now formulate this common attribute in mathematical terms.

In mathematics it is often useful to first introduce the more general and then the more specialized concepts, or in more precise terms, first the less rich and then the more rich structure types. Then all theorems for the less rich are also valid for the more rich structure types. Indeed, everywhere in a mathematical theory where such a general structure is found, the theorems deduced for this structure can be applied. Consider, for example, the structure type "group" and its meaning in many different mathematical theories.

On the basis of physical and mathematical considerations we shall now introduce the structure type "selection procedures."

We begin by introducing a set M. The elements of M shall be used as labels for the microsystems. Therefore we shall loosely refer to M as the set of microsystems (see, for example, [1], §5, §10, and §12 or [2], III, §4). We shall call a subset $\mathscr{S} \subset \mathscr{P}(M)$ a set of selection procedures provided the following axioms are satisfied:

AS 1.1. If $a, b \in \mathscr{S}$, $a \subset b$ then $b \backslash a \in \mathscr{S}$.
AS 1.2. If $a, b \in \mathscr{S}$, then $a \cap b \in \mathscr{S}$.

It is somewhat difficult to make the axioms for such a general concept as a selection procedure plausible to physicists. The following remarks are intended for this purpose.

First, it is evident that a structure $\mathscr{E}$ of "properties" is, on the basis of AE 1 and AE 2, also a structure of "selection procedures." A structure of selection procedures is equivalent to a set of properties (according to AS 1.1 and AS 1.2) if and only if $M \in \mathscr{S}$.

The fact that every set $\mathscr{E}$ of properties is a set of selection procedures may be intuitively expressed as follows: $\mathscr{E}$ consists of the selection procedures which select according to the properties $a \in \mathscr{E}$. If we had introduced the concept of a "property" as a special case of a selection procedure, then we would say that, in physics, there are other methods of "selection" than according to "objective properties."

Mathematically, the distinction between $\mathscr{E}$ and $\mathscr{S}$ appears to be small. For $\mathscr{S}$ we do not require that $M \in \mathscr{S}$. This small distinction permits us to extend the domain of application of selection procedures beyond that of the properties.

In order to make AS 1 more plausible, let us suppose that the selection procedures a, b, etc. are obtained by physical methods. Then a subset a of M represents the set of systems x selected according to the procedure a for some experiment. The set a is, in general, infinite. The set of systems obtained experimentally from the procedure a is always finite, but can be arbitrarily large. Since in principle we do not know how large this number can be, we express this lack of knowledge in the mathematical framework by the expression "infinite" (see [1], §6, §9 and [2], §5, §8). Intuitively speaking, axiom AS 2 states that the set of all x selected according to both selection procedures a and b, that is all $x \in a \cap b$, is a possible selection procedure. If $a \subset b$ we say that the selection procedure a is "finer" than that of b. If $a \subset b$, and if we eliminate (by means of the finer selection procedure a) the systems associated with a, we obtain $b\backslash a$; AS 1 states that $b\backslash a$ is a possible selection procedure.

Have we then not also asserted that all objects $x \in M$ which do not satisfy the selection criterion of a can be "selected" on the basis that they do not satisfy the selection criteria of a?

For "properties" it appears to be meaningful to assert that both a and its complement $M\backslash a$ are properties, because the elements of a differ fundamentally from those of $M\backslash a$ because the latter do not have the property a. For selection procedures, however, it is physically unrealistic to make this assertion. This is the case not only for micro-objects.

For example, let us consider a machine which produces steel spheres (ball bearings). The machine can be considered to be a selection procedure a for steel spheres M. The complementary set $M\backslash a$ is apparently characterized by the fact that the spheres of $M\backslash a$ were not produced by the machine a. For the set a we may make certain (technically important) assertions, while for $M\backslash a$ we may say only that the elements of $M\backslash a$ were not selected according to a.

A similar case exists for the case of a modern electron accelerator. For the "selected" set of electrons we may make important assertions about the experiments for which the electrons are used. What assertions can we make about the electrons which are not prepared by the electron accelerator?

Thus it is meaningful not to require that M be a selection procedure. The addition of the axiom $M \in \mathscr{S}$ to AS 1 and AS 2 would not lead to a contradiction to quantum mechanics. As the above example has shown, the inclusion of the condition $M \in \mathscr{S}$ to the axioms for the structure "selection procedure" is somewhat physically unrealistic. Therefore we will not add the axiom $M \in \mathscr{S}$.

If we add the condition $M \in \mathscr{S}$ to axioms AS 1 and AS 2 then we find that $\mathscr{S}$ will satisfy axioms AE 1 and AE 2 and would therefore be a property structure. In physical terms (that is, on the basis of the mapping principles) an element of $\mathscr{S}$ will not represent an intrinsic "objective" property of an object x but only the "property" that x is selected according to the procedure a. We say that the axiom systems AE 1 and AE 2 or AS 1 and AS 2 alone do not suffice to describe the physical role of the elements of $\mathscr{E}$ and $\mathscr{S}$. Further axioms will be needed in order to formulate the physical structures more precisely.

We shall now state a number of definitions and theorems which we shall need later.

D 2.1. $$\mathscr{S}(a) = \{b \mid b \in \mathscr{S} \text{ and } b \subset a\}.$$

According to AI, §4 and AS 1.1, 2 we find that $\mathscr{S}(a)$ is a Boolean ring of sets with null element $\varnothing$ and unit element a. The set $\mathscr{S}$ itself need not necessarily be a lattice (AI, §1 and §4) since given $a, b \in \mathscr{S}$, $a \cup b$ need not be an element of $\mathscr{S}$.

Th. 2.1. *Given a family of structures of selection procedures* $\{\mathscr{S}_\lambda\}$, *then* $\mathscr{S} = \bigcap_\lambda \mathscr{S}_\lambda$ *is a structure of selection procedures.*

The proof is a simple consequence of AS 1.1, 2.

Th. 2.2. *For each subset Θ of $\mathscr{P}(M)$ there is a smallest structure of selection procedures $\mathscr{S}$ (called the structure of selection procedures generated by Θ) which satisfies $\Theta \subset \mathscr{S}$.*

PROOF. $\mathscr{S}$ is the intersection of all structures of selection procedures $\mathscr{S}_\lambda$ which satisfy $\Theta \subset \mathscr{S}_\lambda$. Since $\mathscr{P}(M)$ is itself a structure of selection procedures, the family $\mathscr{S}_\lambda$ is nonempty.

D 2.2. A set $\mathscr{S}_1$ of selection procedures for which $\mathscr{S}_1 \subset \mathscr{S}$ is said to be *coexistent relative* to $c \in \mathscr{S}$ provided that $\mathscr{S}_1 \subset \mathscr{S}(c)$.

A set $\mathscr{E}$ of properties is a coexistent set of selection procedures relative to M. Every subset of $\mathscr{E}$ is a coexistent set of selection procedures relative to M.

3 Statistical Selection Procedures

Earlier we have presented a mathematical description of the fundamental phenomena associated with the selection of microsystems by means of an apparatus. We shall now consider the mathematical formulation of the second basis of quantum mechanics—statistics. The role of statistics is, of course, not restricted to quantum mechanics. It is, however, an essential component of quantum mechanics. For example, it is possible to resolve the apparent contradiction between the particle and wave descriptions only by the introduction of a statistical viewpoint; see, for example, [2], XI, §1.5–§1.7.

In an experimental context the role of statistics is made manifest by the relative frequency with which a finer selection procedure $b \subset a$ selects relative to a. By this we mean that if we select N systems $x_1, x_2, \ldots, x_N$ according to the selection procedure a, and we obtain N_1 of these systems which also satisfy the selection procedure b, then the relative frequency h is given by $h = N_1/N$. We say that b is statistically dependent on a if the relative frequency h is reproducible. By this we mean that "in physical approximation" we obtain the same relative frequency (in the case of large numbers N of systems) for various experiments involving selection according to a and b. We shall use real numbers to mathematically represent these relative frequencies.

Here we shall not consider the meaning of the expression "in physical approximation" used above. We only note that we do not require that $\alpha = N_1/N$ but only that they are approximately equal $\alpha \approx N_1/N$ (where α is the real number representing the frequency). The nature of this approximation is discussed in [1], §11 where we have placed a particular emphasis on the relationship between theory and experience.

We shall now introduce the concept of a *statistical selection procedure* in order to describe the statistical dependence of selection procedures.

A set $\mathcal{S} \subset \mathcal{P}(M)$ is called a structure of *statistical selection procedures* provided that, in addition to AS 1, the following axiom is satisfied. Let $\mathcal{T} = \{(a, b) \mid a, b \in \mathcal{S}, b \subset a, a \neq \emptyset\}$; let $\lambda: \mathcal{T} \to [0, 1]$.

AS 2.1. If $a_1, a_2 \in \mathcal{S}$, $a_1 \cap a_2 = \emptyset$, $a_1 \cup a_2 \in \mathcal{S}$ then $\lambda(a_1 \cup a_2, a_1) + \lambda(a_1 \cup a_2, a_2) = 1$.

AS 2.2. If $a_1, a_2, a_3 \in \mathcal{S}$, $a_1 \subset a_2 \subset a_3$, $a_2 \neq \emptyset$ then $\lambda(a_1, a_3) = \lambda(a_1, a_2)\lambda(a_2, a_3)$.

AS 2.3. If $a_1, a_2 \in \mathcal{S}, a_2 \subset a_1, a_2 \neq \emptyset$ then $\lambda(a_1, a_2) \neq 0$.

The quantity $\lambda(a, b)$ is called the *probability* of b relative to a and represents the relative frequency (as described above) with which b selects relative to a.

On the basis of this interpretation of $\lambda(a, b)$, the postulates AS 2.1–AS 2.3 are obvious (but not proven; see [1], §5 or [2], III, §4).

If $a_1 \cup a_2$ is a selection procedure then a_1 and a_2 are finer than $a_1 \cup a_2$. If $a_1 \cap a_2 = \emptyset$ then a_1 and a_2 are mutually exclusive. Then if N systems are selected according to $a_1 \cup a_2$ and if N_1 are selected according to a_1, N_2 according to a_2, then we must have $N = N_1 + N_2$, which, after division by N, is what we obtain from AS 2.1. For three selection procedures $a_3 \subset a_2 \subset a_1$ let N_1 systems be selected according to a_1, N_2 of these N_1 systems according to a_2 and N_3 of these N_2 systems according to a_3. Then we find that $N_3/N_1 = (N_3/N_2)(N_2/N_1)$ which is in agreement with AS 2.2.

If $a_1 \supset a_2 \neq \emptyset$ then for N systems, selected according to a_1, a finite number may be selected according to a_2, in agreement with AS 2.3.

We shall now consider a number of simple corollaries of AS 2. Let $a_2 = \emptyset$. From AS 2.1 we obtain

$$\lambda(a_1, a_1) + \lambda(a_1, \emptyset) = 1. \tag{3.1}$$

By multiplication with $\lambda(a_1, a_1)$ we obtain

$$\lambda(a_1, a_1)^2 + \lambda(a_1, a_1)\lambda(a_1, \emptyset) = \lambda(a_1, a_1). \tag{3.2}$$

According to AS 2.2,

$$\lambda(a_1, a_1) + \lambda(a_1, \emptyset) = \lambda(a_1, a_1)$$

and therefore

$$\lambda(a_1, \emptyset) = 0 \tag{3.3}$$

and

$$\lambda(a_1, a_1) = 1. \tag{3.4}$$

If $a_2 \cap a_3 = \emptyset$, and if $a_2 \subset a_1, a_3 \subset a_1$ from AS 2.1 it follows that

$$\lambda(a_2 \cup a_3, a_2) + \lambda(a_2 \cup a_3, a_3) = 1.$$

By multiplication with $\lambda(a_1, a_2 \cup a_3)$ and application of AS 2.2 we obtain

$$\lambda(a_1, a_2 \cup a_3) = \lambda(a_1, a_2) + \lambda(a_1, a_3). \tag{3.5}$$

If $a_3 \subset a_2 \subset a_1$, from (3.5) and $a_2 = a_3 \cup (a_2 \backslash a_3)$ we obtain

$$\lambda(a_1, a_2) = \lambda(a_1, a_3) + \lambda(a_1, a_2 \backslash a_3)$$

and therefore

$$\lambda(a_1, a_2) \geq \lambda(a_1, a_3).$$

D 3.1. A mapping μ of a Boolean ring Σ (see AI, §3) into the unit interval $[0, 1]$ which satisfies $\mu(\varepsilon) = 1$ (where ε is the unit element of Σ) for which the condition $\sigma_1 \wedge \sigma_2 = 0$ implies that $\mu(\sigma_1 \vee \sigma_2) = \mu(\sigma_1) + \mu(\sigma_2)$ is satisfied is called an *additive real measure* on Σ.

From (3.5). AS 2.2, and AS 2.3 it is easy to obtain the following theorem:

Th. 3.1. *$\mu(b) = \lambda(a, b)$ is an additive measure on the Boolean ring $\mathcal{S}(a)$; for $a_3 \subset a_2 \subset a$ we obtain*

$$\lambda(a_2, a_3) = \frac{\mu(a_3)}{\mu(a_2)}.$$

If we fix a, and consider all $b \subset a$, it is sufficient to consider the probability function $\mu(b)$ since, on the Boolean ring $\mathcal{S}(a)$ we obtain all conditional probabilities from the probability function $\mu(b)$.

In addition to the axioms AS 2.1–AS 2.3 we also propose the following:

AS 2.4.1. If $a_\nu \in \mathcal{S}$ is a decreasing sequence for which $\bigcap_\nu a_\nu = \varnothing$, and if there exists an $a \in \mathcal{S}$ for which $a_1 \subset a$ (and thus $a_\nu \subset a$ for all ν) then $\lambda(a, a_\nu) \to 0$.

AS 2.4.2. For every totally ordered subset of $\mathcal{S}$ there exists an upper bound in $\mathcal{S}$.

Axiom AS 2.4.1 is a generalization—in the sense of a mathematical idealization—of the intuitively evident relation (3.3). If for the decreasing sequence we have $a_\nu = \varnothing$ for some ν onward, from (3.3) it follows that AS 2.4.1 is satisfied. In addition, AS 2.4.1 requires that if the sequence of selection procedures a_ν becomes arbitrarily small in the sense of $\bigcap_\nu a_\nu = \varnothing$ then the probabilities $\lambda(a, a_\nu)$ must become arbitrarily small. A situation in which $\lambda(a, a_\nu)$ tends to a nonzero limiting value while, for all practical purposes, there are no more physical systems to be selected by a_ν for large ν is physically unrealistic. Axiom AS 2.4.2 is physically realistic because it asserts that, for a sequence of increasing selection procedures, there is a largest selection procedure. Thus it becomes a substitute for the stronger, but less realistic condition $M \in \mathcal{S}$. Here we refer readers who are interested in these axioms and their implications towards a physically motivated "generalized" probability theory to [1], §11 and [18].[1]

Using axioms AS 2.1–AS 4 we may develop a theory of probability which is similar to that of Kolmogorov (see [18]). We shall not derive any additional results here. Here we shall only note that the probabilistic basis for quantum statistical mechanics differs only slightly from the usual one. If instead, we require that $M \in \mathcal{S}$, then we would find that $\mathcal{S} = \mathcal{S}(M)$ is a Boolean ring, and that the relative probabilities would be completely determined by the probability function $\mu(a) = \lambda(M, a)$. The basis of the statistical selection procedures would then be identical to those of the simplest "classical" probability theory.

We shall now introduce an important definition which we shall need later.

D 3.2. Let $\mathcal{S}$ be a structure of statistical selection procedures. A partition of $a \in \mathcal{S}$, ($a = \bigcup_{i=1}^{n} b_i$, $b_i \neq \varnothing$, $b_i \in \mathcal{S}$, $b_i \cap b_j = \varnothing$, $i \neq j$) is called a *decomposition* of a in the b_i, and a is called a *mixture* of the b_i. The $\lambda(a, b_i)$ are called the *weights* of b_i in a.

Since $\mathcal{S}(a)$ is a Boolean ring, the decomposition of a is nothing other than a disjoint partition of the unit element a of $\mathcal{S}(a)$. With the additive measure $\mu(b)$ defined on $\mathcal{S}(a)$ we obtain $\sum_{i=1}^{n} \mu(b_i) = 1$. The $\mu(b_i)$ describe the "weights" by which the individual components b_i occur in the decomposition. If N systems are selected according to a, and if N_i are selected according to each b_i, then the relation $\mu(b_i) = N_i/N$ must hold in "physical approximation."

4 Physical Systems

We shall now consider the central topic of this chapter: the presentation of a more precise formulation of the concept of a "physical system" and of a more detailed description of the statistical selection procedures used for physical systems. We again begin with a set M, the elements of which will be used to represent "microsystems."

4.1 Preparation Procedures

From the analysis of the experimental process which was presented in I, §2 we found that the "selection" of microsystems takes place in two distinct ways: by preparation and by registration (that is, by selection according to

1. It is possible to add the following axiom to AS 1.1–AS 2.4 without contradicting experience:

AS 1.3. If $a, b \in \mathcal{S}$ and $a \cap b \neq \varnothing$ then $a \cup b \in \mathcal{S}$. We shall not use this axiom in this book.

the result of the interaction of the microsystem with a registration apparatus). Let us begin with the formulation of the mathematical structure describing preparation.

Let $\mathcal{Q} \subset \mathcal{P}(M)$ be a structure for which

APS 1. $\mathcal{Q}$ is a statistical selection procedure.

We shall call $\mathcal{Q}$ a "set of *preparation procedures*." This designation is not a consequence of APS 1, but is shorthand for a mapping principle which maps certain facts onto elements of $\mathcal{Q}$. That is, the elements of $\mathcal{Q}$ will serve as images (see, for example, [2], III, §4 or [1], §5) of certain definite technical facts and processes by which microsystems can be produced in large numbers. Here we do not permit the use of quantum mechanics in their description. The mathematical relation $x \in a$ (where $a \in \mathcal{Q}$) is the translation of the statement: x is obtained by the preparation procedure a. Here it does not make any difference whether this statement is a statement about the past, present, or future (see the discussion in [1], §12).

We shall denote the probability function for $\mathcal{Q}$ (where $\mathcal{Q}$ satisfies APS 1) by $\lambda_{\mathcal{Q}}$. The definitions and theorems of §3 are valid for $\lambda_{\mathcal{Q}}$. Later we shall consider the decomposition (as defined by D 3.2) of preparation procedures.

4.2 Registration Procedures

It is somewhat more difficult to present a mathematical formulation of the registration process. The registration process is characterized by two steps:

(1) Construction and utilization of the registration apparatus.
(2) Selection according to the changes which occur (or do not occur) in the registration apparatus.

Accordingly, we shall define an additional structure on M by means of two subsets of $\mathcal{P}(M)$: $\mathcal{R}$ and $\mathcal{R}_0$. For $\mathcal{R}$ and $\mathcal{R}_0$ we require that the following axioms be satisfied:

APS 2. $\mathcal{R}$ is a selection procedure.

APS 3. $\mathcal{R}_0$ is a statistical selection procedure.

APS 4.1. $\mathcal{R}_0 \subset \mathcal{R}$.

APS 4.2. To each $b \in \mathcal{R}$ there exists a $b_0 \in \mathcal{R}_0$ for which $b \subset b_0$.

In order to develop the physical meaning of axioms APS 2–4 we shall first state what is to be mapped onto the elements of $\mathcal{R}$ and $\mathcal{R}_0$. An element $b_0 \in \mathcal{R}_0$ shall represent the construction and the use of a registration apparatus. This can best be illustrated by an example. Let us consider a counter. Then b_0 will be the set of all microsystems which are registered by the counter. The mathematical relation $x \in b_0$ may be the translation of the following statement: for the registration of x the counter b_0 is applied. The mapping principle may be expressed in more concise form as follows: $\mathcal{R}_0$ is the set of *registration methods*.

For a given microsystem $x \in b_0$ the counter considered above may or may not respond. Let b_+ be the selection procedure for those $x \in b_0$ for which the counter responds. Here we find that $b_+ \subset b_0$. Let b_- denote the set of those x from b_0 for which the counter does not respond. Here we obtain $b_- = b_0 \backslash b_+$; b_+, b_- are elements of $\mathcal{R}$.

Generally we find that $\mathcal{R}$ contains not only the elements of $\mathcal{R}_0$ but also those selection procedures which are finer than the elements of $\mathcal{R}_0$ and are selected according to changes which occur (or do not occur) in the registration apparatus by interaction with the microsystems. This situation is described by axioms APS 4.1 and APS 4.2. The physical interpretation of the elements of $\mathcal{R}$ may be expressed in more concise form as follows: $\mathcal{R}$ is the set of *registration procedures*.

Axioms APS 4.1 and APS 4.2 do not permit us to conclude that the selection procedures from $\mathcal{R}_0$ do not depend on the microsystems but depend only on the apparatus and its "arbitrary" application.

The fact that the registration methods do not depend on the microsystems is, in part, described by APS 3 where we assume that $\mathcal{R}_0$ is a *statistical* selection procedure. Here axiom APS 3 requires that a finer registration method is statistically dependent on a coarser one and is *independent* of the subsequent influence of microsystems. Suppose that $b_0, c_0 \in \mathcal{R}_0$ and that $c_0 \subset b_0$. Then the registration method c_0 satisfies stronger selection criteria (namely that of c_0) than b_0 with respect to the apparatus. In an experiment, this would result in the exclusion of those elements of b_0 which are not also elements of c_0. Thus for c_0 there is a statistical dependence relative to b_0 which depends only on the "finer" selection c_0 of the apparatuses. Experimental physicists will seek to minimize the effect of the "statistical distribution" of the apparatuses because it will interfere with the statistical distributions they wish to measure. Instead of speaking of the statistical dependence of the elements $b_0 \in \mathcal{R}_0$ it is more common to include its influence in the discussion of the so-called experimental errors in the measurement. To an experimental physicist the above statistical distributions of the apparatuses (the effect of which cannot be completely eliminated) will appear as an error in the experimental measurement because he tries, but cannot attain registration methods $b_0 \in \mathcal{R}_0$ for which there are no $c_0 \in \mathcal{R}_0$ for which $c_0 \neq \emptyset$, $c_0 \subset b_0$ and $c_0 \neq b_0$.

We shall denote the probability function for $\mathcal{R}_0$ by $\lambda_{\mathcal{R}_0}$.

We note that axiom APS 2 does not require that $\mathcal{R}$ be a structure of *statistical* selection procedures. This crucial fact permits $\mathcal{R}$ to contain selection procedures which are essentially conditioned by the microsystem. The counter which is characterized by b_0 may or may not respond; thus b_0

can be partitioned into two subsets b_+, b_- where $b_0 = b_+ \cup b_-$, $b_+ \cap b_- = \emptyset$. In nature, however, there are *no* reproducible frequencies $\lambda(b_0, b_+)$. Suppose that we pass N microsystems $x_1, x_2, \ldots, x_N$ through the counter, and that N_+ responses are obtained. By experience we find that N_+/N depends on any circumstances which affect the microsystems. We find that the frequency N_+/N is not reproducible on the basis of the registration procedures alone. On the contrary, counters are used to determine which kinds of microsystems are present. The actual frequency N_+/N depends on the preparation procedures used for the microsystems. Thus we find it necessary to introduce additional postulates which depend on both the preparation procedure and the registration procedure.

4.3 The Dependence of Registration upon Preparation

We shall now consider the following question: Which combinations of $a \in \mathcal{Q}$ and $b_0 \in \mathcal{R}_0$ are physically meaningful? Unfortunately, this problem is not trivial; indeed, in the usual formulation of quantum mechanics it is not even mentioned. In an idealized form, part of this problem can be found in axiomatic field theory as the so-called "causality postulate."

How do we formulate this "combination problem"? In order to formulate this problem in mathematical language we introduce the following definition:

D 4.3.1. Let $a \in \mathcal{Q}$, $b_0 \in \mathcal{R}_0$, $a \neq \emptyset$, $b_0 \neq \emptyset$; we say that a and b_0 *may be combined* provided that if $\tilde{a} \in \mathcal{Q}$, $\tilde{b}_0 \in \mathcal{R}_0$, $\emptyset \neq \tilde{a} \subset a$, $\emptyset \neq \tilde{b}_0 \subset b_0$, we find that $\tilde{a} \cap \tilde{b}_0 \neq \emptyset$.

If a and b_0 may be combined, then clearly $a \cap b_0 \neq \emptyset$. We now define the following set:

$$C = \{(a, b_0) \mid a \in \mathcal{Q}, b_0 \in \mathcal{R}_0 \text{ and } a, b_0 \text{ may be combined}\}. \qquad (4.3.1)$$

The following theorem is an immediate consequence of D 4.3.1.

Th. 4.3.1. *If $\emptyset \neq \tilde{a} \subset a$, $\emptyset \neq \tilde{b}_0 \subset b_0$ and if $a, \tilde{a} \in \mathcal{Q}$; $b_0, \tilde{b}_0 \in \mathcal{R}_0$ and $(a, b_0) \in C$ then we find that $(\tilde{a}, \tilde{b}_0) \in C$.*

Intuitively, the statement $\tilde{a} \cap \tilde{b}_0 \neq \emptyset$ represents the statement that it is physically possible (see [1], §10) to select microsystems according to both $\tilde{a}$ and $\tilde{b}_0$. That is, the microsystems which were prepared according to the preparation procedure $\tilde{a}$ can be used for the registration method $\tilde{b}_0$. Thus the statement that a, b_0 may be combined means that for every finer preparation procedure $\tilde{a} \subset a$ and every finer registration method $\tilde{b}_0 \subset b_0$ there are always systems in $\tilde{a} \cap \tilde{b}_0$.

The physical meaning of the combination problem will be discussed in detail in III, §1. In VII, §1 we shall again return to this subject.

The "largest possible" set C is given by

$$C = \mathcal{Q}' \times \mathcal{R}'_0$$

where

D 4.3.2. $\mathcal{Q}' = \mathcal{Q}\backslash\{\varnothing\}$ and $\mathcal{R}'_0 = \mathcal{R}_0\backslash\{\varnothing\}$.

The condition $C = \mathcal{Q}' \times \mathcal{R}'_0$ is not always realistic. Thus, in the selection of axioms for C we have encountered a physical problem.

Why is the condition $C = \mathcal{Q}' \times \mathcal{R}'_0$ an unrealistic condition for microsystems?

In order to facilitate understanding, we shall now briefly describe an additional structure which is needed in order to describe the relationship between preparation and registration methods. A detailed exposition of this structure will be given in VII, §1. It is concerned with the space–time relationship between these experimental procedures. Since the microsystems are produced by the preparation procedure it makes sense that the micro-objects are registered only after they are produced, and that the preparation apparatus and the registration apparatus do not collide. These remarks show that the question concerning axioms for the set C are not without physical significance. Later we shall see that, under certain circumstances, the combination problem described by C does not play an essential role in the usual formulation of quantum mechanics.

We shall now begin by making a minimal requirement for C.

APS 5.1.1. To each $a \in \mathcal{Q}'$ there exists a $b_0 \in \mathcal{R}'_0$ such that $(a, b_0) \in C$.

This axiom states that it is physically possible (see [1], §10) to apply at least one registration apparatus to the microsystems prepared by a. In III, §1 we will replace APS 5.1.1 by the stronger axiom APS 5.1.2.

We shall now consider a registration procedure $b \in \mathcal{R}$ where $b \neq \varnothing$ (we do not require that $b \in \mathcal{R}_0$); $b \neq \varnothing$ means that there are microsystems which may be selected according to the procedure b. All experience has shown that it is possible to prepare such microsystems. Thus we assert

APS 5.2. To each $b \in \mathcal{R}$, $b \neq \varnothing$ there exists at least one $a \in \mathcal{Q}$ and a $b_0 \in \mathcal{R}_0$ which satisfies $b \subset b_0$, $(a, b_0) \in C$ and $a \cap b \neq \varnothing$.

Axiom APS 5.2 is stronger than APS 4.2.

We shall now consider the mathematical formulation of the statistical dependence of the combined selection procedures. First we define

D 4.3.3.

$$\Theta = \{c \mid c = a \cap b,\ a \in \mathcal{Q},\ b \in \mathcal{R} \text{ and for } a \neq \varnothing \text{ there exists a } b_0 \in \mathcal{R}_0 \text{ which satisfies } b \subset b_0 \text{ and } (a, b_0) \in C\}.$$

Clearly Θ is a subset of $\mathscr{P}(M)$. An element $a \cap b$ of Θ represents the set of all microsystems which are prepared according to a and registered according to b.

From D 4.3.3 and Th. 4.3.1 it follows that:

Th. 4.3.2. *Let* a, $\tilde{a} \in \mathscr{Q}$, $\tilde{a} \subset a$ *and suppose* b, $\tilde{b} \in \mathscr{R}$, $\tilde{b} \subset b$. *If* $a \cap b \in \Theta$ *then* $\tilde{a} \cap \tilde{b} \in \Theta$.

D 4.3.4. Let $\mathscr{S}$ denote the smallest set of selection procedures for which $\Theta \subset \mathscr{S}$ (see Th. 2.2).

In general we find that, neither $\mathscr{Q} \subset \mathscr{S}$ nor $\mathscr{R} \subset \mathscr{S}$ holds!

We shall now formulate the requirement that the combination of preparation and registration procedures leads to reproducible frequencies.

APS 6. We require that $\mathscr{S}$ (as defined by D 4.3.4) shall be a statistical selection procedure.

Let $\lambda_{\mathscr{S}}$ denote the probability function for $\mathscr{S}$.

On the basis of their physical interpretation, $\lambda_{\mathscr{Q}}$, $\lambda_{\mathscr{R}_0}$ and $\lambda_{\mathscr{S}}$ cannot be mutually independent. Experience has shown that, in the combination of preparation procedures $a \in \mathscr{Q}$ and the registration methods $b_0 \in \mathscr{R}_0$ there is no statistical dependence between the selection according to the preparation procedures and the application of the registration method. We formulate the statistical independence of the preparation procedure and the application of registration methods by means of the following axioms:

APS 7. Let $a_1, a_2 \in \mathscr{Q}$, $b_{10}, b_{20} \in \mathscr{R}_0$ where $a_2 \subset a_1$, $b_{10} \subset b_{20}$ and $(a_1, b_{20}) \in C$. Then

APS 7.1. $$\lambda_{\mathscr{S}}(a_1 \cap b_{10}, a_2 \cap b_{10}) = \lambda_{\mathscr{Q}}(a_1, a_2);$$

APS 7.2. $$\lambda_{\mathscr{S}}(a_1 \cap b_{10}, a_1 \cap b_{20}) = \lambda_{\mathscr{R}_0}(b_{10}, b_{20}).$$

Note that if $b \in \mathscr{R}$, $b \notin \mathscr{R}_0$, and if $a_1, a_2 \in \mathscr{Q}$, $a_1 \cap b \neq \varnothing$ then, in general we find that $\lambda_{\mathscr{S}}(a_1 \cap b, a_2 \cap b) \neq \lambda_{\mathscr{Q}}(a_1, a_2)$.

Axiom APS 7.1 expresses the directedness of the interaction of the preparation on the registration apparatus in the following way. The probability associated with the process of the preparation apparatus does not depend on those of the registration apparatus (that is, on b_{10}); see I, §2.

4.4 The Concept of a Physical System

We shall now replace the intuitive notion of a physical system by a well-defined one, one defined in terms of the mathematical structure "preparation and registration procedures" and its physical interpretation (as expressed in terms of the mapping principles). Since the mapping principles are involved in this concept, it is not a derived one, that is, a concept which is defined only by terms and structures deduced in $\mathscr{M}\mathscr{T}_\Sigma$. Such derived concepts will be introduced later, for example, the derived concepts of ensemble, effect, observable (see III and IV).

The expression "physical system" will represent those facts which are to be mapped to elements of the set M together with the well-defined physical processes which are to be mapped to the elements of $\mathscr{Q}$, $\mathscr{R}$, and $\mathscr{R}_0$ whereby the letters $\mathscr{Q}$, $\mathscr{R}$, and $\mathscr{R}_0$ denote not only the sets but also the structures defined by axioms APS 1–7.[1] In less precise terms we say that a physical system is an element of the set M endowed with the structure $\mathscr{Q}$, $\mathscr{R}$, and $\mathscr{R}_0$. In III, §4.1 we shall examine the distinctions which must be made between the concept of a "physical system" and the concept of a "physical object." It is important to note that the concept of a physical system is not necessarily restricted to microsystems; see, for example [1], §12 and [2], XV–XVI and [13].

The facts which we have called "physical systems" have some reality beyond that of the "direct" interpretation of preparation and registration procedures. This additional meaning results from axioms APS 1 and APS 7, which imply that the preparation procedure is independent of the registration procedure. Intuitively, this means that, in the preparation, "something" is produced—that which we have called a physical system—which can be detected by the registration apparatus. A more precise formulation of this state of affairs is presented in [2], XVI and [13].

But the facts which we have called "physical systems" are so closely related to the associated production and detection methods that the typical characterization of physical objects (see §1 and III, §4.1) by objective properties of the systems has not been possible until now.

The statement of facts which are self-existing in the sense that they do not exert any influence on other things is self-contradictory. Such facts are completely inaccessible. Nevertheless, in physics we endeavor to describe, as completely as possible, *portions* of the real world as if these portions were self-existing. The attempt to describe the real world in complete detail would make physics impossible. Physics is possible only because we are able to make structural assertions about portions of the real world without taking into account the structure of the whole world in all its particulars. Only a few "global structures" of the world as a whole are introduced into the description of the physics of its parts, as, for example, the space–time structure (and gravitation, which for sufficiently small regions of space can be neglected due to the existence of local inertial reference systems).

We have made the assumption that the experiments composed of a preparation and a registration procedure can be described as such portions of the world using only space–time as a global structure of the world.

1. Here we shall require the use of axiom APS 5 in its strengthened form as presented in III, §1. We shall use the expression "carriers of physical interaction" to denote the set M when we require only APS 5 in the form presented above.

Is it possible to describe the physical systems as such "self-existing" portions of the world? It would be possible, if the physical systems were physical objects in the sense of §1 and III, §4.1. But since the microsystems are not physical objects (see IV, §8.1), we shall try to eliminate "as far as possible" details of the structure of the preparation and registration apparatuses. This problem is treated in III and IV.

In the case of a microsystem, this problem is not as simple as in the case of a macrosystem because the latter may be described as a physical object. The concept of a physical system described above is too general because it is applicable to both macro- and microsystems. The selection procedures in $\mathscr{S}$ describe a completely conventional "classical statistics" which does not exhibit the "typical" quantum mechanical structure. At what point do we make the transition to "quantum statistics" which appears to be so different? This transition will be made in III, §3 where we introduce axiom AV 4s instead of the postulate that the physical systems are physical objects.

We are able to describe the structure of a physical system only to the extent that the physical system can be prepared and registered. This statement will be formulated in mathematical terms by the following axiom:

APS 8.1.
$$M = \bigcup_{a \in \mathscr{Q}} a.$$

APS 8.2.
$$M = \bigcup_{b \in \mathscr{R}} b.$$

By APS 4.2 we find that APS 8.2 is equivalent to

$$M = \bigcup_{b_0 \in \mathscr{R}_0} b_0.$$

Axiom APS 8 does not describe a profound aspect of reality because the preparation and registration procedures are arbitrary and permit "all possibilities." The axiom states that every physical system interacts with the outside world at least once (APS 8.1) and again (APS 8.2).

4.5 The Structure of Probability Fields for Physical Systems

In this section we shall derive those theorems which may be obtained from the above axioms (that is, without the axioms in III, §1) in order to reduce the problem of the probability structure for the selection procedures of $\mathscr{S}$ to a manageable subset. First, we shall consider how $\mathscr{S}$ is obtained from Θ.

We shall now consider a finite collection $a_i \in \mathscr{Q}$; $i = 1, \ldots, n$ for which $a_i \subset a \in \mathscr{Q}$, $a_i \cap a_j = \varnothing$ for $i \neq j$, and a corresponding collection $b_k \in \mathscr{R}$; $k = 1, \ldots, m$ for which $b_k \subset b \in \mathscr{R}$, $b_k \cap b_l = \varnothing$ for $k \neq l$. Let A be a subset of ordered pairs (i, k), $i = 1, \ldots, n$ and $k = 1, \ldots, m$. From $a \cap b \in \Theta$ and Th. 4.3.2 we obtain $a_i \cap b_k \in \Theta$. Furthermore, we find that $\bigcup_{(i,j) \in A} a_i \cap b_k \subset a \cap b$.

Let Σ denote the set of all $c \subset M$ which may be represented in the form $c = \bigcup_{(i,k)\in A} a_i \cap b_k$ where the a_i, b_k, A satisfy the above conditions. In addition, let Σ' denote the following set:

$$\Sigma' = \left\{ c \mid c = \bigcup_{i=1}^{n} c_i,\ c_i \cap c_k = \varnothing \text{ for } i \neq k,\ c_i \in \Theta \text{ and } c \subset d \in \Theta \right\}. \tag{4.5.1}$$

Then we find that:

Th. 4.5.1. $$\Sigma = \Sigma' = \mathscr{S}.$$

PROOF. Since $\mathscr{S}(a \cap b)$ is a Boolean ring, and since $a_i \cap b_k \in \Theta \subset \mathscr{S}$ and $a_i \cap b_k \subset a \cap b \in \Theta$, we find that $c = \bigcup_{(i,k)\in A} a_i \cap b_k \in \mathscr{S}$. Thus we obtain $\Sigma \subset \mathscr{S}$. We also obtain $\Sigma = \mathscr{S}$ when we prove that Σ is a system of selection procedures, that is, we show that Σ satisfies both conditions AS 1.1 and AS 1.2.

From the definitions of Σ and Σ', from $a_i \cap b_l \in \Theta$ and $a \cap b \in \Theta$ it follows that $\Sigma \subset \Sigma'$. Since $\mathscr{S}(d)$ is a Boolean ring, it follows that $\Sigma' \subset \mathscr{S}$. Thus we need only prove that $\Sigma = \mathscr{S}$.

Let $c = \bigcup_{(i,k)\in A} a_i \cap b_k$ and $\tilde{c} = \bigcup_{(j,l)\in \tilde{A}} \tilde{a}_j \cap \tilde{b}_l$ be two elements of Σ represented in the above form where $i = 1, \ldots, n$; $k = 1, \ldots, m$; $a_i \subset a$, $b_k \subset b$; $j = 1, \ldots, \tilde{n}$, $l = 1, \ldots, \tilde{m}$; $\tilde{a}_j \subset \tilde{a}$, $\tilde{b}_l \subset \tilde{b}$. Then we obtain

$$c \cap \tilde{c} = \bigcup_{\substack{(i,k)\in A \\ (j,l)\in \tilde{A}}} (a_i \cap \tilde{a}_j \cap b_k \cap \tilde{b}_l). \tag{4.5.2}$$

Let $a_{ij} = a_i \cap \tilde{a}_j$, $b_{kl} = b_k \cap \tilde{b}_l$. We find that $a_{ij} \in \mathscr{Q}$, $b_{kl} \in \mathscr{R}$ and $a_{ij} \subset a \cap \tilde{a} \in \mathscr{Q}$, $b_{kl} \subset b \cap \tilde{b} \in \mathscr{R}$. Clearly $a_{ij} \cap a_{kl} = \varnothing$ for $(i,j) \neq (k,l)$; a similar result holds for b_{kl}. Thus, from (4.5.2) and Th. 4.3.2, it follows that $c \cap \tilde{c} \in \Sigma$, that is, Σ satisfies axiom AS 1.2.

In order to show that AS 1.1 is satisfied, we shall assume that $\tilde{c} \subset c$ and compute $c \backslash \tilde{c}$.

Since $\tilde{c} \subset c$, we obtain $\tilde{c} = c \cap \tilde{c}$, from which we obtain, using (4.5.2)

$$c \backslash \tilde{c} = \bigcup_{(i,k)\in A} \left[a_i \cap b_k \Big\backslash \bigcup_{(j,l)\in \tilde{A}} (a_i \cap \tilde{a}_j \cap b_k \cap \tilde{b}_l) \right]. \tag{4.5.3}$$

Since $\mathscr{Q}$ and $\mathscr{R}$ are selection procedures, we obtain $a_{i\tilde{n}+1} \in \mathscr{Q}$, $b_{k\tilde{m}+1} \in \mathscr{R}$, where

$$\begin{aligned} a_{i\tilde{n}+1} &= a_i \Big\backslash \bigcup_{j=1}^{\tilde{n}} (a_i \cap \tilde{a}_j), \\ b_{k\tilde{m}+1} &= b_k \Big\backslash \bigcup_{l=1}^{\tilde{m}} (b_k \cap \tilde{b}_l). \end{aligned} \tag{4.5.4}$$

In addition, we find that

$$\begin{aligned} a_{ij} &\subset a \quad \text{for } i = 1, \ldots, n; j = 1, \ldots, \tilde{n}, \tilde{n}+1, \\ b_{kl} &\subset b \quad \text{for } k = 1, \ldots, m; l = 1, \ldots, \tilde{m}, \tilde{m}+1 \end{aligned} \tag{4.5.5a}$$

and for $j = 1, \ldots, \tilde{n}+1$ we obtain

$$a_{ij} \cap a_{i'j'} = \varnothing \quad \text{for } (i,j) \neq (i',j')$$

and a similar result for the b_{kl}.

From (4.5.4) we conclude that

$$a_i = \bigcup_{j=1}^{\tilde{n}+1} a_{ij}, \qquad b_k = \bigcup_{l=1}^{\tilde{m}+1} b_{kl}$$

and (4.5.5b)

$$a_i \cap b_k = \bigcup_{j=1}^{\tilde{n}+1} \bigcup_{l=1}^{\tilde{m}+1} a_{ij} \cap b_{kl}.$$

Let

$$B = \{(j, l) \mid j = 1, \ldots, \tilde{n} + 1; l = 1, \ldots, \tilde{m} + 1\}.$$

From (4.5.3) we obtain

$$c \backslash \tilde{c} = \bigcup_{(i,k) \in A} \bigcup_{(j,l) \in B \backslash \tilde{A}} (a_{ij} \cap b_{kl}). \tag{4.5.6}$$

From (4.5.6) together with (4.5.5a) and (4.5.5b) we conclude that $c \backslash \tilde{c} \in \Sigma$.

Th. 4.5.2. *The function $\lambda_{\mathscr{S}}$ for $\mathscr{S}$ is uniquely determined by $\lambda_{\mathscr{Q}}$ and by the special values $\lambda_{\mathscr{S}}(a \cap b_0, a \cap b)$ for those $(a, b_0) \in C$ where C is defined by (4.3.1) and $b \in \mathscr{R}$ with $b \subset b_0$.*

PROOF. Since $\mathscr{S} = \Sigma'$ (by Th. 4.5.1) we may express the $c, \tilde{c}$ in $\lambda_{\mathscr{S}}(c, \tilde{c})$ in the form:

$$c = \bigcup_{i=1}^{n} c_i, \qquad \tilde{c} = \bigcup_{k=1}^{m} \tilde{c}_k, \tag{4.5.7}$$

where $c_i \in \Theta$, $\tilde{c}_k \in \Theta$ and $c_i \cap c_k = \varnothing$ for $i \neq k$, $\tilde{c}_k \cap \tilde{c}_l = \varnothing$ for $k \neq l$, where $c \subset d \in \Theta$, $\tilde{c} \subset \tilde{d} \in \Theta$. Since $\tilde{c} \subset c \subset d$ we can set $\tilde{d} = d$ and $\tilde{c}_k \subset c$ without loss of generality.

From (4.5.7) and (3.5) we obtain

$$\lambda_{\mathscr{S}}(c, \tilde{c}) = \sum_{k=1}^{m} \lambda_{\mathscr{S}}(c, \tilde{c}_k). \tag{4.5.8}$$

Since $c \subset d \in \Theta$, from AS 2.2 we obtain

$$\lambda_{\mathscr{S}}(d, c_k) = \lambda_{\mathscr{S}}(d, \check{c}) \lambda_{\mathscr{S}}(c, \tilde{c}_k).$$

Since $c \neq \varnothing$ (otherwise $\lambda_{\mathscr{S}}(c \ldots)$ would not be defined) we find that, by AS 2.3 $\lambda_{\mathscr{S}}(d, c) \neq 0$ and therefore

$$\lambda_{\mathscr{S}}(c, \tilde{c}_k) = \frac{\lambda_{\mathscr{S}}(d, \tilde{c}_k)}{\lambda_{\mathscr{S}}(d, c)}. \tag{4.5.9}$$

Using (3.5) and (4.5.7) we find that

$$\lambda_{\mathscr{S}}(d, c) = \sum_{i=1}^{n} \lambda_{\mathscr{S}}(d, c_i).$$

Thus (4.5.9) becomes

$$\lambda_{\mathscr{S}}(c, \tilde{c}_k) = \frac{\lambda_{\mathscr{S}}(d, \tilde{c}_k)}{\sum_{i=1}^{n} \lambda_{\mathscr{S}}(d, c_i)}. \tag{4.5.10}$$

By substitution of (4.5.10) into (4.5.8) we obtain

$$\lambda_{\mathscr{S}}(c, \tilde{c}) = \frac{\sum_{k=1}^{m} \lambda_{\mathscr{S}}(d, \tilde{c}_k)}{\sum_{i=1}^{n} \lambda_{\mathscr{S}}(d, c_i)}. \tag{4.5.11}$$

Since $d, c_i, \tilde{c}_k \in \Theta$ $\lambda_{\mathscr{S}}$ is determined by the special values $\lambda_{\mathscr{S}}(d, c)$ for $c \subset d$ and $c, d \in \Theta$.

Let $d = a \cap b, a \in \mathscr{Q}, b \in \mathscr{R}$ and $c = \tilde{a} \cap \tilde{b}, \tilde{a} \in \mathscr{Q}, \tilde{b} \in \mathscr{R}$. Then if $c \subset d$ we obtain

$$c = (a \cap \tilde{a}) \cap (b \cap \tilde{b}).$$

That is, if $\bar{a} = a \cap \tilde{a}$ and $\bar{b} = b \cap \tilde{b}$, then $c = \bar{a} \cap \bar{b}$ where $\bar{a} \in \mathscr{Q}, \bar{b} \in \mathscr{R}$ and $\bar{a} \subset a$, $\bar{b} \subset b$.

According to D 4.3.3 there exists a $b_0 \in \mathscr{R}_0$ for which $b \subset b_0$ and $(a, b_0) \in C$; consequently $a \cap b_0 \in \Theta$ and $\bar{b} \subset b \subset b_0$. Since $a \cap b \neq \varnothing$ (otherwise $\lambda_{\mathscr{S}}(d, c) = \lambda_{\mathscr{S}}(a \cap b, \bar{a} \cap \bar{b})$ would not be defined) $a \cap b_0 \neq \varnothing$ and

$$\lambda_{\mathscr{S}}(a \cap b_0, a \cap b) \neq 0.$$

Using the following equations (obtained from AS 2.2)

$$\lambda_{\mathscr{S}}(a \cap b_0, a \cap b)\lambda_{\mathscr{S}}(a \cap b, \bar{a} \cap \bar{b}) = \lambda_{\mathscr{S}}(a \cap b_0, \bar{a} \cap \bar{b})$$

we obtain

$$\lambda_{\mathscr{S}}(a \cap b_0, \bar{a} \cap \bar{b}) = \frac{\lambda_{\mathscr{S}}(a \cap b_0, \bar{a} \cap \bar{b})}{\lambda_{\mathscr{S}}(a \cap b_0, a \cap b)}. \tag{4.5.12}$$

Similarly, from AS 2.2 it follows that

$$\lambda_{\mathscr{S}}(a \cap b_0, \bar{a} \cap \bar{b}) = \lambda_{\mathscr{S}}(a \cap b_0, \bar{a} \cap b_0)\lambda_{\mathscr{S}}(\bar{a} \cap b_0, \bar{a} \cap \bar{b}) \tag{4.5.13}$$

because $a \cap b_0 \neq \varnothing$ whenever $\bar{a} \neq \varnothing$ according to D 4.3.1.

If $\bar{a} \neq \varnothing$, then by (4.5.13) we obtain

$$\lambda_{\mathscr{S}}(a \cap b_0, \bar{a} \cap \bar{b}) = \lambda_{\mathscr{Q}}(a, \bar{a})\lambda_{\mathscr{S}}(\bar{a} \cap b_0, \bar{a} \cap \bar{b});$$

from (4.5.12) we obtain, for $\bar{a} \neq \varnothing$

$$\lambda_{\mathscr{S}}(a \cap b_0, \bar{a} \cap \bar{b}) = \frac{\lambda_{\mathscr{Q}}(a, \bar{a})\lambda_{\mathscr{S}}(\bar{a} \cap b_0, \bar{a} \cap \bar{b})}{\lambda_{\mathscr{S}}(a \cap b_0, a \cap b)}. \tag{4.5.14}$$

For the case in which $\bar{a} = \varnothing$, we obtain

$$\lambda_{\mathscr{S}}(a \cap b, \bar{a} \cap \bar{b}) = 0.$$

According to Th. 4.5.2, in physics it is only necessary to consider those $\lambda_{\mathscr{S}}(a \cap b_0, a \cap b)$ for which $(a, b_0) \in C$, $b \in \mathscr{R}$, $b \subset b_0$; the role of an experiment is to "measure" these special values of the function $\lambda_{\mathscr{S}}$. We now define

D 4.5.1. $\mathscr{F} = \{(b_0, b) \mid b_0 \in \mathscr{R}'_0, b \in \mathscr{R}, b \subset b_0\}$.

$\mathscr{F}$ is called the set of "effect processes" or "questions."

D 4.5.2. $\mathscr{C} = \{(a, f) \mid f = (b_0, b) \in \mathscr{F}$ and $(a, b_0) \in C\}$,

where C is defined by D 4.3.1.

D 4.5.3. On $\mathscr{C}$ we define the real function μ by

$$\mu(a, f) = \mu(a, (b_0, b)) = \lambda_{\mathscr{S}}(a \cap b_0, a \cap b).$$

From Th. 4.5.2 it follows that $\lambda_{\mathscr{S}}$ is determined for all of $\mathscr{S}$ by $\lambda_{\mathscr{Q}}$ and the values of μ on $\mathscr{C}$.

The function μ defined by D 4.5.3 plays a central role in the statistical description of a physical system. We shall now briefly illustrate the nature of the experiments which may be used to test the function μ.

These experiments consist of a preparation procedure a by which a single system $x \in M$ is produced, and upon which a registration method b_0 is applied. As the result of the interaction of the system with the apparatus, a process characterized by b may or may not occur. Let $x_1, \ldots, x_N$ denote N systems which are prepared according to a and to which b_0 is applied; that is, $x_i \in a \cap b_0$ for $i = 1, \ldots, N$. Of these systems, suppose that N_1 systems x_{i_ν}, $\nu = 1, \ldots, N_1$ activate the process characterized by b, that is, $x_{i_\nu} \in b$. According to the physical meaning of $\lambda_{\mathscr{S}}(a \cap b_0, a \cap b)$, if the theory is applicable, we must have approximate agreement of N_1/N with $\mu(a, (b_0, b)) = \mu(a, f) = \lambda_{\mathscr{S}}(a \cap b_0, a \cap b)$.

This basic experimental situation will be repeatedly encountered in the comparison of quantum theory with experiment. We shall not give an account of the numerous examples here. Instead we shall prove a number of theorems about $\mu(a, f)$.

In order to formulate the following theorems we shall now define

D 4.5.4. $\mathbf{1}_{b_0} = (b_0, b_0)$ and $\mathbf{0}_{b_0} = (b_0, \varnothing)$; $\mathbf{1}_{b_0}$ and $\mathbf{0}_{b_0}$ are elements of $\mathscr{F}$.

D 4.5.5. Suppose that $b_0 \in \mathscr{R}_0$ is partitioned by b_i, $i = 1, \ldots, n$, that is, $b_0 = \bigcup_{i=1}^n b_i$ and $b_i \cap b_j = \varnothing$ for $i \neq j$. We call $f_i = (b_0, b_i)$ a (disjoint) partition of $\mathbf{1}_{b_0}$ and write $\mathbf{1}_{b_0} = \bigcup_{i=1}^n f_i$.

The following two theorems describe the central structural properties of the probability function $\lambda_{\mathscr{S}}$.

Th. 4.5.3. *For the function μ the following conditions are satisfied:*

(i) $0 \leq \mu(a, f) \leq 1$.
(ii) *For each $a \in \mathscr{Q}'$ there exists an $f_0 \in \mathscr{F}$ for which $\mu(a, f_0) = 0$.*
(iii) *For each $a \in \mathscr{Q}'$ there exists an $f_1 \in \mathscr{F}$ for which $u(a, f_1) = 1$.*
(iv) *For a decomposition $a = \bigcup_i a_i$ (see D 3.2) the following condition holds for all f for which $(a, f) \in \mathscr{C}$:*

$$\mu\left(\bigcup_i a_i, f\right) = \sum_i \lambda_i \mu(a_i, f),$$

where

$$0 < \lambda_i = \lambda_{\mathscr{Q}}(a, a_i) \leq 1 \quad \textit{and} \quad \sum_i \lambda_i = 1.$$

(v) *Let* $b_{01}, b_{02} \in \mathscr{R}'_0$ *and* $b \subset b_{02} \subset b_{01}$, *let* $f_1 = (b_{01}, b)$, $f_2 = (b_{02}, b)$. *Then the following condition holds for all a for which* $(a, f_1) \in \mathscr{C}$

$$\mu(a, f_1) = \lambda_{\mathscr{R}_0}(b_{01}, b_{02})\mu(a, f_2).$$

(vi) *Let* $(a, b_0) \in C$; *if* $\mathbf{1}_{b_0} = \bigcup_{i=1}^n f_i$ *(see* D 4.5.5) *then* $\sum_{i=1}^n \mu(a, f_i) = 1$.

(vii) *If* $(a, f) \in \mathscr{C}$ *where* $f = (b_0, b)$ *then* $\mu(a, f) = 0$ *if and only if* $a \cap b = \varnothing$.

(viii) *Let* $b_0 = \bigcup_{i=1}^n b_{0i}$ *where* $b_{0i} \in \mathscr{R}'_0$ *and* $b_{0i} \cap b_{0k} = \varnothing$ *for* $i \neq k$ *(that is,* $\bigcup_{i=1}^n b_{0i}$ *is a decomposition of* b_0 *in* $\mathscr{R}_0$*). Then, for* $f = (b_0, b)$ $f_i = (b_{0i}, b_{0i} \cap b)$, *and for every* $a \in \mathscr{Q}'$ *which satisfies* $(a, f) \in \mathscr{C}$ *we obtain*

$$\mu(a, f) = \sum_{i=1}^{n} \lambda_{\mathscr{R}_0}(b_0, b_{0i})\mu(a, f_i)$$

and

$$\sum_{i=1}^{n} \lambda_{\mathscr{R}_0}(b_0, b_{0i}) = 1.$$

(In this case we say that the f_i *describe a decomposition of* f *which is induced by the decomposition of* b_0*).*

PROOF. According to APS 5.1.1, to each $a \in \mathscr{Q}'$ there is at least one $b_0 \in \mathscr{R}'_0$ for which $(a, b_0) \in C$. For such a b_0, from (vi) it follows that (with $n = 1$) $f_1 = \mathbf{1}_{b_0}$, thus proving (iii); similarly, let $f_0 = \mathbf{0}_{b_0}$ $(b_0 = \varnothing)$. From (vii) we obtain (ii). We obtain a proof of (i) from the corresponding property of $\lambda_{\mathscr{S}}$. Thus, it is only necessary to prove (iv)–(viii).

(iv) Since $(a, f) \in \mathscr{C}$ where $f = (b_0, b)$, it follows that $(a, b_0) \in C$. Then by Th. 4.3.1 we obtain $(a_i, b_0) \in C$, that is, $(a_i, f) \in \mathscr{C}$. Thus all the $\mu(a_i, f)$ are defined. Thus, we find that

$$1 = \lambda_{\mathscr{Q}}(a, a) = \lambda_{\mathscr{Q}}\left(a, \bigcup_{i=1}^{n} a_i\right) = \sum_{i=1}^{n} \lambda_{\mathscr{Q}}(a, a_i) = \sum_{i=1}^{n} \lambda_i .$$

Since $a \cap b = \bigcup_i (a_i \cap b)$ we obtain

$$\mu(a, f) = \lambda_{\mathscr{S}}(a \cap b_0, a \cap b) = \sum_{i=1}^{n} \lambda_{\mathscr{S}}(a \cap b_0, a_i \cap b)$$

$$= \sum_{i=1}^{n} \lambda_{\mathscr{S}}(a \cap b_0, a_i \cap b_0)\lambda_{\mathscr{S}}(a \cap b_0, a_i \cap b)$$

$$= \sum_{i=1}^{n} \lambda_{\mathscr{Q}}(a, a_i)\lambda_{\mathscr{S}}(a_i \cap b_0, a_i \cap b) = \sum_{i=1}^{n} \lambda_i \mu(a_i, f).$$

(v) From $(a, f_1) \in \mathscr{C}$ we obtain $(a, b_{01}) \in C$. From Th. 4.3.1 we obtain $(a, b_{02}) \in C$ and therefore $(a, f_2) \in \mathscr{C}$. Thus we obtain

$$\lambda_{\mathscr{S}}(a \cap b_{01}, a \cap b) = \lambda_{\mathscr{S}}(a \cap b_{01}, a \cap b_{02})\lambda_{\mathscr{S}}(a \cap b_{02}, a \cap b)$$

$$= \lambda_{\mathscr{R}_0}(b_{01}, b_{02})\lambda_{\mathscr{S}}(a \cap b_{02}, a \cap b).$$

(vi) Let $b_0 = \bigcup_{i=1}^n b_i$ be a disjoint partition of b_0, and let $(a, b_0) \in C$. Then we find that

$$1 = \lambda_{\mathcal{S}}(a \cap b_0, a \cap b_0) = \sum_{i=1}^n \lambda_{\mathcal{S}}(a \cap b_0, a \cap b_i) = \sum_{i=1}^n \mu(a, f_i).$$

(vii) Let $(a, b_0) \in C$, then if $\mu(a, f) = \lambda_{\mathcal{S}}(a \cap b_0, a \cap b) = 0$ then from AS 2.3 it follows that $a \cap b = \varnothing$. If $a \cap b \neq \varnothing$, then by (3.3) we obtain $\mu(a, f) = \lambda_{\mathcal{S}}(a \cap b_0, a \cap b) \neq 0$.

(viii) If $(a, b_0) \in C$ we obtain

$$\mu(a, f) = \lambda_{\mathcal{S}}(a \cap b_0, a \cap b) = \lambda_{\mathcal{S}}\left(a \cap b_0, \bigcup_{i=1}^n (a \cap b_{0i} \cap b)\right)$$

$$= \sum_{i=1}^n \lambda_{\mathcal{S}}(a \cap b_0, a \cap b_{0i} \cap b).$$

From AS 2.2, Th. 4.3.1 and APS 7.2 we obtain

$$\lambda_{\mathcal{S}}(a \cap b_0, a \cap b_{0i} \cap b) = \lambda_{\mathcal{S}}(a \cap b_0, a \cap b_{0i})\lambda_{\mathcal{S}}(a \cap b_{0i}, a \cap b_{0i} \cap b)$$
$$= \lambda_{\mathcal{R}_0}(b_0, b_{0i})\mu(a, f_i).$$

Since $b_0 = \bigcup_{i=1}^n b_{0i}$ we find that

$$1 = \lambda_{\mathcal{R}_0}(b_0, b_0) = \sum_{i=1}^n \lambda_{\mathcal{R}_0}(b_0, b_{0i}).$$

It is important now to show there is a partial converse to Th. 4.5.3.

Th. 4.5.4. *Given $\lambda_{\mathcal{Q}}$; suppose there is a function $\mu(a,f)$ on $\mathcal{C}$ which satisfies conditions* (i), (iv), (v), (vi), *and* (vii) *of* Th. 4.5.3. *Then on $\mathcal{S}$ there exists one and only one probability function $\lambda_{\mathcal{S}}$ (that is, a function which satisfies* AS 2) *for which $\lambda_{\mathcal{S}}(a \cap b_0, a \cap b) = \mu(a, f)$. This function $\lambda_{\mathcal{S}}$ satisfies the conditions* APS 7 *and conditions* (viii) *of* Th. 4.5.3. *Also, $\lambda_{\mathcal{R}_0}$ is uniquely determined by μ.*

PROOF (Based on a Communication by H. Neumann). We shall use the proof of Th. 4.5.2 in reverse. From the function $\mu: \mathcal{C} \to [0, 1]$ we shall seek a function $\lambda: \mathcal{T}_\Theta \to [0, 1]$ where $\mathcal{T}_\Theta = \{(d, c) \mid d, c \in \Theta \text{ where } c \subset d\}$ which is identical to the restriction of $\lambda_{\mathcal{S}}$ to $\mathcal{T}_\Theta$. Then with the help of $\Sigma' = \mathcal{S}$ we will obtain the function

$$\lambda_{\mathcal{S}}: \mathcal{T}_{\mathcal{S}} \to [0, 1] \quad \text{with } \mathcal{T}_{\mathcal{S}} = \{(d, c) \mid d, c \in \mathcal{S} \text{ where } c \subset d\}.$$

We shall begin with the second step, in order to better establish the conditions which $\lambda: \mathcal{T}_\Theta \to [0, 1]$ must satisfy in order that it may be extended to a probability function $\lambda_{\mathcal{S}}$. We will show that this extension is possible if $\lambda: \mathcal{T}_\Theta \to [0, 1]$ satisfies the following three conditions:

(α) If $c_1, c_2, c_3 \in \Theta$, $c_3 \subset c_2 \subset c_1$, $c_2 \neq \varnothing$ then

$$\lambda(c_1, c_2) = \lambda(c_1, c_2)\lambda(c_2, c_3).$$

(β) If $c, c_i, \bigcup_{i=1}^n c_i \in \Theta$, $c \neq \varnothing$, $\bigcup_{i=1}^n c_i \subset c$, $c_i \cap c_j = \varnothing$ for $i \neq j$ then

$$\lambda\left(c, \bigcup_{i=1}^n c_i\right) = \sum_{i=1}^n \lambda(c, c_i).$$

(γ) If $d, c \in \Theta$, $c \neq \varnothing$ then $\lambda(d, c) \neq 0$.

We now seek to define $\lambda_{\mathscr{S}}: \mathscr{T}_{\mathscr{S}} \to [0, 1]$ by (4.5.11), that is, by the equation

$$\lambda_{\mathscr{S}}(c, \tilde{c}) = \frac{\sum_{k=1}^{m} \lambda(d, \tilde{c}_k)}{\sum_{i=1}^{n} \lambda(d, c_i)}, \tag{4.5.15}$$

where $c \subset d$, $d \in \Theta$ and $c \neq \varnothing$.

In order to use (4.5.15) as a definition, it is necessary that $\lambda(d, c_i) \neq 0$ for at least one value of i; otherwise, by (γ) we would have $c_i = \varnothing$ for all i, and $c = \varnothing$ in contradiction to the hypothesis that $c \neq \varnothing$.

We shall now show that $\lambda_{\mathscr{S}}$, as defined by (4.5.15), is unique. Suppose that, in addition to the decomposition $c = \bigcup_{i=1}^{n} c_i$ and $\tilde{c} = \bigcup_{k=1}^{m} \tilde{c}_k$, there is another decomposition $c = \bigcup_{j=1}^{n'} c'_j$ and $\tilde{c} = \bigcup_{l=1}^{m'} \tilde{c}'_l$. Then we find that

$$c_i = c \cap c_i = \bigcup_{j=1}^{n'} (c_i \cap c'_j) \quad \text{and} \quad \tilde{c}_k = \tilde{c} \cap \tilde{c}_k = \bigcup_{l=1}^{m'} (\tilde{c}_k \cap \tilde{c}'_l).$$

Then from (β) we conclude that

$$\lambda(d, c_i) = \sum_{j=1}^{n'} \lambda(d, c_i \cap c'_j),$$

$$\lambda(d, \tilde{c}_k) = \sum_{l=1}^{m'} \lambda(d, \tilde{c}_k \cap \tilde{c}'_l)$$

and therefore

$$\lambda_{\mathscr{S}}(c, \tilde{c}) = \frac{\sum_{k,l} \lambda(d, \tilde{c}_k \cap \tilde{c}'_l)}{\sum_{i,j} \lambda(d, c_i \cap c'_j)}.$$

This formula is symmetric in the primed and unprimed quantities; in addition, it yields the same value for $\lambda(c, \tilde{c})$ if the primed quantities are used in the right side of the equation (4.5.15). Thus we find that (4.5.15) does not depend on the choice of the decomposition of c and $\tilde{c}$. We must now show that it does not depend on the choice of d.

Let $c \subset d'$ where $d' \in \Theta$. Then $d' \cap d \in \Theta$ and $c \subset d \cap d'$. By (α) we find that

$$\lambda(d, c_i) = \lambda(d, d' \cap d)\lambda(d' \cap d, c_i),$$
$$\lambda(d, \tilde{c}_k) = \lambda(d, d' \cap d)\lambda(d' \cap d, \tilde{c}_k)$$

and, from (4.5.15) we obtain

$$\lambda_{\mathscr{S}}(c, \tilde{c}) = \frac{\sum_{k=1}^{m} \lambda(d' \cap d, \tilde{c}_k)}{\sum_{i=1}^{n} \lambda(d' \cap d, c_i)}.$$

This formula is symmetric in d and d'; thus (4.5.15) is independent of the choice of d.

We shall now show that the function $\lambda_{\mathscr{S}}$ satisfies AS 2.1–AS 2.3.

If $c, \tilde{c} \in \mathscr{S}$, $c \cap \tilde{c} = \varnothing$ and $c \cup \tilde{c} \in \mathscr{S}$ then $c_i \cap \tilde{c}_k = \varnothing$ in (4.5.7) for all i, k. Since $c \cup \tilde{c} \in \mathscr{S} = \Sigma'$, there exists a $d \in \Sigma'$ with $c \cup \tilde{c} \subset d$. Since $c \cup \tilde{c} = [\bigcup_i c_i] \cup [\bigcup_k \tilde{c}_k]$, then by (4.3.15) we obtain

$$\lambda_{\mathscr{S}}(c \cup \tilde{c}, c) = \frac{\sum_i \lambda(d, \tilde{c}_i)}{\sum_i \lambda(d, c_i) + \sum_k \lambda(d, \tilde{c}_k)}$$

and

$$\lambda_{\mathscr{S}}(c \cup \tilde{c}, \tilde{c}) = \frac{\sum_i \lambda(d, \tilde{c}_i)}{\sum_i \lambda(d, c_i) + \sum_k \lambda(d, \tilde{c}_k)}.$$

Thus we find that $\lambda_{\mathscr{S}}(c \cup \tilde{c}, c) + \lambda_{\mathscr{S}}(c \cup \tilde{c}, \tilde{c}) = 1$, or, in other words, AS 2.1 is satisfied.

If $\tilde{\tilde{c}} \subset \tilde{c} \subset c$ and $\tilde{c} \neq \varnothing$ we obtain

$$\lambda(c, \tilde{c})\lambda(\tilde{c}, \tilde{\tilde{c}}) = \frac{\sum_k \lambda(d, \tilde{c}_k)}{\sum_i \lambda(d, c_i)} \frac{\sum_l \lambda(d, \tilde{\tilde{c}}_l)}{\sum_k \lambda(d, \tilde{c}_k)}$$

$$= \frac{\sum_l \lambda(d, \tilde{\tilde{c}}_l)}{\sum_i \lambda(d, c_i)} = \lambda(c, \tilde{\tilde{c}}).$$

Thus we find that AS 2.2 is satisfied.

If $\tilde{c} \neq \varnothing$, then at least one of the $\tilde{c}_k \neq \varnothing$; then according to the definition formula (4.5.15) $\lambda(c, \tilde{c}) \neq \varnothing$, and AS 2.3 is satisfied.

We note that the above result and Th. 4.5.2 show that a function $\lambda: \mathscr{T}_\Theta \to [0, 1]$ can be extended in a unique way to a probability function $\lambda_{\mathscr{S}}: \mathscr{T}_{\mathscr{S}} \to [0, 1]$ if the conditions (α), (β), and (γ) are satisfied for $\lambda: \mathscr{T}_\Theta \to [0, 1]$.

It only remains to show that a function $\mu: \mathscr{C} \to [0, 1]$ which satisfies the conditions (i), (iv), (v), (vi), and (vii) can be extended to a function $\lambda: \mathscr{T}_\Theta \to [0, 1]$ which satisfies (α), (β), and (γ) and APS 7.

We use the formula (4.5.14), that is,

$$\lambda(a \cap b, \bar{a} \cap \bar{b}) = \frac{\lambda_{\mathscr{Q}}(a, \bar{a})\mu(\bar{a}, \bar{f})}{\mu(a, f)}, \tag{4.5.16}$$

where $a \cap b \neq \varnothing$, $f = (b_0, b)$, $\bar{f} = (b_0, \bar{b})$ and $b \subset b_0$ in order to define $\lambda: \mathscr{T}_\Theta \to [0, 1]$.

In order that (4.5.16) be well defined, it is necessary that $\mu(a, f) \neq 0$. From $\mu(a, f) = 0$ it follows that, by (vii) $a \cap b = \varnothing$ in contradiction to the hypothesis $a \cap b \neq \varnothing$. In order to apply (4.5.16), $\mu(a, f)$ must be defined, that is, $\bar{a} \subset a, \bar{a} \neq \varnothing$ and $(\bar{a}, b_0) \in C$.

According to D 4.3.3, $b_0 \in \mathscr{R}_0$ can be chosen so that $(a, b_0) \in C$ because $a \cap b \in \Theta$. Then, by Th 4.2.1 it follows that $(\bar{a}, b_0) \in C$.

In order to show that the function defined by (4.5.16) is unique, we shall assume that $a \cap b = a_1 \cap b_1$ and $\bar{a} \cap \bar{b} = \bar{a}_1 \cap \bar{b}_1$ where $\bar{a}_1 \subset a_1$, $\bar{b}_1 \subset b_1$. We must show that

$$\frac{\lambda_{\mathscr{Q}}(a, \bar{a})\mu(\bar{a}, \bar{f})}{\mu(a, f)} = \frac{\lambda_{\mathscr{Q}}(a_1, \bar{a}_1)\mu(\bar{a}_1, \bar{f}_1)}{\mu(a_1, f_1)}, \tag{4.5.17}$$

where $f_1 = (b_{01}, b_1)$ and $\bar{f}_1 = (b_{01}, \bar{b}_1)$. If $\bar{a} \cap \bar{b} = \bar{a}_1 \cap \bar{b}_1 = \varnothing$, then the right side of (4.5.17) is zero because either $\lambda_{\mathscr{Q}}(a, \bar{a}) = 0$ for $\bar{a} = \varnothing$ or $\mu(\bar{a}, \bar{f}) = 0$ according to (vii); a similar result holds for $\bar{a}_1, \bar{b}_1$. We need, therefore, only consider the case in which $\bar{a} \cap \bar{b} = \bar{a}_1 \cap \bar{b}_1 \neq \varnothing$. We rewrite the left side of (4.5.17) using (vi). We obtain

$$\mu(\bar{a}, (b_0, \bar{b})) = \mu(\bar{a}, (b_0, \bar{b} \cap \bar{b}_1)) + \mu(\bar{a}, (b_0, \bar{b} \backslash (b \cap \bar{b}_1)))$$

and

$$\mu(a, (b_0, b)) = \mu(a, (b_0, b \cap b_1)) + \mu(a, (b_0, b \backslash (b \cap b_1))).$$

Since $a \cap b = a_1 \cap b_1$, $a \cap b = a \cap b \cap b_1$, that is, $a \cap (b \backslash (b \cap b_1)) = \varnothing$. By (vii) $\mu(a, (b_0, b \backslash (b \cap b_1))) = 0$ and $\mu(\bar{a}, (b_0, \bar{b} \backslash (b \cap \bar{b}_1))) = 0$. The left-hand side of (4.5.7) is equal to

$$\frac{\lambda_{\mathscr{Q}}(a, \bar{a})\mu(\bar{a}, (b_0, \bar{b} \cap \bar{b}_1))}{\mu(a, (b_0, b \cap b_1))}.$$

If $b_0 \cap b_{01} \neq \emptyset$, then by (v) we obtain:

$$\mu(a, (b_0, b \cap b_1)) = \lambda_{\mathscr{R}_0}(b_0, b_0 \cap b_{01})\mu(a, (b_0 \cap b_{01}, b \cap b_1)).$$

The condition $b_0 \cap b_{01} \neq \emptyset$ is satisfied because

$$b_0 \cap b_{01} \supset b \cap b_1 \supset a \cap b \cap a_1 \cap b_1 = a \cap b \neq \emptyset.$$

From (v) we also conclude that

$$\mu(\bar{a}, (b_0, \bar{b} \cap \bar{b}_1)) = \lambda_{\mathscr{R}_0}(b_0, b_0 \cap b_{01})\mu(\bar{a}, (b_0 \cap b_{01}, \bar{b} \cap \bar{b}_1)).$$

Using these results, the left-hand side of (4.5.17) is equal to

$$\frac{\lambda_{\mathscr{Q}}(a, \bar{a})\mu(\bar{a}, (b_0 \cap b_{01}, \bar{b} \cap \bar{b}_1))}{\mu(a, (b_0 \cap b_{01}, b \cap b_1)}.$$

According to (iv), for all f for which $(a, f) \in \mathscr{C}$ we obtain

$$\mu(a, f) = \lambda_{\mathscr{Q}}(a, a \cap a_1)\mu(a \cap a_1, f) + \lambda_{\mathscr{Q}}(a, a \backslash a \cap a_1)\mu(a \backslash a \cap a_1, f).$$

Since $a \cap b = a_1 \cap b_1$, it follows that $a \cap b \cap b_1 = (a \cap a_1) \cap (b \cap b_1)$, that is, $(a \backslash a \cap a_1) \cap (b \cap b_1) = \emptyset$. Thus, from (vii) we obtain

$$\mu(a, (b_0 \cap b_{01}, b \cap b_1)) = \lambda_{\mathscr{Q}}(a, a \cap a_1)\mu(a \cap a_1, (b_0 \cap b_{01}, b \cap b_1))$$

and

$$\mu(\bar{a}, (b_0 \cap b_{01}, \bar{b} \cap \bar{b}_1)) = \lambda_{\mathscr{Q}}(\bar{a}, \bar{a} \cap \bar{a}_1)\mu(\bar{a} \cap \bar{a}_1, (b_0 \cap b_{01}, \bar{b} \cap \bar{b}_1)).$$

Thus we obtain the following expression for the left side of (4.3.17):

$$\frac{\lambda_{\mathscr{Q}}(a, \bar{a})\lambda_{\mathscr{Q}}(\bar{a}, \bar{a} \cap \bar{a}_1)\mu(\bar{a} \cap \bar{a}_1, (b_0 \cap b_{01}, \bar{b} \cap \bar{b}_1))}{\lambda_{\mathscr{Q}}(a, a \cap a_1)\mu(a \cap a_1, (b_0 \cap b_{01}, b \cap b_1))}.$$

For $\lambda_{\mathscr{Q}}$ we find that $\lambda_{\mathscr{Q}}(a, \bar{a})\lambda_{\mathscr{Q}}(\bar{a}, \bar{a} \cap \bar{a}_1) = \lambda_{\mathscr{Q}}(a, \bar{a} \cap \bar{a}_1)$ and since $a \cap a_1 \neq \emptyset$, we obtain

$$\lambda_{\mathscr{Q}}(a, \bar{a} \cap \bar{a}_1) = \lambda_{\mathscr{Q}}(a, a \cap a_1)\lambda_{\mathscr{Q}}(a \cap a_1, \bar{a} \cap \bar{a}_1).$$

Thus the left side of (4.5.17) may be rewritten as follows:

$$\frac{\lambda_{\mathscr{Q}}(a \cap a_1, \bar{a} \cap \bar{a}_1)\mu(\bar{a} \cap \bar{a}_1, (b_0 \cap b_{01}, \bar{b} \cap \bar{b}_1))}{\mu(a \cap a_1, (b_0 \cap b_{01}, b \cap b_1))}. \tag{4.5.18}$$

The expression (4.5.18) is symmetric under exchange of quantities having index 1 and those without index 1. The right side of (4.5.17) can be transformed in the same way into an expression which is similar to (4.5.18), proving that (4.5.16) is well defined.

It now remains to show that the function λ defined on $\mathscr{T}_{\Theta}$ satisfies the conditions (α), (β), (γ), and APS 7.

We shall now show that (α) is satisfied:

Let $c_1 = a_1 \cap b_1$, $c_2 = a_2 \cap b_2 \neq \emptyset$, $c_3 = a_3 \cap b_3$ where we suppose that $c_3 \subset c_2 \subset c_1$ (and therefore find that, for example, $c_1 \cap c_2 = c_2$; that is, $c_2 = a_2 \cap b_2 = (a_1 \cap a_2) \cap (b_1 \cap b_2)$. We may therefore assume that $a_3 \subset a_2 \subset a_1$ and that $b_3 \subset b_2 \subset b_1$. According to APS 5.1.1, for a_1 there exists a b_0 such that $(a_1, b_0) \in C$; by Th. 4.3.1 we also find that $(a_2, b_0) \in C$ and (if $a_3 \neq \emptyset$) $(a_3, b_0) \in C$. If $a_3 = \emptyset$ then $c_3 = \emptyset$ and therefore $\lambda(c_1, c_3) = 0$, $\lambda(c_2, c_3) = 0$ and

(α) is satisfied in a trivial way. Let us assume that $a_3 \neq \varnothing$; thus we find that $a_3 \cap b_0 \neq \varnothing$. We shall now assume that $a_3 \neq \varnothing$ and therefore $a_3 \cap b_0 \neq \varnothing$. Since $a_1 \cap b_0 \neq \varnothing$ we may write

$$\lambda(c_1, c_2) = \frac{\lambda_{\mathscr{Q}}(a_1, a_2)\mu(a_2, (b_0, b_2))}{\mu(a_1, (b_0, b_1))},$$

$$\lambda(c_2, c_3) = \frac{\lambda_{\mathscr{Q}}(a_2, a_3)\mu(a_3, (b_0, b_3))}{\mu(a_2, (b_0, b_2))}.$$

Thus we find that

$$\lambda(c_1, c_2)\lambda(c_2, c_3) = \frac{\lambda_{\mathscr{Q}}(a_1, a_2)\lambda_{\mathscr{Q}}(a_2, a_3)\mu(a_3, (b_0, b_3))}{\mu(a_1, (b_0, b_1))}.$$

From $\lambda_{\mathscr{Q}}(a_1, a_2)\lambda_{\mathscr{Q}}(a_2, a_3) = \lambda_{\mathscr{Q}}(a_1, a_3)$ we find that (α) is satisfied.

In order to show that (β) holds, we define $c_i = a_i \cap b_i$ and $\bigcup_{i=1}^n c_i = \tilde{c} = \tilde{a} \cap \tilde{b}$. Since $c_i \subset \tilde{c} \subset c = a \cap b$ we may assume $a_i \subset \tilde{a} \subset a$ and $b_i \subset \tilde{b} \subset b$ as above. The a_i generate a finite Boolean ring in $\mathscr{Q}(\tilde{a})$; we denote the atoms of this finite subring by $\hat{a}_\nu$. Similarly let $\hat{b}_\mu$ denote the atoms of the finite subring obtained from $\mathscr{R}(\tilde{b})$ and the b_i. Then we find that

$$a = \bigcup_{\nu \in A_i} \hat{a}_\nu, \qquad b = \bigcup_{\nu \in B_i} \hat{b}_\nu,$$

where A_i, B_i are the set of indices uniquely associated with a_i and b_i, respectively. We obtain

$$c_i = a_i \cap b_i = \bigcup_{\substack{\nu \in A_i \\ \mu \in B_i}} \hat{a}_\nu \cap \hat{b}_\mu.$$

From $c_i \cap c_j = \varnothing$ for $i \neq j$ it follows that

$$\bigcup_{\substack{\nu \in A_i \cap A'_j \\ \mu \in B_i \cap B'_j}} \hat{a}_\nu \cap \hat{b}_\mu = \varnothing,$$

that is $\hat{a}_\nu \cap \hat{b}_\mu = \varnothing$ for $\nu \in A_i \cap A_j$ and $\mu \in B_i \cap B_j$. Let

$$C_i = \{(\nu, \mu) \mid \nu \in A_i, \mu \in B_i\};$$

then for all pairs $(\nu, \mu) \in C_i \cap C_j$, we find that

$$\hat{a}_\nu \cap \hat{b}_\mu = \varnothing \quad \text{whenever } i \neq j. \tag{4.5.19}$$

We may describe the c_i in terms of the C_i as follows:

$$c_i = \bigcup_{(\nu, \mu) \in C_i} (\hat{a}_\nu \cap \hat{b}_\mu).$$

We obtain

$$\bigcup_{(\nu, \mu) \in \cup C_i} (\hat{a}_\nu \cap \hat{b}_\mu) = \bigcup \tilde{c}_i = c = \tilde{a} \cap \tilde{b} = \left[\bigcup_\nu \hat{a}_\nu\right] \cap \left[\bigcup_\mu \hat{b}_\mu\right]$$
$$= \bigcup_{\nu\mu} (\hat{a}_\nu \cap \hat{b}_\mu).$$

Let $\bar{C}$ denote the set of all index pairs (ν, μ). Then we find that

$$\hat{a}_\nu \cap \hat{b}_\mu = \varnothing \quad \text{for all } (\nu, \mu) \in \bar{C} \backslash \bigcup C_i. \tag{4.5.20}$$

Since by APS 5.1.1 there exists a b_0 for which $(a, b_0) \in C$ by Th. 4.3.1 we find that $(\hat{a}_\nu, b_0) \in C$. For this b_0 we find that

$$\lambda(c, \tilde{c}) = \frac{\lambda_{\mathscr{Q}}(a, \tilde{a})\mu(\tilde{a}, (b_0, b))}{\mu(a, (b_0, b))}.$$

From $b_0 = (b_0 \backslash b) \cup \bigcup_\mu \hat{b}_\mu$ and $b_0 = (b_0 \backslash b) \cup \tilde{b}$ it follows from (vi) that

$$1 = \mu(\tilde{a}, (b_0, b_0 \backslash \tilde{b})) + \sum_\nu \mu(\tilde{a}, (b_0, \hat{b}_\nu)),$$

$$1 = \mu(\tilde{a}, (b, b_0 \backslash \tilde{b})) + \mu(\tilde{a}, (b_0, \tilde{b}))$$

and finally

$$\mu(\tilde{a}, (b_0, \tilde{b})) = \sum_\nu \mu(\tilde{a}, (b_0, \hat{b}_\nu)).$$

From $\tilde{a} = \bigcup_\nu \hat{a}_\nu$, it follows from (iv) that

$$\mu(\tilde{a}, f) = \mu\left(\bigcup_\nu \hat{a}_\nu, f\right) = \sum_\nu \lambda_{\mathscr{Q}}(\tilde{a}, \hat{a}_\nu)\mu(\hat{a}_\nu, f).$$

The preceding result, together with

$$\lambda_{\mathscr{Q}}(a, \tilde{a})\lambda_{\mathscr{Q}}(\tilde{a}, \hat{a}_\nu) = \lambda_{\mathscr{Q}}(a, \hat{a}_\nu)$$

permits us to conclude that

$$\lambda(c, \tilde{c}) = \sum_{\nu\mu} \frac{\lambda_{\mathscr{Q}}(a, \hat{a}_\nu)\mu(\hat{a}_\nu, (b_0, \hat{b}_\mu))}{\mu(a, (b_0, b_\mu))}. \tag{4.5.21}$$

According to (vii), $\mu(\hat{a}_\nu, (b_0, \hat{b}_\mu)) = 0$ for $\hat{a}_\nu \cap \hat{b}_\mu = \varnothing$.

Using (4.5.19) and (4.5.20) we easily obtain

$$\lambda(c, \tilde{c}) = \sum_i \sum_{(\nu, \mu) \in C_i} \frac{\lambda_{\mathscr{Q}}(a, \hat{a}_\nu)\mu(\hat{a}_\nu, (b_0, \hat{b}_\mu))}{\mu(a, (b_0, b))}.$$

If we show that (noting the definition of C_i)

$$\lambda(c, c_i) = \sum_{\substack{\nu \in A_i \\ \mu \in B_i}} \frac{\lambda_{\mathscr{Q}}(a, \hat{a}_\nu)\mu(\hat{a}_\nu, (b_0, \hat{b}_\mu))}{\mu(a, (b_0, b))}, \tag{4.5.22}$$

then we have proven (β). Note that (4.5.22) has the same form as (4.5.21) and is therefore proven.

In order to show that (γ) holds, let $d = a_1 \cap b_1$ and $c = a_2 \cap b_2$ where we assume that $a_2 \subset a_1$, $b_2 \subset b_1$. From

$$\lambda(d, c) = \frac{\lambda_{\mathscr{Q}}(a_1, a_2)\mu(a_2, (b_0, b_2))}{\mu(a_1, (b_0, b_1))}$$

and (vii) it follows that (γ) is satisfied.

We shall now show that APS 7.1 is satisfied. First, we note that

$$\lambda_{\mathscr{S}}(a_1 \cap b_{10}, a_2 \cap b_{10}) = \frac{\lambda_{\mathscr{Q}}(a_1, a_2)\mu(a_2, (b_{10}, b_{10}))}{\mu(a_1, (b_{10}, b_{10})}.$$

According to (iii) (which follows from (vi)) we obtain

$$\mu(a_1, (b_{10}, b_{10})) = \mu(a_2, (b_{20}, b_{20})) = 1.$$

Thus we obtain

$$\lambda_{\mathscr{S}}(a_1 \cap b_{10}, a_2 \cap b_{10}) = \lambda_{\mathscr{Q}}(a_1, a_2).$$

To show that APS 7.2 is satisfied, we note that

$$\begin{aligned}\lambda_{\mathscr{S}}(a_1 \cap b_{10}, a_1 \cap b_{20}) &= \frac{\lambda_{\mathscr{Q}}(a_1, a_1)\mu(a_1, (b_{10}, b_{20}))}{\mu(a_1, (b_{10}, b_{10}))} \\ &= \mu(a_1, (b_{10}, b_{20})).\end{aligned}$$

From (v) we find that

$$\begin{aligned}\mu(a_1, (b_{10}, b_{20})) &= \lambda_{\mathscr{R}_0}(b_{10}, b_{20})\mu(a_1, (b_{20}, b_{20})) \\ &= \lambda_{\mathscr{R}_0}(b_{10}, b_{20}).\end{aligned}$$

Thus we have also shown that $\lambda_{\mathscr{R}_0}$ is determined by μ.

On the basis of the uniqueness proven in Th. 4.5.2, it follows from Th. 4.5.3 that the $\lambda_{\mathscr{S}}$ so defined satisfies condition (viii) of Th. 4.5.3.

Theorem 4.5.4 shows that in order to determine $\lambda_{\mathscr{S}}$ it is sufficient to test the function $\mu(a, f)$.

The conditions (i), (vi), and (vii) are justified (in a trivial way) by the meaning of μ as the mathematical description of a frequency in an experiment, as we have already explained when axiom AS 2 was stated in §3.

Condition (iv) has the following intuitive interpretation: the preparation is not "influenced" by the registration (and therefore is one of the conditions defining the experiment). Otherwise, it would not be possible to conduct a test experiment.

Condition (v) states that a refinement of the registration method will be statistically independent of the preparation process—and is therefore also a condition which defines the experiment.

Conditions (i), (iv), (v), (vi), and (vii) are, for the most part, "rules for correct experimentation" rather than assertions about the physical system under investigation. The complete "information about the physical systems" is to be found in the function $\mu(a, f)$ over $\mathscr{C}$. This fact will justify our future, almost exclusive, interest in this function.

Yet this function μ also is not independent of some of the properties of the preparation and registration apparatuses. In §4.4 we shall attempt, as far as possible, to eliminate the properties of the apparatuses (that is, the properties of the function $\mu(a, f)$ which depend on the structure of the apparatuses). We would like to consider the preparation and registration processes as merely an aid to detect the physical systems and investigate their structure.

The "separation" of the preparation and registration apparatus on one side, and the physical systems on the other side is not only nontrivial, but in the case of microsystems (that is, quantum mechanics) is not possible, at least, in the desired or "expected" sense. The clarification of those concepts, with which we seek to obtain the largest separation, will be the goal of III and IV.

CHAPTER III

Ensembles and Effects

Many of the problems and difficulties encountered in the interpretation of quantum mechanics have the following source: the failure to clearly distinguish between a collection of microsystems obtained by means of a preparation procedure and an ensemble, where the latter is represented by a statistical (density) operator. We shall not describe these misunderstandings here. Instead we shall formulate definitions of the concepts of an "ensemble" and an "effect." An effect is often called a "yes–no" measurement. Here we shall not yet make a distinction between an "effect" and a "decision effect"; such a distinction will be made in §3 and §6. In this book we shall show that the experiments which exhibit paradoxes for the usual interpretation of quantum mechanics can be described in a natural manner.

Here we again state that, in the formulation of quantum mechanics which will be presented here, neither the statistical operators (in particular, the projections onto a vector in Hilbert space, often called a state) nor the self-adjoint operators (which described the so-called "observables") will be used for direct comparison with experience, that is, with experiment. We shall describe the relationship between the mathematical description and experiment exclusively in terms of the preparation and registration procedures and the probability function $\lambda_{\mathscr{S}}$.

In the mapping principles (as described in [1], §5 or [2], III, §4) the fundamental sets are M, $\mathscr{Q}$, $\mathscr{R}_0$, *and* $\mathscr{R}$ and the fundamental relations are $\lambda_{\mathscr{S}}(c, d) = \alpha$, $x \in a$ and $x \in b$. The concepts which will be introduced in III and IV will be *derived* concepts (see [1], §10). The additional axioms which will be introduced in III and IV are additional laws of nature in the sense of [1], §7.3 concerning M, $\mathscr{Q}$, $\mathscr{R}_0$, and $\mathscr{R}$ and $\lambda_{\mathscr{S}}(c, d)$.

In subsequent chapters (for example, in VII) we shall introduce additional relations; these will, however, be related directly to the fundamental sets M, $\mathcal{Q}$, $\mathcal{R}_0$, and $\mathcal{R}$. In addition, it is necessary to describe the apparatuses in more detail than is currently possible using only elements of M, $\mathcal{R}_0$, and $\mathcal{R}$. As we mentioned in II, this problem has not yet been solved, in general, for the case of quantum mechanics. In IV we shall again return to this problem. In IX and XVI we shall make a number of unsystematic special assumptions; in XVII we shall describe a method for the solution of a portion of this problem. A new aspect of this problem will be discussed in [13].

In III and IV we shall redefine the "usual" concepts such as "ensemble," "state," and "observable." These concepts will no longer be dependent on the interpretation problems of quantum mechanics, because we have already stated the relationship between the mathematical description and experiment in II. In III and IV we shall outline (in the sense of [1], §10) certain parts of the *reality domain* of quantum mechanics and consider problems of the structure of microsystems within this framework.

1 Combinations of Preparation and Registration Methods

We shall now continue the discussion of the combination problem which was begun in II, §4.3. For the case in which $C = \mathcal{Q}' \times \mathcal{R}'_0$ axioms APS 5.1 and APS 5.2 become theorems. From $C = \mathcal{Q}' \times \mathcal{R}'_0$ it follows that $\mu(a, f)$ is defined for all $f \in \mathcal{F}$, and that an equivalence relation $a_1 \sim a_2$ is defined on $\mathcal{Q}'$ by the condition

$$\mu(a_1, f) = \mu(a_2, f) \quad \text{for all } f \in \mathcal{F}. \tag{1.1}$$

In our discussion of the combination problem we shall show that for microsystems, physically realistic assumptions guarantee the existence of an equivalence relation $a_1 \sim a_2$ in $\mathcal{Q}'$. Indeed, this will be proven in Th. 1.2.

Assuming that Th. 1.2 holds, we shall introduce the following definition:

D 1.1. Let $\mathcal{K}$ denote the set of all equivalence classes (with respect to the equivalence relation $a_1 \sim a_2$ in $\mathcal{Q}'$). An element of $\mathcal{K}$ is called an *ensemble* or *state*; $\mathcal{K}$ is called the set of ensembles (set of states).

On the basis of the equivalence relation assumed above, we may define a function $\hat{\mu}(w, f)$, $w \in \mathcal{K}$ as follows:

$$\hat{\mu}(w, f) = \mu(a, f) \quad \text{where } a \in w.$$

Then $\hat{\mu}(w, f)$ is defined for all pairs (w, f) for which there exists a $a \in w$ for which $(a, f) \in \mathcal{C}$. Suppose that we may justify, on physical grounds, that $\hat{\mu}$ is defined on all of $\mathcal{K} \times \mathcal{F}$ (that is, axiom APS 5.1.4, which will be introduced below will be satisfied). Then an equivalence relation $f_1 \sim f_2$ on $\mathcal{F}$ will be defined by

$$f_1 \sim f_2 \quad \text{if and only if} \quad \hat{\mu}(w, f_1) = \hat{\mu}(w, f_2) \quad \text{for all } w \in \mathcal{K}.$$

We now introduce the following definition:

D 1.2. Let $\mathscr{L}$ denote the set of all equivalence classes (with respect to the equivalence relation $f_1 \sim f_2$) from $\mathscr{F}$. An element of $\mathscr{L}$ is called an *effect* and $\mathscr{L}$ is called the set of effects.

We obtain the following theorem:

Th. 1.1. *Let $\tilde{\mu}$ be defined over $\mathscr{K} \times \mathscr{L}$, where $\tilde{\mu}(w, g) = \hat{\mu}(w, f)$ where $w \in \mathscr{K}$, $g \in \mathscr{L}$, and $f \in g$. Then $\tilde{\mu}$ satisfies the following conditions:*

(i) $0 \leq \tilde{\mu}(w, g) \leq 1$.
(ii) *If $\tilde{\mu}(w_1, g) \doteq \tilde{\mu}(w_2, g)$ for all $g \in \mathscr{L}$ then $w_1 = w_2$.*
(iii) *If $\tilde{\mu}(w, g_1) = \tilde{\mu}(w, g_2)$ for all $w \in \mathscr{K}$ then $g_1 = g_2$.*
(iv) *There exists a $g_0 \in \mathscr{L}$ such that $\tilde{\mu}(w, g_0) = 0$ for all $w \in \mathscr{K}$.*
(v) *There exists a $g_1 \in \mathscr{L}$ such that $\tilde{\mu}(w, g_1) = 1$ for all $w \in \mathscr{K}$.*

The proof is a simple consequence of II, Th. 4.3.4.

For the following, it is useful to introduce the canonical mappings of $\mathscr{Q}'$ onto $\mathscr{K}$ and $\mathscr{F}$ onto $\mathscr{L}$.

D 1.3. Let φ denote the canonical map which maps the elements $a \in \mathscr{Q}$ to the equivalence classes $w \in \mathscr{K}$ to which a belongs. Let ψ denote the canonical map from $\mathscr{F}$ onto $\mathscr{L}$. Let $f = (b_0, b)$; in addition to $\psi(f)$ we shall also write $\psi(b_0, b)$.

The important concepts "ensemble" and "effect" require the existence of the equivalence relations $a_1 \sim a_2$ in $\mathscr{Q}'$ and $f_1 \sim f_2$ in $\mathscr{F}$. We shall now seek to formulate realistic axioms which will guarantee the existence of such equivalence relations. Here we shall place our emphasis on physical considerations rather than seeking the weakest possible axioms.

The intuitive basis underlying the axioms for C is as follows: In order that a preparation procedure a may be meaningfully combined with a registration method b_0—that is, in order that the microsystem prepared according to the preparation procedure a may be registered according to the method b_0—it is necessary that the registration according to b_0 occur "after" the preparation. The time-ordering of b_0 with respect to a will be discussed in more detail in VII, §1. Here we note that the question whether $a \in \mathscr{Q}$ may be combined with $b_0 \in \mathscr{R}_0$ is not only a question about the physical possibility to position the apparatus for a and b_0, but is also a stipulation that an element $x \in a$ is also an element of b_0 only if the registration of x occurs after x is prepared (see [2], XVI). Only in this way will the transformations which will be introduced in VII, §1 have a simple structure.

We see, therefore, that the equivalence relation $a_1 \sim a_2$ for preparation procedures depends on our stipulation concerning the possible combination of the preparation with registration procedures. $a_1 \sim a_2$ will be satisfied if and only if the effect procedures which may be combined with a_1 and a_2 lead to the same frequencies. Thus the resulting equivalence relation depends strongly on the conditions imposed on the set C!

Let $\mathscr{R}'_{0T}$ denote the set of registration methods $b_0 \neq \emptyset$ which begin "later" than time T. Then we expect that, to each $a \in \mathscr{Q}'$, there exists a T such that $(a, b_0) \in C$ for all $b_0 \in \mathscr{R}'_{0T}$. Obviously, the set of $\mathscr{R}'_{0T}$ for increasing T is a (lower) directed set. We shall now formulate the following axiom:

APS 5.1.2. There exists a directed set (in the sense of inclusion) $\Gamma \subset \mathscr{P}(\mathscr{R}'_0)$ where $\emptyset \notin \Gamma$, so that to each $a \in \mathscr{Q}$ there exists at least one element $\mathscr{R}'_{0\alpha} \in \Gamma$ such that $(a, b_0) \in C$ for all $b_0 \in \mathscr{R}'_{0\alpha}$.

This axiom is, by itself, too weak. From the directedness of Γ it follows that, to every finite set $a_1, a_2, \ldots, a_n \in \mathscr{Q}'$ there exists a common element $\mathscr{R}'_{0\beta} \in \Gamma$ for which $(a_i, b_0) \in C$ for $i = 1, \ldots, n$ and all $b_0 \in \mathscr{R}'_{0\beta}$. It remains an open question whether each element (for example, $\mathscr{R}'_{0\beta}$) contains "sufficiently many" registration methods in order to sufficiently test a preparation procedure. If we consider the set $\mathscr{R}'_{0T}$ described above, our intuition may suggest that it should be possible to test a completely by means of all b_0 from $\mathscr{R}'_{0T}$, that is, if

$$\mu(a_1, (b_0, b)) = \mu(a_2, (b_0, b))$$

for all $b_0 \in \mathscr{R}'_{0T}$ and $b \subset b_0$, then it follows that

$$\mu(a_1, f) = \mu(a_2, f)$$

for all $f \in \mathscr{F}$, providing that both $\mu(a_1, f)$ and $\mu(a_2, f)$ are defined (that is, $(a_1, f) \in \mathscr{C}$ and $(a_2, f) \in \mathscr{C}$)).

For certain macrosystems this assumption will prove to be false (see below). For microsystems and certain "classical" macrosystems (systems of mass-points undergoing conservative forces) the above assumption is successful. We shall now formulate a new axiom (which will be stronger than APS 5.1.2):

APS 5.1.3. Let Γ be a set satisfying APS 5.1.2. Suppose that the following condition is also satisfied: For $a_1, a_2 \in \mathscr{Q}'$ let $\mathscr{R}'_{0\beta}$ be an element of Γ where (a_1, b_0), $(a_2, b_0) \in C$ for all $b_0 \in \mathscr{R}'_{0\beta}$. Suppose that $\mu(a_1, (b_0, b)) = \mu(a_2, (b_0, b))$ is satisfied for all $b_0 \in \mathscr{R}'_{0\beta}$ and all $b \subset b_0$. Then we require that $\mu(a_1, f) = \mu(a_2, f)$ for all $f \in \mathscr{F}$ for which $\mu(a_1, f)$ and $\mu(a_2, f)$ are defined.

Axioms APS 5.1.1–APS 5.1.3 are not mutually independent. We shall now state the relationships among these axioms:

If we assert APS 5.1.2 then we may prove APS 5.1.1 (given in II, §4.3). APS 5.1.3 is clearly a stronger condition than APS 5.1.2. If we assume APS 5.1.3 then APS 5.1.1 and APS 5.1.2 can be proven, and are therefore superfluous.

From APS 5.1.3 we now obtain the following theorem:

Th. 1.2. *Let us define the relation $a_1 \sim a_2$ as follows: If $\mu(a_1,f) = \mu(a_2,f)$ for all $f \in \mathscr{F}$ for which $\mu(a_1,f)$ and $\mu(a_2,f)$ are defined then $a_1 \sim a_2$. The relation $a_1 \sim a_2$ is an equivalence relation.*

PROOF. We need only show that if $a_1 \sim a_2$, $a_2 \sim a_3$ then $a_1 \sim a_3$. According to APS 5.1.2, the directed set Γ contains an element $\mathscr{R}'_{0\beta}$ with

$$(a_1, b_0), (a_2, b_0), (a_3, b_0) \in C \quad \text{for all } b_0 \in \mathscr{R}'_{0\beta}.$$

From $a_1 \sim a_2, a_2 \sim a_3$ we obtain

$$\mu(a_1, (b_0, b)) = \mu(a_2, (b_0, b)) = \mu(a_3, (b_0, b))$$

for all $b \subset b_0$ and $b_0 \in \mathscr{R}'_{0\beta}$. From

$$\mu(a_1, (b_0, b)) = \mu(a_3, (b_0, b))$$

for all $b \subset b_0 \in \mathscr{R}'_{0\beta}$ it follows that, according to APS 5.1.3

$$\mu(a_1,f) = \mu(a_3,f)$$

for all $f \in \mathscr{F}$ for which $\mu(a_1,f)$ and $\mu(a_3,f)$ are defined. Thus we have shown that $a_1 \sim a_3$.

On the basis of Th. 1.2, for $w \in \mathscr{K}$ (where $\mathscr{K}$ is defined by D 1.1) we may define a function $\hat{\mu}$ on a *subset* of $\mathscr{K} \times \mathscr{F}$

$$\hat{\mu}(w,f) = \mu(a,f) \quad \text{where } a \in w.$$

Thus $\hat{\mu}$ is defined for all pairs (w,f) for which there is an $a \in w$ for which $(a,f) \in \mathscr{C}$.

The "intuitive" considerations which have led to the formulation of axiom APS 5.1.3 have now led us to make additional assertions about the set C. We have seen that, for each a, there is a subset $\mathscr{R}'_{0T}$, the elements of which may be combined with a. If we now consider the various a which belong to the same equivalence class, given an $a_1 \in w$ which can be combined with all $b_0 \in \mathscr{R}'_{0T_1}$ (that is the preparation of microsystems is excluded after T_1) it may be possible to find an $a_2 \in w$ which excludes preparations after an earlier time $T_2 (T_2 < T_1)$, but for all effects (b_0, b), where $b_0 \in \mathscr{R}'_{0T_1}$, leads to the same frequencies as a_1. We would like to assert that for each w and each time T it is possible to find an $a \in w$ which may be combined with the registration method $b_0 \in \mathscr{R}'_{0T}$. Together with $\bigcup_T \mathscr{R}'_{0T} = \mathscr{R}'_0$ (where the union is taken over all time).

We come to the postulate:

APS 5.1.4. To each $b_0 \in \mathscr{R}'_0$ in each class $w \in \mathscr{K}$ there exists an $a \in w$ such that $(a, b_0) \in C$. To each pair $a \in \mathscr{Q}', f = \mathscr{F}$ there exists an $f' \in \mathscr{F}$ such that $(a,f') \in \mathscr{C}$ and $\hat{\mu}(w,f) = \hat{\mu}(w,f')$ for all $w \in \mathscr{K}$ for which $\hat{\mu}(w,f)$ and $\hat{\mu}(w,f')$ are defined.

Th. 1.3. *The function $\hat{\mu}$ defined above is defined on $\mathscr{K} \times \mathscr{F}$.*

PROOF. We need to show that, to each $f \in \mathscr{F}$ and $w \in \mathscr{K}$ there is an $a \in w$ for which $(a,f) \in \mathscr{C}$. This is, however, guaranteed by APS 5.1.4 since if $f = (b_0, b)$, there is an $a \in w$ with $(a, b_0) \in C$, that is, $(a,f) \in \mathscr{C}$.

We now obtain the following theorem:

Th. 1.4. *The relation $f_1 \sim f_2$ defined by the condition*

$$\hat{\mu}(w, f_1) = \hat{\mu}(w, f_2) \quad \textit{for all } w \in \mathscr{K}$$

is an equivalence relation. To each pair $a \in \mathscr{Q}'$, $f \in \mathscr{F}$ there exists an $f' \in \mathscr{F}$ such that $(a, f') \in \mathscr{C}$ and $f \sim f'$.

Thus, from D 1.2 we have finally proven Th. 1.1.

In the following (up to and including VI) we shall not consider the precise form of axioms APS 5.1.3 and APS 5.1.4. We shall, however, make use of the equivalence classes of $\mathscr{Q}'$ and $\mathscr{F}$, and the mappings φ and ψ defined by D 5.1.3. The special form of the axioms APS 5.1, 3, 4 have stood the test for the case of microsystems; the considerations presented in VII, §1 depend decisively on these axioms.

The situation is completely different for macrosystems which undergo irreversible processes. Let us select a reference time (which we denote by $t = 0$), and consider only those preparation procedures $a \in \mathscr{Q}$ which are completed before $t = 0$, and the registration procedures which begin after $t = 0$. Then, instead of APS 5.1, 3, 4 we may use the simpler axiom $C = \mathscr{Q}' \times \mathscr{R}'_0$. This is achieved at the expense that the considerations of VII, §1 are no longer applicable, since we may consider only those time translations $t \to t + \tau$ for which $\tau > 0$—a semi-group of time translations!

It is impossible to apply axioms APS 5.1, 3, 4 to irreversible macrosystems because axiom APS 5.1.3 is stronger than APS 5.1.2. The stronger axiom APS 5.1.3 appears to contradict our experience with irreversible macrosystems because the ability to distinguish between the different a_1, a_2 by means of the elements (b_0, b) where $b_0 \in \mathscr{R}'_{0T}$ is reduced as T increases. In particular, if T is sufficiently large, the condition $\mu(a_1, (b_0, b)) = \mu(a_2, (b_0, b))$ for all (b_0, b) for which $b_0 \in \mathscr{R}'_{0T}$ does not imply that $\mu(a_1, f) = \mu(a_2, f)$ for all $f \in \mathscr{F}$ even where $\mu(a_1, f)$ and $\mu(a_2, f)$ are defined. We have noted that APS 5.1.3 cannot be used for irreversible macrosystems. Its usefulness for microsystems, especially for the structures defined in VII, §1 is not self-evident! The distinction between macrosystems and microsystems must be carefully examined if we are to embed a theory of macrosystems into many-body quantum theory. As a result of the above considerations (macrosystems may be prepared only before $t = 0$), the set $\mathscr{Q}_m$ of the preparation procedures for macrosystems can only be a small subset of the set $\mathscr{Q}$ of preparation procedures for a many-body quantum theory; a similar situation holds also for $\mathscr{R}_{0m}$ and $\mathscr{R}_0$ (see [2], XV and [13]).

The explicit form of the axioms—either APS 5.1, 3, 4 or $C = \mathscr{Q}' \times \mathscr{R}'_0$—will not be important in the following—from here to VI inclusive. Here it is important to state that this choice permits us to introduce the concepts of an ensemble and an effect. In the transition from $\mathscr{Q}'$ to $\mathscr{K}$ and from $\mathscr{F}$ to $\mathscr{L}$ we lose (as desired) the special structure of the apparatuses as characterized by $a \in \mathscr{Q}'$ or $b_0 \in \mathscr{R}'_0$. Ensembles and effects then appear to depend more on the systems themselves than upon the preparation and registration apparatus. In this respect, axioms APS 5.1.3 and 4 permit, at least, a partial "separation" of the system from the structure of the apparatuses in the sense of our intention which was expressed in II, §4.4 and at the end of II, §4.5. There remain two problems in the separation described above in defining the concept of ensemble and effect:

(1) In the transition from $\mathcal{Q}'$ to $\mathcal{K}$ and from $\mathcal{F}$ to $\mathcal{L}$ we must ask whether we have lost too much of the structure of the system itself.
(2) On the other hand, we must ask whether we have not included more of the apparatus structure than is necessary.

An ensemble w is characterized only by its statistical distribution. Therefore it is often said that two sets of systems a_1, a_2 for which $\varphi(a_1) = \varphi(a_2) = w$—that is, $a_1 \sim a_2$—are, in reality, equal and differ only in the details of the construction of the two different preparation apparatuses. In other words the two sets a_1, a_2 give the same results for *all* experiments (not only registration). In IV, §5 and §6 and in XVII, §4.4 we shall find that this statement is, in general, not true for microsystems. We find that much too much is lost in the transition from $\mathcal{Q}'$ to $\mathcal{K}$ and from $\mathcal{F}$ to $\mathcal{L}$. In IV we shall seek to recover the loss by introducing the concepts of *observable* and *preparator*.

On the other hand the notion of an effect may perhaps contain too much of the structure of the registration procedure; in general, an effect may also contain the probability that the associated effect process reacts upon a "property" of the microsystem. We would, of course, like to eliminate the "bad" registrations and keep the "good" ones—those which exhibit the "real" properties of the system. Is such a distinction possible?

A search to find a description of the microsystems in themselves, going beyond the concepts of ensembles and effects may proceed in two different directions:

First, by considering the properties or pseudoproperties of systems (see §4), or

Second, by providing a very precise analysis of what we seek to describe by the concepts of observable and preparator (see VI). There we seek to identify those observables which precisely describe the physical systems, and only secondarily describe (if this is inevitable) the measurement (registration) apparatus. We also wish to identify those preparators which describe the structure of the prepared physical systems rather than the structure of the preparation apparatus.

2 Mixtures and Decompositions of Ensembles and Effects

The concept of an "ensemble" (or "state") is uniquely defined by D 1.1 and refers to a number of relatively trivial relationships. The usual intuitive notion of an ensemble (state) does not agree in all details with that defined by D 1.1.

An ensemble $w \in \mathscr{K}$ is not a set of microsystems because w is not a subset of M; w is a subset of $\mathscr{P}(M)$—and is a class of subsets of M. The following fact is of crucial importance in quantum mechanics (but not limited to quantum mechanics). To each class w there is more than one $a \in w$ (see IV, §5 and §6 for details). The substitution of the statement $a \in \mathscr{Q}'$ for the statement $w \in \mathscr{K}$ will quickly result in errors in both logic and intuition. Such errors may be avoided by using the mathematical structure associated with M, $\mathscr{Q}$, $\mathscr{R}_0$, and $\mathscr{R}$ in order to describe microsystems.

For quantum mechanics it is important that, if φ is the map defined in D 1.3, then the condition $\varphi(a_1) = \varphi(a_2)$ does not imply $a_1 = a_2$.

In the following we shall denote the function $\tilde{\mu}: \mathscr{K} \times \mathscr{L} \to \mathbf{R}$ obtained in Th. 1.1 by μ (the same symbol used for the corresponding function on $\mathscr{C}$); it will be clear which function is meant by the arguments. Thus we obtain

$$\mu(a, f) = \mu(\varphi(a), \psi(f)). \tag{2.1}$$

From II, Th. 4.3.4(iv) we obtain the following important theorem on the "decomposition of ensembles."

Th. 2.1. *Let $a = \bigcup_{i=1}^n a_i$ be a decomposition of the preparation procedure a. Then for all $g \in \mathscr{L}$ we obtain*

$$\mu(\varphi(a), g) = \sum_{i=1}^n \lambda_i \mu(\varphi(a_i), g)),$$

where $\lambda_i = \lambda_{\mathscr{Q}}(a, a_i)$, $0 < \lambda_i \leq 1$ and $\sum_{i=1}^n \lambda_i = 1$.

PROOF. According to II, Th. 4.3.4(iv), we find that for $\lambda_i = \lambda_{\mathscr{Q}}(a, a_i)$,

$$\mu(a, f) = \sum_{i=1}^n \lambda_i \mu(a_i, f)$$

for all $f \in \mathscr{F}$ for which $(a, f) \in \mathscr{C}$. Therefore, from Th. 1.3 we obtain

$$\hat{\mu}(\varphi(a), f) = \sum_{i=1}^n \lambda_i \hat{\mu}(\varphi(a_i), f)$$

for all $f \in \mathscr{F}$. Our assertion follows directly from Th. 1.1.

D 2.1. Suppose to $w \in \mathscr{K}$ there is a set of real numbers λ_i, $0 < \lambda_i \leq 1$, $i = 1, \ldots, n$ where $\sum_{i=1}^n \lambda_i = 1$, and a set $w_i \in \mathscr{K}$, $i = 1, \ldots, n$ for which the following condition is satisfied for all $g \in \mathscr{L}$:

$$\mu(w, g) = \sum_{i=1}^n \lambda_i \mu(w_i, g). \tag{2.2}$$

Then (2.2) is called a *decomposition* of w according to the w_i with *weights* λ_i.

We shall now introduce the notion of "mixing" of preparation procedures, a notion which is the inverse of decomposition. We shall describe how we may construct such a mixture w using the apparatuses for the selection of the components w_i.

We may build a new apparatus A using the apparatus A_1 corresponding to the preparation procedure a_1 and the apparatus A_2 corresponding to a_2 in the following way. Suppose we have an apparatus B which randomly generates two states (+) and (−) where (+) occurs with frequency α and (−) with frequency $1 - \alpha$. The arrangement of B, A_1, and A_2 are such that, upon the occurrence of (+) the apparatus A_1 is used, and upon the occurrence of (−) the apparatus A_2 is used. Note that B is also a part of the "large" apparatus A. If (+) occurs in the apparatus A (that is, in its part B), a preparation procedure $a'_1 \subset a$ is determined for which $\lambda_{\mathcal{Q}}(a, a'_1) = \alpha$. Then a'_1 is a "finer" preparation procedure than a, and is selected from a only if (+) occurs. Similarly, if (−) occurs, a preparation procedure a'_2 occurs and $\lambda_{\mathcal{Q}}(a, a'_2) = 1 - \alpha$.

We may be tempted to set $a_1 = a'_1$ and $a_2 = a'_2$. If we do so, we make the following error: the preparation procedures a_1 and a_2 are obtained from the use of a_1 and a_2 independent of the random generator B which controls which of A_1 or A_2 in A is used for the preparation.

From the construction of the apparatus A we find that $a = a'_1 \cup a'_2$ and $a'_1 \cap a'_2 = \emptyset$; that is, a is a mixture of a'_1 and a'_2 with weights $\lambda_{\mathcal{Q}}(a, a'_1) = \alpha$ and $\lambda_{\mathcal{Q}}(a, a'_2) = 1 - \alpha$.

Since the selection according to a'_1 and a_1 is obtained with the same apparatus A_1 (where in the case of a'_1 the apparatus A_1 is only a part of the total apparatus A, but is otherwise unchanged) we expect that $\mathcal{Q}(a_1) = \{a \mid a \in \mathcal{Q},\ a \subset a_1\}$ is isomorphic to $\mathcal{Q}(a'_1) = \{a' \mid a' \in \mathcal{Q},\ a' \subset a_1\}$. That is, there exists an isomorphism i of the Boolean ring $\mathcal{Q}(a_1)$ onton $\mathcal{Q}(a'_1)$ for which

$$\varphi(ia) = \varphi(a) \quad \text{and} \quad \lambda_{\mathcal{Q}}(a_1, a) = \lambda_{\mathcal{Q}}(ia_1, ia) = \lambda_{\mathcal{Q}}(a'_1, ia).$$

Intuitively this isomorphism expresses the fact that a_1 and a'_1 have the same "structure type."

With the above motivation, we now introduce the following definitions:

D 2.2. Two preparation procedures a and $\tilde{a}$ are said to be isomorphic if there is an isomorphism i between the Boolean rings $\mathcal{Q}(a)$ and $\mathcal{Q}(\tilde{a})$ for which

$$\varphi(ia') = \varphi(a') \quad \text{and} \quad \lambda_{\mathcal{Q}}(a, a') = \lambda_{\mathcal{Q}}(ia, ia') \tag{2.3}$$

and if $(a, b_0) \in C$ is equivalent to $(\tilde{a}, b_0) \in C$.

D 2.3. A preparation procedure a is called a *direct mixture* of a_1 and a_2 if there are isomorphic preparation procedures a'_1, a'_2 for which $a'_1 \cap a'_2 = \emptyset$ and $a = a'_1 \cup a'_2$. $\alpha = \lambda_{\mathcal{Q}}(a, a'_1)$ and $1 - \alpha = \lambda_{\mathcal{Q}}(a, a'_2)$ are called the *weights* of a_1 and a_2 in the direct mixture a.

If $\varphi(a_1) \neq \varphi(a_2)$ then the weights α and $1 - \alpha$ in the direct mixture are uniquely determined by $\varphi(a_1)$ and $\varphi(a_2)$. According to II, Th. 4.3(iv) we obtain

$$\mu(a, f) = \alpha\mu(a'_1, f) + (1 - \alpha)\mu(a'_2, f).$$

From D 2.3 we obtain

$$\mu(\varphi(a), g) = \alpha\mu(\varphi(a_1), g) + (1 - \alpha)\mu(\varphi(a_2), g). \tag{2.4}$$

Since $\varphi(a_1) \neq \varphi(a_2)$, there exists a g such that

$$\mu(\varphi(a_1), g) \neq \mu(\varphi(a_2), g).$$

From (2.4) it follows that

$$\alpha = \frac{\mu(\varphi(a), g) - \mu(\varphi(a_2), g)}{\mu(\varphi(a_1), g) - \mu(\varphi(a_2), g)}.$$

The experimental arrangement described above leads us to introduce the following axiom:

AP 1. To each $a_1, a_2 \in \mathcal{Q}'$ and to each rational number α, $0 < \alpha < 1$ there is a direct mixture $a \in \mathcal{Q}'$ of a_1 and a_2 with weight α of a_1 in a.

From AP 1 we obtain the following theorem:

Th. 2.2. *Let $w \in \mathcal{K}$, and let λ_i, $i = 1, \ldots, n$ be rational numbers where $0 < \lambda_i$, $\sum_{i=1}^n \lambda_i = 1$. Then there exists an $a \in \mathcal{Q}'$ and a decomposition $a = \bigcup_{i=1}^n a_i$, $\varphi(a_i) = w_i$ and $\lambda_{\mathcal{Q}}(a, a_i) = \lambda_i$ for which:*

$$\mu(w, g) = \sum_{i=1}^{n} \lambda_i \mu(w_i, g) \quad \textit{for all } g \in \mathcal{L},$$

where $w = \varphi(a)$.

Proof. We use induction on n. For a set of $n + 1$ rational numbers $\lambda_1, \ldots, \lambda_{n+1}$ we consider the set of n rational numbers

$$\alpha_i = \left(\sum_{k=1}^{n} \lambda_k\right)^{-1} \lambda_i \qquad (i = 1, \ldots, n).$$

According to the induction hypothesis, there exists an $\tilde{a} \in \mathcal{Q}'$ with a decomposition $\tilde{a} = \bigcup_{i=1}^n \tilde{a}_i$ with $\varphi(\tilde{a}_i) = w_i$ and $\lambda_{\mathcal{Q}}(\tilde{a}, \tilde{a}_i) = \alpha_i$. Suppose that for w_{n+1} there exists an $\tilde{a}_{n+1}$ with $\varphi(\tilde{a}_{n+1}) = w_{n+1}$. According to APS 1, for $\tilde{a}, \tilde{a}_{n+1}$ there exists a direct mixture a of $\tilde{a}$ and $\tilde{a}_{n+1}$ with weight λ_{n+1} of a_{n+1} in a. By D 2.3 there is an $\hat{a}$ and a a_{n+1} for which $a = \hat{a} \cup a_{n+1}$, $\hat{a} \cap a_{n+1} = \varnothing$, $\lambda_{\mathcal{Q}}(a, \hat{a}) = 1 - \lambda_{n+1}$, $\lambda_{\mathcal{Q}}(a, a_{n+1}) = \lambda_{n+1}$ where $\hat{a}$ is isomorphic to $\tilde{a}$ and a_{n+1} is isomorphic to $\tilde{a}_{n+1}$. From the decomposition $\tilde{a} = \bigcup_{i=1}^n \tilde{a}_i$ and the fact that $\hat{a}$ is isomorphic to $\tilde{a}$, it follows that there is an isomorphic decomposition between $\tilde{a} = \bigcup_{i=1}^n \tilde{a}_i$ and $\hat{a} = \bigcup_{i=1}^{n+1} a_i$. From $a = \hat{a} \cup a_{n+1}$, $\hat{a} \cap a_{n+1} = \varnothing$ it follows that $a = \bigcup_{i=1}^n a_i$ is a decomposition of a. The weights of a_i are for $i \leq n$:

$$\begin{aligned}\lambda_{\mathcal{Q}}(a, a_i) &= \lambda_{\mathcal{Q}}(a, \hat{a})\lambda_{\mathcal{Q}}(\hat{a}, a_i) \\ &= (1 - \lambda_{n+1})\lambda_{\mathcal{Q}}(\tilde{a}, \tilde{a}_i) = (1 - \lambda_{n+1})\alpha_i = \lambda_i,\end{aligned}$$

where the relation (2.3) was used. For $i = n$ we have $\lambda_{\mathcal{Q}}(a, a_{n+1}) = \lambda_{n+1}$. Thus, from (2.3) we obtain $\varphi(a_i) = \varphi(\tilde{a}_i)$ and $\varphi(a_{n+1}) = \varphi(\tilde{a}_{n+1})$. Thus, with the use of Th. 2.1 the theorem is proven.

In AP 1 we have required only that α be a rational number. This has been done with the wish that $\mathcal{Q}$ can be chosen to be a denumerable set (see [1], §9). Th. 2.2 states that to every decomposition (2.2) of w according to the w_i with weights λ_i (λ_i rational) there exists a preparation procedure $a \in \mathcal{Q}$ and a decomposition of a, $a = \bigcup_{i=1}^{n} a_i$ for which $\varphi(a) = w$, $\varphi(a_i) = w_i$ and $\lambda_{\mathcal{Q}}(a, a_i) = \lambda_i$.

It is not difficult to see that AP 1 and Th. 2.2 say little about the structure of microsystems. In fact AP 1 is more of an assertion about the construction of the preparation apparatus. We have introduced AP 1 only because it will illuminate the discussion about the physical assertions of quantum mechanics. We shall now return to our discussion concerning the concepts of preparation and registration procedures and their mixtures and decompositions.

In order to avoid error in connection with AP 1 we find it necessary to make the following remarks. For every decomposition $a = \bigcup_{i=1}^{n} a_i$ we may be tempted to believe that the apparatus A corresponding to a must consist of a random generator B which selects the sub-apparatus A_i (which correspond to the a_i). Such a description is incorrect for the following reason:

Although there may be indications on the apparatus A by which the selection procedures a_i are determined, the total structure of the apparatus may be such that it is impossible to uniquely define the component apparatuses A_i of A. For the case of quantum mechanics it is important to note that there are decompositions $a = \bigcup_{i=1}^{n} a_i$ which do not correspond to a partition of the apparatus A into a random generator and components A_i.

If we replace AP 1 by the following somewhat stronger assertion we shall find ourselves in contradiction with quantum mechanics: Suppose that there is a decomposition of w into w_i according to D 2.1, and suppose that $\varphi(a) = w$. Then there is a decomposition of a, $a = \bigcup_i a_i$ for which $\varphi(a_i) = w_i$ and $\lambda_{\mathcal{Q}}(a, a_i) = \lambda_i$.

In VI, §6 we shall find, even in the case in which the relation (2.2) holds, that there are selection procedures a which satisfy $\varphi(a) = w$ for which there are no decompositions $a = \bigcup_i a_i$ for which $\varphi(a_i) = w_i$. Since AP 1 holds, there must be another selection procedure a' which satisfies $\varphi(a') = w$ and permits a decomposition $a' = \bigcup_i a'_i$ where $\varphi(a'_i) = w_i$ and $\lambda_{\mathcal{Q}}(a', a'_i) = \lambda_i$.

Earlier we have attempted to formulate the concept of an ensemble in such a way as to avoid errors in interpretation. In order to continue this effort it is now necessary to examine the relationships between the concept of an *effect* and other concepts which are frequently used in quantum mechanics.

By analogy with the notion of an ensemble, we emphasize that, according to D 1.2, the expression *effect* denotes classes of effect processes (b_0, b). Then the map defined by D 1.3 maps many effect processes (b_0, b) into an effect. In this mapping, however, information is lost about the effect processes (b_0, b)—see, for example, the description of "coexistent" effects in IV, §1. The effect process is characterized by an "apparatus"—the registration method b_0—

and by the "detection response"—the registration procedure b. Thus it is possible that the two apparatuses $b_0^{(1)}$ and $b_0^{(2)}$ with substantially different technical design will represent the *same* effect $g \in \mathscr{L}$: $\psi(b_0^{(1)}, b^{(1)}) = \psi(b_0^{(2)}, b^{(2)}) = g$.

Let b characterize the "detection response" for the apparatus b_0. Some authors call the pair (b_0, b) a "yes–no measurement." In quantum mechanics the concept of a "yes–no" measurement is usually explained in ordinary language before it is used. Thus it will often be unclear whether the expression "yes–no measurement" should refer to the elements (b_0, b) of $\mathscr{F}$ or to the elements $g = \psi(b_0, b)$ of $\mathscr{L}$. This ambiguity can easily lead to misunderstandings. Often the expression "question" is used instead of "yes–no measurement." Here (b_0, b) is interpreted as a question posed to the micro-object; $x \in b$ corresponds to the answer "yes" and $x \in b_0 \backslash b$ corresponds to "no." Here again it is not clear whether the concept "question" should refer to an element of $\mathscr{F}$ or to an element of $\mathscr{L}$.

The expressions "yes–no measurement" and "question" are also used in a more restricted sense. If care is not taken to see how different authors use these expressions, great confusion will result. We find that the expressions "yes–no measurement," "question," and "proposition" are used for the elements of a *subset* of $\mathscr{L}$, that is, for special effects (which we shall call "decision effects" and define in §3 and §6).

We now find it necessary to define the notions of mixture and decomposition in reference to registration methods and effects—these notions will later prove to be useful.

We shall now consider only the notion of decomposition of registration methods which was defined in II, Th. 4.5.3(viii) (and not the more general decomposition of registration procedures; for the latter, see the discussion on observables presented in IV, §1.4).

The following theorem is an immediate consequence of II, Th. 4.5.3.

Th. 2.3. *Let $b_0 = \bigcup_{i=1}^n b_{0i}$ be a decomposition of the registration method b_0 according to the b_{0i} with weights λ_i. Then for an effect process $f = (b_0, b)$ where $f_i = (b_{0i}, b_{0i} \cap b)$ the following equation holds:*

$$\mu(w, \psi(f)) = \sum_{i=1}^{n} \lambda \mu(w, \psi(f_i)) \tag{2.5}$$

for all $w \in \mathscr{K}$.

D 2.4. Let $g \in \mathscr{L}$. Suppose that there is a set of real numbers λ_i where $0 < \lambda_i \leq 1, \sum_{i=1}^n \lambda_i = 1$ and a set $g_i \in \mathscr{L}$ such that

$$\mu(w, g) = \sum_{i=1}^{n} \lambda_i \mu(w, g_i) \tag{2.6}$$

holds for all $w \in \mathscr{K}$, then (2.6) is called a decomposition of the effect g according to the effects g_i with weights λ_i.

Equation (2.5) represents a decomposition of $g = \psi(b_0, b)$ according to the $g_i = \psi(b_{0i}, b_{0i} \cap b)$ with weights $\lambda_i = \lambda_{\mathscr{R}_0}(b_0, b_{0i})$.

The procedure described earlier for the construction of a preparation apparatus A from a random generator B and two preparation apparatuses A_1 and A_2 can also be applied directly to a similar procedure for registration apparatuses. Let us construct a registration apparatus using a random generator B and two registration apparatuses A_1 and A_2. Since a registration apparatus corresponds to an element of $\mathscr{R}'_0$, from the two registration methods b_{01} and b_{02} we obtain a registration method b_0 having the decomposition $b_0 = b'_{01} \cup b'_{02}$ where b'_{01}, b'_{02} correspond to the apparatuses A_1 and A_2, respectively.

By analogy to D 2.2 and D 2.3 we define:

D 2.5. Two registration methods b_0 and b'_0 are isomorphic if there is an isomorphism i of the Boolean ring $\mathscr{R}(b_0)$ to the Boolean ring $\mathscr{R}(b'_0)$ for which $\psi(ib_0, ib) = \psi(b_0, b)$; i is also an isomorphism of $\mathscr{R}_0(b_0)$ to $\mathscr{R}_0(b'_0)$ and $(a, b_0) \in C$ is equivalent to $(a, b'_0) \in C$. (Here we note that $\mathscr{R}_0(b_0)$ is defined by $\mathscr{R}_0(b_0) = \mathscr{R}_0 \cap \mathscr{R}(b_0)$).

D 2.6. A registration method b_0 is said to be a *direct mixture* of the registration methods b_{01}, b_{02} if there are two registration methods b'_{01}, b'_{02}, where b'_{01} is isomorphic to b_{01}, b'_{02} is isomorphic to b_{02} such that $b'_{01} \cap b'_{02} = \varnothing$, $b_0 = b'_{01} \cup b'_{02}$. $\alpha = \lambda_{\mathscr{R}_0}(b_0, b'_{01})$ and $1 - \alpha = \lambda_{\mathscr{R}_0}(b_0, b'_{02})$ are called the weights of b_{01}, b_{02} in the direct mixture b_0.

From D 2.6 and Th. 2.6 it follows that, for every $b \subset b_0$

$$\begin{aligned}\mu(w, \psi(b_0, b)) = {} & \alpha\mu(w, \psi(b'_{01}, b'_{01} \cap b)) \\ & + (1 - \alpha)\mu(w, \psi(b'_{02}, b'_{02} \cap b)). \end{aligned} \tag{2.7}$$

Let $b_1 \subset b_{01}$, $b_2 \subset b_{02}$. Since b'_{01} and b_{01} are isomorphic, and b'_{02} and b_{02} are isomorphic, there exists a $b'_1 \subset b'_{01}$ and a $b'_2 \subset b'_{02}$ such that b_1, b'_1 and b_2, b'_2 are isomorphic, respectively, and $\psi(b_{01}, b_1) = \psi(b'_{01}, b'_1)$, $\psi(b_{02}, b_2) = \psi(b'_{02}, b'_2)$. If $b = b'_1 \cup b'_2$, then from $b'_{01} \cap b'_{02} = \varnothing$ we obtain $b'_{01} \cap b = b'_1$, $b'_{02} \cap b = b'_2$. From (2.7) it follows that

$$\mu(w, \psi(b_0, b)) = \alpha\mu(w, \psi(b_{01}, b_1)) + (1 - \alpha)\mu(b, \psi(b_{02}, b_2)). \tag{2.8}$$

Here in the set of effects $\psi(b_0, b)$ there is a mixture of the effects $\psi(b_{01}, b_1)$ and $\psi(b_{02}, b_2)$ in the ratio α to $(1 - \alpha)$.

We now introduce the following axiom:

AR 1. To each pair $b_{01}, b_{02} \in \mathscr{R}'_0$ and each rational number α, $0 < \alpha < 1$ there exists a direct mixture $b_0 \in \mathscr{R}'_0$ of $b_{\cdot 1}, b_{02}$ with the weight α of b_{01} in b_0.

From AR 1 we obtain:

Th. 2.4. *Let $g_i \in \mathscr{L}$, $i = 1, \ldots, n$ and let $\lambda_i > 0$, $i = 1, \ldots, n$ be rational numbers for which $\sum_{i=1}^n \lambda_i = 1$. Then there exists a $b_0 \in \mathscr{R}'_0$ and a decomposition $b_0 = \bigcup_{i=1}^n b_{0i}$ for which $b_{0i} \in \mathscr{R}'_0$ and there exists a $b \in \mathscr{R}$, $b \subset b_0$ such that $\psi(b_{0i}, b_{0i} \cap b) = g_i$ and $\lambda_{\mathscr{R}_0}(b_0, b_{0i}) = \lambda_i$. Let $w \in \mathscr{K}$, $g = \psi(b_0, b)$. Then*

$$\mu(w, g) = \sum_{i=1}^{n} \lambda_i \mu(w, g_i).$$

PROOF. The proof of this theorem is analogous to that of Th. 2.2. According to the induction hypothesis there exists a $\tilde{b}_0 = \bigcup_{i=1}^n \tilde{b}_{0_i}$ and a $\tilde{b} \subset \tilde{b}_0$ such that $\psi(\tilde{b}_{0i}, \tilde{b}_{0i} \cap \tilde{b}) = g_i$ and $\lambda_{\mathscr{R}_0}(\tilde{b}_0, \tilde{b}_{0i}) = \alpha_i$. Choose $(\tilde{b}_{0n+1}, \tilde{b}_{n+1})$ such that $\psi(\tilde{b}_{0n+1}, \tilde{b}_{n+1}) = g_{n+1}$. According to AR 1, to $\tilde{b}_0$ and $\tilde{b}_{0n+1}$ there exists a direct mixture b_0 of $\tilde{b}_0$, $\tilde{b}_{0n+1}$ with weight λ_{n+1} of $\tilde{b}_{0n+1}$ in b_0. There also exists a $\hat{b}_0$ which is isomorphic to $\tilde{b}_0$ and a b_{0n+1} isomorphic to $\tilde{b}_{0n+1}$ such that

$$b_0 = \hat{b}_0 \cup b_{0n+1},$$

$$\hat{b}_0 \cap b_{0n+1} = \varnothing,$$

$$\lambda_{\mathscr{R}_0}(b_0, \hat{b}_0) = 1 - \lambda_{n+1},$$

$$\lambda_{\mathscr{R}_0}(b_0, b_{0n+1}) = \lambda_{n+1}.$$

From the isomorphism between $\tilde{b}_0$ and $\hat{b}_0$ (see D 2.5) it follows that there is a decomposition $\hat{b}_0 = \bigcup_{i=1}^n b_{0i}$ which is isomorphic to the decomposition $\tilde{b}_0 = \bigcup_{i=1}^n \tilde{b}_{0i}$ for which

$$\psi(\tilde{b}_0, \tilde{b}_{0i}) = \psi(\hat{b}_0, b_{0i}).$$

From $b_0 = \hat{b}_0 \cup b_{0n+1} = \bigcup_{k=1}^{n+1} b_{0k}$, it follows that, for $k \leq n$

$$\begin{aligned}
\lambda_{\mathscr{R}_0}(b_0, b_{0k}) &= \lambda_{\mathscr{R}_0}(b_0, \hat{b}_0)\lambda_{\mathscr{R}_0}(\hat{b}_0, b_{0k}) \\
&= (1 - \lambda_{n+1})\lambda_{\mathscr{S}}(a \cap \hat{b}_0, a \cap b_{0k}) \\
&= (1 - \lambda_{n+1})\lambda_{\mathscr{S}}(a \cap \tilde{b}_0, a \cap \tilde{b}_{0k}) \\
&= (1 - \lambda_{n+1})\lambda_{\mathscr{R}_0}(\tilde{b}_0, \tilde{b}_{0k}) = (1 - \lambda_{n+1})\alpha_k = \lambda_k.
\end{aligned}$$

Thus it follows that $\lambda_{\mathscr{R}_0}(b_0, b_{0n+1}) = \lambda_{n+1}$. To $\tilde{b} \subset \tilde{b}_0$ there exists a $\hat{b}$ $(\subset \hat{b}_0)$ isomorphic to $\tilde{b}$, for which

$$\lambda_{\mathscr{S}}(a \cap \tilde{b}_0, a \cap \tilde{b}_{0i} \cap \tilde{b}) = \lambda_{\mathscr{S}}(a \cap \hat{b}_0, a \cap b_{0i} \cap \hat{b}).$$

Thus we obtain

$$\begin{aligned}
\lambda_{\mathscr{S}}(a \cap \tilde{b}_0, a \cap \tilde{b}_{0i} \cap \tilde{b}) &= \lambda_{\mathscr{R}_0}(\tilde{b}_0, \tilde{b}_{0i})\lambda_{\mathscr{S}}(a \cap \tilde{b}_{0i}, a \cap \tilde{b}_{0i} \cap \tilde{b}) \\
&= \alpha_i \lambda_{\mathscr{S}}(a \cap \tilde{b}_{0i}, a \cap \tilde{b}_{0i} \cap \tilde{b})
\end{aligned}$$

and

$$\begin{aligned}
\lambda_{\mathscr{S}}(a \cap \hat{b}_0, a \cap b_{0i} \cap \hat{b}) &= \lambda_{\mathscr{R}_0}(\hat{b}_0, b_{0i})\lambda_{\mathscr{S}}(a \cap b_{0i}, a \cap b_{0i} \cap \hat{b}) \\
&= \alpha_i \lambda_{\mathscr{S}}(a \cap b_{0i}, a \cap b_{0i} \cap \hat{b}).
\end{aligned}$$

Thus we find that

$$g_i = \psi(\tilde{b}_{0i}, \tilde{b}_{0i} \cap \tilde{b}) = \psi(b_{0i}, b_{0i} \cap \hat{b}).$$

For b_{n+1} (which is isomorphic to $\tilde{b}_{n+1}$) we obtain

$$g_{n+1} = \psi(\tilde{b}_{0n+1}, \tilde{b}_{n+1}) = \psi(b_{0n+1}, b_{n+1}).$$

From $b = \hat{b} \cup b_{n+1}$, since $b_{n+1} \subset b_{0n+1}$ and $\hat{b} \subset \hat{b}_0$, we obtain

$$b_{0i} \cap b = b_{0i} \cap \hat{b} \quad \text{for } i \leq n,$$

$$b_{0n+1} \cap b = b_{n+1}.$$

Thus, from Th. 2.3 we finally obtain

$$\mu(a, (b_0, b)) = \sum_{i=1}^{n+1} \lambda_i \mu(a, (b_{0i}, b_{0i} \cap b))$$

that is,

$$\mu(\varphi(a), g) = \sum_{i=1}^{n+1} \lambda_i \mu(\varphi(a), g_i)).$$

Th. 2.4 states that for every decomposition (2.6) of g into components g_i with weights λ_i (λ_i rational) there exists a registration method $b_0 \in R'_0$ with decomposition $b_0 = \bigcup_{i=1}^{n} b_{0i}$ and detection response $b_i = b_{0i} \cap b \in \mathscr{R}(b_{0i})$ which satisfies

$$\psi(b_0, \textstyle\bigcup_i b_i) = g, \qquad \psi(b_{0i}, b_i) = g_i \quad \text{and} \quad \lambda_{\mathscr{R}}(b_0, b_{0i}) = \lambda_i.$$

Axioms AR 1 and AP 1 have been introduced primarily for the purpose of aiding the discussion of the physical meaning of certain aspects of quantum mechanics. Here it is not necessary to again remind the reader of the possibility of making incorrect conclusions from axiom AR 1 (these are analogous to those described in the case of preparation procedures).

In closing this section we again state that axioms AP 1 and AR 1 are not only applicable to the case of microsystems, but may be used for *all* systems in physics. Of course, it is possible to choose stronger axioms than AP 1 and AR 1 (see the discussion on coexistent decompositions and coexistent effects in IV, §1 and §5 and the concept of a "physical object" which will be introduced in §4.1) in such a way as to exclude the possibility of describing microsystems.

The interest among theoreticians in the sets $\mathscr{Q}$, $\mathscr{R}_0$, and $\mathscr{R}$ and their physical meaning is divided. In a "classical world" in which the elements of M can be considered to be the set of physical objects, the measurement problems underlying $\mathscr{Q}$, $\mathscr{R}_0$, and $\mathscr{R}$ are generally not of theoretical interest. It is sometimes believed that it should be possible to construct a theory of microsystems without making an inquiry into the physics of the measurement process. In §4 and IV, §8.1 we shall find that this is not possible. We shall eliminate much of the preparation and measurement process if we seek to develop the theory with the aid of the sets $\mathscr{K}$ and $\mathscr{L}$ and the function $\mu: \mathscr{K} \times \mathscr{L} \rightarrow [0, 1]$. In order that we may obtain the most general laws of nature governing the processes of preparation and registration of microsystems (analogous to the first and second laws of thermodynamics) in §3, it is not necessary to use the sets $\mathscr{Q}$, $\mathscr{R}_0$, and $\mathscr{R}$. That is, none of the individual physical structures associated with apparatuses denoted by the elements of $\mathcal{Q}$ and $\mathcal{R}$ will be mapped into the mathematical theory $\mathcal{MT}_\Sigma$. This viewpoint will appear to be sufficient, if not more than sufficient, to those whose interest is in the description of microsystems. We shall consider this form of the theory in the following chapters (up to and including XVI), paying close attention to the difficulties inherent in such a viewpoint.

In XVII we shall seek to introduce additional structure on the sets $\mathcal{Q}$, $\mathcal{R}_0$, and $\mathcal{R}$. The reader who is dissatisfied with the lack of structure on $\mathcal{Q}$, $\mathcal{R}_0$, and $\mathcal{R}$ (that is, the treatment of the preparation and registration apparatuses as "black boxes")—a viewpoint shared by the author—is referred to XVII, XVIII, and [13].

3 General Laws: Preparation and Registration of Microsystems

In this section we shall briefly digress and consider a few fundamental ideas for the case of microsystems. This section may be skipped without loss of continuity. We shall consider how the mathematical description of the set of ensembles $\mathcal{K}$ and the set of effects $\mathcal{L}$ which will be formulated in §5 may be deduced for the case of microsystems from physically motivated axioms. A detailed presentation of this topic can be found in [13].

Because of the "finiteness of physics" (see [1], §9 or [2], III, §8) we shall assume that the sets $\mathcal{M}$, $\mathcal{Q}$, $\mathcal{R}$ are countable. Then $\mathcal{K}$ and $\mathcal{L}$ will be countable.

The following theorem is a consequence of Th. 1.1.

Th. 3.1. *There exists a pair of real Banach spaces $\mathcal{B}$, $\mathcal{B}'$ (where $\mathcal{B}'$ is dual to $\mathcal{B}$) and an embedding of $\mathcal{K}$ in $\mathcal{B}$ and $\mathcal{L}$ in $\mathcal{B}'$ (that is, $\mathcal{K}$, $\mathcal{L}$ can be identified with subsets of $\mathcal{B}$ and $\mathcal{B}'$, respectively) for which the following conditions hold:*

(i) *The canonical bilinear form (w, g) defined for the dual pair $\mathcal{B}$, $\mathcal{B}'$ is identical to $\mu(w, g)$ on $\mathcal{K} \times \mathcal{L}$, that is,*

$$\mu(w, g) = (w, g)|_{\mathcal{K} \times \mathcal{L}}.$$

(ii) *$\mathcal{B}$ is a base-norm space (see* AIII, §6) *with basis K where K is equal to $\overline{\text{co}}\,\mathcal{K}$ (where $\overline{\text{co}}\,\mathcal{K}$ is the norm-closed convex set generated by $\mathcal{K}$). The positive cone $\mathcal{B}_+$ generated by K is closed. From Th. 2.2 it follows that K is the norm closure of $\mathcal{K}$—that is, the norm closure of $\mathcal{K}$ is already convex.*

(iii) *The linear span of $\mathcal{L}$ is $\sigma(\mathcal{B}', \mathcal{B})$ dense in $\mathcal{B}'$ (for the $\sigma(\ldots)$-topology see* AIII, §4).

$\mathcal{B}$ and $\mathcal{B}'$ are also uniquely defined (up to isomorphism) by (ii)–(iii).
Since $\mathcal{K}$ is countable, it follows that $\mathcal{B}$ is separable.

The proof of this theorem can be found in [17] and [13].

We shall denote the dual form (x, y) for $\mathscr{B}, \mathscr{B}'$ by $\mu(x, y)$.

From Th. 3.1(ii) it follows that $\mathscr{B}'$ is an order unit space. Since $0 \leq \mu(w, g) \leq 1$ for $w \in \mathscr{K}$ and $g \in \mathscr{L}$ we also obtain $0 \leq \mu(w, g) \leq 1$ for $w \in K$ and $g \in \mathscr{L}$. This means that $\mathscr{L} \subset [\mathbf{0}, \mathbf{1}]$ where $\mathbf{1}$ is the order unit in $\mathscr{B}'$. Let L denote the $\sigma(\mathscr{B}', \mathscr{B})$ closure of $\mathscr{L}$ in $\mathscr{B}'$. From Th. 2.4 it follows that L is convex. Let $\mathscr{D}$ denote the norm-closure of the linear span of $\mathscr{L}$. Then $\mathbf{1} \in \mathscr{D}$ and $\mathscr{D}$ is a separable Banach subspace of $\mathscr{B}'$ ($\mathscr{D}$ is an order unit space). $\mathscr{D}$ is $\sigma(\mathscr{B}', \mathscr{B})$ dense in $\mathscr{B}'$. K is $\sigma(\mathscr{B}, \mathscr{D})$ precompact and $\sigma(\mathscr{B}', \mathscr{D})$ separable.

Let $\mathscr{D}'$ be the Banach space which is dual to $\mathscr{D}$; $\mathscr{D}'$ is a base-norm space. We may identify the space $\mathscr{B}$ with a subspace of $\mathscr{D}'$. Let $\bar{K}$ denote the $\sigma(\mathscr{D}, \mathscr{D}')$-closure of K in $\mathscr{D}'$. $\bar{K}$ is $\sigma(\mathscr{D}', \mathscr{D})$ compact and L is $\sigma(\mathscr{B}', \mathscr{B})$-compact. For the compact sets $\bar{K}$ and L the Krein–Millman theorem holds (AIII, §4).

The topologies $\sigma(\mathscr{B}', \mathscr{B})$ and $\sigma(\mathscr{D}', \mathscr{D})$ have the following physical interpretation: First, the topologies $\sigma(\mathscr{D}', \mathscr{D})$, $\sigma(\mathscr{D}', L \cap \mathscr{D})$ and $\sigma(\mathscr{D}', \mathscr{L})$ on K (and $\bar{K}$) are identical since $\bar{K}$ is compact. The same is true for the topologies $\sigma(\mathscr{B}', \mathscr{B})$, $\sigma(\mathscr{B}', K)$ and $\sigma(\mathscr{B}', \mathscr{K})$ on L since L is compact. The topologies $\sigma(\mathscr{D}', \mathscr{L})$ on K (or $\mathscr{K}$) and $\sigma(\mathscr{B}', \mathscr{K})$ on L (or $\mathscr{L}$) describe the possibility of "physically" distinguishing among ensembles (effects). We shall now illustrate this for the case of ensembles: From $\mu(w_1, g) = \mu(w_2, g)$ for all $g \in \mathscr{L}$, it follows that $w_1 = w_2$. Experimentally we can only use a finite number of registration apparatuses with a finite number of "detection responses"—that is, a *finite* number of g in order to test whether two ensembles w_1 and w_2 are different. In addition, we can only test to within a finite error whether $\mu(w_1, g) = \mu(w_2, g)$. That is, for finitely many $g_1, g_2, \ldots, g_n$ and finite error $\varepsilon > 0$ we can always test whether

$$|\mu(w_1, g_i) - \mu(w_2, g_i)| < \varepsilon \quad (i = 1, 2, \ldots, n). \tag{3.1}$$

The inequalities (3.1) determine, for different ε, n, and g_i the neighborhood basis for the topology $\sigma(\mathscr{D}', \mathscr{L})$.

We shall now present additional axioms for K and L. The physical intuition upon which these axioms are based can be found in [13] and (in part) in [17]. It is already clear that, on the basis of the maps φ of $\mathscr{Q}'$ in $\mathscr{B}$ and ψ of $\mathscr{F}$ into $\mathscr{B}'$ that the following additional axioms will represent indirect assertions about the sets $\mathscr{Q}$, $\mathscr{R}_0$, and $\mathscr{R}$.

In order to formulate additional axioms, we define the following:

D 3.1.
$$\begin{aligned} K_0(B) &= \{w \mid w \in K \text{ and } \mu(w, g) = 0 \text{ for all } g \in B \subset L\}, \\ K_1(B) &= \{w \mid w \in K \text{ and } \mu(w, g) = 1 \text{ for all } g \in B \subset L\}, \\ L_0(A) &= \{g \mid g \in L \text{ and } \mu(w, g) = 0 \text{ for all } w \in A \subset K\}. \end{aligned}$$

$K_0(B)$ and $K_1(B)$ are closed faces[1] of K, $L_0(A)$ is a $\sigma(\mathscr{B}', \mathscr{B})$-closed face of L. If B consists of only a single element g, instead of $K_0(B)$ we shall write $K_0(g)$ and similarly for K_1 and L_0.

1. For the concept of a face, see §6.

It is easy to verify that the order relation $y_1 < y_2$ in $\mathscr{B}'$ is equivalent to the following relation

$$\mu(w, y_1) \leq \mu(w, y_2) \quad \text{for all } w \in K.$$

We shall now state the first law of measurement as an axiom:

AV 1.1. To each pair $g_1, g_2 \in L$ there exists a $g_3 \in L$ for which $g_3 > g_1$, $g_3 > g_2$ and $K_0(g_1) \cap K_0(g_2) \subset K_0(g_3)$.

AV 1.1 is equivalent to the statement that each $L_0(A)$ has a largest element, which we denote by $eL_0(A)$ (see [17] and [13]).

All elements of K (not only those of $\mathscr{K}$) are called ensembles; all elements of L (not only of $\mathscr{L}$) are called effects. The elements $eL_0(A)$ are called decision effects. We shall denote the set of decision effects by G. Let $\partial_e L$ denote the set of extreme points of L; we obtain $G \subset \partial_e L$ (see [17] and [13]).

For an arbitrary subset $\{A_\alpha\}$ of $\mathscr{P}(K)$ we find that the relationship $L_0(\bigcup_\alpha A_\alpha) = \bigcap_\alpha L_0(A_\alpha)$ is satisfied. Thus we find that the set $\{L_0(A) | A \subset K\}$ is a complete lattice with respect to the partial order $\subset$ of set inclusion. Since the map $L_0(A) \to eL_0(A)$ is an order isomorphism of $\{L_0(A) | A \subset K\}$ onto G, we find that G is a complete lattice with respect to the order induced on $G \subset \mathscr{B}'$ by $\mathscr{B}'$.

For the second law, we propose the following:

AV 1.2s. $$L = [\mathbf{0}, \mathbf{1}].$$

From this axiom, it follows that the set $\{K_0(g), g \in L\}$ coincides with the set of so-called *exposed*[1] faces. We may also show that

$$\sup_{w \in K} \mu(w, e) = 1$$

for all $e \in G$ for which $e \neq 0$. Unfortunately, the following relation AV id cannot be proven. It represents only a minor idealization. We shall introduce it as an axiom.

AV id. To each $e \in G$, $e \neq 0$, there exists a $w \in K$ for which $\mu(w, e) = 1$.

This relation is equivalent to the assertion

$$e \in G \quad \text{implies } \mathbf{1} - e \in G.$$

This relation may be used as an axiom instead of AV id.

Then we would find that the map $e \to e^\perp \overset{\text{def}}{=} \mathbf{1} - e$ is an orthocomplementation in the lattice G and that G is orthomodular. The map $e \to K_1(e)$ is an isomorphism between G and the lattice of exposed faces of K.

1. A face F of K is said to be *exposed* if and only if there exists a $y \in \mathscr{B}'$ for which

$$F = \Big\{w \,|\, w \in K, \mu(w, y) = \sup_{w' \in \mathscr{K}} \mu(w', y)\Big\}.$$

We shall now define the notion of "distance" between two elements e_1, e_2 of G (or the corresponding faces $K_1(e_1)$, $K_1(e_2)$ of K) as follows:

$$\Delta(e_1, e_2) = \max\left(\inf_{w \in K_1(e_1)} \mu(w, \mathbf{1} - e_2), \inf_{w \in K_1(e_2)} \mu(w, \mathbf{1} - e_c) \right).$$

As an additional axiom, we assert the following:

AV 3. If $e_1, e_2, e_3 \in G$ and if $e_2 \leq e_1 \leq e_2 \vee e_3$ and $\Delta(e_1, e_3) \neq 0$ then $e_1 = e_2$.

For the case in which G is a Boolean ring axiom AV 3 will be satisfied as a theorem. "Classical systems" are often defined by the requirement that G be a nonatomic Boolean ring. Instead of the nonatomic condition we shall require that each face of K is infinite dimensional. The requirement that G be a Boolean ring may be replaced by other equivalent assertions (see the general discussion in [13]). In D 4.1.2 we shall define what we mean by the expression "physical object." The assertion that G is a Boolean ring may be replaced by the requirement that the physical systems in M are physical objects (for a proof see [1], §12.3).

The next axiom will permit us to distinguish between microsystems and classical systems. We shall call this axiom the "law of quantization."

AV 4s. Every exposed face of K is the upper bound (the lattice union) of an increasing sequence of exposed and *finite*-dimensional faces.

We extend this axiom by the following assertion:

AV 2f. Every finite-dimensional face of K is exposed.

D 3.2. If axiom AV 1.1, AV 1.2s, AV 2f, AV id, AV 3, and AV 4s are satisfied we then say that M (together with the structure $\mathcal{Q}, \mathcal{R}_0, \mathcal{R}$) is a set of microsystems.

The following important theorem holds (the proof will not be given here; see [17] and [13]):

The relations AV 1.1, AV 1.2s, AV 2f, AV id, AV 3, and AV 4s are equivalent to the condition that the Banach spaces $\mathcal{B}, \mathcal{B}'$ can be identified with the spaces $\mathcal{B}(\mathcal{H}_1, \mathcal{H}_2, \ldots)$, $\mathcal{B}'(\mathcal{H}_1, \mathcal{H}_2, \ldots)$ with K as the basis of $\mathcal{B}(\mathcal{H}_1, \mathcal{H}_2, \ldots)$ and L as the order interval $[\mathbf{0}, \mathbf{1}]$ of $\mathcal{B}'(\mathcal{H}_1, \mathcal{H}_2, \ldots)$ where we assume that the lattice-dimension of the irreducible parts of G is not 2 or 3.

Here $\mathcal{B}(\mathcal{H}_1, \mathcal{H}_2, \ldots)$ and $\mathcal{B}'(\mathcal{H}_1, \mathcal{H}_2, \ldots)$ are understood, as they are defined in AIV, §15 with the generalization that the number fields of the Hilbert spaces $\mathcal{H}$ may be either the set of real numbers $\mathbf{R}$, the set of complex numbers $\mathbf{C}$, or the set of quaternions $\mathbf{Q}$.

There are physical arguments (that is, physical facts—see VIII, §2) which permit us to exclude the cases $\mathbf{R}$ and $\mathbf{Q}$. In §5 we shall assert this "end result"—which is historically obtained by means of the correspondence principle (see, for example [12], [2], XI, §1 and [2], XIII, §3) as an axiom for microsystems. The reader who is willing to accept this "end result" as "axioms for microsystems" and the accompanying structure of the set of ensembles K and the set of effects L as a hypothesis which has been verified will be able to follow the rest of this book without knowledge of this section. Those readers who are interested in the problems described in this section are again referred to references [17] and [13].

In order to dispel scepticism that the axioms which describe the sets M, $\mathcal{Q}$, $\mathcal{R}_0$, $\mathcal{R}$ and the function $\lambda_{\mathcal{S}}$ can restrict the probability function μ over $\mathcal{K} \times \mathcal{L}$ more than that which is permitted by the above theorem (or AQ in §5), it has been shown (in [8]) that for each function μ: $\mathcal{K} \times \mathcal{L} \to [0, 1]$ which satisfies the theorem, it is possible to construct a model consisting of sets M, $\mathcal{Q}$, $\mathcal{R}_0$, $\mathcal{R}$ and a function $\lambda_{\mathcal{S}}$. This does not mean that there is only one such construction possible; we must assume that it is possible to construct many nonisomorphic models M, $\mathcal{Q}$, $\mathcal{R}_0$, $\mathcal{R}$, $\lambda_{\mathcal{S}}$ for a given function μ.

4 Properties and Pseudoproperties

In the discussion of the concept of a physical object which was presented in II, §1 we have left open what we mean by the expression "objective property." In this section we shall seek to clarify this and other questions.

Here we shall seek conceptual clarity. We shall, for the most part, only sketch much of the mathematical content; much of the subject matter of this section does not have a direct bearing on the problems treated in this book.

4.1 Properties and Physical Objects

If, in addition to the axioms APS 1–APS 7 we also add the conditions $M \in \mathcal{Q}$, $M \in \mathcal{R}_0$ (and therefore $M \in \mathcal{R}$) *then* $\mathcal{Q}$, $\mathcal{R}_0$, $\mathcal{R}$ and (since $M = M \cap M \in \Theta$ implies $M \in \mathcal{S}$) $\mathcal{S}$ are Boolean rings of sets. Each of these sets can be considered to be a set of properties. These properties are, however, the opposite of that which we have called "objective" because they refer to the preparation and registration apparatuses rather than to the microsystems. For example, $a \in \mathcal{Q}$ is the "property that the systems $x \in a$ are prepared according to the procedure a." An objective property should refer directly to the microsystem itself and be independent of the preparation and registration process. By this we mean that a set a which is selected by a preparation (and similarly a set b which is selected by a registration) may be divided—according to objective properties—into subsets, and that such a "part" of a can be treated as if it were a fictitious preparation procedure.

We shall now seek to formulate this idea mathematically in terms of the relationship between a set of objective properties and the sets $\mathcal{Q}$ and $\mathcal{R}$ in order to precisely define the concept of an objective property. For this purpose we shall use part of the general treatment which is presented in [1], §12.3.

In addition to the structure defined on M by $\mathcal{Q}$, $\mathcal{R}_0$, $\mathcal{R}$ suppose that a set of properties $\mathcal{E}$ is given (that is, $\mathcal{E} \subset \mathcal{P}(M)$ satisfies AE 1 and AE 2).

D 4.1.1. Let $\bar{\mathcal{Q}}$ be the set of selection procedures generated by the set $\{(a \cap p) \mid a \in \mathcal{Q}, p \in \mathcal{E}\}$.

Since $M \in \mathcal{E}$ we find that $\mathcal{Q} \subset \bar{\mathcal{Q}}$. We assert the following axiom:

AE 3. $\bar{\mathcal{Q}}$ is a statistical selection procedure. For the probability function $\lambda_{\bar{\mathcal{Q}}}$, for $a_1, a_2 \in \mathcal{Q}$ we require that $\lambda_{\bar{\mathcal{Q}}}(a_1, a_2) = \lambda_{\mathcal{Q}}(a_1, a_2)$.
II (4.5.1) providing that the set Θ is replaced by the set $\{(a \cap p) \mid a \in \mathcal{Q}, p \in \mathcal{E}\}$.

We may consider $\bar{\mathcal{Q}}$ to be an extended system of preparation procedures. For example, if $a \cap p$ is the extended preparation procedure which prepares the system according to a and results in the selection of those with property p for further experimentation. Thus $\lambda_{\bar{\mathcal{Q}}}(a, a \cap p)$ is the probability that a system prepared according to a also "exhibits" property p.

In complete analogy to II, Th. 4.5.1 it follows that $\bar{\mathcal{Q}}$ is the set found in II (4.5.1) providing that the set θ is replaced by the set $\{(a \cap p) \mid a \in \mathcal{Q}, p \in \mathcal{E}\}$. In particular, to each $\tilde{a} \in \bar{\mathcal{Q}}$ there exists an $a \in \mathcal{Q}$ for which $\tilde{a} \subset a$.

The extended preparation procedures in $\bar{\mathcal{Q}}$ represent idealized refinements of the preparation procedures in $\mathcal{Q}$. This fact motivates the following extension of II, D 4.3.1:

Let $\tilde{a} \in \bar{\mathcal{Q}}$ and let $b_0 \in \mathcal{R}_0$; we say that $\tilde{a}$ may be combined with b_0 if there exists an $a \in \mathcal{Q}$ such that $\tilde{a} \subset a$ and $(a, b_0) \in C$.

We now define the following as an extension of II, D 4.3.1:

$$\bar{C} = \{(\tilde{a}, b_0) \mid \tilde{a} \in \bar{\mathcal{Q}}, b_0 \in \mathcal{R}_0 \text{ and } \tilde{a} \text{ may be combined with } b_0\}.$$

Here we find that $C \subset \bar{C}$.

By analogy with the sets $\mathcal{Q}'$, Θ, $\mathcal{S}$ defined in II, §4.3 we may also define the sets $\bar{\mathcal{Q}}'$, $\bar{\Theta}$, $\bar{\mathcal{S}}$. We would then obtain $\mathcal{Q}' \subset \bar{\mathcal{Q}}'$, $\mathcal{S} \subset \bar{\mathcal{S}}$.

By analogy with APS 6 we shall now introduce the following axiom:

AE 4.1. $\bar{\mathcal{S}}$ is a statistical selection procedure. For the probability function $\lambda_{\bar{\mathcal{S}}}$ we find that if we replace $\mathcal{Q}$ by $\bar{\mathcal{Q}}$, $\lambda_{\mathcal{S}}$ by $\lambda_{\bar{\mathcal{S}}}$ we find that APS 7.1, 2 holds. In addition, for $c_1, c_2 \in \mathcal{S} \subset \bar{\mathcal{S}}$ we find that

$$\lambda_{\bar{\mathcal{S}}}(c_1, c_2) = \lambda_{\mathcal{S}}(c_1, c_2).$$

From AE 4.1 it follows that: if $(a, b_0) \in C$ and $a \cap p \neq \emptyset$ then $a \cap p \cap b_0 \neq \emptyset$.

For C and $\mathcal{E}$ we require that

AE 4.2. Let $b_0 \in \mathcal{R}'_0$ and $p \in \mathcal{E}$. If $a \cap p = \emptyset$ for all a for which $(a, b_0) \in C$ then $p = \emptyset$.

This axiom expresses the requirement that the combination problem and the relation $p \in \mathscr{E}$ are mutually independent.

D 4.1.2. Let $\mathscr{Q}, \mathscr{R}_0, \mathscr{R}, \mathscr{E}$ be defined on M according to the axioms described above and those axioms given in II, §4.3 and III, §1. Then we shall call $\mathscr{E}$ a set of *virtual* properties—virtual with respect to the structures $\mathscr{Q}, \mathscr{R}_0, \mathscr{R}$.

Let $p_1, p_2 \in \mathscr{E}$ and let $b_0 \in \mathscr{R}_0, b \in \mathscr{R}$ and $b \subset b_0$. Then for $\lambda_{\mathscr{F}}$ we find that

$$\begin{aligned}
&\lambda_{\mathscr{F}}(a \cap p_1 \cap b_0, a \cap p_1 \cap p_2 \cap b)\\
&\quad = \lambda_{\mathscr{F}}(a \cap p_1 \cap b_0, a \cap p_1 \cap p_2 \cap b_0)\\
&\qquad \times \lambda_{\mathscr{F}}(a \cap p_1 \cap p_2 \cap b_0, a \cap p_1 \cap p_2 \cap b)\\
&\quad = \lambda_{\mathscr{Q}}(a \cap p_1, a \cap p_1 \cap p_2)\\
&\qquad \times \lambda_{\mathscr{F}}(a \cap p_1 \cap p_2 \cap b_0, a \cap p_1 \cap p_2 \cap b).
\end{aligned} \tag{4.1.1}$$

Thus we find that $\lambda_{\mathscr{F}}(a \cap p_1 \cap b_0, a \cap p_1 \cap p_2 \cap b)$ is determined by the values of $\lambda_{\mathscr{F}}(\tilde{a} \cap b_0, \tilde{a} \cap b)$ where $\tilde{a} \in \overline{\mathscr{Q}}'$ and $(b_0, b) \in \mathscr{F}$. On the basis of this result, we introduce the following selection structure as a substitute for $\mathscr{R}_0$ and $\mathscr{R}$: $\mathscr{R}_0$ is unchanged; instead of $\mathscr{R}$ we consider the set $\overline{\mathscr{R}}$ of all selection procedures generated by all $b \cap p$ where $b \in \mathscr{R}$ and $p \in \mathscr{E}$. We find that $\mathscr{R}_0 \subset \overline{\mathscr{R}}$, $\mathscr{R} \subset \overline{\mathscr{R}}$, and that the system of selection procedures generated by the $\tilde{a} \cap \tilde{b}$ where $\tilde{a} \in \overline{\mathscr{Q}}$, $\tilde{b} \in \overline{\mathscr{R}}$ is identical to $\overline{\mathscr{S}}$ (from which $\lambda_{\overline{\mathscr{F}}}$ is determined by (4.4.1)). It is easy to show that the system $\overline{\mathscr{R}}$ can be considered to be an extended system of registration procedures, that is, if $\overline{\mathscr{Q}}, \mathscr{R}_0, \mathscr{R}$ satisfy the above axioms, then $\overline{\mathscr{Q}}, \mathscr{R}_0, \overline{\mathscr{R}}$ also do.

Thus the function $\lambda_{\mathscr{F}}(a \cap p_1 \cap b_0, (a \cap p_1) \cap (b \cap p_2))$ takes on the following very intuitive meaning. The "idealized refined" prepared systems in $a = a \cap p_1$ will be registered by the method b_0 in such a way as to permit the use of the "idealized refined" registration method $b = b \cap p_2$. For the "idealized" registration procedure $b = b_0 \cap p_2$ we find that $\lambda_{\mathscr{F}}(a \cap p_1 \cap b_0, a \cap p_1 \cap p_2 \cap b_0) = \lambda_{\mathscr{Q}}(a \cap p_1, a \cap p_1 \cap p_2)$ is equal to the probability (which is independent of b_0) that the systems which are prepared according to $a \mathscr{P} p_1$ "have the property" p_2.

The requirements we have imposed on the set $\mathscr{E}$ of virtual properties together with the structures $\mathscr{Q}, \mathscr{R}_0, \mathscr{R}$ appear to be too weak for us to classify them as "objective," especially since $\mathscr{E}$ is not determined by $\mathscr{Q}, \mathscr{R}_0, \mathscr{R}$. Therefore, we shall call the elements of the set $\mathscr{E}$ which we may add to $\mathscr{Q}, \mathscr{R}_0, \mathscr{R}$ "virtual properties"; when we wish to stress their "virtual" character, we shall refer to them as "hidden properties." The expression "hidden variables" is often used instead of hidden properties. The reason for this name will now be given. We note that, to each $x \in M$, the subsets $\mathscr{E}(x) = \{p \mid p \in \mathscr{E} \text{ and } x \in p\}$ correspond to an ultrafilter $\mathscr{E}(x)$ in the Boolean ring $\mathscr{E}$.

According to a theorem by Stone, each Boolean ring may be described in terms of a set Π in which each ultrafilter corresponds to a single point of Π. To each $x \in M$ there is a point $\pi \in \Pi$; every $x \in M$ which corresponds to the same ultrafilter $\mathscr{E}(x)$ is mapped to the same point π. Π is called the space of hidden variables. In this book we shall not attempt to formulate the problem of hidden variables in mathematical terms. Instead, we shall only attempt to formulate what we would intuitively call "nonhidden" or "measurable" properties.

The following condition appears to be obvious: $p \subset M$ is "measurable" (and is therefore not hidden) if $b_0 \cap p \in \mathscr{R}$ for all $b_0 \in \mathscr{R}_0$. This condition is, however, too strong because we may only be able to register a $p \subset M$ approximately.

In II, §3 (after introducing AS 2.4) we have stated that we may mathematically extend the set of selection procedures by adding "idealized limiting elements." We shall do so now, but not for the general case (see [18]), but only for the case of registration procedures, for which we shall add certain idealized registration procedures.

D 4.1.3. A set $c \subset M$ is called an idealized registration procedure if there is a $b_0 \in \mathscr{R}_0$ for which $c \subset b_0$ and

$$c = \bigcup_{\substack{b \in \mathscr{R} \\ b \subset c}} b, \qquad b_0 \backslash c = \bigcup_{\substack{b \in \mathscr{R} \\ b \subset b_0 \backslash c}} b.$$

Th. 4.1.1. *The map ψ (see D 1.3) of $\mathscr{F}$ into $\mathscr{L}$ (and therefore of $\mathscr{F}$ in L where L is defined in §3 and §5) may be extended to an idealized registration procedure c as follows:*

$$\psi(b_0, c) = \sup_{\substack{b \in \mathscr{R} \\ b \subset c}} \psi(b_0, b) = \inf_{\substack{b \in \mathscr{R} \\ b_0 \supset b \supset c}} \psi(b_0, b). \tag{4.1.2}$$

The function $\lambda_{\mathscr{S}}$ may be extended to c as follows:

$$\lambda_{\mathscr{S}}(a \cap b_0, a \cap c) = \sup_{\substack{b \in \mathscr{R} \\ b \subset c}} \lambda_{\mathscr{S}}(a \cap b_0, a \cap b)$$

$$= \inf_{\substack{b \in \mathscr{R} \\ b_0 \supset b \supset c}} \lambda_{\mathscr{S}}(a \cap b_0, a \cap b). \tag{4.1.3}$$

The properties of the function $\lambda_{\mathscr{S}}$ are preserved when $\mathscr{R}$ is extended to include the set of selection procedures generated by all $b \subset c$ and $b \cap (M \backslash c)$.

PROOF. We shall only sketch the essential part of the proof, namely that

$$\sup_{\substack{b \in \mathscr{R} \\ b \subset c}} \psi(b_0, b) = \inf_{\substack{b \in \mathscr{R} \\ b_0 \supset b \supset c}} \psi(b_0, b).$$

Since the set of the $b \subset c$ is upwardly directed (in the sense of the order relation $\subset$ of set inclusion) the set of the $\psi(b_0, b)$ is also upwardly directed in $\mathscr{B}'$. Since $\psi(b_0, b) \in L$ and L is compact, the sup and inf exist (and are also limits in the $\sigma(\mathscr{B}', \mathscr{B})$ topology) and are in L.

From $b \subset c \subset \tilde{b} \subset b_0$ it follows that $\psi(b_0, b) \leq \psi(b_0, \tilde{b})$ and

$$\sup_{\substack{b \in \mathscr{R} \\ b \subset c}} \psi(b_0, b) \leq \inf_{\substack{\tilde{b} \in \mathscr{R} \\ b_0 \supset \tilde{b} \supset c}} \psi(b_0, \tilde{b}).$$

The condition $c \subset \tilde{b} \subset b_0$ is equivalent to the condition $b_0 \backslash \tilde{b} \subset b_0 \backslash c$. Since

$$b_0 \backslash c = \bigcup_{\substack{b' \in \mathscr{R} \\ b' \subset b \backslash c}} b' = \bigcup_{\substack{\tilde{b} \in \mathscr{R} \\ b_0 \supset \tilde{b} \supset c}} (b_0 \backslash \tilde{b}) = b_0 \Bigg\backslash \bigcap_{\substack{\tilde{b} \in \mathscr{R} \\ b_0 \supset \tilde{b} \supset c}} \tilde{b},$$

it follows that

$$c = \bigcap_{\substack{\tilde{b} \in \mathscr{R} \\ b_0 \supset \tilde{b} \supset c}} \tilde{b}.$$

Thus we obtain

$$\varnothing = c \cap (b_0 \backslash c) = c \cap \left(b_0 \Bigg\backslash \bigcup_{b \subset c} b \right).$$

$$= \bigcap_{\substack{\tilde{b} \in \mathscr{R} \\ b_0 \supset \tilde{b} \supset c}} \tilde{b} \cap \bigcap_{b \subset c} (b_0 \backslash b) = \bigcap_{\substack{\tilde{b}, b \in \mathscr{R} \\ b_0 \supset \tilde{b} \supset c \supset b}} \varphi(\tilde{b} \backslash b).$$

Since the set of b is upwardly directed, and the set of $\tilde{b}$ is downwardly directed, the set of $\tilde{b} \backslash b$ is downwardly directed. By AS 2.4.1 we find that

$$\inf_{\tilde{b}, b} \psi(b_0, \tilde{b} \backslash b) = 0$$

and therefore

$$\sup_{\substack{b \in \mathscr{R} \\ b \subset c}} \psi(b_0, b) = \inf_{\substack{\tilde{b} \in \mathscr{R} \\ b_0 \supset \tilde{b} \supset c}} \psi(b_0, \tilde{b}).$$

D 4.1.4. We say that the set $p \subset M$ may be ideally registered if $b_0 \cap p$ is an idealized registration procedure for each $b_0 \in \mathscr{R}_0$.

Th. 4.1.2. *p may be ideally registered if and only if*

$$p = \bigcup_{\substack{b \in \mathscr{R} \\ b \subset p}} b \quad \text{and} \quad M \backslash p = \bigcup_{\substack{b \in \mathscr{R} \\ b \subset M \backslash p}} b. \tag{4.1.4}$$

Proof. If p may be ideally registered then for each $b_0 \in \mathscr{R}_0$ we obtain:

$$b_0 \cap p = \bigcup_{\substack{b \in \mathscr{R} \\ b \subset b_0 \cap p}} b \quad \text{and} \quad b_0 \cap (M \backslash p) = \bigcup_{\substack{b \in \mathscr{R} \\ b \subset b_0 \backslash (M \backslash p) \\ = b_0 \backslash (b_0 \cap p)}} b. \tag{4.1.5}$$

According to APS 8.2 and APS 4.2 we have $\bigcup_{b_0 \subset \mathscr{R}} b_0 = M$. Thus, from APS 4.2. it follows that

$$p = M \cap p = \bigcup_{b_0 \in \mathscr{R}_0} b_0 \cap p = \bigcup_{b_0 \in \mathscr{R}_0} \bigcup_{\substack{b \in \mathscr{R} \\ b \subset b_0 \cap p}} b = \bigcup_{\substack{b \in \mathscr{R} \\ b \subset p}} b. \tag{4.1.6}$$

In a similar way we obtain

$$M \backslash p = \bigcup_{\substack{b \in \mathscr{R} \\ b \subset M \backslash p}} b. \tag{4.1.7}$$

From (4.1.6) and (4.1.7) it immediately follows that p may be ideally registered.

For a set p which may be ideally registered the effects $\psi(b_0, b_0 \cap p) \in L$ and $\psi(b_0, b \cap p) \in L$ where $b \subset b_0$ are uniquely defined.

Th. 4.1.3. *The set $\mathscr{E}_r$ of all sets which may be ideally registered is a Boolean ring.*

The proof of this theorem is easy, and is left to the reader.

By analogy with the case of registration, we shall define the notion of a set which may be ideally prepared. By analogy to D 4.1.2 we define:

D 4.1.5. We say that a set $p \subset M$ may be ideally prepared if

$$p = \bigcup_{\substack{a \in \mathscr{Q} \\ a \subset p}} a \quad \text{and} \quad M \backslash p = \bigcup_{\substack{a \in \mathscr{Q} \\ a \subset M \backslash p}} a. \tag{4.1.8}$$

The statements made above for sets which may be ideally registered are also valid for sets which may be ideally prepared.

Th. 4.1.4. *The function $\varphi\colon \mathscr{Q}' \to \mathscr{K} \subset K$ has a unique extension to the set of all $a \cap p \neq \varnothing$ where $a \in \mathscr{Q}'$ and p may be ideally prepared. The extension is given by*

$$\varphi(a \cap p) = w\mu(w, \mathbf{1})^{-1},$$

where

$$w = \sup_{\substack{\tilde{a} \in \mathscr{Q} \\ \tilde{a} \subset a \cap p}} \lambda_{\mathscr{Q}}(a, \tilde{a})\varphi(\tilde{a})$$

$$= \inf_{\substack{\tilde{a} \in \mathscr{Q} \\ a \supset \tilde{a} \supset a \cap p}} \lambda_{\mathscr{Q}}(a, \tilde{a})\varphi(\tilde{a}). \tag{4.1.9}$$

Proof. The proof proceeds in a similar manner as Th. 4.1.2, where it is only important to note that sup and inf exist in the sense of the norm in B. For, if $w_1, w_2 \in \check{K}$ (for definition of $\check{K}$, see AIII, §6), and $w_1 \geq w_2$ it follows that $\|w_1 - w_2\| = \mu(w_1 - w_2, \mathbf{1})$.

The following theorem follows directly from Th. 4.1.3.

Th. 4.1.5. *The set $\mathscr{E}_p$ of all sets which may be ideally prepared is a Boolean ring of sets.*

Let $\mathscr{E}_m$ denote the set $\mathscr{E}_r \cap \mathscr{E}_p$, that is, the set of all sets which may be both ideally prepared and ideally registered. Clearly $\mathscr{E}_m$ is a Boolean ring of sets.

Th. 4.1.6. *For each $p \in \mathscr{E}_m$ the map $T_p\colon \mathscr{K} \to \check{K}$ (for definition of $\check{K}$, see* AIII, §6) *which is defined by $\varphi(a) \to \lambda_{\mathscr{Q}}(a, a \cap p)\varphi(a \cap p)$ is norm-continuous and has a unique extension T_p on $\mathscr{B}$ which is linear and norm-continuous.*

This map is uniquely determined by the following equation

$$\mu(w, \psi(b_0, b \cap p)) = \mu(T_p w, \psi(b_0, b)), \tag{4.1.10}$$

which is valid for all $w \in K$ *and all* $(b_0, b) \in \mathscr{F}$. *In addition, the following equations hold:*

$$\begin{aligned} T_{M \setminus p} &= \mathbf{1} - T_p, \\ T_{p_1 \cap p_2} &= T_{p_1} T_{p_2} = T_{p_2} T_{p_1}. \end{aligned} \tag{4.1.11}$$

PROOF. If $w = \varphi(a)$ (that is, $w \in \mathscr{K}$) it follows that:

$$\begin{aligned} \mu(w, \psi(b_0, b \cap p) &= \lambda_{\mathscr{F}}(a \cap b_0, a \cap b \cap p) \\ &= \lambda_{\mathscr{F}}(a \cap b_0, a \cap p \cap b_0) \lambda_{\mathscr{F}}(a \cap b_0 \cap p, a \cap p \cap b) \\ &= \lambda_{\mathscr{Q}}(a, a \cap p) \mu(\varphi(a \cap b), \psi(b_0, b)) \\ &= \mu(T_p \varphi(a), \psi(b_0, b)). \end{aligned}$$

Thus (4.1.10) is proven for $w \in \mathscr{K}$. From (4.10), if $\sum_{i=1}^n \alpha_i w_i = 0$, $w_i \in \mathscr{K}$, it follows that $\sum_{i=1}^n \alpha_i T_p w_i = 0$. Therefore T_p can be uniquely extended as a linear map to all of $\overline{\mathscr{B}}$: T_p: $\overline{\mathscr{B}} \to \overline{\mathscr{B}}$ (where $\overline{\mathscr{B}}$ is the linear span of $\mathscr{K}$ in $\mathscr{B}$). If T_p is norm-continuous, then it can be extended to all of $\mathscr{B}$.

In order to prove that T_p is norm-continuous, we shall assume that co $\mathscr{L}$ is $\sigma(\mathscr{B}', \mathscr{B})$-dense in L (which is the case for quantum mechanics—see §3 or §5). Since $[-\mathbf{1}, \mathbf{1}] = 2L - \mathbf{1}$, for $w_1, w_2 \in \mathscr{K}$ we find that

$$\begin{aligned} \|T_p w_1 - T_p w_2\| &= \sup_{g \in L} \mu(T_p w_1 - T_p w_2, 2g - \mathbf{1}) \\ &\leq |\mu(T_p w_1 - T_p w_2, \mathbf{1})| + 2 \sup_{g \in L} \mu(T_p w_1 - T_p w_2, g). \end{aligned}$$

From (4.1.10), for $w \in K$ we find that, for the special case $b = b_0$

$$\mu(T_p w, \mathbf{1}) = \mu(T_p w, \psi(b_0, b_0)) = \mu(w, \psi(b_0, b_0 \cap p)) \tag{4.1.12}$$

and we obtain:

$$|\mu(T_p w_1 - T_p w_2, \mathbf{1})| = |\mu(w_1 - w_2, \psi(b_0, b_0 \cap p)| \leq \|w_1 - w_2\|.$$

Since co $\mathscr{L}$ is dense in L we obtain

$$\sup_{g \in L} \mu(T_p w_1 - T_p w_2, g) = \sup_{g \in \mathscr{L}} \mu(T_p w_1 - T_p w_2, g).$$

For $g \in \mathscr{L}$ there exists a $(b_0, b) \in \mathscr{F}$ with $\psi(b_0, b) = g$. Therefore we find that

$$\sup_{g \in L} \mu(T_p w_1 - T_p w_2, g) = \sup_{(b_0, b) \in \mathscr{F}} \mu(T_p w_1 - T_p w_2, \psi(b_0, b))$$

and, since

$$\mu(T_p w_1 - T_p w_2, \psi(b_0, b)) = \mu(w_1 - w_2, \psi(b_0, b \cap p)) \leq \|w_1 - w_2\|$$

we finally obtain

$$\|T_p w_1 - T_p w_2\| \leq 3\|w_1 - w_2\|,$$

whereupon we have proven the norm-continuity of T_p. Thus T_p is defined on all of $\mathscr{B}$.

From (4.1.10) and

$$\psi(b_0, b \cap p) + \psi(b_0, b \cap (M \backslash p)) = \psi(b_0, b)$$

it follows that

$$\begin{aligned}\mu(w, (b_0, b \cap (M \backslash p))) &= \mu(w, \psi(b_0, b)) - \mu(w, \psi(b_0, b \cap p)) \\ &= \mu(w, \psi(b_0, b)) - \mu(T_p w, \psi(b_0, b)) \\ &= \mu((\mathbf{1} - T_p)w, \psi(b_0, b)).\end{aligned}$$

Thus we find that

$$T_{M \backslash p} = \mathbf{1} - T_p.$$

Since T_p is defined on all of $\mathscr{B}$, for $w \in \check{K}$ it follows that

$$\mu(w, \psi(b_0, b \cap p)) = \mu(T_p w, \psi(b_0, b)).$$

From

$$\begin{aligned}\psi(b_0, b \cap p_1 \cap p_2) &= \sup_{\substack{\tilde{b} \in \mathscr{R} \\ \tilde{b} \subset p_1 \cap p_2 \cap b}} \psi(b_0, \tilde{b}) \\ &= \sup_{\substack{\bar{b}, \tilde{b} \in \mathscr{R} \\ \bar{b} \subset b_1 \cap p, \tilde{b} \subset b_2 \cap p}} \psi(b_0, \bar{b} \cap \tilde{b})\end{aligned}$$

it follows that

$$\psi(b_0, b \cap p_1 \cap p_2) = \sup_{\substack{\tilde{b} \in \mathscr{R} \\ \tilde{b} \subset p_2 \cap b}} \psi(b_0, p_1 \cap \tilde{b}).$$

Thus we obtain

$$\begin{aligned}\mu(w, \psi(b_0, b \cap p_1 \cap p_2)) &= \sup_{\tilde{b} \subset p_2 \cap b} \mu(w, \psi(b_0, p_1 \cap \tilde{b})) \\ &= \sup_{\tilde{b} \subset p_2 \cap b} \mu(T_{p_1} w, \psi(b_0, \tilde{b})) = \mu(T_{p_1} w, \psi(b_0, b \cap p_2)) \\ &= \mu(T_{p_2} T_{p_1} w, \psi(b_0, b)).\end{aligned}$$

From which we obtain

$$T_{p_1 \cap p_2} = T_{p_1} \cdot T_{p_2} = T_{p_2} \cdot T_{p_1}.$$

D 4.1.6. The set E_m of all subsets $p \subset M$ which may be both ideally registered and ideally prepared is called the set of *objective properties* of M.

We note that D 4.1.6 makes sense becuase $\mathscr{E}_m$ is a Boolean ring.

From (4.1.12) it follows that the map T'_p in B' which is dual to T_p satisfies the equation

$$\chi(p) \underset{\text{def}}{=} \psi(b_0, b_0 \cap p) = T_p \mathbf{1} \tag{4.1.13}$$

and therefore $\psi(b_0, b_0 \cap p)$ is independent of b_0. If $p_1 \cap p_2 = \varnothing$, from (4.1.13) it is easy to show that

$$\chi(p_1 \cup p_2) = \chi(p_1) + \chi(p_2). \tag{4.1.14}$$

If, instead of $\mathscr{E}$ we use $\mathscr{E}_m$, and we use the sets $\overline{\mathscr{Q}}$, $\overline{\mathscr{R}}$, etc. which were defined with the help of $\mathscr{E}$, it follows that AE 3 and AE 4.1.2 are theorems and that

$$\lambda_{\overline{\mathscr{S}}}(a \cap p_1 \cap b_0, a \cap p_1 \cap b \cap p_2) = \frac{\mu(T_{p_1}\varphi(a), \psi(b_0, b \cap p_2)}{\mu(T_{p_1}\varphi(a), \mathbf{1})} \tag{4.1.15}$$

and

$$\lambda_{\overline{\mathscr{Q}}}(a, a \cap p) = \mu(T_p\varphi(a), \mathbf{1}) = \mu(\varphi(a), \chi(p)) \tag{4.1.16}$$

are satisfied. Thus $\mathscr{E}_m$ is, in the sense of D 4.1.12, also a set of virtual properties.

In D 4.1.6 we have called the set $\mathscr{E}_m$ the set of objective properties because we believe that the mathematical structure for $\mathscr{E}_m$ characterizes what we mean intuitively by the expression "objective property." By this we mean that the relations AE 3 and AE 4.1.2 describe the condition that the properties of the system be *independent* of the preparation and registration procedures. The condition that $p \in \mathscr{E}_m$ may be both ideally prepared and registered says that p is not "hidden."

We shall call those physical systems which are completely described by the set $\mathscr{E}_m$ of objective properties *physical objects*. How can we find a mathematically precise definition of the expression "completely described"? The set $\mathscr{E}_m$ can be so "small" that different ensembles $w_1, w_2 \in K$ cannot be distinguished by means of $\mu(w, \chi(p))$. By "completely described" we mean that $\mathscr{E}_m$ is sufficiently "large" that the $w \in K$ can be distinguished by the $p \in \mathscr{E}_m$. In mathematical terms we may express this idea as follows: $\chi(\mathscr{E}_m)$ separates K, that is, if $\mu(w_1, \chi(p)) = \mu(w_2, \chi(p))$ for all $p \in \mathscr{E}_m$ then $w_1 = w_2$. Thus we define:

D 4.1.7 A set M of physical systems is called a set of *physical objects* if $\chi(\mathscr{E}_m)$ separates the $w \in K$.[1]

In IV, §8.1 we shall find that microsystems are not physical objects. In [1], §12.3 we have shown what structures K and L must have in order to describe physical objects.

Intuitively, on the basis of the formulation of the set $\mathscr{E}_m$, it follows that $\mathscr{E}_m$ represents "real physical facts." On the basis of the analysis presented in [1], §10 and in [1], §12.3 we have shown that $\mathscr{E}_m$ does represent a set of real physical facts.

For a set $\mathscr{E}$ of virtual properties (defined according to D 4.1.2) we may suppose that $\mathscr{E} \subset \mathscr{E}_r$. Then we may also define the set $\overline{\mathscr{E}}_p$ of all sets which may be ideally prepared with respect to $\overline{\mathscr{Q}}$ (defined according to D 4.1.1). Then, trivially, we find that $\mathscr{E} \subset \overline{\mathscr{E}}_p$ and that $\mathscr{E} \subset \overline{\mathscr{E}}_p \cap \mathscr{E}_r$. Then $\overline{\mathscr{E}} = \overline{\mathscr{E}}_p \cap \mathscr{E}_r$ is a set of imagined properties which at least may be ideally registered. The set $\overline{\mathscr{E}}$ is, in general, not uniquely determined (there are many such sets possible). Thus $\overline{\mathscr{E}}$ is not a set of real physical facts (in the sense of [1], §10), or in other words—is not a set of physically real and objective properties. If such a set $\overline{\mathscr{E}}$

1. More precisely: if it is a certain hypothesis that $\chi(\mathscr{E}_m)$ separates the $w \in K$ (see [1], §12.3).

is so large that the equivalence classes of $\mathcal{Q}'$ defined by the equivalence relation $a_1 \approx a_2$: $\lambda_{\mathcal{Q}}(a_1, a_1 \cap p) = \lambda_{\mathcal{Q}}(a_2, a_2 \cap p)$ for all $p \in \overline{\mathcal{E}}$ are finer than those determined by the $f \in \mathcal{F}$ then $\overline{\mathcal{E}}$ cannot exist for microsystems (see IV, §8.1).

As we found above, microsystems are not physical objects. This fact has created a scandal. Radical assertions—such as "objective knowledge is impossible"—have been proposed. However, the fact that microsystems are not physical objects is itself objective knowledge.

The fact that microsystems are not physical objects does not mean that each separation of microsystems from the preparation and registration procedures is impossible and that in each experiment the complicated structure of each apparatus must be taken into account. It only means that we must abandon the notion of a microscopic "object," one to which we have been accustomed.

In the following section we shall use a different notion—that of a pseudoproperty—in order to separate the microsystems from the preparation and registration apparatuses.

4.2 Pseudoproperties

In quantum mechanics we find that the role of a Boolean ring $\mathcal{E}$ of sets is replaced by another structure $\mathcal{E}_{ps}$. For $\mathcal{E}_{ps} \subset P(M)$ the following axioms are satisfied:

APE 1. $\mathcal{E}_{ps}$ is a lattice with respect to the order $\subset$ (set inclusion) with largest element M (AI, §4).

APE 2. To each $p \in \mathcal{E}_{ps}$ the set

$$D_p = \{q \mid p \in \mathcal{E}_{ps} \text{ and } q \subset M \backslash p\}$$

has a greatest element, which we denote by p^*. For $p_1 \supset p_2$, $p_1 \neq p_2$ we require that $D_{p_1} \neq D_{p_2}$.

A set $\mathcal{E}_{ps} \subset P(M)$ which satisfies APE 1 and APE 2 is called a structure of species *pseudoproperties*.

Th. 4.2.1. *From* APE 1 *and* APE 2 *it follows that* $\mathcal{E}_{ps}$ *is an orthocomplemented lattice.*

PROOF. We must prove the following three relations (AID 1.2):

(i) If $p \subset q$ then $q^* \subset p^*$.
(ii) $(p^*)^* = p$.
(iii) $p \wedge p^* = \varnothing$.

(i) From $p \subset q$ it follows that $D_q \subset D_p$ and we obtain $q^* \subset p^*$.
(ii) From $p^* \subset M \backslash p$ (and therefore $p \subset M \backslash p^*$) it follows that $p \in D_{p^*}$. Let $q \supset p$ where $q \subset D_{p^*}$. Then we would find that $q \subset M \backslash p^*$; that is, $p^* \subset M \backslash q$ and $p^* \subset D_q$—in other words $D_p \subset D_q$. Since $D_q \subset D_p$ we obtain $D_p = D_q$. From APE 2 we obtain $p = q$. Thus we obtain $(p^*)^* = p$.
(iii) Since $p^* \subset M \backslash p$ we find that $p \cap p^* = \emptyset$, and therefore $p \wedge p^* = \emptyset$ (from which we conclude that $\mathscr{E}_{ps}$ has a least element—the empty set $\emptyset$).

Conditions APE 1 and APE 2 are natural generalizations of AE 1 and AE 2. The latter require that if $p_1, p_2 \in \mathscr{E}$, then $p_1 \cap p_2$ and $p_1 \cup p_2 \in \mathscr{E}$; in addition, if $p \in \mathscr{E}$ then $M \backslash p \in \mathscr{E}$. The former require that to $p_1, p_2 \in \mathscr{E}_{ps}$ there exists a *largest* element $p_3 \in \mathscr{E}_{ps}$ for which $p_3 \subset p_1 \cap p_2$ and a smallest element $p_4 \in \mathscr{E}_{ps}$ for which $p_4 \supset p_1 \cap p_2$; also, to $p \in \mathscr{E}_{ps}$ there exists a largest element $p_5 \in \mathscr{E}_{ps}$ for which $p_5 \subset M \backslash p$. The conditions APE 1 and APE 2 are as strong as possible without requiring that conditions AE 1 and AE 2 be satisfied.

Of course, AE 1 and AE 2 say little about the system under consideration unless we relate the structure $\mathscr{E}$ to the physically motivated structures $\mathscr{Q}$, $\mathscr{R}_0$, and $\mathscr{R}$. The same is true for APE 1 and APE 2. We must now relate the structure $\mathscr{E}_{ps}$ to $\mathscr{Q}$, $\mathscr{R}_0$, and $\mathscr{R}$.

Since we do not intend to develop the notion of "hidden" pseudoproperties, we shall proceed to develop the analogs of equations (4.1.4) and (4.1.8). Of course, we cannot assume these equations directly without obtaining the set $\mathscr{E}_m$ as a result.

From (4.1.4) and (4.1.8) it follows that

$$p = \left[\bigcup_{\substack{a \in \mathscr{Q} \\ a \subset p}} a\right] \cup \left[\bigcup_{\substack{b \in \mathscr{R} \\ b \subset p}} b\right]. \tag{4.2.1}$$

This equation suggests the following generalization: For each subset $c \subset M$ we may define

$$\pi(c) = \left[\bigcup_{\substack{a \in \mathscr{Q} \\ a \subset c}} a\right] \cup \left[\bigcup_{\substack{b \in \mathscr{R} \\ b \subset c}} b\right] \tag{4.2.2}$$

from which it follows that

$$\pi(c) = \left[\bigcup_{\substack{a \in \mathscr{Q} \\ a \subset \pi(c)}} a\right] \cup \left[\bigcup_{\substack{b \in \mathscr{R} \\ b \subset \pi(c)}} b\right], \tag{4.2.3}$$

that is, $\pi(c)$ is an element having the form (4.2.1).

For each set p of the form (4.2.1) we define

$$p^* \stackrel{\text{def}}{=} \pi(M \backslash p). \tag{4.2.4}$$

From (4.2.4) it follows that $(p^*)^* \supset p$ and that $p_1 \supset p_2$ implies $p_1^* \subset p_2^*$. Thus it follows from $(p^*)^* \supset p$ that $[(p^*)^*]^* \subset p^*$. From $(p^*)^* \supset p$, if we replace p with p^* we obtain $[(p^*)^*]^* \supset p^*$. Thus we find that $(p^*)^{**} = p^*$.

Let Π denote the set of all p^* where p is of the form (4.2.1). Then Π is the set of all p of the form (4.2.1) for which $p^{**} = p$.

For $p \in \Pi$ we shall use the following abbreviated notation:

$$p_p = \bigcup_{\substack{a \in \mathscr{Q} \\ a \subset p}} a \quad \text{and} \quad p_r = \bigcup_{\substack{b \in \mathscr{R} \\ b \subset p}} b. \tag{4.2.5}$$

We obtain

$$p_p = \bigcup_{\substack{a \in \mathscr{Q} \\ a \in p_p}} a, \qquad p_r = \bigcup_{\substack{b \in \mathscr{R} \\ b \in p_r}} b \quad \text{and} \quad p = p_p \cup p_r. \tag{4.2.6}$$

Th. 4.2.2. Π *is a structure of species pseudoproperties; in addition, the following equations are satisfied*:

$$\begin{aligned} (p_1 \wedge p_2)_p &= p_{1p} \cap p_{2p}, \\ (p_1 \wedge p_2)_r &= p_{1r} \cap p_{2r}. \end{aligned} \tag{4.2.7}$$

PROOF. If $p_1 \supset p_2$ and $p_1^* = p_2^*$ it follows that $p_1 = p_2$; then, by definition (4.2.4) the relation APE 2 is proven.

According to (4.2.2), $\pi(p_1 \cap p_2)$ is the largest element p of the form (4.2.1) for which $p \subset p_1$ and $p \subset p_2$. It only remains to show that $\pi(p_1 \cap p_2) \in \Pi$, that is $[\pi(p_1 \cap p_2)]^{**} = \pi(p_1 \cap p_2)$. From $\pi(p_1 \cap p_2) \subset p_1$ it follows that $[\pi(p_1 \cap p_2)]^{**} \subset p_1^{**} = p_1$. Similarly it follows that $[\pi(p_1 \cap p_2)]^{**} \subset p_2$; thus we find that $[\pi(p_1 \cap p_2)]^{**} \subset p_1 \cap p_2$. We therefore obtain $[\pi(p_1 \cap p_2)]^{**} \subset \pi(p_1 \cap p_2)$. Since $[\pi(p_1 \cap p_2)]^{**} \supset \pi(p_1 \cap p_2)$ is trivial, we have proven that $\pi(p_1 \cap p_2) \in \Pi$.

We shall now prove that $[\pi(p_1^* \cap p_2^*)]^*$ is the smallest element in Π containing p_1 and p_2. From $p \supset p_1$ and $p \supset p_2$ it follows that $p^* \subset p_1^*$ and $p^* \subset p_2^*$, and therefore we obtain $p^* \subset \pi(p_1^* \cap p_2^*)$ from which we conclude that $[\pi(p_1^* \cap p_2^*)]^* \subset p$.

Thus we find that Π is an orthocomplemented lattice. Using $\wedge$ and $\vee$ for the lattice operations we obtain

$$\begin{aligned} p_1 \wedge p_2 &= \pi(p_1 \cap p_2), \\ p_1 \vee p_2 &= [\pi(p_1^* \cap p_2^*)]^*. \end{aligned}$$

From the definition of π we also find that

$$(p_1 \wedge p_2)_p = \bigcup_{\substack{a \in \mathscr{Q} \\ a \subset p_1 \cap p_2}} a; \qquad (p_1 \wedge p_2)_r = \bigcup_{\substack{b \in \mathscr{R} \\ b \subset p_1 \cap p_2}} b.$$

From these results we obtain

$$(p_1 \wedge p_2)_p \subset p_{1p}, \qquad (p_1 \wedge p_2)_p \subset p_{2p}$$

and

$$(p_1 \wedge p_2)_p \subset p_{1p} \cap p_{2p}.$$

If $a_1 \in \mathscr{Q}$, $a_1 \subset p_1$ and $a_2 \in \mathscr{Q}$, $a_2 \subset p_2$ we find that $a_1 \cap a_2 \in \mathscr{Q}$ and $a_1 \cap a_2 \subset p_1 \cap p_2$ and therefore obtain

$$(p_1 \wedge p_2)_p \supset \bigcup_{\substack{a_1 \in \mathscr{Q} \\ a_1 \subset p_1}} \bigcup_{\substack{a_2 \in \mathscr{Q} \\ a_2 \subset p_2}} a_1 \cap a_2 = p_{1p} \cap p_{2p}.$$

Thus we obtain the first equation in (4.2.7). The second equation is obtained similarly.

Let $A_p = \{\varphi(a) \mid a \in \mathcal{Q}', a \subset p\}$. From D 3.1 it follows from $p \cap p^* = \varnothing$ that $a \cap b = \varnothing$ for all $a \subset p$ and $b \subset p^*$, that is, $\psi(b_0, b) \in L_0(A_p)$ for all $b \subset p^*$. Thus we also obtain $\psi(b_0, b) \in L_0(A_{p^*})$ for all $b \subset p$.

From the definition

$$\psi(b_0, b_0 \cap p_r) = \sup_{b \subset b_0 \cap p_r} \psi(b_0, b)$$

it follows that $\psi(b_0, b_0 \cap p_r) \in L_0(A_{p^*})$. For the case in which $p_r = p_p$ (and thus $p_r = p_p = p$) (that is, in the case of *objective* properties), then for all $a \subset p$, for $a \cap p_r = a$ we obtain

$$\lambda_{\mathcal{F}}(a \cap b_0, a \cap b_0 \cap p_r) = 1$$

from which we conclude that $\psi(b_0, b_0 \cap p_r) \in L_1(A_p)$.[1]

For a *pseudoproperty* we shall *require* that there exists at least one registration method b_0 for which $\psi(b_0, b_0 \cap p_r) \in L_1(A_p)$, from which it follows that

$$L_0(A_{p^*}) \cap L_1(A_p) \neq \varnothing. \tag{4.2.8}$$

The fact that there is no $b \in R$ for which $b \subset p$ and $b \subset p^*$, that is, there is no $\psi(b_0, b)$ for which $\psi(b_0, b) \in L_0(A_p)$ and $\psi(b_0, b) \in L_0(A_{p^*})$ leads us to impose the following additional condition for pseudoproperties:

$$L_0(A_p \cup A_{p^*}) = L_0(A_p) \cap L_0(A_{p^*}) = \varnothing. \tag{4.2.9}$$

We note that (4.2.9) is satisfied for objective properties because, for each $a \in \mathcal{Q}$ the relation $a = (a \cap p) \cup (a \cap p^*)$ holds where $p^* = M \backslash p$ (that is, $\varphi(a) = \lambda\varphi(a \cap p) + (1 - \lambda)\varphi(a \cap p^*)$ holds where $\lambda = \lambda_{\mathcal{Q}}(a, a \cap p)$).

We could not show that the set of all p for which there exists a b_0 for which $\psi(b_0, b_0 \cap p_r) \in L_1(A_p)$ and for which (4.2.9) is satisfied is an orthocomplemented sublattice of Π. For this reason it is necessary to define:

D 4.2.1. An orthocomplemented sub-lattice $\mathscr{E}_{ps}$ of Π which satisfies the conditions

(i) to each $p \in \mathscr{E}_{ps}$ there exists a $b_0 \in \mathscr{R}_0$ such that $\psi(b_0, b_0 \cap p_r) \in L_1(A_p)$, and

(ii) the relation (4.2.9) is satisfied for all $p \in \mathscr{E}_{ps}$,

is called a set of *actual physical pseudoproperties*.

D 4.2.2. A set $\mathscr{E}_{ps}$ of actual physical pseudoproperties is said to be *sufficient* if the set of all $\psi(b_0, b_0 \cap p_r) \in L_1(A_p)$ for $p \in \mathscr{E}_{ps}$ separates the elements of K.

Whether such sets $\mathscr{E}_{ps}$ satisfying D 4.2.1 and D 4.2.2 exist for a given theory cannot be determined in general. For microsystems the existence of such sets is certain—see IV, §8.2.

1. By analogy to D 3.1 we define

$$L_1(A) = \{g \mid g \in L \text{ and } \mu(w, g) = 1 \text{ for all } w \in A \subset K\}.$$

5 Ensembles and Effects in Quantum Mechanics

At the end of §3 we suggested that the "end result" obtained from the fundamental laws of preparation and registration given in §3 may also be formulated as an axiom. This axiom may be motivated by a reasoning process which uses the so-called correspondence principle and certain aspects of the measurement process (see I, XVII, [2], XI, §1 and [2], XIII, §3). We shall now formulate the fundamental axiom of quantum mechanics (which according to §3 may be obtained as a theorem from axioms AV 1.1, AV 1.2s, AV 2f, AV id, AV 3, and AV 4s) as follows:

AQ. There exists an injective mapping β of $\mathscr{K}$ into the basis K of $\mathscr{B}(\mathscr{H}_1, \mathscr{H}_2, \ldots)$ and an injective mapping γ of $\mathscr{L}$ into the order unit interval $[\mathbf{0}, \mathbf{1}]$ in $\mathscr{B}'(\mathscr{H}_1, \mathscr{H}_2, \ldots)$ for which

$$\mu(w, g) = \operatorname{tr}((\beta w)(\gamma g))$$

holds for $w \in \mathscr{K}$, $g \in \mathscr{L}$ where $\beta\mathscr{K}$ is (norm) dense in K and $\gamma\mathscr{L}$ is $\sigma(\mathscr{B}', \mathscr{B})$ dense in $[\mathbf{0}, \mathbf{1}]$.

Both Banach spaces $\mathscr{B}(\mathscr{H}_1, \mathscr{H}_2, \ldots)$ and $\mathscr{B}'(\mathscr{H}_1, \mathscr{H}_2, \ldots)$ and the canonical bilinear form $\operatorname{tr}(uv)$ where $u \in \mathscr{B}(\mathscr{H}_1, \mathscr{H}_2, \ldots)$ and $v \in \mathscr{B}(\mathscr{H}_1, \mathscr{H}_2, \ldots)$ are defined in AIV, §15. In AIV, §15 it will also be shown that $\mathscr{B}'(\mathscr{H}_1, \ldots)$ and $\mathscr{B}(\mathscr{H}_1, \ldots)$ may be identified as subsets of the Banach algebra $\mathscr{A}(\mathscr{H}_1, \ldots)$. $\mathscr{B}(\mathscr{H}_1, \ldots)$ is a base–norm space and $\mathscr{B}'(\mathscr{H}_1, \ldots)$ is its dual Banach space (and therefore is a order unit space). The order unit in $\mathscr{B}'(\mathscr{H}_1, \ldots)$ is the unit operator $\mathbf{1}$ in $\mathscr{A}(\mathscr{H}_1, \ldots)$.

In the following we shall often make use of the fact that $\mathscr{B}(\mathscr{H}_1, \ldots)$ and $\mathscr{B}'(\mathscr{H}_1, \ldots)$ are subsets of $\mathscr{A}(\mathscr{H}_1, \ldots)$. Where products of elements from $\mathscr{B}(\mathscr{H}_1, \ldots)$ and $\mathscr{B}'(\mathscr{H}_1, \ldots)$ occur, they are to be understood in the sense of the algebra $\mathscr{A}(\mathscr{H}_1, \ldots)$. (For the formulation of axiom AQ, see also [2], XIII, §3; the above formulation of axiom AQ is somewhat weaker than that given in [2], XIII, §3. The stronger form in [2], XIII, §3 was chosen for its simplicity).

On the basis of AQ we shall identify $\mathscr{K}$ with $\beta\mathscr{K}$ and $\mathscr{L}$ with $\gamma\mathscr{L}$ and write $\mu(w, g) = \operatorname{tr}(w, g)$. Then the maps φ of $\mathscr{Q}'$ into $\mathscr{K}$ and ψ of $\mathscr{F}$ into $\mathscr{L}$ correspond to maps of $\mathscr{Q}'$ in K and $\mathscr{F}$ into $[\mathbf{0}, \mathbf{1}]$. Instead of $[\mathbf{0}, \mathbf{1}]$ we shall write L. According to AQ the set $\varphi\mathscr{Q}'$ is dense in K and $\psi\mathscr{F}$ is dense in L. We shall call the elements of K ensembles and the elements of L effects. Let $\mathscr{D}$ denote the norm-closed subspaces of $\mathscr{B}'(\mathscr{H}_1, \ldots)$ generated by $\mathscr{F}$. Then $\mathscr{D}$ is a Banach space and $\mathscr{B}(\mathscr{H}_1, \ldots)$ may be identified with a subspace of $\mathscr{D}'$. $\mathscr{D}$ is norm-separable; $\mathscr{B}'$ is not. We shall seek to characterize $\mathscr{D}$ by axioms in VII, §8. Let $\bar{K}$ denote the $\sigma(\mathscr{D}', \mathscr{D})$ closure of K in $\mathscr{D}'$. For the case in which $\mathscr{D}$ is the set of real elements of a C^*-algebra the mathematical situation for $\mathscr{D}, \bar{K} \subset \mathscr{D}$ has been thoroughly researched (see, for example, [30]). Note that the so-called "set of states" in the theory of a C^*-algebra is denoted here by $\bar{K}$ instead of K; for the problem of $\bar{K}$ see VI.

In order to prevent misunderstandings about the physical meaning of axiom AQ we make the following comments: The countable subsets $\mathscr{K}$ of

$K \subset \bar{K}$ and $\mathscr{L}$ of L are, in certain topologies, dense in K or L, respectively. Here $\bar{K}$, K, and L are mathematical extensions or "completions" analogous to the completion of the set of rational numbers by the real numbers. We now pose the "converse" question: Are the "physical sets" $\mathscr{K}$ and $\mathscr{L}$ "special" subsets of K, $\bar{K}$, and L? If so—how? An important tool in the resolution of this question is the description of the physical distinguishability of ensembles and effects which we have described in §3. This fact suggests that, in addition to the sets $\mathscr{K}$ and $\mathscr{L}$, the topologies generated on $\mathscr{K}$ by $\mathscr{L}$ and on $\mathscr{L}$ by $\mathscr{K}$ by means of equations of the form (3.1) will also be important. These topologies are identical to $\sigma(\mathscr{D}', \mathscr{D})$ on K and $\bar{K}$ and to $\sigma(\mathscr{B}', \mathscr{B})$ on L. Note that the assumption that the ensembles may be distinguishable by the topology $\sigma(\mathscr{B}, \mathscr{B}')$ is false because the topology $\sigma(\mathscr{B}, \mathscr{B}')$ is actually "finer" on K than is $\sigma(\mathscr{D}', \mathscr{D})$.

Since two elements of K may be physically distinguishable only in the sense of the topology $\sigma(\mathscr{D}', \mathscr{D})$, no subset of K, that is, no special $\mathscr{K}$ has special physical significance. However, K itself is of special significance as a subset of $\bar{K}$ *only* because the topology generated by K on L (which is identical to $\sigma(\mathscr{B}', \mathscr{B})$) correctly describes the physical distinguishability of effects. We note, however, that every subset $\underline{K}$ of K which is dense (with respect to the norm) in K generates the same topologies on L as does $\mathscr{K}$, K, and $\mathscr{B}$. In summary, we find that $\mathscr{K}$ (as a subset of K) and $\mathscr{L}$ (as a subset of L) are not uniquely determined by the physics; every subset $\underline{K}$ of K which is norm-dense in K (including K itself) and every subset $\underline{L}$ of $L \cap \mathscr{D}$ which is norm-dense in $L \cap \mathscr{D}$ (also $L \cap \mathscr{D}$ itself) may be used as the set of "physical" ensembles or effects in the sense that the topologies generated by $\underline{K}$ on $\underline{L}$ and by $\underline{L}$ on $\underline{K}$ are correct for "physical distinguishability" (see also the general discussion in [1], §10.5).

We may, therefore, choose sets $\underline{K}$ and $\underline{L}$ of the above type on the basis of our mathematical convenience without "changing the physics."

In this book $K \subset \mathscr{B}(\mathscr{H}_1, \ldots)$ and $L \subset \mathscr{B}(\mathscr{H}_1, \ldots)$ will play a central role. (We shall discuss $\mathscr{D}$, $\mathscr{D}'$ only briefly in VI and VIII) because the mathematics for Hilbert spaces is extensively developed, and that the methods developed for the dual spaces $\mathscr{D}$, $\mathscr{D}'$ are, at present, too cumbersome for practical computation. The criterion for mathematical accessibility may change in time as new mathematical methods are developed (recall the development of the Hamilton–Jacobi formulation of mechanics.)

If we introduce the axiom AQ as a substitute for a set of axioms such as that given in §3, we find it desirable to substitute the following definition for D 3.2:

D 5.1. If axiom AQ is satisfied, then we shall call M (together with the structures $\mathscr{Q}$, $\mathscr{R}_0$, and $\mathscr{R}$) a set of *microsystems*.

On the basis of the above formulation of the foundations of quantum mechanics it is clear that the Hilbert space (as a complex vector space) does not directly describe a physical structure. Instead it is a computational tool which permits us to cleverly handle the structure of the convex set K. Since the positive affine functionals on K are identical to the elements of the positive cone of $\mathscr{B}'$ (see AIII, §6), it is the structure of K alone (and the choice of topology $\sigma(\mathscr{D}', \mathscr{D})$ on K) which determine the physical structure of microsystems. The structure of K also contains the so-called "wave character" of microsystems; the Schrödinger wave function—the elements of a Hilbert space—are only mathematical tools for obtaining a better "handle" on this "wave character." Only the elements of K are "physically real" (in the sense of [1], §10 and [2], III, §9). The elements of the Hilbert spaces $\mathscr{H}_i$ are not. The elements of $\mathscr{H}_i$ form only a particular representation basis for the elements of K; a representation analogous to the special coordinates used for the representation of the orbit of mass-points in mechanics.

The role of Hilbert space as a representation tool for physically important quantities will be developed in IX.

6 Decision Effects and Faces of K

In II, Th. 4.5.2(iv) and in III, Th. 2.1 we have examined the notion of a decomposition of preparation procedures and defined the notion of a mixture of ensembles by (2.2). Th. 2.2 states that to each "mixture" of ensembles w_i with weights λ_i there is an analogous "mixture" a of preparation procedures a_i.

The notions of ensemble, mixture, and decomposition may be expressed in simple terms if we use AQ. If we identify the elements $w \in \mathscr{K}$ with the elements $w \in K \subset \mathscr{B}$ it follows that (3.2) is equivalent to

$$w = \sum_{i=1}^{n} \lambda_i w_i, \tag{6.1}$$

where (6.1) is written as an equation among elements of the Banach space $\mathscr{B}$. Equation (6.1) states that w is a convex combination of the w_i. Since the set K is convex, every convex combination of elements $w_i \in K$ is also an element of K.

Let $0 < \lambda < 1$ and let $w_1, w_2 \in K$. Then $w = \lambda w_1 + (1 - \lambda)w_2$ is a "mixture" of w_1, w_2 (in ratio λ to $1 - \lambda$)). If $w \in K$ can be written in the form $w = \lambda w_1 + (1 - \lambda)w_2$ where $w_1, w_2 \in K$, $0 < \lambda < 1$ then the decomposition $w = \lambda w_1 + (1 - \lambda)w_2$ represents w as a "mixture" of the ensembles w_1, w_2.

The norm-closed subsets of K which are invariant under mixtures and decompositions play an important role. A norm-closed subset C of K is invariant under mixtures and decomposition if for $w_1, w_2 \in C$ we obtain $\lambda w_1 + (1 - \lambda)w_2 \in C$ for all λ which satisfy $0 \leq \lambda \leq 1$ and if $w \in C$, $w = \lambda w_1 + (1 - \lambda)w_2$ (where $0 < \lambda < 1$ and $w_1, w_2 \in K$) it follows that $w_1, w_2 \in C$.

The norm-closed subsets of K which are closed under mixtures and decomposition are known as norm-closed "faces" in the mathematical theory of convex sets. The "faces" of K obtain their physical meaning from the concepts of mixture and decomposition. We only consider the norm-closed faces because (from the discussion presented in §5) a subset C of K is not physically distinguishable from its norm-closure.

Of special importance are those faces of K which consist only of a single point (and are therefore norm-closed). Such a face is called an extreme point of K and is characterized as follows: An element w is an extreme point of K if and only if when $w = \lambda w_1 + (1 - \lambda)w_2$, $w_1, w_2 \in K$, and $0 < \lambda < 1$ it follows that $w = w_1 = w_2$. The "physical meaning" of an extreme point of K is that it represents an "irreducible ensemble"—often called a "pure state." We shall not use this expression in this book because we do not wish to create false associations with the notion of a pure state—for example, the false notion that all microsystems $x \in a$ for irreducible $\varphi(a)$ are "identical" while only those $x \in a$ for which $\varphi(a)$ is reducible can be nonidentical. Ontological concepts such as "identical" and "nonidentical" can easily lead to contradictions with experiment.

D 6.1. The set of extreme points of a convex set K will be frequently denoted by $\partial_e K$. The set of norm-closed faces of K will be denoted by $\phi(K)$.

What are the elements of $\partial_e(K)$ and of $\phi(K)$?

Th. 6.1. *Every closed face C of K can be written as $C(w)$ for a suitably chosen $w \in K$ where $C(w)$ is the smallest closed face which contains w.*

PROOF. Since $\mathscr{B}$ is separable, K and C are also separable (see AIV, §15). Thus, in C there is a norm-dense denumerable subset $w_i \in C$. Thus there are real numbers λ_i, $0 < \lambda_i \leq 1$, $\sum_i \lambda_i = 1$ for which $w = \sum_i \lambda_i w_i \in C$ since C is norm-closed and convex. We shall now show that $C = C(w)$. From $w \in C$ it follows that $C(w) \subset C$. From $w = \lambda_k w_k + (1 - \lambda_k)w'_k$ where

$$w'_k = (1 - \lambda_k)^{-1} \sum_{i \neq k} \lambda_i w_i \in K,$$

it follows that $w_k \in C(w)$. Since all w_k belong to $C(w)$ and are dense in C, and $C(w)$ is norm-closed, it follows that $C \subset C(w)$.

D 6.2. The set of elements $e \in L$ for which e is a projection operator in the algebra $\mathscr{A}(\mathscr{H}_1, \ldots)$ (that is, $e^2 = e$ and $e^+ = e$) (see AIV, §15) will be denoted by G.

In the following we shall make use of the concise notion which is presented in AIV, §15: $\mathscr{H} = \bigcup_\gamma \mathscr{H}_\gamma$ is the "sum" of the sets $\mathscr{H}_1, \mathscr{H}_2, \ldots$ etc. Every element of $\mathscr{H}$ is characterized by an index γ and a $\varphi \in \mathscr{H}_\gamma$. $\mathscr{H}$ is *not* a vector space. The elements of $\mathscr{A}(\mathscr{H}_1, \mathscr{H}_2, \ldots)$ can be considered to be operators in $\mathscr{H}$. As "subspaces" of $\mathscr{H}$ we shall denote only those subsets $\mathscr{T}$ of $\mathscr{H}$ for which $\mathscr{T} = \bigcup_\gamma \mathscr{T}_\gamma$ where $\mathscr{T}_\gamma$ are closed linear subspaces of $\mathscr{H}_\gamma$. $\mathscr{T}^1_\gamma \perp \mathscr{T}^2_\gamma$ means that for $\mathscr{T}^1 = \bigcup_\gamma \mathscr{T}^1_\gamma$ and $\mathscr{T}^2_\gamma = \bigcup_\gamma \mathscr{T}^2_\gamma$ the relations $\mathscr{T}^1_\gamma \perp \mathscr{T}^2_\gamma$ hold for all γ. $\mathscr{T}^\perp$ is the subspace $\bigcup_\gamma \mathscr{T}^\perp_\gamma$.

Th. 6.2. *To each face C of K there is a uniquely determined $e \in G$ for which $w \in C$ is equivalent to $w = ewe$. The map so defined of all faces of $\phi(K)$ in G is an order isomorphism of $\phi(K)$ onto G.*

PROOF. According to Th. 6.1, $C = C(\tilde{w})$ where $\tilde{w} = \sum_i \lambda_i w_i$ and $\{w_i\}$ is dense in C. Let $\mathcal{T}$ denote the support of $\tilde{w}$, that is, $\mathcal{T}$ is the space which is orthogonal to the eigenspace of $\tilde{w}$ which has eigenvalue 0 (see AIV, §11). Let e be the projection operator for $\mathcal{T}$. Thus we find that $\tilde{w} = e\tilde{w}e$. The condition $\tilde{w} = e\tilde{w}e$ is equivalent to the condition $(\mathbf{1} - e)\tilde{w}(\mathbf{1} - e) = 0$ (see AIV, §6). Using the decomposition which is found in the proof of Th. 6.1

$$\tilde{w} = \lambda_k w_k + (1 - \lambda_k) w'_k$$

it follows that

$$0 = (\mathbf{1} - e)\tilde{w}(\mathbf{1} - e) = \lambda_k(\mathbf{1} - e)w_k(\mathbf{1} - e) + (1 - \lambda_k)(\mathbf{1} - e)w'_k(\mathbf{1} - e).$$

Since the operators $(\mathbf{1} - e)w_k(\mathbf{1} - e)$ and $(\mathbf{1} - e)w'_k(\mathbf{1} - e)$ are positive operators in $\mathcal{A}(\mathcal{H}, \ldots)$ it follows that $(\mathbf{1} - e)w_k(\mathbf{1} - e) = 0$ and therefore $w_k = ew_ke$. Since $\{w_k\}$ is dense in C, it follows that $w = ewe$ for all $w \in C$.

Let $w \in K$ and $w = ewe$. The eigenvectors (which correspond to nonzero eigenvalues) of w lie in the projection space $\mathcal{T}$ of e. Since C is convex, we obtain $w \in C$ because, if for all elements $\varphi \in \mathcal{T}$ the corresponding projections P_φ lie in C (for P_φ see AIV, §6). By analogy with the proof in Th. 6.1 it can be proven that for every eigenelement φ_α having a nonzero eigenvalue, the corresponding P_{φ_α} lies in C. The $\{\varphi_\alpha\}$ span all of $\mathcal{T}$. From

$$\frac{1}{n}\sum_{i=1}^{n} P_{\varphi_{\alpha_i}} = \frac{1}{n}\sum_{i=1}^{n} P_{\chi_i},$$

where the $\{\chi_i\}$ is an orthogonal system which spans the same subspace as the $\{\varphi_{\alpha_i}\}$ it follows that, P_φ lies in C for each φ from a finite-dimensional subspace $\mathcal{T}$ spanned by elements of $\{\varphi_\alpha\}$. Since each φ from $\mathcal{T}$ may be approximated by a φ' from such a finite-dimensional subspace of $\mathcal{T}$ (and, therefore, P_φ is approximated in the norm-topology by $P_{\varphi'}$ (see AIV, §11) it follows that all P_φ for which $\varphi \in \mathcal{T}$ lie in C.

Thus we have shown that C consists of all w for which $w = ewe$. In this way the element e which corresponds to C is uniquely determined because the set of those w for which $w = ewe$ contain all P_φ for which φ are contained in the projection space for e and the projection spaces for different e are different.

It is easy to show that the converse holds, that, for each $e \in G$ the set of all w for which $w = ewe$ is a norm-closed face of K.

It is easily shown that $\tilde{e} > e$ is equivalent to $\tilde{C} \supset C$. Therefore the correspondence between the elements $e \in G$ and C is an order isomorphism of $\phi(K)$ onto G.

$\phi(K)$ is a complete lattice, since $K \in \phi(K)$ and (as is easily shown) that for every set C of closed faces, $\bigcap_\alpha C_\alpha$ is closed. By the order isomorphism of $\phi(K)$ and G it follows that G is a complete lattice (which also follows directly from AIV, §6 because the set of subspaces of $\mathcal{H}$ is a complete lattice (see AIV, §2)).

Th. 6.3. *The condition that the realtion $w = ewe$ holds for $w \in K$ and $e \in G$ is equivalent to the condition $\mu(w, e) = \mathrm{tr}(we) = 1$, that is, $w \in K_1(e)$ where $K_1(e)$ is defined by* D 3.1.

PROOF. From $w = ewe$ it follows that

$$\mu(w, e) = \mathrm{tr}(we) = \mathrm{tr}(we^2) = \mathrm{tr}(ewe) = \mathrm{tr}(w) = 1.$$

From $\mu(w, e) = 1$ it follows that

$$\mu(w, \mathbf{1} - e) = \mathrm{tr}(w(\mathbf{1} - e)) = 0$$

and

$$\mathrm{tr}(w(\mathbf{1} - e)) = \mathrm{tr}((\mathbf{1} - e)w(\mathbf{1} - e)) = 0.$$

Since $(\mathbf{1} - e)w(\mathbf{1} - e) \geq 0$ it follows that $(\mathbf{1} - e)w(\mathbf{1} - e) = 0$ and $w = ewe$ (see AIV, §6).

Using the notation of D 4.1, from Th. 6.2 and Th. 6.3 we find that every $C \in \phi(K)$ is of the form $C = K_1(e)$ for some $e \in G$. Thus we obtain the first part of the following theorem:

Th. 6.4. *An order isomorphism $\phi(K) \leftrightarrow G$ is defined by $C = K_1(e)$, $C \in \phi(K)$ and $e \in G$. An anti-order isomorphism $\phi(K) \leftrightarrow G$ is defined by $C = K_0(e)$ with $C \in \phi(K)$. In particular,*

$$K_1(e_1) \cap K_1(e_2) = K_1(e_1) \wedge K_1(e_2) = K_1(e_1 \wedge e_2),$$

$$K_1(e_1) \vee K_1(e_2) = K_1(e_1 \vee e_2),$$

$$K_0(e_1) \cap K_0(e_2) = K_0(e_1) \wedge K_0(e_2) = K_0(e_1 \vee e_2), \quad \textit{and}$$

$$K_0(e_1) \vee K_0(e_2) = K_0(e_1 \wedge e_2).$$

PROOF. We need only prove that $e \longrightarrow K_0(e)$ is an anti-isomorphism map of G onto $\phi(K)$. This follows directly from the fact that $e \longrightarrow \mathbf{1} - e$ is an anti-isomorphism map of G onto itself and $K_0(e) = K_1(\mathbf{1} - e)$.

Th. 6.5. *The set $\partial_e K$ is equal to the set of all P_φ.*

PROOF. According to AIV, §11, for each $w \in K$ we have the spectral representation

$$w = \sum_\alpha \lambda_\alpha P_{\varphi_\alpha}. \tag{6.2}$$

From $\mu(w, \mathbf{1}) = 1$ it follows that $\sum_\alpha \lambda_\alpha = 1$. Clearly e can be an extreme point of K only if only one $\lambda_\alpha \neq 0$ in (6.2), that is, w is of the form P_φ for some φ.

P_φ is also an extreme point of K; from $P_\varphi = \lambda w_1 + (1 - \lambda)w_2$, $0 < \lambda < 1$ it follows that, for $e = \mathbf{1} - P_\varphi$

$$0 = \lambda e w_1 e + (1 - \lambda) e w_2 e.$$

Since $ew_{1,2}e \geq 0$ it follows that

$$ew_1 e = 0 \quad \text{and} \quad ew_2 e = 0.$$

From $ew_1e = 0$ it follows that $w_1e = 0$ (see AIV, §6); thus, from $e = \mathbf{1} - P_\varphi$ we obtain $w_1 = \eta P_\varphi$. Since $\mathrm{tr}(w_1) = 1$ it follows that $\eta = 1$. Similarly, it follows that $w_2 = P_\varphi$.

In addition to the faces of K (which are special sets of ensembles), the elements of G (effects) also have a special physical meaning. The physical meaning of the elements of L is established by the map ψ of $\mathscr{F}$ into L (as presented in §1, §3, and §5). In many formulations of quantum mechanics there is an implicit assumption that every experiment $(b_0, b) \in \mathscr{F}$ corresponds to a $\psi(b_0, b) \in G$. In these formulations (b_0, b) is an "observable" having measurement values 0, 1 and must be identified with a projection operator e because only projection operators have eigenvalues 0, 1 (see, for example, [2], XI, §2.1).

Actually, the statement that all $\psi(b_0, b)$ are elements of G is incorrect. On the contrary, an experimental physicist will not know in advance whether $\psi(b_0, b)$ at least approximately corresponds to an $e \in G$ or whether he has only registered an effect $g = \psi(b_0, b) \in L$ (see XVII and [2], XII). We shall now seek to obtain a physically meaningful definition of the notion of a decision effect, that is, one which relates to the probability function $\mu(w, g)$. For this purpose we shall use D 3.1.

For $A \subset K$, $L_0(A)$ is a set of effects for which there will be no detection response for any $w \in A$. $K_0L_0(A)$ is a closed face of K with $A \subset K_0L_0(A)$. For $C = K_0L_0(A)$ we find that $L_0(A) = L_0(C)$. An element $g_1 \in L$ is said to be more "sensitive" than $g_2 \in L$ if $g_1 > g_2$, that is, if $\mu(w, g_1) \geq \mu(w, g_2)$ for all $w \in K$. Then the effect g_1 will respond more frequently than the effect g_2. Is there a most sensitive effect in the set $L_0(A) = L_0(C)$? If such an effect exists, and if it cannot respond to the $w \in A$, it will respond to each $w \in K$ "as frequently as possible."

We shall now use a few examples to clarify this situation. A filter for light may be considered to be a b_0; here b is the set of light-quanta which are absorbed by the filter b_0. If $g = \psi(b_0, b) \in L_0(A)$ then $w \in A$ represents the "light" which passes through the filter without absorption. Consider, for example, the absorption coefficient of the filter b_0 as a function of frequency (see Figure 2). All the light which contains only those frequencies for which the absorption coefficient is zero passes through b_0 without absorption. By "combining" several filters of type b_0 it is possible to construct filters for which the absorption coefficient is approximately 0 or 1 and passes all light which passes through b_0 without absorption. Such a filter represents a type of "maximally sensitive" filter which satisfies the additional condition that all light $w \in A$ passes through it without absorption.

The following definition is motivated by the above example:

D 6.3. If $L_0(A)$ has a maximal element e, then e is called a *decision effect*. We shall denote the largest element of $L_0(A)$ by $eL_0(A)$.

Th. 6.6. *Every $L_0(A)$ has a maximal element $L_0(A)$; the set of decision effects is equal to G (see D 6.2). In addition $G = \partial_e L$; $L_0(A) = L_0K_0(eL_0(A))$.*

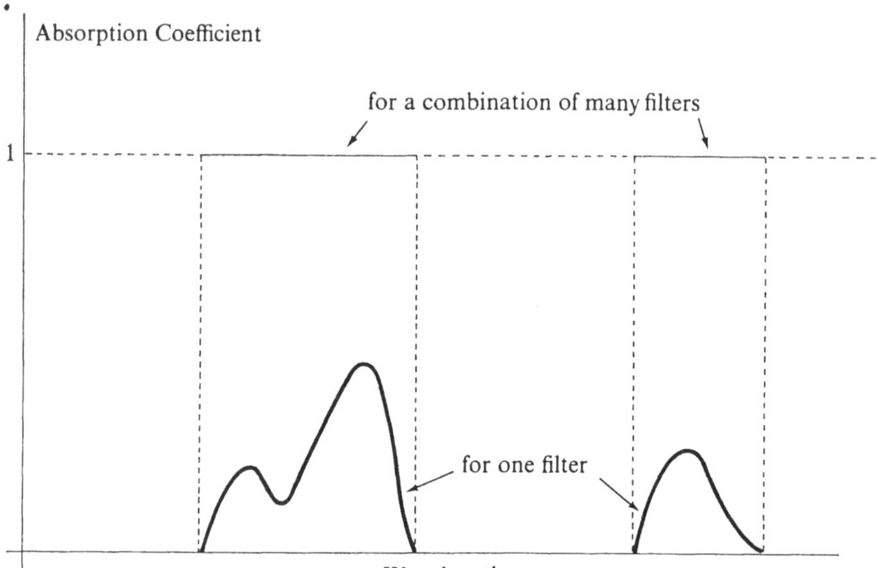

Figure 2

PROOF. As we have seen earlier $L_0(A) = L_0(C)$ where C is a closed face of K. According to Th. 6.4 $C = K_1(\tilde{e})$ where $\tilde{e} \in G$. For $e = \mathbf{1} - \tilde{e}$ we find that $C = K_0(e)$ and we therefore obtain $L_0(A) = L_0 K_0(e)$. Thus we find that $e \in L_0(A)$.

From $g \in L_0(A)$ it follows that $\operatorname{tr}(wg) = 0$ for all $w \in K_0(e)$; in particular, $\operatorname{tr}(P_\varphi g) = 0$ for all $P_\varphi \in K_0(e) = K_1(\tilde{e})$. Since $\tilde{e} = \mathbf{1} - e$, from $P_\varphi \in K_1(\tilde{e})$ it follows that $(\mathbf{1} - e)\varphi = \varphi$. Thus we obtain $0 = \operatorname{tr}(P_\varphi g) = \langle \varphi, g\varphi \rangle$ for all φ for which $(\mathbf{1} - e)\varphi = \varphi$. Since $g \geq 0$ we obtain $g\varphi = 0$ for all φ in the projection space of $\mathbf{1} - e$. From this we conclude that $g(\mathbf{1} - e) = 0$, that is, $g = ge$. This result, together with the adjoint equation $g = eg$ yields $g = ege$. From $g \leq \mathbf{1}$ it follows that $ege \leq e$ and also $g \leq e$. Therefore e is the maximal element of $L_0(A) = L_0 K_0(e)$. Each $e \in G$ is obtained as a maximal element of $L_0(A)$, namely from $L_0(A)$ where $A = K_0(e)$.

Suppose that $e \in G, g_1, g_2 \in L, 0 < \lambda < 1$ and

$$e = \lambda g_1 + (1 - \lambda) g_2.$$

Then it follows that $g_1, g_2 \in L_0 K_0(e)$ and consequently $g_1 \leq e$ and $g_2 \leq e$. If $g_1 \neq e$ (or $g_2 \neq e$) then $\lambda g_1 + (1 - \lambda) g_2 \lneqq e$; therefore $g_1 = g_2 = e$. Thus $G \subset \partial_e L$. For $g \in L$, from $\mathbf{1} - g \leq \mathbf{1}$ it follows that $\mathbf{0} \leq (\mathbf{1} - g)^2 = \mathbf{1} - 2g + g^2 \leq \mathbf{1} - g \leq \mathbf{1}$. We therefore obtain $\mathbf{0} \leq 2g - g^2$ and $\mathbf{0} \leq g^2 \leq \mathbf{1}$. For $g' = 2g - g^2$ and $g'' = g^2$ we find that $g', g'' \in L$ and that $g' = \frac{1}{2}g' + \frac{1}{2}g''$. If $g \in \partial_e L$ then $g' = g'' = g$, that is, $g = g^2$ and $g \in G$. Thus $\partial_e L = G$.

D 6.4. For $e \in G$ we shall let $e^\perp$ denote the element $\mathbf{1} - e$.

Th. 6.7. $e^\perp = \mathbf{1} - e$ *is the projection operator onto the subspace of* $\mathscr{H}$ *which is orthogonal to* $e\mathscr{H}$. *The correspondence* $e \to e^\perp$ *is an orthocomplementation in* *the lattice G. The isomorphism $e \to K_1(e)$ of* Th. 6.4 *defines an analogous orthocomplementation on $\phi(K)$ for which $K_1(e)^\perp = K_1(e^\perp) = K_0(e)$.*

PROOF. The first part of this theorem is a direct consequence of AIV, §6. In order to show that $e \to e^\perp$ is an orthocomplementation it is necessary to show that (see AI, D 1.2):

(i) From $e_1 \leq e_2$ it follows that $e_1^\perp \geq e_2^\perp$.
(ii) $(e^\perp)^\perp = e$.
(iii) $e^\perp \wedge e = \mathbf{0}$.

(i) and (ii) follow directly from $e^\perp = \mathbf{1} - e$. (iii) follows directly from the fact that $e^\perp \wedge e$ is the projection operator on the intersection of the projection spaces of $e^\perp$ and e.

D 6.5. We shall denote the relation $e_1 \leq \mathbf{1} - e_2 = e_2^\perp$ by $e_1 \perp e_2$.

Since $e_1 \leq \mathbf{1} - e_2$ is equivalent to $e_2 \leq \mathbf{1} - e_1$ the relation $e_1 \perp e_2$ is symmetric. If $e_1 \leq e$ and $e_2 \leq e^\perp$ then $e_2 \leq e^\perp \leq e_1^\perp$ and $e_1 \perp e_2$.

Th. 6.8. *If $e_1 \perp e_2$ then $e_1 e_2 = \mathbf{0}$ and $e_1 + e_2 = e_1 \vee e_2$.*

PROOF. Since $e_1 \perp e_2$ and $e_1 e_2 = \mathbf{0}$ are equivalent (see AIV, §6) it easily follows that $(e_1 + e_2)^2 = e_1 + e_2$, that is, $e_1 + e_2 \in G$. It is easy to see that $K_0(e_1 + e_2) = K_0(e_1) \cap K_0(e_2)$; from Th. 6.4 we obtain $K_0(e_1) \cap K_0(e_2) = K_0(e_1 \vee e_2)$. From $K_0(e_1 + e_2) = K_0(e_1 \vee e_2)$ and from Th. 6.4 it follows that $e_1 + e_2 = e_1 \vee e_2$.

D 6.6. An orthocomplemented lattice $\mathscr{V}$ is said to be orthomodular if, for $a, b, c \in \mathscr{V}$, $c \perp b$ and $a \leq c$, the following relation holds:

$$(a \vee b) \wedge c = (a \wedge c) \vee (b \wedge c) = a$$

(see AI, D 2.3).

Th. 6.9. *G is orthomodular.*

PROOF. From $a \wedge c \leq a$, $b \wedge c \leq b$ it follows that $(a \wedge c) \vee (b \wedge c) \leq a \vee b$. From $a \wedge c \leq c$, $b \wedge c \leq c$ it follows that $(a \wedge c) \vee (b \wedge c) \leq c$; thus we obtain

$$(a \wedge c) \vee (b \wedge c) \leq (a \vee b) \wedge c.$$

For G the orthomodularity is equivalent to that found in the case of the subspaces $\imath, \jmath, \ell$ of a Hilbert space. Thus, for $\jmath \perp \ell$ and $\imath \subset \ell$ it is necessary to show that the relation $(\imath \wedge \jmath) \cap \ell \subset \imath$ holds. Since $\imath \perp \jmath$ the vectors in $\imath \vee \jmath$ have the form $\varphi = \varphi_1 + \varphi_2$, where $\varphi_1 \in \imath$ and $\varphi_2 \in \jmath$. Therefore we find that $\varphi_1 \perp \varphi_2$.

Since $\ell \perp \jmath$, for the projection P_ℓ on the subspace ℓ we find that $P_\ell \varphi_2 = 0$. In addition $P_\ell \varphi = P_\ell \varphi_1 = \varphi_1$ (the last equality follows from $\imath \subset \ell$). For each $\varphi \in (\imath \vee \jmath) \cap \ell$ and $\varphi \in \ell$—that is, $P_\ell \varphi = \varphi$ and also $\varphi = \varphi_1$—we obtain $\varphi \in \imath$.

Another proof of Th. 6.9 follows directly from AI, Th. 2.2. (See remarks at the end of AI, §2.)

Th. 6.10. *Decreasing and increasing sequences of elements of G converge in the $\sigma(\mathscr{B}', \mathscr{B})$ topology towards elements of G; an increasing sequence e_ν converges towards $\bigvee_\nu e_\nu$ and a decreasing sequence e_ν converges towards $\bigwedge_\nu e_\nu$.*

PROOF. The convergence of increasing and decreasing sequences in L follows in general from AIII, §6. Let e_ν be a decreasing sequence. From $e_\nu \to g \in L$ it follows that $e_\nu \geq \bigwedge_\nu e_\nu$ and therefore $g \geq \bigwedge_\mu e_\mu$. From $e_\nu \geq e_\mu$ for all $\mu \geq \nu$, in the limit we obtain $e_\nu \geq g$ for all ν. We obtain $K_0(g) \supset K_0(e_\nu)$ for all ν and, from Th. 6.4 we obtain $K_0(g) \supset \bigvee_\nu K_0(e_\nu) = K_0(\bigwedge_\nu e_\nu)$, that is, $g \in L_0K_0(\bigwedge_\nu e_\nu)$. Therefore, from Th. 6.6 we obtain $g \leq \bigwedge_\nu e_\nu$. Thus we find that $g = \bigwedge_\nu e_\nu$.

If the sequence e_ν is increasing, then the sequence $e_\nu^\perp = \mathbf{1} - e_\nu$ is decreasing and $\mathbf{1} - e_\nu \to \bigwedge_\nu e_\nu^\perp$. Therefore we obtain $e_\nu \to \mathbf{1} - \bigwedge_\nu e_\nu^\perp = (\bigwedge_\nu e_\nu^\perp)^\perp = \bigvee_\nu e_\nu$.

CHAPTER IV

Coexistent Effects and Coexistent Decompositions

The structure of the theory of microsystems presented in this book permits us to make a fundamental characterization of the concept of microsystems without making use of the familiar basic concepts of property and observable. The notions of property and pseudoproperty which were introduced in III, §4 are not fundamental concepts but are *derived* concepts—derived from the more fundamental concepts of preparation and registration procedure. In III, §4 we have outlined one possible way in which it is possible to begin to separate the notion of a microsystem from the inherent structure associated with preparation and registration procedures. We shall now consider a second way to accomplish this separation. These two methods will be compared in §8.

We shall now make the difference between the deductive approach presented here and the approaches which are based on the notions of "property" and "observables" more clearly evident, in order that we may be able to more clearly formulate the problems which will be discussed in this chapter.

Many formulations of quantum mechanics begin by making certain specific assumptions about the properties of microsystems: It is assumed that the properties of microsystems may be ascertained by measurement, and may (at least, "after" the measurement) be attributed to the microsystems. Similarly, an observable is considered to be a "measurable quantity" which (again, on the basis of measurement) may be attributed to the microsystems. For instance, it is assumed that, in the measurement of position, we are able to "detect" the position of the microsystem, and that the latter is one of its properties. In other words, it is assumed that the proposition "the microsystem has the measured position" is already meaningful.

We have not used the concepts described above in our formulation of quantum mechanics in II and III because we were not convinced that their meaning can be readily determined. For instance, we do not assume that a proposition such as "the microsystem has the measured position" is already meaningful. Instead, we have only made use of the fundamental concepts—those of preparation (represented by $\mathscr{Q}$) and registration (represented by $\mathscr{R}_0\mathscr{R}$)—concepts which have immediate meaning because they can be explained in terms of "pre-theories" relative to quantum mechanics (for the meaning of the notion of a pre-theory see [1], §5 and [2], §4). It is possible to deduce the concepts of properties and pseudoproperties of microsystems only *after* the development of the theory (see §8).

The microsystems make their presence known by the "response" $b \in \mathscr{R}$ of the "apparatus" $b_0 \in \mathscr{R}_0$. Except for a few general comments made in III, §4, what these responses permit us to conclude about the microsystems themselves remains open. The investigations in III, §4 do not represent a *systematic* path from the preparation and registration methods to the actual structure of microsystems. Instead, they only raise the question whether it is possible to correctly formulate the intuitive notion of a "property" in terms of the structures of preparation and registration procedures. It was not possible to systematically answer the question as to how much the "response" $b \in \mathscr{R}$ of the apparatus $b_0 \in \mathscr{R}_0$ is due to the microsystem $x \in b$ and how much is due to the apparatus b_0. We shall now turn to these problems and questions without (!) prejudging the issue by making arbitrary "assumptions concerning the nature of microsystems." For a realistic clarification it is necessary to obtain all concepts directly from the experiments with microsystems which are described by the structures $\mathscr{Q}$, $\mathscr{R}_0\mathscr{R}$ (see the more general treatment in [1], §10). The physical meaning of quantum mechanics may be obtained then only with the use of elements of the sets $\mathscr{Q}$, $\mathscr{R}_0\mathscr{R}$. We shall not introduce a syntax for assertions about the microsystems themselves; we shall only consider mathematically formulated relations in the form described in §8.3.

1 Coexistent Effects and Observables

The concept of an observable is a very useful concept in quantum mechanics. It will be *derived* from the concepts described by the sets $\mathscr{R}_0\mathscr{R}$. In order to do this, we shall now examine the physical significance of the substructures $\mathscr{R}(b_0)$ of the structures $\mathscr{R}_0\mathscr{R}$.

1.1 Coexistent Registrations

Earlier we have introduced the expression "effect process" to denote a pair (b_0, b) where $b_0 \in \mathscr{R}_0$, $b \in \mathscr{R}$ and $b \subset b_0$. To such a pair the map ψ assigns an effect $\psi(b_0, b) \in L \subset \mathscr{B}'(\mathscr{H}_1, \ldots)$. In III we have only been concerned with the mapping of a single effect process $f = (b_0, b)$ onto an element $g = \psi(b_0, b) \in L$. Here it is important to note that there are, in general, many such $b \subset b_0$ for an apparatus b_0. Let $\mathscr{R}(b_0)$ denote the set of $b \in \mathscr{R}$ for which $b \subset b_0$. In experiments with microsystems we find that, in most cases, the number of registrations for which $b \subset b_0$ is overwhelming. An approximately exhaustive description of those cases cannot be described in a book, no less in a few lines. For the purpose of illustration we shall only consider two typical experiments.

In the first example we shall consider an array of counters in which each microsystem may activate the response of some of them. Let b_0 denote the array of counters. Let us characterize b as follows: three specifically chosen counters will respond; four other specifically chosen counters will not. In this case $\mathscr{R}(b_0)$ will be familiar to those readers who are familiar with modern electronic technology—$\mathscr{R}(b_0)$ is the Boolean switching algebra of the set of possible responses of the counters. For the case in which b_0 consists of a pair of counters we may describe $\mathscr{R}(b_0)$ as follows: b_1 is the registration for which the first counter responds; b_2 is the registration for which the second counter responds. Here $b_1 \cap b_2$ corresponds to the case in which counters 1 and 2 have both responded; $b_0 \backslash b_1$ corresponds to the case in which counter 1 has not responded; similarly for $b_0 \backslash b_2$. Here $(b_0 \backslash b_1) \cap b_2$, $(b_0 \backslash b_2) \cap b_1$, $b_1 \cup b_2$, etc. are described similarly. By including b_0 and $\varnothing$ we find that $\mathscr{R}(b_0)$ has exactly 16 elements.

In the second example we shall consider a cloud chamber (or a bubble chamber) b_0. Here the number of possible registrations is overwhelming. Every condensation droplet (or vapor bubble) may itself be a registration b. the ionization trail (or bubble trail) may also be one. If a magnetic field is applied, then the radius of a circular particle trail will be a "scale value" for certain of these b's. The scale numbers are nothing more than indices which serve to order the overwhelming range of possible registrations $b \subset b_0$. Scale values—even when they are very practical—are of a secondary nature. The primary concept is that of registration possibilities.

These two examples also demonstrate the following fact: The different elements from $\mathscr{R}(b_0)$ need not occur "simultaneously" because $b \subset b_0$ does not necessarily correspond to a "point in time," but instead, may correspond to an extended process, such as a joint response of two counters in an array where one responds later than the other, or, to the entire ionization trail as in a cloud chamber, which persists for a certain length of time. The joint registration of elements $b \in \mathscr{R}(b_0)$ has nothing to do with "simultaneous" registration or "simultaneous" measurement. In quantum theory such expressions as "simultaneously measurable" and "not simultaneously measurable" are often used. These expressions are often misleading and misunderstood. We shall discuss this subject in more detail in VII, XVII, and XVIII. Now we shall state only that the $b \in \mathscr{R}(b_0)$ has nothing to do with "simultaneity." In order to reduce such misunderstanding we shall chose another concept (which we have already introduced in II, D 2.2) for the description of the physical situation represented by $\mathscr{R}(b_0)$.

D 1.1.1. The registration procedures $b \in \mathscr{R}(b_0)$ are said to be *coexistent* with respect to the registration method $b_0 \in \mathscr{R}_0$. The $(b_0, b) \in \mathscr{F}$ which correspond to the same b_0 will be called *coexistent effect processes*.

A subset $A \subset \mathscr{F}$ is a set of coexistent effect processes if and only if all elements in A have the same first component b_0.

How does the structure of coexistent effect processes affected by the mapping ψ of $\mathscr{F}$ into $L \subset \mathscr{B}'(\mathscr{H}_1, \ldots)$?

1.2 Coexistent Effects

If we have a set of coexistent effect processes (b_0, b) then the corresponding set of effects $\psi(b_0, b)$ is a subset of the set of effects $\psi(b_0, b)$ containing all the $b \in \mathscr{R}(b_0)$. For a fixed b_0 the map $\psi(b_0, b)$ defines a map $\psi_0: \mathscr{R}(b_0) \to L$. We shall now investigate the properties of this map.

The following definition is often used:

D 1.2.1. Let F be a map of the Boolean ring Σ (with unit element ε) into the order interval $[0, u]$ of an ordered vector space. Let F satisfy the following conditions:

$$F(\varepsilon) = u,$$

$$F(\sigma_1 \vee \sigma_2) = F(\sigma_1) + F(\sigma_2)$$

for the case in which $\sigma_1 \wedge \sigma_2 = 0$. Then the map F is called an (*additive*) *measure* on Σ.

For the special case in which $[0, u]$ is the interval $[0, 1] \subset \mathbf{R}$ we shall call F (as defined in D 1.2.1) a *real measure* (see II, D 3.1).

Th. 1.2.1. *For fixed b_0 the map $\psi_0(b) = \psi(b_0, b)$ of $\mathscr{R}(b_0)$ into $[0, 1] = L \subset \mathscr{B}'(\mathscr{H}_1, \ldots)$ is an additive measure on $\mathscr{R}(b_0)$.*

PROOF. According to APS 5.1.4, to each $w \in \mathscr{K}$ there exists an $a \in w$ for which $a \cap b_0 \neq \varnothing$. Let $b_1, b_2 \in \mathscr{R}(b_0)$ and suppose that $b_1 \cap b_2 = \varnothing$. We then obtain:

$$\begin{aligned}\lambda_{\mathscr{S}}(a \cap b_0, a \cap (b_1 \cup b_2)) &= \lambda_{\mathscr{S}}(a \cap b_0, (a \cap b_1) \cup (a \cap b_2)) \\ &= \lambda_{\mathscr{S}}(a \cap b_0, a \cap b_1) + \lambda_{\mathscr{S}}(a \cap b_0, a \cap b_2).\end{aligned}$$

We therefore obtain

$$\mu(\varphi(a), \psi(b_0, b_1 \cup b_2)) = \mu(\varphi(a), \psi(b_0, b_1)) + \mu(\varphi(a), \psi(b_0, b_2)).$$

Since, by AQ, $\varphi\mathscr{Q}'$ is dense in K, for all $w \in K$ we obtain

$$\mu(w, \psi(b_0, b_1 \cup b_2)) = \mu(w, \psi(b_0, b_1)) + \mu(w, \psi(b_0, b_2)) \tag{1.2.1}$$

from which we conclude that

$$\psi_0(b_1 \cup b_2) = \psi_0(b_1) + \psi_0(b_2). \tag{1.2.2}$$

Since $\psi(b_0, b_0) = \mathbf{1}$ we obtain $\psi_0(b_0) = \mathbf{1}$.

Equations (1.2.1) and (1.2.2) are equivalent; intuitively they express the additivity of the frequencies for the responses of "b_1 or b_2" for the case in which "b_1 and b_2" cannot both respond. In other words it is a direct consequence of the "switch" from b_1 and b_2 to "b_1 or b_2." In other words these equations are a direct consequence of the intuitively motived equation II (3.5) and the axioms AS 2 for statistical selection procedures. On the basis of previous axioms it follows that there are no restrictive conditions for the map ψ_0 of $\mathscr{R}(b_0)$ in L other than those imposed by equation (1.2.2) and the condition $\psi_0(b_0) = \mathbf{1}$. Thus, it will be possible to impose axiom AOb later. We shall now introduce the following definition which is motivated by Th. 2.1.1.

D 1.2.2. A set $A \subset L$ is called a set of *coexistent* effects if there exists a Boolean ring Σ with additive measure $F\colon \Sigma \to L$ for which $A \subset F\Sigma$.

In order to simplify what follows, we shall introduce the following definition:

D 1.2.3. An additive measure F on Σ is said to be *effective* if $F(\sigma) = 0$ implies that $\sigma = 0$.

We shall now state and prove the following theorem:

Th. 1.2.2. *Let $F\colon \Sigma \to L$ be an additive measure; let $\Sigma_0 = \{\sigma \mid \sigma \in \Sigma$ and $F(\sigma) = 0\}$. Let $\sigma \in \tilde{\sigma}$ where $\tilde{\sigma} \in \Sigma/\Sigma_0$; define $\tilde{F}(\tilde{\sigma}) = F(\sigma)$. Then $\tilde{F}$ is an effective additive measure on the Boolean ring Σ/Σ_0.*

PROOF. First we shall show that Σ_0 is an ideal in Σ: Suppose that $\sigma_1 < \sigma$ and $F(\sigma) = 0$; then from $F(\sigma_1) < F(\sigma)$ it follows that $F(\sigma_1) = 0$. Let $\sigma = \sigma_1 \vee \sigma_2$ and suppose that $F(\sigma_1) = F(\sigma_2) = 0$. Then from $\sigma = \sigma_1 \vee (\sigma_2 \wedge \sigma_1^*)$ and from $\sigma_2 \wedge \sigma_1^* < \sigma_2$ it follows that $F(\sigma) = 0$. Therefore Σ/Σ_0 is a Boolean ring (see AI, end of §3). From $\sigma_1, \sigma_2 \in \tilde{\sigma} \in \Sigma/\Sigma_0$ we obtain $\sigma_1 \wedge \sigma_2^* \in \Sigma_0$ and $\sigma_1^* \wedge \sigma_2 \in \Sigma_0$ from which it follows that

$$\begin{aligned} F(\sigma_1) &= F(\sigma_1 \wedge \sigma_2^*) + F(\sigma_1 \wedge \sigma_2) \\ &= F(\sigma_1^* \wedge \sigma_2) + F(\sigma_1 \wedge \sigma_2) = F(\sigma_2). \end{aligned}$$

Therefore, for $\sigma \in \tilde{\sigma}$ $\tilde{F}(\tilde{\sigma}) = F(\sigma)$ defines a function $\tilde{F}\colon \Sigma/\Sigma_0 \to L$. Let $\tilde{\varepsilon}$ denote the class containing the unit element ε of Σ. Then we obtain $\tilde{F}(\tilde{\varepsilon}) = \mathbf{1}$. If $\tilde{\sigma}_1 \wedge \tilde{\sigma}_2 = 0$, then, for any pair of representatives $\sigma_1 \in \tilde{\sigma}_1$, $\sigma_2 \in \tilde{\sigma}_2$ we find that $\sigma_1 \wedge \sigma_2 \in \Sigma_0$. Hence we find that

$$\begin{aligned} F(\sigma_1 \vee \sigma_2) &= F(\sigma_1) + F(\sigma_1^* \wedge \sigma_2) \\ &= F(\sigma_1) + F(\sigma_1 \wedge \sigma_2) + F(\sigma_1^* \wedge \sigma_2) \\ &= F(\sigma_1) + F(\sigma_2) \end{aligned}$$

and finally

$$\tilde{F}(\tilde{\sigma}_1 \vee \tilde{\sigma}_2) = \tilde{F}(\tilde{\sigma}_1) + \tilde{F}(\tilde{\sigma}_2).$$

From $\tilde{F}(\tilde{\sigma}) = 0$ it follows that $F(\sigma) = 0$ for all $\sigma \in \tilde{\sigma}$, and therefore $\sigma \in \Sigma_0$, that is, $\tilde{\sigma}$ is the null element of Σ/Σ_0.

Therefore, if we wish to investigate coexistent effects Th. 1.2.2 implies that we need only consider those Boolean rings having effective measures. Clearly ψ maps a set of coexistent effect processes into a set of coexistent effects. In addition, it is easy to show that the map $\psi_0: \mathscr{R}(b_0) \to L$ is an effective measure on $\mathscr{R}(b_0)$ because from $\psi_0(b) = 0$ (that is, $\psi(b_0, b) = 0$) it follows that $\lambda_{\mathscr{S}}(a \cap b_0, a \cap b) = 0$ for all $a \in \mathscr{Q}'$ for which $(a, b_0) \in C$ (where C is defined in II (4.3.7)). Thus $a \cap b = \varnothing$ for all $a \in \mathscr{Q}'$ for which $(a, b_0) \in C$. According to APS 5.2 this is possible only if $b = \varnothing$.

Definition D 1.2.2 has the following essential advantage over the special situation $\mathscr{R}_0(b_0) \xrightarrow{\psi_0} L$: We do not have to be concerned with the question whether, given a set A of coexistent effects, there exists a registration method $b_0 \in \mathscr{R}_0$ for which $A \subset \psi_0 \mathscr{R}(b_0)$. In §4 we shall assert that in an "approximate" sense, to each Boolean ring Σ having an effective measure $F: \Sigma \to L$ there is a "realization" described by a $\mathscr{R}(b_0)$; the complication of the "approximate realization" will be eliminated if we consider general effective measures on a Boolean ring.

In AI, §3 we show how to define two operators $\dotplus$ and $\cdot$ for a Boolean ring which satisfy the rules of a commutative algebra. In addition, we show how to define a generalized Boolean ring (without unit element) using $\dotplus$ and $\cdot$. A unit element exists if and only if there exists an ε for which $\varepsilon \cdot a = a$ for all elements a of the Boolean ring.

D 1.2.4. A map F of a generalized Boolean ring Σ into $[\mathbf{0}, \mathbf{1}]$ is called an *additive measure* if, for $\sigma_1, \sigma_2 \in \Sigma$, $\sigma_1 \cdot \sigma_2 = 0$ the following equation is satisfied:

$$F(\sigma_1 \dotplus \sigma_2) = F(\sigma_1) + F(\sigma_2).$$

Note. In the case in which Σ has a unit element ε, we do not require that $F(\varepsilon) = 1$!

Th. 1.2.3. *Let Σ be a generalized Boolean ring and let F be an additive measure on Σ. Then there exists an extension $\tilde{\Sigma}$ of Σ and a (normed) measure $\tilde{F}$ which is a continuation of F on $\tilde{\Sigma}$.*

PROOF. Let $\bar{\Sigma}$ be the two-element Boolean ring which consists of the zero element $\bar{0}$ and the unit element $\bar{\varepsilon}$. Let $\tilde{\Sigma} = \bar{\Sigma} \times \Sigma$ and let $\dotplus$ and $\cdot$ be defined as follows:

$$(\bar{\varepsilon}, \sigma_1) \dotplus (\bar{\varepsilon}, \sigma_2) = (\bar{0}, \sigma_1 \dotplus \sigma_2),$$
$$(\bar{\varepsilon}, \sigma_1) \dotplus (\bar{0}, \sigma_2) = (\bar{\varepsilon}, \sigma_1 \dotplus \sigma_2),$$
$$(\bar{0}, \sigma_1) \dotplus (\bar{0}, \sigma_2) = (\bar{0}, \sigma_1 \dotplus \sigma_2),$$
$$(\bar{\varepsilon}, \sigma_1) \cdot (\bar{\varepsilon}, \sigma_2) = (\bar{\varepsilon}, \sigma_1 \dotplus \sigma_2 \dotplus \sigma_1 \cdot \sigma_2),$$
$$(\bar{\varepsilon}, \sigma_1) \cdot (\bar{0}, \sigma_2) = (\bar{0}, \sigma_2 \dotplus \sigma_1 \cdot \sigma_2),$$
$$(\bar{0}, \sigma_1) \cdot (\bar{0}, \sigma_2) = (\bar{0}, \sigma_1 \cdot \sigma_2).$$

$\tilde{\Sigma}$ is a Boolean ring; Σ may be identified with the subset of all $(\bar{0}, \sigma)$ of $\tilde{\Sigma}$, that is, $\tilde{\Sigma}$ is an extension of Σ. Let

$$\tilde{F}(\bar{0}, \sigma) = F(\sigma), \qquad \tilde{F}(\bar{\varepsilon}, \sigma) = \mathbf{1} - F(\sigma).$$

$\tilde{F}$ is an additive measure on $\tilde{\Sigma}$ which coincides with F on Σ.

From Th. 1.2.3 it follows that a set $A \subset L$ is a set of coexistent effects if there is a generalized Boolean ring Σ and an additive measure F on Σ for which $A \subset F\Sigma$.

We shall now consider the special case in which A consists of a pair of elements g_1, g_2. We then obtain the theorem:

Th. 1.2.4. *The following conditions are equivalent*:

(i) g_1, g_2 *are coexistent.*
(ii) *There exist three elements* $g_1', g_2', g_{12} \in L$ *for which* $g_1 = g_1' + g_{12}$, $g_2 = g_2' + g_{12}$ *and* $g_1' + g_2' + g_{12} = g_1 + g_2' = g_2 + g_1' \in L$.

PROOF. (i) $\Rightarrow$ (ii). Let Σ be a generalized Boolean ring with $g_1, g_2 \in \{F(\sigma) | \sigma \in \Sigma\}$. Here we find that $F(\sigma_1) = g_1$, $F(\sigma_2) = g_2$. Now let us consider the following additional elements in Σ: $\sigma_1 \cdot \sigma_2$, $\sigma_1 \dotplus \sigma_1 \cdot \sigma_2$, $\sigma_2 \dotplus \sigma_1 \cdot \sigma_2$, and $\sigma_1 \vee \sigma_2 = \sigma_1 \dotplus \sigma_2 \dotplus \sigma_1 \cdot \sigma_2$. From the additivity of $F(\sigma)$, and from $F(\sigma_1 \cdot \sigma_2) = g_{12}$, $F(\sigma_1 \dotplus \sigma_1 \cdot \sigma_2) = g_1'$, *and* $F(\sigma_2 \dotplus \sigma_1 \cdot \sigma_2) = g_2'$ it follows that

$$\begin{aligned} g_1 = F(\sigma_1) &= F(\sigma_1 \dotplus \sigma_1 \cdot \sigma_2 \dotplus \sigma_1 \cdot \sigma_2) \\ &= F(\sigma_1 \dotplus \sigma_1 \cdot \sigma_2) + F(\sigma_1 \cdot \sigma_2) = g_1' + g_{12}; \end{aligned}$$

similarly we obtain $g_2 = g_2' + g_{12}$. In addition, we obtain

$$\begin{aligned} F(\sigma_1 \vee \sigma_2) &= F(\sigma_1 \dotplus (\sigma_2 \dotplus \sigma_1 \cdot \sigma_2)) = F(\sigma_1) + F(\sigma_2 \dotplus \sigma_1 \cdot \sigma_2) \\ &= g_1 + g_2' \in L. \end{aligned}$$

(ii) $\Rightarrow$ (i). Let us consider a set Π consisting of three elements (1), (2), and (3), and let $\Sigma = \mathscr{P}(\Pi)$. For the elementary sets $\sigma_1 = \{(1)\}$, $\sigma_2 = \{(2)\}$, $\sigma_3 = \{(3)\}$ we set $F(\sigma_1) = g_1'$, $F(\sigma_2) = g_2'$, $F(\sigma_3) = g_{12}$. It is easy to show that F determines an additive (not necessarily normed) measure satisfying $F(\sigma_1 \dotplus \sigma_3) = g_1$, $F(\sigma_2 \dotplus \sigma_3) = g_2$.

The proof of this theorem is particularly instructive because it shows that g_1, g_2 does not necessarily uniquely determine the remaining values of the measure $F(\sigma)$. For the case in which $\Sigma = \mathscr{P}(\Pi)$ we may obtain different measures because there will be different $g_{12} \in L$ for which $g_1 - g_{12}, g_2 - g_{12}$ and $g_1 + g_2 - g_{12} \in L$. If, for example, $g_1 + g_2 \in L$ then we may choose $g_{12} = 0$. If we can find another $g_{12} \neq 0$, $g_{12} \in L$ for which $g_1 - g_{12}$, $g_2 - g_{12} \in L$, we may then construct two different measures $F(\sigma)$ on $\Sigma = \mathscr{P}(\Pi)$. It is easy to give such an example. Consider three arbitrary elements $\tilde{g}_1, \tilde{g}_2, \tilde{g}_3 \in L$ and choose g_1, g_2 as follows:

$$\begin{aligned} g_1 &= \tfrac{1}{3}\tilde{g}_1 + \tfrac{1}{6}\tilde{g}_3, \\ g_2 &= \tfrac{1}{3}\tilde{g}_2 + \tfrac{1}{6}\tilde{g}_3. \end{aligned} \tag{1.2.3}$$

Then we obtain $g_1 + g_2 = \frac{1}{3}(\tilde{g}_1 + \tilde{g}_2 + \tilde{g}_3) \in L$; alternatively we can set $g_{12} = \frac{1}{6}\tilde{g}_3$.

In the following section we shall return to consider the fact that two effects do not necessarily uniquely determine an effect g_{12} for which "both effects g_1 and g_2 jointly respond."

We shall now mention the following special case of Th. 1.2.4: The following conditions are sufficient for g_1 and g_2 to be coexistent:

(1) $g_1 + g_2 \in L$ (set $g_{12} = 0$).
(2) $g_1 > g_2$ (set $g_{12} = g_2$).

1.3 Commensurable Decision Effects

The notion of a decision effect, which was introduced in III, §6 plays an important role in quantum mechanics. Often the "role" of decision effects is over-estimated—resulting in the exclusion of realistic situations for all of the effects. In order to avoid clouding the issue by introducing "additional hypotheses" or "opinions" we shall proceed step by step to develop the special role of a decision effect. We shall use only the assertions about decision effects which were introduced in III, §6.

We shall now begin with the following obvious definition:

D 1.3.1. A set $A \subset G$ (that is, a set A of decision effects) is said to be *commensurable* if there exists a Boolean ring Σ with additive measure $F: \Sigma \to G$ for which $A \subset F\Sigma$.

The reader should note that, according to this definition, there is a map F of Σ into G. According to D 1.2.2 a set $A \subset G$ is coexistent if there exists a Boolean ring Σ' with additive measure $F': \Sigma' \to L$ for which $A \subset F'\Sigma'$. Thus, a set of commensurable decision effects is also (since $G \subset L$) coexistent. Does the converse hold?

The Boolean ring Σ with measure $F: \Sigma \to G$ represents at least an "idealized" (in the case in which $\Sigma \xrightarrow{F} G$ can only be approximately "realized" by a $\mathscr{R}(b_0) \xrightarrow{\psi_0} L$—see §4) registration method corresponding to decision effects associated with the realizable registrations. It was the practice of theoretical physicists to permit only the use of those measurement methods for which $\psi(f) \in G$. Certainly, this is a particularly interesting special case of a measurement. We note, however, that the general measurement methods for which $\psi(b_0, b)$ are only effects are usable measurements. Indeed, they are the only realistic measurements.

In Th. 1.2.4 we have analyzed the factual content of two coexistent effects. Now we shall consider the special case in which one of the two effects is a decision effect.

Th. 1.3.1. *For $g \in L$, $e \in G$ the following conditions are equivalent:*

(i) *g, e are coexistent.*

(ii) *$g = g_1 + g_2$ where $g_1, g_2 \in L$ and $g_1 \leq e$, $g_2 \leq e^\perp$ (in this partition of g, g_1 and g_2 are uniquely determined).*

(iii) *$e = g'_1 + g'_3$ where $g'_1, g'_3 \in L$ and $g'_1 \leq g$, $g_3 \leq \mathbf{1} - g$ (in this partition of e, g'_1, and g'_3 are uniquely determined and $g'_1 = g_1$ where g_1 is defined in (ii)).*

(iv) *$eg = ge$.*

PROOF. According to Th. 1.2.4 (i) is equivalent to the condition that there exists $g_1, g_2, g_3 \in L$ such that $g = g_1 + g_2$, $e = g_1 + g_3$ and $g_1 + g_2 + g_3 \in L$. From this it follows that $g_1 \leq e$ and $\mathbf{1} \geq g_1 + g_2 + g_3 = g_2 + e$, that is, $g_2 \leq \mathbf{1} - e$; thus we obtain (i) $\Rightarrow$ (ii).

Conversely, if $g = g_1 + g_2$ where $g_1 \leq e$, $g_2 \leq e^\perp$ then $g_3 = e - g_1 \in L$ and $e = g_1 + g_3$ and $g_1 + g_2 + g_3 = g_2 + e \leq e^\perp + e = \mathbf{1}$, that is, $g_1 + g_2 + g_3 \in L$. Thus we have shown that (ii) $\Rightarrow$ (i).

Let $g = g_1 + g_2 = \tilde{g}_1 + \tilde{g}_2$ where $g_1, \tilde{g}_1 \leq e$ and $g_2, \tilde{g}_2 \leq e^\perp$. Then we obtain $g_1 - \tilde{g}_1 = \tilde{g}_2 - g_2$ and $0 \leq e - \tilde{g}_1 \leq e + g_1 - \tilde{g}_1 = e + \tilde{g}_2 - g_2 \leq e + \tilde{g}_2 \leq e + e^\perp \leq \mathbf{1}$. Therefore $g_3 = e + g_1 - \tilde{g}_1 \in L$. Since $g_1 \leq e$ we obtain $K_0(g_1) \supset K_0(e)$; similarly we obtain $K_0(\tilde{g}_1) \supset K_0(e)$. From these results it follows that $K_0(g_3) \supset K_0(e)$, that is, $g_3 \in L_0 K_0(e)$. From $K_0(e) \subset K_0(g_1)$—that is, $g_3 \in L_0 K_0(e)$—and from III, Th. 6.6 it follows that $g_3 \leq e$. From these results it follows that $g_1 \leq \tilde{g}_1$; similarly we may also derive $\tilde{g}_1 \leq g_1$. Therefore we obtain $g_1 = \tilde{g}_1$ and $g_2 = \tilde{g}_2$. Therefore we have proven the uniqueness of the decomposition given in (ii).

From the above decomposition $g = g_1 + g_2$ and $e = g_1 + g_3$ it follows that $g_1 \leq g$ and $g_3 = e - g_1 = e - g + g_2$. Since $g_2 \leq e^\perp$, (see (ii)) it follows that $g_3 \leq e - g + e^\perp = \mathbf{1} - g$. Thus we have shown that (i) $\Rightarrow$ (iii). Conversely, let $e = g'_1 + g'_3$ where $g'_1 \leq g$ and $g'_3 \leq \mathbf{1} - g$ and let $g'_2 = g - g'_1$. Then we obtain $g'_2 = g - e + g'_3 \leq g - e + \mathbf{1} - g = \mathbf{1} - e = e^\perp$. Thus we have shown that (iii) $\Rightarrow$ (ii).

Since we may derive the relation (ii) out of (iii) for the special case $g'_1 = g_1$ it follows that the partition in (iii) is unique.

It now suffices to show that (ii) $\Leftrightarrow$ (iv). From $g_1 \leq e$ it follows that (see AIV, §6) $g_1 = eg_1e$; from $g_2 \leq e^\perp$ it follows that $g_2 = e^\perp g_2 e^\perp$. Therefore, from (ii) we obtain $ge = eg_1e = eg$. Conversely, if $ge = eg$, we define $g_1 = ege$, $g_2 = e^\perp g e^\perp$. Then, since e and g commute, we obtain

$$g = (e + e^\perp)g(e + e^\perp) = ege + e^\perp g e^\perp = g_1 + g_2.$$

In addition, we obtain $g_1 = ege \leq e\mathbf{1}e = e$; similarly we obtain $g_2 \leq e^\perp$. Thus we have shown that (ii) holds.

The commutativity of two coexistent effects does not hold in general! This fact may be verified using example (1.2.1) where $\tilde{g}_1, \tilde{g}_2, \tilde{g}_3$ may be arbitrary elements of L.

Th. 1.3.2. *Let $e_1, e_2 \in G$ be two coexistent decision effects. Suppose that they satisfy the decomposition $e_1 = g_1 + g_2$, $e_2 = g_1 + g_3$ where*

$g_1, g_2, g_3 \in L$ and $g_1 + g_2 + g_3 \in L$ (equivalent to the coexistent requirement). Then g_1, g_2, g_3 are uniquely determined by e_1, e_2 and $g_1 = e_1 \wedge e_2$, $g_2 = e_1 \wedge e_2^\perp$, $g_3 = e_1^\perp \wedge e_2$ and $g_1, g_2, g_3 \in G$. If Σ is a Boolean ring with additive measure $F: \Sigma \to L$ for which $F(\sigma_1) = e_1$, $F(\sigma_2) = e_2$, it follows that $F(\sigma_1 \wedge \sigma_2) = e_1 \wedge e_2$, $F(\sigma_1 \dotplus \sigma_1 \cdot \sigma_2) = e_1 \wedge e_2^\perp$ and $F(\sigma_2 \dotplus \sigma_1 \cdot \sigma_2) = e_2 \wedge e_1^\perp$.

PROOF. According to Th. 1.3.3 (ii) $g_1 \leq e_2$, $g_2 \leq e_2^\perp$, and $g_1 \leq e_1$, $g_3 \leq e_1^\perp$. Therefore it follows that $g_1 \in L_0K_0(e_1)$ and $g_1 \in L_0K_0(e_2)$, that is, $g_1 \in L_0K_0(e_1) \cap L_0K_0(e_2) = L_0(A)$ where $A = K_0(e_1) \cup K_0(e_2)$. According to III, Th. 6.6 $L_0(A) = L_0(C)$ where C is the face generated by A, that is, $C = K_0(e_1) \vee K_0(e_2)$. According to III, Th. 6.4 $K_0(e_1) \vee K_0(e_2) = K_0(e_1 \wedge e_2)$ and therefore $g_1 \in L_0(A) = L_0K_0(e_1 \wedge e_2)$, that is, $g_1 \leq e_1 \wedge e_2$.

Similarly, it follows that $g_2 \leq e_1 \wedge e_2^\perp$, $g_3 \leq e_1^\perp \wedge e_2$. Therefore $e_1 = g_1 + g_2 \leq (e_1 \wedge e_2) + (e_1 \wedge e_2^\perp)$. Since $e_1 \wedge e_2 \leq e_2$ and $e_1 \wedge e_2^\perp \leq e_2^\perp$, we find that $e_1 \wedge e_2 \perp e_1 \wedge e_2^\perp$; thus, by III, Th. 6.8 we obtain $(e_1 \wedge e_2) + (e_1 \wedge e_2^\perp) = (e_1 \wedge e_2) \vee (e_2 \wedge e_2^\perp) \leq e_1$. From $e_1 = g_1 + g_2 \leq (e_1 \wedge e_2) + (e_1 \wedge e_2^\perp) \leq e_1$, it follows that $g_1 = e_1 \wedge e_2$, $g_2 = e_1 \wedge e_2^\perp$ because $g_1 \leq e_1 \wedge e_2$, $g_2 \leq e_2 \wedge e_2^\perp$. Similarly, it follows that $g_3 = e_1^\perp \wedge e_2$. The rest of the theorem follows directly from the proof of Th. 1.2.4.

Th. 1.3.3. *If $F: \Sigma \to G$ is an additive, effective measure on the Boolean ring Σ, then F is an isomorphic map of the Boolean ring Σ onto the Boolean sublattice $F\Sigma$ of G.*

PROOF. According to Th. 1.3.2, for each pair of elements $\sigma_1, \sigma_2 \in \Sigma$ we have $F(\sigma_1 \wedge \sigma_2) = F(\sigma_1) \wedge F(\sigma_2)$. Furthermore, for $\sigma \in \Sigma$ we obtain $\mathbf{1} = F(\varepsilon) = F(\sigma \vee \sigma^*) = F(\sigma) + F(\sigma^*)$: $F(\sigma^*) = \mathbf{1} - F(\sigma) = F(\sigma)^\perp$. Since $\sigma_1 \vee \sigma_2 = (\sigma_1^* \wedge \sigma_2^*)^*$ it follows that $F(\sigma_1 \vee \sigma_2) = F(\sigma_1) \vee F(\sigma_2)$. Thus F is a lattice homomorphism of Σ into G for which $F(\sigma^*) = F(\sigma)^\perp$. In addition, F is injective: From $F(\sigma_1) = F(\sigma_2)$ it follows $F(\sigma_1 \wedge \sigma_2) = F(\sigma_1) \wedge F(\sigma_2) = F(\sigma_1) = F(\sigma_2)$ and from

$$F(\sigma_1) = F((\sigma_1 \wedge \sigma_2) \vee (\sigma_1 \wedge \sigma_2^*)) = F(\sigma_1 \wedge \sigma_2) + F(\sigma_1 \wedge \sigma_2^*)$$

we obtain the relation $F(\sigma_1 \wedge \sigma_2^*) = 0$. Since F is effective, it follows that $\sigma_1 \wedge \sigma_2^* = 0$ and therefore $\sigma_1 = \sigma_1 \wedge \sigma_2$. Similarly we find that $\sigma_2 = \sigma_1 \wedge \sigma_2$, that is, $\sigma_1 = \sigma_2$. Thus F is an isomorphism onto the sublattice $F\Sigma$ of G and $F\Sigma$ is a Boolean sublattice of G.

Th. 1.3.4. *The following six conditions are equivalent for decision effects:*

(i) *e_1, e_2 are coexistent.*
(ii) *e_1, e_2 are commensurable.*
(iii) *The orthocomplemented sublattice Γ of G generated by e_1, e_2 is a Boolean ring.*
(iv) $e_1 = (e_1 \wedge e_2) \vee (e_1 \wedge e_2^\perp)$.
(v) $e_1 e_2 = e_2 e_1$.
(vi) $e_1 e_2 = e_1 \wedge e_2$.

Let $\Sigma \xrightarrow{F} L$ be the Boolean ring defined according to (i) where F is the effective measure. Then there exist two elements $\sigma_1, \sigma_2 \in \Sigma$ for which

$F(\sigma_1) = e_1$, $F(\sigma_2) = e_2$. Let Σ^0 be the Boolean subring of Σ generated by σ_1, σ_2. The restriction of F to Σ^0 is an isomorphism of Σ^0 on Γ (where Γ is defined by (iii)). The identity map i of Γ into itself is an additive measure i on the Boolean ring Γ for which $\Gamma \xrightarrow{i} G$.

PROOF. According to (i) there exists a Boolean ring Σ with an effective measure $F: \Sigma \to L$ (see Th. 1.2.2). According to Th. 1.3.2, there exist two elements σ_1, σ_2 for which $F(\sigma_1) = e_1$, $F(\sigma_2) = e_2$ and the following condition holds: $F(\sigma_1 \wedge \sigma_2) = e_1 \wedge e_2$, $F(\sigma_1 \dotplus \sigma_1 \cdot \sigma_2) = e_1 \wedge e_2^{\perp}$, $F(\sigma_2 \dotplus \sigma_1 \cdot \sigma_2) = e_1 \wedge e_1^{\perp}$. We shall now show that the Boolean subring Σ^0 of Σ generated by σ_1, σ_2 will be mapped by F into G.

Since $\mathbf{1} = F(\varepsilon) = F(\sigma \vee \sigma^*) = F(\sigma) + F(\sigma^*)$ we obtain $F(\sigma^*) = \mathbf{1} - F(\sigma)$. If, in addition, $F(\sigma) \in G$, then we also find $F(\sigma^*) \in G$. It suffices to show that the following eight elements 0, σ_1, σ_2, $\sigma_1 \wedge \sigma_2$, $\sigma_1 \dotplus \sigma_1 \cdot \sigma_2$, $\sigma_2 \dotplus \sigma_1 \cdot \sigma_2$, $\sigma_1 \dotplus \sigma_1 \cdot \sigma_2 \dotplus \sigma_2 \dotplus \sigma_1 \cdot \sigma_2$ and $\sigma_1 \vee \sigma_2 = \sigma_1 \dotplus \sigma_2 \cdot \sigma_2 \dotplus \sigma_2$ may be mapped into G since the remaining eight elements of Σ^0 are complements of the above elements. We obtain:

$$F(\sigma_1 \dotplus \sigma_1 \cdot \sigma_2 \dotplus \sigma_2 \dotplus \sigma_1 \cdot \sigma_2) = F(\sigma_1 \dotplus \sigma_1 \cdot \sigma_2) + F(\sigma_2 \dotplus \sigma_1 \cdot \sigma_2)$$
$$= (e_1 \wedge e_2^{\perp}) + (e_2 \wedge e_1^{\perp}) \in G,$$

since $e_1 \wedge e_2^{\perp} \perp e_2 \wedge e_1^{\perp}$. Similarly, we obtain

$$F(\sigma_1 \vee \sigma_2) = F(\sigma_1 \dotplus \sigma_1 \cdot \sigma_2) + F(\sigma_2) = e_1 \wedge e_2^{\perp} + e_2 \in G,$$

since $e_1 \wedge e_2^{\perp} \perp e_2$.

Therefore the map $\Sigma^0 \xrightarrow{F} G$ is an additive measure on Σ^0 with range in G, whereby (i) $\Rightarrow$ (ii) is proven.

With the application of Th. 1.3.3 to $\Sigma^0 \xrightarrow{F} G$ we find that $\Gamma = F\Sigma^0$ is isomorphic to Σ^0, thereby proving that (ii) $\Rightarrow$ (iii) is proven.

If Γ—as defined in (iii)—is a Boolean ring, then from $e, \tilde{e} \in \Gamma$ and $e \wedge \tilde{e} = 0$ it follows that $\tilde{e} \leq e^{\perp}$ because $\tilde{e} = \tilde{e} \wedge (e \vee e^{\perp}) = (\tilde{e} \wedge e) \vee (\tilde{e} \wedge e^{\perp}) = \tilde{e} \wedge e^{\perp}$. Therefore, by III, Th. 6.8 we find that $e \vee \tilde{e} = e + e^{\perp}$ and that the identity map of Γ onto itself is an additive effective measure. Since $\Gamma \subset G$ we find that e_1, e_2 are also commensurable. That is, we have shown that (iii) $\Rightarrow$ (ii).

As we have shown above after D 1.3.1, (ii) $\Rightarrow$ (i). In Th. 1.3.2 we have shown that (i) $\Rightarrow$ (iv). If (iv) is satisfied, that is, $e_1 = (e_1 \wedge e_2) \vee (e_1 \wedge e_2^{\perp})$, then, by III, Th. 6.8 we obtain $e_1 = g_1 + g_2$ where $g_1 = e_1 \wedge e_2$, $g_2 = e_1 \wedge e_2^{\perp}$. Or, expressed differently we obtain $g_1 \leq e_2, g_1 \leq e_2^{\perp}$. By Th. 1.3.1 it then follows that (iv) $\Rightarrow$ (i).

From Th. 1.3.3 (iv) it follows that (v) $\Leftrightarrow$ (i); (vi) $\Leftrightarrow$ (v) is proven in AIV, §6.

We shall now consider some special cases of commensurable decision effects. If $e_1 \leq e_2$, then, by arguments presented at the end of §1.2 we conclude that e_1, e_2 are coexistent and are therefore commensurable. Similarly, if $e_1 + e_2 \leq \mathbf{1}$ (that is $e_1 + e_2 \in L$) we conclude that e_1, e_2 are coexistent and are therefore commensurable since $e_1 + e_2 \leq \mathbf{1}$ is equivalent to $e_1 \leq \mathbf{1} - e_2 = e_2^{\perp}$, that is, $e_1 \perp e_2$. Hence, from $e_1 \perp e_2$ it follows that e_1 and e_2 are commensurable. If e_1, e_2 are commensurable and $e_1 \wedge e_2 = 0$, it then follows from Th. 1.3.4 (iv) that $e_1 \perp e_2$ and, consequently, by III, Th. 6.8, $e_1 \vee e_2 = e_1 + e_2$.

We shall now develop general criteria for the characterization of commensurable sets of decision effects. For this purpose we shall state and prove the following theorem:

Th. 1.3.5. *Let $A_1 \subset G$, $A_2 \subset G$ and suppose that for each pair $e_1 \in A_1$, and $e_2 \in A_2$ are commensurable. Then $\bigwedge_{e \in A_2} e$ and $\bigvee_{e \in A_2} e$ are commensurable with each $e_1 \in A_1$. Suppose that $e \in G$ is commensurable with each $e_1 \in A_1$, then $e^\perp$ is commensurable with each $e_1 \in A_1$.*

PROOF. The statement that e, e_1 are commensurable is equivalent to the condition that $e_1 = (e_1 \wedge e) \vee (e_1 \wedge e^\perp)$. This relation is symmetric in e and $e^\perp$ because $e = (e^\perp)^\perp$. Thus, $e^\perp, e_1$ are commensurable.

Since e_1, e_2 are commensurable, from Th. 1.3.4 it follows that $e_2 = (e_2 \wedge e_1) \vee (e_2 \wedge e_1^\perp)$ and we therefore obtain

$$\bigvee_{e \in A_2} e = \left[\bigvee_{e \in A_2} (e \wedge e_1)\right] \vee \left[\bigvee_{e \in A_2} (e \wedge e_1^\perp)\right].$$

Since G is orthomodular (see III, D 6.6 and AI, §2) it follows that

$$\left[\bigvee_{e \in A_2} e\right] \wedge e_1 = \left[\left[\bigvee_{e \in A_2} (e \wedge e_1)\right] \vee \left[\bigvee_{e \in A_2} (e \wedge e_1^\perp)\right]\right] \wedge e_1 = \bigvee_{e \in A_2} (e \wedge e_1).$$

A similar result holds if we substitute $e_1^\perp$ for e_1. Thus we obtain

$$\bigvee_{e \in A_2} e = \left[\left(\bigvee_{e \in A_2} e\right) \wedge e_1\right] \vee \left[\left(\bigvee_{e \in A_2} e\right) \wedge e_1^\perp\right].$$

Thus, using Th. 1.3.4(iv) we conclude that $e_1, \bigvee_{e \in A_2} e$ are commensurable. Since $(\bigvee_{e \in A_2} e^\perp)^\perp = \bigwedge_{e \in A_2} e$, the same result follows for $e_1, \bigwedge_{e \in A_2} e$.

Th. 1.3.6. *For $A \subset G$ the following conditions are equivalent:*

(i) *A is coexistent.*
(ii) *A is commensurable.*
(iii) *Each pair $e_1, e_2 \in A$ are coexistent.*
(iv) *Each pair $e_1, e_2 \in A$ are commensurable.*
(v) *The orthocomplemented sublattice Γ_A generated by A is a Boolean ring.*
(vi) *The complete orthocomplemented sublattice $\bar{\Gamma}_A$ generated by A is a Boolean ring.*

PROOF. (iii) ⇒ (iv) is a direct consequence of Th. 1.3.4. (ii) ⇒ (i) and (i) ⇒ (iii) are trivial. If we show that (iv) ⇒ (ii) we will have proven that (i), (ii), (iii), and (iv) are equivalent.

Suppose that (iv) is satisfied. Every element of the sublattice Γ_A described in (v) may be obtained by the finite application of the operators $\wedge$, $\vee$, and $\perp$ to the elements of A. According to Th. 1.3.5 every element of Γ_A is commensurable with every element of A. Again, from Th. 1.3.5 it follows that every pair of elements of Γ_A is commensurable. Thus, from AI, Th. 3.2 it follows that Γ_A is a Boolean ring.

Since $e_1 \wedge e_2 = 0$ and e_1, e_2 are commensurable (see the discussion preceding Th. 1.3.5) it follows that $e_1 \vee e_2 = e_1 + e_2$ and that the identity map of Γ_A is an additive measure. Therefore, since $A \subset \Gamma_A$, the set A is a set of commensurable decision effects. Thus we have proven (iv) ⇒ (ii), (iv) ⇒ (v), and (v) ⇒ (ii).

(v) ⇒ (vi). By Zorn's lemma it follows that the set Ξ of Boolean subrings Γ' for which $A \subset \Gamma' \subset G$ contains a maximal element Γ_m. Γ_m must be a complete Boolean ring, otherwise, there would be a subset $B \subset \Gamma_m$ for which the element $\tilde{e} = \bigvee_{e \in B} e \in G$ would not lie in Γ_m. Then, by Th. 1.3.5 and by (ii) ⇔ (iv) we would find that $\{\Gamma_m, \tilde{e}\}$ would be a set of commensurable decision effects which, according to (v), would be contained in a Boolean ring. This contradicts the fact that Γ_m is maximal. $\bar{\Gamma}_A$ must be a sublattice of Γ_m and must therefore be a Boolean ring.

(vi) ⇒ (v) is trivial.

Th. 1.3.7. *Let e_ν be a sequence of commensurable decision effects which converges in the $\sigma(\mathscr{B}', \mathscr{B})$ topology towards $e \in G$. Then e is commensurable with all $\bar{e} \in G$ which are commensurable with the e_ν. In addition, it is possible to choose a subsequence e_{ν_k} such that $e = \bigwedge_{n=1}^{\infty} \tilde{e}_n$ where $\tilde{e}_n = \bigvee_{k=n}^{\infty} e_{\nu_k}$. (From this result we obtain the following special case: Every complete Boolean subring of G is $\sigma(\mathscr{B}', \mathscr{B})$ closed in G.)*

Proof. From III, Th. 6.10 $\tilde{e}_n \to \tilde{e}$ in the $\sigma(\mathscr{B}', \mathscr{B})$ topology where $\tilde{e} = \bigwedge_{n=1}^{\infty} \tilde{e}_n$. Since $\tilde{e}_n \geq e_{\nu_k}$ for all $k \geq n$, in the limit $k \to \infty$ we find that $\tilde{e}_n \geq e$ and therefore find that $\tilde{e} \geq e$. If we show that, in the $\sigma(\mathscr{B}', \mathscr{B})$ topology $\tilde{e}_n \to e$ then from $\tilde{e}_n \geq \tilde{e} \geq e$ we would obtain $e = \tilde{e}$.

First we shall show that e is commensurable with all the $\bar{e} \in G$ which are commensurable with the e_ν. If $\bar{e}$ is commensurable with the e_ν then, by Th. 1.3.4(iv) and III, Th. 6.8 we obtain

$$\bar{e} = (e_\nu \wedge \bar{e}) + (\bar{e} \wedge e_\nu^\perp),$$

$$e_\nu = (e_\nu \wedge \bar{e}) + (e_\nu \wedge \bar{e}^\perp).$$

Since L is $\sigma(\mathscr{B}', \mathscr{B})$-compact (see AIII, §4 and §6), we may select a subsequence e_{ν_α} from e_ν for which $e_{\nu_\alpha} \wedge \bar{e} \to g_1 \in L$, $e_{\nu_\alpha} \wedge \bar{e}^\perp \to g_2 \in L$, $\bar{e} \wedge e_{\nu_\alpha}^\perp \to g_3 \in L$, and therefore $\bar{e} = g_1 + g_3$, $e = g_1 + g_2$. From $e_{\nu_\alpha} \wedge \bar{e} \leq e_{\nu_\alpha}$ and from $e_{\nu_\alpha} \wedge \bar{e} \leq \bar{e}$ it follows that $g_1 \leq e$ and $g_3 \leq \bar{e}$. Hence it follows that $K_0(g_1) \supset K_0(e)$ and $K_0(g_1) \supset K_0(\bar{e})$. Therefore, from III, Th. 6.4 it follows that

$$K_0(g_1) \supset K_0(e) \vee K_0(\bar{e}) = K_0(e \wedge \bar{e}).$$

According to III, Th. 6.6 it follows that $g_1 \leq e \wedge \bar{e}$. Similarly we may prove that $g_2 \leq e \wedge \bar{e}^\perp$ and $g_3 \leq \bar{e} \wedge e^\perp$. From these results we obtain

$$g_1 + g_2 + g_3 \leq [(e \wedge e^\perp) \vee (e \wedge \bar{e}^\perp)] + (\bar{e} \wedge e^\perp).$$

Since $(e \wedge \bar{e}) \vee (e \wedge \bar{e}^\perp) \leq e$ and $\bar{e} \wedge e^\perp < e^\perp$ we obtain

$$g_1 + g_2 + g_3 \leq e + e^\perp = \mathbf{1}.$$

Therefore, by Th. 1.2.4, e and $\bar{e}$ are coexistent and, by Th. 1.3.4 are commensurable.

Since we have assumed that the e_ν are all commensurable, e is therefore commensurable with all e_ν, and, by Th. 1.3.5, is also commensurable with $\tilde{e}$ and the $\tilde{e}_n$.

Since e is commensurable with the $\tilde{e}_n = \bigvee_{k=n}^{\infty} e_{\nu_k}$ and with the e_{ν_k} we therefore obtain (see the proof of Th. 1.3.5)

$$\tilde{e}_n \wedge e^\perp = \bigvee_{k=1}^{\infty} (e_{\nu_k} \wedge e^\perp).$$

Since the $e_{\nu_k} \wedge e^{\perp}$ are commensurable, we find that

$$\tilde{e}_n \wedge e^{\perp} \leq \sum_{k=n}^{\infty} (e_{\nu_k} \wedge e^{\perp}).$$

According to AIII, §4 the $\sigma(\mathscr{B}', \mathscr{B})$-topology on L may be characterized by a norm $\|\ldots\|_\sigma$. For this norm we find that

$$\|\tilde{e}_n \wedge e^{\perp}\|_\sigma \leq \sum_{k=n}^{\infty} \|e_{\nu_k} \wedge e^{\perp}\|_\sigma.$$

Since $\tilde{e}_n \geq e$ we find that $\tilde{e}_n - e = \tilde{e}_n \wedge e^{\perp}$ and we therefore obtain

$$\|\tilde{e}_n - e\|_\sigma \leq \sum_{k=n}^{\infty} \|e_{\nu_k} \wedge e^{\perp}\|_\sigma.$$

Since L is compact, we may choose a subsequence e_{ν_k} from the e_ν such that $e_{\nu_k} \wedge e^{\perp} \to g \in L$ in the $\sigma(\mathscr{B}', \mathscr{B})$-topology. From $e_{\nu_k} \wedge e^{\perp} \leq e_{\nu_k}$ it follows that, in the limit $g \leq e$; from $e_{\nu_k} \wedge e^{\perp} \leq e^{\perp}$ it follows that $g \leq e^{\perp}$.

Hence, from the above we may conclude that $g \leq e^{\perp} \wedge e$, that is, $e_{\nu_k} \wedge e^{\perp} \to 0$.

Thus it is possible to select a subsequence such that $\|e_{\nu_k} \wedge e^{\perp}\|_\sigma \leq (\frac{1}{2})^k$ from which we conclude that $\|\tilde{e}_n - e\|_\sigma \xrightarrow{n} 0$ and $\tilde{e}_n \to e$.

D 1.3.2. The set of all $e \in G$ which are commensurable with all the elements of G is called the *center* Z of G.

Th. 1.3.8. *Z is the set of all $e = (E_1, E_2, \ldots)$ for which the E_ν in H_ν are either* **0** *or* **1**.

PROOF. According to Th. 1.3.4 Z is the set of all $e = (E_1, E_2, \ldots)$ which commute with all the other $e \in G$, which proves the assertion.

According to Th. 1.3.6(vi) it follows that Z is a complete Boolean subring of G (which can be easily proven directly). Z is atomic, with atoms $q_i = (\mathbf{0}, \mathbf{0}, \ldots, \mathbf{1}_i, \ldots)$ where the components are equal to **0** except for the ith position.

The fact that Z is atomic is a consequence of AV 4s from III, §3. The atomic character of Z is characteristic for microsystems.

1.4 Observables

In §1.1 and §1.2 we have implicitly presented the structure upon which the observable concept will be based. The concept of an observable is none other than an abstract idealization of the structure represented by the map $\mathscr{R}(b_0) \xrightarrow{\psi_0} L$. Before we proceed to give a precise definition of the notion of an observable we shall make a number of preliminary remarks in order to reduce the possibility of misunderstandings. Many readers will find it somewhat surprising that we shall say little about "measurement values," "measurement scales" and the like when we introduce the notion of an observable. Is a "measurement" necessarily quantitative? In response to this question, we emphasize that it is important for the reader to put aside the notion that the *essential* aspect of physics is that of *quantitative* measurement. Otherwise, the reader will not obtain a correct understanding of the methods of theoretical physics. Parameterization of registration procedures can be very "convenient," "practical," and "useful," but it has no *fundamental* meaning in reference to the mapping of physical reality by means of the mathematical structures. This becomes evident in the fact that it is possible, in principle, to record all measurements digitally and store them in a computer. In addition, the structure of a Boolean ring (for example, of $\mathscr{R}(b_0)$) becomes more tractable if we do not insist on imposing a more or less arbitrary parameterization of the Boolean ring. Finally, the abstract structure of a Boolean ring is more transparent than the "usual" parametric formulation.

At this point we could define an observable directly in terms of $\mathscr{R}(b_0)$ and the map $\psi_0: \mathscr{R}(b_0) \to L$. This approach will, however, lead to a number of mathematically inconvenient structures. Instead, we shall proceed in an abstract manner in analogy to the definition of coexistent effects. The first approach would be to define an observable by means of a Boolean ring Σ and an effective measure $F: \Sigma \to L$ (see [2], XIII, D 5.6; in [2] we have presented this definition. There we did not find it necessary to discuss the process of completion of Σ). For this approach, in order to make the concept more realistic physically, and to simplify the mathematical treatment, we shall impose a number of additional conditions upon the notion of an observable.

Mathematically, in order to formulate several theorems more simply, it is always very convenient to make the sets complete (relative to the uniform structure). We shall apply such a completion to the ring Σ.

Th. 1.4.1. *Let $w \in K$; the sets*

$$N_{w,\varepsilon} = \{(\sigma_1, \sigma_2) \mid \sigma_1, \sigma_2 \in \Sigma, \mu(w, F(\sigma_1 \dotplus \sigma_2)) < \varepsilon\}$$

form a fundamental system of sets for the uniform structure U_g (see AII, §2) *of the Boolean ring Σ with effective measure $F: \Sigma \to L$. U_g separates Σ because F is effective. U_g is metrizable, with metric $d(\sigma_1, \sigma_2) = \mu(w_0, F(\sigma_1 + \sigma_2))$ where we choose an "effective" w_0 (such an effective w_0 exists according to* III, Th. 6.1), *that is, a w_0 for which $C(w_0) = K$.*

Proof. From $w_0 \in K$ we will have proved the theorem if we can show that $d(\sigma_1, \sigma_2)$ is a metric, and that, to each pair w, ε there exists an ε' for which $\mu(w_0, g) < \varepsilon'$ implies that $\mu(w, g) < \varepsilon$ for all $g \in L$.

Since $F(\sigma_1 \dotplus \sigma_2) + F(\sigma_2 \dotplus \sigma_3) = F(\sigma_1 \dotplus \sigma_3) + F((\sigma_1 \dotplus \sigma_3) \cdot (\sigma_2 \dotplus \sigma_3))$ we obtain $F(\sigma_1 \dotplus \sigma_3) \leq F(\sigma_1 \dotplus \sigma_2) + F(\sigma_2 \dotplus \sigma_3)$. Then we find that $d(\sigma_1, \sigma_2)$ satisfies the triangle inequality. From $d(\sigma_1, \sigma_2) = 0$ it follows that $\mu(w_0, F(\sigma_1 \dotplus \sigma_2)) = 0$ and we therefore obtain $\mu(w_0, F(\sigma_1 \dotplus \sigma_2)) = 0$ for all $w \in C(w_0) = K$, that is, $F(\sigma_1 \dotplus \sigma_2) = 0$. Since F is effective, it follows that $\sigma_1 \dotplus \sigma_2 = 0$, that is $\sigma_1 = \sigma_2$. Thus $d(\sigma_1, \sigma_2)$ is a metric.

We shall now show that we may choose a *special* effective $w_0 \in K$ such that, to each pair w, ε there exists an ε' for which $\mu(w_0, g) < \varepsilon'$ implies $\mu(w, g) < \varepsilon$ for all $g \in L$. For this purpose, we introduce a denumerable subset $\{w_\nu\}$ which is dense in K and define $w_0 = \sum_{\nu=1}^{\infty} \lambda_\nu w_\nu, \lambda_\nu > 0, \sum_{\nu=1}^{\infty} \lambda_\nu = 1$.

Then, for $w \in K, w_\rho \in \{w_\nu\}$ we obtain

$$\begin{aligned}\mu(w, g) &\leq |\mu(w - w_\rho, g)| + \mu(w_\rho, g) \\ &\leq \|w - w_\rho\| + \mu(w_\rho, g).\end{aligned}$$

Since $\mu(w, g) = \sum_{\nu=1}^{\infty} \lambda_\nu \mu(w_\nu, g)$, we find that

$$\lambda_\rho \mu(w_\rho, g) \leq \mu(w_0, g)$$

and we obtain

$$\mu(w, g) \leq \|w - w_\rho\| + \lambda_\rho^{-1} \mu(w_\rho, g).$$

If we now choose w_ρ so that $\|w - w_\rho\| < \varepsilon/2$ and choose $\varepsilon' < \lambda_\rho(\varepsilon/2)$ then we obtain $\mu(w, g) < \varepsilon$.

From Th. 1.4.3 and Th. 2.1.11 it follows that the metrics

$$d_1(\sigma_1, \sigma_2) = \mu(w_1, F(\sigma_1 \dot{+} \sigma_2))$$

and

$$d_2(\sigma_1, \sigma_2) = \mu(w_2, F(\sigma_1 \dot{+} \sigma_2))$$

are equivalent for effective pairs $w_1, w_2 \in K$.

Here it is important to note that the following theorems (up to Th. 2.1.11) only require that there exists a single w_0 for which the metric defined in Th. 1.4.1 generates the uniform structure U_g!

Let Σ_g denote Σ endowed with the uniform structure U_g defined in Th. 1.4.1.

Th. 1.4.2. *The map $F: \Sigma_g \to L$ (where L is endowed with the uniform structure defined by $\sigma(\mathscr{B}', \mathscr{B})$) and the maps $(\sigma_1, \sigma_2) \to \sigma_1 \dot{+} \sigma_2$ and $(\sigma_1, \sigma_2) \to \sigma_1 \cdot \sigma_2$ of $\Sigma_g \times \Sigma_g$ into Σ_g are uniformly continuous.*

PROOF. The uniform structure on L determined by $\sigma(\mathscr{B}', \mathscr{B})$ is the initial structure for the maps $L \xrightarrow{\mu(w, g)} [0, 1] \subset \mathbf{R}$ for $w \in K$ because every $y \in \mathscr{B}$ may be expressed in the form $y = \alpha w_1 - \beta w_2$ where $w_1, w_2 \in K$. If the composite map $\Sigma_g \to L \xrightarrow{\mu(w, g)} [0, 1]$ is uniformly continuous, that is, the map $\mu(w, F(\sigma))$ is uniformly continuous for each w, then the map F will be uniformly continuous (see AII, §2). From

$$F(\sigma_1) + F(\sigma_2 \dot{+} \sigma_1 \cdot \sigma_2) = F(\sigma_1 \vee \sigma_2) = F(\sigma_2) + F(\sigma_1 \dot{+} \sigma_1 \cdot \sigma_2)$$

and from

$$F(\sigma_1 \dot{+} \sigma_2) = F(\sigma_1 \dot{+} \sigma_1 \cdot \sigma_2) + F(\sigma_2 \dot{+} \sigma_1 \cdot \sigma_2)$$

it follows that

$$|\mu(w, F(\sigma_1) - F(\sigma_2))| \leq \mu(w, F(\sigma_1 \dot{+} \sigma_2)),$$

thus proving the uniform continuity of $\mu(w, F(\sigma))$.

From

$$F(\sigma_1 \dotplus \tilde{\sigma}_1 \dotplus \sigma_2 \dotplus \tilde{\sigma}_2) \leq F(\sigma_1 \dotplus \tilde{\sigma}_1) + F(\sigma_2 \dotplus \tilde{\sigma}_2)$$

it is easy to show that $(\sigma_1, \sigma_2) \to \sigma_1 \dotplus \sigma_2$ is uniformly continuous. From

$$\begin{aligned} F(\sigma_1 \dotplus \tilde{\sigma}_1) + F(\sigma_2 \dotplus \tilde{\sigma}_2) &\geq 2F((\sigma_1 \dotplus \tilde{\sigma}_1)\cdot(\sigma_2 \dotplus \tilde{\sigma}_2)) \\ &= 2F(\sigma_1 \cdot \sigma_2 \dotplus \sigma_1 \cdot \tilde{\sigma}_2 \dotplus \tilde{\sigma}_1 \cdot \sigma_2 \dotplus \tilde{\sigma}_1 \cdot \tilde{\sigma}_2) \\ &= 2F(\sigma_1 \cdot \sigma_2 \dotplus \tilde{\sigma}_1 \cdot \tilde{\sigma}_2) + 2F(\sigma_1 \cdot \tilde{\sigma}_2 \dotplus \tilde{\sigma}_1 \cdot \sigma_2) \end{aligned}$$

it follows that

$$F(\sigma_1 \cdot \sigma_2 \dotplus \tilde{\sigma}_1 \cdot \tilde{\sigma}_2) \leq \tfrac{1}{2}[F(\sigma_1 \dotplus \tilde{\sigma}_1) + F(\sigma_2 \dotplus \tilde{\sigma}_2)],$$

thus proving that the map $(\sigma_1, \sigma_2) \to \sigma_1 \cdot \sigma_2$ is uniformly continuous.

Th. 1.4.3. *The uniform completion $\hat{\Sigma}_g$ of Σ_g is a (lattice-theoretically) complete Boolean ring. The additive measure F may be extended on $\hat{\Sigma}_g$ to an additive measure $\hat{\Sigma}_g \to L$. If F is effective on Σ_g then its extension is effective on $\hat{\Sigma}_g$. For each subset $\Gamma \subset \hat{\Sigma}_g$ a denumerable subset $\sigma_\nu \in \Gamma$ can be so chosen such that $\bigvee_{\sigma\in\Gamma} \sigma = \bigvee_\nu \sigma_\nu$ (and similarly for $\bigwedge$). In addition $\bigvee_{\nu=1}^{n} \sigma_\nu \to \bigvee_{\nu=1}^{\infty} \sigma_\nu$. Both F and $m(\sigma) = \mu(w, F(\sigma))$, are also σ-additive measures on $\hat{\Sigma}_g$.*

PROOF. From Th. 1.4.2 and from the general theorems about the extension of uniformly continuous maps we find that $\hat{\Sigma}_g$ is a Boolean ring and that the map $\Sigma_g \xrightarrow{F} L$ may be extended as a uniformly continuous map because L is $\sigma(\mathscr{B}', \mathscr{B})$-complete since it is $\sigma(\mathscr{B}', \mathscr{B})$-compact. Similarly, it is easy to show that the extension of F is an additive measure on $\hat{\Sigma}_g$ since $\sigma_1, \sigma_2 \in \hat{\Sigma}_g$ may be approximated arbitrarily well by $\tilde{\sigma}_1, \tilde{\sigma}_2 \in \Sigma_g$; if $\sigma_1 \cdot \sigma_2 = 0$ then $\tilde{\sigma}_1 \cdot \tilde{\sigma}_2$ will approximate the null-element arbitrarily well. Similarly it follows that F is effective on $\hat{\Sigma}_g$ when it is effective on Σ_g.

In order to prove the lattice-theoretical completeness of $\hat{\Sigma}_g$ it suffices to show that, for a $\Gamma \subset \hat{\Sigma}_g$ the upper bound $\bigvee_{\sigma\in\Gamma} \sigma$ exists because if $\tilde{\Gamma}$ is the set of all $\tilde{\sigma}$ for which $\tilde{\sigma} \leq \sigma$ for all $\sigma \in \Gamma$ then we obtain $\bigwedge_{\sigma\in\Gamma} \sigma = \bigvee_{\tilde{\sigma}\in\tilde{\Gamma}} \tilde{\sigma}$.

Let ϕ denote the set of all *finite* subsets of Γ. Then $\bigvee_{\sigma\in\Gamma} \sigma = \bigvee_{\varphi\in\phi} \sigma_\varphi$ where $\sigma_\varphi = \bigvee_{\sigma\in\varphi} \sigma$. Since $\hat{\Sigma}_g$ is a Boolean ring $\sigma_\varphi \in \hat{\Sigma}_g$ the set of σ_φ, $\varphi \in \phi$ is a directed subset of $\hat{\Sigma}_g$. From the additivity of the measure $m_0(\sigma) = \mu(w_0, F(\sigma))$ where w_0 is defined in Th. 1.4.1, we may conclude that the set of numbers $m_0(\sigma_\varphi)$ is upwardly directed; since $m_0(\sigma_\varphi) \leq 1$ it is also bounded. The set of $m_0(\sigma_\varphi)$ therefore has an upper bound which we denote by α.

If $\varphi_1, \varphi_2 \in \phi$ and if $\varphi \in \phi$, $\varphi \subset \varphi_1$, $\varphi \subset \varphi_2$ then

$$\begin{aligned} m_0(\sigma_{\varphi_1} \dotplus \sigma_{\varphi_2}) &= m_0(\sigma_{\varphi_1} \dotplus \sigma_\varphi \dotplus \sigma_\varphi \dotplus \sigma_{\varphi_2}) \\ &\leq m_0(\sigma_{\varphi_1} \dotplus \sigma_\varphi) + m_0(\sigma_\varphi \dotplus \sigma_{\varphi_2}). \end{aligned}$$

Furthermore, since $\sigma_\varphi < \sigma_{\varphi_1}$, $\sigma_\varphi < \sigma_{\varphi_2}$ we find that

$$m_0(\sigma_{\varphi_1} \dotplus \sigma_\varphi) = m_0(\sigma_{\varphi_1}) - m_0(\sigma_\varphi) \leq \alpha - m_0(\sigma_\varphi);$$

a similar expression holds for σ_{φ_2}. Thus we obtain $m_0(\sigma_{\varphi_1} \dotplus \sigma_{\varphi_2}) \leq 2[\alpha - m_0(\sigma_\varphi)]$. Using $d(\ldots)$ from Th. 1.4.1 we obtain

$$d(\sigma_{\varphi_1}, \sigma_{\varphi_2}) \leq 2[\alpha - m_0(\sigma_\varphi)].$$

Since α is the upper limit of all $m_0(\sigma_\varphi)$, the directed set σ_φ, $\varphi \in \phi$ converges towards an element $\sigma_\phi \in \hat{\Sigma}_g$.

Since $\sigma_\varphi \cdot \sigma_{\varphi_1} = \sigma_\varphi$ for $\varphi_1 \supset \varphi$ we obtain $\sigma_\varphi \cdot \sigma_\phi = \sigma_\varphi$, that is, $\sigma_\phi \geq \sigma_\varphi$ for all $\varphi \in \phi$. If $\sigma > \sigma_\varphi$ for all $\varphi \in \phi$, then from $\sigma \cdot \sigma_\varphi = \sigma_\varphi$ it follows that, in the limit $\sigma \cdot \sigma_\phi = \sigma_\phi$ and also $\sigma_\phi < \sigma$. Therefore $\sigma_\phi = \bigvee_{\varphi \in \phi} \sigma_\varphi = \bigvee_{\sigma \in \Gamma} \sigma$.

As a special case of the above result, for a denumerable set $\{\sigma_\nu\}$ we obtain

$$\tilde{\sigma}_n = \bigvee_{\nu=1}^{n} \sigma_\nu \xrightarrow{n} \bigvee_{\nu=1}^{\infty} \sigma_\nu ;$$

since $\tilde{\sigma}_n$ is an increasing sequence, bounded from above, we therefore obtain

$$\tilde{\sigma}_n \longrightarrow \bigvee_{m=1}^{\infty} \tilde{\sigma}_m = \bigvee_{\nu=1}^{\infty} \sigma_\nu .$$

We shall now show that, in general, it is possible to choose a denumerable subset $\{\sigma_\mu\}$ from Γ such that $\bigvee_{\sigma \in \Gamma} \sigma = \bigvee_{\mu=1}^{\infty} \sigma_\mu$. First we can choose a denumerable sequence φ_ν from ϕ such that $d(\sigma_{\varphi_2}, \sigma_\phi) \to 0$. For $\tilde{\sigma} = \bigvee_\nu \sigma_{\varphi_\nu}$ we find that $\sigma_{\varphi_\nu} < \tilde{\sigma} < \sigma_\phi = \bigvee_{\varphi \in \phi} \sigma_\varphi$. Therefore, as above, since $\sigma_{\varphi_\nu} \to \sigma_\phi$, it follows that $\sigma_\phi < \tilde{\sigma} < \sigma_\phi$; that is, $\tilde{\sigma} = \sigma_\phi$. Thus, using the *denumerable* set $\gamma = \bigcup_\nu \varphi_\nu \subset \Gamma$: we obtain $\bigvee_{\sigma \in \Gamma} \sigma = \sigma_\phi = \tilde{\sigma} = \bigvee_\nu \sigma_{\varphi_\nu} = \bigvee_{\sigma \in \gamma} \sigma$.

F is σ-additive if for a denumerable sequence $\sigma_\nu \in \hat{\Sigma}_g$ for which $\sigma_\nu \cdot \sigma_\mu = 0$ for $\nu \neq \mu$

$$F\left(\bigvee_\nu \sigma_\nu\right) = \sum_\nu F(\sigma_\nu).$$

Let $\tilde{\sigma}_m = \bigvee_{\nu=1}^{m} \sigma_\nu$; then we obtain $F(\tilde{\sigma}_m) = \sum_{\nu=1}^{m} F(\sigma_\nu)$. We must therefore show that $F(\tilde{\sigma}_m)$ converges towards $F(\bigvee_\nu \sigma_\nu)$ in the $\sigma(\mathscr{B}', \mathscr{B})$-topology. Above we have already shown that $\tilde{\sigma}_m \xrightarrow{m} \bigvee_\nu \sigma_\nu$; from the continuity of F this assertion is proven. From the σ-additivity of F it easily follows that $m(\sigma) = \mu(w, F(\sigma))$ is also σ-additive.

Th. 1.4.4. *Let Σ be a (lattice-theoretically) complete Boolean ring, let m_0 be a σ-additive effective measure $\Sigma \xrightarrow{m_0} [0, 1] \subset \mathbf{R}$ and let U_g be the uniform structure induced by $d(\sigma_1, \sigma_2) = m_0(\sigma_1 \dotplus \sigma_2)$. Then Σ is U_g-complete. Then, given a convergent sequence $\sigma_\nu \to \sigma$ a subsequence σ_{ν_i} can be chosen such that the decreasing sequence $\tilde{\sigma}_m = \bigvee_{i=m}^{\infty} \sigma_{\nu_i}$ converges to σ and $\sigma = \bigwedge_m \tilde{\sigma}_m$.*

PROOF. Σ is U_g-complete if every Cauchy sequence σ_ν converges to a limit $\sigma \in \Sigma$. From σ_ν we select a (yet to be determined) subsequence σ_{ν_k} and (since Σ is lattice-theoretically complete) construct $\tilde{\sigma}_n = \bigvee_{k=n}^{\infty} \sigma_{\nu_n}$ and $\sigma = \bigwedge_{n=1}^{\infty} \tilde{\sigma}_n$. We will now show that $\sigma_\nu \to \sigma$. Since σ_ν is a Cauchy sequence, it suffices to show that, for a subsequence $\sigma_{\nu_k} \to \sigma$, that is, $d(\sigma_{\nu_k}, \sigma) \to 0$.

Suppose that m_0 is σ-additive and that $\hat{\sigma}_\mu$ is a decreasing sequence which satisfies $\bigwedge_\mu \hat{\sigma}_\mu = 0$. We claim that $m_0(\hat{\sigma}_\mu) \to 0$. From

$$\hat{\sigma}_1 = (\hat{\sigma}_1 \dotplus \hat{\sigma}_2) \vee (\hat{\sigma}_2 \dotplus \hat{\sigma}_3) \vee \cdots \vee (\hat{\sigma}_{m-1} \dotplus \hat{\sigma}_m) \vee \cdots$$

it follows that, on the basis of σ-additivity, that

$$m_0(\hat{\sigma}_1 \dotplus \hat{\sigma}_2) + m_0(\hat{\sigma}_2 \dotplus \hat{\sigma}_3) + \cdots + m_0(\hat{\sigma}_{m-1} \dotplus \hat{\sigma}_m) \xrightarrow{m} m_0(\hat{\sigma}_1).$$

Since $m_0(\sigma_\mu \dotplus \sigma_{\mu+1}) = m_0(\sigma_\mu) - m_0(\sigma_{\mu+1})$ the left-hand side of the above equation is equal to $m_0(\hat{\sigma}_1) - m_0(\hat{\sigma}_m)$. Therefore it follows that $m_0(\hat{\sigma}_m) \to 0$.

From $\sigma = \bigwedge_{n=1}^{\infty} \tilde{\sigma}_n$ it follows that $\hat{\sigma}_n = \tilde{\sigma}_n \dot{+} \sigma$ is a decreasing sequence for which $\bigwedge_{n=1}^{\infty} \hat{\sigma}_n = 0$. Therefore $d(\tilde{\sigma}_n, \sigma) = m_0(\tilde{\sigma}_n \dot{+} \sigma) \to 0$. From $\tilde{\sigma}_m = \sigma_{\nu_m} \vee (\bigvee_{i=m+1}^{\infty} \sigma_{\nu_i})$ it follows that

$$\tilde{\sigma}_m \dot{+} \sigma_{\nu_m} < \bigvee_{i=m+1}^{\infty} (\sigma_{\nu_i} \dot{+} \sigma_{\nu_m}) = \bigvee_{i=m+1}^{\infty} \left[\sigma_{\nu_i} \dot{+} \sigma_{\nu_m} \dot{+} (\sigma_{\nu_i} \dot{+} \sigma_{\nu_m}) \cdot \bigvee_{j=m+1}^{i-1} (\sigma_{\nu_j} \dot{+} \sigma_{\nu_m}) \right].$$

Thus, since m_0 is σ-additive, it follows that

$$d(\tilde{\sigma}_m, \sigma_{\nu_m}) \leq \sum_{i=m+1}^{\infty} m_0 \left[\sigma_{\nu_i} \dot{+} \sigma_{\nu_m} \dot{+} (\sigma_{\nu_i} \dot{+} \sigma_{\nu_m}) \cdot \bigvee_{j=m+1}^{i-1} (\sigma_{\nu_j} \dot{+} \sigma_{\nu_m}) \right]$$

$$\leq \sum_{i=m+1}^{\infty} m_0(\sigma_{\nu_i} \dot{+} \sigma_{\nu_m}) = \sum_{i=m+1}^{\infty} d(\sigma_{\nu_i}, \sigma_{\nu_m}).$$

Since σ_ν is a Cauchy sequence, we may select a subsequence such that $d(\sigma_{\nu_k}, \sigma_\mu) < 1/2^k$ for $\mu > \nu_k$. Thus we obtain

$$d(\tilde{\sigma}_m, \sigma_{\nu_m}) \leq \sum_{k=m+1}^{\infty} \frac{1}{2^k} \xrightarrow{m} 0.$$

From $d(\sigma_{\nu_m}, \sigma) \leq d(\sigma_{\nu_m}, \tilde{\sigma}_m) + d(\tilde{\sigma}_m, \sigma)$ it follows that $d(\sigma_{\nu_m}, \sigma) \to 0$ and therefore also $\sigma_\nu \to \sigma$. Since $\sigma \in \Sigma$, Σ is U_g-complete.

In reference to Th. 1.4.3 it is important to note that the following situation is possible: For a subset $\Gamma \subset \Sigma_g$ there exists an upper bound in Σ_g which is usually denoted by $\bigvee_{\sigma \in \Gamma} \sigma$. We shall, however, denote *this* upper bound *in* Σ_g by $\bigvee_{\sigma \in \Gamma}^{\Sigma} \sigma$. The same subset Γ, as a subset of $\hat{\Sigma}_g$ has, according to Th. 1.4.3, an upper bound in $\hat{\Sigma}_g$, which in Th. 1.4.3 was denoted by $\bigvee_{\sigma \in \Gamma} \sigma$. Note that it is possible that

$$\bigvee_{\sigma \in \Gamma}^{\Sigma} \sigma \neq \bigvee_{\sigma \in \Gamma} \sigma.$$

In this sense it is possible that F is σ-additive in $\hat{\Sigma}_g$, but is not σ-additive in Σ_g!

The map $\Sigma \xrightarrow{F} L$ represents an idealization of the situation $\mathscr{R}(b_0) \xrightarrow{\psi_0} L$. For Σ, as an idealization of $\mathscr{R}(b_0)$, the uniform structure U_g which was introduced in Th. 1.4.1 also appears to be physically meaningful. U_g distinguishes two elements σ_1, σ_2 from Σ with the aid of *finitely* many $w \in K$ by means of the probabilities $\mu(w, F(\sigma_1 \dot{+} \sigma_2))$ where $\sigma_1 \dot{+} \sigma_2$ is that "registration" in which *only one* of the two registrations σ_1, σ_2 have responded. It is premature, however, to identify U_g with the uniform structure of the "physical imprecision" associated with Σ (this uniform structure of "physical imprecision" is treated in a general setting in [1], §6 and §9) because of the following unexpected property of U_g:

In general $\hat{\Sigma}_g$ is not compact—as we would expect (according to [1], §9) for a uniform structure of physical imprecision. If we postulate that Σ is denumerable, then $\hat{\Sigma}_g$ will be separable. $\mathscr{R}(b_0)$ is denumerable, in agreement with the assumptions made at the beginning of III, §3 about M, $\mathscr{Q}$, $\mathscr{R}$.

Since $\hat{\Sigma}_g$ need not be compact, we would like to obtain another uniform structure which would provide a better map of the *physical* imprecision of the distinguishability of the elements of Σ than that provided by U_g.

For the special case $\Sigma \xrightarrow{F} G$ the Boolean ring Σ is isomorphic to the image of Σ in G. The uniform structure which describes the physical precision in G is that generated by $\sigma(\mathscr{B}', \mathscr{B})$. The latter may be transferred to Σ as the initial structure which corresponds to the map F. This result would suggest that it is reasonable to use the initial uniform structure on Σ (which is generated by the map $\Sigma \xrightarrow{F} L$) as the uniform structure of physical imprecision. But the initial uniform structure does not, in general, separate Σ; for $\Sigma = \mathscr{R}(b_0)$, $F = \psi_0$ we obtain (by means of $F(\sigma_1) = F(\sigma_2)$) the same equivalence classes of elements (b_0, b) which we have considered in III, §1.

The uniform structure U_g permits us to compare each σ_1 with all the other σ's. By analogy to the fact that we can test a σ_1 only with a finite number of $w \in K$ we may only compare a σ_1 with a finite number of the $\sigma \in \Sigma$. This fact leads us to suggest that we adopt the initial uniform structure generated by the maps $\sigma \longrightarrow F_{\tilde{\sigma}}(\sigma) = F(\sigma \cdot \tilde{\sigma})$—that is, the weakest uniform structure for which the maps $F_{\tilde{\sigma}}$ are uniformly continuous for each $\tilde{\sigma}$ (see AII, §2)—as the uniform structure U_p of physical imprecision. Since the uniform structure generated by $\sigma(\mathscr{B}', \mathscr{B})$ is identical to the initial structure defined by the maps $L \xrightarrow{\mu(w,g)} [0, 1] \subset \mathbf{R}$ for all $w \in K$, U_p is equal to the initial structure which is generated by the maps:

$$\Sigma \xrightarrow{\mu(w, F(\tilde{\sigma}, \sigma))} [0, 1] \quad \text{for all } w \in K, \tilde{\sigma} \in \Sigma.$$

We therefore "test" a σ by means of U_p using a finite number of the $\tilde{\sigma} \in \Sigma$ and a finite number of the $w \in K$ (see the general treatment presented in [1], §9).

According to Th. 1.4.2 the maps $\Sigma \xrightarrow{F(\sigma \cdot \tilde{\sigma})} L$ (for fixed $\tilde{\sigma}$) are uniformly continuous. Therefore U_g is finer than U_p. If F is effective (which we shall always assume—see Th. 1.2.2) then U_p will also separate Σ because if $F(\sigma_1 \cdot \tilde{\sigma}) = F(\sigma_2 \cdot \tilde{\sigma})$ for all $\tilde{\sigma} \in \Sigma$ then, for $\tilde{\sigma} = \sigma_2$ we obtain $F(\sigma_2) = F(\sigma_1 \cdot \sigma_2)$. Thus, from $F(\sigma_2) = F(\sigma_2 \dotplus \sigma_1 \cdot \sigma_2) + F(\sigma_1 \cdot \sigma_2)$ it follows that $F(\sigma_2 \dotplus \sigma_1 \cdot \sigma_2) = 0$, that is, $\sigma_2 \dotplus \sigma_1 \cdot \sigma_2 = 0$ and hence $\sigma_2 = \sigma_1 \cdot \sigma_2$. Thus we may derive (for $\tilde{\sigma} = \sigma_1$) $\sigma_1 = \sigma_1 \cdot \sigma_2$. Therefore we obtain $\sigma_1 = \sigma_2$.

The map $F(\sigma \cdot \tilde{\sigma})$ may be extended to all of $\hat{\Sigma}_g$ as a uniformly continuous map by means of the extension of F upon $\hat{\Sigma}_g$ in such a way that U_p will be defined in all of $\hat{\Sigma}_g$. The uniform structure of U_p is coarser than that of U_g and is the initial structure for the maps $\hat{\Sigma}_g \xrightarrow{F} L$ where $F_{\tilde{\sigma}}(\sigma) = F(\sigma \cdot \tilde{\sigma})$ for all $\tilde{\sigma} \in \Sigma_g$. Would we obtain a finer initial structure on $\hat{\Sigma}_g$ if we admit all $\tilde{\sigma} \in \hat{\Sigma}_g$? In fact we obtain the same structure U_p as the initial structure for the maps $\Sigma_g \xrightarrow{F} L$ for all $\tilde{\sigma} \in \hat{\Sigma}_g$. This becomes evident from the following estimate for $\hat{\sigma} \in \hat{\Sigma}_g$ and $\tilde{\sigma} \in \Sigma_g$, which is obtained by analogy with the proof of Th. 1.4.2.

$$\begin{aligned} |\mu(w, F(\sigma_1 \cdot \hat{\sigma}) - F(\sigma_2 \cdot \hat{\sigma}))| \leq{} & |\mu(w, F(\sigma_1 \cdot \hat{\sigma}) - F(\sigma_2 \cdot \tilde{\sigma}))| \\ & + |\mu(w, F(\sigma_1 \cdot \tilde{\sigma}) - F(\sigma_2 \cdot \tilde{\sigma}))| \\ & + |\mu(w, F(\sigma_2 \cdot \tilde{\sigma}) - F(\sigma_2 \cdot \hat{\sigma}))|; \end{aligned}$$

$$\begin{aligned} |\mu(w, F(\sigma \cdot \hat{\sigma}) - F(\sigma \cdot \tilde{\sigma}))| &\leq \mu(w, F(\sigma \cdot \hat{\sigma} \dotplus \sigma \cdot \tilde{\sigma})) \\ &= \mu(w, F(\sigma \cdot (\hat{\sigma} \dotplus \tilde{\sigma}))) \leq \mu(w, F(\hat{\sigma} \dotplus \tilde{\sigma}) \end{aligned}$$

and we therefore obtain

$$|\mu(w, F(\sigma_1 \cdot \hat{\sigma}) - F(\sigma_2 \cdot \hat{\sigma}))| \leq 2\mu(w, F(\hat{\sigma} \dotplus \tilde{\sigma})) + |\mu(w, F(\sigma_1 \cdot \tilde{\sigma}) - F(\sigma_2 \cdot \tilde{\sigma}))|.$$

The previous discussion shows that if we impose the condition $\Sigma_g = \hat{\Sigma}_g$ we do not lose any physical generality, but we do gain mathematical simplicity. We define:

D 1.4.1. Let Σ be a Boolean ring with additive measure $F: \Sigma \longrightarrow L$ such that Σ is complete with respect to the uniform structure U_g defined in Th. 1.4.1 (or equivalently by Th. 1.4.3 and Th. 1.4.4, where F is σ-additive and Σ is lattice theoretically complete). We define the uniform structure of "physical imprecision" U_p on Σ as the initial uniform structure generated by the maps $\Sigma \xrightarrow{F_{\tilde{\sigma}}} L$, $F_{\tilde{\sigma}}(\sigma) = F(\sigma \cdot \tilde{\sigma})$, $\tilde{\sigma} \in \Sigma$ and L is endowed with the uniform structure generated by $\sigma(\mathscr{B}', \mathscr{B})$. We shall let Σ_p denote Σ together with the uniform structure U_p (Σ_g is therefore complete; Σ_p need not necessarily be complete).

Th. 1.4.5. *Σ_p is precompact (see* AII, §2). *The topologies generated by U_p and U_g are identical.*

PROOF. Since L is $\sigma(\mathscr{B}', \mathscr{B})$-compact, Σ_p is precompact (see AII, §2). Since U_g is finer than U_p, it is only necessary to show that the identity map $\Sigma_p \longrightarrow \Sigma_g$ is continuous (not necessarily uniformly continuous!) in order to prove that the topologies generated by U_p and U_g are identical. Now, for fixed σ we have:

$$d(\sigma, \sigma_1) = \mu(w_0, F(\sigma \dotplus \sigma_1)) = \mu(w_0, F(\sigma) + F(\sigma_1) - 2F(\sigma \cdot \sigma_1))$$
$$\leq 2|\mu(w_0, F(\sigma) - F(\sigma \cdot \sigma_1)| + |\mu(w_0, F(\sigma_1) - F(\sigma)|.$$

Therefore, for the case in which $\tilde{\sigma}_1 = \sigma$, $\tilde{\sigma}_2 = \varepsilon$ (ε = unit element) we have

$$d(\sigma, \sigma_1) \leq 2|\mu(w_0, F(\sigma \cdot \tilde{\sigma}_1) - F(\sigma_1 \cdot \tilde{\sigma}_1))| + |\mu(w_0, F(\sigma \cdot \tilde{\sigma}_2) - F(\sigma_1 \cdot \tilde{\sigma}_2))|$$

from which we conclude that the identity map $\Sigma_p \longrightarrow \Sigma_g$ is continuous.

According to Th. 1.4.5 the structure Σ_g, Σ_p is of the form which has been discussed more generally in [1], §9. U_p is somewhat "more physical" than U_g. Σ_p is precompact (totally bounded), but the structure of this Boolean ring may not necessarily be extended to the completion $\hat{\Sigma}_p$. We note that U_g is characterized by the fact that $\hat{\Sigma}_g = \Sigma_g$ is a complete Boolean ring; in order to obtain a Boolean ring for completion it is methodologically desirable to introduce U_g and not only U_p.

Since we wish to describe an idealization of the situation $\mathscr{R}(b_0) \xrightarrow{\psi_0} L$ by $\Sigma \xrightarrow{F} L$ and since $\mathscr{R}(b_0)$ is denumerable, it seems reasonable to introduce the notion of an observable in the following way:

D 1.4.2. By an *observable* we mean a pair of objects (Σ, F) where Σ is a Boolean ring and F is an additive effective measure $F: \Sigma \longrightarrow L$ for which Σ is complete and *separable* with respect to the uniform structure U_g (defined by Th. 1.4.1 using F). We shall denote the observable (Σ, F) also by $\Sigma \xrightarrow{F} L$.

According to D 1.4.2, $\Sigma_1 \xrightarrow{F_1} L$ and $\Sigma_2 \xrightarrow{F_2} L$ are considered to be the "same" observable if there is an isomorphism $\Sigma_1 \xrightarrow{i} \Sigma_2$ of the Boolean rings for which $F_1(\sigma) = F_2(i\sigma)$.

Th. 1.4.6. *$\Sigma \xrightarrow{F} L$ is an observable if and only if Σ is lattice theoretically complete, F is σ-additive and there exists a denumerable Boolean sublattice Σ_a of Σ whose (lattice theoretical) completion in Σ is equal to Σ.*

PROOF. According to Th. 1.4.3 Σ (of D 1.4.2) is lattice theoretically complete and F is σ-additive. Since Σ is U_g-separable (according to D 1.4.2) there exists a denumerable subset $A \subset \Sigma$ which is U_g-dense in Σ. The Boolean ring Σ_A generated by A is denumerable. The lattice theoretical completion of Σ_A in Σ is a (lattice theoretically) complete Boolean ring $\tilde{\Sigma}_A \subset \Sigma$. According to Th. 1.4.4 $\tilde{\Sigma}_A$ is U_g-complete. Since $\tilde{\Sigma}_A$ is U_g-dense in Σ and since Σ is U_g-complete $\tilde{\Sigma}_A = \Sigma$.

Conversely, if Σ is lattice theoretically complete and F is σ-additive, then by Th. 1.4.4 Σ is U_g-complete. We shall now show that Σ_a is U_g-dense in Σ. Let $\hat{\Sigma}_{ag}$ be the completion of Σ_a with respect to U_g. Since Σ is U_g-complete, $\hat{\Sigma}_{ag} \subset \Sigma$. Note that $\hat{\Sigma}_{ag}$ is, according to Th. 1.4.3, lattice theoretically complete; therefore the lattice theoretical completion Σ of Σ_a lies in $\hat{\Sigma}_{ag}$, that is $\Sigma \subset \hat{\Sigma}_{ag}$. Therefore $\hat{\Sigma}_{ag} = \Sigma$ and Σ_a is U_g-dense in Σ.

In previous theorems we have only assumed that F maps the Boolean ring Σ into L. For the special case in which $\Sigma \xrightarrow{F} G$, then it is, in principle, conceivable that the extension of the map F to the completion $\hat{\Sigma}_g$ will yield points which are not elements of G. The following theorem rules out this possibility.

Th. 1.4.7. *Suppose that, for a Boolean ring Σ, an additive measure $F: \Sigma \to G$ is given. Then, for the extension of $F: \hat{\Sigma}_g \to L$ $F\hat{\Sigma}_g$ is contained in the $\sigma(\mathscr{B}', \mathscr{B})$-closure of $F\Sigma$ in G.*

PROOF. According to Th. 1.4.1 U_g is metrizable. Thus it suffices to show that, for a sequence $\sigma_\nu \in \Sigma$ satisfying $d(\sigma_\nu, \sigma) \to 0$ for $\sigma \in \hat{\Sigma}_g$, that if $F(\sigma_\nu) \in G$ then $F(\sigma) \in G$. Since $F(\sigma_\nu) \to F(\sigma)$ in the $\sigma(\mathscr{B}', \mathscr{B})$-topology, then we have also proven that $F\Sigma_g$ is in the $\sigma(\mathscr{B}', \mathscr{B})$ closure of $F\Sigma$ in G.

Since $F(\sigma_\nu)$ converges towards $F(\sigma)$, in order to prove that $F(\sigma) \in G$ it suffices to consider a subsequence σ_{ν_i}; where the latter is chosen such that (using the notation of Th. 1.4.4)

$$\tilde{\sigma}_m = \bigvee_{i=m}^{\infty} \sigma_{\nu_i} \to \sigma.$$

Then we obtain $F(\tilde{\sigma}_m) \to F(\sigma)$.

According to Th. 1.4.3 we have

$$\tilde{\sigma}_{m,N} = \bigvee_{i=m}^{N} \sigma_{\nu_i} \xrightarrow[N]{} \tilde{\sigma}_m$$

we therefore obtain

$$F(\tilde{\sigma}_{m,N}) \xrightarrow[N]{} F(\tilde{\sigma}_m).$$

Since $\tilde{\sigma}_{m,N} \in \Sigma$ we obtain $F(\tilde{\sigma}_{m,N}) \in G$. Therefore, by III, Th. 6.10 $F(\tilde{\sigma}_m) \in G$. $F(\tilde{\sigma}_m)$ is a decreasing sequence; therefore, by III, Th. 6.10 we obtain $F(\sigma) \in G$.

Th. 1.4.7 shows that in the case in which $\Sigma \xrightarrow{F} G$ we may assume that, Σ_g is complete without any loss of generality. Thus we define:

D 1.4.3. A *decision observable* is a pair (Σ, F) where Σ is a Boolean ring and F is an additive effective measure $F: \Sigma \to G$ for which Σ is complete with respect to the uniform structure U_g. (We do *not* require that Σ is U_g-separable; this fact is a result of Th. 1.4.8.)

Th. 1.4.8. *For a decision observable Σ_g is always separable. F is an isomorphism onto the image set $F\Sigma \subset G$; for each decision observable Σ may be identified with a U_g-closed Boolean sublattice of G. Each (lattice theoretically) complete Boolean sublattice of G is also U_g-complete, and conversely, so that decision observables may be identified with (lattice theoretically) complete Boolean sublattices of G. The uniform structure U_p defined in* D 1.4.1 *is identical to the uniform structure on Σ generated by $\sigma(\mathscr{B}', \mathscr{B})$. The topology on Σ generated by U_g is identical to the $\sigma(\mathscr{B}', \mathscr{B})$ topology on Σ (therefore Σ is $\sigma(\mathscr{B}', \mathscr{B})$ closed in G since Σ is U_g-complete (see* Th. 1.3.7).

PROOF. On the basis of Th. 1.3.3 Σ may be identified with $F\Sigma$. According to Th. 1.4.4 the U_g-complete Boolean sublattices of G and the lattice-theoretically complete sublattices of G are the same. Therefore it is only necessary to show that every complete Boolean sublattice of G is U_g-separable and that the uniform structure U_p is identical to that generated by $\sigma(\mathscr{B}', \mathscr{B})$.

Let Σ be a complete Boolean sublattice of G. In order to prove that Σ is U_g-separable, we recall that $\mathscr{B}'$ is $\sigma(\mathscr{B}', \mathscr{B})$ separable (see AIII, §4). Therefore Σ is $\sigma(\mathscr{B}', \mathscr{B})$-separable. Therefore there exists a countable subset $A \subset \Sigma$ which is $\sigma(\mathscr{B}', \mathscr{B})$-dense in Σ. The Boolean subring Σ_A (where Σ_A is generated by A—according to Th. 1.3.6(v)) is denumerable; let $\bar{\Sigma}_A$ denote the complete Boolean subring generated by A according to Th. 1.3.6(vi)). Since Σ is complete, we obtain $\Sigma_A \subset \bar{\Sigma}_A \subset \Sigma$.

Since A is $\sigma(\mathscr{B}', \mathscr{B})$-dense in Σ and since $\bar{\Sigma}_A$ and Σ are, according to Th. 1.3.7, $\sigma(\mathscr{B}', \mathscr{B})$-closed in G, we obtain $\bar{\Sigma}_A = \Sigma$. Since the lattice theoretical completion $\bar{\Sigma}_A$ of Σ_A is equal to Σ, then, according to Th. 1.4.6, Σ is U_g-separable.

Since Σ is identified with a subset of G, the map F becomes the identity map and the structure U_p (defined by D 1.4.1) is the initial structure generated by the maps $F_{\tilde{e}}(e) = e \wedge \tilde{e}$ (for all $\tilde{e} \in \Sigma$) of Σ into G. For $\tilde{e} = 1$ we obtain the identity map; thus U_p is finer than the uniform structure defined by $\sigma(\mathscr{B}', \mathscr{B})$. If, however, the maps $F_{\tilde{e}}$ (for fixed $\tilde{e}$) are uniformly continuous with respect to $\sigma(\mathscr{B}', \mathscr{B})$ then U_p is identical to the uniform structure generated by $\sigma(\mathscr{B}', \mathscr{B})$.

The fact that $F_{\tilde{e}}$ is uniformly continuous is a consequence of

$$\mu(w, \tilde{e} \wedge e_1 - \tilde{e} \wedge e_2) = \mu(w', e_1 - e_2), \tag{1.4.1}$$

where w' depends on w and $\tilde{e}$, as we wished to prove.

Since $\tilde{e}, e_1, e_2$ are commensurable, then, by Th. 1.3.4(v) and (vi) we obtain $\tilde{e} \wedge e_1 = \tilde{e}e_1 = e_1\tilde{e} = \tilde{e}e_1\tilde{e}$ and, correspondingly, $\tilde{e} \wedge e_2 = \tilde{e}e_2 = e_2\tilde{e} = \tilde{e}e_2\tilde{e}$. For $w' = \tilde{e}w\tilde{e}$, from $\mu(w, g) = \mathrm{tr}(wg)$ we obtain $\mu(w, \tilde{e}(e_1 - e_2)\tilde{e}) = \mu(\tilde{e}w\tilde{e}, e_1 - e_2)$ thus proving (1.4.1).

The fact that U_p is identical to the uniform structure defined by $\sigma(\mathscr{B}', \mathscr{B})$ on $\Sigma \subset G$ is only an assurance that we do not provide two different uniform structures for physical imprecision, because, as a subset of G, Σ already has the uniform structure of physical imprecision defined by $\sigma(\mathscr{B}', \mathscr{B})$ (see III, §3).

Th. 1.4.8 permits us to characterize the decision observables entirely by complete Boolean subrings Σ of G. In this way the Boolean operations in Σ immediately represent the switching-algebra of an idealized registration apparatus—the measurement apparatus for the observable. However, this "simplification" which naively identifies the measurements of decision observables with subsets Σ of decision effects makes the explanation of the measurement problem more difficult.

This difficulty is increased by the fact that (as we will see in §2.5) the "usual" concept of an observable is identical with what we called a decision observable and that we are accustomed to view only these decision observables. We note, however, that our notion of an observable is more general and realistic as a deduced concept; since it is a deduced concept we do not need to make vague statements such as "an observable is something that can be measured." Its physical meaning may be obtained from its definition as an idealization of the situation $\mathscr{R}(b_0) \xrightarrow{\psi_0} L$ to which we have already given a physical interpretation—a conceptual joining of the registration procedures $b \in \mathscr{R}(b_0)$ and the set of possible frequencies $\lambda_{\mathscr{S}}(a \cap b_0, a \cap b)$ for the various preparation procedures $a \in \mathscr{Q}'$. In this description, the problem of finding a "measurement method" $b_0 \in \mathscr{R}_0$ which permits us to at least approximately "measure" the observable $\Sigma \xrightarrow{F} L$ need no longer be assigned to a domain between theory and experiment which may only be described by "words" and not by a theory. Instead, it becomes a question which must be treated in a theory which is perhaps more comprehensive than quantum mechanics because the physical interpretation of quantum mechanics does not depend upon the concept of an observable but depends instead on the actual physical methods of preparation and registration. In §4 we shall begin with a step by step discussion of the problem of the measurement of an observable. This topic will be treated in more detail in XVII and XVIII. Nevertheless, the solution of the measurement problem concerning the macroscopic signals of a macroscopic apparatus will not be treated in this book. This problem has been solved in [13], X where the compatibility of an extrapolated quantum mechanics for "many particles" with the macroscopic description of the macroscopic measurement and preparation apparatuses is demonstrated. Such a macroscopic description of the apparatuses was used as a starting point for the foundation of quantum mechanics presented in II.

2 Structures in the Class of Observables

By introducing the concept of an observable we seek to eliminate a portion of the structure associated with a particular apparatus from the registration process. For registration methods only the abstract structure of a Boolean ring together with an additive measure $F: \Sigma \to L$ remains. Without any additional analysis it is already evident that the concept of an observable already contains too much of the structure of the registration methods. We should therefore seek to eliminate "unnecessary" and "bad" registration methods. Is it possible that such an elimination procedure will lead us to the concept of a decision observable? If this is the case, then the decision observables are the "true" measurements of the microsystems and exhibit the real structure of the microsystems.

These and similar questions make it necessary for us to examine the concept of an observable more closely, as we shall do in this section. The reader who is not interested in a deeper analysis of the concept of an observable may omit this section. This is possible because, in §1 we have already introduced a number of important theorems which have a close connection with the analysis of §2.

2.1 The Spaces $\mathscr{B}(\Sigma)$ and $\mathscr{B}'(\Sigma)$

Let Σ be a Boolean ring and let $m_0: \Sigma \to [0, 1] \subset \mathbf{R}$ be an additive real effective measure. Σ may be complete with respect to U_g (that is, Σ together with the uniform structure defined by the metric $d(\sigma_1, \sigma_2) = m_0(\sigma_1 \dotplus \sigma_2)$). Let w_0 be defined as in Th. 1.4.1; then, according to §1.4 we may choose m_0 as follows:

$$m_0(\sigma) = \mu(w_0, F(\sigma)).$$

According to Th. 1.4.4 a (lattice theoretically) complete Boolean ring Σ with a σ-additive effective measure is U_g-complete.

We shall now recapitulate some of the results obtained in §1.

Th. 2.1.1. *To each set $\Gamma \subset \Sigma$ there is a denumerable subset $\{\sigma_\nu\} \subset \Gamma$ for which $\bigvee_{\sigma \subset \Gamma} \sigma = \bigvee_\nu \sigma_\nu$. $\tilde{\sigma}_n = \bigvee_{\nu=1}^n \sigma_\nu$ is an increasing sequence for which $\bigvee_n \tilde{\sigma}_n = \bigvee_\nu \sigma_\nu = \bigvee_{\sigma \in \Gamma} \sigma$ and for which $\tilde{\sigma}_n \to \bigvee_{\sigma \in \Gamma} \sigma$ (in the topology generated by U_g). For every increasing sequence σ_ν we obtain $\sigma_\nu \to \bigvee_\nu \sigma_\nu$. A similar result holds for $\bigwedge_{\sigma \in \Gamma} \sigma$.*

Proof. See Th. 1.4.3.

Th. 2.1.2. *Let σ_ν be a convergent sequence (in the topology generated by U_g) and suppose $\sigma_\nu \to \sigma$. Then we may choose a subsequence σ_{ν_i} such that, for $\tilde{\sigma}_m = \bigvee_{i=m}^\infty \sigma_{\nu_i}$ the following relationships hold:*

$$\tilde{\sigma}_m \to \sigma, \qquad \sigma = \bigwedge_m \tilde{\sigma}_m.$$

Proof. See Th. 1.4.4.

D 2.1.1. A real function $x(\sigma)$ over Σ is said to be a signed additive (or signed σ-additive) measure over Σ if there exists a real number c such that $|x(\sigma)| < c$ for all σ and if, for $\sigma = \bigvee_\nu \sigma_\nu, \sigma_\nu \wedge \sigma_\mu = 0$ for $\nu \neq \mu$ the equation

$$x(\sigma) = \sum_\nu x(\sigma_\nu) \tag{2.1.1}$$

holds for finitely (or countably) many σ_ν.

All theorems about signed σ-additive measures will also hold for σ-additive measures.

Th. 2.1.3. *Let x be a signed, additive measure over Σ. Then the following conditions are equivalent:*

(i) *x is σ-additive.*
(ii) *If σ_n is a decreasing sequence for which $\bigwedge_n \sigma_n = 0$ then $x(\sigma_n) \to 0$.*
(iii) *If σ_n is a decreasing sequence for which $\bigwedge_n \sigma_n = \sigma$ then $x(\sigma_n) \to x(\sigma)$.*
(iv) *If σ_n is an increasing sequence for which $\bigvee_n \sigma_n = \sigma$ then $x(\sigma_n) \to x(\sigma)$.*

PROOF. (i) $\Rightarrow$ (iv). Assume that σ_n is defined according to (iv). Let $\sigma_0 = 0$; then $\sigma = \bigvee_{n=0}^\infty (\sigma_{n+1} \dotplus \sigma_n)$. Then by (i) it follows that

$$\begin{aligned} x(\sigma) &= \lim_{m\to\infty} \sum_{n=0}^m [x(\sigma_{n+1}) - x(\sigma_n)] \\ &= \lim_{m\to\infty} x(\sigma_{m+1}). \end{aligned}$$

(iv) $\Rightarrow$ (iii). Let σ_n be defined according to (iii). Then σ_n^* is a sequence for which $\bigvee_n \sigma_n^* = \sigma^*$. According to (iv) it follows that $x(\sigma_n^*) \to x(\sigma^*)$. Since $x(\sigma^*) + x(\sigma) = x(\varepsilon)$ it follows $x(\sigma_n) \to x(\sigma)$.

(iii) $\Rightarrow$ (ii) is trivial, since (ii) is a special case of (iii).

(ii) $\Rightarrow$ (i). Let $\sigma = \bigvee_{\nu=1}^\infty \sigma_\nu$ where $\sigma_\nu \wedge \sigma_\mu = 0$ for $\nu \neq \mu$. The sequence $\tilde{\sigma}_n = \sigma \dotplus \bigvee_{\nu=1}^n \sigma_\nu$ is a decreasing sequence, for which, according to (ii)

$$x\left(\sigma \dotplus \bigvee_{\nu=1}^n \sigma_\nu\right) \xrightarrow{n} 0.$$

Since

$$\begin{aligned} x\left(\sigma \dotplus \bigvee_{\nu=1}^m \sigma_\nu\right) &= x(\sigma) - x\left(\bigvee_{\nu=1}^n \sigma_\nu\right) \\ &= x(\sigma) - \sum_{\nu=1}^n x(\sigma_\nu), \end{aligned}$$

it follows that $x(\sigma) = \sum_{\nu=1}^\infty x(\sigma_\nu)$.

Th. 2.1.4. *Let m be an additive measure over Σ. Then the following conditions are equivalent:*

(i) *m is σ-additive,*
(ii) *m is continuous.*

PROOF. According to Th. 2.1.3, the condition that m is σ-additive can be replaced by one of the other conditions (ii)–(iv) in Th. 2.1.3.

(ii) $\Rightarrow$ (i). For an increasing sequence σ_n for which $\bigvee_n \sigma_n = \sigma$, from Th. 2.1.1 it follows that $\sigma_n \rightarrow \sigma$; therefore, from (ii) we obtain $m(\sigma_n) \rightarrow m(\sigma)$. Thus, from Th. 2.1.3(iv) we have proven (i).

(i) $\Rightarrow$ (ii). Let σ_ν be a convergent sequence for which $\sigma_\nu \rightarrow \sigma$. Since $0 \leq m(\sigma_\nu) \leq 1$ the set $m(\sigma_\nu)$ is convergent if and only if it has only one accumulation point. If α is any accumulation point we may choose a subsequence σ_{ν_k} such that $m(\sigma_{\nu_k}) \rightarrow \alpha$. According to Th. 2.1.2 a subsequence of σ_{ν_k} can be chosen (for simplicity we shall also use σ_{ν_k} to denote the subsequence) such that, if $\tilde{\sigma}_m = \bigvee_{i=m}^{\infty} \sigma_{\nu_i}$ then $\sigma = \bigwedge_m \tilde{\sigma}_m$. Since m is σ-additive, from Th. 2.1.3(iii) it follows that $m(\tilde{\sigma}_m) \rightarrow m(\sigma)$. Since $\tilde{\sigma}_m > \sigma_{\nu_m}$ we obtain $m(\tilde{\sigma}_m) \geq m(\sigma_{\nu_m}) \rightarrow \alpha$. Therefore $m(\sigma) \geq \alpha$.

From the subsequence σ_{ν_k}, since $\sigma_{\nu_k}^* \rightarrow \sigma^*$ according to Th. 2.1.2, a subsequence can be chosen (which we shall denote by σ_{ν_k}) such that

$$\sigma^* = \bigwedge_m \hat{\sigma}_m \quad \text{with } \hat{\sigma}_m = \bigvee_{i=m}^{\infty} \sigma_{\nu_i}^* .$$

Thus, we obtain

$$\sigma = \bigvee_m \hat{\sigma}_m^* \quad \text{where } \hat{\sigma}_m^* = \bigwedge_{i=m}^{\infty} \sigma_{\nu_i} .$$

Since m is σ-additive, from Th. 2.1.3(iv) we obtain

$$m(\hat{\sigma}_m^*) \longrightarrow m(\sigma).$$

Since $m(\hat{\sigma}_m^*) \leq m(\sigma_{\nu_m})$ and $m(\sigma_{\nu_m}) \rightarrow \alpha$ it follows that $m(\sigma) \leq \alpha$. Therefore $\alpha = m(\sigma)$.

Th. 2.1.5. *To each σ-additive measure m there exists one and only one $\sigma_s \in \Sigma$ for which $m(\sigma_s) = 1$ and $m(\sigma) \neq 0$ for all σ for which $0 \neq \sigma < \sigma_s$.*

PROOF. Let $\Gamma = \{\sigma \mid \sigma \in \Sigma \text{ and } m(\sigma) = 1\}$. Then, from Th. 2.1.1 it follows that there is a countable subset $\{\sigma_\nu\}$ of Γ such that

$$\sigma_s \overset{\text{def}}{=} \bigwedge_{\sigma \in \Gamma} \sigma = \bigwedge_\nu \sigma_\nu .$$

We shall now show that $\tilde{\sigma}_n = \bigwedge_{\nu=1}^{n} \sigma_\nu \in \Gamma$. We need only show that if $\sigma_1, \sigma_2 \in \Gamma$ then it follows that $\sigma_1 \wedge \sigma_2 \in \Gamma$. From $m(\sigma_2) = 1$ it follows that $m(\sigma_2^*) = 0$ and therefore $m(\sigma_1 \wedge \sigma_2^*) = 0$.

From $1 = m(\sigma_1) = m(\sigma_1 \wedge \sigma_2^*) + m(\sigma_1 \wedge \sigma_2)$, it follows that $m(\sigma_1 \wedge \sigma_2) = 1$, that is $\sigma_1 \wedge \sigma_2 \in \Gamma$. Since $\sigma_s = \bigwedge_n \tilde{\sigma}_n$ from $m(\tilde{\sigma}_n) = 1$ and Th. 2.1.3(iii) it follows that $m(\sigma_s) = 1$. Thus Γ contains a least element, namely σ_s.

Suppose that $\sigma \leq \sigma_s$ where $m(\sigma) = 0$; then it follows that $m(\sigma_s \dotplus \sigma) = m(\sigma_s) - m(\sigma) = 1$ and we therefore obtain $\sigma_s \dotplus \sigma \in \Gamma$. Since σ_s is the smallest element of Γ and since $\sigma_s \dotplus \sigma \leq \sigma_s$ it follows that $\sigma_s \dotplus \sigma = \sigma_s$, that is, $\sigma = 0$.

From $m(\hat{\sigma}) = 1$ it follows that $\hat{\sigma} \in \Gamma$ and therefore $\hat{\sigma} > \sigma_s$. Let $\sigma = \hat{\sigma} \dotplus \sigma_s$. Then we obtain $m(\sigma) = m(\hat{\sigma}) - m(\sigma_s) = 0$. If, in addition, $\hat{\sigma} \neq \sigma_s$ then there exists a σ, $0 \neq \sigma < \hat{\sigma}$ with $m(\sigma) = 0$. Thus we have proven the uniqueness of σ_s.

D 2.1.2. We shall call σ_s (as defined in Th. 2.1.5) the support of m.

Th. 2.1.6. *To each signed σ-additive measure x on Σ there is exactly one partition $\varepsilon = \sigma_+ \vee \sigma_- \vee \sigma_0$ (that is, $\sigma_+ \wedge \sigma_- = \sigma_+ \wedge \sigma_0 = \sigma_- \wedge \sigma_0 = 0$) for which $x(\sigma) > 0$ for all σ for which $0 \neq \sigma \subset \sigma_+$, and $x(\sigma) < 0$ for all σ for which $0 \neq \sigma < \sigma_-$ and $x(\sigma) = 0$ for all $\sigma < \sigma_0$.*

PROOF. Let $A = \{\sigma \mid x(\tilde{\sigma}) < 0 \text{ for all } \tilde{\sigma}, 0 \neq \tilde{\sigma} < \sigma\}$; let $\sigma_- = \bigvee_{\sigma \in A} \sigma$. We shall now show that $\sigma_- \in A$: Let $\tilde{\sigma}$ satisfy $0 \neq \tilde{\sigma} < \sigma_-$. Then there must be a $\sigma_1 \in A$ for which $\sigma_1 \wedge \tilde{\sigma} \neq 0$; thus we obtain $x(\sigma_1 \wedge \tilde{\sigma}) < 0$. Since $\tilde{\sigma} < \sigma_-$ we obtain $\tilde{\sigma} = \tilde{\sigma} \wedge \sigma_- = \bigvee_{\sigma \in A}(\sigma \wedge \tilde{\sigma})$. According to Th. 2.1.1 there exists a countable subset $\{\sigma_\nu\} \subset A$ for which $\tilde{\sigma} = \bigvee_\nu (\sigma_\nu \wedge \tilde{\sigma})$. We may choose σ_1 such that $m(\sigma_1 \wedge \tilde{\sigma}) < 0$. We may rewrite $\hat{\sigma}_n = \bigvee_{\nu=1}^n (\sigma_\nu \wedge \tilde{\sigma})$ recursively in the form:

$$\hat{\sigma}_n = (\sigma_1 \wedge \tilde{\sigma}) \vee \tilde{\sigma}_2 \vee \cdots \vee \tilde{\sigma}_n,$$

where

$$\tilde{\sigma}_m = (\sigma_m \wedge \tilde{\sigma}) \dotplus (\sigma_m \wedge \tilde{\sigma}) \wedge \left[\bigvee_{\nu=1}^{m-1} (\sigma_\nu \wedge \tilde{\sigma})\right]$$

from which it follows that

$$x(\hat{\sigma}_n) = x(\sigma_1 \wedge \tilde{\sigma}) + \sum_{\nu=2}^{n} x(\tilde{\sigma}_\nu).$$

From $\tilde{\sigma}_\nu < \sigma_\nu$ it follows that $x(\tilde{\sigma}_\nu) \leq 0$. Therefore $x(\hat{\sigma}_n) \leq x(\sigma_1 \wedge \tilde{\sigma})$. From Th. 2.1.3(iv) it follows that $x(\tilde{\sigma}) \leq x(\sigma_1 \wedge \tilde{\sigma}) < 0$. Thus we obtain $\sigma_- \in A$. Therefore A contains a greatest element σ_- and $A = \{\sigma \mid 0 \neq \sigma < \sigma_-\}$. We shall now show that $x(\sigma) \geq 0$ for all $\sigma < \sigma_-^*$. Let $\sigma' < \sigma_-^*$ and let $x(\sigma') < 0$. Since $\sigma' < \sigma_-^*$ we find that $\sigma' \notin A$, from which it follows that there exists a σ_1 such that $0 \neq \sigma_1 < \sigma'$ for which $x(\sigma_1) \geq 0$. Let n_1 be the smallest positive integer for which there exists a $\sigma_1 < \sigma'$ for which $x(\sigma_1) > 1/n_1$. Then

$$x(\sigma' \dotplus \sigma_1) = x(\sigma') - x(\sigma_1) < 0 \quad \text{and} \quad \sigma' \dotplus \sigma_1 < \sigma' < \sigma_-^*.$$

We find a similar situation holds for $\sigma' \dotplus \sigma_1$: Let n_2 be the smallest positive integer for which there exists a σ_2 for which $\sigma_2 < \sigma' \dotplus \sigma_1$ and $x(\sigma_2) > 1/n_2$.

Continuing in a similar fashion, from $\hat{\sigma} = \sigma' \dotplus \bigvee_\nu \sigma_\nu$ we obtain

$$x(\hat{\sigma}) = x(\sigma') - \sum_\nu x(\sigma_\nu) < x(\sigma') - \sum_\nu \frac{1}{n_\nu} < 0.$$

Thus we obtain $1/n_\nu \to 0$. If $\tilde{\sigma} < \hat{\sigma}$, then from the construction of σ_ν it follows that $x(\tilde{\sigma}) \leq 0$. Thus $\tilde{m}(\sigma) = (1/x(\hat{\sigma}))x(\hat{\sigma} \wedge \sigma)$ is a positive σ-additive measure. On the support σ_s of $\tilde{m}$ we find that $\sigma_s < \hat{\sigma}$; since $\tilde{m}(\sigma) > 0$ for all σ for which $0 \neq \sigma < \sigma_s$ we obtain $\sigma_s \in A$ and therefore $\sigma_s < \sigma_-$ in contradiction to $\sigma_s < \hat{\sigma} < \sigma' < \sigma_-^*$. Thus $m_+(\sigma) = (1/x(\sigma_-^*))x(\sigma_-^* \wedge \sigma)$ is a positive σ-additive measure for which $x(\sigma_-^*) > 0$. Let σ_+ be the support of m_+. Then $\sigma_+ < \sigma_-^*$. For all σ for which $0 \neq \sigma < \sigma_+$ we obtain $x(\sigma) > 0$. If $\sigma < \sigma_0 \overset{\text{def}}{=} (\sigma_+ \vee \sigma_-)^*$ then $x(\sigma) = x(\sigma \wedge (\sigma_+ \vee \sigma_-)^*) = x(\sigma \wedge \sigma_+^* \wedge \sigma_-^*) = x(\sigma_-^*)m_+(\sigma \wedge \sigma_+^*) = 0$ because σ_+ is the support of m_+.

Suppose that, in addition to $\varepsilon = \sigma_+ \vee \sigma_- \vee \sigma_0$ there exists a second partition of the same type $\varepsilon = \sigma_+^1 \vee \sigma_-^1 \vee \sigma_0^1$. Then we would obtain $\sigma_-^1 \in A$ and therefore $\sigma_-^1 < \sigma_-$. Since, for all σ for which $\sigma < \sigma_+^1 \vee \sigma_0$, and therefore for all σ for which $\sigma < (\sigma_+^1 \vee \sigma_0) \wedge \sigma_- x(\sigma) = 0$ we must obtain $(\sigma_+^1 \vee \sigma_0^1) \wedge \sigma_- = 0$. Thus we obtain $\sigma_-^1 = \sigma_-$ and therefore $\sigma_+ \vee \sigma_0 = \sigma_+^1 \vee \sigma_0^1$; from the uniqueness of the support we therefore obtain $\sigma_+ = \sigma_+^1$ and $\sigma_0 = \sigma_0^1$.

Th. 2.1.7. *Every signed σ-additive measure x on Σ may be written in the form* $x = \alpha m_1 - \beta m_2$ *where* $\alpha, \beta \geq 0$ *and* m_1, m_2 *are σ-additive measures such that the supports* σ_{s1}, σ_{s2} *of* m_1, m_2 *satisfy the relation* $\sigma_{s1} \wedge \sigma_{s2} = 0$; α, β, m_1, *and* m_2 *are uniquely determined.*

PROOF. From the partition $\Sigma = \sigma_+ \vee \sigma_- \vee \sigma_0$, from Th. 2.1.6 we obtain the following result: Let

$$m_1(\sigma) = \frac{1}{x(\sigma_+)} x(\sigma_+ \wedge \sigma), \qquad m_2(\sigma) = \frac{1}{x(\sigma_-)} x(\sigma_- \wedge \sigma),$$

$$\alpha = x(\sigma_+), \qquad \beta = -\, x(\sigma_-),$$

we obtain the desired partition and $\sigma_{s1} = \sigma_+, \sigma_{s2} = \sigma_-$.

Conversely, suppose that we are given a partition which satisfies Th. 2.1.7. Since, for all σ for which $\sigma < \sigma_{s2} x(\sigma) < 0$, $\sigma_{s2} < \sigma_-$ is obtained from Th. 2.1.6. Since, for all σ for which $\sigma < \sigma_- \wedge \sigma_{s2}^*$, $x(\sigma) = \alpha m_1(\sigma) \geq 0$, we must have $\sigma_- \wedge \sigma_{s2}^* = 0$. Therefore $\sigma_{s2} = \sigma_-$. Thus we obtain $\alpha m_1(\sigma) = x(\sigma_-^* \wedge \sigma)$ and therefore obtain $\sigma_{s1} = \sigma_+$.

D 2.1.3. Let $\mathscr{B}(\Sigma)$ denote the set of all signed σ-additive measures on Σ; let $K(\Sigma)$ denote the set of all σ-additive measures on Σ.

For every finite set $\{x_\nu\} \subset \mathscr{B}(\Sigma)$ the sum $x = \sum_\nu \alpha_\nu x_\nu$ is a signed σ-additive measure. This result, together with Th. 2.1.7, yields the following theorem:

Th. 2.1.8. *$\mathscr{B}(\Sigma)$ is a linear vector space. The linear hull of $K(\Sigma)$ is $\mathscr{B}(\Sigma)$.*

Th. 2.1.9. *The absolute convex set generated by* $K(\Sigma)$*—the set* $V = \bigcup_{0 \leq \lambda \leq 1} [\lambda K(\Sigma) - (1 - \lambda) K(\Sigma)]$*—defines a norm in* $\mathscr{B}(\Sigma)$ *by means of its Minkowski functional.*[1] *For this norm* $\|x\| = \alpha + \beta$ *where* α, β *are uniquely defined by* Th. 2.1.7. *With the positive cone* $\mathscr{B}_+(\Sigma)$ *given by*

$$\mathscr{B}_+(\Sigma) = \{x \mid x(\sigma) \geq 0 \text{ for all } \sigma \in \Sigma\}$$

$\mathscr{B}(\Sigma)$ becomes an ordered vector space. For $x > 0$ we obtain $\|x\| = x(\varepsilon)$; for $x \gneq 0$ we obtain $x = x(\varepsilon) m$ where $m = (1/x(\varepsilon))x \in K(\Sigma)$. Therefore $K(\Sigma)$ is the base for the cone $\mathscr{B}_+(\Sigma)$ and $\mathscr{B}(\Sigma)$ is a base-norm space (see AIII, §6).

PROOF. It suffices to show that the Minkowski functional for V is equal to $\alpha + \beta$ where α, β are obtained from Th. 2.1.7, because V is then absorbing and the Minkowski functional is equal to zero only if $\alpha = \beta = 0$, that is, $x = 0$.

The Minkowski functional for V is equal to $\alpha + \beta$ since, if $x = \alpha m_1 - \beta m_2 = \lambda[\mu \tilde{m}_1 - (1 - \mu)\tilde{m}_2]$ where α, β, m_1, m_2, are defined in Th. 2.1.7 and $0 \leq \mu \leq 1$, $\tilde{m}_1, \tilde{m}_2 \in K(\Sigma)$, it follows that $\lambda \geq \alpha + \beta$. Let σ_{s1}, σ_{s2} be the supports of m_1 and m_2, respectively. Then it follows that

$$\alpha + \beta = x(\sigma_{s1}) - x(\sigma_{s2}) = \lambda[\mu(\tilde{m}_1(\sigma_{s1}) - \tilde{m}_1(\sigma_{s2})) - (1 - \mu)(\tilde{m}_2(\sigma_{s1}) - \tilde{m}_2(\sigma_{s2}))].$$

1. If V is convex and if $x \in V$ implies also $-x \in V$, the Minkowski functional is defined by $p(x) = \inf\{\lambda \mid \lambda^{-1} x \in V\}$.

Since $\tilde{m}_1(\sigma_{s1}) + \tilde{m}_1(\sigma_{s2}) = \tilde{m}_1(\sigma_{s1} \vee \sigma_{s2}) \leq \tilde{m}_1(\varepsilon) = 1$ and since $\tilde{m}_2(\sigma_{s1}) + \tilde{m}_2(\sigma_{s2}) \leq 1$ it follows that

$$|\mu(\tilde{m}_1(\sigma_{s1}) - \tilde{m}_1(\sigma_{s2})) - (1 - \mu)(\tilde{m}_2(\sigma_{s1}) - \tilde{m}_2(\sigma_{s2}))| \leq 1.$$

Thus $\lambda \geq \alpha + \beta$. The remainder of the theorem is easy to prove.

Th. 2.1.10. *$\mathscr{B}(\Sigma)$ is a Banach space.*

PROOF. Let x_n be a Cauchy sequence in $\mathscr{B}(\Sigma)$. Let $x = \alpha m_1 - \beta m_2$ (where α, β, m_1, m_2 are given by Th. 2.1.7). It then follows that $|x(\sigma)| \leq \alpha m_1(\sigma) + \beta m_2(\sigma) \leq \alpha + \beta = \|x\|$. Therefore the real numbers $x_n(\sigma)$, for fixed σ, form a Cauchy sequence. From $x_n(\sigma) \to x(\sigma)$ a real function x is defined on Σ which (as may easily be shown) is additive. From

$$\begin{aligned} |x_n(\sigma)| &\leq |x_n(\sigma) - x_m(\sigma)| + |x_m(\sigma)| \\ &\leq \|x_n - x_m\| + \|x_m\| \end{aligned}$$

it follows that there exists a real number c for which $|x_n(\sigma)| < c$ for all n and σ. Therefore $|x(\sigma)| < c$, that is, x is a signed additive measure on Σ for which $|x(\sigma)| < c$ for all $\sigma \in \Sigma$.

We shall now show that x is also σ-additive, and is therefore an element of $\mathscr{B}(\Sigma)$.

According to Th. 2.1.3 x is σ-additive if, for any decreasing sequence σ_ν satisfying $\bigwedge_\nu \sigma_\nu = 0$, it follows that $x(\sigma_\nu) \to 0$. In the following equation

$$\begin{aligned} |x(\sigma_\nu)| &\leq |x(\sigma_\nu) - x_n(\sigma_\nu)| + |x_n(\sigma_\nu) - x_m(\sigma_\nu)| + |x_m(\sigma_\nu)| \\ &\leq |x(\sigma_\nu) - x_n(\sigma_\nu)| + \|x_n - x_m\| + |x_m(\sigma_\nu)| \end{aligned}$$

we may choose N such that, for $n, m > N$, $\|x_n - x_m\| < \varepsilon$. Next, we hold $m > N$ fixed, and choose ν so large that $|x_m(\sigma_\nu)| < \varepsilon$($x_m$ is σ-additive). Then, for fixed ν we choose n so large that $|x(\sigma_\nu) - x_n(\sigma_\nu)| < \varepsilon$. Therefore, we obtain $|x(\sigma_\nu)| \to 0$.

Let σ_+^n, σ_-^n be the elements which, according to Th. 2.1.6, correspond to the signed measures $(x_n - x)$. Then by Th. 2.1.9 and Th. 2.1.7 we obtain

$$\begin{aligned} \|x_n - x\| &= x_n(\sigma_+^n) - x(\sigma_+^n) - [x_n(\sigma_-^n) - x(\sigma_-^n)] \\ &\leq |x_n(\sigma_+^n) - x_m(\sigma_+^n)| + |x_m(\sigma_+^n) - x(\sigma_+^n)| \\ &\quad + |x_n(\sigma_-^n) - x_m(\sigma_-^n)| + |x_m(\sigma_-^n) - x(\sigma_-^n)| \\ &\leq 2\|x_n - x_m\| + |x_m(\sigma_+^n) - x(\sigma_+^n)| + |x_m(\sigma_-^n) - x(\sigma_-^n)|. \end{aligned}$$

We now choose N such that $\|x_n - x_m\| < \varepsilon/2$ for all $n, m > N$. Then, holding n fixed and choosing m so large that

$$|x_m(\sigma_+^n) - x(\sigma_+^n)| < \varepsilon,$$

$$|x_m(\sigma_-^n) - x(\sigma_-^n)| < \varepsilon.$$

From this, it follows that $\|x_n - x\| \to 0$.

Th. 2.1.11. *Let m be an effective σ-additive measure. Then*

$$K(\Sigma) = \overline{\mathrm{co}}\{\hat{m}_\sigma \mid \hat{m}_\sigma(\tilde{\sigma}) = m(\sigma)^{-1} m(\sigma \wedge \tilde{\sigma}), \sigma \in \Sigma, \sigma \neq 0\},$$

where $\overline{\mathrm{co}}\{\ldots\}$ is the norm-closure of $\mathrm{co}\{\ldots\}$, that is, the convex subset of $K(\Sigma)$ generated by all the measures $\hat{m}_\sigma$ is dense (in norm) in $K(\Sigma)$.

Let m_1, m_2 be two effective σ-additive measures. Then the corresponding metrics $d_1(\sigma_1, \sigma_2) = m_1(\sigma_1 \dotplus \sigma_2)$ and $d_2(\sigma_1, \sigma_2) = m_2(\sigma_1 \dotplus \sigma_2)$ are equivalent, that is, the uniform structure U_g of Σ will be generated by $d(\sigma_1, \sigma_2) = m(\sigma_1 \dotplus \sigma_2)$ where m is any effective σ-additive measure.

PROOF. For $\tilde{m} \in K(\Sigma)$ and for m as defined in Th. 2.1.11, the $x_\lambda = \tilde{m} - \lambda m$, for real numbers $\lambda \geq 0$ are elements of $\mathscr{B}(\Sigma)$. For each x_λ we may define $\sigma_-(\lambda)$ according to Th. 2.1.6; clearly $\lambda_1 \geq \lambda_2$ implies $\sigma_-(\lambda_1) > \sigma_-(\lambda_2)$. We shall now show that $\bigvee_{\lambda \geq 0} \sigma_-(\lambda) = \varepsilon$. Suppose not; then, for $\hat{\sigma} = [\bigvee_{\lambda \geq 0} \sigma_-(\lambda)]^*$ we would obtain $\hat{\sigma} < \sigma_-(\lambda)^*$ for all $\lambda \geq 0$, that is, $x_\lambda(\hat{\sigma}) \geq 0$ for all $\lambda \geq 0$. Therefore we would obtain $\tilde{m}(\hat{\sigma}) \geq \lambda m(\hat{\sigma})$ for all $\lambda \geq 0$.

Since $m(\hat{\sigma}) \neq 0$, this contradicts the fact that $\tilde{m}(\hat{\sigma}) \leq 1$. Let $\delta > 0$; then define $\lambda_n = n\delta$ for $n = 0, 1, 2, \ldots$ and $\sigma_n = \sigma_-(n\delta) \dotplus \sigma_-((n-1)\delta)$. Thus we obtain $\sigma_n \wedge \sigma_m = 0$ for $n \neq m$ and $\bigvee_{n=1}^\infty \sigma_n = \varepsilon$.

Let $\hat{m}_n(\sigma) = m(\sigma_n)^{-1} m(\sigma \wedge \sigma_n)$; we define

$$x_{\delta, N} = \sum_{n=1}^{N} (n-1)\delta m(\sigma_n)\hat{m}_n.$$

Clearly $x_{\delta, N} \in \mathscr{B}_+(\Sigma)$ and $\hat{m}_n \in K(\Sigma)$. Since

$$\tilde{m}(\sigma) = \sum_{n=1}^{\infty} \tilde{m}(\sigma \wedge \sigma_n)$$

it follows that

$$\tilde{m}(\sigma) - x_{\delta, N}(\sigma) = \sum_{n=1}^{N} [\tilde{m}(\sigma \wedge \sigma_n) - (n-1)\delta m(\sigma \wedge \sigma_n)] + \sum_{n=N+1}^{\infty} \tilde{m}(\sigma \wedge \sigma_n).$$

We obtain $\sigma \wedge \sigma_n < \sigma_-(n\delta)$ and $\sigma \wedge \sigma_n < \sigma_-((n-1)\delta)^*$; from which it follows that

$$\tilde{m}(\sigma \wedge \sigma_n) - n\delta m(\sigma \wedge \sigma_n) \leq 0,$$
$$\tilde{m}(\sigma \wedge \sigma_n) - (n-1)\delta m(\sigma \wedge \sigma_n) \geq 0.$$

From which we obtain

$$0 \leq \tilde{m}(\sigma \wedge \sigma_n) - (n-1)\delta m(\sigma \wedge \sigma_n) \leq \delta m(\sigma \wedge \sigma_n)$$

and

$$0 \leq \tilde{m}(\sigma) - x_{\delta, N}(\sigma) \leq \delta \sum_{n=1}^{N} m(\sigma \wedge \sigma_n) + \sum_{n=N+1}^{\infty} \tilde{m}(\sigma \wedge \sigma_n)$$
$$\leq \delta m(\sigma) + \sum_{n=N+1}^{\infty} \tilde{m}(\sigma \wedge \sigma_n).$$

Since $0 \leq \tilde{m}(\sigma) - x_{\delta,N}(\sigma)$ we obtain $\|\tilde{m} - x_{\delta,N}\| = \tilde{m}(\varepsilon) - x_{\delta,N}(\varepsilon)$ (see Th. 1.2.9). It follows that

$$\|\tilde{m} - x_{\delta, N}\| \leq \delta + \sum_{n=N+1}^{\infty} \tilde{m}(\sigma_n).$$

We may now choose $\delta < \varepsilon/4$, and N so large that $\sum_{n=N+1}^{\infty} \tilde{m}(\sigma_n) < \varepsilon/4$ where the latter is possible because of the convergence of the series

$$\sum_{n=1}^{\infty} \tilde{m}(\sigma_n) = \tilde{m}(\varepsilon) = 1.$$

Therefore we obtain $\|\tilde{m} - x_{\delta,N}\| < \varepsilon/2$. Since $\|\tilde{m}\| = 1$ it follows that $1 - \varepsilon/2 \leq \|x_{\delta N}\| \leq 1 + \varepsilon/2$ and

$$\left\| \tilde{m} - \frac{x_{\delta,N}}{\|x_{\delta,N}\|} \right\| < \varepsilon.$$

Since $x_{\delta,N} > 0$, $m_{\delta,N} = x_{\delta,N}/\|x_{\delta,N}\|$ is an element of $K(\Sigma)$ and $m_{\delta,N}$ is a convex linear combination of the $\tilde{m}_n$.

Suppose that $\tilde{m}$ is effective. We shall now show that the metrics $d(\sigma_1, \sigma_2) = m(\sigma_1 \dotplus \sigma_2)$ and $\tilde{d}(\sigma_1, \sigma_2) = \tilde{m}(\sigma_1 \dotplus \sigma_2)$ are equivalent. By symmetry between m and $\tilde{m}$ we need only show that, for each $\tilde{\varepsilon} > 0$ there exists an $\varepsilon > 0$ such that $d(\sigma_1, \sigma_2) < \varepsilon$ implies that $\tilde{d}(\sigma_1, \sigma_2) < \tilde{\varepsilon}$. For $x_{\delta,N}$, as defined above, we obtain

$$\begin{aligned} \tilde{m}(\sigma_1 \dotplus \sigma_2) &= \tilde{m}(\sigma_1 \dotplus \sigma_2) - x_{\delta,N}(\sigma_1 \dotplus \sigma_2) + x_{\delta,N}(\sigma_1 \dotplus \sigma_2) \\ &\leq \|\tilde{m} - x_{\delta,N}\| + \sum_{n=1}^{N} (n-1)\delta m(\sigma_n \cdot \sigma_1 \dotplus \sigma_n \cdot \sigma_2) \\ &\leq \|\tilde{m} - x_{\delta,N}\| + m(\sigma_1 \dotplus \sigma_2) \sum_{n=1}^{N} (n-1)\delta. \end{aligned}$$

We now choose $x_{\delta,N}$ (and therefore choose δ, N) such that $\|\tilde{m} - x_{\delta,N}\| < \tilde{\varepsilon}/2$. Then we choose $\varepsilon < \tilde{\varepsilon}/2 \sum_{n=1}^{N} (n-1)\delta$. Then, for $d(\sigma_1, \sigma_2) < \varepsilon$ it follows that $\tilde{d}(\sigma_1, \sigma_2) < \tilde{\varepsilon}$.

Th. 2.1.12. *$\mathscr{B}(\Sigma)$ is separable if Σ is separable.*

PROOF. Suppose $\{\sigma_\nu\}$ is a countable set which is dense in Σ, that is, to each $\tilde{\sigma} \in \Sigma$, for arbitrary $\varepsilon > 0$ there exists a σ_ν such that $d(\tilde{\sigma}, \sigma_\nu) = m_0(\tilde{\sigma} \dotplus \sigma_\nu) < \varepsilon$. Let $\hat{m}_{\tilde{\sigma}}(\sigma) = m_0(\tilde{\sigma})^{-1} m_0(\tilde{\sigma} \wedge \sigma)$ where m_0 is the effective measure (see the beginning of §2.1). According to Th. 2.1.8 and Th. 2.1.11 it suffices to show that, to each $\hat{m}_{\tilde{\sigma}}$ there exists a $\hat{m}_{\sigma_\nu}$ for which $\|\hat{m}_{\tilde{\sigma}} - \hat{m}_{\sigma_\nu}\|$ is arbitrarily small. This fact is a consequence of the following estimates:

$$\begin{aligned} \|\hat{m}_{\tilde{\sigma}} - \hat{m}_{\sigma_\nu}\| &\leq \frac{1}{m_0(\tilde{\sigma})} \|m_0(\tilde{\sigma} \wedge \sigma) - m_0(\sigma_\nu \wedge \sigma)\| \\ &\quad + \left| \frac{1}{m_0(\tilde{\sigma})} - \frac{1}{m_0(\sigma_\nu)} \right| \|m_0(\sigma_\nu \wedge \sigma)\| \\ &\leq \frac{1}{m_0(\tilde{\sigma})} \|m_0(\tilde{\sigma} \wedge \sigma) - m_0(\sigma_\nu \wedge \sigma)\| \\ &\quad + \left| \frac{1}{m_0(\tilde{\sigma})} - \frac{1}{m_0(\sigma_\nu)} \right|. \end{aligned}$$

From

$$\begin{aligned} m_0(\tilde{\sigma} \wedge \sigma) - m_0(\sigma_\nu \wedge \sigma) &= m_0(\tilde{\sigma} \wedge \sigma_\nu \wedge \sigma) + m_0((\tilde{\sigma} \dotplus \tilde{\sigma} \wedge \sigma_\nu) \wedge \sigma) \\ &\quad - m_0(\tilde{\sigma} \wedge \sigma_\nu \wedge \sigma) - m_0((\sigma_\nu \dotplus \tilde{\sigma} \wedge \sigma_\nu) \wedge \sigma) \end{aligned}$$

it follows that

$$\|m_0(\tilde{\sigma} - \sigma) - m_0(\sigma_v \wedge \sigma)\| \leq \|m_0((\tilde{\sigma} \dotplus \tilde{\sigma} \wedge \sigma_v) \wedge \sigma)\| + \|m_0((\sigma_v \dotplus \tilde{\sigma} \wedge \sigma_v) \wedge \sigma)\|$$
$$= m_0(\tilde{\sigma} \dotplus \tilde{\sigma} \cdot \sigma_v) + m_0(\sigma_v \dotplus \tilde{\sigma} \cdot \sigma_v)$$
$$= d(\tilde{\sigma}, \sigma_v).$$

Therefore we obtain

$$\|\hat{m}_{\sigma_v} - \hat{m}_{\tilde{\sigma}}\| \leq \frac{1}{m_0(\tilde{\sigma})} d(\tilde{\sigma}, \sigma_v) + \left|\frac{1}{m_0(\tilde{\sigma})} - \frac{1}{m_0(\sigma_v)}\right|.$$

We shall now examine the properties of the dual Banach space $\mathscr{B}'(\Sigma)$ corresponding to $\mathscr{B}(\Sigma)$.

Let $l \in \mathscr{B}'(\Sigma)$, that is, l is a bounded linear functional over $\mathscr{B}(\Sigma)$. The norm of l is given by

$$\|l\| = \sup_{\|x\| \leq 1} |l(x)|.$$

It is easy to show that

$$\|l\| = \sup\{l(m) \mid m \in K(\Sigma)\}.$$

The positive cone $\mathscr{B}'_+(\Sigma)$ in $\mathscr{B}'(\Sigma)$ is defined as the set of all l for which $l(m) \geq 0$ for $m \in K(\Sigma)$.

The unit sphere in $\mathscr{B}'(\Sigma)$ may then be described by the order interval $[-\mathbf{1}, \mathbf{1}]$ where $\mathbf{1}$ is the functional for which $\mathbf{1}(m) = 1$ for all $m \in K(\Sigma)$. Therefore $\mathbf{1}(x) = \alpha - \beta$ (where α, β are defined in Th. 2.1.7).

The unit sphere $[-\mathbf{1}, \mathbf{1}]$ is $\sigma(\mathscr{B}', \mathscr{B})$-compact. Thus it satisfies the Krein–Milman theorem (see AIII, §4) which says that a convex set is spanned by its set of extreme points $\partial_e[-\mathbf{1}, \mathbf{1}]$. The unit sphere is isomorphic to the set $L(\Sigma) \stackrel{\text{def}}{=} [\mathbf{0}, \mathbf{1}]$. What are the elements of $\partial_e L(\Sigma)$?

For fixed $\sigma \in \Sigma$, $l_\sigma(m) = m(\sigma)$ is a bounded linear functional which satisfies $l_\sigma \in L(\Sigma)$. We shall now show that $\partial_e L(\Sigma)$ is precisely the set of all l_σ where $\sigma \in \Sigma$.

Th. 2.1.13. $l_\sigma \in \partial_e L(\Sigma), \sigma_1 \neq \sigma_2 \Rightarrow l_{\sigma_1} \neq l_{\sigma_2}$.

PROOF. Let $l_{\sigma_1} = \lambda l + (1 - \lambda)l'$ where $0 < \lambda < 1$. Then, for all m for which $m(\sigma_1) = 1$ it follows that $l(m) = 1$; for all m satisfying $m(\sigma_1) = 0$ it follows that $l(m) = 0$. Each m can be written as follows:

$$m = m(\sigma_1)m_{\sigma_1} + m(\sigma_1^*)m_{\sigma_1^*},$$

where

$$m_{\tilde{\sigma}}(\sigma) = \frac{m(\sigma \wedge \tilde{\sigma})}{m(\tilde{\sigma})}.$$

Therefore, since $l(m_{\sigma_1}) = 1$ and $l(m_{\sigma_1^*}) = 0$ it follows that

$$\begin{aligned} l(m) &= m(\sigma_1)l(m_{\sigma_1}) + m(\sigma_1^*)l(m_{\sigma_1^*}) \\ &= m(\sigma_1) = l_{\sigma_1}(m). \end{aligned}$$

$$\begin{aligned} l_{\sigma_1} = l_{\sigma_2} &\Rightarrow m_0(\sigma_1)^{-1}m_0(\sigma_1 \wedge \sigma_1) = m_0(\sigma_1)^{-1}m_0(\sigma_1 \wedge \sigma_2) \\ &\Rightarrow m_0(\sigma_1 \dotplus \sigma_1 \wedge \sigma_2) = 0 \Rightarrow \sigma_1 = \sigma_1 \wedge \sigma_2 \end{aligned}$$

and similarly $\sigma_2 = \sigma_1 \wedge \sigma_2$, that is, $\sigma_1 = \sigma_2$.

The fact that each element of $\partial_e L(\Sigma)$ is of the form l_σ is obtained as a corollary of the important "spectral theorem" for $l \in \mathscr{B}'(\Sigma)$. In order to prove this fact we introduce the following definition and notation:

Let the canonical dual form for $\mathscr{B}$, $\mathscr{B}'$ be denoted by (x, y). Therefore every linear functional $y \in \mathscr{B}'$ may be expressed as a function over $\mathscr{B}$ in terms of (x, y) for fixed y. Let

$$\begin{aligned} s(y \mid \sigma) &= \sup\{(m, y) \mid m \in K(\Sigma) \text{ and } m(\sigma) = 1\}, \\ \Gamma_\alpha(y) &= \{\sigma \mid s(y \mid \sigma) \leq \alpha\}, \\ \sigma(y \leq \alpha) &= \bigvee_{\sigma \in \Gamma_\alpha(y)} \sigma. \end{aligned} \tag{2.1.2}$$

We claim that $\sigma(y \leq \alpha) \in \Gamma_\alpha(y)$, that is, $\sigma(y \leq \alpha)$ is the largest element of $\Gamma_\alpha(y)$.

In order to show this result, we shall first show that, if $\sigma_1, \sigma_2 \in \Gamma_\alpha$ then $\sigma_1 \vee \sigma_2 \in \Gamma_\alpha$.

The following is obvious:

$$\begin{aligned} s(y \mid \sigma_1 \vee \sigma_2) \geq s(y, \sigma_1), \\ s(y \mid \sigma_1 \vee \sigma_2) \geq s(y, \sigma_2). \end{aligned} \tag{2.1.3}$$

Let $\tilde{\sigma}_1 = \sigma_1 \dotplus \sigma_1 \cdot \sigma_2 \leq \sigma_1$; we then obtain $\sigma_1 \vee \sigma_2 = \tilde{\sigma}_1 \vee \sigma_2$ and $\tilde{\sigma}_1 \wedge \sigma_2 = 0$ from which we conclude that

$$m(\sigma_1 \vee \sigma_2) = m(\tilde{\sigma}_1) + m(\sigma_2).$$

Suppose that $m(\sigma_1 \vee \sigma_2) = 1$. If $m(\tilde{\sigma}_1) = 1$ (or $m(\tilde{\sigma}_1) = 0$ and therefore $m(\sigma_2) = 1$) it follows that $(m, y) \leq s(y \mid \sigma_1)$ (or $\leq s(y \mid \sigma_2)$). If $m(\tilde{\sigma}_1) \neq 1, \neq 0$, then, setting $\lambda = m(\tilde{\sigma}_1)$ we obtain:

$$m_1(\sigma) = \frac{1}{\lambda} m(\tilde{\sigma}_1 \wedge \sigma), \qquad m_2(\sigma) = \frac{1}{1 - \lambda} m(\sigma_2 \wedge \sigma);$$

m_1, m_2 are elements of $K(\Sigma)$ and $m_1(\tilde{\sigma}_1) = 1$, $m_2(\tilde{\sigma}_2) = 1$. Since $m(\sigma_1 \vee \sigma_2) = 1$ we obtain

$$m(\sigma) = m((\sigma_1 \vee \sigma_2) \wedge \sigma) = \lambda m_1(\sigma) + (1 - \lambda)m_2(\sigma).$$

From this result it follows that

$$\begin{aligned} (m, y) &= \lambda(m_1, y) + (1 - \lambda)(m_2, y) \\ &\leq \lambda s(y \mid \sigma_1) + (1 - \lambda)s(y \mid \sigma_2). \end{aligned}$$

This result also holds (as we have seen above) for $\lambda = 0$ and $\lambda = 1$. Therefore we obtain:

$$s(y \mid \sigma_1 \vee \sigma_2) \leq \sup_{0 \leq \lambda \leq 1} [\lambda s(y \mid \sigma_1) + (1 - \lambda) s(y \mid \sigma_2)]$$
$$= \max[s(y \mid \sigma_1), s(y \mid \sigma_2)].$$

Combining this result with (2.1.3) we obtain:

$$s(y \mid \sigma_1 \vee \sigma_2) = \max[s(y \mid \sigma_1), s(y \mid \sigma_2)]$$

thus proving that, if $\sigma_1, \sigma_2 \in \Gamma_\alpha$ then $\sigma_1 \vee \sigma_2 \in \Gamma_\alpha$. Therefore we have shown that, for $\sigma_\nu \in \Gamma_\alpha$ we obtain

$$\bigvee_{\nu=1}^{m} \sigma_\nu \in \Gamma_\alpha .$$

This result, together with Th. 2.1.1, shows that there is an increasing sequence $\sigma_n \in \Gamma_\alpha$ such that $\sigma(y \leq \alpha) = \bigvee_n \sigma_n$. Let $m \in K(\Sigma)$ satisfy $m(\sigma(y \leq \alpha)) = 1$. Then, from Th. 2.1.3(iv) it follows that

$$m(\sigma_n) \longrightarrow 1. \tag{2.1.4}$$

Clearly, we may have only a finitely many $m(\sigma_n) = 0$. These can be omitted, and we may require that $m(\sigma_n) \neq 0$. Then

$$m_n(\sigma) = \frac{m(\sigma_n \wedge \sigma)}{m(\sigma_n)}$$

is therefore an element of $K(\Sigma)$ which satisfies $m_n(\sigma_n) = 1$. For the complement σ_n^* of σ_n we find that

$$m_n^*(\sigma) = \frac{m(\sigma_n^* \wedge \sigma)}{m(\sigma_n^*)}$$

is an element of $K(\Sigma)$ if $m(\sigma_n) \neq 1$.

If $m(\sigma_n) = 1$ then $(m, y) \leq s(y \mid \sigma_n) \leq \alpha$. Let $m(\sigma_n) \neq 1$ for all n. Then from

$$m = m(\sigma_n) m_n + (1 - m(\sigma_n)) m_n^*$$

it follows that

$$(m, y) = m(\sigma_n)(m_n, y) + (1 - m(\sigma_n))(m_n^*, y)$$
$$\leq m(\sigma_n) s(y \mid \sigma_n) + (1 - m(\sigma_n)) \|y\|$$
$$\leq m(\sigma_n)\alpha + (1 - m(\sigma_n)) \|y\|.$$

Therefore, together with (2.1.4) it follows that (again, as in the case in which $m(\sigma_n) = 1$) $(m, y) \leq \alpha$. For an $m \in K(\Sigma)$ for which $m(\sigma(y \leq \alpha)) = 1$ we also obtain $(m, y) \leq \alpha$, that is, $\sigma(y \leq \alpha) \in \Gamma_\alpha$.

Thus, we obtain

$$\alpha_1 \geq \alpha_2 \Rightarrow \sigma(y \leq \alpha_1) > \sigma(y \leq \alpha_2). \tag{2.1.5}$$

Let $\sigma^*(y \le \alpha)$ denote the complement of $\sigma(y \le \alpha)$. We will now show that the relation $(m, y) > \alpha$ is satisfied for all $m \in K(\Sigma)$ which satisfy $m(\sigma^*(y \le \alpha)) = 1$.

Suppose that there exists an $\overline{m} \in K(\Sigma)$ which satisfies $\overline{m}(\sigma^*(y \le \alpha)) = 1$ and $(\overline{m}, y) \le \alpha$. Let $\tilde{y} = y - \alpha\mathbf{1}$; then $(\overline{m}, y) \le \alpha$ is equivalent to $(\overline{m}, \tilde{y}) \le 0$. Let $\bar{\sigma}$ be the support of $\overline{m}$ (see D 2.1.2). Then $\bar{\sigma} < \sigma^*(y \le \alpha)$, $\overline{m}(\bar{\sigma}) = 1$ and $\overline{m}(\sigma) \ne 0$ for all σ satisfying $0 \ne \sigma \le \bar{\sigma}$.

Let n_1 be the smallest positive integer for which there exists a σ_1 satisfying $\sigma_1 < \bar{\sigma}$ and $(x_{\sigma_1}, \tilde{y}) \ge 1/n_1$, where $x_{\sigma_1}(\sigma) = \overline{m}(\sigma_1 \wedge \sigma)$. For $\bar{\sigma} \dotplus \sigma_1$ we may make the same assumption as we made for $\bar{\sigma}$: Let n_2 be the smallest positive integer for which there exists a $\sigma_2 < \bar{\sigma} \dotplus \sigma_1$ such that $(x_{\sigma_2}, \tilde{y}) \ge 1/n_2$. Proceeding in this fashion we obtain a sequence σ_ν for which $\sigma_\nu \wedge \sigma_\mu = 0$ for $\nu \ne \mu$ such that $\bigvee_\nu \sigma_\nu = \bar{\sigma}$.

Let $\tilde{\sigma} = \bar{\sigma} \dotplus \bigvee_\nu \sigma_\nu$; we will now show that $\tilde{\sigma} \ne 0$. From $\bar{\sigma} = \tilde{\sigma} \vee \bigvee_\nu \sigma_\nu$ it follows that

$$\overline{m}(\sigma) = \overline{m}(\bar{\sigma} \wedge \sigma) = \overline{m}(\tilde{\sigma} \wedge \sigma) + \sum_\nu \overline{m}(\sigma_\nu \wedge \sigma). \tag{2.1.6}$$

Since

$$\left\| \overline{m} - x_{\tilde{\sigma}} - \sum_{\nu=1}^{N} x_{\sigma_\nu} \right\| = \overline{m}(\varepsilon) - \overline{m}(\tilde{\sigma}) - \sum_{\nu=1}^{N} \overline{m}(\sigma_\nu)$$

in the norm-topology of $\mathscr{B}(\Sigma)$ we obtain

$$\overline{m} = x_{\tilde{\sigma}} + \sum_{\nu=1}^{\infty} x_{\sigma_\nu} \tag{2.1.7}$$

and therefore obtain

$$\begin{aligned}(\overline{m}, \tilde{y}) &= (x_{\tilde{\sigma}}, \tilde{y}) + \sum (x_{\sigma_\nu}, \tilde{y}) \\ &\ge (x_{\tilde{\sigma}}, \tilde{y}) + \sum_\nu^\nu \frac{1}{n_\nu},\end{aligned} \tag{2.1.8}$$

If $\tilde{\sigma} = 0$ (and therefore $x_{\tilde{\sigma}} = 0$) then we would obtain $(\overline{m}, \tilde{y}) > 0$ in contradiction to $(\overline{m}, \tilde{y}) \le 0$. Therefore $\tilde{\sigma} \ne 0$ and $\overline{m}(\tilde{\sigma}) \ne 0$ (since $\tilde{\sigma} \le \bar{\sigma}$ and $\bar{\sigma}$ is the support of $\overline{m}$). From (2.1.8) it follows that $\sum_\nu 1/n_\nu$ is convergent and therefore $1/n_\nu \to 0$.

We must have $(x_\sigma, \tilde{y}) \le 0$ for all $\sigma < \tilde{\sigma}$, otherwise there would be a $\hat{\sigma} < \tilde{\sigma}$ for which $(x_{\hat{\sigma}}, \tilde{y}) = \varepsilon > 0$, contradicting the definition of n_ν and $1/n_\nu \to 0$.

For all σ satisfying $0 \ne \sigma < \tilde{\sigma}$ we find that $x_{\tilde{\sigma}}(\sigma) = \overline{m}(\tilde{\sigma} \wedge \sigma) = \overline{m}(\sigma) \ne 0$. The measure $\overline{m}_{\tilde{\sigma}} \in K(\Sigma)$ given by

$$\overline{m}_{\tilde{\sigma}}(\sigma) = \frac{\overline{m}(\tilde{\sigma} \wedge \sigma)}{\overline{m}(\tilde{\sigma})}$$

is therefore "effective" in $\tilde{\sigma}$, that is, $\overline{m}_{\tilde{\sigma}}(\sigma) \ne 0$ for all σ for which $0 \ne \sigma < \tilde{\sigma}$. Thus, in the same manner as in Th. 2.1.11 it follows that $\text{co}\{\overline{m}_{\hat{\sigma}} \mid \hat{\sigma} \le \tilde{\sigma}\}$ is norm-dense in the set of all m for which $m(\tilde{\sigma}) = 1$. Therefore, for all m

satisfying $m(\tilde{\sigma}) = 1$ we obtain $(m, \tilde{y}) \leq 0$, that is, $(m, y) = (m, \tilde{y} + \alpha 1) = (m, \tilde{y}) + \alpha = \leq \alpha$; or, in other words $\tilde{\sigma} \in \Gamma_\alpha$ and therefore $\tilde{\sigma} < \sigma(y \leq \alpha)$ in contradiction to $\tilde{\sigma} < \bar{\sigma} < \sigma^*(y \leq \alpha)$.

Therefore, we have shown that $(m, y) > \alpha$ is satisfied for all m satisfying $m(\sigma^*(y \leq \alpha)) = 1$.

We see, therefore, that $\sigma(y \leq \alpha) = 0$ for all $\alpha < -\|y\|$ and $\sigma(y \leq \alpha) = \varepsilon$ for $\alpha \geq \|y\|$.

D 2.1.4. The totally ordered subset $\{\sigma(y \leq \alpha)\} \subset \Sigma$ is called the *spectral family* of $y \in \mathscr{B}'(\Sigma)$.

Th. 2.1.14. *The spectral family $\sigma(y \leq \alpha)$ is right-continuous, that is, $\sigma(y \leq \alpha + \varepsilon) \to \sigma(y \leq \alpha)$ for $\varepsilon > 0$ and $\varepsilon \to 0$.*

PROOF. According to Th. 2.1.1 it suffices to show that, for $\hat{\sigma} = \bigwedge_{\varepsilon > 0} \sigma(y \leq \alpha + \varepsilon)$ the relation $\hat{\sigma} = \sigma(y \leq \alpha)$ holds. Immediately it follows that $\sigma(y \leq \alpha) < \hat{\sigma}$. On the other hand

$$s(y \mid \hat{\sigma}) = \sup\{(m, y) \mid m \in K(\Sigma) \text{ and } m(\hat{\sigma}) = 1\}.$$

Since $\hat{\sigma} \leq \sigma(y \leq \alpha + \varepsilon)$ from $(m, y) \leq \alpha + \varepsilon$ for each $\varepsilon > 0$ we obtain $(m, y) \leq \alpha$ for each $m \in K(\Sigma)$ and $m(\hat{\sigma}) = 1$, that is, $\hat{\sigma} < \sigma(y \leq \alpha)$.

D 2.1.5. A map $\mathbf{R} \xrightarrow{\rho} \Sigma$ satisfying the conditions

$$\alpha_2 > \alpha_1 \Rightarrow \rho(\alpha_2) > \rho(\alpha_1); \qquad \rho(\alpha) = 0 \quad \text{for } \alpha \leq -c,$$

$\rho(\alpha) = \varepsilon$ for $\alpha \geq c$ for some c, and satisfying $\rho(\alpha+) = \rho(\alpha)$ (ρ is right-continuous) is called a *(generalized) spectral family*.

Th. 2.1.15. (Spectral Theorem). *For each spectral family there is a linear bounded functional $l \in \mathscr{B}'(\Sigma)$ defined by*

$$l(m) = \int_{-c}^{+c} \alpha \, dm(\rho(\alpha)), \qquad m \in K(\Sigma) \tag{2.1.9a}$$

such that

$$\sigma(l \leq \alpha) = \rho(\alpha). \tag{2.1.10}$$

For the special case in which $y \in \mathscr{B}'(\Sigma)$, $\rho(\alpha) = \sigma(y \leq \alpha)$ then l in (2.1.9a) *is equal to y. We may write* (2.1.9a) *(l_σ is defined in* Th. 2.1.13) *as follows:*

$$l = \int_{-c}^{+c} \alpha \, dl_{\rho(\alpha)}, \tag{2.1.9b}$$

where the integral is to be considered to be a limit in the norm of $\mathscr{B}'(\Sigma)$.

PROOF. $l(m)$ is a bounded linear functional if it is affine and bounded on $K(\Sigma)$; this result follows directly from (2.1.9a).

Let $m(\rho(\alpha_1)) = 1$; then $m(\rho(\alpha)) = 1$ for $\alpha \geq \alpha_1$. From (2.1.9a) it follows that $l(m) = (m, l) \leq \alpha_1$, that is, $\rho(\alpha_1) \leq \sigma(l \leq \alpha_1)$.

Now let $m(\sigma(l \leq \alpha_1)) = 1$. From (2.1.9a) it follows that

$$(m, l) = \int_{-c}^{+c} \alpha \, dm(\rho(\alpha) \wedge \sigma(l \leq \alpha_1)). \tag{2.1.11}$$

Suppose that $\rho(\alpha_1) \neq \sigma(l \leq \alpha_1)$; then there exists a $\delta > 0$ such that $\rho(\alpha_1 + \delta) \neq \sigma(l \leq \alpha_1)$ since ρ is continuous from the right. Therefore $\rho(\alpha_1 + \delta) \wedge \sigma(l \leq \alpha_1) \neq \sigma(l \leq \alpha_1)$ or $\rho(\alpha_1 + \delta) > \sigma(l \leq \alpha_1)$. If $\rho(\alpha_1 + \delta) > \sigma(l \leq \alpha_1)$ for all $\delta > 0$ it would follow that $\rho(\alpha_1 +) = \rho(\alpha_1) = \sigma(l \leq \alpha_1)$. If there exists a δ such that $\rho(\alpha_1 + \delta) \wedge \sigma(l \leq \alpha_1) \neq \sigma(l \leq \alpha_1)$ then $\tilde{\sigma} = \sigma(l \leq \alpha_1) \dot{+} \rho(\alpha_1 + \delta) \wedge \sigma(l \leq \alpha_1) \neq 0$ and $\tilde{\sigma} \leq \sigma(l \leq \alpha_1)$. Let us choose a $m \in K(\Sigma)$ such that $m(\tilde{\sigma}) = 1$. Then it follows that

$$m(\rho(\alpha) \wedge \sigma(l \leq \alpha_1)) = 0 \quad \text{for } \alpha < \alpha_1 + \delta,$$
$$m(\rho(c) \wedge \sigma(l \leq \alpha_1)) = m(\sigma(l \leq \alpha_1)) = 1$$

and, therefore, from (2.1.11) $l(m) \geq \alpha_1 + \delta$ in contradiction to the fact that $\tilde{\sigma} \leq \sigma(l \leq \alpha_1)$.

If $\alpha_1 > \alpha_2$, then, from the definition of $\sigma(y \leq \alpha)$ we obtain the inequality

$$\begin{aligned} \alpha_2 m(\sigma(y \leq \alpha_1) \dot{+} \sigma(y \leq \alpha_2)) &\leq (x_{\sigma(y \leq \alpha_1) \dot{+} \sigma(y \leq \alpha_2)}, y) \\ &\leq \alpha_1 m(\sigma(y \leq \alpha_1) \dot{+} \sigma(y \leq \alpha_2)), \end{aligned} \tag{2.1.12}$$

where x_σ is defined by $x_\sigma(\tilde{\sigma}) = m(\sigma \wedge \tilde{\sigma})$.

We may rewrite (2.1.12) in the form

$$\begin{aligned} \alpha_2[m(\sigma(y \leq \alpha_1)) - m(\sigma(y \leq \alpha_2))] &\leq (x_{\ldots}, y) \\ &\leq \alpha_1[m(\sigma(y \leq \alpha_1)) - m(\sigma(y \leq \alpha_2))]. \end{aligned}$$

For a partition of the real axis for which $\alpha_{n+1} > \alpha_n$ for

$$\Delta_n m = m(\sigma(y \leq \alpha_{n+1})) - m(\sigma(y \leq \alpha_n))$$

and

$$m = \sum_n x_{\sigma(y \leq \alpha_{n+1}) \dot{+} \sigma(y \leq \alpha_n)}$$

we obtain the inequality

$$\sum_n \alpha_n \Delta_n m \leq (m, y) \leq \sum_n \alpha_{n+1} \Delta_n m. \tag{2.1.13a}$$

With $\Delta_n l = l_{\sigma(y \leq \alpha_{n+1})} - l_{\sigma(y \leq \alpha_n)}$ this inequality can be written in the form ($<$ is in the order in $\mathscr{B}'(\Sigma)$!)

$$\sum_n \alpha_n \Delta_n l < y \sum_n \alpha_{n+1} \Delta_n l \tag{2.1.13b}$$

because (2.1.13a) holds for all $m \in K(\Sigma)$.

Thus it follows that the left- and right-hand sides of (2.1.13b) converge (with respect to the norm of $\mathscr{B}'(\Sigma)$)—as the maximum interval length tends zero—towards the same limit

$$y = \int \alpha \, dl_{\sigma(y \leq \alpha)}, \tag{2.1.14}$$

Th. 2.1.16. $\partial_e L(\Sigma) = \{l_\sigma \mid \sigma \in \Sigma\}$, *where* $l_\sigma(m) = m(\sigma)$.

PROOF. On the basis of Th. 2.1.13 it is only necessary to show that there are no other extreme points of L other than the l_σ. If $0 \leq y \leq 1$, then, from the spectral theorem it follows that:

$$(m, y) = \int_{-\delta}^{1} \alpha \, dm(\sigma \leq \alpha), \tag{2.1.15}$$

where $\delta > 0$ can be arbitrarily chosen. We define two elements $y_1, y_2 \in L(\Sigma)$ by means of the following equations:

$$(m, y_1) = \int_{-\delta}^{1} (2\alpha - \alpha^2) \, dm(\sigma(y \leq \alpha)) \tag{2.1.16}$$

and

$$(m, y_2) = \int_{-\delta}^{1} \alpha^2 \, dm(\sigma(y \leq \alpha)) \tag{2.1.17}$$

for which $y = \frac{1}{2}y_1 + \frac{1}{2}y_2$; y can be an extreme point of $L(\Sigma)$ if and only if $y_1 = y_2 = y$, that is, if $\sigma(y \leq \alpha) = \sigma(y_1 \leq \alpha) = \sigma(y_2 \leq \alpha)$ for all α. On the basis of (2.1.15)–(2.1.17) this is the case only if $\sigma(y \leq \alpha) = \sigma(y = 0)$ for all α for which $0 \leq \alpha < 1$. Thus

$$(m, y) = m(\varepsilon \dotplus \sigma(y = 0)) = l_{\varepsilon \dotplus \sigma(y=0)}. \tag{2.1.18}$$

Th. 2.1.17. *The map* $\sigma \longrightarrow l_\sigma$ *is an order isomorphism of* Σ *onto* $\partial_e L(\Sigma)$.

PROOF. From Th. 2.1.16 and Th. 2.1.13 it is obvious that the mapping is a bijection. Thus, it is clear that $\sigma_1 > \sigma_2 \Leftrightarrow l_{\sigma_1} > l_{\sigma_2}$, where $l_{\sigma_1} > l_{\sigma_2}$ is the order relation in $\mathscr{B}'(\Sigma)$, that is, $(m, l_{\sigma_1}) = m(\sigma_1) \geq (m, l_{\sigma_2}) = m(\sigma_2)$ for all $m \in K(\Sigma)$.

On the basis of Th. 2.1.17 we may identify Σ with $\partial_e L(\Sigma)$ and write $(m, l_\sigma) = (m, \sigma)$.

Since we may identify a representation of a Boolean ring with a measure space (see [31] and §2.5) and the elements of $\mathscr{B}'(\Sigma)$ with a measurable, essentially bounded function (where two functions which differ only on a set of zero measure are considered to be the "same" element of $\mathscr{B}'(\Sigma)$) the abstract space $\mathscr{B}'(\Sigma)$ is sometimes called the space of measurable functions. We will not, however, make use of this representation of Σ and of this notion of $\mathscr{B}'(\Sigma)$.

We shall now state (without proof) an interesting theorem which is concerned with the axioms introduced in III.

Th. 2.1.18. *If the lattice G which was introduced in* III, §3 *is distributive then it follows that (without making use of* AV 3 *and* AV Vs *in* III, §3) *the following isomorphisms hold for the Banach spaces defined in* III, §3:

$$\mathscr{B} \longrightarrow \mathscr{B}(\Sigma), \qquad \mathscr{B}' \longrightarrow \mathscr{B}'(\Sigma) \quad \textit{where } G \longrightarrow \partial_e L(\Sigma)$$

(and therefore $G \longrightarrow \Sigma$), $L \longrightarrow L(\Sigma)$, $K \longrightarrow K(\Sigma)$ *for a suitable chosen* Σ. *For a proof, see* [13], VII, §5.3.

2.2 Mixture Morphisms Corresponding to an Observable

According to Th. 1.4.3, for each observable $\Sigma \xrightarrow{F} L$, there exists a σ-additive measure m defined by $m(\sigma) = \mu(w, F(\sigma))$ for each $w \in K$; for w_0 (from Th. 1.4.1) $m_0(\sigma) = \mu(w_0, F(\sigma))$ is effective. In this way we clearly obtain a map of K (as a subset of $\mathscr{B}(\mathscr{H}_1, \mathscr{H}_2, \ldots)$) into $K(\Sigma)$ (see also [17], p.380) as follows:

$$\begin{aligned} K &\longrightarrow K(\Sigma), \\ w &\longrightarrow \mu(w, F(\sigma)). \end{aligned} \tag{2.2.1}$$

We shall now find it useful to investigate such maps. In V, §4.1 these maps will be studied in a more general setting. The discussion presented in this section may serve to motivate the studies presented in V. We will therefore use several general results from V here. From V, D 4.1.1 it is easy to show that the map (2.2.1) is a mixture morphism (abbreviation: mi-morphism). According to V, Th. 4.1.2 such a mi-morphism uniquely defines a map S of $\mathscr{B}(\mathscr{H}_1, \ldots)$ into $\mathscr{B}(\Sigma)$ which is norm-continuous.

Earlier we have called a $w \in K(\mathscr{H}_1, \ldots)$ effective if $C(w) = K$.

Th. 2.2.1 (H. Neumann [21]). *The mi-morphism S defined by* (2.2.1) *transforms effective ensembles $w \in K(\mathscr{H}_1, \ldots)$ into effective measures $m \in K(\Sigma)$. The adjoint map S' of $\mathscr{B}'(\Sigma)$ into $\mathscr{B}'(\mathscr{H}_1, \ldots)$ maps (according to* V, Th. 4.1.3) *$L(\Sigma)$ into $L(\mathscr{H}_1, \ldots)$. The restriction of S' upon $\partial_e L(\Sigma)$ is identical to the vector measure $F: \Sigma \to L$ providing that we identify Σ with $\partial_e L(\Sigma)$ (according to* Th. 2.1.17). *S is uniquely defined by the restriction of S' upon $\partial_e L(\Sigma)$.*

PROOF. Let w be effective. From $m(\sigma) = \mu(w, F(\sigma)) = 0$ it follows that $F(\sigma) = 0$; if $F(\sigma) = 0$ then $\sigma = 0$ since F was assumed to be effective.

Since $\sigma \in \partial_e L(\Sigma) = \Sigma$ and if $Sw = m$ it follows that $\mu(w, S'\sigma) = (Sw, \sigma) = (m, \sigma) = m(\sigma)$ and, according to (2.2.1) $\mu(w, S'\sigma) = \mu(w, F(\sigma))$ for all $w \in K$. Thus we obtain $S'\sigma = F(\sigma)$.

Suppose that we have two mi-morphisms S_1 and S_2 for which S_1' and S_2' are identical on $\partial_e L(\Sigma)$. Then, for all $w \in K(\mathscr{H}_1, \ldots)$ we obtain $0 = \mu(w, (S_1' - S_2')\sigma) = (S_1 w - S_2 w, \sigma)$ for all $\sigma \in \partial_e L(\Sigma)$. According to either the Krein–Milman theorem or the spectral theorem (Th. 2.1.15) $L(\Sigma) = \overline{\text{co}}\partial_e L(\Sigma)$. Thus it follows that $(S_1 w - S_2 w, h) = 0$ for all $h \in L(\Sigma)$; since $L(\Sigma)$ spans all of $\mathscr{B}'(\Sigma)$, the same result holds also for all $h \in \mathscr{B}'(\Sigma)$. It follows that $S_1 w = S_2 w$ for all $w \in K(\mathscr{H}_1, \ldots)$ from which we conclude that $S_1 = S_2$.

Th. 2.2.2 (H. Neumann [21]). *If S is a mi-morphism of $K(\mathscr{H}_1, \ldots)$ into $K(\Sigma)$ which maps effective ensembles into effective measures, then the restriction of the adjoint map S' onto $\partial_e L(\Sigma)$ defines a σ-additive effective measure F on Σ such that $F(\sigma) = S'\sigma$ and $F(\varepsilon) = 1$ (where ε is the unit element of Σ). In addition, Σ_g is complete if U_g is the uniform structure determined by F.*

PROOF. According to V, Th. 4.1.3 S' maps the set $L(\Sigma)$ into $L(\mathscr{H}_1, \ldots)$. We note that if $\sigma_1 \wedge \sigma_2 = 0$ we obtain $m(\sigma_1 \vee \sigma_2) = m(\sigma_1) + m(\sigma_2)$, from which it follows that $\sigma_1 \vee \sigma_2 = \sigma_1 + \sigma_2$, where the σ_i are elements of $\partial_e L(\Sigma) \subset \mathscr{B}'(\Sigma)$. Thus it follows that $S'(\sigma_1 \vee \sigma_2) = S'\sigma_1 + S'\sigma_2$; thus $S'\sigma$ is an additive measure on Σ. If we define $F(\sigma) = S'\sigma$, then, for $\sigma \in \partial_e L(\Sigma)$, $m = Sw$ we obtain $m(\sigma) = (m, \sigma) = (Sw, \sigma) = \mu(w, S'\sigma) = \mu(w, F(\sigma))$.

For $\sigma = \varepsilon$ we find that $1 = m(\varepsilon) = \mu(w, F(\varepsilon))$ for all $w \in K(\mathscr{H}_1, \ldots)$ and we therefore obtain $F(\varepsilon) = \mathbf{1}$.

For w_0 defined by Th. 1.4.1 we find that Sw_0 is an effective σ-additive measure m_0 on Σ for which U_g is the uniform structure generated by $d(\sigma_1, \sigma_2) = m_0(\sigma_1 \dotplus \sigma_2)$. Thus, by Th. 1.4.4, Σ is U_g-complete.

Th. 2.2.1 and Th. 2.2.2 show that it is possible to uniquely characterize σ-additive measures $F: \Sigma \to L$ on complete Boolean rings Σ by mi-morphisms $S: K(\mathscr{H}_1, \ldots) \to K(\Sigma)$ which transform effective ensembles w into effective measures. $S': L(\Sigma) \to L(\mathscr{H}_1, \ldots)$ represents a type of extension of the map $F: \Sigma \to L(\mathscr{H}_1, \ldots)$ because we may identify Σ with $\partial_e L(\Sigma)$ and S' is equal to F on $\partial_e L(\Sigma)$. Since S' is linear, S' maps the convex set $L(\Sigma)$ (which is generated by $\partial_e L(\Sigma)$) onto a convex subset of $L(\mathscr{H}_1, \ldots)$. What is the physical interpretation of these convex subsets?

2.3 The Kernel of an Observable; Mixture of Effects for an Observable

In III, §2 we discussed the use of the concept of a random generator in the formulation of the concept of the direct mixture of two registration methods. We shall now consider the special case in which $b_{01} = b_{02}$. We begin by defining the following special case of III, D 2.6.

D 2.3.1. A b_0 is said to be a direct mixture of the same $\tilde{b}_0$ if there exist two registration methods b'_{01}, b'_{02} such that b'_{01} is isomorphic to $\tilde{b}_0$, b'_{02} is isomorphic to $\tilde{b}_0$ such that $b'_{01} \cap b'_{02} = \emptyset$, $b_0 = b'_{01} \cup b'_{02}$. We call $\alpha = \lambda_{\mathscr{R}_0}(b_0, b'_{01})$ and $1 - \alpha$ the weights of b'_{01} and b'_{02} in the direct mixture b_0.

The direct mixture of $\tilde{b}_0$ with itself is nothing other than an extension of the Boolean rings $\mathscr{R}(\tilde{b}_0)$ to a Boolean ring $\mathscr{R}(b_0)$ with the aid of a random generator such that all convex combinations of effects $\psi(\tilde{b}_0, \tilde{b})$ appear in the ratio α to $1 - \alpha$. According to III (2.8), to each pair $\tilde{b}_1, \tilde{b}_2 \subset \tilde{b}_0$ there exists a $b \subset b_0$ such that

$$\psi(b_0, b) = \alpha\psi(\tilde{b}_0, \tilde{b}_1) + (1 - \alpha)\psi(\tilde{b}_0, \tilde{b}_2). \tag{2.3.1}$$

From III, Th. 2.4 we obtain the following special case:

Th. 2.3.1. *Let $\tilde{b}_0$ be a registration method, and let $\lambda_i > 0$ be a set of rational numbers such that $\sum_{i=1}^n \lambda_i = 1$. Then there exists a registration method b_0 such that, in $\mathscr{R}(b_0)$ there exists a b for every series of $\tilde{b}_i \in \mathscr{R}(\tilde{b}_0)$ for which*

$$\psi(b_0, b) = \sum_{i=1}^{n} \lambda_i \psi(\tilde{b}_0, \tilde{b}_i). \tag{2.3.2}$$

It is not difficult to obtain mixtures of effects from the extension of the ring $\mathcal{R}(b_0)$ by means of a random generator. This procedure apparently does not lead to new information. In Th. 2.3.3 we shall show that we may obtain an abstract version of Th. 2.3.1

The converse question is more interesting: Let $\Sigma \xrightarrow{F} L$ be an observable. Suppose there exist three elements $\sigma, \sigma_1, \sigma_2$ which satisfy the following equation:

$$F(\sigma) = \alpha F(\sigma_1) + (1 - \alpha)F(\sigma_2)$$

(which is an abstract version of (2.3.1)). Is it possible to "shrink" Σ in such a way as to eliminate σ?

We have placed these descriptions of mixtures of effects in the foreground in order to make it easier for the reader to visualize the physics underlying the following abstract discussion about convex combinations and to understand the desire to seek "small" observables.

We shall begin the abstract discussion with a definition:

D 2.3.2. Let $F: \Sigma \rightarrow L$ be an additive measure on the Boolean ring Σ. (We do not require that $\Sigma = \hat{\Sigma}_g$ or that $\hat{\Sigma}_g$ be separable.) Let $\overline{\text{co}}(F\Sigma)$ denote the $\sigma(\mathscr{B}', \mathscr{B})$-convex closure of the set $F\Sigma$; we shall call it the *convex range* of the measure F. If $\Sigma \xrightarrow{F} L$ is an observable, then we shall call $\overline{\text{co}}(F\Sigma)$ the *convex range* of the observable.

According to the previous discussion, in order to make an "economical" measurement, it is not necessary to observe a "response" of a σ for which there exists a pair σ_1, σ_2 satisfying (2.3.3). Conversely, if we are interested in the physical meaning of mixtures of effects, by using direct mixtures we may introduce arbitrarily many convex combinations of the form (2.3.3). It is possible to prove the following: To each $\Sigma \xrightarrow{F} L$ there exists an extension Σ' of Σ for which, if $\Sigma' \xrightarrow{F'} L'$ then $\overline{\text{co}}(F\Sigma) = F'\Sigma'$ (see [21]); this result has the following intuitive meaning: we may introduce arbitrary convex combinations (corresponding to the physical notion of a direct mixture given in Th. 2.3.1) if we are willing to "enlarge" the Boolean ring.

Since L is $\sigma(\mathscr{B}', \mathscr{B})$-compact, $\overline{\text{co}}(F\Sigma)$ is, as a closed subset of L, also $\sigma(\mathscr{B}', \mathscr{B})$-compact. According to the Krein–Milman theorem $\overline{\text{co}}(F\Sigma)$ is generated by its set of extreme points $\partial_e \overline{\text{co}}(F\Sigma)$. Physically this set $\partial_e \overline{\text{co}}(F\Sigma)$ is the essential set of effects for the observable $\Sigma \xrightarrow{F} L$. We therefore define:

D 2.3.3. $\partial_e \overline{\text{co}}(F\Sigma)$ is called the extreme kernel of the range of the measure $F: \Sigma \rightarrow L$. If $\Sigma \xrightarrow{F} L$ is an observable, then it is called extreme kernel of the observable.

D 2.3.4. Two observables $\Sigma_1 \xrightarrow{F_1} L$ and $\Sigma_2 \xrightarrow{F_2} L$ are said to be convex equivalent if $\overline{\text{co}}(F_1\Sigma_1) = \overline{\text{co}}(F_2\Sigma_2)$.

On the basis of the Krein–Milman theorem the following assertions are equivalent:

(i) $\Sigma_1 \xrightarrow{F_1} L$ and $\Sigma_2 \xrightarrow{F_2} L$ are convex equivalent.
(ii) $\Sigma_1 \xrightarrow{F_1} L$ and $\Sigma_2 \xrightarrow{F_2} L$ have the same extremal kernel.

Physically, in order that we do not have to measure anything more than is necessary, it is desirable to make an observable "as small as possible." Mathematically, this corresponds to the question whether there is a "smallest" range of an observable, given a particular convex range. The answer is based upon the following theorems:

Th. 2.3.2. *Let $F: \Sigma \to L$ be an effective measure on the Boolean ring Σ (we do not require that $\Sigma = \hat{\Sigma}_g$ or that Σ_g is separable). Then $\overline{\text{co}}(F\Sigma) = \overline{\text{co}}(F\hat{\Sigma}_g)$.*

PROOF. From Th. 1.4.3 it follows that, since F is continuous (according to Th. 1.4.2), $F\hat{\Sigma}_g \subset \overline{F\Sigma}$, from which the assertion follows.

Th. 2.3.3. *To each additive measure $F: \Sigma \to L$ there exists a convex equivalent observable $\Sigma_1 \xrightarrow{F_1} L$. If desirable, this observable may be chosen such that $\overline{\text{co}}(F\Sigma) = \overline{F_1\Sigma_1}$ (where $\bar{A}$ is the $\sigma(\mathscr{B}', \mathscr{B})$-closure of A).*

PROOF. Since L is $\sigma(\mathscr{B}', \mathscr{B})$-separable, we may choose a countable subset $\{\sigma_\nu\} \subset \Sigma$ such that $\{F(\sigma_\nu)\}$ is dense in $F\Sigma$. Thus $\overline{\text{co}}(F\Sigma) = \overline{\text{co}}(\{F(\sigma_\nu)\})$. The σ_ν generate a countable Boolean subring Σ' of Σ such that $\overline{\text{co}}(F\Sigma) = \overline{\text{co}}(F\Sigma')$. $\Sigma_1 = \hat{\Sigma}'_g$ is separable, hence the extension F_1 of F (as a function over Σ') onto $\hat{\Sigma}'_g$ defines an observable $\Sigma_1 \xrightarrow{F_1} L$. According to Th. 2.3.2 $\overline{\text{co}}(F_1\Sigma_1) = \overline{\text{co}}(F\Sigma')$; therefore $\Sigma_1 \xrightarrow{F_1} L$ is convex equivalent to $\Sigma \xrightarrow{F} L$.

If we wish to obtain $\overline{\text{co}}(F\Sigma) = \overline{F_1\Sigma_1}$ we then extend $F: \Sigma \to L$ to $F'': \Sigma'' \to L$ for which $\overline{\text{co}}(F\Sigma) = F''\Sigma''$ (see above and [21]). Then, as above, we may construct the observable $\Sigma_1 \xrightarrow{F_1} L$ by means of a countable subset $\{\sigma_\nu\} \subset \Sigma''$ for which $\{F''(\sigma_\nu)\}$ is dense in $F''\Sigma''$.

In order to investigate the convex range we may always assume that $\Sigma = \hat{\Sigma}_g$, that is, that Σ is complete. This makes it possible to use the results of §2.2. We then obtain

Th. 2.3.4 (H. Neumann [21]). *Let $F: \Sigma \to L$ be an effective additive measure, and let $\Sigma = \hat{\Sigma}_g$. Let $S: K(\mathscr{H}_1, \ldots) \to K(\Sigma)$ be the mi-morphism defined by* (2.2.1) *and let S' be the dual map (see* Th. 2.2.1) *$S': L(\Sigma) \to L(\mathscr{H}_1, \ldots)$. Then the following conditions are satisfied:*

(i) $S'L(\Sigma) = \overline{\text{co}}(F\Sigma)$.
(ii) *To each extreme point $g_e \in \partial_e \overline{\text{co}}(F\Sigma) = \partial_e S'L(\Sigma)$ there is one and only one $\sigma \in \Sigma$ for which $g_e = F(\sigma)$.*
(iii) $G \cap \overline{\text{co}}(F\Sigma) = G \cap \partial_e \overline{\text{co}}(F\Sigma) = G \cap (F\Sigma)$.

PROOF. Since S' is continuous with respect to the $\sigma(\mathscr{B}', \mathscr{B})$ topologies and $L(\Sigma)$ is compact, we find that $S'L(\Sigma)$ is compact. Since $L(\Sigma)$ is convex, $S'L(\Sigma)$ is convex. Then, since Th. 2.2.1 Σ may be identified with $\partial_e L(\Sigma)$, we find that $F\Sigma \subset S'L(\Sigma)$

and therefore obtain $\overline{\mathrm{co}}(F\Sigma) \subset S'L(\Sigma)$. Then, by the Krein–Milman theorem $L(\Sigma) = \overline{\mathrm{co}}\partial_e L(\Sigma)$ and F is continuous, $S'L(\Sigma) = S'\,\overline{\mathrm{co}}\partial_e L(\Sigma) = \overline{\mathrm{co}}S'\partial_e L(\Sigma) = \overline{\mathrm{co}}(F\Sigma)$ whereby we have proven (i).

Let $g_e \in \partial_e\,\overline{\mathrm{co}}(F\Sigma) = \partial_e S'L(\Sigma)$. The set $S'^{-1}g_e \cap L(\Sigma)$, that is, the set of all $f \in L(\Sigma)$ with $S'f = g_e$ forms a closed face of $L(\Sigma)$, as we will show: $S'^{-1}g_e$ is the inverse image of a closed set and is therefore closed. Therefore $S'^{-1}g_e \cap L(\Sigma)$ is also closed. It is easy to verify that $S'^{-1}g_e \cap L(\Sigma)$ is convex. Let $f \in S'^{-1}g_e \cap L(\Sigma)$, $f_1, f_2 \in L(\Sigma)$ and suppose that

$$f = \lambda f_1 + (1 - \lambda)f_2 \quad \text{where } 0 < \lambda < 1.$$

Then

$$S'f = g_e = \lambda S'f_1 + (1 - \lambda)S'f_2.$$

Since g_e is an extreme point, $S'f_1 = S'f_2 = g_e$, that is, $f_1, f_2 \in S'^{-1}g_e \cap L(\Sigma)$.

According to the Krein–Milman theorem the face $S'^{-1}g_e \cap L(\Sigma)$ contains an extreme point of $L(\Sigma)$ which, according to Th. 2.2.1 may be identified with a $\sigma \in \Sigma$. Therefore there exists a $\sigma \in \Sigma$ for which $F(\sigma) = g_e$.

Suppose that, for $\sigma_1, \sigma_2 \in \Sigma$ we have $F(\sigma_1) = F(\sigma_2) = g_e$. Since F is additive, it follows that

$$g_e = F(\sigma_1) = F(\sigma_1 \dot{+} \sigma_1 \cdot \sigma_2) + F(\sigma_1 \cdot \sigma_2),$$

$$g_e = F(\sigma_2) = F(\sigma_2 \dot{+} \sigma_1 \cdot \sigma_2) + F(\sigma_1 \cdot \sigma_2).$$

We therefore obtain

$$\begin{aligned} &\tfrac{1}{2}F(\sigma_1 \vee \sigma_2) + \tfrac{1}{2}F(\sigma_1 \wedge \sigma_2) \\ &\quad = \tfrac{1}{2}[F(\sigma_1 \dot{+} \sigma_1 \cdot \sigma_2) + F(\sigma_1 \cdot \sigma_2) + F(\sigma_2 \dot{+} \sigma_1 \cdot \sigma_2)] + \tfrac{1}{2}F(\sigma_1 \cdot \sigma_2) \\ &\quad = g_e. \end{aligned}$$

Since g_e is an extreme point, the following condition must be satisfied:

$$F(\sigma_1 \vee \sigma_2) = F(\sigma_1 \wedge \sigma_2) = g_e.$$

Thus it follows that

$$F(\sigma_1 \dot{+} \sigma_2) = F(\sigma_1 \vee \sigma_2) - F(\sigma_1 \wedge \sigma_2) = 0.$$

Since F is effective, it follows that $\sigma_1 \dot{+} \sigma_2 = 0$, that is, $\sigma_1 = \sigma_2$. Since $G \subset \partial_e L(\mathscr{H}_1, \ldots)$ (see III, Th. 6.6) we obtain $G \cap \overline{\mathrm{co}}(F\Sigma) = G \cap \partial_e\,\overline{\mathrm{co}}(F\Sigma)$. Then, since by (ii) $\partial_e\,\overline{\mathrm{co}}(F\Sigma) \subset F\Sigma$ we obtain

$$G \cap \partial_e\,\overline{\mathrm{co}}(F\Sigma) \subset G \cap F\Sigma \subset G \cap \overline{\mathrm{co}}(F\Sigma).$$

D 2.3.5. Let $F\colon \Sigma \to L$ where Σ is U_g-complete. Then the subset

$$N = \{\sigma \mid \sigma \in \Sigma \text{ and } F(\sigma) \in \partial_e\,\overline{\mathrm{co}}(F\Sigma)\}$$

is called the *kernel* of the measure F. If $\Sigma \xrightarrow{F} L$ is an observable, then N is called the kernel of the observable $\Sigma \xrightarrow{F} L$.

According to Th. 2.3.4 the map $\sigma \to F(\sigma)$ is a bijection of N onto $\partial_e\,\overline{\mathrm{co}}(F\Sigma)$.

For the experimental technique of measurement the kernel N of an observable is the essential component. For example, in order to determine the frequencies $\mu(w, F(\sigma))$ for all $\sigma \in \Sigma$ we need only measure the "responses" corresponding to the "indications" $\sigma \in N$. In §2.4 we shall consider this topic in more detail.

From Th. 2.3.4(iii) we find that we do not introduce any new additional decision effects into $F\Sigma$ by taking the closure $\overline{\text{co}}$ of the domain $F\Sigma$ (providing, of course, that Σ is U_g-closed). For the special case in which $F\Sigma \subset G$ it follows that, since $\Sigma \xrightarrow{F} G$ is, by definition D 1.4.3, a decision observable, that

$$F\Sigma = G \cap F\Sigma = G \cap \overline{F\Sigma} = G \cap \overline{\text{co}}(F\Sigma)$$

and therefore

$$F\Sigma \text{ is } \sigma(\mathscr{B}', \mathscr{B})\text{-closed in } G.$$

Thus we obtain an alternative proof of the portion of Th. 1.3.7 which is enclosed by brackets; see also Th. 1.4.8. Therefore, the kernel N of a decision observable is identical to Σ, and we may therefore identify Σ, and therefore N with $F\Sigma$. According to Th. 1.4.8 $N = \Sigma$ is U_g-separable.

We shall now show that the above result holds in general.

Th. 2.3.5. *The kernel N of a measure $F: \Sigma \to L$ (where Σ is complete in U_g) is U_g-separable (in general Σ is not U_g-separable—see* [11]*). There exists an observable $\Sigma_1 \xrightarrow{F_1} L$ where $\Sigma_1 \subset \Sigma$ and $F_1 = F_{|\Sigma_1}$ for which the kernel $N_1 = N$ (where N is the kernel of $F: \Sigma \to L$). The complete Boolean ring $\Sigma_2 \subset \Sigma$ generated by N is separable; therefore $\Sigma_2 \xrightarrow{F_2} L$ where $F_2 = F_{|\Sigma_2}$ is an observable. If $\partial_e \overline{\text{co}}(F\Sigma) \subset G$, then $\partial_e \overline{\text{co}}(F\Sigma)$ is a complete Boolean sublattice of G and $\Sigma_2 \xrightarrow{F_2} L$ is an isomorphism $\Sigma_2 \xrightarrow{F_2} \partial_e \overline{\text{co}}(F\Sigma)$.*

PROOF. If we construct Σ_1 as in the proof of Th. 2.3.3, then it follows that $\Sigma_1 \subset \Sigma$ and that $\Sigma_1 \xrightarrow{F} L$ is an observable (that is, Σ_1 is U_g-separable). According to Th. 2.3.3, for this observable we have $\overline{\text{co}}(F\Sigma) = \overline{\text{co}}(F\Sigma_1)$ and therefore $\partial_e \overline{\text{co}}(F\Sigma) = \partial_e \overline{\text{co}}(F\Sigma_1)$. Since Σ_1 also satisfies the assumptions of Th. 2.3.4, $F(\sigma)$ is, according to Th. 2.3.4(ii) a bijection of $N_1 \subset \Sigma_1$ upon $\partial_e \overline{\text{co}}(F\Sigma_1)$. Since $F(\sigma)$ is also a bijection of $N \subset \Sigma$ onto $\partial_e \overline{\text{co}}(F\Sigma)$ it follows that $N_1 = N$. Since Σ_1 is U_g-separable, we find that $N = N_1$, as a subset of Σ_1 is U_g-separable. Thus Σ_2 is also separable. From Th. 1.4.7 it follows that the last part of Th. 2.3.5 holds.

According to Th. 2.3.5 the complete Boolean ring Σ_2 generated by the kernel N (which exists according to Th. 2.3.5) yields the smallest possible observable for which nothing is lost with respect to the measure $F: \Sigma \to L$ since $\partial_e \overline{\text{co}}(F_2\Sigma_2) = \partial_e \overline{\text{co}}(F\Sigma)$. Thus, as far as the physics of the situation is concerned, it suffices to only consider those observables $\Sigma \xrightarrow{F} L$ for which Σ is the complete Boolean ring generated by the kernel N.

D 2.3.6. Let N be the kernel corresponding to the observable $\Sigma \xrightarrow{F} L$. If the complete Boolean ring generated by N is equal to Σ, then we shall call $\Sigma \xrightarrow{F} L$ a *kernel observable*.

Since the kernel N of a decision observable is identical to Σ, every decision observable is a kernel observable.

D 2.3.7. Let $\Sigma \xrightarrow{F} L$ be an observable. The observable $\Sigma_2 \xrightarrow{F_2} L$ generated by the kernel N of $\Sigma \xrightarrow{F} L$ is called the *associated kernel observable* corresponding to $\Sigma \xrightarrow{F} L$.

For a given observable, for measurement purposes it is sufficient to consider only the associated kernel observable and the associated measurements. If we consider only the class of kernel observables, we have taken an additional step towards our goal of describing "measurements" in terms of the structure of microsystems: the exclusion of mixtures of effects from the measurement. Have we actually attained this goal? Suppose that the relation $\partial_e \overline{\text{co}}(F\Sigma) \subset G$ holds for an observable $\Sigma \xrightarrow{F} L$. Then the kernel is a complete Boolean ring (see Th. 1.4.8), and the kernel observable is therefore a decision observable $N \xrightarrow{F_2} G$, where $F_2 = F_{|N}$. It would suffice to consider only decision observables if all kernel observables would be decision observables.

If this would really be the case, then we could say that the decision effects are the "true" measurements of the structure of microsystems. Unfortunately, as we shall find in the remainder of this chapter, this is not the case.

In this section we have already seen that it is sufficient to consider only observables—in particular, kernel observables, instead of more general measures $F: \Sigma \to L$.

To close this section, we shall now state the following theorem, which is a direct consequence of the previous theorems:

Th. 2.3.6. *A set $A \subset L$ is a set of coexistent effects if and only if there exists an observable (which we may also assert is a kernel observable) $\Sigma \xrightarrow{F} L$ for which $A \subset \overline{\text{co}}(F\Sigma)$.*

2.4 Mixtures and Decompositions of Observables

We shall now, according to the implications of the previous section, only consider observables—as described by U_g-separable Boolean rings.

We shall now seek to define, in mathematical terms, the following intuitive idea: A registration method b_{01} registers "more" than a second b_{02}. For this purpose, we again consider, as an idealization of $\mathscr{R}(b_{01})$ and $\mathscr{R}(b_{02})$, two observables $\Sigma_1 \xrightarrow{F_1} L$ and $\Sigma_2 \xrightarrow{F_2} L$ and the following possibility: Suppose that we are given a homomorphism h of a Boolean ring Σ_1 into the Boolean ring Σ_2 for which the following diagram is commutative:

$$\begin{array}{ccc} \Sigma_1 & \xrightarrow{h} & \Sigma_2 \\ & {}_{F_1}\searrow \quad \swarrow_{F_2} & \\ & L & \end{array} \tag{2.4.1}$$

A homomorphism h is an isomorphism of Σ_1 onto the Boolean subring $h\Sigma_1$ of Σ_2 if and only if no other element of Σ_1 except the zero element is mapped to the zero element of Σ_2. If this is the case, then from $h(\sigma_1) = h(\sigma_2)$ it follows that $h(\sigma_1 \dotplus \sigma_2) = h(\sigma_1) \dotplus h(\sigma_2) = 0$ and $\sigma_1 \dotplus \sigma_2 = 0$ and that $\sigma_1 = \sigma_2$. h^{-1} exists, and is also a homomorphism, where the latter can be proven in the following way. From $\sigma_3 = h^{-1}[h(\sigma_1) \vee h(\sigma_2)]$ and $h(\sigma_1 \vee \sigma_2) = h(\sigma_1) \vee h(\sigma_2)$ it follows that $\sigma_3 = (\sigma_1 \vee \sigma_2)$ and so on.

If (2.4.1) is commutative, and if $\Sigma_1 \xrightarrow{F_1} L$, $\Sigma_2 \xrightarrow{F_2} L$ are observables, that is, F_1, F_2 are effective, then if $h(\sigma) = 0$ it follows that $\sigma = 0$: According to (2.4.1) it follows that if $h(\sigma) = 0$ then $F_1(\sigma) = F_2(\sigma) = F_2(0) = 0$; therefore $\sigma = 0$. h is an isomorphism of Σ_1 onto $h\Sigma_1 \subset \Sigma_2$.

We will prove that, in addition, h is not only an isomorphism of the Boolean ring Σ_1 onto $h\Sigma_1 \subset \Sigma_2$, but is also an isomorphism of the uniform structures of the two rings. These uniform structures are generated by

$$d_1(\sigma_1, \sigma_2) = \mu(w_0, F_1(\sigma_1 \dotplus \sigma_2))$$

and

$$d_2(\sigma_1, \sigma_2) = \mu(w_0, F_2(\sigma_1 \dotplus \sigma_2)),$$

where w_0 is defined in Th. 1.4.1

Since $F_2(h(\sigma_1) \dotplus h(\sigma_2)) = F_2 h(\sigma_1 \dotplus \sigma_2) = F_1(\sigma_1 \dotplus \sigma_2)$ we find that

$$d_1(\sigma_1, \sigma_2) = d_2(h(\sigma_1), h(\sigma_2)).$$

Thus $h\Sigma_1$ is U_g-complete, as is Σ_1, and is, according to Th. 1.4.3, a complete Boolean sublattice of Σ_2 and h is also an isomorphism of the complete Boolean lattices Σ_1 and $h\Sigma_1$. On the basis of (2.4.1) we may identify $\Sigma_1 \xrightarrow{F_1} L$ with $h\Sigma_1 \xrightarrow{F_2} L$. We shall not, however, do this in the following.

D 2.4.1. Let (Σ_1, F_1) and (Σ_2, F_2) be a pair of observables. We shall say that $\Sigma_2 \xrightarrow{F_2} L$ is *more extensive* than $\Sigma_1 \xrightarrow{F_1} L$ (written $(\Sigma_1, F_1) \prec (\Sigma_2, F_2)$) if there is a homomorphism h (of the type described above) for which the diagram (2.4.1) is commutative.

This definition is the precise form of the following statement: $\Sigma_2 \xrightarrow{F_2} L$ "measures more" than $\Sigma_1 \xrightarrow{F_2} L$.

The relation $\prec$ is a pre-order: Suppose that $(\Sigma_1, F_1) \prec (\Sigma_2, F_2)$ and $(\Sigma_2, F_2) \prec (\Sigma_3, F_3)$. Then, from the commutative diagram

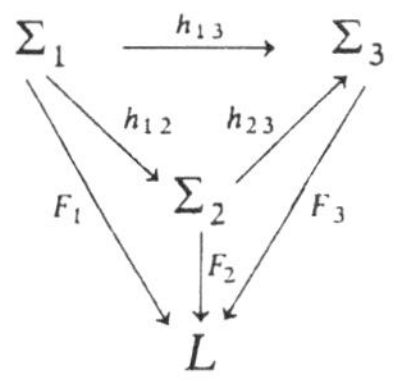

(2.4.2)

it immediately follows that $(\Sigma_1, F_1) \prec (\Sigma_3, F_3)$.

D 2.4.2. A pair of observables (Σ_1, F_1) and (Σ_2, F_2) are said to be *equivalent* if $(\Sigma_1, F_1) \prec (\Sigma_2, F_2)$ and $(\Sigma_2, F_2) \prec (\Sigma_1, F_1)$.

It is easy to verify the fact that if $(\Sigma_1, F_1) \prec (\Sigma_2, F_2)$ then the following relationships hold: $F_1\Sigma_1 \subset F_2\Sigma_2$ and $\overline{\text{co}}(F_1\Sigma_1) \subset \overline{\text{co}}(F_2\Sigma_2)$. Thus, if two observables are equivalent, then they are convex equivalent (see D 2.3.4).

If (Σ_1, F_1) is the kernel observable associated with the observable (Σ, F) then $(\Sigma_1, F_1) \prec (\Sigma, F)$ and (Σ_1, F_1) is convex-equivalent to (Σ, F). Let $(\Sigma_2, F_2) \prec (\Sigma, F)$ and let (Σ_2, F_2) be convex-equivalent to (Σ, F). Then, from (2.4.1) it follows that

$$\overline{\text{co}}(Fh\Sigma_2) = \overline{\text{co}}(F_2\Sigma_2) = \overline{\text{co}}(F\Sigma)$$

and therefore

$$\partial_e\,\overline{\text{co}}(Fh\Sigma_2) = \partial_e\,\overline{\text{co}}(F\Sigma).$$

From this result, and as a consequence of Th. 2.3.4 it follows that the kernels N_1 and N of both observables obey the relation $hN_1 = N$. If (Σ_1, F_1) is the kernel observable associated with (Σ, F) then $h\Sigma_1 \subset \Sigma_2$. In this sense the kernel observable associated with (Σ, F) is the "smallest" observable (Σ_2, F_2) which is convex-equivalent to (Σ, F) and which satisfies $(\Sigma_2, F_2) \prec (\Sigma, F)$. This result exhibits the exceptional status of kernel observables.

The physical description of the registration process motivates the need to study an additional relationship between observables:

Let $b_0 = \bigcup_{i=1}^n b_{0i}$ be a decomposition of the registration method b_0. Then, according to III, Th. 2.3 we obtain

$$\psi(b_0, b) = \sum_{i=1}^{n} \lambda_i \psi(b_{0i}, b_{0i} \cap b), \tag{2.4.3}$$

where $\lambda_i = \lambda_{\mathcal{R}_0}(b_0, b_{0i})$. We may also obtain λ_i by means of the function $\mu(w, g)$ as follows:

$$\lambda_{\mathcal{R}_0}(b_0, b_{0i}) = \lambda_{\mathcal{S}}(a \cap b_0, a \cap b_{0i})$$

for any a for which $a \cap b_0 \neq \varnothing$, that is,

$$\lambda_i = \lambda_{\mathcal{R}_0}(b_0, b_{0i}) = \mu(w, \psi(b_0, b_{0i}))$$

holds for all w, that is, $\mu(w, \psi(b_0, b_{0i})) = \lambda_i$ is independent of w. Thus it follows that

$$\psi(b_0, b_{0i}) = \lambda_i \mathbf{1}. \tag{2.4.4}$$

In equation (2.4.4) we meet, for the first time, an example of a general result (which will be discussed in XVII, §2.3) that we will not obtain any information about a microsystem from an effect of the form $\lambda\mathbf{1}$. This is self-evident in the case of (2.4.4) since b_{0i}, as well as b_0 are, as elements of $\mathcal{R}_0$, statistically independent of the preparation; (2.4.4) is a direct consequence of APS 7.2.

According to the discussions of III, §2 we may make arbitrary direct mixtures of registration methods. The direct mixtures are of great importance for the interpretation of quantum mechanics as well as for the selection of axioms in the sense of III, §3. In III, §2 we have already noted that, in general, an experimental physicist does not create mixtures of registration methods in an experiment. In general, as far as possible, he makes decompositions. In making direct mixtures we lose almost nothing in the way of experimental results (see III, §2) providing the weights λ_i are not too small. We note, however, that the use of direct mixtures makes experiments more difficult without making even a small improvement with respect to the results. On the other hand, the decomposition of registration procedures, if carried out experimentally by improving the tolerances of the apparatus represents a refinement of the results, because it then becomes possible to measure the effect processes $(b_{0i}, b_{0i} \cap b)$ instead of only (b_0, b) by means of the improved apparatus (see III, §2).

If it is possible to measure the effects $\psi(b_{0i}, b_{0i} \cap b)$ then the measurement of the mixtures $\psi(b_0, b)$ is not interesting because no new information is obtained. Thus it is reasonable to seek, in mathematical terms, a solution of the problem of decompositions, making the transition from the "special" situation $\mathscr{R}(b_0) \xrightarrow{\psi_0} L$ to the idealized case $\Sigma \xrightarrow{F} L$.

For this purpose we see that the decomposition described by (2.4.3) may be characterized in the two following different ways:

(1) We may consider $\psi_0(b) = \psi(b_0, b)$ and $\psi_{0i}(b) = \psi(b_{0i}, b_{0i} \cap b)$ to be additive measures on the same Boolean ring $\mathscr{R}(b_0)$. Then (2.4.3) describes the decomposition of measures on $\mathscr{R}(b_0)$ as follows:

$$\psi_0(b) = \sum_i \lambda_i \psi_{0i}(b). \tag{2.4.5}$$

(2) Let us consider the additive measures $\psi_0 : \mathscr{R}(b_0) \to L$ and $\tilde{\psi}_{0i} = \psi(b_{0i}, b_{0i} \cap b) : \mathscr{R}(b_{0i}) \to L$. Let j_i be the injective map of the subset $\mathscr{R}(b_{0i})$ into $\mathscr{R}(b_0)$. Then the following diagram is commutative:

$$\begin{array}{ccc} \mathscr{R}(b_{0i}) & \xrightarrow{j_i} & \mathscr{R}(b_0) \\ & {\scriptstyle \lambda_i \tilde{\psi}_{0i}} \searrow \quad \swarrow {\scriptstyle \psi_0} & \\ & L & \end{array}$$

These two characterizations suggest the following two idealizations concerning observables:

1′. We begin with a *single* observable $\Sigma \xrightarrow{F} L$ and consider a decomposition of F into additive measures $\hat{F}_i : \Sigma \to L$ of the form

$$F(\sigma) = \sum_{i=1}^{m} \lambda_i \hat{F}_i(\sigma) \quad \text{where } \lambda_i > 0, \sum_{i=1}^{n} \lambda_i = 1. \tag{2.4.6}$$

The $\hat{F}_i$ need not be effective measures. It is easy to show that the uniform structure U_g where the latter is defined by the metric

$$d(\sigma_1, \sigma_2) = \mu(w_0, F_i(\sigma_1 \dotplus \sigma_2))$$

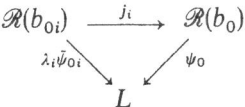

(where w_0 is defined in Th. 1.4.1) is finer than the uniform structures U_{g_i} where the latter are defined by the metrics

$$d_i(\sigma_1, \sigma_2) = \mu(w_0, \hat{F}_i(\sigma_1 \dotplus \sigma_2)).$$

Since the $\hat{F}_i$ are uniformly continuous with respect to the U_{g_i} (see Th. 1.4.2), they are also uniformly continuous with respect to U_g and, as is the case of F, they are also σ-additive (because σ-additivity is equivalent to the condition that $\hat{F}_i(\sigma_\nu) \to 0$ for decreasing sequences which satisfy $\sigma_\nu \to 0$—see Th. 2.1.3).

2′. Let $\Sigma \xrightarrow{F} L$ and $\Sigma_i \xrightarrow{F_i} L$ ($i = 1, 2, \ldots$) be observables. In addition, suppose that there exist isomorphic maps h_i of Σ_i onto an interval $[0, \eta_i]$ of Σ where $\eta_i \wedge \eta_j = 0$ for $i \neq j$ and $\bigvee_i \eta_i = \dot{\varepsilon}$ for which the following diagrams commute:

$$\begin{array}{ccc} \Sigma_i & \xrightarrow{h_i} & \Sigma \\ & {\scriptstyle \lambda_i F_i} \searrow \quad \swarrow {\scriptstyle F} & \\ & L & \end{array} \tag{2.4.7}$$

We then say that the observable $\Sigma \xrightarrow{F} L$ is a mixture of the observables $\Sigma_i \xrightarrow{F_i} L$ with weights λ_i. From the diagram (2.4.7) it follows that $Fh_i(\varepsilon) = \lambda_i \mathbf{1}$, that is, $F(\eta_i) = \lambda_i \mathbf{1}$ and therefore $\lambda_i > 0$. Since the η_i are a decomposition of ε, we obtain $\mathbf{1} = F(\varepsilon) = \sum_i F(\eta_i) = \sum_i \lambda_i \mathbf{1}$, that is, $\sum_i \lambda_i = 1$.

Condition 2′ idealizes in a most precise way the decomposition of a $b_0 \in \mathscr{R}_0$ into different $b_{0i} \in \mathscr{R}_0$ as described by condition 2. From 2′ it follows that 1′ holds as follows: Let

$$\hat{F}_i(\sigma) = \frac{1}{\lambda_i} F_i h_i^{-1}(\eta_i \wedge \sigma). \tag{2.4.8}$$

Since $\eta_i \wedge \sigma < \eta_i$ (2.4.8) defines an additive measure over Σ. From $\sigma = \bigvee_i (\eta_i \wedge \sigma)$ it follows that $F(\sigma) = \sum_i F(\eta_i \wedge \sigma)$. According to the diagram (2.4.7) and definition (2.4.8) it follows that $F(\sigma) = \sum_i \lambda_i \hat{F}_i(\sigma)$—which is a decomposition of the form (2.4.6).

The question whether a structure of the form 1′ exists is mathematically simpler than the question whether a structure of the form 2′ exists. 1′ is somewhat weaker than 2′ (we have already seen that 1′ follows directly from 2′), but physically is not much weaker, as we shall see with the help of the following theorem:

Th. 2.4.1. *Suppose that, for the measure F of an observable $\Sigma \xrightarrow{F} L$ there exists a partition of the form (2.4.6). Then there exists an observable $(\tilde{\Sigma}, \tilde{F}) \succ (\Sigma, F)$, a homomorphism $\Sigma \xrightarrow{h} \tilde{\Sigma}$ and a partition of $\tilde{\varepsilon}_n$ of $\tilde{\Sigma}$ of the form*

$$\tilde{\varepsilon} = \bigvee_{i=1}^{n} \tilde{\eta}_i \quad \textit{where } \tilde{\eta}_i \wedge \tilde{\eta}_j = 0 \textit{ for } i \neq j,$$

such that the following diagrams commute:

$$\begin{array}{ccc} \Sigma & \xrightarrow{\tilde{h}_i} & \tilde{\Sigma} \\ & {\scriptstyle \lambda_i \hat{F}_i} \searrow \quad \swarrow {\scriptstyle \tilde{F}} & \\ & L & \end{array} \qquad (i = 1, 2, \ldots, n), \tag{2.4.9}$$

where $\tilde{h}_i(\sigma) = h(\sigma) \wedge \tilde{\eta}_i$.

PROOF. As we have already seen (in 1′ and the subsequent discussion) the $\hat{F}_i$ are σ-additive measures over Σ. The sets

$$J_i = \{\sigma \mid \sigma \in \Sigma \text{ and } \hat{F}_i(\sigma) = 0\}$$

are complete Boolean subrings of Σ; the factor rings $\Sigma_i = \Sigma/J_i$ are complete Boolean rings. A σ-additive effective measure $\Sigma_i \xrightarrow{F_i} L$ is defined by

$$F_i(\rho) = \hat{F}_i(\sigma) \quad \text{for any } \sigma \in \rho \in \Sigma_i \tag{2.4.10}$$

Therefore we find that the $\Sigma_i \xrightarrow{F_i} L$ are observables.

Using the $\Sigma_i = \Sigma/J_i$ we construct the product $\tilde{\Sigma}$ of the Σ_i. The elements of $\tilde{\Sigma}$ are therefore the n-tuples $(\sigma_1, \sigma_2, \ldots, \sigma_n)$ with the ordering $(\sigma_1, \sigma_2, \ldots) < (\tilde{\sigma}_1, \tilde{\sigma}_2, \ldots)$ if and only if $\sigma_i < \tilde{\sigma}_i$ $(i = 1, 2, \ldots, n)$ (see AI, §3). Thus $\tilde{\Sigma}$ is complete. Using the F_i defined by (2.4.10) we define the following measure $\tilde{F}$ on $\tilde{\Sigma}$:

$$\tilde{F}(\sigma_1, \ldots, \sigma_n) = \sum_i \lambda_i F_i(\sigma_i).$$

This measure is, as in the case of the F_i, σ-additive. Therefore $\tilde{\Sigma} \xrightarrow{\tilde{F}} L$ is an observable. Let $\tilde{\eta}_i = (0, 0, \ldots, \varepsilon_i, \ldots)$. Then we obtain the decomposition $\tilde{\varepsilon} = \bigvee_i \tilde{\eta}_i$ of the unit element $\tilde{\varepsilon}$ of $\tilde{\Sigma}$ and

$$\begin{aligned} \tilde{F}(\sigma_1, \ldots, \sigma_n) &= \sum_{i=1}^{n} \tilde{F}(\tilde{\eta}_i \wedge (\sigma_1, \ldots, \sigma_n)) \\ &= \sum_{i=1}^{n} \tilde{F}(0, 0, \ldots, \sigma_i, \ldots) \\ &= \sum_{i=1}^{n} \lambda_i F_i(\sigma_i). \end{aligned}$$

The map $\sigma_i \rightarrow (0, 0, \ldots, \sigma_i, 0, \ldots)$ is a homomorphism h_i in the sense of 2′ for which the diagram

$$\begin{array}{ccc} \Sigma_i & \xrightarrow{h_i} & \tilde{\Sigma} \\ & {\scriptstyle \lambda_i F_i} \searrow \quad \swarrow {\scriptstyle \tilde{F}} & \\ & L & \end{array}$$

is commutative. Thus we have a structure of type 2′ between Σ_i and $\tilde{\Sigma}$.

Using the canonical homomorphism $\Sigma \xrightarrow{k_i} \Sigma/J_i$, from $h_i k_i$ we obtain a homomorphism of Σ into $\tilde{\Sigma}$ for which the diagram

$$\begin{array}{ccc} \Sigma & \xrightarrow{h_i k_i} & \tilde{\Sigma} \\ & {\scriptstyle \lambda_i \hat{F}_i} \searrow \quad \swarrow {\scriptstyle \tilde{F}} & \\ & L & \end{array} \tag{2.4.11}$$

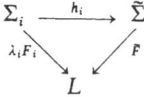

is commutative. Let $h(\sigma) = \bigvee_{i=1}^{n} h_i k_i(\sigma)$; h then defines a homomorphism $\Sigma \xrightarrow{h} \tilde{\Sigma}$ which satisfies the conditions of Th. 2.4.1. The fact that h is a homomorphism follows directly from

$$h(\sigma) = (h_1 k_1(\sigma), h_2 k_2(\sigma), \ldots, h_n k_n(\sigma)).$$

Thus, we obtain

$$\tilde{h}_i(\sigma) = h(\sigma) \wedge \tilde{\eta}_i = h_i k_i(\sigma)$$

from which it follows that (2.4.11) transforms into (2.4.9).

It is easy to verify that condition 2′ is equivalent to the representation of the lattice $\tilde{\Sigma}$ as a product which was used in the proof of Th. 2.4.1. Using the η_i in 2′, for each $\sigma \in \Sigma$ we may form the decomposition $\sigma = \bigvee_{i=1}^{n} (\sigma \wedge \eta_i)$, which may be rewritten in "component form"

$$\sigma = (\sigma \wedge \eta_1, \sigma \wedge \eta_2, \ldots, \sigma \wedge \eta_n).$$

The h_i in (2.4.7) are nothing other than isomorphisms by which we may identify the Σ_i with the Boolean ring of the ith component.

We may also represent condition 2′ as follows: Σ permits a "representation" as a product of the Σ_i as follows:

$$\sigma = (\sigma_1, \sigma_2, \ldots, \sigma_n) \quad \text{where } \sigma_i \in \Sigma_i,$$

where $F(\sigma)$ may be represented in terms of the $F_i\colon \Sigma_i \to L$ in the following form:

$$F(\sigma) = \sum_{i=1}^{n} \lambda_i F_i(\sigma_i).$$

From the diagram (2.4.9) it follows that

$$\tilde{F}(\eta_i) = \tilde{F}(h(\varepsilon) \wedge \tilde{\eta}_i) = \tilde{F}h_i(\varepsilon) = \lambda_i \tilde{F}_i(\varepsilon) = \lambda_i \mathbf{1}.$$

The observable $\tilde{\Sigma} \xrightarrow{\tilde{F}} L$ in Th. 2.4.1 may be represented as a product

$$\tilde{\sigma} = (\tilde{\sigma} \wedge \tilde{\eta}_1, \tilde{\sigma} \wedge \tilde{\eta}_2, \ldots, \tilde{\sigma} \wedge \tilde{\eta}_n)$$

and

$$\tilde{F}(\tilde{\sigma}) = \sum_{i=1}^{n} \lambda_i \tilde{F}_i(\tilde{\sigma} \wedge \tilde{\eta}_i),$$

where the $\tilde{F}_i$ are defined by

$$\tilde{F}_i(\tilde{\sigma} \wedge \tilde{\eta}_i) = \frac{1}{\lambda_i} \tilde{F}(\tilde{\sigma} \wedge \tilde{\eta}_i).$$

Therefore the observable $\tilde{\Sigma} \xrightarrow{\tilde{F}} L$ may be decomposed according to the "refined registration methods" $\tilde{\eta}_i$. Th. 2.4.1 states that if the measure F for the observable $\Sigma \xrightarrow{F} L$ satisfies the decomposition (2.4.6), then by the map h this observable can be identified with a portion (namely with $h\Sigma \subset \tilde{\Sigma}$) of the observable $\tilde{\Sigma} \xrightarrow{\tilde{F}} L$ because the decomposition of $\tilde{\Sigma} \xrightarrow{\tilde{F}} L$ according to the refined registration procedures $\tilde{\eta}_i$ leads to a partition of the restriction of the measure $\tilde{F}$ on $h\Sigma$ (which is identical to F on Σ) of the form (2.4.6). Therefore a decomposition of the form (2.4.6) can always be considered to be a decomposition of a "more extensive" observable according to a set of refined registration methods.

Hence, in experimental situations it is desirable to seek those observables for which the measure F permits no proper decompositions of the form (2.4.6). For this purpose we shall now introduce the following definition:

D 2.4.3. An observable $\Sigma \xrightarrow{F} L$ is said to be *irreducible* if, for any decomposition of F in the following form

$$F(\sigma) = \lambda F_1(\sigma) + (1 - \lambda)F(\sigma), \tag{2.4.12}$$

where $0 < \lambda < 1$ and F_1, F_2 are additive measures $F_1, F_2 \colon \Sigma \to L$ it follows that $F_1 = F_2 = F$.

It immediately follows that every decision observable is irreducible.

In complete analogy to the set $K(\Sigma)$ which was defined in Th. 2.1.1 we shall now introduce a set $K(\Sigma, L)$ as the set of all σ-additive measures $F\colon \Sigma \to L$ on the complete Boolean ring Σ. Then $K(\Sigma, L)$ is a convex set; the measure F for an irreducible observable $\Sigma \xrightarrow{F} L$ is clearly an extreme point of $K(\Sigma, L)$. On the basis of (2.2.1), Th. 2.2.1, and Th. 2.2.2 (see the remarks following Th. 2.2.2) we may also identify $K(\Sigma, L)$ with the set of all mi-morphisms (not only those which transform effective w into effective measures) $K(\mathscr{H}_1, \ldots) \to K(\Sigma)$. An investigation of the mathematical structure of $K(\Sigma, L)$ remains yet to be done; in particular, the structure of $\partial_e K(\Sigma, L)$ is not yet known. In each case we find that all F for which $\Sigma \xrightarrow{F} G$ belongs to $\partial_e K(\Sigma, L)$.

An experimental physicist must strive to measure irreducible observables (or at least "approximately irreducible" observables). We shall not describe what we mean here by "approximately irreducible" observables. For this purpose it is necessary to introduce a uniform structure of physical imprecision in $K(\Sigma, L)$ (for the general structure see [1], §9) with respect to which $K(\Sigma, L)$ is precompact (totally bounded).

We shall now consider the following question: Is irreducibility preserved in the transition to a more extensive observable? This need not be the case; in the following theorem we shall find that if the more extensive observable is reducible, then each component is more extensive than that of the irreducible observable.

Th. 2.4.2. *Let $\Sigma \xrightarrow{F} L$ be irreducible and let $\tilde{\Sigma} \xrightarrow{\tilde{F}} L$ be more extensive than $\Sigma \xrightarrow{F} L$ and reducible in the sense of 2′. Then every component $\Sigma_i \xrightarrow{F_i} L$ is also more extensive than $\Sigma \xrightarrow{F} L$.*

PROOF. According to 2′ and diagram (2.4.7) we may identify Σ_i with $[0, \eta_i] \subset \tilde{\Sigma}$ and F_i with $(1/\lambda_i)\tilde{F}$. Since $\tilde{\Sigma} \xrightarrow{\tilde{F}} L$ is more extensive than $\Sigma \xrightarrow{F} L$ there exists a homomorphism h such that

$$\begin{array}{ccc} \Sigma & \xrightarrow{h} & \tilde{\Sigma} \\ & {\scriptstyle F}\searrow \quad \swarrow{\scriptstyle \tilde{F}} & \\ & L & \end{array} \tag{2.4.13}$$

The map $h_i(\sigma) = h(\sigma) \wedge \eta_i$ is a homomorphism $\Sigma \xrightarrow{h_i} \Sigma_i = [0, \eta_i] \subset \tilde{\Sigma}$. On the basis of the identification of $[0, \eta_i]$ with Σ_i we obtain

$$\begin{aligned} \tilde{F}h_i(\sigma) &= \tilde{F}(h(\sigma) \wedge \eta_i) = \lambda_i F_i(h(\sigma) \wedge \eta_i) \\ &= \lambda_i F_i h_i(\sigma). \end{aligned} \tag{2.4.14}$$

On the other hand, on the basis of (2.4.13) we find that

$$F(\sigma) = \tilde{F}h(\sigma) = \sum_i \tilde{F}(h(\sigma) \wedge \eta_i).$$

Since $F(\sigma)$ is irreducible, it follows that

$$\tilde{F}(h(\sigma) \wedge \eta_i) = \lambda_i F(\sigma).$$

Equation (2.4.14) transforms into

$$F(\sigma) = F_i h_i(\sigma). \tag{2.4.15}$$

Equation (2.4.15) says that the diagram

$$\begin{array}{ccc} \Sigma & \xrightarrow{h_i} & \Sigma_i \\ & {\scriptstyle F}\searrow \quad \swarrow{\scriptstyle F_i} & \\ & L & \end{array}$$

is commutative, and thus we find that $\Sigma_i \xrightarrow{F_i} L$ is more extensive than $\Sigma \xrightarrow{F} L$.

Therefore Th. 2.4.2 guarantees that, in the transition from an irreducible observable to a more extensive observable (at least approximately) we *may* make use of irreducible observables. In IX, §1 we shall study such transitions for decision observables which are always irreducible.

If we combine our wish for irreducible observables with the wish for kernel observables, then the "desired" measurements would be irreducible kernel observables. The irreducible kernel observable characterizes, so to say, the "optimum apparatus without unnecessary redundancy" which we would seek to approximate in an experiment.

We may also wish to determine whether it is reasonable to interpret the irreducible kernel-observables as "measurements of the structure of microsystems." Are all irreducible kernel observables also decision observables?

According to the remarks following D 2.3.6 and the remarks following D 2.4.3 every decision observable is an irreducible kernel observable.

If all irreducible kernel observables were also decision observables, then we could justifiably assert that every measurement made by a sufficiently refined apparatus—neglecting unnecessary mixtures of registrations—is traceable to measurement of decision observables. It should be mentioned that this is indeed the case for "classical" physical systems. However, in the measurement of microsystems there are irreducible kernel observables which are not decision observables. We shall now exhibit an example:

Let φ, χ be two normalized orthogonal vectors in one of the Hilbert spaces $\mathscr{H}_i$, for example, $\mathscr{H}_1$. Let us consider the following effects:

$$\begin{aligned} g_1 &= \alpha(P_\varphi, 0, 0, \ldots), \\ g_2 &= \alpha(P_\psi, 0, 0, \ldots), \end{aligned} \tag{2.4.16}$$

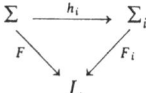

where

$$\psi = \frac{(\varphi + \chi)}{\sqrt{2}} \quad \text{and} \quad \alpha = 2 - \sqrt{2}.$$

Later we shall show that $g_1 + g_2 \leq \mathbf{1}$—see below. Thus g_1, g_2 are coexistent. Let Σ be the Boolean ring consisting of the subsets of a set of three elements. We assign the following measures of the atoms η_1, η_2, η_3 of Σ as follows:

$$\eta_1 \rightarrow g_1,$$

$$\eta_2 \rightarrow g_2,$$

$$\eta_3 \rightarrow \mathbf{1} - g_1 - g_2.$$

In this way we determine an observable $\Sigma \xrightarrow{F} L$ for which the remaining three elements of Σ (excluding 0 and ε) correspond to

$$\eta_1 \cup \eta_2 \rightarrow g_1 + g_2,$$

$$\eta_1 \cup \eta_3 \rightarrow \mathbf{1} - g_2,$$

$$\eta_2 \cup \eta_3 \rightarrow \mathbf{1} - g_1.$$

$\Sigma \xrightarrow{F} L$ is a kernel observable with $N = \Sigma$ since $\overline{\text{co}}(F\Sigma)$ is a three-dimensional parallelepiped: the four points $0, g_1, g_2, g_1 + g_2$ generate a two-dimensional parallelogram. If we add $\mathbf{1} - g_1 - g_2$ to these four points we obtain the remaining four points of the parallelepiped $\overline{\text{co}}(F\Sigma)$.

Now we need only show that $\Sigma \xrightarrow{F} L$ is irreducible. Let $0 < \lambda < 1$, $F(\sigma) = \lambda F_1(\sigma) + (1 - \lambda)F_2(\sigma)$. For $\sigma = \eta_1$ we obtain the following equations for the components in $\mathscr{H}_1$ (the other components are 0):

$$\alpha P_\varphi = \lambda F_1(\eta_1) + (1 - \lambda)F_2(\eta_1). \tag{2.4.17}$$

For each vector $\tau \perp \varphi$ it follows from $F_1(\eta_1) > 0$, $F_2(\eta_1) > 0$ that $\langle \tau, F_1(\eta_1)\tau \rangle = 0$. Thus, since $F_1(\eta_1) > 0$ we must obtain $F_1(\eta_1)\tau = 0$. Therefore $F_1(\eta_1) = \varepsilon_1 P_\varphi$. In the same way it follows that $F_1(\eta_2) = \varepsilon_2 P_\psi$. If we then show that $\varepsilon_1 = \varepsilon_2 = \alpha$, we would then easily obtain $F_1(\sigma) = F(\sigma)$ for all $\sigma \in \Sigma$, that is, $\Sigma \xrightarrow{F} L$ is irreducible.

To this purpose we shall consider all $\varepsilon_1, \varepsilon_2$ for which

$$\eta_1 \rightarrow \varepsilon_1(P_\varphi, 0, \ldots),$$

$$\eta_2 \rightarrow \varepsilon_2(P_\psi, 0, \ldots)$$

determine an additive measure $\Sigma \rightarrow L$. This is the case only if $\varepsilon_1 P_\varphi + \varepsilon_2 P_\psi \leq \mathbf{1}$ (as an operator in $\mathscr{H}_1$).

Let us consider the operator $\varepsilon_1 P_\varphi + \varepsilon_2 P_\psi$. It is easy to determine the $\varepsilon_1, \varepsilon_2$ for which this operator takes on 1 as an eigenvalue: $\varepsilon_1, \varepsilon_2$ must satisfy the condition

$$\varepsilon_1 + \varepsilon_2 - \tfrac{1}{2}\varepsilon_1\varepsilon_2 - 1 = 0. \tag{2.4.18}$$

(For the case in which $\varepsilon_1 = \varepsilon_2$ we obtain the special values $\varepsilon_1 = \varepsilon_2 = \alpha = 2 - \sqrt{2}$.) In the following diagram (Figure 3) we illustrate the domain in the $(\varepsilon_1, \varepsilon_2)$ plane for which $\varepsilon_1 P_\varphi > 0$, $\varepsilon_2 P_\psi > 0$ and $\varepsilon_1 P_\varphi + \varepsilon_2 P_\psi < \mathbf{1}$ by means of shading. It is bounded by the curve (2.4.18). This domain is convex, and $\varepsilon_1 = \varepsilon_2 = \alpha$ is an extreme point. Thus, by (2.4.17) it follows that $\varepsilon_1 = \varepsilon_2 = \alpha$.

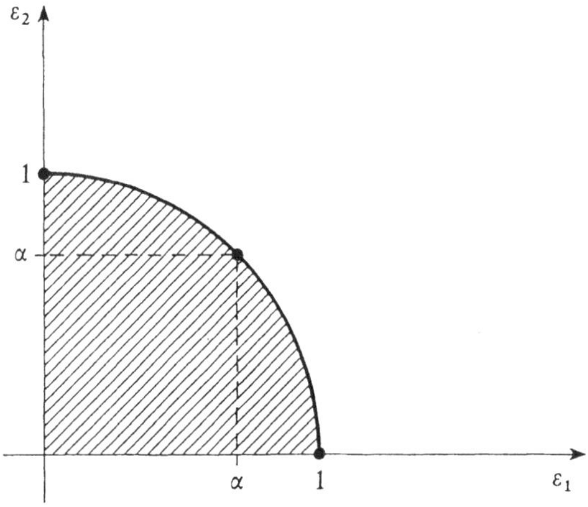

Figure 3

The reader should show that the observable constructed in (2.4.16) is reducible for $\alpha = \frac{1}{2}$, and, in the sense of Th. 2.4.1, can be described as a mixture of two decision observables.

According to Th. 2.4.2, the observable constructed using $\alpha = 2 - \sqrt{2}$ can also not be "decomposed" if we would make a transition into a more extensive observable. If we had an apparatus which measures the observable characterized by (2.4.16) we would need only two yes–no responses, namely, η_1 and η_2. These responses are mutually exclusive, that is, they never occur together. The maximum frequency (for an ensemble) for the response of η_1 is *not* equal to 1 but $\alpha = 2 - \sqrt{2}$; similarly for η_2. There are, however, ensembles for which we "always" obtain at least one of the two responses η_1, η_2 since $\alpha(P_\varphi + P_\psi)$ has the eigenvalue 1! The observable $\Sigma \xrightarrow{F} L$ cannot be improved. It is very instructive to clarify the notions of coexistence and observable for this example, and to realize that observables which are not decision observables are not necessarily the result of "poor" or "unskilled" experimentation.

In §3 we shall find that the observable described above is complementary to the decision observables $(P_\varphi, 0, \ldots)$ and $(P_\psi, 0, \ldots)$; we shall discuss this fact in more detail in §3.

2.5 Measurement Scales for Observables

In the definition of an observable we have not discussed the notion of a measurement scale. In practical applications measurement scales often (but not always) play an important role. They are always somewhat arbitrary, but they are also very practical. In the case of a well-developed theory, a well-chosen measurement scale can both be very practical and involve some of the physical structures of the theory; see, for example, the observables defined by the infinitesimal transformations of a group (see VII); here a particular scale is preferred. Rectangular coordinates represent particular selection of scales which are often preferred to arbitrary coordinates in three dimensions. Since we have not developed a detailed formulation of quantum mechanics until this chapter, it is not possible to develop specific measurement scales in this chapter. This fact must be emphasized, otherwise the reader may think that we have already formulated the selection of a measurement scale in mathematical terms. What do we mean by a measurement scale?

According to §2.1 and §2.2, to an observable $\Sigma \xrightarrow{F} L$ there exists a space $\mathscr{B}(\Sigma)$ together with the set $K(\Sigma)$ of σ-additive measures over Σ. The set Σ symbolizes the set of registrations $b \in \mathscr{R}(b_0)$. In general, by a measurement scale we mean a sequence of numbers such that the fact that the measurement values fall within an interval of the scale can be represented by one of the $b \subset b_0$. We shall now replace this intuitive idea by a mathematically correct definition.

D 2.5.1. An element $y \in \mathscr{B}'(\Sigma)$ is called a *measurement scale* for the complete Boolean ring Σ.

The expression "random variable" is frequently used instead of "measurement scale." We choose not to use the former expression because the term "random" only obscures the meaning, since it can be assigned with the meaning of pure chance. In physics the concept of pure chance, as a *fundamental concept*, is without meaning.

We wish to clarify the fact that D 2.5.1 corresponds to the intuitive meaning of a measurement scale described above. At first y is an affine functional on $K(\Sigma)$. In §2.1 we have already said that we may identify $\mathscr{B}'(\Sigma)$ with the set of measurable functions. But, as we mentioned in §2.1, we shall not take this mathematical route to obtain a representation of a Boolean ring.

The spectral family defined in D 2.1.4 makes the following definition plausible:

D 2.5.2. Let $\alpha_1 < \alpha_2$; we define $\sigma(y \leq \alpha_2) \dotplus \sigma(y \leq \alpha_1)$ to be the registration for which the scale value of y lies in the interval $(\alpha_1, \alpha_2]$.

This definition will be justified by the following facts (let $m \in K(\Sigma)$):

$$m(\sigma(y \leq \alpha_2) \dotplus \sigma(y \leq \alpha_1)) = 1 \Rightarrow \alpha_1 \leq (m, y) \leq \alpha_2,$$

$$m(\sigma^*(y \leq \alpha_2)) = 1 \Rightarrow (m, y) > \alpha_2, \qquad m(\sigma(y \leq \alpha_1)) = 1 \Rightarrow (m, y) \leq \alpha_1.$$

D 2.5.3. Let $\Sigma(y)$ denote the complete Boolean ring generated by $\{\sigma(y \le \alpha) \mid \alpha \in \mathbf{R}\}$. We shall call $\Sigma(y)$ the ring of registrations generated by the measurement scale y.

Therefore it "suffices" to measure the probabilities for all registrations $\sigma(y \le \alpha)$ in order to determine (in principle) the probabilities for all $\sigma \in \Sigma(y)$. Thus, for the purpose of measurement it is "economical" to introduce measurement scales. In the following we shall consider this topic in more detail.

The following concepts will be important in the discussion which follows:

D 2.5.4. The set

$$\mathrm{Sp}(y) = \{\alpha \mid \sigma(y \le \alpha + \varepsilon) \dotplus \sigma(y \le \alpha - \varepsilon) \ne 0 \text{ for all } \varepsilon > 0\}$$

is called the *spectrum* of the scale y.

The set

$$\mathrm{Sp}_d(y) = \{\alpha \mid \sigma(y \le \alpha) \dotplus \sigma(y \le \alpha -) \ne 0\}$$

is called the *discrete spectrum* of the scale y.

The set

$$\mathrm{Sp}_c(y) = \{\alpha \mid [\sigma(y \le \alpha + \varepsilon) \dotplus \sigma(y \le \alpha)] \vee [\sigma(y \le \alpha - \varepsilon) \dotplus \sigma(y \le \alpha -)] \ne 0 \text{ for all } \varepsilon > 0\}$$

is called the *continuous spectrum* of the scale y.

It is easy to see that $\mathrm{Sp}(y)$ is a closed subset of $\mathbf{R}$.

Since there always exists an interval about $\alpha \notin \mathrm{Sp}(y)$ for which $\sigma(y \le \alpha + \varepsilon) \dotplus \sigma(y \le \alpha - \varepsilon) = 0$, the registration of a scale value in $(\alpha - \varepsilon, \alpha + \varepsilon]$ is "impossible" (see D 2.5.2). Therefore we shall call $\mathrm{Sp}(y)$ the set of "possible measurement values" for the scale y.

Those readers who are already accustomed to the intricacies of quantum mechanics will at first be somewhat surprised to find that such an essential concept as the "spectrum of possible measurement values" apparently depends only on the arbitrary choice of a Boolean ring *and* a scale y. The spectrum of a (conventional) quantum mechanical observable (for example, the energy) exhibits the typical quantum mechanical structure—namely, the structure of discrete measurement values. The following remarks are in order:

(1) The Boolean ring Σ of an observable $\Sigma \xrightarrow{F} L$ is not at all "arbitrary." For example, according to Th. 1.4.8, for a decision observable F we may identify Σ with the set of decision effects $F\Sigma \subset G$.

(2) The choice of a scale y for a given Σ appears to be arbitrary. If, however, we seek to use a scale y for which $\Sigma(y)$ is "as large as possible," for example, $\Sigma(y) = \Sigma$, then the scale y exhibits the structure of Σ.

One purpose (but not the only one) for the introduction of a measurement scale is to manage only with the set $\{\sigma(y \leq \alpha)\}$ instead of the complete Boolean ring. That is, the introduction of a measurement scale corresponds to the "need" to "measure only as much as is necessary," as we have already described above.

(3) It is possible that the choice of a measurement scale for a decision observable does not correspond to the question of the introduction of the most "practical" information about a measurement apparatus. Instead, if the decision observable is defined as an infinitesimal transformation, then the physical meaning of the scale is obtained from the transformations. We may express this fact as follows: The Boolean algebra of the decision observable is generated by the spectral family $E(\lambda)$ of a one-parameter unitary group given by

$$U_\tau = \int e^{i\lambda\tau}\, dE(\lambda) = e^{iA\tau} \quad \text{where } A = \int \lambda\, dE(\lambda).$$

Thus the λ-scale is *determined* by U_τ. Observables which are defined in terms of infinitesimal transformations will be discussed in VII and VIII.

If $\Sigma \xrightarrow{F} L$ is an observable, and y is a measurement scale, then $\Sigma(y) \xrightarrow{F} L$ will be an observable for which $(\Sigma(y), F) \prec (\Sigma, F)$ holds (see D 2.4.1).

D 2.5.5. If $\Sigma \xrightarrow{F} L$ is an observable, and if y is a measurement scale, then we shall call $\Sigma(y) \xrightarrow{F} L$ the *partial observable* generated by the scale y.

We may expand the above concept to a finite collection of scales as follows: Let $\Sigma(y_1, y_2, \ldots, y_n)$ denote the Boolean ring generated by the $\sigma(y_i \leq \alpha)$, $i = 1, 2, \ldots, n$. Let $\Sigma(y_1, y_2, \ldots, y_n) \xrightarrow{F} L$ denote the partial observables generated by the scales $y_1, y_2, \ldots, y_n$.

In physics it is customary to select a set of "practical" scales in order to obtain $\Sigma(y_1, y_2, \ldots, y_n) = \Sigma$.

Here we shall not consider the case in which $\Sigma(y)$ is "smaller" than Σ. Such cases can be treated using the methods discussed in §2.1 and §2.2. Instead, we shall consider only the case in which $\Sigma(y_1, \ldots, y_n) = \Sigma$. This is sufficient (see Th. 2.5.6) because, for separable Boolean rings Σ there always exists a scale y for which $\Sigma(y) = \Sigma$, and in §2.2 we have seen that we may restrict ourselves to observables—that is, to separable Boolean rings.

Th. 2.5.1. *To each separable complete Boolean ring Σ there exists a totally ordered subset Δ such that the smallest complete Boolean ring containing Δ is equal to Σ.*

PROOF. Since Σ is separable, Σ contains a countably dense subset Γ. Using the elements γ_ν of Γ we may recursively define the following totally ordered subsets of Σ:

$$\Delta_1: 0 < \gamma_1;$$

$$\Delta_2: 0 < \gamma_1 \cdot \gamma_2 < \gamma_1 < \gamma_1 \dotplus (\gamma_2 \dotplus \gamma_1) \cdot \gamma_2;$$

and with

$$\Delta_n: 0 < \pi_1 < \pi_2 < \cdots < \pi_\nu < \pi_{\nu+1} < \cdots < \pi_m$$

we set

$$\begin{aligned}\Delta_{n+1}: 0 < \pi_1 \cdot \gamma_{n+1} &< \pi_1 < \pi_1 \dotplus (\pi_2 \dotplus \pi_1) \cdot \gamma_{n+1} < \pi_2 < \cdots \\ &< \pi_\nu < \pi_\nu \dotplus (\pi_{\nu+1} \dotplus \pi_\nu) \cdot \gamma_{n+1} < \pi_{\nu+1} < \cdots \\ &< \pi_m < \pi_m \dotplus (\gamma_{n+1} \dotplus \pi_m) \cdot \gamma_{n+1}.\end{aligned}$$

We therefore obtain

$$\begin{aligned}\gamma_{n+1} = {} & \pi_1 \cdot \gamma_{n+1} \dotplus \pi_1 \dotplus [\pi_1 \dotplus (\pi_2 \dotplus \pi_1) \cdot \gamma_{n+1}] \dotplus \pi_2 \\ & \dotplus [\pi_2 \dotplus (\pi_3 \dotplus \pi_2) \cdot \gamma_{n+1}] \dotplus \cdots \dotplus \pi_m \dotplus [\pi_m \dotplus (\gamma_{n+1} \dotplus \pi_m) \cdot \gamma_{n+1}].\end{aligned}$$

Thus, recursively, we find that the Boolean ring generated by the Δ_n contains all of the elements $\gamma_1, \gamma_2, \ldots, \gamma_n$. Clearly the Boolean ring generated by $\gamma_1, \gamma_2, \ldots, \gamma_n$ contains the set Δ_n.

Since $\Delta_{n+1} \supset \Delta_n$ and since each Δ_n contains a finite number of elements, the set $\Delta = \bigcup_n \Delta_n$ is countable. Thus we see that Δ is totally ordered, because any two elements of Δ, for sufficiently high n, are elements of a Δ_n. The Boolean ring Σ_Δ generated by Δ is therefore also countable. Since $\Gamma \subset \Sigma_\Delta$ and Γ is dense in Σ, Σ_Δ is dense in Σ, that is, the complete Boolean ring generated by Δ is equal to Σ.

Th. 2.5.2. *To each separable complete Boolean ring Σ there exists a totally ordered and closed subset $\bar{\Delta}$ such that the smallest complete Boolean ring containing $\bar{\Delta}$ is equal to Σ.*

PROOF. On the basis of Th. 2.5.1 it suffices to show that the closure $\bar{\Delta}$ of Δ is totally ordered.

Let σ_ν, σ'_μ be two sequences for which $\sigma_\nu, \sigma'_\mu \in \Delta$ and $\sigma_\nu \to \sigma, \sigma'_\mu \to \sigma'$. We must show that either $\sigma < \sigma'$ or $\sigma' < \sigma$. If there exists an N such that, for all $\nu, \mu > N$, the relation $\sigma_\nu < \sigma'_\mu$ is satisfied, then $\sigma < \sigma'$. If there exists an N such that, for all $\nu, \mu > N$ the relation $\sigma_\nu > \sigma'_\mu$ is satisfied, then $\sigma > \sigma'$. If no such N exists, then, for each N there are two pairs $\sigma_{\nu_1}, \sigma'_{\mu_1}$ and $\sigma_{\nu_2}, \sigma'_{\mu_2}$ for which $\sigma_{\nu_1} < \sigma'_{\mu_1}$ and $\sigma_{\nu_2} > \sigma'_{\mu_2}$. Thus we obtain four subsequences $\sigma_{\nu_1}, \sigma_{\nu_2}, \sigma'_{\mu_1}, \sigma'_{\mu_2}$. From $\sigma_{\nu_n} < \sigma'_{\mu_n}$ it follows that, in the limit $\sigma < \sigma'$. Similarly, from $\sigma_{\nu_2} > \sigma'_{\mu_2}$ it follows that, in the limit, $\sigma > \sigma'$. Therefore we obtain $\sigma = \sigma'$.

Th. 2.5.3. *Let Δ be a totally ordered subset of Σ for which $\varepsilon \in \Delta$. Let Σ_Δ be the smallest Boolean subring of Σ containing Δ. Then Σ_Δ is the set of all elements of the form $\sigma = \sum_{\nu=1}^{n} \dotplus\, \sigma_\nu$, where $\sigma_\nu \in \Delta$; the representation of σ as a sum of elements of Δ is unique.*

PROOF. Since the product of finitely many elements of Δ is again an element of Δ (namely, the smallest of these sets), every $\sigma \in \Sigma_\Delta$ is of the form $\sigma = \sum_{\nu=1}^{n} \dot{+}\, \sigma_\nu$, where $\sigma_\nu \in \Delta$.

We will now show that this representation is unique. Let

$$\sum_{\nu=1}^{n} \dot{+}\, \sigma_\nu = \sum_{\mu=1}^{m} \dot{+}\, \sigma'_\mu.$$

Then we obtain

$$\sum_{\nu=1}^{n} \dot{+}\, \sigma_\nu \dot{+} \sum_{\mu=1}^{m} \dot{+}\, \sigma'_\mu = 0.$$

In the last sum let us suppose that the quantities are ordered (Δ is totally ordered). Then we obtain a sum of the form $\sum_{\rho=1}^{n+m} \dot{+}\, \sigma''_\rho = 0$ with $\sigma''_{\rho+1} < \sigma''_\rho$. We may write this sum as follows

$$(\sigma''_1 \dot{+} \sigma''_2) \dot{+} (\sigma''_3 \dot{+} \sigma''_4) \dot{+} \cdots = 0 \tag{2.5.1}$$

For the $\eta_n = \sigma''_{2n-1} \dot{+} \sigma''_{2n}$ $(n = 1, 2, \ldots)$ we obtain $\eta_n \cdot \eta_m = 0$ for $n \neq m$. Thus (2.5.1) takes on the following form

$$\sum_n \dot{+}\, \eta_n = 0. \tag{2.5.2}$$

If we multiply (2.5.2) by η_m, then it follows that $\eta_m = 0$, that is, $\sigma''_{2n-1} \dot{+} \sigma''_{2n} = 0$ for all n and therefore

$$\sigma''_{2n-1} = \sigma''_{2n} \quad \text{for all } n. \tag{2.5.3}$$

Since all σ_ν are different, and the σ'_μ are different, equation (2.5.3) can hold only if σ_ν and σ'_μ are pairwise identical.

Th. 2.5.4. *Let Δ be a totally ordered and closed subset of a complete Boolean ring Σ. Then Δ is compact, and will be mapped by an effective σ-additive measure m homomorphically to a closed subset ω of* $[0, 1]$.

PROOF. Let $\sigma_1, \sigma_2 \in \Delta$, $\sigma_1 \neq \sigma_2$. Then either $\sigma_1 > \sigma_2$ or $\sigma_1 < \sigma_2$; thus it follows that $m(\sigma_1 \dot{+} \sigma_2) = |m(\sigma_1) - m(\sigma_2)|$ and, since m is effective, $m(\sigma_1 \dot{+} \sigma_2) \neq 0$. Since m is σ-additive and effective, then, according to Th. 2.1.11 the uniform structure U_g on Σ (and therefore also on Δ) is defined by the metric $d(\sigma_1, \sigma_2) = m(\sigma_1 \dot{+} \sigma_2)$. Since $m(\sigma_1 \dot{+} \sigma_2) = |m(\sigma_1) - m(\sigma_2)|$, the uniform structure U_g on Δ is equal to the initial uniform structure for the map $m\colon \Delta \to [0, 1]$. Since the image ω of Δ is precompact (totally bounded) (AII, §2), Δ is precompact (totally bounded). However, since Δ is U_g-complete (because it is closed in Σ), Δ is compact, and therefore the image ω of Δ is a compact subset of $[0, 1]$. Thus $\sigma \to m(\sigma)$ is a homomorphic map of Δ onto ω.

It is always possible to include the null and unit elements in Δ. In the following we shall assume that they have been included.

We note that Th. 2.5.3 holds, and that ω contains 0 and 1. It is not difficult to construct a spectral family from a totally ordered closed set Δ in Σ. To this purpose, we begin by introducing the map $\Delta \to \omega \subset [0, 1]$ which we obtained from Th. 2.5.4.

Th. 2.5.5. *Let $\tau(\lambda) = \sup\{\alpha \mid 0 \leq \alpha \leq \lambda$ and $\alpha \in \omega\}$; $\tau(\lambda)$ is an upper continuous map defined on the interval $[0, 1]$ which increases from 0 to 1. The map $\sigma(\lambda)$, defined by $\sigma \in \Delta$ for which $m(\sigma) = \tau(\lambda)$ is a spectral family for which $\sigma(0) = 0$ and $\sigma(1) = \varepsilon$. The set $\{\sigma(\lambda)\}$ is equal to Δ.*

PROOF. From the definition it immediately follows that $\tau(\lambda)$ is increasing and that $\tau(0) = 0$, $\tau(1) = 1$ since $0 \in \omega$ and $1 \in \omega$. From $\tau(\lambda) \leq \lambda$ it follows that $\tau(\lambda + \varepsilon) \leq \lambda + \varepsilon$. Therefore, if $\lambda \in \omega$ it follows that $\tau(\lambda+) = \lambda = \tau(\lambda)$. If $\lambda \notin \omega$, then the closed subset $\{\alpha \mid \alpha \geq \lambda$ and $\alpha \in \omega\}$ has an infimum β such that $\beta \in \omega$ since ω is closed. We obtain $\beta > \lambda$ since $\lambda \notin \omega$. For an ε satisfying $0 < \varepsilon < \beta - \lambda$ it follows that $\tau(\lambda + \varepsilon) = \tau(\lambda)$ and therefore $\tau(\lambda+) = \tau(\lambda)$. Therefore $\tau(\lambda)$ is continuous from above.

Since the values of the function $\tau(\lambda)$ lie in ω, and Δ is homeomorphically mapped onto ω by m, $m^{-1}\tau$ is a map of $[0, 1]$ in Δ which we shall denote by $\sigma(\lambda)$. Therefore $\sigma(\lambda)$ is uniquely defined by $m(\sigma(\lambda)) = \tau(\lambda)$. Thus we obtain $\sigma(0) = 0$ and $\sigma(1) = \varepsilon$. Since m^{-1} is increasing and upper continuous, $\sigma(\lambda) = m^{-1}\tau$ is therefore a spectral family.

Since, for $\lambda \in \omega$, the relation $\tau(\lambda) = \lambda$ holds, we obtain $\{\sigma(\lambda)\} = \Delta$.

Th. 2.5.6. *There exists a measurement scale y such that $\Sigma(y) = \Sigma$.*

PROOF. Choose the totally ordered and closed set $\Delta = \bar{\Delta}$ according to Th. 2.5.9. Then, according to Th. 2.5.5 we obtain a spectral family $\{\sigma(\lambda)\}$ where Σ is the smallest possible Boolean ring containing $\{\sigma(\lambda)\}$. The linear functional y obtained from the spectral family by (2.1.9) therefore satisfies the relation $\Sigma(y) = \Sigma$ (on the basis of Th. 2.1.15).

Th. 2.5.1–2.5.6 show that it is sufficient to consider the spectral family $\sigma(\lambda)$ and the corresponding functionals $y \in \mathscr{B}'(\Sigma)$.

A spectral family $\sigma(\lambda)$ is a totally ordered subset of Σ but is not necessarily closed. We obtain the closure $\overline{\{\sigma(\lambda)\}}$ of $\{\sigma(\lambda)\}$ by adding all of the limit points $\sigma(\lambda-)$ to $\{\sigma(\lambda)\}$. Theorem 2.5.3 is also applicable to nonclosed totally ordered subsets of Σ. Let $\Delta = \{\sigma(\lambda)\}$, that is, Δ is a spectral family, and y is its corresponding unique measurement scale. It is easy to obtain a representation of Σ_Δ in terms of subsets of the real axis.

Th. 2.5.7. *Let Δ be the spectral family of y and let $\mathrm{Sp}(y)$ be the spectrum of y. Then the bijection $\sigma(\lambda) \leftrightarrow (-\infty, \lambda] \cap \mathrm{Sp}(y)$ of Δ into subsets of the real axis can be uniquely extended to an isomorphism of Σ_Δ to the Boolean ring of sets P_I generated by the set*

$$\{(-\infty, \lambda] \cap \mathrm{Sp}(y) \mid \lambda \in \mathbf{R}\}.$$

PROOF. The proof is obtained directly from Th. 2.5.3. It is only necessary to order the elements in the sum $\sum_{\nu=1}^{n} \dotplus \sigma(\lambda_\nu)$ such that $\lambda_{\nu+1} < \lambda_\nu$. Then we consider

$$\sum_{\nu=1}^{n} \dotplus \sigma(\lambda_\nu) = [\sigma(\lambda_1) \dotplus \sigma(\lambda_2)] \vee [\sigma(\lambda_3) \dotplus \sigma(\lambda_4)] \cdots.$$

Each bracket, for example, $[\sigma(\lambda_3) \dotplus \sigma(\lambda_4)]$ corresponds to an interval $(\lambda_4, \lambda_3] \cap \mathrm{Sp}(y)$. The elements of P_I are unions of finitely many such intervals $(\alpha, \beta] \cap \mathrm{Sp}(y)$.

In practice, a representation of $\Sigma(y)$ in terms of subsets of $\mathbf{R}$ which is not one-to-one is more commonly used. Such a representation is obtained as follows:

For $\lambda_2 < \lambda_1$ let the interval $I = (\lambda_2, \lambda_1]$ correspond to the element $\sigma(I) = \sigma(\lambda_1) \dotplus \sigma(\lambda_2)$ of Σ. Let k be an arbitrary subset of $\mathbf{R}$. We define a covering u of k by a denumerable set of such intervals I_ν for which $k \subset \bigcup_\nu I_\nu$. To each subset of $k \subset \mathbf{R}$ there is a corresponding element $\bar{\sigma}(k)$ of Σ defined by $\bar{\sigma}(k) = \bigwedge_u \bigvee_\nu \sigma(I_\nu)$. If u is a covering of k and v is a covering of $\mathbf{R} - k$ then $u \cup v$ is a covering of all of $\mathbf{R}$. It is easy to show that, for every covering of $\mathbf{R}$ the relation $\bigvee_\nu \sigma(I_\nu) = \varepsilon$ holds. Thus it follows that $\bar{\sigma}(k) \vee \bar{\sigma}(\mathbf{R} - k) = \varepsilon$. We shall call a set "measurable" if $\bar{\sigma}(k) \wedge \bar{\sigma}(\mathbf{R} - k) = 0$. Let P denote the set of all "measurable" subsets of $\mathbf{R}$. We shall show that P is a σ-complete Boolean ring of sets. In general P need not be a complete Boolean ring! A Boolean ring P of sets is said to be σ-complete if the union and intersection of countably many sets in P is an element of P.

Th. 2.5.8. *The set P described above is a σ-complete Boolean ring of sets for which $\mathbf{R} \in P$. The map $\sigma(k) = \bar{\sigma}(k)$: $P \longrightarrow \Sigma(y)$ is a surjective σ-homomorphism of P onto $\Sigma(y)$, that is, a homomorphism in which P and $\Sigma(y)$ are σ-complete Boolean rings. Let J be the kernel of this map, that is, $J = \{k \mid k \in P, \sigma(k) = 0\}$. Then the mapping $P \to \Sigma(y)$ may be expressed in canonical form as follows:*

$$P \longrightarrow P/J \longrightarrow \Sigma(y),$$

where $P/J \longrightarrow \Sigma(y)$ is an isomorphism. In particular, P/J is also a complete Boolean ring. Every interval $I = (\lambda_2, \lambda_1]$ is an element of P; in particular, we obtain $\sigma(I) = \sigma(\lambda_1) \dotplus \sigma(\lambda_2)$.

PROOF. In order to prove that P is a σ-complete Boolean ring of sets it suffices to show that if $k \in P$ then $\mathbf{R} - k \in P$; in addition, if $\{k_\nu\}$, $k_\nu \in P$ is a countable set, $\bigcup_\nu k_\nu \in P$ because $\bigcap_\nu k_\nu = \mathbf{R} - \bigcup_\nu (\mathbf{R} - k_\nu)$.

If $k \in P$, then, from the definition it follows that $\bar{\sigma}(k) \wedge \bar{\sigma}(\mathbf{R} - k) = 0$ and therefore $\mathbf{R} - k \in P$ and $\sigma(\mathbf{R} - k) = \sigma^*(k)$.

We will now show that, if $k_\nu \in P$ then

$$\bigcup_\nu k_\nu \in P \quad \text{and} \quad \sigma\left(\bigcup_\nu k_\nu\right) = \bigvee_\nu \sigma(k_\nu).$$

Since $\mathbf{R} - \bigcup_\nu k_\nu = \bigcap_\nu (\mathbf{R} - k_\nu)$, from the definition of $\bar{\sigma}$ it easily follows that $\bar{\sigma}(\mathbf{R} - \bigcup_\nu k_\nu) \le \bar{\sigma}(\mathbf{R} - k_\nu)$ for all ν, that is, $\bar{\sigma}(\mathbf{R} - \bigcup_\nu k_\nu) \le \bigwedge_\nu \bar{\sigma}(\mathbf{R} - k_\nu) = \bigwedge_\nu \sigma(\mathbf{R} - k_\nu) = \bigwedge_\nu \sigma^*(k_\nu)$. Since $\bar{\sigma}(\bigcup_\nu k_\nu) \cup \bar{\sigma}(\mathbf{R} - \bigcup_\nu k_\nu) = \varepsilon$ we obtain $\bar{\sigma}^*(\mathbf{R} - \bigcup_\nu k_\nu) \le \bar{\sigma}(\bigcup_\nu k_\nu)$, that is, $\bar{\sigma}(\bigcup_\nu k_\nu) \ge [\bigwedge_\nu \sigma^*(k_\nu)]^* = \bigvee_\nu \sigma(k_\nu)$. If we show that $\bar{\sigma}(\bigcup_\nu k_\nu) \le \bigvee_\nu \sigma(k_\nu)$ then we obtain $\bar{\sigma}(\bigcup_\nu k_\nu) \wedge \bar{\sigma}(\mathbf{R} - \bigcup_\nu k_\nu) = 0$, that is, $\bigcup_\nu k_\nu \in P$ and $\sigma(\bigcup_\nu k_\nu) = \bigvee_\nu \sigma(k_\nu)$.

Let U_ν be a covering of k_ν. Then $V = \bigcup_\nu U_\nu$ is a covering of $\bigcup_\nu k_\nu$ because V is, like the U_ν, a countable set of intervals. Therefore we obtain

$$\bar{\sigma}\left(\bigcup_\nu k_\nu\right) \le \bigwedge_{V = \bigcup_\nu U_\nu} \bigvee_{\nu, \mu} \sigma(I_{\mu\nu}),$$

where $I_{\mu\nu}$ are, for fixed ν, the intervals for U_ν and $\bigwedge$ is taken over all V of the form $V = \bigcup_\nu U_\nu$. We will now show that the U_ν may be chosen such that $\bigvee_\nu \bigvee_\mu \sigma(I_{\mu\nu})$ will be "arbitrarily close" to $\bigvee_\nu \sigma(k_\nu)$, that is, with the metric d from Σ, for every $\varepsilon > 0$ we may make

$$d\left(\bigvee_\nu \sigma(k_\nu), \bigvee_{\nu,\mu} \sigma(I_{\mu\nu})\right) < \varepsilon$$

for a suitable choice of U_ν.

For this purpose we shall show some relations for a subset k: If $U_1 = \{I_\nu^{(1)}\}$ and $U_2 = \{I_\mu^{(2)}\}$ are two coverings of k, it then follows immediately that $V = \{I_\nu^{(1)} \cap I_\mu^{(2)}\}$ is a covering of k. From this result we obtain

$$\left[\bigvee_\nu \sigma(I_\nu^{(1)})\right] \wedge \left[\bigvee_\mu \sigma(I_\mu^{(2)})\right] = \bigvee_{\nu,\mu} (\sigma(I_\nu^{(1)})) \wedge \sigma(I_\mu^{(2)})$$
$$= \bigvee_{\nu\mu} \sigma(I_\nu^{(1)} \cap I_\mu^{(2)}).$$

For a covering U we shall use the following abbreviation:

$$\sigma_U = \bigvee_{I_\nu \in U} \sigma(I_\nu).$$

From which we obtain $\sigma_V = \sigma_{U_1} \wedge \sigma_{U_2}$.

Similarly, for a finite number of coverings $U_1, U_2, \ldots, U_n$ there exists a covering V such that $\sigma_V = \bigwedge_{\nu=1}^n \sigma_{U_\nu}$.

According to Th. 2.1.1, from $\bar{\sigma}(k) = \bigwedge_U \sigma_U$ it follows that there is a sequence U_ν of coverings such that for $\tilde{\sigma}_n = \bigwedge_{\nu=1}^n \sigma_{U_\nu}$ the relationship $\bar{\sigma}(k) = \bigwedge_n \tilde{\sigma}_n$ and $\tilde{\sigma}_n \to \bar{\sigma}(k)$ hold. Since, to U_ν, $\nu = 1, 2, \ldots, n$ there exists a covering V_n such that $\sigma_{V_n} = \bigwedge_{\nu=1}^n \sigma_{U_\nu}$, there exists a sequence of coverings V_n for which σ_{V_n} is decreasing and converges towards $\bar{\sigma}(k)$, and satisfies $\bar{\sigma}(k) = \bigwedge_n \sigma_{V_n}$.

For a given $\varepsilon > 0$ we may choose such a covering U_ν for each k_ν such that

$$d(\sigma_{U_\nu}, \sigma(k_\nu)) = m_0(\sigma_{U_\nu} \dotplus \sigma(k_\nu)) < \frac{\varepsilon}{2^\nu}.$$

Since

$$d\left(\bigvee_{\nu,\mu} \sigma(I_{\mu\nu}), \bigvee_\nu \sigma(k_\nu)\right) = m_0\left(\bigvee_\nu \sigma_{U_\nu} \dotplus \bigvee_\nu \sigma(k_\nu)\right)$$

and

$$\bigvee_\nu \sigma_{U_\nu} \dotplus \bigvee_\nu \sigma(k_\nu) = \left[\bigvee_\nu \sigma_{U_\nu}\right] \wedge \left[\bigwedge_\mu \sigma^*(k_\mu)\right]$$
$$= \bigvee_\nu \left[\sigma_{U_\nu} \wedge \bigwedge_\mu \sigma^*(k_\mu)\right] < \bigvee_\nu [\sigma_{U_\nu} \wedge \sigma^*(k_\nu)]$$
$$= \bigvee_\nu [\sigma_{U_\nu} \dotplus \sigma(k_\nu)]$$

we obtain

$$d\left(\bigvee_{\nu,\mu} \sigma(I_{\mu\nu}), \bigvee_\nu \sigma(k_\nu)\right) \leq \sum_\nu m_0(\sigma_{U_\nu} \dotplus \sigma(k_\nu))$$
$$< \varepsilon \sum_{\nu=1}^{\infty} \frac{1}{2^\nu} = \varepsilon.$$

Thus it follows that

$$\bigwedge_{V=\bigcup_\nu U_\nu} \bigvee_{\nu,\mu} \sigma(I_{\mu\nu}) = \bigvee_\nu \sigma(k_\nu)$$

from which we have proven that

$$\bar{\sigma}\left(\bigcup_\nu k_\nu\right) < \bigvee_\nu \sigma(k_\nu).$$

With this result we have also proven that the map $P \to \Sigma(y)$ is a σ-homomorphism. The fact that the map $P \to \Sigma(y)$ is surjective follows from Th. 2.1.1 and Th. 2.1.2 because the elements of $\Sigma(y)$ may be obtained from Σ_Δ by the joint $\vee$ and meet $\wedge$ of denumerably many elements. $P/J \to \Sigma(y)$ is therefore an order isomorphism and therefore P/J is isomorphic to $\Sigma(y)$, that is, P/J is a complete Boolean ring.

It is easy to see that $I = (\lambda_1, \lambda_2]$ is an element of P and that $\sigma(I) = \sigma(\lambda_1) \dotplus \sigma(\lambda_2)$.

If $\Sigma(y)$ is equal to Σ then we have obtained a representation of Σ in terms of subsets of scale values of the measurement scale y.

For $k \in P$ $\sigma(k)$ is the "registration" for which the scale value of y lies in the set k. It directly follows that $\mathbf{R}\backslash \mathrm{Sp}(y) \in J$, that is, only scale values of the spectrum of y will be registered.

It is easy to verify that the discrete spectrum $\mathrm{Sp_d}$ is the set of those values λ for which $\sigma(k) \neq 0$ with k as the set consisting only of a single point λ. The set of these $\sigma(k)$ is equal to the set of atoms in $\Sigma(y)$.

It is not difficult to extend Th. 2.5.8 to the case of finitely many scales $y_1, y_2, \ldots, y_n$ and $\Sigma(y_1, y_2, \ldots, y_n)$. P is then a σ-complete Boolean ring of sets in the n-dimensional space $\mathbf{R}^n$. In physics we frequently find that we choose n scales (instead of one) for which $\Sigma(y_1, y_2, \ldots, y_n) = \Sigma$ because n scales often prove to be both practical and theoretically useful (in specific situations which we will discuss later, for example, in VII and VIII).

Here we have treated the problem of measurement scales in terms of a structure which may be "added" to the structure of a complete Boolean ring Σ. Thus we see that the measurement scale is not necessarily involved in the concept of an observable. In this way it is clear that the measurement scale is nothing other than a preferred method which permits an overview of Σ (or, in particular, of $\mathscr{R}(b_0)$). $\mathscr{R}(b_0)$ is of primary interest for experiment; the measurement scale for the apparatus b_0 is only a very practical tool, or may have an additional physical meaning (see the discussion of (3) above) concerning transformations.

In D 2.5.4 we have defined different components of the spectrum $\mathrm{Sp}(y)$ for the measurement scale y. If m is a σ-additive effective measure on Σ then for $k \in P$ $\mu(k) = m(\sigma(k))$ defines a σ-additive measure μ on P for which the sets of μ-measure 0 coincide with the elements of J. $(\mathbf{R}, P, \mu)$ is then a so-called measure space. We may separate $\mathrm{Sp}(y)$ into three disjoint components: $k_d = \mathrm{Sp_d}(y)$, k_{cc} and k_{sc} where $\mu_d(k) = \mu(k \cap k_d)$ is a discrete measure, $\mu_{cc}(k) = \mu(k \cap k_{cc})$ is absolutely continuous with respect to Lebesque measure, and $\mu_{sc}(k) = \mu(k \cap k_{sc})$ is singular with respect to Lebesque measure, that is, the Lebesque measure of k_{sc} is 0. This decomposition of $\mathrm{Sp}_d(y)$ does not depend on the σ-additive effective measure m on Σ.

Let $\{\sigma(\lambda)\}$ be a spectral family. Then, by the map $F: \Sigma \to L$ which corresponds to an observable, there exists a totally ordered subset $\{F(\lambda)\}$ of L defined by $F(\lambda) = F(\sigma(\lambda))$. Therefore $F(\lambda)$ is an increasing function $\mathbf{R} \to L$ which is continuous from above $F(\lambda+) = F(\lambda)$. We shall then call $F(\lambda)$ a spectral family of effects. Earlier we have constructed a σ-complete Boolean ring of sets P using the spectral family $\{\sigma(\lambda)\}$, and we have seen that P/J is isomorphic to Σ. The construction of P is, however, possible without knowledge of $\sigma(\lambda)$, if we make use of the spectral family of effects $\{F(\lambda)\}$.

We may proceed by using any totally ordered subset $l \subset L$. The closure $\bar{l}$ of L in the $\sigma(\mathscr{B}', \mathscr{B})$-topology is totally ordered. This fact may be proven analogously to the proof of Th. 2.5.2. If W is an effective ensemble from $K(\mathscr{H}_1, \ldots)$ then $\bar{l}$ will be mapped injectively into $[0, 1]$ by the map $F \to \mu(W, F)$ in complete analogy to Th. 2.5.4. By introducing the parameter τ from Th. 2.5.5 we may obtain a spectral family of effects from l; thus, it suffices to start with such a spectral family of effects.

If we are given a spectral family of effects $\{F(\lambda)\}$ then we may use the methods leading to Th. 2.5.8 to obtain an analogous result.

For the interval $I = (\lambda_2, \lambda_1]$ we define $F(I) = F(\lambda_1) - F(\lambda_2)$. Therefore we obtain $F(I) \in L$. Furthermore, let $\bar{F}(k) = \inf_U (\Sigma_\nu F(I_\nu))$. This infimum exists because the set of $F_U = \sum_\nu F(I_\nu)$ is a lower-directed set (here we have used the following result from Th. 2.5.8: if $U_1 = \{I_\nu^{(1)}\}$ and $U_2 = \{I_\mu^{(2)}\}$ are coverings of k, then $V = \{I_\nu^{(1)} \cap I_\mu^{(2)}\}$ is a covering of k). We obtain $\bar{F}(k) + \bar{F}(\mathbf{R}\backslash k) > \mathbf{1}$. A set k is said to be measurable if $\bar{F}(\mathbf{R}\backslash k) = \mathbf{1} - \bar{F}(k)$. Let P denote the set of measurable k. Clearly P is a σ-complete Boolean ring and $k \to \bar{F}(k)$ is a σ-additive measure. Let J be the set of all k having $\bar{F}$-measure 0. Then $\Sigma_0 = P/J$ is a complete Boolean ring and Σ_0 together with the map $F: \Sigma_0 \to L$ defined by $\bar{F}: P \to P/J \xrightarrow{F} L$ is an observable.

For the special case in which $F(\lambda)$ is determined by $F(\sigma(\lambda))$ then P and J are identical to the sets P, J of Th. 2.5.8. That is, by using Σ_0 we may recover the Boolean ring Σ and the mapping $F: \Sigma \to L$ from which we obtain the spectral families $\{\sigma(\lambda)\}$ and $\{F(\lambda) = F(\sigma(\lambda))\}$. In this sense each totally ordered subset of L uniquely defines an observable, and each observable may be obtained in this way.

Therefore it is not surprising that we often use a spectral family of effects instead of the total observable. The spectral family of effects not only determines the observable but also determines the measurement scale y of the observable by means of (2.1.9b) where $\Sigma_0 = P/J$ replaces Σ and $\rho(\alpha) \in \Sigma_0$ is the class of elements of P which belong to the interval $(-\infty, \alpha]$. For $k \in P$ we find that $\mu(W, F(k))$ is equal to the probability of obtaining a scale value of y from the set k in the case in which the observable $\Sigma_0 \xrightarrow{F} L$ was measured and the ensemble W was prepared. Thus it is not yet clear how we may obtain apparatuses which will measure the desired observable $\Sigma \to L$ and to prepare the desired ensemble W. These problems will be discussed first in §4 and §7 and then later in XVII and XVIII.

For decision observables the above relationships between Σ, the spectral family $\{\sigma(\lambda)\}$ and the measures $E: \Sigma \to G$ is somewhat simplified. According to Th. 1.4.8 Σ may be directly identified with a subset of G such that all results from Th. 2.5.1–Th. 2.5.8 are applicable to decision observables providing that we consider Σ as a subset of G. In particular, the map $P \to \Sigma(y)$ is also a map of P into G. Therefore spectral families of decision effects uniquely determine a decision observable with scale, and Σ may be chosen as the complete orthocomplemented sublattice of G generated by the spectral family $\{E(\lambda)\}$ where the identity map of Σ onto itself is the measure $\Sigma \to G$ of the observable. The scale y corresponding to the spectral set is determined by

$$y = \int \lambda \, dl_{E(\lambda)}.$$

For decision observables the maps S and S' have the special properties described in §2.2.

Th. 2.5.9. *For a decision observable S' is injective and $SK(\mathscr{H}_1, \ldots) = K(\Sigma)$. If we identify Σ with a subset of G, then we obtain S'^{-1} from the spectral representation of the operator $A = Sy \in \mathscr{B}'(\mathscr{H}_1, \ldots)$:*

$$A = \int \lambda \, dE(\lambda), \tag{2.5.4}$$

where $(S'^{-1})A = y$ has the form

$$y = \int \lambda \, dl_{E(\lambda)}. \tag{2.5.5}$$

Proof. We begin by noting that equations (2.5.4) and (2.5.5) are valid as a limit (in norm) in $\mathscr{B}'(\mathscr{H}_1, \ldots)$ and $\mathscr{B}'(\Sigma)$, respectively. According to Th. 2.1.15 each $y \in \mathscr{B}'(\Sigma)$ may be written in the form (2.5.5). Since $S'E = E$ for $E \in \Sigma \subset G$ and since S' is continuous in norm, it follows that $S'y$ is equal to A. Since both spectral representations (2.5.4) and (2.5.5) are unique (for the operator A, see AIV, §8 and §15 and for y, Th. 2.1.15) we therefore find that S' is injective and $S'^{-1}(A)$ (where A is defined for $E(\lambda) \in \Sigma \subset G$ by (2.5.4)) is equal to y by (2.5.5).

$SK(\mathscr{H}_1, \ldots) = K(\Sigma)$ follows from Th. 2.1.11 as follows: each $\tilde{m} \in K(\Sigma)$ may according to Th. 2.1.11, be approximated (in norm) arbitrarily well by $x_{\delta, N}$. For an arbitrary effective $w \in K(\mathscr{H}_1, \ldots)$ $m = Sw$ is an effective $m \in K(\Sigma)$. For this m $x_{\delta, N}$ is of the form

$$x_{\delta, N}(\sigma) = \sum_{n=1}^{N} (n - 1)\delta m(\sigma \wedge \sigma_n),$$

where we now consider σ, σ_n to be elements of $\Sigma \subset G$. Thus $m(\sigma \wedge \sigma_n) = m(\sigma\sigma_n)$, where $\sigma\sigma_n$ are products of the "operators" σ, σ_n of G. From $m(\sigma) = \mu(w, \sigma) = \operatorname{tr}(w\sigma)$ it follows that (since σ commutes with σ_n where $\sigma, \sigma_n \in \Sigma$!):

$$m(\sigma \wedge \sigma_n) = \operatorname{tr}(w\sigma\sigma_n) = \operatorname{tr}(w\sigma\sigma_n^2) = \operatorname{tr}(w\sigma_n\sigma\sigma_w)$$
$$= \operatorname{tr}(\sigma_n w \sigma_n \sigma).$$

Thus we obtain

$$x_{\delta,N} = \sum_{n=1}^{N} (n-1)\delta S(\sigma_n w \sigma_n)$$
$$= S\left(\sum_{n=1}^{N} (n-1)\delta\sigma_n w \sigma_n\right).$$

According to the proof of Th. 2.1.11 $\|x_{\delta,N}\|^{-1}x_{\delta,N} \in K(\Sigma)$ converges in norm to $\tilde{m}$. Thus $\|z_{\delta,N}\|^{-1}z_{\delta,N} \in K(\mathscr{H}_1, \ldots)$ with $z_{\delta,N} = \sum_{n=1}^{N} (n-1)\delta\sigma_n w \sigma_n$ converges in norm towards a $\tilde{w} \in K(\mathscr{H}_1, \ldots)$. Thus we obtain $\tilde{m} = S\tilde{w}$.

According to Th. 2.5.9 we may therefore identify the space $\mathscr{B}'(\Sigma)$ with the subspace of $\mathscr{B}'(\mathscr{H}_1, \ldots)$ of all operators of the form (2.5.4) where $E(\lambda)$ is a spectral family of the complete Boolean sublattice Σ of G. $\mathscr{B}'(\Sigma)$ is an abelian algebra. $K(\Sigma)$ arises from the partition of $K(\mathscr{H}_1, \ldots)$ into equivalence classes where all the $w \in K(\mathscr{H}_1, \ldots)$ belongs to the same class if the w cannot be distinguished by means of $\mathrm{tr}(we)$ where $e \in \Sigma$, that is, the linear forms $\mathrm{tr}(wg)$ agree on the subspace $B'(\Sigma)$ of $B'(\mathscr{H}_1, \ldots)$.

If, for one or more scales $y_1, \ldots, y_n$ we have $\Sigma(y_1, \ldots, y_n) = \Sigma$ (according to Th. 2.5.6 we may always choose a scale y such that $\Sigma(y) = \Sigma$) then the corresponding operators $A_1, \ldots, A_n$ and the decision observables and the scales $y_1, \ldots, y_n$ are uniquely determined. Therefore we may uniquely characterize a decision observable with measurement scales by a finite number of commuting operators $A_1, \ldots, A_n \in B'(\mathscr{H}_1, \ldots)$. The spectral families of $A_1, \ldots, A_n$ generate the corresponding complete Boolean ring $\Sigma \subset G$.

We therefore define

D 2.5.6. Let $A \in B'(\mathscr{H}_1, \ldots)$; the decision observable and its corresponding scale which is uniquely determined by A is called a *scale observable*.

We shall often use the expression "A is a scale observable" or, more briefly, "A is an observable." Thus we have explained the connection between the usual language of quantum mechanics in which the self-adjoint operators are called observables. In this explanation we expect that, at least, "in principle" misunderstandings are impossible.

The brief characterization of a decision observable having a scale by a single operator $A \in B'(\mathscr{H}_1, \ldots)$ is not applicable to more *general* observables. In such cases, to each scale $y \in \mathscr{B}'(\Sigma)$ there corresponds an operator $A = S'y \in \mathscr{B}'(\mathscr{H}_1, \ldots)$ but the operator A does not permit us to reconstruct the Boolean ring and the scale y!

Since, for fixed Σ, we may introduce different measurement scales $y \in \mathscr{B}'(\Sigma)$, the question arises as to what scale we obtain when we replace λ by a new scale value $f(\lambda)$ where $f(\lambda)$ is a real function. Let us consider the set

$$k(\alpha) = \{a \mid f(\lambda) \leq \alpha\}. \tag{2.5.6}$$

For a registration the question: is $f(\lambda) \leq \alpha$? is meaningful only if the set $k(\alpha) \in P$, that is, $k(\alpha)$ is measurable. Then the registration $f(\lambda) \leq \alpha$ can be associated with the element $\sigma(\dot{k}(\alpha)) \in \Sigma$. In this way we may obtain the precise meaning of the "renaming" of the scale λ by $f(\lambda)$, and we find that it is meaningful to consider only these $f(\lambda)$ for which the sets $k(\alpha)$ of (2.5.6) are measurable, that is, are elements of P. Such functions are usually called measurable functions.

The element $\hat{y}$ corresponding to the renaming of the scale defined by $f(\lambda)$ is given by

$$\hat{y} = \int \alpha \, d\sigma(k(\alpha)). \tag{2.5.7}$$

Since $\sigma(k(\alpha))$ may be obtained from the $\sigma(y \leq \beta)$ we may therefore write (2.5.7) as follows:

$$\hat{y} = \int f(\lambda) \, d\sigma(y \leq \lambda), \tag{2.5.8}$$

where (2.5.8) is defined in terms of (2.5.7). Instead of (2.5.8) we sometimes use the abbreviation $\hat{y} = f(y)$. Since the integrals (2.5.7) and (2.5.8) exist as limits in the norm, $f(\lambda)$ must be essentially bounded, that is, a c can be found such that the sets $\mathbf{R}$ and

$$\{\lambda \,|\, -c \leq f(\lambda) \leq c\}$$

differ only by a set of measure zero.

Th. 2.5.10. *Let y be a scale for which $\Sigma(y) = \Sigma$. Then, to each $\hat{y} = \mathscr{B}'(\Sigma)$ there exists a measurable, essentially bounded $f(\lambda)$ for which $\hat{y} = f(y)$.*

Proof. We seek a $f(\lambda)$ for which $\sigma(\hat{y} \leq \alpha) = \sigma(k(\alpha))$ where $k(\alpha)$ satisfies (2.5.6). Let α be a rational number. For this purpose we choose an arbitrary $\tilde{k}(\alpha)$ from the class of subsets of P which is, according to Th. 2.5.8, in 1:1 correspondence with $\sigma(\hat{y} \leq \alpha)$. For rational $\beta < \alpha$, since $\sigma(\hat{y} \leq \beta) < \sigma(\hat{y} \leq \alpha)$, we find that the set

$$j(\alpha, \beta) = \tilde{k}(\beta) \dotplus \tilde{k}(\alpha) \cap \tilde{k}(\beta)$$

is a set of measure zero. Therefore, since the set of rational β which satisfies $\beta < \alpha$ is denumerable, we find that the set $\bigcup_{\beta < \alpha} j(\alpha, \beta)$ has measure zero. From this result it follows that the set

$$\hat{k}(\alpha) = \tilde{k}(\alpha) \cup \bigcup_{\beta < \alpha} j(\alpha, \beta)$$

belongs to the same class as $\tilde{k}(\alpha)$ For the $\hat{k}(\alpha)$ we find that $\alpha_1 \geq \alpha_2 \Rightarrow \hat{k}(\alpha_1) \supset \hat{k}(\alpha_2)$. Thus, for rational α we have obtained a set of $\hat{k}(\alpha) \in P$ which increases with α and satisfies $\sigma(\hat{k}(\alpha)) = \sigma(\hat{y} \leq \alpha)$. Since the set of $\alpha > \beta$ is denumerable, for *each* β a set $k(\beta) = \bigcap_{\alpha > \beta} \hat{k}(\alpha) \in P$ is determined.

From Th. 2.5.8 it follows that

$$\sigma(k(\beta)) = \bigwedge_{\alpha > \beta} \sigma(\hat{k}(\alpha)) = \bigwedge_{\alpha > \beta} \sigma(\hat{y} \leq \alpha) = \sigma(\hat{y} \leq \beta),$$

where we obtain the last expression from the fact that the spectral set $\sigma(\hat{y} \leq \beta)$ is upper continuous.

Thus the desired function $f(\lambda)$ is obtained as follows:

$$f(\lambda) = \inf\{\beta \mid \lambda \in k(\beta)\}.$$

This theorem may be transformed from the scale y into an analogous theorem about its corresponding scale observable.

Th. 2.5.11. *If a decision observable is completely determined by the scale observable A (that is, the spectral family of A generates all of $\Sigma \subset G$) then all of the scale observables corresponding to the decision observable A are functions $f(A)$ of A.*

PROOF. By analogy to (2.5.6) and (2.5.8) $f(A)$ is defined by

$$f(A) = \int \alpha \, dE(k(\alpha)) = \int f(\alpha) \, dE(\alpha),$$

where $E(\alpha)$ is the spectral family of A. The theorem follows directly from the map S' of $\mathscr{B}'(\Sigma)$ into $\mathscr{B}'(\mathscr{H}_1, \ldots)$.

The fact that the scale values of the observable $f(A)$ are obtained from the scale values of A by means of the real function f is not a mysterious result of the correspondence principle, where the latter provided the basis for the initial development of quantum mechanics more than 50 years ago. It is, instead, a consequence of the definition of a scale observable, that is, of the mapping S' of a scale y onto an operator $A = S'y$ from $\mathscr{B}'(\mathscr{H}_1, \ldots)$. Clearly the concept of a measurement scale is not a quantum mechanical concept, but arises from the registration methods described by $\mathscr{R}(b_0)$ and idealized by means of Σ. Thus we have clarified and explained the usual, more or less, intuitively based methods entirely on the basis of the comparison of theory and experience in terms of $a' \in \mathscr{Q}'$, $b_0 \in \mathscr{R}_0$, $b \in \mathscr{R}$.

From this viewpoint it is no longer remarkable that one and the same operator A can represent the following different things: a scale observable, an effect (if $0 \leq A \leq 1$) or even an ensemble (in the case in which $0 \leq A$ and $\mathrm{tr}(A) = 1$). A mathematical term does not, in itself, have any physical meaning. The physical meaning is obtained when we, in addition, state what it represents. Here it is hoped that the formulation presented above will eliminate many of these "apparent" problems.

3 Coexistent and Complementary Observables

If we apply a registration method b_0, then by $\mathscr{R}(b_0) \xrightarrow{\psi_0} L$ (more precisely the U_g-completion of $\mathscr{R}(b_0)$) an observable is determined. In an experiment it is possible that we may only be interested in a Boolean subring Σ_1 of $\mathscr{R}(b_0)$; then $\Sigma_1 \xrightarrow{\psi_0} L$ is, so to say, a type of partial observable of $\mathscr{R}(b_0) \xrightarrow{\psi_0} L$. If $\Sigma_2 \xrightarrow{\psi_0} L$ is a second such partial observable, then $\Sigma_1 \xrightarrow{\psi_0} L$ and $\Sigma_2 \xrightarrow{\psi_0} L$ will "both" be measured by the registration method b_0. Thus we are led to the following general definition.

D 3.1. Two observables $\Sigma_1 \xrightarrow{F_1} L$ and $\Sigma_2 \xrightarrow{F_2} L$ are said to be *coexistent* if there exists an observable $\Sigma \xrightarrow{F} L$ and two homomorphisms h_1, h_2 such that the following diagram is commutative:

$$\begin{array}{ccccc} \Sigma_1 & \xrightarrow{h_1} & \Sigma & \xleftarrow{h_2} & \Sigma_2 \\ & {\scriptstyle F_1}\searrow & \downarrow{\scriptstyle F} & \swarrow{\scriptstyle F_2} & \\ & & L & & \end{array}$$

In §2.4 we have shown that we may identify Σ_1 with $h\Sigma_1 \subset \Sigma$ and F_1 with $F_{|h\Sigma_1}$; and similarly for Σ_2. From D 3.1 it follows that, in particular, the set

$$\{F \mid F = F_1(\sigma_1) \text{ for } \sigma_1 \in \Sigma_1 \text{ or } F = F(\sigma_2) \text{ for } \sigma_2 \in \Sigma_2\}$$

is a set of coexistent effects.

If $\Sigma_1 \xrightarrow{F_1} G$ and $\Sigma_2 \xrightarrow{F_2} G$ are decision observables, then, according to Th. 1.4.8 we may identify Σ_1 and Σ_2 with Boolean sublattices of G. $\Sigma_1 \xrightarrow{F_1} G$ and $\Sigma_2 \xrightarrow{F_2} G$ may be coexistent only if $\{\Sigma_1, \Sigma_2\}$ is a subset of a complete Boolean ring $\Sigma \subset G$. For such a Σ the diagram in D 3.1 is trivially satisfied.

D 3.2. Two coexistent decision observables are also said to be *commensurable*.

The above results therefore show that:

Th. 3.1. *Two decision observables are commensurable only if their combined images form a set of commensurable decision effects.*

From Th. 2.5.9 and Th. 1.3.4(v) we obtain the following extension:

Th. 3.2. *Two decision observables with scales are commensurable only if the scale observables commute.*

Thus we have obtained the joining of the well-known and common characterization of commensurable observables without the need to make use of the usual long-winded discussion of what is meant by joint measurements. Here we have intentionally avoided the use of the expression "simultaneous measurement." We shall return to this problem and its accompanying misconceptions in XVII, §2 and XVIII, §4. At this point we have only defined the notion of coexistence and commensurability of observables in D 3.1 and D 3.2, respectively, only in the form of an idealization. We must also look into the question concerning the realization of these idealized definitions; this will be done in §4.

If two observables are coexistent, then, according to D 3.1 there exists an observable $\Sigma \xrightarrow{F} L$ for which $F_1\Sigma_1 \subset F\Sigma$ and $F_2\Sigma_2 \subset F\Sigma$. We shall also use a concept which will characterize the "extreme case" of noncoexistence between two observables. For this purpose we shall define the subset

$$\Xi = \{(\lambda_1 \mathbf{1}, \lambda_2 \mathbf{1}, \ldots) \mid 0 \le \lambda_i \le 1\}$$

of L. Obviously the elements of Ξ are those which are coexistent relative to each element of L.

D 3.3. Two observables $\Sigma_1 \xrightarrow{F_1} L$, $\Sigma_2 \xrightarrow{F_2} L$ are said to be complementary if

$$F_1\Sigma_1 \not\subset \Xi, \qquad F_2\Sigma_2 \not\subset \Xi$$

and, for each observable $\Sigma \xrightarrow{F} L$ it follows that

$$F_1\Sigma_1 \cap F\Sigma \subset \Xi \quad \text{or} \quad F_2\Sigma_2 \cap F\Sigma \subset \Xi.$$

It is not difficult to see that two decision observables are complementary only if, for $e_1 \in F_1\Sigma_1$ and $e_2 \in F_2\Sigma_2$ and e_1, e_2 commensurable, then it follows that either e_1 or $e_2 \in Z$ where Z is the center of G (the definition of Z is given in D 1.3.2).

4 Realizations of Observables

We have simplified the analysis of the structure of observables by using the idealized version of the definition of an observable $\Sigma \xrightarrow{F} L$ instead of the realistic one $\mathscr{R}(b_0) \xrightarrow{\psi_0} L$. We must now ask whether it is possible to realize (in the sense of the construction of a measurement apparatus) the observable $\Sigma \xrightarrow{F} L$, that is, whether there exists a $b_0 \in \mathscr{R}_0$ for which the following diagram commutes:

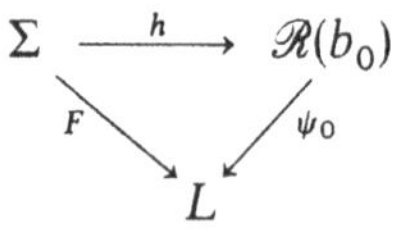

where h is an isomorphism. The requirement that, to each observable $\Sigma \xrightarrow{F} L$ there exists a $\mathscr{R}(b_0)$ such that the above diagram is satisfied, is clearly too strong since we have assumed that $\mathscr{Q}$, $\mathscr{R}$ are denumerable (see III, §3).

The following axiom governing "approximate" realizations may be added to the previous axioms:

AOb. To each observable $\Sigma \xrightarrow{F} L$ and each *finite* Boolean subring $\tilde{\Sigma}$ of Σ and each $\sigma(\mathscr{B}', \mathscr{B})$-neighborhood U of $0 \in \mathscr{B}'(\mathscr{H}_1, \ldots)$ there exists a $\mathscr{R}(b_0)$ and a homomorphism $\tilde{\Sigma} \xrightarrow{h} \mathscr{R}(b_0)$ such that

$$\psi_0 h(\sigma) - F(\sigma) \in U$$

for all $\sigma \in \tilde{\Sigma}$ where ψ_0 is the measure $\mathscr{R}(b_0) \xrightarrow{\psi_0} L$.

If we do not explicitly require axiom AOb, then, in any case, we may consider AOb to be a "certain" hypothesis in the sense of [1], §10.1; we shall not attempt to establish this result here. The proof that AOb is a "certain" hypothesis implies that the addition of axiom AOb does not lead to any contradictions in the mathematical theory. As we have explained in

[1], §10.4, it is only a matter of taste whether AOb is added as an axiom, except, if on the basis of experience there is strong evidence that, in nature, there are strong barriers to the possibility that apparatuses satisfying AOb can actually be constructed for all observables $\Sigma \xrightarrow{F} L$ (see [1], §10). But we have no indication of such barriers for the case of microsystems. However, AOb is wrong in the case of an extrapolated quantum mechanics for "many particles" (see [13], X).

Axiom AOb expresses the following statement: Each observable can be measured "approximately" (note that $\sigma(\mathscr{B}', \mathscr{B})$ determines the structure of physical imprecision in L—see III, §3, and in general [1], §9).

The structure described by AOb by which we may "approximately" measure an observable is essential for the application of quantum mechanics. Most of the important observables in quantum mechanics, for example, position, momentum, and angular momentum can only be approximately measured. An noteworthy experiment illustrating this fact is given by the Stern–Gerlach experiment in which the angular momentum of an atom is measured. In fact, the angular momentum is only approximately measured, because the procedure represents a measurement for angular momentum only for a subset of the ensembles (a subset for which the neighborhood U can be specified). This situation may be more clearly understood in the presentation of the theory of this experiment in [2], XI, §7.2. The angular momentum will be measured only for such ensembles which pass through the magnetic field in a particular way. For ensembles which, for example, do not pass through the apparatus, the apparatus does not make any measurements of angular momentum!

The definition of coexistence of two observables given in D 3.1 has, on the basis of AOb, the following consequence: there exists a measurement method $b_0 \in \mathscr{R}_0$ by which it is possible to (at least approximately) make a joint measurement of both of them. This can be clearly inferred from the diagram in D 1.3.1 as follows.

Let $\tilde{\Sigma}_1$ be a finite Boolean subring of Σ_1, and similarly, let $\tilde{\Sigma}_2$ be such for Σ_2. Then $h_1\tilde{\Sigma}_1$ and $h_2\tilde{\Sigma}_2$ generate a finite Boolean subring $\tilde{\Sigma}$ in Σ, to which, according to AOb, given a neighborhood U there exists a $\mathscr{R}(b_0)$ satisfying $h\tilde{\Sigma} \subset \mathscr{R}(b_0)$ and $\psi h(\sigma) - F(\sigma) \in U$. From this result it follows that $\tilde{\Sigma}_1 \xrightarrow{hh_1} \mathscr{R}(b_0)$, $\tilde{\Sigma}_2 \xrightarrow{hh_2} \mathscr{R}(b_0)$ and $\psi_0 hh_1(\sigma) - F(\sigma) \in U$ for all σ from $\tilde{\Sigma}_1$ and $\psi_0 hh_2(\sigma) - F_2(\sigma) \in U$ for all σ from $\tilde{\Sigma}_2$. Thus b_0 is an approximate measurement method for $\Sigma_1 \xrightarrow{F_1} L$ as well as for $\Sigma_2 \xrightarrow{F_2} L$, that is, the coexistent observables $\Sigma_1 \xrightarrow{F_1} L$ and $\Sigma_2 \xrightarrow{F_2} L$ can be (approximately) jointly measured by using the *single* method b_0.

Of course, the converse of this situation does not hold. It is indeed possible to make joint approximate measurements of two noncoexistent (even complementary) observables. Naturally both cannot be measured with arbitrary accuracy. For example, it is even possible that an apparatus will be able to measure both position and momentum in which the errors of measurement of position Δp and that of momentum Δp satisfy $\Delta p \Delta q < 1/2$

for a certain *sub*set of ensembles. This can occur only for certain subsets of ensembles. This fact has led to much unnecessary confusion. The source of this confusion is the false interpretation that such an attribution of position and momentum to single microsystems is forbidden by the Heisenberg uncertainty relations (see §8.2).

Axiom AOb states that it is "physically possible" (see [1], §10.4) that every observable may be approximately measured. However, AOb does not show how we may find the appropriate measurement b_0. The theory presented here cannot be used for this purpose because it does not contain any mathematical description of the technical construction of the "apparatus" $b_0 \in \mathcal{R}_0$. In XVII we shall undertake a partial step in the direction of this "construction problem" for the apparatus b_0. In XVIII we shall be introduced into some of the deep problems found in this area. In [13], X these problems are solved in principle.

The fact that the theory presented here does not yet provide a description of the physical structure for the apparatus b_0 gives rise to another deficiency of quantum mechanics. We shall now provide a brief description of this deficiency.

Suppose that we are given a detailed description of the physical structure for a given apparatus (for example, a particle counter). Then, using the mapping axioms given in [1], §5 we may express this fact by identifying the apparatus with a particular b_0. Apparatuses for which the internal physical structure is actually different will correspond to different b_0's.

According to the definition III, D 1.3 of the map ψ should be obtained directly from knowledge of the internal physical structure of the apparatus. But only the existence of this map is assured in the theory. It remains open which $g = \psi(b_0, b)$ corresponds to a given definite experimental effect process. In this situation we may choose to "guess" which $g \in L$ corresponds (at least, approximately) to a given (b_0, b) and to "add" the resulting guesses to the theory as axioms. In XI and XVI we will use this approach and the defects of this approach will become very evident. In XVII we shall describe a method by which it may be possible to make some progress in the area. A valid solution of the problem of obtaining an approximate determination of the map ψ is, however, not attained in XVII. A general method for finding ψ will be given in [13], X.

5 Coexistent Decompositions of Ensembles

In other formulations of quantum mechanics the question whether it is possible to make joint measurements of pairs of observables has played an important role. The question whether it is possible to make *joint preparations* is, however, either usually ignored or it is treated as a minor part of the measurement problem. Thus, in the literature, a primary emphasis is placed upon the discussion (in a somewhat limited way) of idealized measurements of the first kind. These idealized measurements are concerned with preparations of a well-defined form (see XVII, §5). We may clarify many of the fundamental questions in quantum mechanics by separating the question about the possibility of making joint preparations from the problem of registration. This question can be posed in a natural manner using the methodology presented here. To do so only requires that we begin with the methods of III, §2.

Using III, Th. 2.1 and the identification of $\varphi(a)$ with the elements of K according to AQ, for a decomposition $a = \sum_{i=1}^{n} a_i$ of a preparation procedure a we obtain

$$\varphi(a) = \sum_{i=1}^{n} \lambda_i \varphi(a_i), \tag{5.1}$$

where $\lambda_i = \lambda_{\mathcal{Q}}(a, a_i)$, $0 \leq \lambda_i \leq 1$ and $\sum_{i=1}^{n} \lambda_i = 1$.

If we have two decompositions of the *same* $a \in \mathcal{Q}'$

$$a = \bigcup_{i=1}^{n} a_i = \bigcup_{k=1}^{m} \tilde{a}_k$$

we then obtain

$$\varphi(a) = \sum_{i=1}^{n} \lambda_i \varphi(a_i) = \sum_{k=1}^{m} \tilde{\lambda}_k \varphi(a_k).$$

We may create a new decomposition from the above decompositions as follows:

$$a = \bigcup_{i,k}{}' (a_i \cap \tilde{a}_k). \tag{5.2}$$

Here we shall use the prime ′ to indicate that we shall not take the union (or summation) over those (i, k) for which $a_i \cap \tilde{a}_k = \varnothing$. Thus, from (5.2) we obtain

$$\varphi(a) = \sum_{i,k}{}' \tilde{\lambda}_i^k \varphi(a_i \cap \tilde{a}_k), \tag{5.3}$$

where

$$\lambda_{ik} = \lambda_{\mathcal{Q}}(a, a_i \cap \tilde{a}_k).$$

In addition, we find that

$$\varphi(a_i) = \sum_{k}{}' \lambda_k^i \varphi(a_i \cap \tilde{a}_k)$$

and

$$\varphi(\tilde{a}_k) = \sum_{i}{}' \lambda_i^k \varphi(a_i \cap \tilde{a}_k),$$

where

$$\lambda_k^i = \lambda_{ik}\left(\sum_k \lambda_{ik}\right)^{-1}, \qquad \tilde{\lambda}_i^k = \lambda_{ik}\left(\sum_i \lambda_{ik}\right)^{-1}.$$

The decomposition may be expressed in a very simple way if, in addition to φ, we introduce the maps

$$\varphi_a(\tilde{a}) = \lambda_{\mathcal{Q}}(a, \tilde{a})\varphi(\tilde{a})$$

of $\mathcal{Q}(a) = \{\tilde{a} \mid \tilde{a} \in \mathcal{Q}, \tilde{a} \subset a\}$ into $\check{K}$.

Th. 5.1. *The map φ_a with $\varphi_a(\tilde{a}) = \lambda_{\mathcal{Q}}(a, \tilde{a})\varphi(\tilde{a})$ of $\mathcal{Q}(a)$ in $\check{K}$ is an additive measure over the Boolean ring $\mathcal{Q}(a)$ which satisfies $\varphi_a(a) = \varphi(a) \in K$.*

PROOF. Let $\tilde{a} = \tilde{a}_1 \cup \tilde{a}_2$ where $\tilde{a}_1 \cap \tilde{a}_2 = \varnothing$; Let $\tilde{a}_3 = a \backslash \tilde{a}$. Then it follows that $a = \bigcup_{i=1}^3 \tilde{a}_i$ is a decomposition for which according to (5.1) $\varphi(a) = \varphi_a(\tilde{a}_1) + \varphi_a(\tilde{a}_2) + \varphi_a(\tilde{a}_3)$. In addition, $a = \tilde{a} \cup \tilde{a}_3$ is a decomposition, so that we obtain $\varphi(a) = \varphi_a(\tilde{a}) + \varphi_a(\tilde{a}_3)$.

D 5.1. A $\tilde{w} \in \check{K}$ satisfying $\tilde{w} < w \in K$ is called a *mixture component* or *component* of w.

If $\tilde{w}$ is a component of w, then $w - \tilde{w}$ also is, because $0 < w - \tilde{w} < w$; $w = \tilde{w} + (w - \tilde{w})$ is therefore a decomposition of w into the components $\tilde{w}$ and $(w - \tilde{w})$.

Two decompositions of the same preparation procedure $a = \bigcup_{i=1}^n a_i = \bigcup_{k=1}^m \tilde{a}_k$ generate two decompositions of the ensemble $\varphi(a)$ as follows:

$$\varphi(a) = \sum_{i=1}^{n} \varphi_a(a_i) = \sum_{k=1}^{m} \varphi_a(\tilde{a}_k),$$

where the components $\varphi_a(a_i)$, $\varphi_a(\tilde{a}_k)$ lie in the range of the additive measure φ_a on the Boolean ring $\mathcal{Q}(a)$.

This result suggests the following definition:

D 5.2. Two decompositions of an ensemble

$$w = \sum_{i=1}^{n} w_i = \sum_{k=1}^{m} \tilde{w}_k \quad \text{where } w_i, \tilde{w}_k \in \check{K}$$

are said to be *coexistent* if there exists a Boolean ring Σ and an additive measure $\Sigma \xrightarrow{W} \check{K}$ for which $W(\varepsilon) = w$ and $w_i, \tilde{w}_k \in W\Sigma$.

Two decompositions of a preparation procedure a result in coexistent decompositions of $\varphi(a)$. In §7 we shall return to the problem of the realization of coexistent decompositions.

D 5.3. A set $A \subset \check{K}$ is called a set of *coexistent components* of w if there exists a Boolean ring Σ and an additive measure $\Sigma \xrightarrow{W} \check{K}$ such that $W(\varepsilon) = w$ and $A \subset W\Sigma$.

By analogy with the case of coexistent effects we need only consider effective measures. Here it is reasonable to consider an idealization of $\Sigma \xrightarrow{W} \check{K}$

obtained by mathematical completion. This possibility follows directly from the following theorem.

Th. 5.2. *Let W be an effective additive measure on the Boolean ring $\Sigma \xrightarrow{W} \check{K}$ which satisfies $W(\varepsilon) = w \in K$. Then $m_0(\sigma) = \|W(\sigma)\| = \mu(W(\sigma), \mathbf{1})$ is an effective additive measure satisfying $m_0(\varepsilon) = 1$, and $d(\sigma_1, \sigma_2) = m_0(\sigma_1 \dotplus \sigma_2)$ is a metric in Σ for which W is uniformly continuous as a map in the Banach space $\mathscr{B}$.*

PROOF. From $\sigma = \sigma_1 \vee \sigma_2$ and $\sigma_1 \wedge \sigma_2 = 0$ it follows that $W(\sigma) = W(\sigma_1) + W(\sigma_2)$ and $m_0(\sigma) = \mu(W(\sigma), \mathbf{1}) = \mu(W(\sigma_1), \mathbf{1}) + \mu(W(\sigma_2), \mathbf{1}) = m_0(\sigma_1) + m_0(\sigma_2)$. Since $W(\varepsilon) \in K$ we obtain $m_0(\varepsilon) = \mu(W(\varepsilon), \mathbf{1}) = 1$. m_0 is effective, because if $m_0(\sigma) = 0$ it follows that $\mu(W(\sigma), 1) = 0$ and therefore $\|W(\sigma)\| = 0$, that is, $W(\sigma) = 0$; since W is, by assumption, effective, we obtain $\sigma = 0$.

From

$$W(\sigma_1) + W(\sigma_2 \dotplus \sigma_1 \cdot \sigma_2) = W(\sigma_1 \vee \sigma_2)$$
$$= W(\sigma_2) + W(\sigma_1 \dotplus \sigma_1 \cdot \sigma_2)$$

and

$$W(\sigma_1 \dotplus \sigma_2) = W(\sigma_1 \dotplus \sigma_1 \cdot \sigma_2) + W(\sigma_2 \dotplus \sigma_1 \cdot \sigma_2)$$

it follows that, for all $g \in L$:

$$|\mu(W(\sigma_1) - W(\sigma_2), g))| \leq \mu(W(\sigma_1 \dotplus \sigma_2), g)$$
$$\leq \mu(W(\sigma_1 \dotplus \sigma_2), \mathbf{1}) = d(\sigma_1, \sigma_2).$$

On the other hand, we have

$$\|W(\sigma_1) - W(\sigma_2)\| = \sup_{-\mathbf{1} \leq y \leq \mathbf{1}} \mu(W(\sigma_1) - W(\sigma_2), y).$$

Since $[-\mathbf{1}, \mathbf{1}] = 2L - \mathbf{1}$, for $y = 2g - \mathbf{1}$ we obtain

$$\mu(W(\sigma_1) - W(\sigma_2), y) = 2\mu(W(\sigma_1) - W(\sigma_2), g)$$
$$- \mu(W(\sigma_1) - W(\sigma_2), \mathbf{1})$$

and thus we find that

$$\|W(\sigma_1) - W(\sigma_2)\| \leq 3 \sup_{g \in L} |\mu(W(\sigma_1) - W(\sigma_2), g)|$$

Therefore we obtain

$$\|W(\sigma_1) - W(\sigma_2)\| \leq 3d(\sigma_1, \sigma_2).$$

From Th. 5.2 it follows that, as is the case in §1.4, it is possible to complete Σ and extend W on the completion of Σ. Then W becomes a σ-additive measure on the completion of Σ. If Σ is a (lattice-theoretically) complete Boolean ring, and if W is a σ-additive measure then Σ is complete with respect to the metric

$$d(\sigma_1, \sigma_2) = \mu(W(\sigma_1 \dotplus \sigma_2), \mathbf{1}).$$

We define the following analog of an observable:

D 5.4. A Boolean ring Σ with effective measure $W: \Sigma \to \breve{K}$ for which $W(\varepsilon) = w \in K$ and for which Σ (in the metric determined by W) is complete and separable is called a *preparator* of w.

It follows that two decompositions of an ensemble

$$w = \sum_{i=1}^{n} w_i = \sum_{k=1}^{m} \tilde{w}_k$$

are coexistent only if there exists a preparator $\Sigma \xrightarrow{W} \breve{K}$ such that $w_i, \tilde{w}_k \in W\Sigma$.

The mathematical similarity between preparator and observable runs much deeper, and depends upon the following theorem:

Th. 5.3. *For fixed $w \in K$ each $\tilde{w} \in \breve{K}$ satisfying $\tilde{w} < w$ can be written in the form $\tilde{w} = w^{1/2} g w^{1/2}$, where $g \in L$. Let $g \in L$; then, for $w \in K$: $w^{1/2} g w^{1/2} \in \breve{K}$. The correspondence $g \to w^{1/2} g w^{1/2}$ is an order-isomorphism of $[0, e]$ onto $[0, w] \subset \breve{K}$ where e is the decision effect satisfying $K_1(e) = C(w)$.*

PROOF. From $\tilde{w} = w^{1/2} g w^{1/2}$ and from $\mathbf{0} < g < \mathbf{1}$ it follows that $\langle \varphi, \tilde{w}\varphi \rangle = \langle w^{1/2}\varphi, g w^{1/2}\varphi \rangle \leq \langle w^{1/2}\varphi, w^{1/2}\varphi \rangle = \langle \varphi, w\varphi \rangle$, that is, $\tilde{w} \leq w$. Since $\langle w^{1/2}\varphi, g w^{1/2}\varphi \rangle \geq 0$ we obtain $\tilde{w} > 0$.

Let $0 < \tilde{w} < w$; we shall now consider the support $\mathfrak{r}$ of w. The domain of definition of the operator $w^{-1/2}$ is dense in $\mathfrak{r}$ because, if $w = \sum_\nu w_\nu P_{\varphi_\nu}$ and $w_\nu \neq 0$, then the φ_ν form a complete orthonormal basis for $\mathfrak{r}$, and all vectors of the form $\sum_{\nu=1}^{N} \varphi_\nu a_\nu$ lie in the domain of definition of $w^{-1/2}$. If we then write $A = \tilde{w}^{1/2} w^{-1/2}$, then A is defined in a dense set in $\mathfrak{r}$; in addition, since $\tilde{w} < w$ we obtain

$$\begin{aligned} \|A\varphi\|^2 &= \langle A\varphi, A\varphi \rangle = \langle w^{-1/2}\varphi, \tilde{w} w^{-1/2}\varphi \rangle \\ &\leq \langle w^{-1/2}\varphi, w w^{-1/2}\varphi \rangle = \langle \varphi, \varphi \rangle, \end{aligned}$$

that is, $\|A\| \leq 1$. Clearly A may be extended to all of $\mathfrak{r}$. Therefore, for all $\varphi \in \mathfrak{r}$ we obtain

$$A w^{1/2}\varphi = \tilde{w} w^{-1/2} w^{1/2}\varphi = \tilde{w}^{1/2}\varphi$$

and thus we find that

$$\tilde{w}^{1/2} = A w^{1/2}.$$

Thus we obtain

$$\tilde{w} = \tilde{w}^{1/2}\tilde{w}^{1/2} = (\tilde{w}^{1/2})^{+}\tilde{w}^{1/2} = w^{1/2} A^{+} A w^{1/2}.$$

If we define g by $A^{+}A = g$ we then obtain $\tilde{w} = w^{1/2} g w^{1/2}$. Since (see AIV, §3 and §4) $\|A^{+}\| = \|A\|$ and $\|A^{+}A\| \leq \|A^{+}\| \, \|A\|$ we therefore find that $\|g\| \leq 1$ and therefore $\mathbf{0} \leq g \leq \mathbf{1}$.

From the operator equation in $\mathfrak{r}$: $w^{1/2} g_1 w^{1/2} = w^{1/2} g_2 w^{1/2}$ it follows that $\langle w^{1/2}\varphi, (g_1 - g_2) w^{1/2}\varphi \rangle = 0$ for all $w^{1/2}\varphi$. Since $w^{1/2}\mathfrak{r}$ is dense in $\mathfrak{r}$ we obtain $\langle \psi, (g_1 - g_2)\psi \rangle = 0$ for all $\psi \in \mathfrak{r}$ and thus we find that $g_1 - g_2 = 0$ is satisfied as an operator equation in $\mathfrak{r}$.

Th. 5.4. *Let $\Sigma \xrightarrow{W} \check{K}$ be an additive measure with $W(\varepsilon) = w \in K$. Then an additive measure η is defined in terms of the bijection χ and the following diagram as follows:*

$$\tilde{w} \xrightarrow{\chi} g \quad \text{where } \tilde{w} = w^{1/2} g w^{1/2},$$

where $[0, w] \xrightarrow{\chi} [0, e]$, where $e \in G$ and $K_1(e) = C(w)$

$$\begin{array}{ccc} \Sigma & \xrightarrow{\eta} & [0, e] \\ & {\scriptstyle W}\searrow \quad \nearrow{\scriptstyle \chi} & \\ & [0, w] & \end{array}$$

The uniform structure U_g defined by η is the same as that one defined by W.

PROOF. The fact that η is an additive measure follows directly from Th. 5.3. U_g is determined by the metric

$$d_\eta(\sigma_1, \sigma_2) = \mu(w_0, \eta(\sigma_1 \dotplus \sigma_2))$$

for a w_0 for which $C(w_0) = K$ (see Th. 1.4.1). For example, we may choose w_0 as follows:

$$w_0 = \tfrac{1}{2}w + \tfrac{1}{2}\hat{w},$$

where w is defined in Th. 5.4 and $\hat{w} \in K_0(e)$ and $C(\hat{w}) = K_0(e)$. Thus, since $0 < \eta(\sigma) < e$ we obtain

$$d_\eta(\sigma_1, \sigma_2) = \tfrac{1}{2}\mu(w, \eta(\sigma_1 \dotplus \sigma_2)).$$

On the other hand, the metric for W is given by

$$\begin{aligned} d_W(\sigma_1, \sigma_2) &= \mu(W(\sigma_1 \dotplus \sigma_2), \mathbf{1}) \\ &= \mu(w^{1/2}\eta(\sigma_1 \dotplus \sigma_2)w^{1/2}, \mathbf{1}) \\ &= \mu(w, \eta(\sigma_1 \dotplus \sigma_2)). \end{aligned}$$

According to Th. 5.4, if we admit, as observables, the $\Sigma \xrightarrow{\eta} L$ for which $\eta(\varepsilon) = e \in G$ and $e \neq 1$, then we have established a 1:1 correspondence between the preparators $\Sigma \xrightarrow{W} K$ and the observables $\Sigma \xrightarrow{\eta} L$. This correspondence permits us to, by analogy, transform all theorems about observables to theorems about preparators. We leave the proof of this result as an exercise for the reader. We may, without difficulty also define the following concepts for preparators: a preparator for an ensemble w is more extensive than another for the same ensemble, and: two preparators for the same ensemble are coexistent (see also §6). These concepts are completely analogous to those defined for observables. We shall now explain the distinction between preparator and observable. The distinction depends on the physical meaning and, mathematically upon the structure of the measure

$$W(\sigma) = w^{1/2}\eta(\sigma)w^{1/2}$$

which is somewhat different than $\eta(\sigma)$.

The desire not to consider "unnecessary" convex combinations of the $\eta(\sigma)$ is analogous to the desire not to consider unnecessary mixtures of the $W(\sigma)$;

here the case of decompositions is more interesting physically than that of mixtures. We shall therefore seek preparators for which $\Sigma \xrightarrow{\eta} L$ corresponds to an kernel observable.

We find, however, that the decomposition of observables described in §2.4 has an alternative interpretation when applied to η. Suppose, for example, that

$$\eta(\sigma) = \lambda\eta_1(\sigma) + (1 - \lambda)\eta_2(\sigma) \tag{5.4}$$

holds for all $\sigma \in \Sigma$. Then it follows that

$$W(\sigma) = \lambda W_1(\sigma) + (1 - \lambda)W_2(\sigma). \tag{5.5}$$

Equation (5.5) represents a decomposition of each ensemble $W(\sigma)$. We note, however, that this decomposition fulfils on the basis of (5.5), an additional condition. Since $\eta(\varepsilon) = e = \lambda\eta_1(\varepsilon) + (1 - \lambda)\eta_2(\varepsilon)$ we find that $\eta_1(\varepsilon) = \eta_2(\varepsilon) = e$ and we therefore obtain $W_1(\varepsilon) = W_2(\varepsilon) = w$, that is, the ensemble w underlying the preparator is not decomposed by (5.5).

A decomposition which does not, in general, satisfy this condition may be obtained, for example, from a partition $\varepsilon = \sigma_1 \vee \sigma_2, \sigma_1 \wedge \sigma_2 = 0$ of the form

$$\begin{aligned} W(\sigma) &= W(\sigma_1 \wedge \sigma) + W(\sigma_2 \wedge \sigma) \\ &= \lambda W_1(\sigma) + (1 - \lambda)W_2(\sigma), \end{aligned}$$

where $\lambda = \mu(W(\sigma_1), \mathbf{1})$, $W_1(\sigma) = \lambda^{-1}W(\sigma_1 \wedge \sigma)$ and

$$W_2(\sigma) = (1 - \lambda)^{-1}W(\sigma_2 \wedge \sigma).$$

In these cases, we find that, in general $W_1(\varepsilon) = \lambda^{-1}W(\sigma_1)$ is not equal to $W(\varepsilon) = w$. If, in (5.5) we have $W_1(\sigma) = W_2(\sigma) = W(\varepsilon) = w$, then, according to Th. 5.4 it follows that $\eta_1(\sigma), \eta_2(\sigma)$ are uniquely determined with $\eta_1(\varepsilon) = \eta_2(\varepsilon) = e$ and must also satisfy (5.4).

Decompositions of the form (5.5) for which $W_1(\varepsilon) = W_2(\varepsilon) = W(\varepsilon) = w$ are called decompositions of the preparator. One of the arguments presented in §2.4 for observables is not applicable because, in the set of preparation procedures there is no special subset of the type $\mathscr{R}_0$ in $\mathscr{R}$. Despite this fact, we may still deduce from the decompositions of preparators in a way analogous to the case of observables the goal to realize experimentally as far as possible irreducible preparators.

If $\Sigma \xrightarrow{\eta} L$ is a decision observable (with the exception that we do not require that $\eta(\varepsilon) = e = 1$) then the corresponding preparator is irreducible. The preparator for which $\Sigma \xrightarrow{\eta} L$ is a decision observable plays a theoretically distinguished role.

D 5.5. A preparator $\Sigma \xrightarrow{W} \check{K}$ is called a *decision preparator* if the corresponding observable $\Sigma \xrightarrow{\eta} L$ is a decision observable.

According to Th. 1.4.8, for a decision observable we may identify Σ with a Boolean ring $\eta\Sigma$. A decision preparator is uniquely defined by a Boolean subring Σ of G (possibly with an $e \neq 1$ as unit element; e is the support of w) together with the map

$$W(\tilde{e}) = w^{1/2}\tilde{e}w^{1/2} \tag{5.6}$$

of Σ into $\breve{K}$ for $\tilde{e} \in \Sigma$. Since e is the support of w, we find that $W(e) = w^{1/2}ew^{1/2} = w^{1/2}w^{1/2} = w$ (as we would expect).

In the desire to eliminate the lack of symmetry between preparator and observable, we will often find the following statement: the ensemble $w = (\mathbf{1}, \mathbf{1}, \ldots) = \mathbf{1}$ corresponds to the case of "complete ignorance," and the fact that w is not an element of K is taken only for a matter of mathematical "inconvenience."

For $w = \mathbf{1}$ equation (5.6) may be formally transformed into $W(\tilde{e}) = \tilde{e}$ and we may come to the false conclusion that a preparator is nothing other than a decomposition of $w = \mathbf{1}$ with respect to a (decision) observable. According to the formulation of quantum mechanics developed here we cannot dismiss w in (5.6); w is not a measure of knowledge or ignorance, but is only a mathematical symbol $w = \varphi(a)$ corresponding to an apparatus (!) a by means of the map φ of $\mathcal{Q}'$ in K. Such an apparatus a does not exist for the case in which $\varphi(a)$ is approximately equal to $\mathbf{1}$. On the contrary, we may *approximately* represent each $w = \sum_\nu w_\nu P_{\varphi_\nu}$ as a *finite* sum $\sum_{\nu=1}^N \tilde{w}_\nu P_{\varphi_\nu}$, where $\tilde{w}_\nu = w_\nu(\sum_{\mu=1}^N w_\mu)^{-1}$. In physical terms each ensemble which we may obtain from a preparation procedure has, for all practical purposes, a finite-dimensional support. This is a very important aspect of the structure of microsystems, and is a direct consequence of axiom AV 4s.

The following important question is often discussed: For a given observable $\Sigma \xrightarrow{F} L$ is it possible to find a preparator of w for which w can be decomposed in such a way that there is no dispersion with respect to the observable?

Let $\Sigma \xrightarrow{F} L$ be an observable, and let $\tilde{\Sigma} \xrightarrow{W} \breve{K}$ be a preparator. Suppose that, to each $\sigma \in \Sigma$ there exists a $\tilde{\sigma} \in \tilde{\Sigma}$ for which

$$\mu(W(\tilde{\sigma}), 1 - F(\sigma)) = 0 \quad \text{and} \quad \mu(W(\varepsilon \dotplus \tilde{\sigma}), F(\sigma)) = 0, \tag{5.7}$$

that is, the mixture component $W(\tilde{\sigma})$ triggers the response for the effect $F(\sigma)$ with certainty and the mixture component $W(\varepsilon \dotplus \tilde{\sigma})$ does not trigger the response for $F(\sigma)$ with certainty.

D 5.6. A preparator $\tilde{\Sigma} \xrightarrow{W} \breve{K}$ is said to be *dispersion-free* with respect to the observable $\Sigma \xrightarrow{F} L$ if to each $\sigma \in \Sigma$ there exists a $\tilde{\sigma} \in \tilde{\Sigma}$ for which (5.7) is satisfied.

What relationships must be satisfied between a preparator and an observable in order that the preparator be dispersion-free with respect to an observable?

Let e_w denote the support of $w = W(\varepsilon)$. Then, since $W(\tilde{\sigma}) \leq W(\varepsilon)$ we obtain $W(\tilde{\sigma}) = e_w W(\tilde{\sigma}) e_w$ for all $\tilde{\sigma}$ and therefore find that

$$\mu(W(\tilde{\sigma}), g) = \mu(e_w W(\tilde{\sigma}) e_w, g) = \mu(W(\tilde{\sigma}), e_w g e_w)$$

for all $W(\tilde{\sigma})$ and $g \in L$. An additive measure $\Sigma \xrightarrow{\hat{F}} L$ (which is not necessarily effective) is defined by $\hat{F}(\sigma) = e_w F(\sigma) e_w$. Eq. (5.7) is satisfied for $\hat{F}(\sigma)$ as well as for $F(\sigma)$:

$$\mu(W(\tilde{\sigma}), 1 - \hat{F}(\sigma)) = \mu(W(\tilde{\sigma}), e_w - \hat{F}(\sigma)) = 0$$

and (5.8)

$$\mu(W(\varepsilon \dotplus \tilde{\sigma}), \hat{F}(\sigma)) = 0.$$

Let $\lambda = \mu(W(\tilde{\sigma}), \mathbf{1})$; then from (5.8) it follows that $\lambda^{-1} W(\tilde{\sigma}) \in K_1(\hat{F}(\sigma))$ and $(1 - \lambda)^{-1} W(\varepsilon \dotplus \tilde{\sigma}) \in K_0(\hat{F}(\sigma))$. For $e(\sigma)$ and $e_0(\sigma) \in G$ and $K_1(\hat{F}(\sigma)) = K_1(e(\sigma))$, $K_0(\hat{F}(\sigma)) = K_0(e_0(\sigma))$ it follows that

$$e(\sigma) < \hat{F}(\sigma) < e_0(\sigma). \tag{5.9}$$

Since $w = W(\tilde{\sigma}) + W(\varepsilon \dotplus \tilde{\sigma})$ it follows that

$$C(w) = C(\lambda^{-1} W(\tilde{\sigma})) \vee C((1 - \lambda)^{-1} W(\varepsilon \dotplus \tilde{\sigma})),$$

that is,

$$K_1(e_w) \subset K_1(e(\sigma)) \vee K_0(e_0(\sigma))$$

from which it follows that

$$\begin{aligned} K_1(e_w) &\subset K_1(e(\sigma)) \vee K_1(e_0(\sigma)^\perp) \\ &= K_1(e(\sigma) \vee e_0(\sigma)^\perp). \end{aligned}$$

Thus it follows that

$$e_w < e(\sigma) \vee e_0(\sigma)^\perp.$$

Since, according to (5.9) $e(\sigma) < e_0(\sigma)$ we obtain

$$\begin{aligned} e(\sigma) \vee e_0(\sigma)^\perp &= e(\sigma) + e_0(\sigma)^\perp \\ &= e(\sigma) + \mathbf{1} - e_0(\sigma) = (e_0(\sigma) - e(\sigma))^\perp. \end{aligned}$$

Therefore $e_w \perp e_0(\sigma) - e(\sigma)$.

Thus, according to (5.9) we have $\hat{F}(\sigma) = e(\sigma) + (\hat{F}(\sigma) - e(\sigma))$, where $0 < \hat{F}(\sigma) - e(\sigma) < e_0(\sigma) - e(\sigma)$. According to the definition of $\hat{F}$ we obtain $e(\sigma) < e_w$. Therefore we obtain $\hat{F}(\sigma) - e(\sigma) = e_w(\hat{F}(\sigma) - e(\sigma))e_w < e_w(e_0(\sigma) - e(\sigma))e_w = 0$ and hence we find that $\hat{F}(\sigma) = e(\sigma)$. Therefore the measure $\hat{F}(\sigma)$ is a projection-valued measure $\Sigma \xrightarrow{\hat{F}} G$ for $\hat{F}(\varepsilon) = e_w$.

From $e_w F(\sigma) e_w = e(\sigma) \in G$, it follows that, for a $\varphi \in e(\sigma)\mathscr{H}_i$ such that $\|\varphi\| = 1$:

$$\begin{aligned} F(\sigma)\varphi &= (\mathbf{1} - e_w)F(\sigma)\varphi + e_w F(\sigma)\varphi \\ &= (\mathbf{1} - e_w)F(\sigma)\varphi + e_w F(\sigma) e_w \varphi \\ &= (\mathbf{1} - e_w)F(\sigma)\varphi + e(\sigma)\varphi \\ &= (\mathbf{1} - e_w)F(\sigma)\varphi + \varphi. \end{aligned}$$

Since $(\mathbf{1} - e_w)F(\sigma)\varphi \perp \varphi = e_w\varphi$ it follows that

$$\|F(\sigma)\varphi\|^2 = \|(\mathbf{1} - e_w)F(\sigma)\varphi\|^2 + 1.$$

Since $0 < F(\sigma) < 1$ we obtain $\|F(\sigma)\| \leq 1$ and therefore $\|F(\sigma)\varphi\|^2 \leq 1$. Therefore we find that $(\mathbf{1} - e_w)F(\sigma)\varphi = 0$ for all $\varphi \in e(\sigma)\mathscr{H}_i$, and therefore $F(\sigma)\varphi = \varphi$, that is,

$$F(\sigma)e(\sigma) = e(\sigma). \tag{5.10}$$

For $\varphi \in (e_w - e(\sigma))(\mathscr{H}_1, \mathscr{H}_2, \ldots)$ we find that

$$\begin{aligned}\langle\varphi, F(\sigma)\varphi\rangle &= \langle e_w\varphi, F(\sigma)e_w\varphi\rangle \\ &= \langle\varphi, e_wF(\sigma)e_w\varphi\rangle = \langle\varphi, e(\sigma)\varphi\rangle = 0\end{aligned}$$

and (since $F(\sigma) > 0$) $F(\sigma)\varphi = 0$, that is,

$$F(\sigma)(e_w - e(\sigma)) = 0. \tag{5.11}$$

From (5.10) and (5.11) it follows that

$$F(\sigma)e_w = e(\sigma)$$

and therefore

$$(\mathbf{1} - e_w)F(\sigma)e_w = 0, \qquad e_wF(\sigma) = e(\sigma)$$

and (5.12)

$$F(\sigma) = e(\sigma) + (\mathbf{1} - e_w)F(\sigma)(\mathbf{1} - e_w).$$

The measure $F(\sigma)$ may therefore be expressed as a sum of two terms, one equal to $e(\sigma)$ and the other term is orthogonal to the support of w. In order that a preparator of w be dispersion-free with respect to the observable $\Sigma \xrightarrow{F} L$ $F(\sigma)$ must therefore be, relative to w, a measure of decision effects $\Sigma \xrightarrow{\hat{F}} G$. For every $\Sigma \xrightarrow{\hat{F}} G$ does there exist a preparator of w which is dispersion-free with respect to $\Sigma \xrightarrow{\hat{F}} G$?

For $\hat{F}(\sigma) = e(\sigma)$ equation (5.8) is equivalent to

$$w^{1/2}\eta(\tilde{\sigma})w^{1/2}(e_w - e(\sigma)) = 0$$

and

$$w^{1/2}(1 - \eta(\tilde{\sigma}))w^{1/2}e(\sigma) = 0,$$

that is, equivalent to

$$we(\sigma) = w^{1/2}\eta(\tilde{\sigma})w^{1/2}e(\sigma) = w^{1/2}\eta(\tilde{\sigma})w^{1/2} \tag{5.13}$$

from which we obtain the adjoint equation:

$$e(\sigma)w = w^{1/2}\eta(\tilde{\sigma})w^{1/2} = we(\sigma).$$

Therefore we find that $e(\sigma)$ commutes with w and with $w^{1/2}$, so that, from (5.13), it follows that

$$w^{1/2}\eta(\tilde{\sigma})w^{1/2} = w^{1/2}e(\sigma)w^{1/2}, \tag{5.14}$$

Since $e(\sigma) < e_w$, from Th. 5.3 it follows that $e(\sigma) = \eta(\tilde{\sigma})$, that is, $\eta\tilde{\Sigma} \supset \hat{F}\Sigma$. Instead of $\tilde{\Sigma} \xrightarrow{\eta} L$ we may consider the Boolean subring $\hat{F}\Sigma$ of G and determine $\hat{F}\Sigma \xrightarrow{\hat{W}} \breve{K}$ by

$$\hat{W}(\hat{F}(\sigma)) = w^{1/2}\hat{F}(\sigma)w^{1/2}. \tag{5.15}$$

For a preparator of w, instead of $\tilde{\Sigma} \xrightarrow{W} \breve{K}$ we may use $(\hat{F}\Sigma) \xrightarrow{\hat{W}} \breve{K}$ where $\hat{W}$ is defined by (5.15). A preparator can be dispersion-free with respect to $\Sigma \xrightarrow{\hat{F}} L$ only if (5.15) results in dispersion-free ensembles, that is, if the $\hat{F}(\sigma)$ commute with w for all σ. Then we would obtain

$$\mu(w^{1/2}e(\sigma)w^{1/2}, \mathbf{1} - e(\sigma)) = 0$$

and

$$\mu(w^{1/2}(\mathbf{1} - e(\sigma))w^{1/2}, e(\sigma)) = 0.$$

Therefore, if and only if w commutes with the *decision* observable determined by $\Sigma \xrightarrow{\hat{F}} [0, e_w]$, there exists exactly one preparator of w which is dispersion-free with respect to $\Sigma \xrightarrow{\hat{F}} [0, e_w]$.

On the other hand, to each decision observable $\Sigma \xrightarrow{\hat{F}} [0, e_w]$ there exists a $w \in K$ (with support e_w) which commutes with the observable providing that the decision observable has only a discrete spectrum. According to D 2.5.4 and Th. 1.4.8 this result is equivalent to the condition that the Boolean subring $\hat{F}\Sigma$ of G is atomic. This result follows from the fact that w has only discrete eigenvalues and that each eigenspace has finite dimension.

Therefore, there are no preparators which are dispersion-free with respect to an observable which have a continuous spectrum. This fact plays an important role in the investigation of the so-called "measurements of the first kind." This is a typical aspect of "quantum structures," that is, of the structure of microsystems since, in classical theories, to any decision observable there exist dispersion-free preparators. Thus it is understandable that some physicists have not only sought to weaken the distinction between observables and preparators, but have also sought (by means of "tricks") the "elimination" of the typical "quantum mechanical structures." For the case of microsystems the preparation and registration processes are no longer "parallel" as is the case for macrosystems. Therefore the implications for the so-called measurements of the first kind appear to be highly idealized (see XVII, §5). However, on the basis of II we shall find that it is not necessary to require the existence of measurements of the first kind as a basis for quantum mechanics.

6 Complementary Decompositions of Ensembles

In complete analogy to the definition of the notion of coexistent observables we introduce the following definition:

D 6.1. A preparator $\Sigma \xrightarrow{W} \breve{K}$ of the ensemble w is said to be *more extensive* than another preparator $\Sigma_1 \xrightarrow{W_1} \breve{K}$ for the same ensemble if there exists a homomorphism h for which the following diagram commutes:

$$\begin{array}{ccc} \Sigma_1 & \xrightarrow{h} & \Sigma \\ & {}_{W_1}\searrow \quad \swarrow_{W} & \\ & \breve{K} & \end{array}$$

D 6.2. Two preparators $\Sigma_1 \xrightarrow{W_1} \breve{K}$, $\Sigma_2 \xrightarrow{W_2} \breve{K}$ for the same ensemble w are said to be *coexistent* if there exists a preparator $\Sigma \xrightarrow{W} \breve{K}$ for the same ensemble which is more extensive than the preparators $\Sigma_1 \xrightarrow{W_1} \breve{K}$ and $\Sigma_2 \xrightarrow{W_2} \breve{K}$.

Therefore we find that, for two coexistent preparators, there exists a preparator $\Sigma \xrightarrow{W} \breve{K}$ and two homomorphisms h_1, h_2 such that the following diagram commutes:

$$\begin{array}{ccccc} \Sigma_1 & \xrightarrow{h_1} & \Sigma & \xleftarrow{h_2} & \Sigma_2 \\ & {}_{W_1}\searrow & \downarrow{\scriptstyle W} & \swarrow_{W_2} & \\ & & \breve{K} & & \end{array} \tag{6.1}$$

We shall now provide a physical interpretation of this result: $\Sigma \xrightarrow{W} \breve{K}$ represents (in idealized form) a preparation procedure a for which Σ_1, Σ_2 may be considered to be Boolean subrings of $\mathscr{Q}(a)$, that is, to be parts of a single preparation procedure and its possible decompositions.

For the observables which (formally) correspond to the preparators there exists a diagram analogous to (6.1) as follows:

$$\begin{array}{ccccc} \Sigma_1 & \xrightarrow{h_1} & \Sigma & \xleftarrow{h_2} & \Sigma_2 \\ & {}_{\eta_1}\searrow & \downarrow{\scriptstyle \eta} & \swarrow_{\eta_2} & \\ & & L & & \end{array} \tag{6.2}$$

where $\eta(\varepsilon) = \eta_1(\varepsilon) = \eta_2(\varepsilon)$ is the support of w.

In the sense of diagram (6.2) the corresponding observables are, by D 3.1, coexistent observables.

The situation for two preparators is substantially different than is the case of observables relative to the problem of "perfect" noncoexistence. Such "perfect" noncoexistent preparators play an important role in those problems of quantum mechanics which are associated with an epistomology (theory of cognition).

We shall now characterize (in idealized form) the following situation: For $w \in K$ suppose that we are given two decompositions of w:

$$w = \sum_{i=1}^{n} w_i = \sum_{k=1}^{m} \tilde{w}_k, \tag{6.3}$$

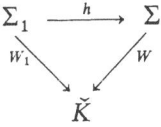

where $w_i, \tilde{w}_k \in \breve{K}$. Suppose that there exists a pair of preparation procedures a and $\tilde{a}$ for which $\varphi(a) = \varphi(\tilde{a}) = w$, and that $a, \tilde{a}$ may be decomposed as follows: $a = \bigcup_{i=1}^{n} a_i$, $\tilde{a} = \bigcup_{k=1}^{m} \tilde{a}_k$, where $\varphi_a(a_i) = w_i$, $\varphi_{\tilde{a}}(\tilde{a}_k) = \tilde{w}_k$. If it follows that $a \cap \tilde{a} = \varnothing$ then the preparation procedures a and $\tilde{a}$ have nothing in common—it is not possible to prepare a single microsystem which could be considered to be prepared according to both a and $\tilde{a}$. The preparation procedures $a, \tilde{a}$ are mutually exclusive, even though a and $\tilde{a}$ result in the preparation of the same (!) ensemble $w = \varphi(a) = \varphi(\tilde{a})$.

Suppose that $a \cap \tilde{a} \neq \varnothing$; then there exists a common Boolean subring $\mathcal{Q}(a \cap \tilde{a})$ of $\mathcal{Q}(a)$ and $\mathcal{Q}(\tilde{a})$. We shall now try to express, in a mathematically idealized form, the fact that two preparators do not have such a "common part."

From the preparator $\Sigma \xrightarrow{W} \check{K}$ of an ensemble w we may easily obtain new preparators as follows: Let $[0, \eta]$ be an interval from Σ. For $\lambda = \mu(W(\eta), \mathbf{1})$ we find that $[0, \eta] \xrightarrow{\lambda^{-1}W} \check{K}$ is a preparator of the ensemble $w_0 = \lambda^{-1}W(\eta)$. We shall call the preparator obtained in this way the preparator canonically determined by $[0, \eta]$.

D 6.3. Let $\Sigma_1 \xrightarrow{W_1} \check{K}$ and $\Sigma_2 \xrightarrow{W_2} \check{K}$ be two preparators for the *same* ensemble w. These preparators are said to be *disjoint* if, for any pair of intervals $[0, \eta_1] \subset \Sigma_1$ and $[0, \eta_2] \subset \Sigma_2$ the following condition cannot be satisfied: Let $\lambda_1 = \mu(W_1(\eta_1), \mathbf{1})$, $\lambda_2 = \mu(W_2(\eta_2), \mathbf{1})$, then $\lambda_1^{-1}W_1(\eta_1) = \lambda_2^{-1}W_2(\eta_2)$ and the preparators canonically defined by $[0, \eta_1]$ and $[0, \eta_2]$ are coexistent.

Let Σ_1, Σ_2 denote the closures of $\mathcal{Q}(a)$ and $\mathcal{Q}(\tilde{a})$, respectively. Suppose that $\varphi(a) = \varphi(\tilde{a}) = w$, $W_1 = \varphi_a$, $W_2 = \varphi_{\tilde{a}}$. Then if $\mathcal{Q}(a) \xrightarrow{\varphi_a} \check{K}$ and $\mathcal{Q}(\tilde{a}) \xrightarrow{\varphi_{\tilde{a}}} \check{K}$ (more precisely, their completions) are disjoint, then we find that $a \cap \tilde{a} = \varnothing$ because the intervals $[\varnothing, a \cap \tilde{a}] \subset Q(a)$ and $[\varnothing, a \cap \tilde{a}] \subset Q(\tilde{a})$ canonically determine the same preparator, and therefore are (trivially) coexistent.

D 6.4. Let $\Sigma_1 \xrightarrow{W_1} \check{K}$ and $\Sigma_2 \xrightarrow{W_2} \check{K}$ be a pair of preparators. We say that they are *complementary* if any preparator which is more extensive than $\Sigma_1 \xrightarrow{W_1} \check{K}$ is disjoint from any preparator which is more extensive than $\Sigma_2 \xrightarrow{W_2} \check{K}$.

D 6.5. Two decompositions of an ensemble

$$w = \sum_{i=1}^{n} w_i = \sum_{k=1}^{m} \tilde{w}_k$$

are said to be *complementary* if each pair of preparators $\Sigma_1 \xrightarrow{W_1} \check{K}$, $\Sigma_2 \xrightarrow{W_2} \check{K}$ for which $w_i \in W_1\Sigma_1$, $\tilde{w}_k \in W_2\Sigma_2$ are complementary.

How may we determine whether two decompositions are complementary?

Th. 6.1. *Two decompositions of an ensemble*

$$w = \sum_{i=1}^{n} w_i = \sum_{k=1}^{m} \tilde{w}_k$$

are complementary only if, for each pair $w_i, \tilde{w}_k$ *the following condition is satisfied:*

$$w_0 \in \check{K}, \qquad w_0 < w_i, \qquad w_0 < \tilde{w}_k \Rightarrow w_0 = 0.$$

PROOF. We shall assume that the decompositions are not complementary. Then there exists a pair of preparators $\Sigma_1 \xrightarrow{W_1} \check{K}$, $\Sigma_2 \xrightarrow{W_2} \check{K}$ for which $w_i \in W_1\Sigma_1$, $\tilde{w}_k \in W_2\Sigma_2$ and a pair of coexistent intervals $[0, \eta_1] \subset \Sigma_1$, $[0, \eta_2] \subset \Sigma_2$. For $\sigma_i \in \Sigma_1$ satisfying $W_1(\sigma_i) = w_i$ and for $\tilde{\sigma}_k \in \Sigma_2$ satisfying $W_2(\tilde{\sigma}_k) = \tilde{w}_k$ we obtain the following diagram:

$$\begin{array}{ccccc} [0, \eta_1] & \xrightarrow{h_1} & \Sigma & \xleftarrow{h_2} & [0, \eta_2] \\ & {\scriptstyle \lambda_1^{-1}W_1} \searrow & \downarrow {\scriptstyle W} & \swarrow {\scriptstyle \lambda_2^{-1}W_2} & \\ & & \check{K} & & \end{array}$$

Let $\tau_i = h_1(\sigma_i \wedge \eta_1)$ and $\rho_k = h_2(\tilde{\sigma}_k \wedge \eta_2)$. From this diagram it follows that

$$Wh_1(\sigma_i \wedge \eta_1) = \lambda_1^{-1}W_1(\sigma_i \wedge \eta_1) = W(\tau_i),$$
$$Wh_2(\tilde{\sigma}_k \wedge \eta_2) = \lambda_2^{-1}W_2(\tilde{\sigma}_k \wedge \eta_2) = W(\rho_k).$$

From $\bigvee_i \sigma_i = \varepsilon$ it follows that $\bigvee_i (\sigma_i \wedge \eta_1) = \eta_1$ and we obtain $\bigvee_i h(\sigma_i \wedge \eta_1) = \bigvee_i \tau_i = h_1(\eta_1) = \varepsilon$. Similarly we obtain $\bigvee_k \rho_k = \varepsilon$ and find that $\bigvee_{i,k} (\tau_i \wedge \rho_k) = \varepsilon$. Thus we obtain

$$W(\varepsilon) = \lambda_1^{-1}W_1(\eta_1) = \lambda_2^{-1}W_2(\eta_2) = \sum_{i,k} W(\tau_i \wedge \rho_k).$$

Therefore we cannot have all $W(\tau_i \wedge \rho_k) = 0$. Suppose that $W(\tau_1 \wedge \rho_1) \neq 0$. Then we obtain $w_1 = W_1(\sigma_1) > W_1(\sigma_1 \wedge \eta_1) = \lambda_1 W(\tau_1) > \lambda_1 W(\tau_1 \wedge \rho_1)$ and similarly $\tilde{w}_1 > \lambda_2 W(\tau_1 \wedge \rho_1)$. Therefore there exists a $w_0 \in \check{K}$, $w_0 \neq 0$ for which $w_0 < w_1$, $w_0 < \tilde{w}_1$.

In order to prove the converse, we assume that there exists a $w_0 \in \check{K}$ for which $w_0 \neq 0$, $w_0 < w_1$, $w_0 < \tilde{w}_1$. We then introduce the following Boolean rings: Let Σ_1 denote the set of all subsets of a set of $(n + 1)$ elements which we shall denote by $0, 1, \ldots, n$. The one element subsets are the atoms of Σ_1. For these atoms $\alpha_0, \alpha_1, \ldots, \alpha_n$ we define W_1 as follows:

$$\begin{aligned} W_1(\alpha_0) &= w_0, \\ W_1(\alpha_1) &= w_1 - w_0, \\ W_1(\alpha_2) &= w_2, \\ &\vdots \\ W_1(\alpha_n) &= w_n. \end{aligned}$$

In the same way we define $\Sigma_2 \xrightarrow{W_2} \check{K}$ with atoms $\beta_0, \beta_1, \ldots, \beta_m$:

$$\begin{aligned} W_2(\beta_0) &= w_0, \\ W_2(\beta_1) &= \tilde{w}_1 - w_0, \\ W_2(\beta_2) &= \tilde{w}_2, \\ &\vdots \\ W_2(\beta_m) &= \tilde{w}_m. \end{aligned}$$

The interval $[0, \beta_0]$ is then isomorphic to the interval $[0, \alpha_0]$. Let Σ denote the Boolean ring consisting only of two elements, the zero and unit elements.

We then obtain the following diagram:

$$\begin{array}{ccccc} [0, \alpha_0] & \longrightarrow & \Sigma & \longleftarrow & [0, \beta_0] \\ & \searrow^{\lambda_1^{-1} W_1} & \downarrow W & {}^{\lambda_2^{-1} W_2}\swarrow & \\ & & \check{K} & & \end{array}$$

where $W(\varepsilon) = \lambda_{1,2}^{-1} w_0$ and $\lambda_{1,2} = \mu(w_0, \mathbf{1})$.

Let e_1, e_2 denote the supports of $w_1 \in \check{K}$ and $w_2 \in \check{K}$, respectively. If $e_1 \wedge e_2 = 0$ then there does not exist a $w_0 \in \check{K}$ such that $w_0 < w_1$ and $w_0 < w_2$, because, since $w_0 < w_1$ and $w_0 < w_2$ the relations $e_0 < e_1$ and $e_0 < e_2$ must hold for the support e_0 of w_0. The converse is not true in general: Even if $e_1 \wedge e_2 \neq 0$ it is possible that for $w_0 < w_1, w_0 < w_2, w_0 \in K$ that the condition $w_0 = 0$ holds. We shall now give a simple example: For a Hilbert space $\mathscr{H}$ and a complete orthonormal basis φ_ν in $\mathscr{H}$ let $w_1 = 3 \sum_{\nu=1}^{\infty} (1/4^\nu) P_{\varphi_\nu}, w_2 = P_\varphi$, where $\varphi = \sum_{\nu=1}^{\infty} (1/2^{\nu/2}) \varphi_\nu$.

Then the supports of w_1 and w_2 are $\mathbf{1}$ and P_φ, respectively, and we find that $\mathbf{1} \wedge P_\varphi = P_\varphi \neq 0$. A $w_0 \in \check{K}$ for which $w_0 < w_1$ and $w_0 < w_2$ must therefore have P_φ as support, that is, must have the form λP_φ for $\lambda > 0$. For $\lambda P_\varphi < w_1$ it follows that, for all $\psi \in \mathscr{H}$

$$\lambda \langle \psi, P_\varphi \psi \rangle = \lambda |\langle \varphi, \psi \rangle|^2 \leq \langle \psi, w_1 \psi \rangle.$$

In particular, for all φ_μ we must have

$$\lambda |\langle \varphi, \varphi_\mu \rangle|^2 = \lambda \frac{1}{2^\mu} \leq \langle \varphi_\mu, w_1 \varphi_\mu \rangle = \frac{3}{4^\mu},$$

that is, $\lambda \leq 3/2^\mu$ for all μ, that is, $\lambda = 0$.

Each $w \in K$ may be expressed as a mixture of the extreme points of K, a result which follows from the spectral decomposition theorem for operators of trace class (see AIV, §9 and §11). Thus it follows that

$$w = \sum_\nu w_\nu P_{\varphi_\nu}, \qquad w_\nu > 0, \tag{6.4}$$

where the φ_ν $(\nu = 1, \ldots)$ are pairwise orthonormal vectors from $\mathscr{H} = \bigcup_i \mathscr{H}_i$.

We will now show that any $w \in K$ for which the spectral representation (6.4) containing a sum of two P_{φ_ν} where φ_ν are from the same $\mathscr{H}_i$ has complementary decompositions. It suffices to show this for one Hilbert space only, that is, for $\mathscr{H} = \mathscr{H}_1$. We shall show that for $\mathscr{H} = \mathscr{H}_1$ a $w \in K(\mathscr{H})$ has complementary decompositions if w is not an extreme point of $K(\mathscr{H})$.

According to our previous discussion, two decompositions

$$w = \sum_\nu \alpha_\nu P_{\varphi_\nu} = \sum_\mu \beta_\mu P_{\psi_\mu} \tag{6.5}$$

are complementary if each P_{ψ_μ} is different from each P_{φ_ν}. Then $P_{\varphi_\nu} \wedge P_{\psi_\mu} = 0$ for all pairs ν, μ. We need only show that each w which is not an extreme point (that is $w \neq P_\varphi$) has two decompositions (6.5).

We shall first prove this result for the case in which the support of w is two-dimensional.

Therefore, let $w = \lambda P_\varphi + (1 - \lambda)P_\psi$, where $\varphi \perp \psi$ and $0 < \lambda < 1$. Let χ be another vector in the plane spanned by $\{\varphi, \psi\}$. We claim there exists another decomposition of the form:

$$w = \mu P_\chi + (1 - \mu)P_\eta$$

for some $0 < \mu < 1$ and $\eta \in \mathscr{H}$.

For $\chi = \varphi a + \psi b$, $|a|^2 + |b|^2 = 1$ we need only set

$$\mu = \frac{\lambda(1 - \lambda)}{\lambda|b|^2 + (1 - \lambda)|a|^2},$$

$$\eta = \frac{\varphi\lambda\bar{b} - \psi(1 - \lambda)\bar{a}}{[\lambda^2|b|^2 + (1 - \lambda^2)|a|^2]^{1/2}}.$$

Let $a \neq 0, b \neq 0$; then the decompositions

$$w = \lambda P_\varphi + (1 - \lambda)P_\psi = \mu P_\chi + (1 - \mu)P_\eta$$

are complementary.

If the support of w is three-dimensional, that is, $= \lambda_1 P_{\varphi_1} + \lambda_2 P_{\varphi_2} + \lambda_3 P_{\varphi_3}$ where $\langle \varphi_i, \varphi_k \rangle = 0$ for $i \neq k$, we may write

$$w = (\lambda_1 + \rho)\left(\frac{\lambda_1}{\lambda_1 + \rho}P_{\varphi_1} + \frac{\rho}{\lambda_1 + \rho}P_{\varphi_2}\right)$$

$$+ (\lambda_3 + \lambda_2 - \rho)\left(\frac{\lambda_2 - \rho}{\lambda_2 + \lambda_3 - \rho}P_{\varphi_2} + \frac{\lambda_3}{\lambda_3 + \lambda_2 - \rho}P_{\varphi_3}\right),$$

where $0 < \rho < \lambda_2$. We may then apply the result proven above to each of the ensembles

$$w_1 = \frac{\lambda_1}{\lambda_1 + \rho}P_{\varphi_1} + \frac{\rho}{\lambda_1 + \rho}P_{\varphi_2}, \quad w_2 = \frac{\lambda_2 - \rho}{\lambda_3 + \lambda_2 - \rho}P_{\varphi_2} + \frac{\lambda_3}{\lambda_3 + \lambda_2 - \rho}P_{\varphi_3}.$$

We then obtain four vectors, χ_1, η_1 in the plane spanned by $\{\varphi_1, \varphi_2\}$ and χ_2, η_2 in the plane spanned by $\{\varphi_2, \varphi_3\}$, and a decomposition of the form

$$w = \mu_1 P_{\chi_1} + \mu_1' P_{\eta_1} + \mu_2 P_{\chi_2} + \mu_2' P_{\eta_2}$$

which is complementary to the decomposition

$$w = \lambda_1 P_{\varphi_1} + \lambda_2 P_{\varphi_2} + \lambda_3 P_{\varphi_3},$$

Finally, if $w = \sum_\nu \lambda_\nu P_{\varphi_\nu}$, where $\langle \varphi_\nu, \varphi_\mu \rangle = 0$ for $\nu \neq \mu$, we may apply the above proof to the parts $\lambda_1 P_{\varphi_1} + \lambda_2 P_{\varphi_2}, \lambda_3 P_{\varphi_3} + \lambda_4 P_{\varphi_4}, \ldots$; if we obtain an odd number of terms in the sum, then we may apply the result obtained for the three-dimensional case to the final three terms.

Thus we find that complementary decompositions are not unusual.

The structure of complementary decompositions has lead to a continuing discussion about the relationship between epistomology (theory of cognition) and quantum theory. Especially many are inclined to reject quantum theory. It is easy to see that the arguments used to reject quantum theory depend on an impermissible identification of the preparation procedure $a \in \mathcal{Q}'$ with the ensemble $\varphi(a) \in K$. Such an automatic identification appears to be natural in the usual formulation of quantum mechanics because the usual formulation takes into account only the set K but not the sets M or $\mathcal{Q}$, nor the canonical mapping φ of $\mathcal{Q}'$ into K.

7 Realizations of Decompositions

If, for a preparator $\Sigma \xrightarrow{W} \check{K}$ there exists a preparation procedure $a \in \mathcal{Q}'$ and a homomorphism h of Σ into $\mathcal{Q}(a)$ such that the following diagram is commutative:

$$\begin{array}{ccc} \Sigma & \xrightarrow{h} & \mathcal{Q}(a) \\ & {\scriptstyle W}\searrow \quad \swarrow{\scriptstyle \varphi_a} & \\ & \check{K} & \end{array}$$

then we may identify Σ with the Boolean subring $h\Sigma$ of $\mathcal{Q}(a)$. If this is the case we may call the preparation procedure a (together with $\mathcal{Q}(a)$ and φ_a) a realization of the preparator $\Sigma \xrightarrow{W} \check{K}$. The requirement that to each preparator there exists a realization in this sense is too strong (see also §4). In complete analogy to §4 we now impose the following requirement:

APr. To each preparator $\Sigma \xrightarrow{W} \check{K}$ and each finite Boolean subring $\tilde{\Sigma}$ of Σ and each $\sigma(\mathcal{B}, \mathcal{D})$ neighborhood U of 0 there exists a $\mathcal{Q}(a) \xrightarrow{\varphi_a} \check{K}$ and a homomorphism $\tilde{\Sigma} \xrightarrow{h} \mathcal{Q}(a)$ such that $\varphi_a h(\sigma) - W(\sigma) \in U$ for all $\sigma \in \tilde{\Sigma}$. (For the space $\mathcal{D}$ see III, §3.)

APr means that we may realize (in physical approximation) each preparator and, therefore, each decomposition of a $w \in K$. In other words, it is "physically possible" to (approximately) realize each preparator. We note that APr does not, however, provide us with any information about how we may, in an actual situation, build the apparatus. Such information cannot be obtained from the theory presented here because it does not contain a mathematical description of the structure of the apparatus. In XVIII we shall take a number of small steps in the direction of the investigation of the construction problem, in which we shall consider "transpreparation" processes.

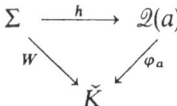

The problems concerning the maps ψ which were described in §4 are practically the same for the map $\varphi\colon \mathcal{Q}' \to K$. If the construction of a preparation apparatus is known to us, we still do not have any theoretical tools by which we may compute the elements $\varphi(a) \in K$ corresponding to the preparation procedures a. By analogy to the case of registration procedures (see §4) we may guess, on the basis of a known classical theory, together with the use of a correspondence principle, the ensemble $w = \varphi(a)$ and add this result formally as an axiom. At first we have no other choice, even though this procedure may not be very satisfactory (see, for example, XI–XVI).

The development of experimental methods for preparation and registration has up to now, taken place in a manner similar to that described above using classical theories—statistical theories—and quantum mechanics. This first step allows us to "estimate" the $\varphi(a)$ and $\psi(b_0, b)$. Then, by varying the experiment, that is, by using different combinations of preparation and registration procedures, we seek to improve the values of $\varphi(a)$ and $\psi(b_0, b)$. Therefore, in physics it is not the case that there are *no* adequate means of determining the functional relationships between the preparation and registration apparatuses and their corresponding $\varphi(a)$ and $\psi(b_0, b)$. The only failure is that there is no comprehensive and systematic theory of the macroscopic preparation and measurement apparatuses.

In XVII we shall show that there is at least one realistic theoretical route by which, using measurement collisions, measurement transformations, transpreparations, and starting with poorly known values of $\varphi(a)$ before collision and $\psi(b_0, b)$ after collision we may obtain theoretically computed and very precise values of $\varphi(\tilde{a})$ and $\psi(\tilde{b}_0, \tilde{b})$ for new preparation procedures $\tilde{a}$ or new effect processes $(\tilde{b}_0, \tilde{b})$, respectively. It may be that the method described in XVII for the "improvement" is the only "practical" possibility for experimentation, and that the "desired" comprehensive theory (see [13], X) is more a requirement for epistomology (see XVIII).

8 Objective Properties and Pseudoproperties of Microsystems

In the literature of quantum mechanics the discussion about *properties* of microsystems, about *propositions* about microsystems, and about problems of logic in connection with such propositions about microsystems has taken on vast proportions. In this book we cannot attempt to provide an overview of all these topics. Here we shall only discuss the problem about the properties of microsystems in terms of the formulation presented in III, §4.

8.1 Objective Properties of Microsystems and Superselection Rules

We shall now seek to describe the structure $\mathscr{E}_m$ of objective properties which was defined in III, D 4.1.6.

From III (4.1.11) it follows that, for each $g \in L$ together with the mapping T_p' (which is dual to T_p) of L into itself (from V, §4.1 we find that T_p is an operation; thus, for T_p' it follows that V, Th. 4.1.3 holds):

$$g = T_p' g + T_{M\setminus p}' g. \tag{8.1.1}$$

From III (4.1.13) it follows that

$$\chi(p) = T_p'\mathbf{1} = T_p'g + T_p'(\mathbf{1} - g). \tag{8.1.2}$$

In addition we find that

$$\begin{aligned} T_p'g + T_{M\setminus p}'g + T_p'(\mathbf{1} - g) &= T_{M\setminus p}'g + T_p'\mathbf{1} \\ &\leq T_{M\setminus p}'\mathbf{1} + T_{p'}\mathbf{1} = \mathbf{1}. \end{aligned} \tag{8.1.3}$$

From (8.1.1), (8.1.2), (8.1.3), and Th. 1.2.4 it follows that g and $\chi(p)$ are coexistent. Therefore $\chi(p)$ is coexistent with each $g \in L$, and is therefore coexist with each $e \in G$. Thus, according to Th. 1.3.4 $\chi(p)$ commutes with each $e \in G$. Therefore $\chi(p)$ has the following form:

$$T_p'\mathbf{1} = \chi(p) = (\lambda_1\mathbf{1}, \lambda_2\mathbf{1}, \ldots). \tag{8.1.4}$$

Since $T_p'g + T_{M\setminus p}'g = g$, for each $g \in L$ we obtain $T_p'g \leq g$ and find that

$$T_p'(0, 0, \ldots, g_i, \ldots) = (0, 0, \ldots, g_i', 0, \ldots),$$

where $g_i' \leq g_i$. Since T_p' is linear, from the previous result and (8.1.4) it follows that

$$\mathrm{T}_p'(0, 0, \ldots, \mathbf{1}, \ldots) = (0, 0, \ldots, \lambda_i\mathbf{1}, 0, \ldots). \tag{8.1.5}$$

From which we find that

$$T_p'^2(0, 0, \ldots, \mathbf{1}, \ldots) = (0, 0, \ldots, \lambda_i^2\mathbf{1}, \ldots)$$

and (8.1.6)

$$T_p'^2\mathbf{1} = (\lambda_1^2\mathbf{1}, \ldots, \lambda_i^2\mathbf{1}, \ldots).$$

From III (4.1.11) we obtain the following special case: $T_p^2 = T_{p\cap p} = T_p$; from this result, and from (8.1.6) and (8.1.4) it follows that either $\lambda_i = 0$ or 1. Therefore $\chi(p)$ is an element of the center Z (see Th. 1.3.8) of G. The map $\mathscr{E}_m \xrightarrow{\chi} L$ therefore maps into Z as follows:

$$\mathscr{E}_m \xrightarrow{\chi} Z. \tag{8.1.7}$$

Z is a complete Boolean ring. χ must then be an isomorphism of $\mathscr{E}_m$ onto a Boolean subring of Z since the following condition is satisfied: if $p \neq 0$ then we must have $\chi(p) \neq 0$, since if $\lambda_{\mathscr{Q}}(a, a \cap p) = \mu(\varphi(a), \chi(p)) = 0$ we would obtain $a \cap p = \varnothing$ for all $a \in \mathscr{Q}'$ and therefore $p \cap \bigcup_{a \in \mathscr{Q}} a = \varnothing$, and, from APS 8.1 we would obtain $p = \varnothing$.

It is a certain hypothesis in the sense of [1], §10.1 that the map (8.1.7) is surjective. Since we do not wish to discuss the hypotheses which are described in [1], §10.1, we shall formulate, as an axiom, the condition that (8.1.7) is surjective. Then it follows that χ is an isomorphism between $\mathscr{E}_m$ and Z.

According to III (4.1.8), for $p \in \mathscr{E}_m$ we obtain

$$p = \bigcup_{\substack{a \in \mathscr{Q} \\ a \subset p}} a.$$

From $a \subset p$ it follows that $\lambda_{\mathscr{Q}}(a, a \cap p) = \lambda_{\mathscr{Q}}(a, a) = 1$; conversely, from $\lambda_{\mathscr{Q}}(a, a \cap p) = 1$, it follows that $\lambda_{\mathscr{Q}}(a, a \cap (M \setminus p)) = 0$ since $\lambda_{\mathscr{Q}}(a, a \cap p) + \lambda_{\mathscr{Q}}(a, a \cap (M \setminus p)) = 1$. Thus it follows that $a \cap (M \setminus p) = \varnothing$ and therefore $a \subset p$, where the latter is equivalent to the condition that $\mu(\varphi(a), \chi(p)) = 1$, that is, $\varphi(a) \in K_1(\chi(p))$. Therefore, for each $e \in Z$ an objective property is defined by

$$e \to p(e) = \bigcup_{\substack{a \in \mathscr{Q} \\ \varphi(a) \in K_1(e)}} a, \tag{8.1.8}$$

where

$$\chi(p(e)) = e.$$

Thus we find that (8.1.8) is the inverse mapping of (8.1.7).

Thus it is understandable that we may sometimes say (somewhat incorrectly) that Z itself is a collection of objective properties of microsystems.

Since $\mu(w_1, e) = \mu(w_2, e)$ for all $e \in Z$ does not imply that $w_1 = w_2$, it follows that the microsystems (as described in terms of III, D 4.1.7) are *not* physical objects.

Of special importance are the atoms of the center Z (see the end of §1.3), because we could always assume that each microsystem x has one and only one property which is characterized by an atom of Z. Let Z_A denote the set of atoms of Z. Then we obtain

$$M = \bigcup_{e \in Z_A} p(e), \tag{8.1.9}$$

where $p(e)$ is defined by (8.1.8).
If $e_1, e_2 \in Z_A$ and if $e_1 \neq e_2$ we then find that

$$p(e_1) \cap p(e_2) = \varnothing. \tag{8.1.10}$$

In physics it is customary to use "names" to denote objective properties $p(e)$, $e \in Z_A$—such as "$p(e_1)$ is the set of electrons," "$p(e_2)$ is the set of hydrogen atoms" or "$p(e_3)$ is the set of helium atoms," etc. The introduction of these "names" requires a more extensive characterization of the individual elements of Z_A than we have previously seen, that is, there must be additional axioms which are needed to describe how (that is, by means of which apparatus) we may produce or prepare the individual $e \in Z_A$. An experimental physicist can easily provide us with several apparatuses by which we may "produce" or detect, for example, electrons or hydrogen atoms.

Although we are unable, at this stage of the development of the theory, to assign characteristic names to the atoms of the center Z_A, we now find it desirable to introduce the concept of a system type.

D 8.1.1. The elements $e \in Z_A$ are called *system types*; for $e \in Z_A$ we shall call the set $p(e)$ the set of systems of type e.

Using (8.1.9) and (8.1.10), for each $a \in \mathcal{Q}'$ we obtain the following decomposition:

$$a = \bigcup_{e \in Z_A} (a \cap p(e)). \tag{8.1.11}$$

Thus we obtain

$$\varphi(a) = \sum_e \lambda(e)\varphi(a \cap p(e)), \tag{8.1.12}$$

where $\lambda(e) = \operatorname{tr}(\varphi(a)e)$. Equation (8.1.12) is nothing other than a decomposition of each $w \in K$ into components with respect to the different $\mathcal{H}_\nu$ as follows: Let $w = (W_1, W_2, \ldots)$ and $w_\nu = (0, 0, \ldots, W_\nu, \ldots)$. Then $w = \sum_\nu w_\nu$. This composition of w is uniquely determined by the condition $w_\nu \operatorname{tr}(w_\nu)^{-1} \in K_1(e_\nu)$. This decomposition is coexistent with all other decompositions of w. Thus it is understandable that in physics we are, in most cases, concerned only with the individual system types. This is evident from the fact that each registration procedure b may be decomposed as follows

$$b = \bigcup_{e \in Z_A} (b \wedge p(e)) \tag{8.1.13}$$

from which it follows that

$$\psi(b_0, b) = \sum_{e \in Z_A} \psi(b_0, b \cap p(e)). \tag{8.1.14}$$

Equation (8.1.14) is nothing other than a decomposition of each effect $F = (F_1, F_2, \ldots)$ into its components $F_\nu \leq e_\nu$. Here, for a given system type only the probability $\operatorname{tr}(W_\nu F_\nu)$ is of interest.

The set $\mathcal{E}_m$ of objective properties, such as the set Z_A of system types are clearly related to the concept of "super selection rules." We may (in a formal mathematical sense) construct a Hilbert space $\mathcal{H} = \sum_i \oplus \mathcal{H}_i$ from a direct sum of the individual Hilbert spaces $\mathcal{H}_i$, where the latter, according to AIV, §15, determine the algebra $\mathcal{A}(\mathcal{H}_1, \mathcal{H}_2, \ldots)$. Then $\mathcal{A}(\mathcal{H}_1, \mathcal{H}_2, \ldots)$ will be identified with a subalgebra of $\mathcal{L}(\mathcal{H})$. This subalgebra is characterized by the condition that all A in $\mathcal{A}(\mathcal{H}_1, \mathcal{H}_2, \ldots)$ commute with the projections P_i onto the individual subspaces $\mathcal{H}_i$ of H (see AIV, §15). The P_i are, however, the atoms of the center Z. Each P_i is then a "super selection rule" since the P_i commute with all "actual" decision observables in $\mathcal{A}(\mathcal{H}_1, \mathcal{H}_2, \ldots)$ and are therefore, in all circumstances, invariant quantities. Invariant "under all circumstances" is only an imprecise formulation of what we have previously called "objective properties."

In closing we shall now ask whether there exists a set $\overline{\mathcal{E}} = \overline{\mathcal{E}}_p \cap \mathcal{E}_r$ (as defined at the end of III, §4.1) which is so large that the equivalence relation on $\mathcal{Q}'$ defined by $a_1 \approx a_2$: $\{\lambda_{\bar{\mathcal{Q}}}(a_1, a_1 \cap p) = \lambda_{\bar{\mathcal{Q}}}(a_2, a_2 \cap p)$ for all $p \in \overline{\mathcal{E}}\}$ is finer than that defined by the $f \in \mathcal{F}$. From [1], §12.3 it follows that such a set does not exist, because we may apply the results of that section upon $\overline{\mathcal{Q}}$, $\mathcal{R}_0$, $\mathcal{R}$ and obtain the following result.

There exists a complete Boolean ring Σ and a pair of maps $\bar{\mathscr{Q}} \xrightarrow{\bar{\varphi}} K(\Sigma)$ and $\mathscr{F} \xrightarrow{\bar{\psi}} L(\Sigma)$ where $\bar{\varphi}\bar{\mathscr{Q}}$ is dense in $K(\Sigma)$ and $\bar{\psi}\mathscr{F}$ is dense in $L(\Sigma)$. Since $\bar{\mathscr{Q}} \supset \mathscr{Q}$, a σ-continuous and surjective map $L(\Sigma) \rightarrow L$ is defined by the maps $\mathscr{F} \xrightarrow{\bar{\psi}} L(\Sigma)$ and $\mathscr{F} \xrightarrow{\psi} L$.

According to §2 this would mean that *all* effects would be coexistent—in particular, G would be a Boolean ring—in contradiction to our axioms for microsystems.

8.2 Pseudoproperties of Microsystems

In §8.1 we have found that microsystems are not micro-objects—that is, they do not have a sufficient set of objective properties. We shall now consider the analogous question concerning the physically realizable pseudoproperties of microsystems (see III, D 4.2.1). Let $\mathscr{E}_{ps}$ denote the set of physically realizable pseudoproperties. For $p \in \mathscr{E}_{ps}$ we obtain the following result from III, §4.2:

Let $K_0 L_0(A_p) = K_1(e) = K_0(e^{\perp})$ where $e \in G$, and let

$$K_0 L_0(A_{p^*}) = K_1(e^*) = K_0(e^{*\perp}) \quad \text{where } e^* \in G.$$

Then, according to III (4.2.8) there exists a $g \in L_0(A_{p^*})$ and $g \in L_1(A_p)$ for which

$$g \leq e^{*\perp} = \mathbf{1} - e^* \quad \text{and} \quad g \geq e. \tag{8.2.1}$$

From which it follows that

$$e \leq e^{*\perp} \quad \text{that is } e \perp e^* \quad \text{and} \quad e + e^* \leq \mathbf{1}. \tag{8.2.2}$$

From III (4.2.9) it follows that

$$\begin{aligned}\varnothing &= L_0(A_p) \cap L_0(A_{p^*}) = L_0 K_0(e^{*\perp}) \cap L_0 K_0(e^{\perp}) \\ &= L_0 K_0(e^{*\perp} \wedge e^{\perp})\end{aligned}$$

and therefore $e^{*\perp} \wedge e^{\perp} = 0$, that is,

$$e^* \vee e = \mathbf{1}. \tag{8.2.3}$$

From (8.2.2) and (8.2.3) we obtain

$$e + e^* = \mathbf{1} \quad \text{that is} \quad e^* = e^{\perp}. \tag{8.2.4}$$

From (8.2.1) it follows that there exists a particular $b_0 \in \mathscr{R}_0$ such that

$$g = e = \psi(b_0, b_0 \cap p_r).$$

From $A_p \subset K_1(e)$ it follows that for all $a \subset p$

$$\varphi(a) \in K_1(e). \tag{8.2.5}$$

From $\psi(b_0, b) \in L_0(A_{p^*})$ it follows that, for all $b \subset p$

$$\psi(b_0, b) \leq e^{*\perp} = e. \tag{8.2.6}$$

Conversely, if $\varphi(a) \in K_1(e)$, then it follows that

$$0 = \mu(\varphi(a), e^{\perp}) \geq \mu(\varphi(a), \psi(b_0, b))$$

for all $b \subset p^*$. Therefore we obtain

$$\lambda_{\mathscr{S}}(a \cap b_0, a \cap b) = 0$$

and

$$a \cap b = \varnothing \quad \text{for all } b \subset p^* \quad \text{that is} \quad a \cap (p^*)_r = \varnothing. \tag{8.2.7}$$

Suppose that $a \cap (p^*)_p \neq \varnothing$; then there exists an $\tilde{a} \subset p^*$ such that $a \cap \tilde{a} \neq \varnothing$. For $a' = a \cap \tilde{a}$, since $a' \subset a$ we clearly obtain $\varphi(a') \in K_1(e)$. Since $a' \subset \tilde{a} \subset p^*$ we obtain

$$\varphi(a') \in A_{p^*} \subset K_0(e^{*\perp}) = K_0(e).$$

Thus we are led to a contradiction, from which it follows that

$$a \cap (p^*)_p = \varnothing. \tag{8.2.8}$$

From (8.2.7) and (8.2.8) it follows that $a \cap p^* = \varnothing$, that is, $a \subset M \backslash p^*$ and therefore $a \subset p$.

Therefore we may strengthen (8.2.5) as follows:

$$a \subset p \Leftrightarrow \varphi(a) \in K_1(e). \tag{8.2.9}$$

Suppose that $\psi(b_0, b) \leq e$. Then, for $\varphi(a) \in A_{p^*} \subset K_0(e)$ we obtain

$$0 = \mu(\varphi(a), \psi(b_0, b)) = \lambda_{\mathscr{S}}(a \cap b_0, a \cap b)$$

that is $a \cap b = \varnothing$ for all $a \subset p^*$, from which it follows that

$$b \cap (p^*)_p = \varnothing. \tag{8.2.10}$$

Suppose that $b \cap (p^*)_r \neq \varnothing$; then there would be a $\tilde{b} \subset p^*$ for which $b' = \tilde{b} \cap p \neq \varnothing$. Since $b' \subset b$ we obtain $\psi(b_0, b') \leq \psi(b_0, b) \leq e$. Since $b' \subset \tilde{b}$ there exists a $\tilde{b}_0$ for which $\psi(\tilde{b}_0, b') \leq \psi(\tilde{b}_0, \tilde{b}) \in L_0(A_p) = L_0 K_0(e^{\perp})$ and we therefore obtain $\psi(\tilde{b}_0, b') \leq \psi(\tilde{b}_0, b) \leq e^{\perp}$. From $\psi(b_0, b') \leq e$ and $\psi(b_0, b') \leq e^{\perp}$ it follows that $\psi(b_0, b') = 0$ in contradiction to $b' \neq \varnothing$. Therefore we obtain

$$b \cap (p^*)_r = \varnothing. \tag{8.2.11}$$

From (8.2.10) and (8.2.11) it follows that $b \cap p^* = \varnothing$, that is, $b \subset M \backslash p^*$ and we obtain $b \subset p$.

Therefore we may strengthen (8.2.6) as follows:

$$b \subset p \Leftrightarrow \psi(b_0, b) \leq e. \tag{8.2.12}$$

From (8.2.9) it follows that

$$p_r = \bigcup_{b \in \mathscr{R}_e} b \tag{8.2.13}$$

where

$$\mathscr{R}_e = \{b \mid b \in \mathscr{R} \text{ and there exists a } b_0 \in \mathscr{R}_0 \text{ for which } \psi(b_0, b) \leq e\}.$$

From (8.2.12) we obtain

$$p_p = \bigcup_{a \in \mathcal{Q}_e} a \quad \text{where } \mathcal{Q}_e = \{a \mid a \in \mathcal{Q}' \text{ and } \varphi(a) \in K_1(e)\}. \tag{8.2.14}$$

Finally we obtain

$$p = \left[\bigcup_{a \in \mathcal{Q}_e} a\right] \cup \left[\bigcup_{b \in \mathcal{R}_e} b\right]. \tag{8.2.15}$$

To each physically realizable pseudoproperty p there exists a corresponding $e \in G$ which is obtained from the equation $K_0 L_0(A_p) = K_1(e)$ and defines a map $\mathscr{E}_{ps} \to G$. This map is injective, because the image $e \in G$ of p is, according to (8.2.15), uniquely determined.

We will now show that, for an arbitrary $e \in G$, the corresponding p (which we shall denote by $p(e)$) given by (8.2.15) is an element of Π (where the latter is defined by III, §4.2).

By using the method of proof of (8.2.7)–(8.2.11) it may be shown that, if $a \in \mathcal{Q}_e$, then, for all $\tilde{a} \in \mathcal{Q}_{e^\perp}$ and all $\tilde{b} \in \mathcal{R}_{e^\perp}$ then it follows that $a \cap \tilde{a} = \varnothing$, $a \cap \tilde{b} = \varnothing$, that is, $a \cap p(e^\perp) = \varnothing$. Similarly, for $b \in \mathcal{R}_e$ it follows that $b \cap p(e^\perp) = \varnothing$. Therefore $p(e) \cap p(e^\perp) = \varnothing$. This is equivalent to $p(e^\perp) \subset p(e)^*$ where $p(e)^* = \pi(M \backslash p(e))$ where $\pi(c)$ is defined by III (4.2.2).

From $\tilde{a} \subset p(e)$ it follows that $\tilde{a} \cap p(e^\perp) = \varnothing$ and hence $\tilde{a} \cap b = \varnothing$ for all $b \in \mathcal{R}_{e^\perp}$, that is, $\mu(\varphi(\tilde{a}), \psi(b_0, b)) = 0$ for all $\psi(b_0, b) \leq e^\perp$. If

$$\sup_{b \in \mathcal{R}_{e^\perp}} \psi(b_0, b) = e^\perp \tag{8.2.16}$$

then it follows that $\mu(\varphi(\tilde{a}), e^\perp) = 0$ and therefore $\tilde{a} \in Q_e$.

From $\tilde{b} \subset p(e)$ it follows that $\tilde{b} \cap p(e^\perp) = \varnothing$ and hence $\tilde{b} \cap a = \varnothing$ for all $a \in Q_{e^\perp}$, that is, $\mu(\varphi(a), \psi(\tilde{b}_0, \tilde{b})) = 0$ for all $a \in Q_{e^\perp}$.
Let $A_e = \{\varphi(a) \mid a \in \mathcal{Q}', \varphi(a) \in K_1(e)\}$; if

$$L_0(A_{e^\perp}) = L_0 K_1(e^\perp) = L_0 K_0(e), \tag{8.2.17}$$

that is, if the face generated by A_e is equal to $K_1(e)$ then it follows that $\psi(\tilde{b}_0, \tilde{b}) \in L_0 K_0(e)$ and hence $\psi(\tilde{b}_0, \tilde{b}) \leq e$, that is, $\tilde{b} \in \mathcal{R}_e$.

Thus we obtain

$$p(e) = \left[\bigcup_{\substack{a \in \mathcal{Q} \\ a \subset p(e)}} a\right] \cup \left[\bigcup_{\substack{b \in \mathcal{R} \\ b \subset p(e)}} b\right]$$

that is $p(e)$ has the form III (4.2.2).

At the same time we have also proven that if $\tilde{a} \subset M \backslash p(e^\perp)$ and $\tilde{b} \subset M \backslash p(e^\perp)$ it also follows that $\tilde{a} \subset p(e)$ and $\tilde{b} \subset p(e)$. Thus, from $p(e^\perp) \subset p(e)^*$ (see above) and from $\tilde{a} \subset M \backslash p(e)^*$, $\tilde{b} \subset p(e)^*$ we also obtain $\tilde{a} \subset p(e)$ and $\tilde{b} \subset p(e)$, that is, $p(e) = p(e)^{**}$. Therefore $p(e)$ is an element of Π. Furthermore it follows that $p(e)^* = p(e^\perp)$ since (8.2.16) and (8.2.17) are also satisfied if we replace e by $e^\perp$.

From $K_1(e_1) \cap K_1(e_2) = K_1(e_1 \wedge e_2)$ it follows that $\mathcal{Q}_{e_1} \cap \mathcal{Q}_{e_2} = \mathcal{Q}_{e_1 \wedge e_2}$. From $g \leq e_1$ and $g \leq e_2$ it follows that $g \leq e_1 \wedge e_2$ and $\mathcal{R}_{e_1} \cap \mathcal{R}_{e_2} = \mathcal{R}_{e_1 \wedge e_2}$, Thus $p(e_1) \wedge p(e_2) = p(e_1 \wedge e_2)$.

Let $e \in G$; then condition (8.2.16), that is,

$$\sup_{b \in \mathscr{R}_e} \psi(b_0, b) = e$$

is satisfied, if (in correspondence to the assumption from III, D 4.2.1) to e there exists a $b_0 \in \mathscr{R}_0$ for which

$$\psi(b_0, b_0 \cap p(e)_r) = e. \tag{8.2.18}$$

The condition (8.2.17) is satisfied for an $e \in G$, that is, $L_0(A_e) = L_0 K_1(e)$, if there exists an $a \in Q'$ for which

$$C(\varphi(a)) = K_1(e), \tag{8.2.19}$$

where $C(\varphi(a))$ is the face generated by $\varphi(a)$ (see also III, §3).

If the elements e of an orthocomplemented sublattice G_p of G satisfy conditions (8.2.18) and (8.2.19) then $\mathscr{E}_{ps} = \{p(e) \mid e \in G_p\}$ is an orthocomplemented sublattice of Π which is isomorphic to G_p. It is a certain hypothesis that such a sublattice G_p of G exists which is $\sigma(\mathscr{B}', \mathscr{B})$ dense in G.

Since, as in §8.1, we shall not discuss certain hypotheses, here we shall assert the existence of such a sublattice G_p as an axiom.

For such a G_p, $\mathscr{E}_{ps} = \{p(e) \mid e \in G\}$ is a sufficient system (see III, D 4.2.2) of physically realizable pseudoproperties.

From (8.1.18) and $\chi(M \backslash p(e)) = p(e^\perp)$ (where χ is defined by §8.1) it follows that we may always assume that $Z \subset G_p$. We therefore assume that $Z \subset G_p$. Thus we obtain

$$\mathscr{E}_m = \{p(e) \mid e \in Z\} \subset \mathscr{E}_{ps}.$$

We may characterize $\mathscr{E}_m$ as a subset of $\mathscr{E}_{ps}$ as follows:

$$\mathscr{E}_m = \{p \mid p \in E_{ps} \text{ and } p^* = M \backslash p\}.$$

Proof. For $e \in Z$, with $p(e)$ defined by (8.1.8), we find that $\chi(M \backslash p(e)) = p(e^\perp)$ and therefore $p(e)^* = M \backslash p(e)$. Conversely, suppose that $p(e) \in \mathscr{E}_{ps}$ and $p(e)^* = M \backslash p(e)$, but that $e \in Z$ is not satisfied. Since G_p is dense in G there exists an $\tilde{e} \in G_p$ which is not commensurable with e. From $(\tilde{e} \wedge e) \vee (\tilde{e} \wedge e^\perp) \leq \tilde{e}$ and since e and $\tilde{e}$ are not commensurable, it follows that $e_1 = \tilde{e} - [(\tilde{e} \wedge e) \vee (\tilde{e} \wedge e^\perp)] \neq \varnothing$. Since G_p is an orthocomplemented lattice we also obtain $e_1 \in G_p$. In addition, we find that $e_1 \wedge e = 0$ and $e_1 \wedge e^\perp = 0$.

Since $(p(e_1) \wedge p(e))_r = p(e_1)_r \cap p(e)_r$ and $p(e_1) \wedge p(e) = p(e_1 \wedge e) = p(0) = \varnothing$ we therefore find that $p(e_1)_r \cap p(e)_r = \varnothing$. Similarly we obtain $p(e_1)_r \cap p(e^\perp)_r = \varnothing$, $p(e_1)_p \cap p(e)_p = \varnothing$ and $p(e_1)_p \cap p(e^\perp)_p = \varnothing$. Therefore we finally obtain $p(e_1) \cap p(e) = \varnothing$ and $p(e_1) \cap p(e^\perp) = \varnothing$. Since $p(e^\perp) = p(e)^* = M \backslash p(e)$ it follows that $p(e_1) = \varnothing$ in contradiction to $e_1 \neq 0$. Therefore, for $e \notin Z$ we find that $p(e)^* \neq M \backslash p(e)$.

Since G_p is dense in G, in addition to designating the set $\mathscr{E}_{ps} = \{p(e) \mid e \in G_p\}$ as a set of pseudoproperties of microsystems, we shall also refer to G itself (somewhat imprecisely) as a set of pseudoproperties. The last designation is, however, often misunderstood. We have introduced the set $\mathscr{E}_{ps}$ in order to be more precise. We may reduce difficulty in interpretation if we translate "$x \in p$" for $p \in \mathscr{E}_{ps}$ into normal language by the expression "x has the pseudoproperty p."

8.3 Logic of Decision Effects?

Now that we have discussed the meaning of decision effects in various circumstances we shall now briefly discuss a number of expressions which frequently lead to misunderstandings in quantum mechanics.

Since the set G is, in some respects, analogous to the set of properties of a classical system, it is common to refer to an element $e \in G$ as a property (and not more precisely as a pseudoproperty). Since we are not accustomed to describe individual microsystems mathematically (as we do in this book) we frequently express the above assertion in ordinary language as follows: "a particular microsystem has the property e." In seeking to give such statements a verifiable meaning it was recognized that if the assertion "the microsystem has the property e_1 and has the property e_2" is replaced by the assertion "the microsystem has the property $e_1 \wedge e_2$" then doubts about the validity of the usual two-valued logic arise. A similar case exists if the negation of the proposition "the microsystem has the property e" is replaced by the proposition "the microsystem has the property $e^\perp$." In this way a nonstandard logic of propositions was derived in which the logical operations are (in a sense) parallel to the lattice operations in G. This parallelism may be expressed as follows:

$$\begin{aligned} \text{and} &\leftrightarrow \wedge \\ \text{or} &\leftrightarrow \vee, \\ \text{not} &\leftrightarrow \perp. \end{aligned} \tag{8.3.1}$$

The lattice G is often called a quantum logic—even in the case in which the relationships (8.3.1) are not strictly required or "believed."

In the discussion of propositions concerning the properties of microsystems an alternative possibility proceeds from the idea that, in quantum mechanics, it is not possible to formulate so-called "objective" propositions of the form "the microsystem has the property e." Instead, it is suggested that we may only formulate "subjective" propositions, such as, "I know that the microsystem under consideration has the property e." Here there exists two different types of negation: "I do not know that the microsystem has the property e" and "I know that the microsystem has the property $e^\perp$." Here "I do not know . . ." can be considered to be an imprecise form of the proposition "I do not for certain know whether the microsystem in question has the property e or $e^\perp$; there is a probability α (possibly subjective) that the microsystem has the property e (and $1 - \alpha$ for the property $e^\perp$). In this way attempts have been made to develop a "probability logic."

The notion of a "probability logic" will not be discussed here (see, for example, [16] and [9]). Instead, we shall continue the development of the fundamental description of individual microsystems in mathematical terms, terms which are in some correspondence to the more intuitive ideas formulated above.

Instead of G we shall consider the dense subset G_p of G which was introduced at the end of §8.2, and the corresponding set of physically realizable pseudoproperties $\mathscr{E}_{ps}$ together with the isomorphic map $e \longrightarrow p(e)$.

Let $e \in G_p$; we shall now express the relationship

$$x \in p(e) \tag{8.3.2}$$

in ordinary language as follows: "The microsystem x has the pseudoproperty e." Here it is important to note that the ordinary language formulation should not be construed to mean *anything other* than the relation described by (8.3.2). In this formulation (8.3.2) is primarily a relation in the mathematical description of a *physical* theory, that is, it has a *physical interpretation*. We should not, however, make the mistake of using the ordinary language description of (8.3.2) as an alternative interpretation in *addition* to that already given in II.

We shall now proceed as follows: According to the methodology presented in [1] certain relationships in a mathematical theory $\mathscr{MT}$ (as a part of a physical theory $\mathscr{PT}$) may be considered to be a representation of real physical facts. Here it is important to emphasize that, in *addition* to the mathematical formulation of $\mathscr{MT}$ there does not exist another type of "proposition" formulated in $\mathscr{PT}$. The mathematical formulation (for example, (8.3.2)) is, on the basis of the physical interpretation of the fundamental sets—for example, $\mathscr{Q}, \mathscr{R}, \mathscr{R}_0$ and the real function $\lambda_{\mathscr{S}}$ in II—considered to be an assertion about reality. Thus the logic used in $\mathscr{PT}$ is only that of the mathematical theory. In this way we obtain, in a natural way, certain mathematical assertions of a real character (see [1], §10 or [2], III, §9). We shall now describe this situation using (8.3.2) without making use of the general formulation presented in [1], §10.

In the mathematical framework we cannot assign the "values" *true* or *false* to the relation (8.3.2). The meaning and importance of relations such as (8.3.2) in a mathematical theory is much more complicated and requires a more precise analysis. Here we shall impose all of the axioms previously introduced and also those introduced in subsequent chapters (for example, VII).

We shall begin by providing a logical analysis of relations of the form (8.3.2). Here, by the expression "logical analysis" we mean an analysis in the sense of a mathematical theory.

We may logically "combine" two relations of the form (8.3.2) by means of a logical conjunction "and" as follows:

$$x \in p(e_1) \quad \text{and} \quad x \in p(e_2), \tag{8.3.3}$$

where (8.3.3) is equivalent to the relation

$$x \in p(e_1) \cap p(e_2). \tag{8.3.4}$$

We note that, in general

$$p(e_1) \cap p(e_2) \neq p(e_1 \wedge e_2). \tag{8.3.5}$$

We note that

$$p(e) = p(e)_p \cup p(e)_r$$

and

$$\begin{aligned} p(e_1) \cap p(e_2) &= (p(e_1)_p \cup p(e_1)_r) \cap (p(e_2)_p \cup p(e_2)_r) \\ &= (p(e_1)_p \cap p(e_2)_p) \cup (p(e_1)_p \cap p(e_2)_r) \\ &\quad \cup (p(e_1)_r \cap p(e_2)_p) \cup (p(e_1)_r \cap p(e_2)_r) \\ &= p(e_1 \wedge e_2)_p \cup p(e_1 \wedge e_2)_r \\ &\quad \cup (p(e_1)_p \cap p(e_2)_r) \cup (p(e_1)_r \cap p(e_2)_p) \\ &= p(e_1 \wedge e_2) \cup (p(e_1)_p \cap p(e_2)_r) \cup (p(e_1)_r \cap p(e_2)_p). \end{aligned}$$

For the special case in which $e_1 \wedge e_2 = 0$ we find that

$$p(e_1) \cap p(e_2) = (p(e_1)_p \cap p(e_2)_r) \cup (p(e_1)_r \cap p(e_2)_p)$$

need not be empty! For example, $p(e_1)_p \cap p(e_2)_r \neq \varnothing$ if there exists a preparation procedure a for which $\varphi(a) \in K_1(e_1)$ and $\mu(\varphi(a), e_2) \neq 0$ because, for $\psi(b_0, b_0 \cap p(e_2)_r) = e_2$ (see (8.2.18)) we must then have

$$\lambda_{\mathscr{S}}(a \cap b_0, a \cap b_0 \cap p(e_2)_r) \neq 0,$$

that is, $a \cap p(e_2)_r \neq \varnothing$.

For example, let e_1, e_2 be the following decision effects: e_1: the momentum lies within a compact region W of momentum space, e_2: the position lies within a compact region V of position space (see VII, §4). Then the set (8.3.4) is nonempty because it is possible to prepare microsystems from $p(e_1)_p$ (that is, with momentum in W) which may be registered according to $p(e_2)_r$ (with position in V). If we express $x \in p(e_1)$, $x \in p(e_2)$ as follows: "x has momentum in W" and "x has position in V," respectively, then (8.3.4) is equivalent to the logical conjunction (8.3.3) which says that "x has momentum in W and position in V." For sufficiently small domains W and V the latter statement may appear to contradict the Heisenberg uncertainty relation. We shall now find that this contradiction is only an apparent one.

The following objection is frequently made—particularly in the case of position and momentum: "after" the registration of the position (that is, for the elements of $p(e_2)_r$) the momentum has been changed. This is indeed the case. This fact is, however, irrelevant to the interpretation of quantum mechanics. Quantum mechanics makes assertions concerning the interaction between the preparation apparatus and the registration apparatus which results from microsystems. Therefore all such assertions are concerned with the microsystems "between" preparation and registration; here (8.3.3) represents an assertion which is both correct and important. (In XVII, §4 we shall examine the "trans-preparation" process and obtain a number of conclusions concerning the "passage" of microsystems through the registration apparatus. We note, however, that these special processes—special relative to the more general process of preparation—are not needed for the interpretation of quantum mechanics.)

We shall now show by an example that the "strength" of the disturbance of a system which occurs during the registration process is not an important issue in the interpretation of quantum mechanics. Let us consider a "classical" system, for example, a bullet which is fired by a gun (as the preparation apparatus) and produces a hole in a target (the registration apparatus). No one will object to the use of the expressions "position of the bullet" and "momentum of the bullet" in the description of bullet immediately *before* it is "registered" by the target. Here the "strength" of the influence of the target is unimportant, for example, the bullet may become embedded into the target if, for example, the target is a metal plate.

The real distinction between macro- and microsystems lies in the structure of the convex set K which, for macrosystems, is completely different than that for microsystems (see, for example, the remarks in III, §3).

The following claim is often made: the "classical" mode of description is made possible whenever the disturbance of the measurement can be neglected. This claim is false, and avoids the actual problem, making an understanding of the problem more difficult.

Indeed, it is correct to say that *every* registration disturbs—for both the case of a classical system and that of a quantum mechanical system. Whether a system need be described classically or quantum mechanically has nothing to do with the "strength" of the disturbance in the registration process. Without disturbance there would hardly be any systems because without disturbances we cannot prepare and register, that is, cannot "extract" the system from its surroundings and make observations; without such interaction we cannot talk about systems at all.

For classical systems the set $\mathscr{E}_m$ of objective properties is "sufficiently large" that we may interpret the preparation and registration procedures by means of the objective properties (this topic was outlined in III, §4), even though the interaction during registration may be very large! (see the remarks in III, §4 and the discussion in [1], §12).

We note that (8.3.5) holds in general. This fact has led to much unnecessary speculation. It is not difficult to see that the expression "and" is used in *different* senses in $e_1 \wedge e_2$ and $p(e_1) \cap p(e_2)$, respectively. The mathematical formulation used here does not allow the usage of such imprecise language.

The same can be examined if we consider the negation of (8.3.2)

$$x \notin p(e). \tag{8.3.6}$$

(8.3.6) is equivalent to

$$x \in M \backslash p(e). \tag{8.3.7}$$

In general we find that

$$M\backslash p(e) \neq p(e^{\perp}). \tag{8.3.8}$$

In §8.2 we have seen that

$$M\backslash p(e) = p(e^{\perp})$$

if and only if $e \in Z$. Thus, from (8.3.7) we find that, in general, the two statements "x does not have the pseudoproperty e" and "x has the pseudoproperty $e^{\perp}$" are *not* equivalent. Thus, in "ordinary" language we may easily encounter the following difficulty: Suppose, for example, that e is the decision effect that the position lies in the domain V. Then, from the statement "x does not lie in V" we may easily (and incorrectly!) conclude that "the position of x lies in V'" where V' is the complement of the set V—$e^{\perp}$ is the decision effect for the statement "the position of x lies in V'."

In ordinary language it is evident that we may easily arrive at contradictions with logic. The mathematical language of relation (8.3.2) does not permit such confusion.

Let $\mathscr{E}_{pb}$ be the smallest Boolean ring of sets which is generated by $\mathscr{E}_{ps}$. $\mathscr{E}_{pb}$ contains (in the sense of equivalences (8.3.3), (8.3.4) and (8.3.6), (8.3.7)) all possible "logical relations" of the pseudoproperties in $\mathscr{E}_{ps}$. The Boolean ring $\mathscr{E}_{pb}$ is a reflection of the "ordinary" logic.This fact demonstrates that the question about the properties of a microsystem is not a question of logic because the elements of $\mathscr{E}_{pb}$ may be called properties of the microsystem, and $\mathscr{E}_{pb}$ contains "sufficiently many" properties since $\mathscr{E}_{ps}$ already contains sufficiently many pseudoproperties. The problem is therefore not associated with the construction of a Boolean ring of sets $\mathscr{E} \subset \mathscr{P}(M)$ but with the question posed in III, §1 about "*objective* properties." If we define the term "objective" in the sense of III, §4.1 then $\mathscr{E}_{pb}$ is clearly not a set of "objective" properties because the set of objective properties is (according to §8.1 and §8.2) the subset $\mathscr{E}_m$ of $\mathscr{E}_{ps}$. Clearly $\mathscr{E}_m$ is not a sufficient set of properties.

Since $\mathscr{E}_m$ is a Boolean ring of sets, the "logical operations" do not force us to leave the set $\mathscr{E}_m$, and we may, without hesitation, speak about these objective properties in ordinary language as, for example, "x is an electron." Using the following properties, we shall show that the elements of $\mathscr{E}_{pb}$ are not "objective" properties. Let $p = p(e)$. Then the following element

$$M\backslash p = (M\backslash p_p) \cap (M\backslash p_r)$$

belongs to $\mathscr{E}_{pb}$. Let us consider two preparation procedures a_1 and a_2 which belong to the same ensemble, that is, $\varphi(a_1) = \varphi(a_2)$. Suppose that $\mu(\varphi(a_1), e) \neq 0, \neq 1$. According to §6 it is possible that a_1 can be complementary to a_2 (see the EPR paradox in XVII, §4.4). Let a_2 have a decomposition for which $\tilde{a}_2 \subset a_2$ and $\varphi(\tilde{a}_2) \in K_1(e)$. a_1 may not have such a decomposition. Therefore it follows that $a_1 \cap p_p = \varnothing$, $a_2 \cap p_p \neq \varnothing$. Similarly according to §3 there may exist a pair of registration procedures b_0^1 and b_0^2 for which $b_0^1 \cap p_r = \varnothing$ and $\psi(b_0^2, b_0^2 \cap p_r) = e$; b_0^1 is a registration procedure for which the corresponding observable is not coexistent with any effect $g < e$.

Thus we obtain

$$a_1 \cap b_0^1 \subset M \backslash p,$$

that is, for all systems $x \in a_1 \cap b_0^1$ we obtain $x \notin p$. On the other hand, from $\mu(\varphi(a_1), e) = \mu(\varphi(a_1), \psi(b_0^2, b_0^2 \cap p_r)) \neq 0$ we obtain $a_1 \cap b_0^2 \cap p_r \neq \varnothing$, that is, in a_1 there exists a system x for which $x \in p$. Therefore, by the "application of the registration method b_0^1" alone (that is, without any selection according to a registration b!) we have therefore selected the "property" $M \backslash p$ from the systems of type a_1, although a_1 also contains elements of p. The "property" $M \backslash p$ is therefore not "objective" because it depends upon the application of a registration *method*.

Here we again make the remark that we do not use the designation "objective property" in a meaning opposite to that of "subjective," but (as we have already expressed in an intuitive way in II, §1) in the sense of "independent from the methods of preparation and registration," that is, objective in the sense of the properties ascribed to the systems. The opposite of "objective" properties is therefore "relative" properties, not subjective meaning and knowledge about properties.

The desire for the intuitive idea that although the microsystems are emitted by the preparation apparatus, they exist independently after the emission, that is, *no* interaction exists after emission between the microsystems and the preparation apparatus, and later, that the microsystems on the basis of their inherent structure—that is, their "objective properties" act upon the registration apparatus—is so compelling that many of us may wish to adopt the "hidden" properties hypothesis to retain this idea.

In the following discussion we shall ignore properties which are "completely hidden" in the sense that they have *nothing* to do with the preparation and registration processes. The above idea may yet have an additional meaning: It is perhaps not possible to construct a sufficiently good preparation procedure in order to produce systems with definite specified objective properties. Otherwise, the registration process may be deficient in this respect—the objective properties may only be partially registered. It is in *this* sense that we shall attempt to mathematically formulate the notion of "hidden objective properties." This notion will be somewhat different than that described in III, §4.1.

In addition to the structure previously introduced, we shall consider an additional structure $\mathscr{E}_h$ for which $\mathscr{E}_h \subset \mathscr{P}(M)$ where $\mathscr{E}_h$ is a Boolean ring of sets. The elements of $\mathscr{E}_h$ will represent the hidden properties.

For $p = p(e)$, $e \in G_p$ there exists a registration method b_0 for which $\psi(b_0, b_0 \cap p_r) = e$ (see §8.2). Then we obtain

$$\lambda_{\mathscr{S}}(a \cap p_r \cap b_0, a \cap p_p \cap b_0 \cap p_r) = 1,$$

that is, $a \cap p_p \cap b_0 \subset p_r$. Since the relation $(b_0 \cap p_r) \cup (b_0 \cap p_r^*) = b_0$ is satisfied for this b_0, for each system registered according to b_0 either $b_0 \cap p_r$ or $b_0 \cap p_r^*$ will "respond." We shall attempt to interpret this situation in the following way: $p = p_p \cup p_r$ possibly does not include *all* systems which have the same objective property which is made evident by the response $b_0 \cap p_r$ during the registration process b_0. Suppose there exists a $\varepsilon \in \mathscr{E}_h$ for which $\varepsilon \supset p_r$ such that for systems in ε the registration response is always obtained for $b_0 \cap p_r$ and therefore a response is never observed for p_r^*, that is, $\varepsilon \cap p_r^* = \emptyset$. Since the systems in $a \cap p_r$ are such that $b_0 \cap p_r$ responds with certainty, for all a we should find that $a \cap p_p \subset \varepsilon$ and therefore $p_p \subset \varepsilon$. Since the systems in $a \cap p_p^*$ are such that $b_0 \cap p_r$ does not respond with certainty we should find that $\varepsilon \cap p_p^* = \emptyset$. We then find that $\varepsilon \supset p_r$ and $\varepsilon \supset p_p$ are equivalent to $\varepsilon \supset p_p \cup p_r = p$; $\varepsilon \cap p_r^* = \emptyset$ and $\varepsilon \cap p_p^* = \emptyset$ is equivalent to $\varepsilon \cap (p_p^* \cap p_r^*) = \varepsilon \cap p^* = \emptyset$.

These speculations suggest that the following axiom is desirable:

AH 1. To each $p \in \mathscr{E}_{ps}$ there exists a $\varepsilon \in \mathscr{E}_h$ for which $\varepsilon \supset p$ and $\varepsilon \cap p^* = \emptyset$. Here $\varepsilon \cap p^* = \emptyset$ is equivalent to $M \backslash \varepsilon \supset p^*$. Since $\mathscr{E}_{ps} = pG_p$ we obtain: To each $e \in G_p$ there exists a $\varepsilon \in \mathscr{E}_h$ for which $\varepsilon \supset p(e)$ and $M \backslash \varepsilon \supset p(e^\perp)$.

In practical terms axiom AH 1 is not very restrictive because it is satisfied by $\mathscr{E}_h = \mathscr{E}_{ps}$ itself for the case in which $\varepsilon = p(e)$.

The following idea is more restrictive: Suppose that e_1, e_2 are two decision effects in G_p for which $e_1 \perp e_2$. Then we may think of a registration method b_0 for which $\psi(b_0, b_0 \cap p_{1r}) = e_1$, $\psi(b_0, b_0 \cap p_{2r}) = e_2$ where $p_1 = p(e_1)$, $p_2 = p(e_2)$. Then from $p_{1r} \cap p_{2r} = \emptyset$ and

$$b_0 = (b_0 \cap p_{1r}) \cup (b_0 \cap p_{2r}) \cup [b_0 \backslash (b_0 \cap (p_{1r} \cup p_{2r}))]$$

we obtain

$$\psi(b_0, b_0 \backslash (b_0 \cap (p_{1r} \cup p_{2r}))) = e_3 = \mathbf{1} - (e_1 + e_2)$$

and therefore

$$b_0 \backslash (b_0 \cap (p_{1r} \cup p_{2r})) = b_0 \cap p_{3r},$$

where $p_3 = p(e_3)$. This registration method b_0 permits us to separate the systems according to p_{1r}, p_{2r}, p_{3r}. For three objective properties $\varepsilon_1, \varepsilon_2, \varepsilon_3$ for which exactly one of $b_0 \cap p_{1r}$, $b_0 \cap p_{2r}$, or $b_0 \cap p_{3r}$ will respond with certainty we should therefore have $\varepsilon_1 \cup \varepsilon_2 \cup \varepsilon_3 = M$ because, for each system, in every case one of the "responses" $b_0 \cap p_{1r}, b_0 \cap p_{2r}, b_0 \cap p_{3r}$ must occur. In this way we are led to the following axiom:

AH 2. For $e_1, e_2, e_3 \in G_p$ and $e_1 + e_2 + e_3 = \mathbf{1}$, $\varepsilon_1, \varepsilon_2, \varepsilon_3 \in \mathscr{E}_h$ where $\varepsilon_i \supset p(e_i)$ $(i = 1, 2, 3)$ we require that $\varepsilon_1 \cup \varepsilon_2 \cup \varepsilon_3 = M$.

Let us define a map $\phi: G_p \to \mathscr{P}(M)$ as follows:

$$\phi(e) = \bigcap_{\substack{\varepsilon \in \mathscr{E}_h \\ \varepsilon \supset p(e)}} \varepsilon.$$

Then it easily follows that

$$e_1 < e_2 \Rightarrow \phi(e_1) \subset \phi(e_2) \tag{8.3.9}$$

and, from AH 1 it follows that:

$$\phi(e^{\perp}) \subset M \backslash \phi(e). \tag{8.3.10}$$

If $e_1, e_2, e_3 \in G_p$ and $e_1 + e_2 + e_3 = \mathbf{1}$, then, from AH 2 it follows that

$$\phi(e_1) \cup \phi(e_2) \cup \phi(e_3) = M. \tag{8.3.11}$$

If we assume that if $e \in G_p$ then $[0, e] \cap G_p$ is dense in $[0, e]$ we may then prove that the existence of a map ϕ satisfying (8.3.9), (8.2.10), and (8.3.11) leads to contradictions; for the case in which $G_p = G$ an elementary proof can be found in [2], XVIII). We shall not present discussions here of any weaker hypothesis for "hidden properties." Instead, we shall make an analysis of the structure $\mathscr{E}_{ps}$ of "physically real" pseudoproperties and their physical interpretation.

For (8.2.4) it follows that p_p depends upon the preparation procedure because $a = (a \cap p_p) \cup (a \backslash a \cap p_p)$ is a decomposition of a which must be coexistent with all possible decompositions in $\mathscr{Q}(a)$. Similarly, from (8.2.3) it follows that p_r depends on the registration method since $b_0 = (b_0 \cap p_r) \cup (b_0 \backslash b_0 \cap p_r)$ must be coexistent with all the other registrations in $\mathscr{R}(b_0)$. The discussions in previous sections about the structure of preparators and observables are therefore applicable to the sets $a \cap p_p$ and $b_0 \cap p_r$, respectively, as substructures. The elements of $\mathscr{E}_{ps}$ represent only a part of the structures of preparators and observables. The analysis of the preparators and the observables has shown that if we consider only $\mathscr{E}_{ps}$ then we lose part of the general structure of the interaction transfer mechanism from the preparation to the registration systems, where this mechanism is independent of the special technical construction of the preparation and registration apparatuses as described by the corresponding irreducible kernal observables or preparators, respectively. We cannot speak of microsystems except in the *context* of preparation and registration even if we neglect as many of the "accidental" properties of the preparation and registration apparatuses as possible. We have also found that even the structure described by the pseudoproperties in $\mathscr{E}_{ps}$ actually refers back to the preparation and registration processes. Only the "objective" properties in $\mathscr{E}_m$ can be separated from the preparation and registration procedures. $\mathscr{E}_m$ is not sufficient. Therefore it would be necessary to attempt to demonstrate the subjectivity of every assertion about nature. Such attempts would greatly exceed the real procedures in physics and contradict our original intentions to base the description of microsystems in terms of preparation and registration procedures which can be described in an "objective" form. In II we have argued that the interpretation of quantum mechanics depends only on the mode of description of the preparation and registration processes, a description which is given already before any knowledge of microsystems and quantum mechanics. This is specially important if we wish to consider the "physical possibility" of assertions of the form (8.3.2), that is, the question whether it is possible to realize situations in which the assertion (8.3.2) is true.

The structure of assertions of the form (8.3.2) for an unspecified x and an unspecified e is meaningful only if we are interested in studying the logical operations, as we have done earlier. We are, however, also interested in assertions of the form (8.3.2) in those cases in which x and e are particular specified elements. In order to emphasize this definiteness we shall modify the notion of (8.3.2) by using the subscript 1 as follows:

$$x_1 \in p(e_1). \tag{8.3.12}$$

By requiring that x_1 and e_1 be definite elements we mean that x_1 and e_1 are already *defined* before we use the relation (8.3.12) in our mathematical framework. For x_1 this may be achieved by requiring that x_1 is already a label for an actual system in the context of an actual experiment which has been carried out and the result of which is written down in the mathematical framework before we may use (8.3.12). In [1] the mathematical scheme of the theory in which previous experimental results have been incorporated is denoted by $\mathcal{MTA}$. In brief we say that x_1 is a definite element if x_1 is already a label appearing in $\mathcal{MTA}$ before we add the relation (8.3.12) to $\mathcal{MTA}$.

e_1 is a definite element if, for instance, it is defined as the decision effect to find the position of the system in a given (that is, in a technically determined) region $\mathscr{V}$ in the laboratory system, a decision effect which will be defined in VII, §4.

It is possible to add the relation (8.3.12) to the mathematical scheme $\mathcal{MTA}$ and determine whether

(1) (8.3.12) may be derived as a theorem in $\mathcal{MTA}$.
(2) The negation of (8.3.12) may be derived as a theorem in $\mathcal{MTA}$.
(3) Either (8.3.12) or its negation may be added to $\mathcal{MTA}$ without producing a contradiction (naturally both cannot).

In case (1) we say that "x_1 actually has the pseudoproperty e_1." In case (2) we say that "x_1 actually does not have the pseudoproperty e_1." Earlier we have seen that case (2) is *not* equivalent to the statement that "x_1 actually has the pseudoproperty $e_1^{\perp}$," except in the case in which $e_1 \in Z$—then e_1 would be an objective property. In case (3) we say that "x_1 may possibly have the property e_1" or ". . . not have . . . ." Every mathematician knows that case (3) can occur in a mathematical theory such as $\mathcal{MTA}$. The existence of case (3) has nothing to do with a "new" mathematical logic. The existence of case (3) is possible using "normal" mathematics and "normal" logic. Only some physicists have difficulty with case (3) and have the opinion that it may be necessary to introduce a new logic in physics in order to interpret case (3).

The source of these difficulties lies in the fact that we always (often unconsciously) assume that the elements of $\mathscr{E}_{ps}$ are like the inherent properties of the system—so that for $x \in M$ and $p \in \mathscr{E}_{ps}$ only $x \in p$ or $x \notin p$ will be true. For the elements $p \in \mathscr{E}_m$ we may, in fact, make such a claim, but not for the elements $p \in \mathscr{E}_{ps}$. Case (3) is essential! Case (3) is possible only because the elements $p \in \mathscr{E}_{ps}$ are not "objective" properties but, more generally, are assertions about the microsystems relative to the preparation and registration processes. For this reason it is important to examine case (3) in more detail.

We shall assume that the system x_1 is prepared according to some preparation procedure a_1. In this way we may introduce the relation $x_1 \in a_1$ into the mathematical theory as an experimental result. On the basis of the construction of the preparation apparatus corresponding to a_1 (the corresponding information cannot be represented in terms of the theory described here—see XVII) and on the basis of additional experiments we may determine $\varphi(a_1)$. If $\varphi(a_1)$ is determined by experiment then the value of $\mu(\varphi(a_1), e_1)$ is determined by the theory.

(A) Suppose $\mu(\varphi(a_1). e_1) = 1$. In this case we obtain $a_1 \in \mathcal{Q}_{e_1}$ and obtain (8.3.12) as a theorem.

(B) Suppose that $\mu(\varphi(a_1), e_1) = 0$. In this case $\mu(\varphi(a_1), e_1^\perp) = 1$, that is, $x_1 \in p(e_1^\perp)$ and $x_1 \notin p(e_1)$ are theorems.

Therefore we find that (A) corresponds to case (1) and (B) corresponds to case (2).

(C) Suppose that $\mu(\varphi(a_1), e_1) \neq 1$ and $\neq 0$. To comment on this case the actual experimental situation is essential since we are not dealing with an imaginary microsystem. x_1 is an actual microsystem. How was the preparation of the system carried out? Has the system x_1 already been registered? Has a registration method already been applied to x_1? ... etc. ...?

We shall first consider the case in which a registration method has not yet been applied. Then the *only* experimental facts are $x_1 \in a_1$ and (possibly) for some of the $\tilde{a} \subset a_1$ we have also obtained $x_1 \in \tilde{a}$. Since the relationship $x_1 \in \tilde{a}$ can be experimentally verified only for finitely many $\tilde{a}_\nu$ we find that

$$\hat{a} = \bigcap_\nu \tilde{a}_\nu \tag{8.3.13}$$

is an element of $\mathcal{Q}(a_1)$ and we obtain $x_1 \in \hat{a}$. If $\mu(\varphi(\hat{a}), e_1) = 1$ or if $\mu(\varphi(\hat{a}), e_1) = 0$ we obtain cases (A) and (B), respectively. Here it is important to remark that in the case in which not all the $x_1 \in \tilde{a}_\nu$ are included in the "protocol" of the experiment (for example, it is possible that some of these relations have been overlooked) in reality there exists such an $\hat{a}$ satisfying (8.3.13) even thoughwe may fail to take it into account in $\mathcal{MTA}$. Thus it is possible that, in the case $\mu(\varphi(\hat{a}), e_1) = 1$, that "$x_1$ really has the pseudo-property $p(e_1)$" although we may be unaware of this fact. Here we find that the $a \in \mathcal{Q}(a_1)$ to which the system x belongs is determined by the Boolean ring $\mathcal{Q}(a_1)$ even in the case in which an experimental result $x \in a$ for an $a \in \mathcal{Q}(a_1)$ has been overlooked. The incompleteness associated with the introduction of the experimental results into a mathematical theory (that is, the incompleteness of the axioms denoted by $\mathcal{MAP}$ set down in [1], §5 and [2], III, §4) may permit "possibilities" which nature does not allow. For example, it is possible that the relation $\mu(\varphi(\tilde{a}), e_1) = 1$ is satisfied for a particular experimental $\tilde{a}$ but that we have only observed that $x \in a_1$ (where $a_1 \supset \tilde{a}$) where $\mu(\varphi(a_1), e_1) < 1$;

then the "possibility" remains that $x_1 \notin p(e_1)$ while, in reality $x_1 \in p(e_1)$. This "possibility" remains only if we have overlooked the fact that x_1 actually satisfies the finer preparation procedure $\tilde{a} \subset a_1$.

We are not interested in the possibilities which are caused by the incompleteness of the axioms $\mathscr{MAP}$.

Therefore we shall now assume that the experimental situation is described by the relationship $x_1 \in \hat{a}$, where $\hat{a}$ is defined by (8.3.13) and that $\mu(\varphi(\hat{a}), e_1)$ has been calculated, and found not to be equal to either 0 or 1.

Then the relationship (8.3.12) or its negation can be introduced into the theory without producing a contradiction. In such a case, possibility (3) holds—"it is possible that x_1 has the pseudoproperty $p(e_1)$." What are the possibilities engendered in this case?

Since we have assumed that $x_1 \in \hat{a}$ is the best possible observation (that is, we have excluded the possibility of an incomplete "protocol"), since $\mu(\varphi(\hat{a}), e_1) \neq 1$, we must have $\hat{a} \cap p_{1p} = \varnothing$ (where $p_1 = p(e_1)$) otherwise in $Q(a_1)$ there would be selection procedures which are finer than $\hat{a}$ which we have overlooked. By assuming that $\hat{a} \cap p_1 = \varnothing$ we will simplify the following discussions.

From $\hat{a} \cap p_{1p} = \varnothing$ and $x_1 \in \hat{a}$ it follows that $x_1 \notin p_{1p}$. Then $x_1 \in p_1 = p(e_1)$ is equivalent to $x_1 \in p_{1r}$. Since $\mu(\varphi(\hat{a}), e_1) \neq 0$ we may add the relation $x_1 \in p_{1r}$ without contradiction. It would be false, however, to say that x_1 has the pseudoproperty p_1 with probability $\mu(\varphi(\hat{a}), e_1)$ because the probability that $x_1 \in p_1$ is realized in an experiment depends on the following:

(a) The registration method b_0 which we apply to x_1 can be chosen arbitrarily. Thus the "possibility" that $x_1 \in p_1$ may be obtained from $x_1 \in \hat{a}$ depends upon the possibility of the arbitrary(!) choice of b_0. For the choice of b_0 there are no probabilities. Here b_0 is freely "available" (for this concept see [1], §11 and §12).

Suppose that a particular choice $b_0^{(1)}$ is made, that is, as an additional experimental fact $x_1 \in b_0^{(1)}$ is observed. By the introduction of the relation $x_1 \in b_0^{(1)}$ it becomes necessary to alter our assessment of the relation (8.3.12) as follows:

(a1) $b_0^{(1)}$ has been chosen such that $b_0^{(1)} \cap p_{1r} = \varnothing$. Such a choice has occurred if the observable corresponding to $\mathscr{R}(b_0^{(1)}) \xrightarrow{\psi_0} L$ is complementary to the observable $\{0, e_1, e_1^{\perp}, \mathbf{1}\}$ (see §3). Then, by the choice(!) of such a registration method $b_0^{(1)}$, from $x_1 \in b_0^{(1)}$ we obtain the following result as a theorem in $\mathscr{MTA}$: $x_1 \notin p_{1r}$, and, since $x_1 \notin p_{1p}$ we also obtain $x_1 \notin p_1$. Therefore we obtain the case that "x_1 actually does *not* have the pseudoproperty p_1."

In this case (a1) it is easy to see that the elements $p \in \mathscr{E}_{ps}$ cannot represent properties of "isolated" microsystems because they are unconditionally connected with the possibilities inherent in the preparation and registration processes because the application of a registration method(!) can make it impossible that $x_1 \notin p_1$.

(a2) Let $b_0^{(1)}$ be chosen such that $b_0^{(1)} \cap p_{1r} \neq \varnothing$. Here a registration has not yet taken place. Then $x_1 \in b_0^{(1)}$ does not prohibit the addition of the relation $x_1 \in p_{1r}$. In this case we have "x_1 may possibly have the pseudoproperty p_1."

The "possibilities" for $x_1 \in p_1$ cannot be further influenced because there exist reproducible frequencies $\mu(\varphi(\hat{a}), \psi(b_0^{(1)}, b))$ for the various registrations $b \subset b_0^{(1)}$. Therefore, with a degree of justification, we may call

$$\mu(\varphi(\hat{a}), \psi(b_0^{(1)}, b_0^{(1)} \cap p_{1r}))$$

the "probability for $x_1 \in p_1$." Since $\psi(b_0^{(1)}, b_0^{(1)} \cap p_{1r})$ is, in general, smaller than e_1, it is possible that

$$\mu(\varphi(\hat{a}), \psi(b_0^{(1)}, b_0^{(1)} \cap p_{1r})) \leqq \mu(\varphi(\hat{a}), e_1)!$$

Indeed, it is possible that

$$\mu(\varphi(\hat{a}), \psi(b_0^{(1)}, b_0^{(1)} \cap p_{1r})) = 0$$

even in the case in which $b_0^{(1)} \cap p_{1r} \neq \varnothing$.

If $\mu(\varphi(\hat{a}), \psi(b_0^{(1)} \cap p_{1r})) = 0$ then "x_1 does not actually have the pseudoproperty p_1." If $\mu(\varphi(\hat{a}), \psi(b_0^{(1)}, b_0^{(1)} \cap p_{1r}) \neq 0$ then "x_1 may possibly have the pseudoproperty p_1."

(a2m) According to APE 3.4 we may choose $b_0^{(1)}$ such that

$$b_0^{(1)} = (b_0^{(1)} \cap p_{1r}) \cup (b_0^{(1)} \cap p_{1r}^*).$$

Then we obtain $\psi(b_0^{(1)}, b_0^{(1)} \cap p_{1r}) = e_1$. If $b_0^{(1)}$ is chosen in this way, then $x_1 \in p_1$ is possible with the maximal probability $\mu(\varphi(\hat{a}), e_1)$.

In case (a2) and in the special case (a2m), providing that $\mu(\varphi(\hat{a}), \psi(b_0^{(1)}, b_0^{(1)} \cap p_{1r})) \neq 0)$, there exists a registration $b_+ \subset b_0^{(1)} \cap p_{1r}$. If the experiment is carried out, then either $x_1 \in b_0^{(1)} \cap p_{1r}$ or $x_1 \in b_0^{(1)} \backslash b_0^{(1)} \cap p_{1r}$ will be observed upon registration from the registration apparatus.

For the experimental result $x_1 \in b_0^{(1)} \cap p_{1r}$ we therefore obtain (8.3.12) as a theorem: "x_1 has the pseudoproperty p_1" and the possible pseudoproperty p_1 of x_1 has been realized. Otherwise, if we experimentally obtained the result $x_1 \in b_0^{(1)} \backslash (b_0^{(1)} \cap p_{1r})$ then we obtain the statement $x_1 \notin p_1$ as a theorem, that is, "x_1 does not actually have the pseudoproperty p_1." Of course, this does not mean that the relation $x_1 \in p_1^* = p(e_1^\perp)$ must be satisfied. If $b_0^{(1)}$ had been chosen according to the case (a2m) then, from the experimental result

$$x_1 \in b_0^{(1)} \backslash (b_0^{(1)} \cap p_{1r})$$

we would indeed obtain $x_1 \in p_1^* = p(e_1^\perp)$ as a theorem, that is, "x_1 actually has the pseudoproperty p_1^*."

The analysis presented above shows the complicated structure of case (3) in which $x_1 \in p_1$ and $x_1 \notin p_1$ can be introduced without contradiction. This analysis shows the essential point that the knowledge of the subject does not play a role, and that, at most the incompleteness of the "protocol of an experiment" (as, for example, stored in a computer) may leave additional "possibilities" open, and that the latter are actually established by the experiment if they are not established by the protocol. This analysis shows that quantum mechanics permits the description of all experimental situations even for individual microsystems without, as we have found, the need for the introduction of a new logic, providing that we do not consider mathematical logic to be unusual because of the occurrence of case (3).

The fact that we do not need to introduce a new logic does not mean that it is not possible to introduce new language together with a new logic. For example, it is possible to express relations such as (8.3.2) in a new language which expresses more of the ontological character of such expressions than does the formal mathematical language. For example, it is possible to formulate expressions like (8.3.2) as follows: "The microsystem x has the pseudoproperty e" (see the discussion following (8.3.2)). This "new" linguistic formulation can be considered to be interpreted by means of (8.3.2). Using this and similar expressions as "elementary propositions" it is possible to obtain new propositions by means of logical operations; these new logical operations need not be the logical operations of $\mathscr{MT}$. Instead, they may be introduced on the basis of a dialog (see [16]), in which the verification of a proposition corresponds to what we have described above by: The proposition is physically verified on the basis of an experiment (see [1], §10.4).

According to the previous discussion in §8.3 we are now in the position to correctly explain the physical meaning of many of the famous quantum mechanical facts. We will first consider the uncertainty relation, often called the Heisenberg Uncertainty Relation. It has played a great role in the conceptual development of quantum mechanics, but is often loaded with considerable historical ballast; we shall now present a review of this topic.

We shall begin by briefly proving the following mathematical theorem:

Let $e_1, e_2 \in G$, let $\alpha_1 = \operatorname{tr}(we_1)$, $\alpha_2 = \operatorname{tr}(we_2)$. Then there exist e_1, e_2 such that for each $w \in K$ at least one of the following two relations is false:

$$\begin{aligned} \operatorname{tr}(w(e_1 - \alpha_1\mathbf{1})^2) &= \alpha_1 - \alpha_1^2 \leq \tfrac{1}{16}, \\ \operatorname{tr}(w(e_2 - \alpha_2\mathbf{1})^2) &= \alpha_2 - \alpha_2^2 \leq \tfrac{1}{16}. \end{aligned} \tag{8.3.14}$$

First we shall present the proof for the case of a single Hilbert space, that is, for $\mathscr{B} = \mathscr{B}(\mathscr{H})$. Let us consider a complete orthonormal basis which we divide into two sets φ_ν $(\nu = 1, 2, \ldots)$ and ψ_μ $(\mu = 1, 2, \ldots)$. Let

$$e_1 = \sum_\nu P_{\varphi_\nu} \quad \text{and} \quad \mathbf{1} - e_1 = e_1^\perp = \sum_\nu P_{\psi_\nu}.$$

Let

$$\chi_\nu = \frac{1}{\sqrt{2}}(\varphi_\nu + \psi_\nu) \quad \text{and} \quad \eta_\nu = \frac{1}{\sqrt{2}}(\varphi_\nu - \psi_\nu).$$

Let us assume that the relation $\alpha_1 - \alpha_1^2 \leq \frac{1}{16}$ is satisfied for w. Then it follows that either $\alpha_1 \leq \frac{1}{8}$ or $(1 - \alpha_1) \leq \frac{1}{8}$, that is, either $\operatorname{tr}(we_1) \leq \frac{1}{8}$ or $\operatorname{tr}(we_1^\perp) \leq \frac{1}{8}$. We shall only consider the case $\operatorname{tr}(we_1) \leq \frac{1}{8}$; the proof of the other is similar. From $\operatorname{tr}(we_1) \leq \frac{1}{8}$ it follows that

$$\begin{aligned}
\tfrac{1}{8} > \operatorname{tr}(we_1) &= \sum_\nu \|\sqrt{w}\varphi_\nu\|^2 > \tfrac{1}{2}\sum_\nu (\|\sqrt{w}\chi_\nu\| - \|\sqrt{w}\eta_\nu\|)^2 \\
&\geq \tfrac{1}{2}\sum_\nu \|\sqrt{w}\chi_\nu\|^2 + \tfrac{1}{2}\sum_\nu \|\sqrt{w}\eta_\nu\|^2 - \sum_\nu \|\sqrt{w}\chi_\nu\|\,\|\sqrt{w}\eta_\nu\| \\
&\geq \tfrac{1}{2}\operatorname{tr}(we_2) + \tfrac{1}{2}\operatorname{tr}(we_2^\perp) \\
&\quad - \left(\sum_\nu \|\sqrt{w}\chi_\nu\|^2\right)^{1/2} \left(\sum_\nu \|\sqrt{w}\eta_\nu\|^2\right)^{1/2} \\
&\geq \tfrac{1}{2} - \operatorname{tr}(we_2)^{1/2}\operatorname{tr}(w(\mathbf{1} - e_2))^{1/2} = \tfrac{1}{2} - \alpha_2^{1/2}(1 - \alpha_2)^{1/2}
\end{aligned}$$

and we obtain

$$\alpha_2 - \alpha_2^2 = \alpha_2(1 - \alpha_2) \geq (\tfrac{1}{2} - \tfrac{1}{8})^2 = (\tfrac{3}{8})^2 > (\tfrac{2}{8})^2 = \tfrac{1}{16}.$$

The proof may be extended to the case of more than one Hilbert space. Since G_p is dense in G, the theorem also applies for two suitable $e_1, e_2 \in G_p$.

Since at least one of the relations (8.3.14) must fail for every w, it represents an uncertainty relationship between the two decision effects e_1 and e_2. We have stated the case for a pair of decision effects in order to show that the uncertainty relations have *nothing* to do with the accuracy of measurements!

In order to discover the physical meaning of the relations (8.3.14) we shall now rewrite them in terms of preparation and registration procedures. For $p(e_1) = p_1$ and $p(e_2) = p_2$ let two registration procedures $b_0^{(1)}$ and $b_0^{(2)}$ be chosen such that $\psi(b_0^{(1)}, b_0^{(1)} \cap p_{1r}) = e_1$ and $\psi(b_0^{(2)}, b_0^{(2)} \cap p_{2r}) = e_2$ are satisfied.

For each preparation procedure a from

$$\lambda_{\mathscr{S}}(a \cap b_0^{(1)}, a \cap b_0^{(1)} \cap p_{1r}) \leq \tfrac{1}{8}$$

or

$$\lambda_{\mathscr{S}}(a \cap b_0^{(1)}, a \cap b_0^{(1)} \cap p_{1r}^*) \leq \tfrac{1}{8}$$

it follows that

$$\lambda_{\mathscr{S}}(a \cap b_0^{(2)}, a \cap b_0^{(2)} \cap p_{2r}) \geq \tfrac{1}{8}$$

and

$$\lambda_{\mathscr{S}}(a \cap b_0^{(2)}, a \cap b_0^{(2)} \cap p_{2r}^*) \geq \tfrac{1}{8}$$

must be satisfied.

$$\lambda_{\mathscr{S}}(a \cap b_0^{(1)}, a \cap b_0^{(1)} \cap p_{1r}) \leq \tfrac{1}{8}$$

states that, for the registration $b_0^{(1)} \cap p_{1r}$ at most $\frac{1}{8}$ of the systems prepared according to the procedure a will respond. Correspondingly,

$$\lambda_{\mathscr{S}}(a \cap b_0^{(1)}, a \cap b_0^{(1)} \cap p_{1r}^*) \leq \tfrac{1}{8}$$

states that, in the registration $b_0^{(1)} \cap p_{1r}$ more than $\frac{7}{8}$ of all systems prepared according to a will respond.

A "high" value for the response or nonresponse of $b_0^{(1)} \cap p_{1r}$ for the systems obtain from the preparation procedure a leads automatically to a "low" value as well for the response as for the nonresponse of $b_0^{(2)} \cap p_{2r}$, that is, the frequency of response must lie between $\frac{1}{8}$ and $\frac{7}{8}$.

The uncertainty relations expressed by the relation (8.3.14) express the following statement for possible preparation apparatuses: It is impossible to experimentally produce a preparation procedure for which the frequencies α_1 and α_2 satisfy both $\alpha_1 - \alpha_1^2 \leq \frac{1}{16}$ and $\alpha_2 - \alpha_2^2 \leq \frac{1}{16}$. This relation says nothing about the possibility of building registration apparatuses. On the contrary, we have assumed that we have made two experiments $a \cap b_0^{(1)}$ and $a \cap b_0^{(2)}$ where we permit $b_0^{(1)} \cap b_0^{(2)} = \varnothing$. From the uncertainty relations we cannot say anything about the possibility of obtaining joint measurements of e_1 and e_2. In fact, if e_1 and e_2 satisfy (8.3.14) then they are not commensurable in the sense of D 1.3.1. Then, if $\psi(b_0^{(1)}, b_0^{(1)} \cap p_{1r}) = e_1$ and if $\psi(b_0^{(2)}, b_0^{(2)} \cap p_{2r}) = e_2$ and if $b_0^{(1)} \cap b_0^{(2)} = b_0 \neq \varnothing$ it follows that:

$$\psi(b_0^{(1)}, b_0^{(1)} \cap p_{1r}) = \lambda_{\mathscr{R}_0}(b_0^{(1)}, b_0)\psi(b_0, b_0 \cap p_{1r}) = e_1$$

and

$$\psi(b_0^{(2)}, b_0^{(2)} \cap p_{2r}) = \lambda_{\mathscr{R}_0}(b_0^{(2)}, b_0)\psi(b_0, b_0 \cap p_{2r}) = e_2.$$

From

$$1 > \psi(b_0, b_0 \cap p_{1r}) > \frac{e_1}{\lambda_{\mathscr{R}_0}(b_0^{(1)}, b_0)}$$

it follows that $\lambda_{\mathscr{R}_0}(b_0^{(1)}, b_0) = 1$, that is, $b_0 = b_0^{(1)} = b_0^{(2)}$ resulting in the fact that e_1 and e_2 are coexistent, in contradiction to the fact that e_1 and e_2 do not commute. Therefore we *must* have $b_0^{(1)} \cap b_0^{(2)} = \varnothing$, that is, the two registration methods must be mutually exclusive.

The above clarification of this concept is necessary because we are accustomed to intuitively make more or less correct conclusions using the uncertainty relations. These uncertainty relations are usually formulated for scale observables which, according to D 2.5.6, are always decision observables.

Suppose that A and B are two such scale observables for which $A, B \in \mathscr{B}'(\mathscr{H}_1, \mathscr{H}_2, \ldots)$. A and B are therefore "bounded" operators (see AIV, §15). The "dispersion" of measurement values of A in the ensemble w is defined by

$$\mathrm{Str}(A) = \mathrm{tr}(wA'^2), \tag{8.3.15}$$

where $A' = A - \mathbf{1}\,\mathrm{tr}(wA)$.

Here $\mathrm{tr}(wA)$ is the so-called expectation value of A in the ensemble w, that is, it is approximately the experimental mean value $\bar{\alpha} = (1/N) \sum_{\nu=1}^{N} \alpha_\nu$ of the measurement results α_ν for a large number N of repeated experiments. $\mathrm{Str}(A)$ is the mean of the square of the deviation, that is, it is experimentally the approximate value of the mean value $(1/N) \sum_{\nu=1}^{N} (\alpha_\nu - \bar{\alpha})^2$. We may make the physical meaning more clear with the aid of a preparation procedure a for which $\varphi(a) = w$ and a registration method b_0 for which $\mathscr{R}(b_0) \xrightarrow{\psi_0} L$ represents a very good approximation for the scale decision observable A (for the realization of an observable, see §4).

For $\Delta A = \sqrt{\mathrm{Str}(A)}$ and the corresponding equation (8.3.15) for an observable B it follows that:

$$(\Delta A)(\Delta B) = \sqrt{\mathrm{tr}(wA'^2)}\sqrt{\mathrm{tr}(wB'^2)}.$$

Let $D = A' + i\alpha B'$ where α is real. Then D^+D is self-adjoint and $D^+D > 0$. Thus it follows that

$$\mathrm{tr}(wD^+D) \geq 0$$

and, for all α we have

$$(\Delta A)^2 + \alpha^2(\Delta B)^2 + \alpha\, \mathrm{tr}(wC) \geq 0, \tag{8.3.16}$$

where

$$C = i(A'B' - B'A') = i(AB - BA).$$

In order that (8.3.16) is satisfied, it is necessary and sufficient that the minimum of (8.3.16) with respect to α is non-negative; we therefore obtain

$$(\Delta A)(\Delta B) \geq (\tfrac{1}{2})|\mathrm{tr}(wC)|. \tag{8.3.17}$$

(If $\Delta B = 0$ we exchange A with B in the derivation; if both ΔA and $\Delta B = 0$ then from (8.3.16) it follows that $\mathrm{tr}(wC) = 0$ from which (8.3.17) is satisfied).

If equality holds in (8.3.17), then it follows that there exists an α for which $\mathrm{tr}(wD^+D) = \mathrm{tr}(D\sqrt{w}\sqrt{w}D) = 0$. This is equivalent to the condition that $D\sqrt{w} = 0$ and also $Dw = 0$. If there exists an α such that $Dw = 0$ then we obtain equality in (8.3.17). For $w = \sum_\nu w_\nu P_{\varphi_\nu}$ (according to AIV, §11) we obtain $Dw = 0$ is equivalent to the statement: $D\varphi_\nu = 0$ for all φ_ν (for which $w_\nu \neq 0$). This result follows from

$$0 = \mathrm{tr}(D^+Dw) = \sum_\nu w_\nu\, \mathrm{tr}(D^+DP_{\varphi_\nu}) = \sum_\nu w_\nu \|DP_{\varphi_\nu}\|^2.$$

Thus we have seen that the "uncertainty relations" (8.3.16) and (8.3.17) are determined essentially by the noncommutativity of the operators A and B, and, according to §3, are determined by the fact that A and B are not commensurable. The fact that A and B are not commensurable does not directly appear in the derivation of the uncertainty relations, but only indirectly, in terms of the mathematical structure of noncommutivity.

The best known case of the uncertainty relation (8.3.16) is the case in which A represents the position observable Q, and B represents the corresponding momentum observable P; P and Q will be described in detail in VII, §4. For P and Q (see VII (4.22) we obtain

$$PQ - QP = -i\mathbf{1},$$

that is, $C = -i\mathbf{1}$; therefore (8.3.16) takes the form

$$(\Delta P)(\Delta Q) \geq \tfrac{1}{2}, \tag{8.3.18}$$

this is the famous Heisenburg uncertainty relation which has played an important role in the evolution of quantum mechanics.

(Since P and Q are not bounded observables, the above proof does not apply. Let $w = \sum_\nu w_\nu P_{\varphi_\nu}$. If for $w_\nu \neq 0$ all the φ_ν lie in the domain of definition of the operators P^2 and Q^2, then the above derivation will be valid. If, for example, one of the φ_ν, say φ_1, does not lie in the domain of definition of P^2 (or Q^2) then $\text{tr}(wP_{\varphi_1})$ need not exist; we may then consider the operator $P' - \bar{p}\mathbf{1}$ with arbitrary values of $\bar{p}$ and obtain $\text{Str}(P)$ is infinite. Since $\text{Str}(Q)$ cannot be equal to zero, equation (8.3.18) will still be satisfied.)

The following conclusions are often made in connection with the Heisenberg uncertainty relation:

"The position and momentum of a particle cannot be simultaneously determined with arbitrary precision. The measurement uncertainties ΔP and ΔQ for a simultaneous measurement must satisfy (8.3.18)."

Or, somewhat more concisely:

"Position and momentum cannot be simultaneously measured to arbitrary accuracy."

These assertions are half-truths, and are not valid conclusions of equation (8.3.18).

First, the expression "time" does not appear in either (8.3.18) or (8.3.16). The meaning of the expression "simultaneous" in this context is unclear. If we conclude that position and momentum may be measured to arbitrary accuracy at different times, that conclusion will be false (see VII, §6 and XVII). As we have already seen, (8.3.16) is only indirectly concerned with the fact that A and B are not commensurable. As we have seen in §1–§4 the concept of commensurability is defined in an entirely different manner than is the uncertainty principle.

Again, it is important to note that ΔA and ΔB have nothing to do with measurement imprecisions; on the contrary, in the derivation of (8.3.16) it was assumed that they were measured with "ideal precision." What does it mean to measure an observable "imprecisely?"

Such a concept does not appear in this chapter. Have we overlooked part of the structure of the registration process? No, this is not the case. It was essential that, in the introduction of the concept of a registration procedure $b \in \mathcal{R}$ that the registration is precise, whether $x \in b$ or $x \notin b$. Therefore there are no "imprecise" registrations. What then does an experimental physicist mean by the expressions "measurement error," "measurement imprecisions," etc.? He is making a comparison between a "real" observable $\mathscr{R}(b_0) \xrightarrow{\psi_0} L$ and the desired observable $\Sigma \xrightarrow{F} L$; we have already discussed this problem in §4. Here the "measurement error" is understood to be the "difference" between $\mathscr{R}(b_0) \xrightarrow{\psi_0} L$ and $\Sigma \xrightarrow{F} L$ which is characterized in AOb (see §4) by the differences $\psi_0 h(\sigma) - F(\sigma)$. If $\Sigma \xrightarrow{F} L$ is a scale decision observable, then the experimental physicist seeks to describe the difference between his real apparatus, as described by $\mathscr{R}(b_0) \xrightarrow{\psi_0} L$ and a "real scale," by "errors" between the real scale of the apparatus and the ideal scale of the scale observable he wished to measure. This "error" depends upon the ensemble used in the experiment. This subject was discussed in §4 in connection with axiom AOb. The discussion of errors is a typical experimental problem because it is not related to the theoretical postulates of AOb but is concerned with the construction of the real experimental apparatus.

Thus, when we say that we can only make imprecise joint measurements of both P and Q, the assertion is made that, for a real apparatus b_0 having two scales x and y, the corresponding partial observables of $\mathscr{R}(b_0) \xrightarrow{\psi_0} L$ may only approximately measure the ideal observables P and Q with errors, where the errors are "somewhat similar" to those described by (8.3.18). To analyze these "errors" more precisely is somewhat more difficult than may at first appear. At least we know that the well-defined quantities ΔP and ΔQ in (8.3.18) are not measurement errors.

Although ΔP and ΔQ are often falsely interpreted as measurement errors, it is probable that a relationship which is similar to (8.3.18) will be obtained if we define a suitable notion of a measurement error. This is certainly not surprising. However, the derivation of a relation which is similar to (8.3.18) for the "measurement errors" of "approximate P" and "approximate Q" is much more difficult and, in all probability, cannot in general be carried out. For this reason we shall not proceed further in this direction (see [19]).

Certain related problems, such as the problem of a sequence of measurements or the problem posed in III, §1 (that the registration must occur "after" the preparation), and that the registration must take place in another region of space than the preparation cannot be clarified using only the axioms presented here. This is due in part to the fact that we have not built into the theory described here a description of space and time. Such a clarification is extremely desirable, since the role of space and time in quantum mechanics is of great importance. Such a clarification is possible only after we have investigated the transformation properties of preparation and registration procedures in V–VIII.

CHAPTER V

Transformations of Registration and Preparation Procedures. Transformations of Effects and Ensembles

In II–IV we have been primarily motivated by physical considerations. In this chapter we shall be concerned with questions which play an important role in all mathematical theories—the definition and examination of the role of morphisms. The concepts presented in II–IV were motivated by physical considerations; here again the mathematical structure will reflect the physical situation. In this book we shall not investigate the underlying general mathematical problem itself because we are interested in the physical significance of the morphisms. We have already encountered this problem in previous chapters and have already anticipated some of the applications of morphisms. In VII we shall describe another important application; in XVII we shall become familiar with additional important physical examples of morphisms.

1 Morphisms for Selection Procedures

At first it may appear to be desirable to consider mappings of the set M into itself. In classical physics we are accustomed to studying transformations of state space (for example, phase space in classical mechanics). Here we note that we may not identify the state space of classical theories with the set of physical systems. For example, it is possible that many systems have the same state. In classical theories the notion of the transformation of systems is, in general, physically vague. In quantum mechanics, except for the maps considered in XVII, §4.1, we also find that there are no physically interesting maps of M into itself. For these reasons we shall not consider maps of the set M into itself.

On the other hand, the mapping of selection procedures is of substantial interest. Let $\mathcal{S}_1$ and $\mathcal{S}_2$ be two systems of selection procedures on the sets M_1 and M_2.

D 1.1. Let $\mathcal{S}_1 \xrightarrow{h} \mathcal{S}_2$ be a map satisfying the following conditions:

$$h(a \cap b) = h(a) \cap h(b),$$

$$h(a \backslash b) = h(a) \backslash h(b) \quad \text{for } a \supset b.$$

Such a map will be called an sp-morphism.

For $b \subset a$ (hence $a \cap b = b$) from $h(a \cap b) = h(a) \cap h(b)$ it follows that $h(a) \cap h(b) = h(b)$, that is, $h(a) \supset h(b)$. Thus h preserves the order relation and, if the first requirement is satisfied, the second requirement of D 1.1 is meaningful because $b \subset a$ implies $h(b) \subset h(a)$. If $a, b \in \mathcal{S}_1$, $a \cup b \in \mathcal{S}_1$ then from $h(a) \subset h(a \cup b)$ and $h(b) \subset h(a \cap b)$ we obtain $h(a) \cup h(b) \subset h(a \cup b)$; on the other hand, from $h(a \cup b \backslash a) = h(a \cup b) \backslash h(a)$, $a \cup b \backslash a \subset b$ it follows that $h(a \cup b) \backslash h(a) \subset h(b)$ and that

$$h(a) \cup [h(a \cup b) \backslash h(a)] = h(a \cup b) \subset h(a) \cup h(b).$$

Therefore we obtain $h(a \cup b) = h(a) \cup h(b)$.

In accord with the usual terminology we shall say that a bijective mapping h is an sp-isomorphism if both h and h^{-1} are sp-morphisms.

Th. 1.1. *Let N_1 be a subset of M_1, let $N_2 = M_1 \backslash N_1$, and let $\mathcal{S}_1$ be a system of selection procedures. Then the set*

$$\mathcal{S}_2 = \{b \mid b = a \cap N_2 \text{ and } a \in \mathcal{S}_1\}$$

is a system of selection procedures and the mapping

$$h(a) = a \cap N_2$$

is an sp-*morphism.*

PROOF. The proof is simple and left to the reader.

Th. 1.2. *If h is an* sp-*morphism and if*

$$\mathcal{I} = \{a \mid a \in \mathcal{S}_1 \text{ and } h(a) = \varnothing\}$$

then $\mathcal{I}$ satisfies the following properties:

(1) $a \in \mathcal{I}, b \in \mathcal{S}_1$ *and* $b \subset a \Rightarrow b \in \mathcal{I}$.
(2) $a_1, a_2 \in \mathcal{I}, a_1 \cup a_2 \in \mathcal{S}_1 \Rightarrow a_1 \cup a_2 \in \mathcal{I}$.

PROOF. (1) Since h is order preserving, from $h(b) \subset h(a) = \varnothing$ it follows that $h(b) = \varnothing$.

(2) We obtain $h(a_1 \cup a_2) = h(a_1) \cup h(a_2) = \varnothing \cup \varnothing = \varnothing$.

D 1.2. A subset $\mathscr{I} \subset \mathscr{S}$ is said to be an ideal in $\mathscr{S}$ providing that the following conditions are satisfied:

(1) $a \in \mathscr{I}, b \in \mathscr{S}$ and $b \subset a \Rightarrow b \in \mathscr{I}$.
(2) $a_1, a_2 \in \mathscr{I}, a_1 \cup a_2 \in \mathscr{S} \Rightarrow a_1 \cup a_2 \in \mathscr{I}$.

Th. 1.3. *If h is an sp-morphism $\mathscr{S}_1 \xrightarrow{h} \mathscr{S}_2$ then, by means of the ideal $\mathscr{I}$ described in Th. 1.2 it is possible to decompose the map h as follows:*

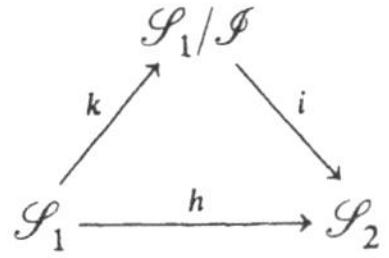

where i is an injection.

Proof. From $h(a_1) = h(a_2)$ it follows that $h(a_1 \cap a_2) = h(a_1) \cap h(a_2) = h(a_1) = h(a_2)$ and we obtain $h(a_1 \backslash a_1 \cap a_2) = h(a_1) \backslash h(a_1 \cap a_2) = \emptyset$, $h(a_2 \backslash a_1 \cap a_2) = \emptyset$, that is, $a_1 \backslash a_1 \cap a_2, a_2 \backslash a_1 \cap a_2 \in \mathscr{I}$. An equivalence relation $a_1 \sim a_2$ is defined by $a_1 \backslash a_1 \cap a_2, a_2 \backslash a_1 \cap a_2 \in \mathscr{I}$. We obtain the following identity:

$$a_1 \backslash (a_1 \cap a_3) = [(a_1 \backslash a_1 \cap a_2) \backslash (a_1 \backslash a_1 \cap a_2) \cap a_3] \cup [a_1 \cap (a_2 \backslash a_2 \cap a_3)].$$

From $a_1 \backslash a_1 \cap a_2 \in \mathscr{I}$ and $a_2 \backslash a_2 \cap a_3 \in \mathscr{I}$ it follows that $a_1 \backslash a_1 \cap a_3 \in \mathscr{I}$. Similarly, from $a_3 \backslash a_3 \cap a_2 \in \mathscr{I}$ and $a_2 \backslash a_2 \cap a_1 \in \mathscr{I}$ it follows that $a_3 \backslash a_3 \cap a_1 \in \mathscr{I}$.

2 Morphisms of Statistical Selection Procedures

For many applications the probability function obeys certain laws under sp-morphisms. We shall now formulate these laws.

If $a \in \mathscr{S}$, h is an sp-morphism and $\mathscr{I}$ is the ideal defined in Th. 1.2, then the set of all $a \cap \tilde{a}$ for which $\tilde{a} \in \mathscr{I}$ is a subset of $\mathscr{I}$ which we shall denote by $\mathscr{I}$(a). We therefore obtain $\mathscr{I}(a) = \mathscr{S}(a) \cap \mathscr{I}$. For $\mathscr{I}(a)$ we therefore obtain

$$a_1 \subset a_2, a_2 \in \mathscr{I}(a) \Rightarrow a_1 \in \mathscr{I}(a),$$

$$a_1, a_2 \in \mathscr{I}(a) \Rightarrow a_1 \cup a_2 \in \mathscr{I}(a)$$

since if $a_1, a_2 \in \mathscr{I}(a)$ then it follows that $a_1 \cup a_2 \in \mathscr{I}$ since $a_1 \cup a_2 \subset a$!

D 2.1. An ideal $\mathscr{I}$ is said to be closed with respect to a statistical selection procedure $\mathscr{S}$ if $\sup_{\tilde{a} \in \mathscr{I}(a)} \lambda(a, \tilde{a}) = 1$ implies the relation $a \in \mathscr{I}$.

The condition $\sup_{\tilde{a} \in \mathscr{I}(\tilde{a})} \lambda(a, \tilde{a}) = 1$ means that there exists an $\tilde{a} \in \mathscr{I}$ for which the probability for the selection $a \backslash \tilde{a}$ is, for all practical purposes, equal to zero.

D 2.2. An sp-morphism h of a statistical selection procedure $\mathcal{S}_1$ in a statistical selection procedure $\mathcal{S}_2$ is said to be an ssp-morphism if the ideal $\mathcal{I}$ is closed and if, for $a_1 \subset a_2$ the following condition holds:

$$\lambda_2(h(a_1), h(a_2)) = \frac{\alpha_2}{\alpha_1} \lambda_1(a_1, a_2),$$

where α_1, α_2 are defined as follows:

$$\alpha_i = 1 - \sup_{\tilde{a} \in \mathcal{I}(a_i)} \lambda_1(a_i, \tilde{a}) = \inf_{\tilde{a} \in \mathcal{I}(a_i)} \lambda_1(a_i, a_i \backslash \tilde{a}).$$

Since the ideal $\mathcal{I}$ is closed, it follows that the condition $h(a) = \varnothing$ is equivalent to the condition

$$\alpha = 1 - \sup_{\tilde{a} \in \mathcal{I}(a)} \lambda_1(a, \tilde{a}) = 0.$$

For $h(a_1) \neq \varnothing$ and therefore $\alpha_1 \neq 0$; therefore the condition given for λ_1, λ_2 is well defined.

If an ssp-homomorphism is an sp-isomorphism then it is also an ssp-isomorphism, since for each $a \neq \varnothing$ we obtain $\alpha = 1$ and therefore

$$\lambda_2(h(a_1), h(a_2)) = \lambda_1(a_1, a_2).$$

Conversely, if $\alpha = 1$ for all $a \neq \varnothing$, then it follows that $\lambda_1(a, \tilde{a}) = 0$ for all $\tilde{a} \in \mathcal{I}(a)$. If $a \in \mathcal{I}$ and if $a \neq \varnothing$ then $a \in \mathcal{I}(a)$ and therefore $\lambda_1(a, a) = 1$ in contradiction to $\lambda_1(a, \tilde{a}) = 0$ for all $\tilde{a} \in \mathcal{I}(a)$. Therefore $\mathcal{I}$ contains only the null set, that is, the ssp-homomorphism is an ssp-isomorphism.

D 2.3. A subset $\mathcal{S}_1 \subset \mathcal{S}$ of a selection procedure $\mathcal{S}$ is called a separated part of $\mathcal{S}$, if $\mathcal{S}_1$ is a selection procedure and if, for each pair of elements $a_1 \in \mathcal{S}_1, a_2 \in \mathcal{S} \backslash \mathcal{S}_1$ the intersection $a_1 \cap a_2 = \varnothing$.

It is easy to see that if $\mathcal{S}_1$ is a separate part of $\mathcal{S}$ then $\mathcal{S}_2 = \mathcal{S} \backslash \mathcal{S}_1$ is also a separate part.

Th. 2.1. *Let h be an ssp-morphism of $\mathcal{S}_1$ into $\mathcal{S}_2$. If the relation $\lambda_2(h(a_1), h(a_2)) = \lambda_1(a_1, a_2)$ is satisfied for the case $a_1 \supset a_2$ and $h(a_1) \neq 0$, then $\mathcal{I}$ is a separate part of $\mathcal{S}_1$ and h is an ssp-isomorphism of $\mathcal{S}' = \mathcal{S}_1 \backslash \mathcal{I}$ onto a partial selection procedure $h\mathcal{S}_1 = \mathcal{S}'_2 \subset \mathcal{S}_2$.*

PROOF. Let $a \notin \mathcal{I}$ and $\tilde{a} \in \mathcal{I}$. If $a \cap \tilde{a} \neq \varnothing$ then, since $h(a \cap \tilde{a}) = 0$ we obtain

$$0 = \lambda_2(h(a), h(a \cap \tilde{a})) = \lambda_1(a, a \cap \tilde{a}) \neq 0$$

which is a contradiction. Therefore $\mathcal{I}$ is a separate part of $\mathcal{S}_1$.

For $a \in \mathcal{S}'_1$ then $h(a) \neq \varnothing$ for $a \neq \varnothing$ and h is an sp-isomorphism of $\mathcal{S}'_1$ onto $\mathcal{S}'_2 = h\mathcal{S}'_1 = h\mathcal{S}_1$. As a result the probabilities are invariant—therefore h is an ssp-isomorphism of $\mathcal{S}'_1$ onto $\mathcal{S}'_2$.

3 Morphisms of Preparation and Registration Procedures

We now turn from the general case to the case of the preparation and registration procedures which are important for quantum mechanics.

We shall now assume that we are given $M_1, \mathcal{Q}_1, \mathcal{R}_{01}, \mathcal{R}_1$ and $M_2, \mathcal{Q}_2, \mathcal{R}_{02}, \mathcal{R}_2$.

D 3.1. An ssp-morphism of $\mathcal{Q}_1$ into $\mathcal{Q}_2$ where $\mathcal{Q}_1$ and $\mathcal{Q}_2$ are statistical selection procedures is called a preparation morphism (abbreviated p-morphism). By analogy with D 2.2 we define $\alpha(a) = \inf_{\tilde{a} \in \mathcal{I}(a)} \lambda_{\mathcal{Q}_1}(a, a \backslash \tilde{a})$, where $\mathcal{I}(a)$ is defined in §2

The p-morphisms and the p-automorphisms shall play a particularly important role.

D 3.2. A p-morphism h will be said to be recording-invariant (r-invariant) if $\varphi_1(a_1) = \varphi_1(a_2)$ implies that $\varphi_2(h(a_1)) = \varphi_2(h(a_2))$ and $\alpha(a_1) = \alpha(a_2)$; here φ_1 and φ_2 are defined according to III, D 3.1.

Th. 3.1. *For an* r-*invariant* p-*morphism h a map*

$$\varphi_1 \mathcal{Q}'_1 = \mathcal{K}_1 \xrightarrow{S} \check{K}_2$$

is defined by

$$S\varphi_1(a) = \alpha(a)\varphi_2(h(a)).$$

For a decomposition $a = \bigcup_{\nu=1}^n a_\nu$ *it follows that*

$$S\left(\sum_\nu \lambda_\nu \varphi_1(a_\nu)\right) = \sum_\nu \lambda_\nu S\varphi_1(a_\nu),$$

where $\lambda_\nu = \lambda_{\mathcal{Q}_1}(a, a_\nu)$. *If h is a* p-*isomorphism, then* $\alpha(a) = 1$ *and* $\varphi_1 \mathcal{Q}'_1 = \mathcal{K}_1 \xrightarrow{S} \mathcal{K}_2 \subset K_2$.

PROOF. Clearly S is well defined. From $a = \bigcup_{\nu=1}^n a_\nu$ it follows that $h(a) = \bigcup_{\nu=1}^n h(a_\nu)$. Since $a_\nu \cap a_\mu = \varnothing$ for $\nu \neq \mu$ we also obtain $h(a_\nu) \cap h(a_\mu) = \varnothing$. Therefore we obtain

$$\varphi_2(h(a)) = \sum_\nu \lambda_{\mathcal{Q}_2}(h(a), h(a_\nu))\varphi_2(h(a_\nu)),$$

and

$$S\varphi_1(a) = \alpha(a)\varphi_2(h(a)) = \alpha(a) \sum_\nu \lambda_{\mathcal{Q}_2}(h(a), h(a_\nu))\varphi_2(h(a_\nu)).$$

According to D 2.2 we have

$$\alpha(a)\lambda_{\mathcal{Q}_2}(h(a), h(a_\nu)) = \alpha(a_\nu)\lambda_{\mathcal{Q}_1}(a, a_\nu)$$

and we obtain

$$S\varphi_1(a) = \sum_\nu \lambda_{\mathscr{Q}_1}(a, a_\nu)\alpha(a_\nu)\varphi_2(h(a_\nu))$$
$$= \sum_\nu \lambda_{\mathscr{Q}_1}(a, a_\nu)S\varphi_1(a_\nu).$$

From $\varphi_1(a) = \sum_\nu \lambda_{\mathscr{Q}_1}(a, a_\nu)\varphi_1(a_\nu)$ (according to III, Th. 3.1 and the equivalence of III (3.2) and III (6.1)) we obtain the desired result.

The map S defined in Th. 3.1 is (according to III, Th. 3.2) a rational affine map of the rational, convex set $\varphi_1\mathscr{Q}_1 = \mathscr{K}_1$ (which is dense in K_1) into K_2.

D 3.3. We shall call an sp-morphism h of $\mathscr{R}_1$ into $\mathscr{R}_2$ for which the restriction to $\mathscr{R}_{01}$ is an ssp-isomorphism $\mathscr{R}_{01}$ onto $\mathscr{R}_{02}$ a recording morphism (abbreviation: r-morphism).

From an r-morphism h we may easily obtain a (canonical) map of effect processes; this map we shall also denote by h:

D 3.4. For $(b_0, b) \in \mathscr{F}_1$ *we define the map* $\mathscr{F}_1 \xrightarrow{h} \mathscr{F}_2$ by $h(b_0, b) = (h(b_0), h(b))$.

D 3.5. A r-morphism h is said to be preparation-invariant (p-invariant if $\psi_1(f_1) = \psi_1(f_2)$ implies that $\psi_2(h(f_1)) = \psi_2(h(f_2))$. ψ_1 and ψ_2 correspond to III, D 3.1.

Th. 3.2. *For a p-invariant r-morphism h a map $\psi\mathscr{F}_1 \xrightarrow{T} L_2$ is defined by $T\psi_1(f) = \psi_2(h(f))$ which satisfies $T\mathbf{1} = \mathbf{1}$. For the decomposition of the unit effect* **1** *(see* III, D 4.5.5)

$$\mathbf{1}_{b_0} = \bigcup_i f_i$$

we obtain

$$\mathbf{1} = \sum_i \psi_1(f_i) \quad \textit{and} \quad \mathbf{1} = \sum_i T\psi_1(f_i).$$

PROOF. The proof is similar to the proof of Th. 3.1.

According to III, Th. 2.4 T is a rational affine map of the rational convex set $\psi\mathscr{F}_1$ in L_2.

In many physical applications an r-morphism is not only p-invariant but, in addition, if f_1, f_2 are hardly distinguishable by testing with preparation procedures (even if $\psi(f_1) \neq \psi(f_2)$) then the same is true for the images $h(f_1)$ and $h(f_2)$. We therefore define:

D 3.6. An r-morphism h is said to be preparation-continuous (p-continuous) if to each $\varepsilon > 0$ and $a \in \mathscr{Q}_2'$ there exists a $\delta > 0$ and a finite number of $a_i \in \mathscr{Q}_1'$ such that

$$|\mu_2(\varphi_2(a), \psi_2(h(f)) - \mu_2(\varphi_2(a), \psi_2(h(\hat{f}))| < \varepsilon$$

whenever

$$|\mu_1(\varphi_1(a_i), \psi_1(f)) - \mu_1(\varphi_1(a_i), \psi_1(\hat{f}))| < \delta$$

for all a_i. An r-isomorphism is said to be a p-continuous r-isomorphism if both h and h^{-1} are p-continuous.

It is easy to see that a p-continuous r-morphism is also p-invariant.

An analogous definition can also be made for p-morphisms:

D 3.7. A p-morphism h is said to be recording-continuous (r-continuous) if to each $\varepsilon > 0$ and $f \in \mathscr{F}_2$ there exists a $\delta > 0$ and a finite number of $f_i \in \mathscr{F}_1$ such that

$$|\mu_2(\alpha(a)\varphi_2(h(a)), \psi_2(f)) - \mu_2(\alpha(\hat{a})\varphi_2(h(\hat{a})), \psi_2(f))| < \varepsilon$$

whenever

$$|\mu_1(\varphi_1(a), \psi_1(f_i)) - \mu_1(\varphi_1(\hat{a}), \psi_1(f_i))| < \delta$$

for all f_i. A p-isomorphism is said to be an r-continuous p-isomorphism if both h and h^{-1} are r-continuous.

Again it is easy to see that an r-continuous p-morphism is also r-invariant.

D 3.8. A p-isomorphism $h: \mathscr{Q}_1 \to \mathscr{Q}_2$ is said to be dual to an r-isomorphism $h': \mathscr{R}_2 \to \mathscr{R}_1$ if $(h(a), b_0) \in C_2$ is equivalent to $(a, h'(b_0)) \in C_1$ and if

$$\mu_2(h(a), (b_0, b)) = \mu_1(a, h'(b_0, b)). \tag{3.1}$$

Here C_1 and C_2 are defined by analogy with C in II (4.3.1).

Th. 3.3. *If a* p-*isomorphism* $h: \mathscr{Q}_1 \to \mathscr{Q}_2$ *and* r-*isomorphism* $h': \mathscr{R}_2 \to \mathscr{R}_1$ *are dual, then* h^{-1} *and* h'^{-1} *are also dual.*

Proof. From $(h^{-1}(a), b_0) \in C_1$ it follows that, for $a' = h^{-1}(a)$ and $b'_0 = h'^{-1}(b_0)$ that $(a', h'(b'_0)) \in C_1$ and, according to D 3.8 $(h(a'), b'_0) \in C_2$, that is, $(a, h'^{-1}(b_0)) \in C_2$. In this way it follows from $(a, h'^{-1}(b_0)) \in C_2$ that $(h^{-1}(a), b_0) \in C_1$.

From (3.1) it follows that for $a' = h^{-1}(a)$ and $b'_0 = h'^{-1}(b_0)$, $b' = h'^{-1}(b)$

$$\begin{aligned}\mu_1(h^{-1}(a), (b_0, b)) &= \mu_1(a', h'(b'_0, b'))\\ &= \mu_2(h(a'), (b'_0, b')) = \mu_2(a, h'^{-1}(b_0, b)).\end{aligned}$$

Th. 3.4 (H. Neumann). *If a* p-*isomorphism h and an* r-*isomorphism h' are dual, then h is an* r-*continuous* p-*isomorphism and h' is a* p-*continuous* r-*isomorphism.*

Proof. According to Th. 3.3 we need only show that h is r-continuous and that h' is p-continuous.

Since h is a p-isomorphism, we find that $\alpha(a) = 1$. According to D 3.7 it suffices to show that

$$\mu_2(\varphi_2(h(a)), \psi_2(b_0, b)) = \mu_1(\varphi_1(a), \psi_1(h'(b_0, b)). \tag{3.2}$$

First we shall show that h is r-invariant. Suppose that $a' \sim a$ and that $h(a') \nsim h(a)$. Then there exists a (b_0, b) such that $(h(a'), b_0) \in C_2$, $(h(a), b_0) \in C_2$ and

$$\mu_2(h(a'), (b_0, b)) \neq \mu_2(h(a'), (b_0, b)).$$

Thus it follows that $(a', h'(b_0)) \in C_1$ and $(a, h'(b_0)) \in C_1$ and $\mu_1(a', h'(b_0, b)) \neq \mu_1(a, h'(b_0, b))$ which contradicts $a' \sim a$.

According to APS 5.1.4 there exists an $a' \in \varphi_1(a)$ satisfying $(a', h'(b_0)) \in C_1$. Then, according to D 3.8

$$\begin{aligned}\mu_1(\varphi_1(a), \psi_1(h'(b_0, b)) &= \mu_1(a', h'(b_0, b)) \\ &= \mu_2(h(a'), (b_0, b)) = \mu_2(\varphi_2(h(a'), \psi_2(b_0, b)).\end{aligned}$$

Since h is r-invariant, from $a' \sim a$ it follows that $h(a') \sim h(a)$, that is, $\varphi_2(h(a')) = \varphi_2(h(a))$.

According to D 3.6 the relation (3.2) suffices to show that h' is p-continuous (observe that h' is a map $\mathscr{R}_2 \rightarrow \mathscr{R}_1$ and not $\mathscr{R}_1 \rightarrow \mathscr{R}_2$ as in D 3.6!).

4 Morphisms of Ensembles and Effects

Since r-invariant p-morphisms and p-invariant r-morphisms always occur in applications, it is understandable that our emphasis in the investigation of morphisms in quantum mechanics will be concerned with morphisms of ensembles and effects.

4.1 Morphisms of Ensembles

D 4.1.1. An affine mapping S of K_1 into K_2 is called a mixture morphism (mi-morphism).

D 4.1.2. An affine map S of $\check{K}_1$ into $\check{K}_2$ is called an operation.

Th. 4.1.1. *A rational affine and norm-continuous map S of a (rational affine) set $\mathscr{K}_1$ which is dense in K_1 into $\check{K}_2$ may be uniquely extended to an operation $\check{K}_1$ in $\check{K}_2$.*

PROOF. Since S is norm-continuous, S may be extended as an affine mapping onto $K_1 = \overline{\text{co}}\, K_1$ and therefore onto the whole space $\mathscr{B}_1$.

Since a $w \in \check{K}_1$ may be written in the form $\lambda\tilde{w}$ where $0 \leq \lambda \leq 1$, $\tilde{w} \in K_1$ and S is affine in K_1, S may be extended onto all of $\check{K}_1$ by means of the equation $S(\lambda\tilde{w}) = \lambda S(\tilde{w})$. Thus this extension of S is an operation $\check{K}_1 \xrightarrow{S} \check{K}_2$.

Th. 4.1.2. *An operation S of $\check{K}_1$ into $\check{K}_2$ may be uniquely extended as a linear mapping of $\mathscr{B}_1$ in $\mathscr{B}_2$ with norm $\|S\| < 1$. Every mixture morphism $K_1 \xrightarrow{S} K_2$ has a unique extension as a linear map of $\mathscr{B}_1$ in $\mathscr{B}_2$ for which $\|S\| = 1$; in particular, every mixture morphism can be extended in this way to an operation.*

Every positive norm-continuous linear map $\mathscr{B}_1 \xrightarrow{S} \mathscr{B}_2$ with $\|S\| \leq 1$ (restricted to $\check{K}_1$) is an operation. Every positive linear map $\mathscr{B}_1 \xrightarrow{S} \mathscr{B}_2$ is norm-continuous and $\|S\|^{-1}S$ is an operation.

PROOF. Since $\mathscr{B}_1$ is spanned by K_1, S can be extended to $\mathscr{B}_1$. For $w \in K_1$, since $Sw_1 \in \check{K}_2$ the relation $\|Sw_1\| \leq 1$ and for $x = \alpha w_1 - \beta w_2$ and $\|x\| = \alpha + \beta$ we may conclude that $\|Sx\| \leq \|x\|$, it follows that S is norm-continuous and $\|S\| \leq 1$. In this way we find that every positive map satisfying $\|S\| < 1$ is an operation, because $\check{K}$ is the intersection of the unit sphere with the positive cone.

Since every positive linear map is norm-continuous, this result holds in general (see AIII, §6).

Thus we see that a bijective mixture morphism $K_1 \xrightarrow{S} K_2$ is a mixture isomorphism, that is, S^{-1} is a mixture morphism. Thus it follows that a bijective operation $\check{K}_1 \xrightarrow{S} \check{K}_2$ is a mixture isomorphism, that is, $K_1 \xrightarrow{S} K_2$.

Th. 4.1.3. *To each operation S there exists a dual map S' of $\mathscr{B}'_2$ in $\mathscr{B}'_1$ for which $L_2 \xrightarrow{S'} L_1$. S' is $\sigma(\mathscr{B}'_2, \mathscr{B}_2)$–$\sigma(\mathscr{B}'_1, \mathscr{B}_1)$ continuous; S is a mixture morphism if and only if $S'\mathbf{1} = \mathbf{1}$.*

PROOF. The fact that S' exists and is $\sigma(\mathscr{B}'_2, \mathscr{B}_2)$–$\sigma(\mathscr{B}'_1, \mathscr{B}_1)$-continuous follows from Th. 4.1.2, that is, from the fact that S is norm-continuous (see AIII, §5). From $K_1 \xrightarrow{S} K_2$ it follows that $\mu_2(Sw, \mathbf{1}) = 1 = \mu_1(w, S'\mathbf{1})$ holds for all $w \in K_1$, and we therefore obtain $S'\mathbf{1} = \mathbf{1}$. If $S'\mathbf{1} = \mathbf{1}$ then, for all $w \in K_1$ it follows that $\mu_2(Sw, \mathbf{1}) = \mu_2(w, S'\mathbf{1}) = \mu_1(w, \mathbf{1}) = 1$ and we therefore obtain $Sw \in K_2$.

Th. 4.1.4. *For an* mi-*morphism S the following statements are equivalent:*

(i) *S is a mixture isomorphism.*
(ii) *$K_1 \xrightarrow{S} K_2$ is injective and SK_1 is dense in norm in K_2.*
(iii) *$L_2 \xrightarrow{S} L_1$ is bijective.*
(iv) *S' is an isomorphic map of the Banach spaces.*
(v) *S is an isomorphic map of the Banach spaces.*

PROOF. (i) ⇒ (ii) trivial.

(ii) ⇒ (iii). Since SK_1 is norm-dense in K_2 it follows that $(S\mathscr{B}_1)^{\perp} = 0$ which is equivalent to the condition that S' is injective. Since SK_1 is norm-dense in K_2 it follows that

$$\|S'y\| = \sup_{w \in K_1} |\mu_1(w, S'y)| = \sup_{w \in K_1} |\mu_2(Sw, y)| = \sup_{\tilde{w} \in K_2} |\mu_2(\tilde{w}, y)| = \|y\|$$

that is S' is norm-preserving. Thus it follows that $S'\mathscr{B}'_2$ is a norm-closed subspace of $\mathscr{B}'_1$.

We shall now show that if $K_1 \xrightarrow{S} K_2$ is injective then $\mathscr{B}_1 \xrightarrow{S} \mathscr{B}_2$ is also injective: Every $x \in \mathscr{B}_1$ can be written in the form $x = \alpha w_1 - \beta w_2$, where $\alpha, \beta \geq 0$ and $w_1, w_2 \in K_1$. Then from $0 = Sx = \alpha Sw_1 - \beta Sw_2$ and from $Sw_1, Sw_2 \in K_2$ it follows that $\alpha = \beta$ and that $Sw_1 = Sw_2$. Thus it follows that $w_1 = w_2$ and finally $x = 0$. Since $\mathscr{B}_1 \xrightarrow{S} \mathscr{B}_2$ is injective, it follows that $S'\mathscr{B}'_2$ is σ-dense in $\mathscr{B}'_1$. Since the unit sphere $[-\mathbf{1}, \mathbf{1}]$ is σ-compact, the set $A = S'[-\mathbf{1}, \mathbf{1}]$ is therefore compact and convex. Since S' preserves the norm we obtain $A = S'\mathscr{B}'_2 \cap [-\mathbf{1}, \mathbf{1}]$ and $S'\mathscr{B}'_2$ is therefore also σ-closed (see AIII, §4) and $S'\mathscr{B}'_2 = \mathscr{B}'_1$ and $S'[-\mathbf{1}, \mathbf{1}] = [-\mathbf{1}, \mathbf{1}]$. Since, according to Th. 4.1.4 $S'\mathbf{1} = \mathbf{1}$ and $L = \frac{1}{2}(\mathbf{1} + [-\mathbf{1}, \mathbf{1}])$ it follows that $S'L_2 = L_1$ and (iii) is satisfied.

(iii) ⇒ (iv). If S' is a bijective map of L_2 onto L_1 then (since, according to Th. 4.1.3 $S'\mathbf{1} = \mathbf{1}$ and $[-\mathbf{1}, \mathbf{1}] = 2L - \mathbf{1}$) it follows that S' is also a bijective map of the unit spheres onto each other. Hence (iv) holds.

(iv) ⇒ (v). We shall now show that $(S')^{-1}$ is σ-continuous: $(S')^{-1}$ as the inverse of the σ-continuous bijective map of the σ-compact unit sphere is σ-continuous on the unit sphere and is therefore σ-continuous everywhere (see AIII, §5). Thus it follows that S^{-1} exists and satisfies $(S^{-1})' = (S')^{-1}$. Since S' maps the unit sphere bijectively, we obtain $\|Sx\| = \|x\|$, from which we have proven (v).

(v) ⇒ (i). The existence of $(S')^{-1}$ and $(S')^{-1} = (S^{-1})'$ is clear. Then, according to Th. 4.1.3 $S'\mathbf{1} = \mathbf{1}$ and we obtain $(S')^{-1}\mathbf{1} = \mathbf{1}$. Let $w \in K_2$; since $SK_1 \subset K_2$ from $\|S^{-1}w\| = \|w\| = 1$ it follows that $\mu_1(S^{-1}w, \mathbf{1}) = \mu_2(w, \mathbf{1}) = 1$, we obtain $S^{-1}K_2 \subset K_1$.

D 4.1.3. An operation (or mixture morphism) is said to be $\mathscr{D}$-continuous when, as a map $K_1 \to \check{K}_2$ is continuous with respect to the topologies $\sigma(K_1, \mathscr{D}_1)$, $\sigma(K_2, \mathscr{D}_2)$; the spaces $\mathscr{D}_i$ are defined in III, §4 and §5.

As a mapping $K_1 \to K_2$ a $\mathscr{D}$-continuous mixture morphism may be naturally extended as a $\mathscr{D}$-continuous operation.

Th. 4.1.5. *Every $\mathscr{D}$-continuous rational affine map $\mathscr{K}_1 \xrightarrow{S} \mathscr{K}_2$ can be extended to a mapping $\mathscr{B}_1 \xrightarrow{S} \mathscr{B}_2$ which is norm-continuous and, as a mapping $K_1 \xrightarrow{S} \check{K}_2$ is also $\mathscr{D}$-continuous.*

Proof. By definition, the norm in $\mathscr{D}'$ is equal to

$$\|x\| = \sup\{\mu(x, y) \mid \|y\| \leq 1, y \in \mathscr{D}\}.$$

Since $\mathscr{D} \cap L$ is $\sigma(\mathscr{B}', \mathscr{B})$-dense in L (see III, §5) and since the set $\{y \mid \|y\| \leq 1, y \in \mathscr{D}\}$is $\sigma(\mathscr{B}', \mathscr{B})$-dense in the unit sphere of $\mathscr{B}'$, for elements of $\mathscr{B} \subset \mathscr{D}'$, the norm in $\mathscr{D}'$ is identical to the norm in B.

We shall now show that the unit sphere of $\mathscr{D}'$ is equal to the convex set generated by $\bar{K}^\sigma \cup (-\bar{K}^\sigma)$.

The unit sphere in $\mathscr{D}'$ is the set which is bipolar to the set $K \cup (-K)$ in the dual pair $(\mathscr{D}', \mathscr{D})$. Therefore the unit sphere of $\mathscr{D}'$ is equal to the $\sigma(\mathscr{D}', \mathscr{D})$-closed convex set generated by $\bar{K}^\sigma \cup (-\bar{K}^\sigma)$. Since $\bar{K}^\sigma$ and $(-\bar{K}^\sigma)$ are $\sigma(\mathscr{D}', \mathscr{D})$-compact (as bounded closed sets in $\mathscr{D}'$; see AIII, §4) the convex set generated by $\bar{K}^\sigma \cup (-\bar{K}^\sigma)$ is already compact, and is therefore $\sigma(\mathscr{D}', \mathscr{D})$-closed, that is, equal to the unit sphere of $\mathscr{D}'$.

Let $\mathbf{1} \in \mathscr{D}$; since the relation $\mu(w, \mathbf{1}) = 1$ for all $w \in \bar{K}^\sigma$ follows from the same relation for all $w \in K$ it follows that $\|w\| = 1$ for all $w \in \bar{K}^\sigma$.

Since the closures $\bar{K}_1^\sigma$ of K_1 in $\mathscr{D}_1$ and $\bar{K}_2^\sigma$ of $\check{K}_2$ in $\mathscr{D}_2$ are compact, S may be extended as a $\mathscr{D}$-continuous map $\bar{K}_1^\sigma \xrightarrow{S} \bar{K}_2^\sigma$, where this extension becomes an affine map. Since $\mathscr{B}_1$ is the linear span of K_1, S may be extended to a map $\mathscr{B}_1 \to \mathscr{D}'_2$.

For $x \in \mathscr{B}_1$ and $x = \alpha w_1 - \beta w_2$, where $w_1, w_2 \in K_1$ and $\|x\| = \alpha + \beta$ it follows that

$$\|Sx\| \leq \alpha\|Sw_1\| + \beta\|Sw_2\| \leq \alpha + \beta$$

because the norm $\|w'\| \leq 1$ for elements $w' \in \bar{\check{K}}_2^\sigma$. S is therefore norm-continuous as a map $\mathscr{B}_2 \to \mathscr{D}'_2$.

Since K_1 is norm-complete and $\mathscr{K}_1$ is norm-dense in K_1 and $\check{K}_2$ is norm-complete, and from $\mathscr{K}_1 \xrightarrow{S} \check{K}_2$ we therefore obtain $K_1 \xrightarrow{S} \check{K}_2$, that is, S is an operation. According to Th. 4.1.1 and Th. 4.1.2 it follows that S is a norm-continuous map $\mathscr{B}_1 \to \mathscr{B}_2$.

Since $\bar{K}_1^\sigma \xrightarrow{S} \bar{\check{K}}_2^\sigma$ is $\mathscr{D}$-continuous, the map $K_1 \xrightarrow{S} \check{K}_2$ is $\mathscr{D}$-continuous.

Th. 4.1.6. *For a $\mathscr{D}$-continuous affine map $K_1 \xrightarrow{S} \check{K}_2$ the extension of S on $\mathscr{B}_1$ is continuous in the topologies $\sigma(\mathscr{B}_2, \mathscr{D}_1)$ and $\sigma(\mathscr{B}_2, \mathscr{D}_2)$ and, equivalently, S' maps the subspace $\mathscr{D}_2$ of $\mathscr{B}'_2$ in $\mathscr{D}_1$.*

PROOF. First we shall show that each element $x \in \mathscr{D}'$ may be written in the form $x = \alpha w_1 - \beta w_2$, where $w_1, w_2 \in \bar{K}^\sigma$ and $\|x\| = \alpha + \beta$. If $x' = \|x\|^{-1}x$ then x' is in the unit sphere of $\mathscr{D}'$, and, since $\text{co}(\bar{K}^\sigma \cup (-\bar{K}^\sigma))$ is the unit sphere in $\mathscr{D}'$, is therefore of the form $x' = \lambda w_1 - (1-\lambda)w_2$, where $0 \leq \lambda \leq 1$ and $w_1, w_2 \in \bar{K}^\sigma$. Thus it follows that $x = \alpha w_1 - \beta w_2$ with $\alpha = \|x\|\lambda$, $\beta = \|x\|(1-\lambda)$, that is, $\alpha + \beta = \|x\|$. Thus it follows that the set of all $\alpha w_1 - \beta w_2$, where $w_1, w_2 \in K$ is $\sigma(\mathscr{D}', \mathscr{D})$ dense in $\mathscr{D}'$, that is, $\mathscr{B}$ is $\sigma(\mathscr{D}', \mathscr{D})$ dense in $\mathscr{D}'$.

We may define a $\sigma(\mathscr{D}'_1, \mathscr{D}_2)$-continuous affine function on K_1 by means of the equation $l(w) = \mu_2(Sw, y)$ (for fixed $y \in \mathscr{D}_2$) because S is continuous in the topologies $\sigma(\mathscr{D}'_1, \mathscr{D}_1)$ on K_1 and $\sigma(\mathscr{D}'_2, \mathscr{D}_2)$ on K_2. In this way we may obtain an extension of $l(w)$ as a σ-continuous affine function on all of $\bar{K}^\sigma$.

Since each $x \in \mathscr{D}'_1$ is of the form $x = \alpha w_1 - \beta w_2$, where $\|x\| = \alpha + \beta$ and $w_1, w_2 \in \bar{K}_1^\sigma$, l may be extended on all of $\mathscr{D}'$ as a norm-continuous linear form because

$$\begin{aligned}|l(x)| &\leq \alpha|l(w_1)| + \beta|l(w_2)| \\ &\leq (\alpha + \beta) \sup_{w \in \bar{K}_1^\sigma} |l(w)| = \|x\| \sup_{w \in \bar{K}_1^\sigma} |l(w)|.\end{aligned}$$

We shall now show that l is, in addition, $\sigma(\mathscr{D}'_1, \mathscr{D}_1)$-continuous as a linear form over $\mathscr{D}'_1$.

Here l is $\sigma(\mathscr{D}'_1, \mathscr{D}_1)$-continuous if l is $\sigma(\mathscr{D}'_1, \mathscr{D}_1)$-continuous on the unit sphere of $\mathscr{D}'_1$ (see AIII, §4). Thus we need only show that $(l(x)| < \delta$ for $\|x\| \leq 1$ and a suitable $\sigma(\mathscr{D}'_1, \mathscr{D}_1)$-neighborhood of 0.

We may write x in the form $x = \alpha w_1 - \beta w_2$ where $\alpha + \beta = \|x\| \leq 1$ and $w_1, w_2 \in \bar{K}_1^\sigma$. Thus it follows that

$$x = \tfrac{1}{2}(\alpha + \beta)(w_1 - w_2) + (\alpha - \beta)(\tfrac{1}{2}w_1 + \tfrac{1}{2}w_2).$$

Since $\mu_1(w, \mathbf{1}) = 1$ for all $w \in \bar{K}_1^\sigma$, it follows that

$$x = \tfrac{1}{2}\|x\|(w_1 - w_2) + \mu_1(x, \mathbf{1})(\tfrac{1}{2}w_1 + \tfrac{1}{2}w_2).$$

Thus

$$|l(x)| \leq \tfrac{1}{2}\|x\| l(w_1 - w_2) + \lambda|\mu(x, \mathbf{1})|,$$

where $\lambda = \sup_{w \in \bar{K}_1^\sigma} |l(w)|$.

We shall now choose the $\sigma(\mathscr{D}'_1, \mathscr{D}_1)$ neighborhood of 0 as follows: Since l is $\sigma(\mathscr{D}'_1, \mathscr{D}_1)$-continuous in $\bar{K}^{\sigma}_1$, there are finitely many $y_i \in \mathscr{D}_1$ for which $|l(w - \tilde{w})| < \delta/2$ providing that $|\mu(w - \tilde{w}, y_i)| \leq 1$. For $\varepsilon = \delta/2\lambda$ we define $\hat{y}_i = (1/\varepsilon)y_i$. We choose the $\sigma(\mathscr{D}'_1, \mathscr{D}_1)$ neighborhoods of 0 as follows: $|\mu(x, \hat{y}_i)| < \frac{1}{4}$ for all i and $|\mu(x, \mathbf{1})| < \sigma/4$, where σ is the smallest of the numbers $\|\hat{y}_i\|$ and ε. We will now prove that $|l(x)| < \delta$ if $\|x\| \leq 1$ and if x lies in the specified neighborhood of 0:

(a) If $\|x\| \leq \varepsilon$, then since $|l(w_1 - w_2)| < 2\lambda$ it follows that

$$|l(x)| \leq \varepsilon\lambda + \frac{\sigma}{4}\lambda \leq 2\varepsilon\lambda = \delta.$$

(b) If $\|x\| \geq \varepsilon$ then, since $\|x\| < 1$, it follows that

$$|l(x)| \leq \tfrac{1}{2}|l(w_1 - w_2)| + \tfrac{1}{4}\sigma\lambda \leq \tfrac{1}{2}|l(w_1 - w_2)| + \varepsilon\lambda$$

and from

$$\mu(x, \hat{y}_i) = \tfrac{1}{2}\|x\|\mu(w_1 - w_2, \hat{y}_i) + \mu(x, \mathbf{1})\tfrac{1}{2}\mu(w_1 + w_2, \hat{y}_i)$$

it follows that

$$\begin{aligned}|\mu(w_1 - w_2, \hat{y}_i)| &\leq \frac{2}{\|x\|}|\mu(x, \hat{y}_i)| + |\mu(x, \mathbf{1})|\frac{1}{\|x\|}|\mu(w_1 + w_2, \hat{y}_i)| \\ &\leq \frac{2}{\varepsilon}|\mu(x, \hat{y}_i)| + \frac{1}{4}\sigma\frac{1}{\varepsilon}2\|\hat{y}_i\| < \frac{1}{\varepsilon}\frac{1}{2} + \frac{1}{2}\frac{1}{\varepsilon} = \frac{1}{\varepsilon},\end{aligned}$$

that is, $|\mu(w_1 - w_2, y_i)| \leq 1$. Therefore we obtain $|l(w_1 - w_2)| < \delta/2$ and

$$|l(x)| \leq \frac{\delta}{2} + \varepsilon\lambda = \delta.$$

Thus we have shown that l is $\sigma(\mathscr{D}'_1, \mathscr{D}_1)$-continuous in all of $\mathscr{D}'_1$, and that there exists a $\tilde{y} \in \mathscr{D}_1$ for which

$$l(x) = \mu_2(Sx, y) = \mu_1(x, \tilde{y}) = \mu_1(x, S'y)$$

from which it follows that $S'y \in \mathscr{D}_1$ for $y \in \mathscr{D}_2$, that is, $S'\mathscr{D}_2 \subset \mathscr{D}_1$. This is equivalent to the condition that S is, as a map $\mathscr{D}'_1 \to \mathscr{D}'_2$, continuous in the topologies $\sigma(\mathscr{D}'_1, \mathscr{D}_1)$ and $\sigma(\mathscr{D}'_2, \mathscr{D}_2)$.

Since $\mathscr{B}_1 \subset \mathscr{D}'_1$ and $\mathscr{B}_2 \subset \mathscr{D}'_2$, the same statement holds for the map $S: \mathscr{B}_1 \to \mathscr{B}_2$.

Th. 4.1.7. *Let h be an* r*-continuous* p*-morphism. A rational affine and $\mathscr{D}$-continuous map $\mathscr{K}_1 \xrightarrow{S} \check{K}_2$ is defined by the equation $S\varphi_1(a) = \alpha(a)\varphi_2(h(a))$. This map may be extended to an affine $\mathscr{D}$-continuous map $K_1 \xrightarrow{S} \check{K}_2$. The continuation S is, as a map $\mathscr{B}_1 \xrightarrow{S} \mathscr{B}_2$, both norm-continuous and $\mathscr{D}$-continuous, that is, $S'\mathscr{D}_2 \subset \mathscr{D}_1$. If h is a* p*-isomorphism, then $K_1 \xrightarrow{S} K_2$ is bijective.*

PROOF. From Th. 3.1 and from the conclusions following Th. 3.1 it follows that $\mathscr{K}_1 \xrightarrow{S} \check{K}_2$ is a rational affine map. From the fact that h is r-continuous and the fact that the topologies $\sigma(K, \mathscr{D})$ and $\sigma(\check{K}, \mathscr{L})$ are (with $\mathscr{L} = \psi\mathscr{F}$) identical (since $\mathscr{L}$ is norm-dense in $L \cap \mathscr{D}$) it follows that $\mathscr{K}_1 \xrightarrow{S} \check{K}_2$ is $\mathscr{D}$-continuous. According to Th. 4.1.5 it follows that $K_1 \xrightarrow{S} \check{K}_2$ is affine and $\mathscr{D}$-continuous and that $\mathscr{B}_1 \xrightarrow{S} \mathscr{B}_2$ is norm-continuous. According to Th. 4.1.6 it follows that $\mathscr{B}_1 \xrightarrow{S} \mathscr{B}_2$ is also $\mathscr{D}$-continuous and $S'\mathscr{D}_2 \subset \mathscr{D}_1$.

If h is a p-isomorphism, then, according to Th. 3.1 $\mathscr{K}_1 \xrightarrow{S} K_2$. Since h is a p-isomorphism, we obtain $S\mathscr{K}_1 = \mathscr{K}_2$. For h^{-1} there is an $\tilde{S}$ with $\tilde{S}\mathscr{K}_2 = \mathscr{K}_1$ and $S\tilde{S} = \mathbf{1}$, $\tilde{S}S = \mathbf{1}$ on $\mathscr{K}_2$ (resp. $\mathscr{K}_1$). Since S and $\tilde{S}$ are norm-continuous, and $\mathscr{K}_1, \mathscr{K}_2$ norm-dense in K_1 and K_2, respectively, then for the extensions we obtain $K_1 \xrightarrow{S} K_2$, $K_2 \xrightarrow{\tilde{S}} K_1$ and $S\tilde{S} = \mathbf{1}$, $\tilde{S}S = \mathbf{1}$.

4.2 Morphisms of Effects

D 4.2.1. Let T be a mapping of L_1 into L_2; if the relation

$$T(g_1 + g_2) = Tg_1 + Tg_2$$

is satisfied for all $g_1, g_2, g_1 + g_2 \in L_1$, then we shall call the map T an effect morphism. T is said to be $\mathscr{B}$-continuous if it is continuous as a map $L_1 \longrightarrow L_2$ in the topologies $\sigma(\mathscr{B}'_1, \mathscr{B}_1)$, $\sigma(\mathscr{B}'_2, \mathscr{B}_2)$.

Th. 4.2.1. *A mapping T of L_1 into L_2 which satisfies the relation $T(g_1 + g_2) = Tg_1 + Tg_2$ for $g_1, g_2, g_1 + g_2 \in L_1$ has a unique extension to $\mathscr{B}'_1$ and satisfies $\|T\| < 1$. T is positive.*

PROOF. From $g_1, g_2 \in L_1$ and $g_2 > g_1$ it follows that $g_2 = g_1 + (g_2 - g_1)$ and that $Tg_2 = Tg_1 + T(g_2 - g_1)$, that is, $Tg_2 > Tg_1$ and $T(g_2 - g_1) = Tg_2 - Tg_1$. For $g \in L_1$ and $ng \in L_1$ (for integer values of n) by induction, we obtain $T(ng) = nTg$. For integer m and $g_1 \in L_1$ we obtain $g = g_1/m \in L$ and $g_1 = mg$; therefore we find that $Tg_1 = mTg$, that is, $T(g_1/m) = (Tg_2)/m$. Thus, for $ng/m \in L_1$ we obtain $T(ng/m) = nT(g/m) = nT(g)/m$. Thus, for all rational numbers λ, for $g \in L_1$, $\lambda g \in L_1$ we obtain $T(\lambda g) = \lambda Tg$. If λ is irrational, then, to each $\varepsilon > 0$ we may choose two rational numbers λ_1, λ_2 such that $\lambda_1 < \lambda < \lambda_2$ and $\lambda_2 - \lambda_1 < \varepsilon$. Let $\lambda_2 g \in L_1$. Since $\lambda_1 g < \lambda g < \lambda_2 g$ we obtain $\lambda_1 Tg < T(\lambda g) < \lambda_2 Tg$. Since ε was arbitrary it follows that $T(\lambda g) = \lambda Tg$. If λ is irrational and $\lambda g \in L_1$, but $(\lambda + \varepsilon)g \notin L_1$ for each $\varepsilon > 0$, we then choose $g_1 = g - \delta g$, where $0 < \delta < 1$. Then, for irrational numbers λ we also obtain $T(\lambda g_1) = \lambda Tg_1$ and $T(\lambda\delta g) = \lambda T(\delta g)$, because, for sufficiently small $\varepsilon > 0$ we always obtain $(\lambda + \varepsilon)g_1 \in L_1$ and $(\lambda + \varepsilon)\delta g \in L_1$. Since $g = g_1 + \delta g$ we obtain $Tg = Tg_1 + T(\delta g)$ and, since $\lambda g = \lambda g + \lambda\delta g$ we also obtain $T(\lambda g) = T(\lambda g_1) + T(\lambda\delta g) = \lambda Tg_1 + \lambda T(\delta g) = \lambda Tg$.

Therefore T is an affine mapping of L_1 into L_2. Since $\mathscr{B}'_1$ is the linear span of L_1 (for example, each $y \in B'_1$ may be written in the form $y = \alpha g - \beta\mathbf{1}$, where $g \in L$), T may be extended as a linear map from $\mathscr{B}'_1$ into $\mathscr{B}'_2$.

Since the unit sphere of $\mathscr{B}'_1$ is equal to $2L - \mathbf{1}$ (and similarly for $\mathscr{B}'_2$) and since $L_1 \xrightarrow{T} L_2$ we find that T maps the unit sphere of $\mathscr{B}'_1$ into the unit sphere of $\mathscr{B}'_2$, that is, $\|T\| \leq 1$.

Each $y \in B'_{1+}$ is of the form λg, where $\lambda > 0$, $g \in L_1$; therefore T is a positive mapping.

Th. 4.2.2. *If T is a $\mathscr{B}$-continuous effect morphism, then the extension of the map T onto $\mathscr{B}'_1$ is also $\sigma(\mathscr{B}'_1, \mathscr{B}_1)$–$\sigma(\mathscr{B}'_2, \mathscr{B}_2)$-continuous.*

PROOF. Since the unit sphere of $\mathscr{B}'_1$ is equal to $2L_1 - \mathbf{1}$ the map T is σ-continuous on the unit sphere. Then, according to AIII, §5, T is σ-continuous in all of $\mathscr{B}'_1$.

Th. 4.2.3. *If T is a linear positive mapping of $\mathscr{B}'_1$ into $\mathscr{B}'_2$ then T is norm-continuous. T maps L_1 into L_2 if and only if $T\mathbf{1} \in L_2$.*

PROOF. The norm-continuity of T is a result of general theorems (see AIII, §6). We may easily prove this result directly. For $y \in \mathscr{B}'_1$ we define

$$\beta = -\inf\{\mu(w, y) \mid w \in K_1\}$$

providing that the inf is less than zero, otherwise set $\beta = 0$. We also define

$$\alpha = \sup\{\mu(w, y) \mid w \in K_1\}.$$

Then $\alpha > -\beta$ and

$$\|y\| = \sup\{\mu(w, y) \mid w \in K_1\} = \max\{|\alpha|, \beta\}.$$

If we set

$$g = \frac{1}{\alpha + \beta}(y + \beta\mathbf{1})$$

then $0 \leq \mu(w, g) \leq 1$ for all $w \in K_1$, that is, $g \in L_1$. Since T is a positive map we obtain $0 \leq Tg \leq T\mathbf{1}$ and obtain

$$\begin{aligned} Ty &= T((\alpha + \beta)g - \beta\mathbf{1}) = (\alpha + \beta)Tg - \beta T\mathbf{1} \\ &\leq (\alpha + \beta)T\mathbf{1} - \beta T\mathbf{1} = \alpha T\mathbf{1} \end{aligned}$$

and

$$Ty \geq -\beta T\mathbf{1}.$$

Thus it follows that $\|Ty\| \leq \|T\mathbf{1}\| \, \|y\|$, that is, T is norm-continuous and (since $\mathbf{1}$ is an element of the unit sphere) we finally obtain $\|T\| = \|T\mathbf{1}\|$.

If $L_1 \xrightarrow{T} L_2$ we obtain $T\mathbf{1} \in L_2$. Conversely, if $T\mathbf{1} \in L_2$ then, for $g_1 \in L_1$, that is, $0 \leq g_1 \leq \mathbf{1}$ (since T is positive) and we obtain $0 \leq Tg_1 < T\mathbf{1}$. Since $T\mathbf{1} \in L_2$ we obtain $T\mathbf{1} \leq \mathbf{1}$ and therefore obtain $Tg_1 \in L_2$.

For $\|T\| = \|T\mathbf{1}\|$ it follows that for each positive mapping T the map $\|T\|^{-1}T$ maps the set L_1 into L_2.

To each $\sigma(\mathscr{B}'_1, \mathscr{B}_2)$–$\sigma(\mathscr{B}'_2, \mathscr{B}_2)$-continuous mapping T of $\mathscr{B}'_1$ into $\mathscr{B}'_2$ there exists an adjoint mapping T' of $\mathscr{B}_2$ into $\mathscr{B}_1$ (see AIII, §5) for which

$$\mu_2(x, Ty) = \mu_1(T'x, y).$$

If T maps L_1 into L_2 then, for $w \in K_2$ we obtain:

$$0 \leq \mu_2(w, Ty) \leq 1$$

and also

$$0 \leq \mu_1(T'w, y) \leq 1$$

for all $y \in L_1$, that is, $T'w \in \check{K}_1$. T' is therefore an operation. The $\mathscr{B}$-continuous effect morphisms and the operations therefore uniquely correspond to adjoint mappings, as we have already seen in §4.1. In particular, T' is a mixture morphism if and only if $T\mathbf{1} = \mathbf{1}$.

In closing this section we shall now give criteria for the $\mathscr{B}$-continuity of an effect morphism.

Th. 4.2.4. *An effect morphism T is $\mathscr{B}$-continuous if each $\sigma(\mathscr{B}', \mathscr{B})$-convergent sequence g_ν in L_1 (therefore $g_\nu \to g \in L_1$) satisfies $Tg_\nu \to Tg$ in the $\sigma(\mathscr{B}'_2, \mathscr{B}_2)$ topology.*

PROOF. The $\sigma(\mathscr{B}', \mathscr{B})$-topology is metrizable (see AIII,§4) and it is therefore only necessary to consider sequences.

Th. 4.2.5. *An effect morphism T is $\mathscr{B}$-continuous if and only if, for each decreasing sequence of decision effects e_ν satisfying $\bigwedge_\nu e_\nu = 0$ the relation $Te_\nu \to 0$ in the $\sigma(\mathscr{B}'_2, \mathscr{B}_2)$-topology.*

PROOF. Since e_ν is decreasing it follows that Te_ν is decreasing in L_2 and therefore converges in the $\sigma(\mathscr{B}'_2, \mathscr{B}_2)$ topology: $Te_\nu \to g \in L_2$. Thus the hypothesis of the theorem is equivalent to the condition that $\bigwedge_\nu e_\nu = 0$ implies that $g = 0$.

If e_ν is a increasing sequence satisfying $\bigvee_\nu e_\nu = e$, it follows that $e - e_\nu$ is a decreasing sequence, and that $T(e - e_\nu) = Te - Te_\nu \to 0$, that is, Te_ν converges to Te. If e_n is a sequence of pairwise orthogonal decision effects satisfying $\sum_{n=1}^{\infty} e_n = e$, then $e_\nu = \sum_{n=1}^{\nu} e_n$ is an increasing sequence satisfying $\bigvee_\nu e_\nu = e$. Therefore, in the $\sigma(\mathscr{B}'_2, \mathscr{B}_2)$ topology $Te_\nu = \sum_{n=1}^{\nu} Te_n \to Te$, that is, $T(\sum_{n=1}^{\infty} e_n) = \sum_{n=1}^{\infty} Te_n$. Thus, for each $w_2 \in K$ we obtain

$$\mu_2\left(w_2, T\sum_{n=1}^{\infty} e_n\right) = \sum_{n=1}^{\infty} \mu_2(w_2, Te_n).$$

Thus $m(e) = \mu_2(w_2, Te)$ is a σ-additive measure on the lattice G. According to Gleason's theorem (see AIV, §12) there exists a uniquely determined $w_1 \in K_1$ for which

$$m(e) = \mu_2(w_2, Te) = \mu_1(w_1, e).$$

A mapping $K_2 \xrightarrow{S} K_1$ is defined by $w_2 \to w_1$ where it is clearly evident that it is affine, that is, is a mixture morphism. We need only show that $S' = T$, that is,

$$\mu_2(w, Tg) = \mu_2(w, S'g)$$

for all $w \in K_2$ and all $g \in L_1$. We then have $\mu_1(Sw, e) = \mu_2(w, Te)$ and therefore $\mu_2(w, Te) = \mu_2(w, S'e)$ for all $w \in K_2$ and for all $e \in G_1$.

Thus we obtain

$$\mu_2(w, Tx) = \mu_2(w, S'x)$$

for all x in the linear space spanned by G_1. Since T is norm-continuous, the same is true for all x in the norm-closed subspaces of $\mathscr{B}'_1$ spanned by G_1. According to the spectral representation theorem (see AIV, §8) the norm-closed subspace of $\mathscr{B}'_1$ spanned by G_1 is $\mathscr{B}'_1$ itself.

The converse is easy to see. If e_ν is a decreasing sequence of decision effects, then e_ν converges towards e in the $\sigma(\mathscr{B}'_1, \mathscr{B}_1)$ topology. If $\bigwedge_\nu e_\nu = 0$, from III, Th. 6.10 or IV, Th. 1.3.7 we find $e = 0$, that is, $e_\nu \to 0$. Therefore from the $\mathscr{B}$-continuity of T it follows that $Te_\nu \to 0$.

Th. 4.2.6. *A* p-*continuous* r-*morphism* h *determines a* $\mathscr{B}$-*continuous effect morphism* $T: \mathscr{D}_1 \to \mathscr{D}_2$ *by means of the map* $\psi\mathscr{F}_1 \xrightarrow{T} L_2$ *where the latter is given by* Th. 3.2. T' *is a* $\mathscr{D}$-*continuous mixture morphism.*

Proof. According to D 3.6 the map $\psi\mathscr{F}_1 \xrightarrow{T} L_2$ is $\mathscr{B}$-continuous since $\varphi\mathscr{Q}'$ is dense (in norm) in K. Thus T can be uniquely extended onto the σ-compact set L_1 in which $\psi\mathscr{F}_1$ is σ-dense because L_2 is σ-compact. It follows that $L_1 \xrightarrow{T} L_2$. Since $\psi\mathscr{F}_1 \xrightarrow{T} L_2$ is, according to the remark following Th. 3.2, a rational affine map, $L_1 \xrightarrow{T} L_2$ is an affine map. Therefore T may be extended as a linear map $\mathscr{B}'_1 \xrightarrow{T} \mathscr{B}'_2$. Thus, according to D 4.2.1, T is a $\mathscr{B}$-continuous effect morphism. According to Th. 3.2 $T\mathbf{1} = \mathbf{1}$. T' is therefore a mixture morphism.

Since T is norm-continuous, and since $\psi\mathscr{F}_1 \xrightarrow{T} \psi\mathscr{F}_2 \subset \mathscr{D}_2 \cap L_2$ and since $\psi\mathscr{F}_1$ is norm-dense in $\mathscr{D}_1 \cap L_1$ we obtain $\mathscr{D}_1 \cap L_1 \xrightarrow{T} \mathscr{D}_2 \cap L_2$. Since $\mathscr{D} \cap L$ spans the space $\mathscr{D}$ we obtain $\mathscr{D}_1 \xrightarrow{T} \mathscr{D}_2$. Therefore T' is $\mathscr{D}$-continuous.

Th. 4.2.7. *Let* h *be a* p-*isomorphism and let* h' *be a* r-*isomorphism*; *let* h *be dual to* h' (D 3.8). *Then the maps* $S\varphi_1(a) = \varphi_2(h(a))$ *and* $T\psi_2(f) = \psi_1(h'(f))$ *determine a* $\mathscr{D}$-*continuous linear map* $\mathscr{B}_1 \xrightarrow{S} \mathscr{B}_2$, *where* $K_1 \xrightarrow{S} K_2$ *is bijective and a* $\mathscr{B}$-*continuous linear map* $\mathscr{B}'_2 \xrightarrow{T} \mathscr{B}'_1$, *where* $L_2 \xrightarrow{T} L_1$ *is bijective,* $T = S'$. *In addition,* $T\mathscr{D}_2 = \mathscr{D}_1$.

The proof follows directly from Th. 3.4, Th. 4.1.7, Th. 4.2.6, and Th. 4.1.4.

4.3 Coexistent Operations and Coexistent Effect Morphisms

The norm-continuous maps S of $\mathscr{B}_1$ into $\mathscr{B}_2$ form a Banach algebra $\mathscr{A}(\mathscr{B}_1, \mathscr{B}_2)$ and therefore form a Banach space (see AIII, §6) with the norm

$$\begin{aligned}\|S\| &= \sup\{\|Sx\| \mid x \in \mathscr{B}_1, \|x\| \leq 1\}\\ &= \sup\{\|Sw\| \mid w \in K_1\}.\end{aligned}$$

In $\mathscr{A}(\mathscr{B}_1, \mathscr{B}_2)$ a positive cone $\mathscr{A}_+(\mathscr{B}_1, \mathscr{B}_2)$ is defined by $S \geq 0$ or equivalently $\{Sx \geq 0 \text{ for all } x \geq 0\}$.

$\mathscr{A}(\mathscr{B}_1, \mathscr{B}_2)$ is not only complete with respect to the norm topology but also with respect to the topology of simple convergence, that is, for a sequence S_n with $S_n w$ which is norm-convergent in $\mathscr{B}_2$ for all $w \in K_1$ there exists an $S \in \mathscr{A}(\mathscr{B}_1, \mathscr{B}_2)$ satisfying $\|S_n w - Sw\| \to 0$ for all $w \in K_1$ and therefore $\|S_n x - Sx\| \to 0$ for all $x \in \mathscr{B}_1$.

Th. 4.3.1. *A map* $S \in \mathscr{A}(\mathscr{B}_2, \mathscr{B}_2)$ *is an operation if and only if* S *is positive and* $\|S\| \leq 1$.

PROOF. From $\mu(Sw, \mathbf{1}) = \|Sw\|$ for $w \in K_1$ and S positive and $\|S\| = \sup\{\|Sw\| \mid w \in K_1\}$ it follows that $\|Sw\| \leq 1$, that is, $Sw \in \check{K}_2$.

D 4.3.1. We shall denote the set of operations—that is, the intersection of $\mathscr{A}_+(\mathscr{B}_1, \mathscr{B}_2)$ with the unit sphere—by Π.

D 4.3.2. An additive mapping of a Boolean ring $\Sigma \xrightarrow{\chi} \Pi$ for which $\chi(\varepsilon)$ is a mixture morphism (where ε is the unit element of Σ) will be called an operation measure.

For an effective ensemble w_0 we may define a uniform structure in Σ by means of the metric

$$d(\sigma_1, \sigma_2) = \mu(\chi(\sigma_1 \dotplus \sigma_2)w_0, \mathbf{1}), \tag{4.3.1}$$

which is identical to that defined in IV, Th. 5.2 provided that we set

$$W(\sigma) = \chi(\sigma)w_0 \tag{4.3.2}$$

or is identical with that defined by IV, Th. 1.4.1 if we rewrite (4.3.1) in the form

$$d(\sigma_1, \sigma_2) = \mu(w_0, \chi'(\sigma_1 \dotplus \sigma_2)\mathbf{1}) \tag{4.3.3}$$

and set $F(\sigma) = \chi'(\sigma)\mathbf{1}$.

Th. 4.3.2. *The mapping $\Sigma \xrightarrow{\chi} \Pi$ is uniformly continuous with respect to the metric in Σ and the uniform structure of simple convergence in Π.*

PROOF. According to IV, Th. 5.2 for each fixed w the map $\Sigma \xrightarrow{\chi w} \check{K}_2$ is uniformly continuous with respect to the norm topology in $\check{K}_2$. This implies that the map $\Sigma \xrightarrow{\chi} \Pi$ is uniformly continuous with respect to the uniform structure of simple convergence in Π.

From Th. 4.3.2 and the fact that Π is complete in the topology of simple convergence, it follows directly from IV, Th. 1.4.3 that Σ can be completed and that χ can be extended on the completion..

Th. 4.3.3. *$\Sigma \xrightarrow{\chi} \Pi$ together with a complete and separable Σ determines for each $w \in K_1$ a preparator of the ensemble $\chi(\varepsilon)w$ by means of the map $\Sigma \xrightarrow{\chi w} \check{K}_2$.*

PROOF. The proof is a simple corollary of IV, D 5.4 and the preceding results.

The map $\Sigma \xrightarrow{\chi} \Pi$ is uniquely defined by the maps $\Sigma \xrightarrow{\chi w} \check{K}_2$ for all $w \in K_1$. We therefore define:

D 4.3.3. We shall call an additive measure $\Sigma \xrightarrow{\chi} \Pi$ on a complete separable Boolean ring Σ a trans-preparator.

Th. 4.3.4. *Let Σ be complete and separable, and let an additive measure χ be defined by $\Sigma \xrightarrow{\chi} \Pi$. Then to each $\chi(\sigma)$ (as a map $\mathscr{B}_1 \xrightarrow{\chi(\sigma)} \mathscr{B}_2$) the adjoint map $\chi'(\sigma)$ (as a map $\mathscr{B}'_2 \xrightarrow{\chi'(\sigma)} \mathscr{B}'_1$, we may define an additive map $\Sigma \xrightarrow{\chi'} P$ (where P is the set of $\mathscr{B}$-continuous effect morphisms). For each $g \in L_2$ an additive measure is defined by $\Sigma \xrightarrow{\chi' g} L_1$. For $g = \mathbf{1}$ the map $\Sigma \xrightarrow{\chi' \mathbf{1}} L$ is an observable.*

PROOF. The proof follows simply, noting that $\chi'(\varepsilon)\mathbf{1} = \mathbf{1}$ because $\chi'(\varepsilon)w \in K_2$.

D 4.3.4. The map $\Sigma \xrightarrow{\chi'} P$ which is conjugate to a trans-preparator $\Sigma \xrightarrow{\chi} \Pi$ is called the adjoint effect transformer to $\Sigma \xrightarrow{\chi} \Pi$. The observable $\Sigma \xrightarrow{\chi' \mathbf{1}} L_1$ is said to be the observable associated with the trans-preparator.

D 4.3.5. A set $\tilde{\mathscr{A}}$ of operations is said to be coexistent if there exists a Boolean ring Σ and an additive measure $\Sigma \xrightarrow{\chi} \Pi$ for which $\tilde{\mathscr{A}} \subset \chi\Sigma$.

We shall not carry out further analysis of the trans-prepartors in a manner which is similar to the analysis of observables and preparators which was presented in IV. In XVII, §4 we shall present a number of applications in order to become familiar with the concepts presented in §4.3.

5 Isomorphisms and Automorphisms of Ensembles and Effects

In quantum mechanics automorphisms of effects and ensembles play an important role. For this reason we must present a careful and precise explanation of the structure of these maps. In addition, we shall investigate the question of the possibility of the extension of bijective maps, for example, isomorphisms of decision effects onto Banach spaces, that is, to effect morphisms.

D 5.1. If T is a bijective map of L_1 onto L_2 which satisfies the following condition:

$$\{g_1, g_2 \in L_1, g_1 \leq g_2 \Leftrightarrow Tg_1 \leq Tg_2\}$$

then we say that T is an order isomorphism of L_1 onto L_2.

In D 5.1 we therefore consider the structure type: L as an ordered set.

Since $\mathbf{1}$ is the largest and $\mathbf{0}$ is the smallest element of L_1 (and correspondingly for L_2) it follows that $T\mathbf{1} = \mathbf{1}$ and $T\mathbf{0} = \mathbf{0}$.

Since $\mathbf{1} - g$ corresponds to the case in which g does not respond, the following comprehensive structure type is suggested: L is an ordered set with a dual automorphism $g \to g^* = \mathbf{1} - g$. Thus we obtain the following restricted set of isomorphsims satisfying D 5.1:

D 5.2. An order isomorphism T of L_1 onto L_2 is called a $*$-isomorphism if $T(\mathbf{1} - g) = \mathbf{1} - Tg$ for all $g \in L_1$. The same considerations can be repeated if we consider the subset G of L as an ordered set (as a lattice) and, in the second case, as an ordered set with the dual automorphism $e \longrightarrow e^{\perp}$. The isomorphisms of G as an ordered set are known as lattice isomorphisms.

D 5.3. A lattice isomorphism T of G_1 onto G_2 is said to be an $\perp$-isomorphism of G_1 onto G_2 providing $T(e^{\perp}) = (Te)^{\perp}$.

We may also consider the following structure type: L together with the map $(g_1, g_2) \longrightarrow g_1 + g_2$ where $g_1 + g_2 < \mathbf{1}$.

According to D 4.2.1 an isomorphism of L with respect to this structure type is called an effect isomorphism. If an effect isomorphism $L_1 \rightarrow L_2$ is $\mathscr{B}$-continuous, then so is the inverse $L_2 \longrightarrow L_1$ since L_1, L_2 are compact (see §4.2).

If $L_1 = L_2$ or if $G_1 = G_2$ we then use the expression automorphism instead of isomorphism.

Th. 5.1. *Every effect isomorphism T of L_1 onto L_2 is an $*$-isomorphism and T may be uniquely extended as a linear mapping of $\mathscr{B}'_1$ onto $\mathscr{B}'_2$; then T will be a norm preserving isomorphism of the Banach spaces $\mathscr{B}'_1$ and $\mathscr{B}'_2$. If T is $\mathscr{B}$-continuous then T' is a mixture isomorphism of K_2 onto K_1 and an isomorphism of the Banach spaces $\mathscr{B}_2, \mathscr{B}_1$ and T^{-1} is also $\mathscr{B}$-continuous. If T is $\mathscr{B}$-continuous then the restriction of T onto G_1 is a $\perp$-isomorphism of G_1 onto G_2.*

PROOF. From $g + (\mathbf{1} - g) = \mathbf{1}$ it follows that $Tg + T(\mathbf{1} - g) = T\mathbf{1}$. From Th. 4.2.1 it follows that T is positive and preserves the order. Therefore $T\mathbf{1} = \mathbf{1}$ and $T(\mathbf{1} - g) = \mathbf{1} - Tg$, that is, T is a $*$-isomorphism. From Th. 4.2.1 it follows that the map T may be extended on $\mathscr{B}'_1$ as a bijective mapping from $\mathscr{B}'_1$ to $\mathscr{B}'_2$; in addition, since $\|Ty\| \leq \|y\|$ and since $\|T^{-1}\tilde{y}\| \leq \|\tilde{y}\|$ we find that T preserves the norm, and it follows that T is an isomorphism of the Banach spaces.

If T is $\mathscr{B}$-continuous, then, according to §4.2, T' is a mixture morphism of K_2 into K_1. According to Th. 4.1.4 T' is a mixture isomorphism and that $\mathscr{B}_2 \xrightarrow{T'} \mathscr{B}_1$ is an isomorphism of the Banach spaces. Since $(T^{-1})' = (T')^{-1}$ it follows that T' is $\mathscr{B}$-continuous.

Let $e \in G_1$. Then, from $\mu(w, Te) = \mu(T'w, e)$ it follows that $w \in K_0(Te) \Leftrightarrow T'w \in K_0(e)$. Therefore there exists an $e' \in G_2$ such that $K_0(Te) = K_0(e')$ and $Te < e'$. From this result it follows that $e \leq T^{-1}e'$. From $w \in K_0(e') = K_0(Te)$ it follows that $\mu(w, e') = 0$ and that $\mu(w, TT^{-1}e') = \mu(T'w, T^{-1}e') = 0$, that is, $T^{-1}e' \in L_0K_0(e)$ and therefore $T^{-1}e' \leq e$. Therefore $T^{-1}e' = e$, that is, $Te = e' \in G_2$.

The last property may be proved more easily if we use the fact that G is the set of extreme points of L (see III, Th. 6.6).

From $T(\lambda g + (1 - \lambda)g) = \lambda Tg + (1 - \lambda)Tg$ it follows that, under a bijective linear mapping of L_1 onto L_2 the extreme points of L_1 are bijectively mapped onto the extreme points of L_2. Since $T(\mathbf{1} - e) = \mathbf{1} - Te$ it follows that $(Te)^{\perp} = T(e^{\perp})$.

We may obtain a stronger result than Th. 5.1 if we use the fact that for $g \in L$ the spectral representation

$$g = \int_0^1 \lambda \, de(\lambda) \tag{5.1}$$

holds (see AIV, §8).

Th. 5.2. *The restriction of a $*$-isomorphism $L_1 \xrightarrow{T} L_2$ to G_1 is a $\perp$-isomorphism $G_1 \to G_2$.*

PROOF. Let $e \in G_1$; suppose that $Te = g \in L_2$. Then we obtain $T(\mathbf{1} - e) = \mathbf{1} - g$. Since $g' < e$ and $g' < \mathbf{1} - e = e^{\perp}$ implies that $g' = 0$ it follows that, except for the null element, there exists no element in L_2 which is smaller than g and $\mathbf{1} - g$. From (5.1) it follows that this is the case if and only if the spectrum of $e(\lambda)$ contains only 0 and 1, that is, $g \in G_2$. Therefore TG_1 is mapped isomorphically onto G_2. Since $Te^{\perp} = T(\mathbf{1} - e) = \mathbf{1} - Te = (Te)^{\perp}$. T is a $\perp$-isomorphism of G_1 onto G_2.

We now have the following situation: An effect isomorphism of L_1 onto L_2 is also a $*$-isomorphism of L_1 onto L_2.

Every $*$-isomorphism restricted to G_1 becomes a $\perp$-isomorphism of the lattices G_1, G_2.

We now turn to the converse problem: When is it possible to extend a $\perp$-isomorphism of G_1 onto G_2 to a $*$-isomorphism of $L_1 \to L_2$? When is a $*$-isomorphism of $L_1 \to L_2$ an effect isomorphism? . . . etc.

Th. 5.3. *Let T be a mapping of H_1 into G_2 where H_1 is a $\sigma(\mathscr{B}'_1, \mathscr{B}_1)$-dense subset of G_1. To each $w \in K_2$ there may exist an $x' \in \mathscr{B}_1$ such that $\mu_2(w, Te) = \mu_1(x'_1, e)$ for all $e \in H_1$. Then T has a unique extension as a linear $\sigma(\mathscr{B}'_1, \mathscr{B}_1)$-$\sigma(\mathscr{B}'_2, \mathscr{B}_2)$-continuous map of $\mathscr{B}'_1$ into $\mathscr{B}'_2$. In addition $TL_1 \subset L_2$, that is, T is a $\mathscr{B}$-continuous effect morphism.*

PROOF. By hypothesis $\mu_2(w, Te)$ is, as a function defined on H_1, the restriction of a linear $\sigma(\mathscr{B}'_1, \mathscr{B}_1)$-continuous functional over $\mathscr{B}'_1$. x' is uniquely determined because H_1 is $\sigma(\mathscr{B}'_1, \mathscr{B}_1)$-dense in G_1 and $\mathscr{B}'_1$ is the $\sigma(\mathscr{B}'_1, \mathscr{B}_1)$-closure of the linear span of G_1.

From $0 \leq \mu_2(w, Te) \leq 1$ it follows that $0 \leq \mu_1(x', e) \leq 1$ on H_1 and thus $0 \leq \mu_1(x', g) \leq 1$ on the $\sigma(\mathscr{B}'_1, \mathscr{B}^1)$-closed convex set L_1 which is generated by G_1 and therefore also by H_1. Thus it follows that $x' = \lambda w'$ where $0 \leq \lambda \leq 1$ and $w' \in K_1$, that is, $x' \in \check{K}_1$.

It is easy to see that an operation S is defined as a mapping of K_2 into $\check{K}_1$ by $w \to x'$, and that this map may be extended to a map of $\mathscr{B}_2$ into $\mathscr{B}_1$. The dual map S' corresponding to S is identical on H_2 to the previously defined map T. Hence we find that S' is the extension of T on $\mathscr{B}'_1$, and that this extension is unique because T is $\sigma(\mathscr{B}'_1, \mathscr{B}_1)$-$\sigma(\mathscr{B}'_2, \mathscr{B}_2)$-continuous and that the space spanned by H_1 is $\sigma(\mathscr{B}'_1, \mathscr{B}_1)$-dense in $\mathscr{B}'_1$! Therefore S is equal to T', the map which is dual to T. Since T' is an operation, T is itself an effect morphism.

Th. 5.4. *If, for the map $T = S'$ which was defined in* Th. 5.3, *the set G_2 lies in the $\sigma(\mathscr{B}'_2, \mathscr{B}_2)$-closure of TG_1, then T is surjective as a mapping of L_1 onto L_2 and of $\mathscr{B}'_1$ onto $\mathscr{B}'_2$ and, in addition, T' is a mixture morphism $K_2 \to K_1$ and T' is an injective map of $\mathscr{B}_2$ into $\mathscr{B}_1$.*

PROOF. Since $G_1 \subset L_1$ and $TG_1 \subset TL_1$ and since TL_1 is $\sigma(\mathscr{B}'_2, \mathscr{B}_2)$-compact (because T is $\mathscr{B}$-continuous and L_1 is $\sigma(\mathscr{B}'_1, \mathscr{B}_1)$-compact!) the $\sigma(\mathscr{B}'_1, \mathscr{B}_1)$-closure of TG_1 lies in TL_1. Then, according to our hypothesis, $G_2 \subset TL_1$. Since TL_1 is convex and compact, and since $L_2 = \overline{\text{co}}\, G_2$ we therefore obtain $TL_1 = L_2$. Since

L_2 spans all of $\mathscr{B}'_2$ we obtain $T\mathscr{B}'_1 = \mathscr{B}'_2$. Since T preserves the order and $TL_1 = L_2$ we obtain $T\mathbf{1} = \mathbf{1}$. Thus it follows that $\mu(T'w, \mathbf{1}) = \mu(w, T\mathbf{1}) = \mu(w, \mathbf{1}) = 1$, that is, $TK_2 \subset K_1$. T' is injective since $T\mathscr{B}'_1 = \mathscr{B}'_2$.

If, in addition, T is injective onto L_1, then T is a $\mathscr{B}$-continuous effect isomorphism and T' is a mixture isomorphism.

Is it possible to determine whether the extension of T onto L_1 is injective merely by looking at the map T only on G_1?

Th. 5.5. *If we assume the hypothesis of* Th. 5.3 *and the assumptions $TG_1 \subset G_2$ then T is injective onto $\mathscr{B}'_1$ if and only if, for $e \in G$ and $e \neq 0$ it follows that $Te \neq 0$.*

PROOF. Since $y = \alpha g - \beta\mathbf{1}$, for each $y \in \mathscr{B}'_1$ we obtain

$$y = \int_{-\beta}^{\alpha-\beta} \lambda \, de(\lambda)$$

from which it follows that

$$Ty = \int_{-\beta}^{\alpha-\beta} \lambda \, d(Te(\lambda)).$$

Since T preserves the order, $Te(\lambda)$ is, for increasing λ, an increasing sequence of elements of G. Since $Te \neq 0$ for $e \neq 0$ it follows that $Te(\lambda_1) - Te(\lambda_2) \neq 0$ for $e(\lambda_1) - e(\lambda_2) \neq 0$ and we therefore obtain $Ty \neq 0$ for $y \neq 0$.

Th. 5.6. *Let e_ν be a pairwise orthogonal set of elements of G_1. Then, for a $\perp$-isomorphism T (and for any lattice homomorphism satisfying $T(e^\perp) = (Te)^\perp$) of G_1 into G_2 the Te_ν are pairwise orthogonal and $T(\sum_\nu e_\nu) = \sum_\nu Te_\nu$.*

PROOF. From $e_1 \perp e_2$—that is, $e_1 \leq e_2^\perp$—it follows that $Te_1 \leq T(e_2^\perp) = (Te_2)^\perp$, that is, $Te_1 \perp Te_2$. Since the Te_ν are pairwise orthogonal and since T is a lattice homomorphism it follows that $T(\sum_\nu e_\nu) = T(\bigvee_\nu e_\nu) = \bigvee_\nu (Te_\nu) = \sum_\nu Te_\nu$.

Th. 5.7. *Let T be a lattice homomorphism of G_1 into G_2 for which $Te^\perp = (Te)^\perp$ (for example, a $\perp$-isomorphism of G_1 onto G_2). Then, by applying Gleason's theorem (see* AIV, §12*) to G_1 it follows that T may be uniquely extended as a linear $\sigma(\mathscr{B}'_1, \mathscr{B}_1)$-$\sigma(\mathscr{B}'_2, \mathscr{B}_2)$-continuous map $T\colon \mathscr{B}'_1 \to \mathscr{B}'_2$. The extension of T is a $\mathscr{B}$-continuous effect morphism.*

If TG_1 is $\sigma(\mathscr{B}'_2, \mathscr{B}_2)$-dense in G_2 then T is a surjective map of L_1 onto L_2 and of $\mathscr{B}'_1$ onto $\mathscr{B}'_2$ and T' is injective as a map of $\mathscr{B}_2$ into $\mathscr{B}_1$.

If $Te \neq 0$ for $e \neq 0$, $e \in G_1$ then T is a $\mathscr{B}$-continuous effect isomorphism.

PROOF. For all $w \in K_2$ $\mu_2(w, Te)$ is a positive function over G_1, which, according to Th. 5.6 satisfies $\mu_2(w, T(\sum_\nu e_\nu)) = \sum_\nu \mu_2(w, Te_\nu)$ and $\mu_2(w, T\mathbf{1}) = 1$. Then, according to Gleason's theorem there exists an $x \in \mathscr{B}_1$ for which $\mu(w, Te) = \mu(x, e)$. The remainder of the theorem follows directly from Th. 5.3, Th. 5.4, and Th. 5.5.

Th. 5.8. *The restriction of an order isomorphism $L_1 \xrightarrow{T} L_2$ onto G_1 results in a lattice isomorphism of G_1 onto G_2.*

PROOF. Since both T and T^{-1} are order isomorphisms, we need only show that $TG_1 \subset G_2$.

Let $A(G)$ denote the set of atoms of the lattice G. Let $p \in G_1$ be an atom of G_1. We will show that $Tp \in A(G_2)$. We note that all $g \in L_1$ which satisfy $g \leq p$ are of the form λp, where $0 \leq \lambda \leq 1$. The set $\{g \mid g \leq p\}$ is therefore totally ordered, and so is the set $\{g' \mid g' \leq Tp\}$ since T is an order isomorphism of L_1 onto L_2. This is the case only if $Tp = \alpha p'$, where $p' \in A(G)$. Suppose that $\alpha \neq 1$, then $T^{-1}p' \geq p$ and $T^{-1}p' \neq p$ and $T^{-1}p' = \beta p''$, where $p'' \in A(G_1)$, which is impossible. Therefore $\alpha = 1$, that is, $Tp \in A(G_2)$. Therefore T generates a bijective mapping of $A(G_1)$ onto $A(G_2)$.

If $e \in G_1$ then for every $p \in A(G_1)$ for which $p \leq e$ we obtain $Tp \leq Te$. If $g \geq e_\lambda$ for $\lambda \in \Lambda$ we obtain $K_1(g) \supset K_1(e_\lambda)$ and $K_1(g) \supset \bigcup_{\lambda \in \Lambda} K_1(e_\lambda) = K_1(\bigvee_\lambda e_\lambda)$ from which it follows that $g \geq \bigvee_{\lambda \in \Lambda} e_\lambda$. Therefore we obtain:

$$Te \geq \bigvee_{\substack{p \leq e \\ p \in A(G_1)}} Tp = e' \in G_2,$$

a similar result holds for T^{-1}:

$$T^{-1}e' \geq \bigvee_{\substack{q \in e' \\ q \in A(G_2)}} T^{-1}q = e'' \in G_1.$$

Since $Tp \in A(G_2)$ and $Tp \leq e'$ holds for all $p \leq e$, it follows that all $p \in A(G)$ for which $p \leq e$ are elements of the set of all $T^{-1}q$ (where $q \leq e'$, $q \in A(G_2)$), that is,

$$e'' \geq \bigvee_{\substack{p \leq e \\ p \in A(G_1)}} p = e.$$

Since $Te > e'$ we obtain $e > T^{-1}e' = e''$; therefore we obtain $e = e''$ and $Te = Te'' \leq e'$, and we finally obtain $Te = e'$.

Th. 5.9. *If, in addition to the hypothesis of* Th. 5.8 *we also assume that the* Tg, $T(\mathbf{1} - g)$ *are coexistent for all* $g \in L_1$, *then the restriction of* T *on* $G_1 \to G_2$ *is a* $\perp$*-isomorphism.*

PROOF. From $\{Tg, T(\mathbf{1} - g)\}$ is coexistent, for $g = e \in G_1$ it follows that $\{Te, T(e^\perp)\}$ are coexistent and are therefore commensurable. Since $e \wedge e^\perp = 0$ we also obtain $(Te) \wedge (Te^\perp) = 0$ and also $T(e^\perp) \perp Te$. Since $e \vee e^\perp = \mathbf{1}$ we obtain $Te \vee Te^\perp = \mathbf{1}$ and finally $T(e^\perp) = (Te)^\perp$.

Th. 5.10. *Let* G_1, G_2 *be two atomic lattices, and let* $A(G_1)$, $A(G_2)$ *denote their sets of atoms, respectively. Let* T *be a bijective mapping of* $A(G_1)$ *onto* $A(G_2)$ *for which both* T *and* T^{-1} *maps orthogonal atoms into orthogonal atoms, that is,* T *is an isomorphic map of* $A(G_1)$ *onto* $A(G_2)$ *with respect to the species of structure determined by* $\perp$. *Then* T *may be uniquely extended to a* $\perp$*-isomorphism from* $G_1 \to G_2$.

PROOF. For each $e \in G_1$ the set of all atoms for which $p \leq e$ is uniquely determined and satisfies

$$e = \bigvee_{\substack{p \leq e \\ p \in A(G_1)}} p.$$

For each order isomorphic mapping $\tilde{T}$ of G_1 onto G_2 the following equation holds:

$$\tilde{T}e = \bigvee_{\substack{p \in e \\ p \in A(G_1)}} (Tp).$$

For this reason we define the extension of T (first defined only on $A(G_1)$) as follows:

$$Te = \bigvee_{\substack{p \leq e \\ p \in A(G_1)}} (Tp).$$

Then it follows that $e_1 \leq e_2$ implies that $Te_1 \leq Te_2$.

For the set of atoms $q \leq e^\perp$ it follows that $Te^\perp = \bigvee_{q \leq e^\perp} (Tq)$. Since $p \perp q$ it follows that $Tp \perp Tq$; we therefore obtain $Te^\perp \perp Te$. Since there exists a complete system of pairwise orthogonal p_ν, q_μ satisfying $p_\nu \leq e$ and $q_\mu < e^\perp$ and $(\bigvee_\nu p_\nu) \vee (\bigvee_\mu q_\mu) = \mathbf{1}$, we find that

$$\left(\bigvee_\nu Tp_\nu\right) \vee \left(\bigvee_\mu Tq_\mu\right) = \mathbf{1}$$

otherwise there would be an atom r which would be orthogonal to all Tp_ν, Tq_μ so that $T^{-1}r$ would be orthogonal to all p_ν, q_μ in contradiction to the condition $(\bigvee_\nu p_\nu) \vee (\bigvee_\mu q_\mu) = \mathbf{1}$. From $(\bigvee_\nu Tp_\nu) \vee (\bigvee_\mu Tq_\mu) = \mathbf{1}$ and $\bigvee_\nu Tp_\nu \leq Te$ and $\bigvee_\mu Tq_\mu \leq Te^\perp$ it then follows that $Te^\perp = (Te)^\perp$, $\bigvee_\nu Tp_\nu = Te$ and $\bigvee_\mu Tq_\mu = Te^\perp$.

Suppose there exists an atom r for which $r \leq Te$. Then we also obtain $r \perp Te^\perp$, that is, $r \perp$ to all Tq for which $q \leq e^\perp$. In this way $T^{-1}r \perp$ to all $q \leq e^\perp$ and we obtain $T^{-1}r \leq e$. That is, each atom $r \leq Te$ may be obtained as the image Tp of an atom $p \leq e$. Thus $Te_1 = Te_2$ implies that $e_1 = e_2$ and

$$e = \bigvee_{\substack{r \leq Te \\ r \in A(G_2)}} T^{-1}r$$

is proven.

Since the procedure presented above can also be applied to the extension of the map T^{-1} of $A(G_2)$ to $A(G_1)$ and

$$e = \bigvee_{\substack{r \leq Te \\ r \in A(G_2)}} T^{-1}r$$

holds, it follows that T is a bijective map from G_1 onto G_2.

Th. 5.11. *Let T be an order automorphism of L onto itself (for example, an effect automorphism). Therefore we obtain $TG = G$. Let q_ν denote the atoms of the center Z of G (see* IV, *end of* §1.3). *For this T there exists a bijective mapping p of the integers (a permutation) so that $T = \sum_\nu T_\nu$ such that $T_\mu q_\nu = \delta_{\mu\nu} q_{p(\nu)}$ is satisfied. The T_ν are order isomorphisms of L_ν onto $L_{p(\nu)}$ where the L_ν are equal to* $[\mathbf{0}, \mathbf{1}]$ *in $\mathscr{B}'(\mathscr{H}_\nu)$; $\mathscr{B}'(\mathscr{H}_\nu)$ is therefore isomorphic to the subspace of all operators of the form* $(0, 0, \ldots, A_\nu, \ldots)$ *in $\mathscr{B}'(\mathscr{H}_1, \mathscr{H}_2, \ldots)$. Each order automorphism T may, in this case, therefore be represented as a "direct sum" of order isomorphisms $L_\nu \xrightarrow{T_\nu} L_{p(\nu)}$ for the irreducible $L_\nu, L_{p(\nu)}$. If T is an $*$-automorphism, then the T_ν are $*$-isomorphisms. If T is an effect automorphism, then the T_ν are effect isomorphisms. If T is $\mathscr{B}$-continuous, then the T_ν are also $\mathscr{B}$-continuous.*

PROOF. Let G_ν be the sublattice of all $e \in G$ for which $e \leq q_\nu$. Each $e \in G$ may be uniquely written in the form $e = \sum_\nu e_\nu$ where $e_\nu \in G_\nu$. The following sums are to be understood in this way. From $e \leq q_\nu$ it follows that $Te < Tq_\nu = \sum_\mu (Tq_\nu)_\mu$. For $Te = \sum_\mu (Te)_\mu$ it follows that $(Te)_\mu < (Tq_\nu)_\mu$ for all $e \in G_\nu$. Therefore T generates an order isomorphic map of G_ν on the lattice $\tilde{G}$ of all $e' = \sum_\mu e_\mu$ where $e_\mu < (Tq_\nu)_\mu$. If $(Tq_\nu)_\mu \neq 0$ for more than one μ then the lattice $\tilde{G}$ would be reducible, in contradiction to the assumption that G_ν is irreducible. Therefore there exists, for each ν precisely one $\mu = p(\nu)$ for which $(Tq_\nu)_\mu \neq 0$. Since the same is also true for T^{-1} we must have $(Tq_\nu) = q_\mu$ in addition to $(Tq_\nu) < q_\mu$ and $p(\nu)$ must be a bijective map of the set of the ν. Therefore we obtain $Tq_\nu = q_{p(\nu)}$.

Since $g \in L$ and $g \leq q_\nu$ it follows that $Tg \leq Tq_\nu = q_{p(\nu)}$; therefore T generates an order isomorphism map T_ν of L_ν onto $L_{p(\nu)}$. Therefore it remains only to show that (for $g = \sum_\nu g_\nu$ and $g_\nu < q_\nu$) $Tg = \sum_\nu Tg_\nu$ in order to set $T = \sum_\nu T_\nu$ where we obtain $T_\nu g_\mu = 0$ for $\nu \neq \mu$.

Since $Tg = \sum_\mu (Tg)_\mu$ we must show that $(Tg)_{p(\nu)} = Tg_\nu$. Since $(Tg)_\mu < q_\mu$ we find that $T^{-1}(Tg)_{p(\nu)} = g'_\nu < q_\nu$ where $Tg'_\nu = (Tg)_{p(\nu)}$. It remains to show that $g'_\nu = g_\nu$. Thus from $(Tg)_{p(\nu)} < Tg$ it follows that $g'_\nu = T^{-1}(Tg)_{p(\nu)} < g$. From $g_\nu < g$ it follows that $Tg_\nu < Tg$ and since $Tg_\nu \leq q_{p(\nu)}$ we obtain $Tg_\nu < (Tg)_{p(\nu)}$, that is, $g_\nu < T^1(Tg)_{p(\nu)} = g'_\nu$; therefore $g'_\nu = g_\nu$.

If T is a $*$-automorphism, then $T(\mathbf{1} - g) = \mathbf{1} - Tg$. If $g < q_\nu$ and therefore $g' = q_\nu - g < q_\nu$ then $\mathbf{1} - g = q_\nu - g + \sum_{\mu \neq \nu} q_\mu$. Thus it follows that

$$T(\mathbf{1} - g) = T_\nu(q_\nu - g) + \sum_{\mu \neq \nu} q_{p(\mu)} = \mathbf{1} - Tg = \sum_\rho q_\rho - Tg$$
$$= \sum_\mu q_{p(\mu)} - Tg = \sum_{\mu \neq \nu} q_{p(\mu)} + q_{p(\nu)} - Tg.$$

Since $g < q_\nu$ we obtain $Tg < q_{p(\nu)}$ and therefore obtain $T(q_\nu - g) = q_{p(\nu)} - Tg$, that is, T_ν is a $*$-isomorphism since q_ν or $q_{p(\nu)}$ are the unit elements of L_ν or $L_{p(\nu)}$.

From $g = g_1 + g_2$ it follows that $g_\nu = g_{1\nu} + g_{2\nu}$. Thus it follows that, for an effect automorphism $Tg = Tg_1 + Tg_2$ and therefore

$$\sum_\nu T_\nu g = \sum_\nu T_\nu g_1 + \sum_\nu T_\nu g_2,$$

that is,

$$\sum_\nu T_\nu g_\nu = \sum_\nu T_\nu g_{1\nu} + \sum_\nu T_\nu g_{2\nu}$$

and we obtain (since the partition is unique) $T_\nu g_\nu = T_\nu g_{1\nu} + T_2 g_{2\nu}$, that is, T_ν is an effect isomorphism.

The $\mathscr{B}$-continuity of T_ν follows from that of T, in which T is applied to such g having only a single component $g = g_\nu$ which is different from 0.

On the basis of Th. 5.11 we are therefore interested in $*$-isomorphism or effect isomorphism between two irreducible systems L_1 and L_2, that is, $L_1 \subset \mathscr{B}'(\mathscr{H}_1)$, $L_2 \subset \mathscr{B}'(\mathscr{H}_2)$.

Since G is a Gleason lattice, and the spectral representation theorem holds for $\mathscr{B}'$, each $*$-automorphism of G may be continued to an automorphism T of the Banach space $\mathscr{B}'$ which is $\sigma(\mathscr{B}', \mathscr{B})$-continuous, and the adjoint mapping T' is an automorphism of $\mathscr{B}$. We obtain $T = \sum_\nu T_\nu$ and $T' = \sum_\nu T'_\nu$ where the T_ν are isomorphisms of $\mathscr{B}'_\nu$ on $\mathscr{B}'_{p(\nu)}$ and T'_ν are the adjoint isomorphisms of $\mathscr{B}_{p(\nu)}$ and $\mathscr{B}_\nu(\mathscr{B}_\mu = \mathscr{B}(\mathscr{H}_\mu), \mathscr{B}'_\mu = \mathscr{B}'(\mathscr{H}_\mu))$.

From the preceding theorems we obtain the following special case:

Th. 5.12. *Each $\perp$-isomorphism T of $G_1 \subset \mathscr{B}'(\mathscr{H}_1)$ onto $G_2 \subset \mathscr{B}'(\mathscr{H}_2)$ may be uniquely extended to a $\mathscr{B}$-continuous effect isomorphism T where the latter is also an isomorphism of the Banach space $\mathscr{B}'_1 = \mathscr{B}'(\mathscr{H}_1)$ onto $\mathscr{B}'_2 = \mathscr{B}'(\mathscr{H}_2)$. The adjoint map T' is an isomorphism of the Banach space $\mathscr{B}_2$ onto $\mathscr{B}_1$ where T' is a mixture isomorphism $K_2 \to K_1$.*

For $\varphi \in \mathscr{H}$ and $\|\varphi\| = 1$ let P_φ be the projection operator $Pf = \varphi\langle\varphi, f\rangle$. Then P_φ may be considered to be an element of $K \subset \mathscr{B}(\mathscr{H})$ as well as of $G \subset \mathscr{B}'(\mathscr{H})$. The following theorem is to be understood in this sense where $T: \mathscr{B}'_1 \to \mathscr{B}'_2$, $T': \mathscr{B}_2 \to \mathscr{B}_1$ and $T'P_\varphi$ means $P_\varphi \in K_2 \subset \mathscr{B}_2 = \mathscr{B}(\mathscr{H}_2)$ and TP_ψ means $P_\psi \in G_1 \subset \mathscr{B}'_1 = \mathscr{B}'(\mathscr{H}_1)$.

Th. 5.13. *Let T' be a mixture isomorphism. Then $T'P_\varphi = T^{-1}P_\varphi$. Since TP_ψ is an atom, $TP_\psi = P_{\psi'}$ where ψ' is determined by T and ψ up to a factor $e^{i\alpha}$(α is real).*

PROOF. $\mathrm{tr}((T'P_\varphi)e) = \mathrm{tr}(P_\varphi Te)$. For $e = T^{-1}P_\varphi = P_\psi$; therefore it follows that $\mathrm{tr}((T'P_\varphi)P_\psi) = 1$. For $e = \mathbf{1} - P_\psi$ we obtain $Te = \mathbf{1} - TP_\psi = \mathbf{1} - P_\varphi$ and therefore $\mathrm{tr}((T'P_\varphi)(\mathbf{1} - P_\psi)) = 0$. Thus we obtain $T'P_\varphi = T^{-1}P_\varphi$.

If $e_1 \perp e_2$ then $Te_1 \perp Te_2$ and $T^{-1}e_1 \perp T^{-1}e_2$. Therefore we obtain:

Th. 5.14. *For $T'P_{\varphi_\nu} = P_{\psi_\nu}$ and pairwise orthogonal φ_ν, the φ_ν are pairwise orthogonal. With $TP_{\psi_\nu} = P_{\varphi_\nu}$ and pairwise orthogonal ψ_ν the φ_ν are also pairwise orthogonal.*

Th. 5.15. *From $w = \sum_\nu \lambda_\nu P_{\varphi_\nu}$ (each $w \in K$ may be written in this form with pairwise orthogonal φ_ν) and $T'P_{\varphi_\nu} = T^{-1}P_{\varphi_\nu} = P_{\psi_\nu}$ it follows that*

$$T'w = \sum_\nu \lambda_\nu P_{\psi_\nu}.$$

PROOF. The proof follows directly from the fact that T' is linear and norm-continuous as a mixture morphism.

Th. 5.16. *Let $T'P_{\varphi_1} = P_{\psi_1}$, $T'P_{\varphi_2} = P_{\psi_2}$ (that is, $P_{\varphi_1} = TP_{\psi_1}$, $P_{\varphi_2} = TP_{\psi_2}$). Then $|\langle\varphi_1, \varphi_2\rangle|^2 = |\langle\psi_1, \psi_2\rangle|^2$.*

PROOF. $\mathrm{tr}(P_{\psi_1}P_{\psi_2}) = \mathrm{tr}((T'P_{\varphi_1})P_{\psi_2}) = \mathrm{tr}(P_{\varphi_1}(TP_{\psi_2})) = \mathrm{tr}(P_{\varphi_1}P_{\varphi_2})$.

Th. 5.17. *Let T be an isomorphism of $A(G_1)$ onto $A(G_2)$ (see* Th. 5.10*). Then T may be uniquely extended as a $\perp$-isomorphism of $G_1 \to G_2$. For $P_{\varphi_1} = TP_{\psi_1}$ and $P_{\varphi_2} = TP_{\psi_2}$ we obtain*

$$|\langle\varphi_1, \varphi_2\rangle|^2 = |\langle\psi_1, \psi_2\rangle|^2.$$

The proof is a direct consequence of Th. 5.10, Th. 5.12, and Th. 5.16.

Th. 5.18. *To each isomorphism T of $A(G_1)$ onto $A(G_2)$ there exists an isomorphism or anti-isomorphism U of $\mathscr{H}_1$ onto $\mathscr{H}_2$ which satisfies*

$$TP_\psi = P_\varphi \quad \text{and} \quad \varphi = U\psi; \qquad TP_\psi = UP_\varphi U^{-1},$$

where

$$\psi \in \mathscr{H}_1, \qquad \varphi \in \mathscr{H}_2.$$

PROOF. Let ψ_ν be a complete orthonormal basis for $\mathscr{H}_1$. A complete orthonormal basis φ_ν for $\mathscr{H}_2$ is defined by $TP_{\psi_\nu} = P_{\varphi_\nu}$ where the φ_ν are uniquely determined except for factors of the form $e^{i\alpha_\nu}$.

For $\psi_0 = \sum_{\nu=1}^{\infty} (1/2^{\nu/2})\psi_\nu$ and $TP_{\psi_0} = P_{\varphi_0}$ and $\varphi_0 = \sum_\nu x_\nu \varphi_\nu$ it follows that since $|\langle\varphi_0, \varphi_\nu\rangle|^2 = |\langle\psi_0, \psi_\nu\rangle|^2$ that $|x_\nu|^2 = 1/2^\nu$. The factors $e^{i\alpha_\nu}$ for the φ_ν are arbitrary, and may therefore be chosen such that $\varphi_0 = \sum_\nu (1/2^{\nu/2})\varphi_\nu$.

We will now set $\psi = \sum_\nu a_\nu \psi_\nu$ and set $\varphi = \sum_\nu b_\nu \varphi_\nu$ where $TP_\psi = P_\varphi$. We will now show that all $b_\nu = a_\nu$ or all $b_\nu = \bar{a}_\nu$.

From $|\langle\varphi_\nu, \varphi\rangle|^2 = |\langle\psi_\nu, \psi\rangle|^2$ it follows that $|a_\nu| = |b_\nu|$. Since T may be extended as a $\perp$-isomorphism on all of G_1 (by Th. 5.17) we shall consider the map T of the following projection operators: P_ψ, $P = \sum'_\nu P_{\psi_\nu}$ (where the $'$ means that we perform the sum only on a certain subset N' of the natural numbers) and $P^\perp$. We obtain $(P_\psi \vee P^\perp) \wedge P = P_\chi$ where $\chi = P\psi \|P\psi\|^{-1}$. For $Q = \sum'_\nu P_{\varphi\nu} = TP$ we obtain

$$(P_\varphi \vee Q^\perp) \wedge Q = TP_\chi = P_{\chi'},$$

where

$$\chi' = Q\varphi \|Q\varphi\|^{-1}.$$

That is, if ψ is mapped onto φ then every partial sum $\sum'_\nu a_\nu \psi_\nu$ is mapped onto $\sum'_\nu b_\nu \psi_\nu$ up to a normalization factor, which is, since $|a_\nu| = |b_\nu|$ identical: $\|P\psi\| = \|Q\varphi\|$.

From

$$|\langle\psi_0, P\psi\rangle| \, \|P\psi\|^{-1} = |\langle\varphi_0, Q\varphi\rangle| \, \|Q\varphi\|^{-1}$$

and since $\|P\psi\| = \|Q\varphi\|$ it follows that

$$\left|\sum_\nu{}' a_\nu \frac{1}{2^{\nu/2}}\right| = \left|\sum_\nu{}' b_\nu \frac{1}{2^{\nu/2}}\right| \tag{5.2}$$

for every $\sum'$ over a subset N' of the ν. Since a factor in φ and ψ are arbitrary, and $|a_\nu| = |b_\nu|$ we may choose $a_1 = b_1$ real and nonzero (in the case in which $a_1 = b_1 = 0$ we choose a different $\nu \neq 1$). From (5.2) we obtain especially:

$$\left|\frac{1}{2^{1/2}} a_1 + \frac{1}{2^{\nu/2}} a_\nu\right| = \left|\frac{1}{2^{1/2}} a_1 + \frac{1}{2^{\nu/2}} b_\nu\right|$$

for all ν. Thus it follows that either $a_\nu = b_\nu$ or $a_\nu = \bar{b}_\nu$. For each two different ν, μ from (5.2) it follows that

$$\left|\frac{1}{2^{\nu/2}} a_\nu + \frac{1}{2^{\mu/2}} a_\mu\right| = \left|\frac{1}{2^{\nu/2}} b_\nu + \frac{1}{2^{\mu/2}} b_\mu\right|.$$

If a_ν is not real, and $b_\nu = a_\nu$ then it follows that $b_\mu = a_\mu$; if a_ν is not real and if $b_\nu = \bar{a}_\nu$, then it follows that $b_\mu = \bar{a}_\mu$. Thus we can either have all $b_\nu = a_\nu$ or all $b_\nu = \bar{a}_\nu$.

We must distinguish between two different cases:

(1) $\psi_1 + i\psi_2$ is mapped onto $\varphi_1 + i\varphi_2$.

(2) $\psi_1 + i\psi_2$ is mapped onto $\varphi_1 - i\varphi_2$.

Case (1). Since every partial sum is mapped into a partial sum, it follows that $\psi_1 + i\psi_2 + \varphi_\nu \to \varphi_1 + i\psi_2 + \varphi_\nu$ and therefore we obtain $\psi_\nu + i\psi_2 \to \varphi_\nu + i\varphi_2$. From this result we obtain $\psi_\nu + i\psi_2 + i\psi_\mu \to \varphi_\nu + i\varphi_2 + i\varphi_\mu$ and also $\psi_\nu + i\psi_\mu \to \varphi_\nu + i\varphi_\mu$ for arbitrary pairs ν, μ.

From $\psi_\nu + i\psi_\mu \to \varphi_\nu + i\varphi_\mu$ it follows that (for real a, b)

$$\psi_\nu + i\psi_\mu + (a + ib)\psi_\rho \to \varphi_\nu + i\varphi_\mu + (a + ib)\varphi_\rho$$

and thus we obtain $\psi_\nu + (a + ib)\psi_\rho \to \varphi_\nu + (a + ib)\varphi_\rho$. Let $\sum_\nu c_\nu\psi_\nu$ be an arbitrary vector. We shall assume that c_{ν_1} is real and nonzero. Since $c_{\nu_1}\psi_{\nu_1} + c_\rho\psi_\rho \to c_{\nu_1}\varphi_{\nu_1} + c_\rho\psi_\rho$ for all ρ it follows that $\sum_\nu c_\nu\psi_\nu \to \sum_\nu c_\nu\varphi_\nu$.

Case (2). In the same way as Case (1) it is easy to show that if

$$\psi_\nu + i\psi_\mu \to \varphi_\nu - i\varphi_\mu$$

for all pairs, then $\psi_\nu + (a + ib)\psi_\rho \to \varphi_\nu + (a - ib)\varphi_\rho$ and therefore

$$\sum_\nu c_\nu\psi_\nu \to \sum_\nu \bar{c}_\nu\varphi_\nu.$$

An isomorphic map U of $\mathscr{H}_1$ onto $\mathscr{H}_2$ is defined by $U(\sum_\nu c_\nu\psi_\nu) = \sum_\nu c_\nu\varphi_\nu$; an anti-isomorphic map U is defined by $U(\sum_\nu c_\nu\psi_\nu) = \sum \bar{c}_\nu\varphi_\nu$ (see AIV, §13). In both cases $TP_\psi = P_\varphi$ where $\varphi = U\psi$. For an $f \in \mathscr{H}_2$ we obtain

$$(TP_\psi)f = P_\varphi f = \varphi\langle\varphi, f\rangle = U\psi\langle U\psi, f\rangle.$$

For $f = Ug$ we have $\langle U\psi, Ug\rangle = \langle\psi, g\rangle$ if U is an isomorphism (or $\langle U\psi, Ug\rangle = \langle g, \psi\rangle$ in the case in which U is an anti-isomorphism). Thus it follows that $(TP_\psi)f = U\psi\langle\psi, g\rangle$ (or $= (U\psi)\langle g, \psi\rangle$) and therefore $(TP_\psi)f = UP_\psi U^{-1}f$ (or $= (U\psi)\langle g, \psi\rangle = U[\psi\langle\psi, g\rangle] = UP_\psi U^{-1}f$), where $TP_\psi = UP_\psi U^{-1}$ is proven for both cases. In the last bracket we used the fact that, for an anti-isomorphic map $U(\alpha\psi) = \bar{\alpha}U\psi$.

From Th. 5.17 it follows that:

Th. 5.19. *The mapping T extended to all of G_1 (or the $\perp$-isomorphism T of G_1 onto G_2) has the form $Te = UeU^{-1}$ where U is either an isomorphism or anti-isomorphism of $\mathscr{H}_1$ onto $\mathscr{H}_2$.*

Th. 5.20. *Every isomorphism or anti-isomorphism U of $\mathscr{H}_1$ onto $\mathscr{H}_2$ generates a $\perp$-isomorphism of G_1 onto G_2 by means of the equation $Te = UeU^{-1}$. U is determined up to a phase factor $e^{i\alpha}$ by T.*

PROOF. It is easy to see that $e \to U_1eU_1^{-1}$ is a $\perp$-isomorphism of G_1 onto G_2. Let $U_1eU_1^{-1} = U_2eU_2^{-1}$. Then we obtain $U_2^{-1}U_1e = eU_2^{-1}U_1$. The unitary map $U_2^{-1}U_1$ of $\mathscr{H}_1$ onto itself commutes with all $e \in G_1$, and we therefore obtain $U_2^{-1}U_1 = e^{i\alpha}\mathbf{1}$ from which we obtain $U_1 = U_2e^{i\alpha}\mathbf{1} = e^{\pm i\alpha}U_2$ (where + or − depends on whether U_1, U_2 are isomorphisms or anti-isomorphisms.

Th. 5.21. *The $\perp$-isomorphism $Te = UeU^{-1}$ of G_1 onto G_2 may be extended to a $\mathscr{B}$-continuous effect isomorphism of L_1 upon L_2 by means of the equation $Tg = UgU^{-1}$. The uniquely determined isomorphism of $\mathscr{B}'_1$ onto $\mathscr{B}'_2$ described in* Th. 5.7 *is given by $Ty = UyU^{-1}$. $T'x = U^{-1}xU$ is the isomorphism of $\mathscr{B}_2$ onto $\mathscr{B}_1$ and is a mixture isomorphism of K_2 onto K_1.*

PROOF. The fact that $Ty = UyU^{-1}$ is an effect isomorphism and is an isomorphism of $\mathscr{B}'_1$ onto $\mathscr{B}'_2$ is clear. $\mathrm{tr}((T'w)e) = \mathrm{tr}(w(Te)) = \mathrm{tr}(wUeU^{-1})$; if U is unitary, it then follows that $\mathrm{tr}((T'w)e) = \mathrm{tr}(U^{-1}wUe)$ and, therefore, $T'w = U^{-1}wU$. Since $\mathrm{tr}(w(Te))$ is real, from AIV, §13 it follows that, for an anti-isomorphism $\mathrm{tr}(wUeU^{-1}) = \mathrm{tr}(U^{-1}wUe)$ and we therefore obtain $T'w = U^{-1}wU$.

Th. 5.22. *If the restriction of a $*$-automorphism $L \xrightarrow{\tilde{T}} L$ where $L \subset \mathscr{B}'(\mathscr{H})$ onto G satisfies $\tilde{T}e = e$ then $\tilde{T}g = g$ is satisfied, that is, $\tilde{T}$ is the identity map of L onto itself.*

PROOF. Since $\tilde{T}p = p$ for all $p \in A(G)$, it follows that, by analogy with the proof of Th. 5.8, $\tilde{T}(\lambda p) = \tau(\lambda, p)p$ for $0 \leq \lambda \leq 1$. We obtain $\tau(0, p) = 0$, $\tau(1, p) = 1$ and $\tau(\lambda, p)$ increases monotonically with λ.

Let p_ν be finitely many pairwise orthogonal elements of $A(G)$, it follows that, for all $g = \sum_\nu \lambda_\nu p_\nu$ where $0 < \lambda_\nu \leq 1$, that g^{-1} (where g^{-1} is the reciprocal operator to g) exists in the subspace $(\sum_\nu p_\nu)\mathscr{H}$. We now consider all atoms $P_\varphi \leq \sum_\nu p_\nu$, that is, all φ for which $\varphi \in (\sum_\nu p_\nu)\mathscr{H}$. We seek all values of λ, $0 \leq \lambda \leq 1$ for which $g \geq \lambda P_\varphi$. For the maximal $\bar{\lambda}$ for which $g \geq \bar{\lambda} P_\varphi$ there exists a χ (for which $\sum_\nu p_\nu \chi = \chi$) with $g\chi = \bar{\lambda} P_\varphi \chi$. With $\eta = g\chi$ (that is, $\chi = g^{-1}\eta$) we obtain $P_\varphi \eta = \eta$ and $\bar{\lambda}^{-1}\eta = P_\varphi g^{-1} P_\varphi \eta$ and we therefore obtain $\bar{\lambda}^{-1} = \langle \varphi, g^{-1}\varphi \rangle$. From this result it follows that for $g = \sum_{\nu \neq \mu} p_\nu + \beta p_\mu$ the following relationship holds:

$$\beta = \bar{\lambda}\beta(1 - \|p_\mu \varphi\|^2) + \bar{\lambda}\|p_\mu \varphi\|^2$$

and we therefore obtain

$$\|p_\mu \varphi\|^2 = \beta(1 - \bar{\lambda})\bar{\lambda}^{-1}(1 - \beta)^{-1}.$$

Since $g = \sum_{\nu \neq \mu} p_\nu + \beta p_\mu$ for all g for which $\sum_{\nu \neq \nu} p_\nu \leq g \leq \sum_\nu p_\nu$ we therefore obtain

$$\sum_{\nu \neq \mu} p_\mu \leq \tilde{T}g \leq \sum_\nu p_\nu,$$

that is,

$$\tilde{T}g = \sum_{\nu \neq \mu} p_\nu + \sigma_\mu(\beta) p_\mu .$$

$\tau(\bar{\lambda}, P_\varphi)$ must be the maximal value of a τ for which $\tilde{T}g \geq \tau P_\varphi$. From $\tilde{T}g \geq \tau(\bar{\lambda}, P_\varphi)P_\varphi$ where $\tau(\bar{\lambda}, P_\varphi)$ is the maximal τ it follows that $\tau(\bar{\lambda}, P_\varphi) = \langle \varphi, (\tilde{T}g)^{-1}\varphi \rangle$ and we obtain

$$\sigma_\mu(\beta) = \tau(\bar{\lambda}, P_\varphi)\sigma_\mu(\beta)(1 - \|p_\mu \varphi\|^2) + \tau(\bar{\lambda}, P_\varphi)\|p_\mu \varphi\|^2,$$

that is,

$$\|p_\mu \varphi\|^2 = \frac{\sigma_\mu(\beta)(1 - \tau(\bar{\lambda}, P_\varphi))}{\tau(\bar{\lambda}, P_\varphi)(1 - \sigma_\mu(\beta))}.$$

If we insert the above value of $\|p_\mu \varphi\|^2$, it follows that

$$\frac{\sigma_\mu(\beta)(1-\beta)}{\beta(1-\sigma_\mu(\beta))} = \frac{\tau(\bar{\lambda}, P_\varphi)(1-\bar{\lambda})}{\bar{\lambda}(1-\tau(\bar{\lambda}, P_\varphi))}. \tag{5.3}$$

In this formula we shall hold p_μ fixed, while we consider φ to vary in $(\sum_\nu p_\nu)\mathscr{H}$; we may set the value

$$\beta = \frac{\bar{\lambda}\|p_\mu \varphi\|^2}{1-\bar{\lambda}(1-\|p_\mu \varphi\|^2)}$$

in the above equation. The left-hand side depends therefore only on $\bar{\lambda}$ and $\|p_\mu \varphi\|^2$; therefore the right side can depend only on $\bar{\lambda}$ and $\|p_\mu \varphi\|^2$. If the Hilbert space $\mathscr{H}$ is more than two-dimensional then the right-hand side is independent of all other components $\|p_\nu \varphi\|^2$ $(\nu \neq \mu)$. Since the definition of $\tau(\bar{\lambda}, P_\varphi)$ does not depend on the choice of p_μ, it follows that the right side is independent of each $\|p_\lambda \varphi\|^2$ for every arbitrary atom p_λ in G and is therefore independent of P_φ (for all $\varphi \in (\sum_\nu p_\nu)\mathscr{H}$).

On the other hand we have

$$\bar{\lambda} = \frac{\beta}{\beta(1-\|p_\mu \varphi\|^2) + \|p_\mu \varphi\|^2}.$$

If in (5.3) we hold β on the left-hand side fixed, then the expression on the left side is a constant, while $\bar{\lambda}$ on the right-hand side (which is independent of P_φ) may vary, and for fixed $\beta \neq 0$, may vary between 0 and β. The right side depends only on $\bar{\lambda}$ and is therefore constant in the interval $1 > \bar{\lambda} > \beta$. Since β is arbitrary in $1 \geq \beta \geq 0$ it follows that the right side is constant for $1 \geq \bar{\lambda} > 0$ (for $\bar{\lambda} = 0$ we obtain $\tau(0, P_\varphi) = 0$). Since the left side of (5.3) is positive, there exists a constant $a > 0$ such that

$$\tau(\lambda, P_\varphi) = \frac{a\lambda}{1+\lambda a - \lambda}$$

(this is also the case for $\lambda = 0$!).

$\tau(\lambda, P_\varphi)$ is therefore independent of P_φ for all $\varphi \in (\sum_\nu p_\nu)\mathscr{H}$; $\tau(\lambda, P_\varphi) = \tau(\lambda)$.

Since each pair φ_1, φ_2 together with a $\varphi \in (\sum_\nu p_\nu)\mathscr{H}$ lies in a finite-dimensional subspace of $\mathscr{H}$ we therefore obtain

$$\tilde{T}(\lambda P_\varphi) = \tau(\lambda) P_\varphi \quad \text{for all } \varphi \in \mathscr{H}.$$

If $e \in G$ and if $g = \alpha e$, where $0 \leq \alpha \leq 1$, then g is uniquely determined by $g \geq \alpha P_\varphi$ for all $\varphi \in e\mathscr{H}$ and $g \ngeq (\alpha + \varepsilon)P_\varphi$ for $\varepsilon > 0$. Therefore $\tilde{T}g > \tau(\alpha)P_\varphi$ for all $\varphi \in e\mathscr{H}$ and $\tilde{T}g \ngeq \tau(\alpha + \varepsilon)P_\varphi$. Since $g \leq e$ we obtain $\tilde{T}g \leq e$ and therefore obtain $\tilde{T}(\alpha e) = \tau(\alpha)e$.

Let $g = (\mathbf{1} - e) + \beta e$; then, for all φ $g \geq \bar{\lambda}P_\varphi$, where $\bar{\lambda}^{-1} = \langle \varphi, g^{-1}\varphi \rangle = (1 - \|e\varphi\|^2) + \beta^{-1}\|e\varphi\|^2$ and we obtain $\tilde{T}g \geq \tau(\bar{\lambda})P_\varphi$. Thus it follows that $\tilde{T}g = (\mathbf{1} - e) + \tau(\beta)e$ because for $\tilde{T}g = (\mathbf{1} - e) + \tau(\beta)e$ it follows that $\tau(\bar{\lambda})^{-1} = (1 - \|e\varphi\|^2) + \tau(\beta)^{-1}\|e\varphi\|^2$, which, for the above value of λ and the form of the function τ leads to an identity.

Since $\tilde{T}$ is a $*$-automorphism of L, we must have $\tilde{T}(\mathbf{1} - \alpha e) = \mathbf{1} - \tau(\alpha)e$. Since $\mathbf{1} - \alpha e = (\mathbf{1} - e) + (1 - \alpha)e$, it follows that $\tilde{T}(\mathbf{1} - \alpha e) = (\mathbf{1} - e) + \tau(1 - \alpha)e$, that is, $1 - \tau(\alpha) = \tau(1 - \alpha)$ from which we obtain $a = 1$ and $\tau(\lambda) = \lambda$.

Since each $g \in L$ is determined by the maximal $\bar{\lambda}$ for which $g \geq \bar{\lambda} P_\varphi$ for all φ we obtain $\tilde{T}g = g$.

Th. 5.23. *Each $*$-isomorphism T of L_1 onto L_2 (where $L_1 \subset \mathscr{B}'(\mathscr{H}_1)$, $L_2 \subset \mathscr{B}'(\mathscr{H}_2)$) has the form $Tg = UgU^{-1}$ with an isomorphic or anti-isomorphic map U of $\mathscr{H}_1$ onto $\mathscr{H}_2$.*

PROOF. The restriction of T onto G_1 is a $\perp$-isomorphism. On G_1 we therefore obtain $Te = UeU^{-1}$ with either an isomorphism or anti-isomorphism map U of $\mathscr{H}_1$ onto $\mathscr{H}_2$. $\tilde{T}g = U^{-1}(Tg)U$ is a $*$-automorphism of L onto itself with $\tilde{T}e = e$ for $e \in G$. According to Th. 5.2.2 T is the identity map on all of L; therefore $Tg = UgU^{-1}$.

Here we have closed the circle: all $*$-isomorphisms of L_1 onto L_2 are determined in the above way. We have formulated these theorems such that we can begin the circle at any point, for example, with the isomorphic maps of $A(G_1)$ onto $A(G_2)$ and with Th. 5.17 or with all $\perp$-isomorphisms of G_1 onto G_2.

Th. 5.23 is not valid for reducible L_1, L_2, so that, in general each $*$-isomorphism is not necessarily an effect isomorphism. If $\mathscr{H}_1 = \mathscr{H}_2$ and $L_1 = L_2, G_1 = G_2$, we may only replace "iso" with "auto" in all the preceding theorems. An automorphism U of a Hilbert space onto itself will also be called a unitary mapping; similarly an anti-automorphism map U will be called an anti-unitary map.

Before closing this chapter we shall consider the $*$-automorphisms T of the whole reducible system $L \subset \mathscr{B}'(\mathscr{H}_1, \mathscr{H}_2, \ldots)$. Each $y \in \mathscr{B}'(\mathscr{H}_1, \mathscr{H}_2, \ldots)$ can be considered to be an operator $y = \sum_\nu y_\nu$ where each y_ν operates in $\mathscr{H}_\nu$.

Let $U_{p(\nu)\nu}$ denote the isomorphism (or anti-isomorphism) of $\mathscr{H}_\nu$ onto $\mathscr{H}_{p(\nu)}$. Then we obtain

$$Ty = \sum_\nu U_{p(\nu)\nu} y_\nu U_{p(\nu)\nu}^{-1}. \tag{5.4}$$

If we embed $\mathscr{H} = \bigcup_\nu \mathscr{H}_\nu$ into the Hilbert space

$$\bar{\mathscr{H}} = \sum_\nu \oplus \mathscr{H}_\nu$$

then we may consider a $y \in \mathscr{B}'(\mathscr{H}_1, \mathscr{H}_2, \ldots)$ to be an operator having the form

$$y\varphi = \sum_\nu y_\nu \varphi_\nu,$$

where

$$\varphi = \sum_\nu \varphi_\nu \quad \text{and} \quad \varphi_\nu \in \mathscr{H}_\nu.$$

If the $U_{p(\nu)\nu}$ are all isomorphisms, then a unitary operator is defined by

$$U\varphi = \sum_\nu U_{p(\nu)\nu} \varphi_\nu \tag{5.5}$$

and we obtain

$$Ty = UyU^{-1}. \tag{5.6}$$

In general (that is, if not all the $U_{p(\nu)\nu}$ are isomorphisms or are anti-isomorphisms) then the mapping defined by (5.5) is neither unitary or anti-unitary.

The map T' which is adjoint to T is given by

$$T'x = \sum_{\nu} U^{-1}_{p(\nu)\nu} x_{p(\nu)} U_{p(\nu)\nu} \tag{5.7}$$

or, according to (5.5), is given by

$$T'x = U^{-1}xU. \tag{5.8}$$

Here it is important to note that U is not only defined except for a factor $e^{i\alpha}$ by (5.5) and (5.6) but that with

$$\tilde{U}\varphi = \sum_{\nu} e^{i\alpha\nu} U_{p(\nu)\nu}\varphi_{\nu} \tag{5.9}$$

we have also

$$Ty = \tilde{U}y\tilde{U}^{-1} = UyU^{-1}. \tag{5.10}$$

With the proof of the preceding theorems we have, at the same time also investigated the structure of mixture isomorphisms because, according to §4.1 and §4.2, the mixture isomorphisms S and $\mathscr{B}$-continuous effect isomorphisms T correspond uniquely to the equations $S' = T$ and $T' = S$. Therefore we do not need to derive any additional theorems for mixture isomorphisms.

In deriving the structure of mixture isomorphisms we have, after all, not needed all the preceding theorems. These theorems have served to show "how few" assumptions about the maps $L_1 \to L_2$ or $G_1 \to G_2$ or $A(G_1) \to A(G_2)$ were already sufficient to determine the $\mathscr{B}$-continuous effect isomorphisms. If we had begun with the mixture isomorphism S, then we would only need the following theorems for $T = S'$:

(1) The restriction of T onto G is a $\perp$-isomorphism $G_1 \to G_2$ (see Th. 5.1).
(2) T maps $A(G_1)$ bijectively onto $A(G_2)$ and, correspondingly, T^{-1} maps $A(G_2)$ onto $A(G_1)$ where orthogonal atoms are mapped onto orthogonal atoms (this result follows directly from 1).
(3) Th. 5.11.
(4) Th. 5.13, Th. 5.14, Th. 5.15, and Th. 5.16.
(5) Th. 5.18.
(6) $T'w = Sw = U^{-1}wU$ follows directly from Th. 5.21 and Th. 5.7.

Th. 5.22 and 5.23 were not needed.

With $T' = S$ and for $x \in \mathscr{B}(\mathscr{H}_1, \mathscr{H}_2, \ldots)$, from (5.7) we finally obtain

$$Sx = \sum_{\nu} U^{-1}_{p(\nu)\nu} x_{p(\nu)} U_{p(\nu)\nu}. \tag{5.11}$$

A mixture isomorphism is $\mathscr{D}$-continuous if $T = S'$ transforms the space $\mathscr{D}$ into itself. If we assume that the center Z of G is a subset of $\mathscr{D}$ then the components y_ν of a $y \in \mathscr{D}$ form a Banach subspace $\mathscr{D}_\nu$ of $\mathscr{B}'(\mathscr{H}_\nu)$. Then S is $\mathscr{D}$-continuous only if $U_{p(\nu)\nu} y_\nu U^{-1}_{p(\nu)\nu} \in \mathscr{D}_{p(\nu)}$ for $y_\nu \in \mathscr{D}_\nu$.

Th. 5.24. *Every* p*-continuous* r*-isomorphism h determines (by means of the map $\psi\mathscr{F}_1 \xrightarrow{T} L_2$ defined in* Th. 3.2) *a $\mathscr{B}$-continuous effect isomorphism T which maps $\mathscr{D}_1$ isomorphically onto $\mathscr{D}_2$ (as Banach spaces). T' is a $\mathscr{D}$-continuous mixture isomorphism.*

Proof. According to Th. 4.2.6 we need only show that $L_1 \xrightarrow{T} L_2$ is bijective and that $\mathscr{D}_1 \xrightarrow{T} \mathscr{D}_2$ is bijective. Since $\psi\mathscr{F}_1$ is σ-dense in L_1 and $\psi\mathscr{F}_2$ is σ-dense in L_2 it follows from $\psi\mathscr{F}_1 \xrightarrow{T} \psi\mathscr{F}_2$ that $L_1 \xrightarrow{T} L_2$ is surjective. For the map $\tilde{T}$ corresponding to h^{-1} on $\psi\mathscr{F}_1$ and $\psi\mathscr{F}_2$ we obtain $T\tilde{T} = \mathbf{1}$ and $\tilde{T}T = \mathbf{1}$. Since $L_2 \xrightarrow{\tilde{T}} L_1$ is surjective, it follows that $L_1 \xrightarrow{T} L_2$ is injective. Since $\mathscr{D}_2 \xrightarrow{\tilde{T}} \mathscr{D}_1$ and $\tilde{T} = T^{-1}$ we find that $\mathscr{D}_1 \xrightarrow{T} \mathscr{D}_2$ is bijective.

CHAPTER VI

Representation of Groups by Means of Effect Automorphisms and Mixture Automorphisms

In the previous chapter we have seen that there is a one-to-one relationship between the $\mathscr{B}$-continuous effect automorphisms and mixture automorphisms which is defined by the adjoint maps. In this chapter we shall investigate the representation of groups by means of $\mathscr{B}$-continuous effect automorphisms. If we make the transition from the $\mathscr{B}$-continuous effect automorphisms to the mixture automorphisms, then we obtain the corresponding "adjoint" representation. Thereby, to each representation of a group by a mixture automorphism we obtain a corresponding "adjoint" representation in terms of a $\mathscr{B}$-continuous effect automorphism. Here we shall only consider representations by means of $\mathscr{B}$-continuous effect automorphisms because the representation of a group by means of mixture automorphisms would only result in an unnecessary repetition of the results derived here.

1 Homomorphic Maps of a Group $\mathscr{G}$ in the Group $\mathscr{A}$ of $\mathscr{B}$-continuous Effect Automorphisms

From the fact that the product of two effect automorphisms is an effect automorphism and that both the identity and the inverse of a $\mathscr{B}$-continuous effect automorphism is again such, it easily follows that the set of $\mathscr{B}$-continuous effect automorphisms form a group. By the expression: the representation of a group $\mathscr{G}$ by means of $\mathscr{B}$-continuous effect automorphisms we mean a (group)-homomorphism of $\mathscr{G}$ into the group $\mathscr{A}$ of $\mathscr{B}$-continuous effect automorphisms.

If we are given a map of a set M into the set $\mathscr{A}$, then we may easily construct a mapping $M \times \mathscr{B}' \to \mathscr{B}'$ in the following way: For $a \in M$ and $a \to T$ $(T \in \mathscr{A})$ we set $(a, y) \to Ty$. For convenience we shall often use the abbreviation ay for the map Ty of (a, y) providing that in the particular circumstance there exists a particular fixed map $a \to T$. In addition, for the image set $\{ay \mid a \in M\}$ we will use the abbreviation My; similarly, for $\{ay \mid a \in M, y \in \mathscr{B}'\}$ we will use the abbreviation $M\mathscr{B}'$; for $\{ay \mid a \in M, y \in L\} = ML$.

Therefore, for $a, b \in \mathscr{G}$, for a representation of a group $\mathscr{G}$ we obtain $a(by) = (ab)y$ for all $y \in L$ (and therefore all $y \in \mathscr{B}'$). Thus it follows (e the unit element of $\mathscr{G}$) that $a(ey - y) = 0$; since a is an effect automorphism it follows that $ey = y$. From $a^{-1}a = e$ it then follows that $a^{-1}ay = y$, that is, a^{-1} is the inverse effect automorphism relative to a.

1.1 Generation of a Representation of $\mathscr{G}$ in $\mathscr{A}$ by Means of a Representation of $\mathscr{G}$ by r-Automorphisms

In applications we frequently find that a representation of a group $\mathscr{G}$ is given in terms of r-automorphisms. We will therefore assume that to each element $a \in \mathscr{G}$ there corresponds an r-automorphism $\mathscr{F} \xrightarrow{a} \mathscr{F}$ which satisfies $a_1(a_2 f) = (a_1 a_2) f$ (for all $f \in \mathscr{F}$). Thus, for $a_2 = e$ we obtain $a(ef) = af$. Since a is injective, it follows that $ef = f$, that is, e is the identity map in $\mathscr{F}$. From $(a^{-1}a)f = ef = f$ it follows that a^{-1} is the inverse r-automorphism which corresponds to a.

If, for all $a \in G$, the maps $\mathscr{F} \xrightarrow{a} \mathscr{F}$ are p-continuous, then a and also a^{-1} are, according to V, D 3.6, p-continuous r-automorphisms.

If we are given a representation of $\mathscr{G}$ in terms of p-continuous r-automorphisms, then according to V, Th. 5.24, there exists a representation of $\mathscr{G}$ by means of $\mathscr{B}$-continuous effect automorphisms, that is, a representation of $\mathscr{G}$ in $\mathscr{A}$. In addition, according to Th. 5.2.4, for all a in $\mathscr{G}$ we obtain $a\mathscr{D} = \mathscr{D}$, that is, $\mathscr{G}$ leaves $\mathscr{D}$ invariant.

If to each r-automorphism $a \in \mathscr{G}$ there exists a dual p-automorphism a' then, according to V, Th. 3.4, all a are p-continuous. For the representation of $\mathscr{G}$ in $\mathscr{A}$ obtained in this way the mixture automorphisms which correspond to the p-automorphisms are precisely the adjoint maps corresponding to the representation elements of $\mathscr{G}$ in $\mathscr{A}$.

In the following we shall only consider topological groups $\mathscr{G}$ (since we may consider finite and countably infinite groups to be topological groups under the discrete topology, the following considerations are also valid for such groups).

The topology of a topological group is uniquely determined by the neighborhood filter of the unit element e (see AV, §10.1). A topological group is given if the neighborhood filter of the unit element satisfies the conditions below; these conditions will be interpreted "physically." By this we mean that the neighborhood filter of the unit element should (ideally)) relate physical imprecision to the group elements which will be physically interpreted as being distinguishable from the unit element (see [1], §5 and §9). Therefore it is "physically" reasonable to postulate the following conditions:

TG 1. To each neighborhood U of the unit element there exists a neighborhood V for which $VV \subset U$.

TG 2. To each neighborhood U of the unit element there exists a neighborhood V for which $V^{-1} \subset U$.

VV is defined as the set of all ab for which $a, b \in V$. TG 1 therefore says that the product of two elements a, b is not distinguishable from the unit element (with the imprecision represented by the set U) if both elements are "near enough" to the unit element.

Similarly TG 2 says that if a is not "distinguishable" from the unit element then a^{-1} is also not "distinguishable" from the unit element.

TG 3. To each $a \in \mathcal{G}$ and each neighborhood U of the unit element there exists another neighborhood V for which $V \subset aUa^{-1}$, that is, $a^{-1}Va \subset U$.

TG 3 says that, if a group element b cannot be distinguished from the unit element (with imprecision defined by V), then if we consider $a^{-1}ba$, that is, we apply b "at the location a" and then transform back by a we then obtain an element which is not distinguishable from the unit element.

A neighborhood filter of the unit element (or a basis for such a neighborhood filter) which satisfies TG 1–TG 3 makes $\mathcal{G}$ into a topological group for which the neighborhood basis of an element $a \in \mathcal{G}$ is given by aU where U is the neighborhood of the unit element (AV, §10.1).

With the neighborhood filter of the unit element there are two (eventually coinciding) uniform structures defined on $\mathcal{G}$ which are compatible with the group operations.

A right-handed (or left-handed) uniform structure is defined by the family of sets $\{(a, b) \mid ba^{-1} \in U$ (or $a^{-1}b \in U$, respectively)$\}$ where U is a neighborhood of the unit element. The topologies associated with the left- and right-handed uniform structures are identical with the original topology of the group.

If a group $\mathcal{G}$ is complete with respect to the right-handed uniform structure, then it is also complete with respect to the left-handed uniform structure, and conversely (AV, §10.2). We may complete $\mathcal{G}$ as a group if the map $a \longrightarrow a^{-1}$ transforms a Cauchy filter of a right-handed uniform structure into a Cauchy filter of the right-handed uniform structure (AV, §10.2). We will assume that this is the case for all physically meaningful groups. Then, instead of $\mathcal{G}$ we may use its completion. For this reason we shall assume that all groups which have a "physical interpretation" are complete. (Finite and denumerable groups with discrete topologies are complete.) Every locally compact group is complete (AV, §10.2).

In addition to the so-called "finiteness of physics" condition (see [1], §9) we shall require that physically interpretable groups also satisfy the following condition:

$\mathscr{G}$ is separable, and the neighborhood filter of the unit element has a denumerable basis. Then it follows that the uniform structure of $\mathscr{G}$ is metrizable (see AV, §10.2). Then $\mathscr{G}$ is a Baire space (see AII, §3) since $\mathscr{G}$ is complete.

Thus we assume that "physical" groups $\mathscr{G}$ will always be, as topologically metrizable, complete and separable groups relative to the corresponding unique right- and left-handed uniform structures.

As we mentioned above, finite and denumerable groups with discrete topologies satisfy the above requirements.

We have provided a simple physical interpretation for the neighborhood filter of the unit element and derived two uniform structures from the compatibility condition; we have also found that we may consider the group to be complete with respect to this uniform structure. This does not, however, mean that one of these uniform structures describes the physical imprecision for the whole group, that is, the physical imprecision which permits us to compare two arbitrary group elements. Physically it is an important distinction whether we are able to only consider elements $b_2 \in \mathscr{G}$ which are neighbors of a fixed element b_1 or we have a procedure by which we may compare arbitrary pairs (b_1, b_2).

The fact that we have considered the set aUa^{-1} where U is an arbitrary neighborhood of the unit element does not contradict the assumption that only elements in the vicinity of the unit element may be compared. From the above symmetry assumption we may at most conclude that we cannot better distinguish elements of U if we perform a translation of the unit to the location a. Our assumption about the uniform structure of physical imprecision for the group implies the opposite: We cannot compare a pair of arbitrary elements as well as we can for two elements in the vicinity of the unit elements. As we have explained in [1], §9, for the uniform structure of physical imprecision it is desirable that the set under consideration (in this case $\mathscr{G}$) is precompact.

For a "physical" group we shall now assume that, in addition, there exists an additional uniform structure ph of physical imprecision on $\mathscr{G}$ which is weaker than the above and for which $\mathscr{G}_{ph}$ (that is, $\mathscr{G}$ together with the uniform structure ph) is precompact and metrizable and generates the same topology, that is, the topology of $\mathscr{G}$ which is compatible with the group operations. In addition, for *fixed* a the maps $b \longrightarrow ab$ and $b \longrightarrow ba$ should be ph-uniformly continuous.

Since the topologies of $\mathscr{G}$ and $\mathscr{G}_{ph}$ are identical, $\mathscr{G}_{ph}$ is separable.

In AV, §10.2 we have proven that such a structure ph always exists for $\mathscr{G}$. If $\mathscr{G}$ is compact, then ph, and both the left- and right-handed uniform structures of $\mathscr{G}$ are identical because a compact space has only a single uniform structure. If $\mathscr{G}$ is not compact, then the uniform structure ph is not uniquely determined by $\mathscr{G}$ and by the above requirements. Then *ph* requires an additional physical structure which is not determined by the neighborhood filter of the unit element.

The completion $\hat{\mathscr{G}}_{ph}$ of $\mathscr{G}_{ph}$ is compact and is called the physical compactification of $\mathscr{G}$. In general $\hat{\mathscr{G}}_{ph}$ is not a group. If $\mathscr{G}$ is compact, then $\mathscr{G}$ is its own compactification. Although $\hat{\mathscr{G}}_{ph}$ is not necessarily a group, the maps $b \to ab$ and $b \to ba$ may be, for fixed $a \in \mathscr{G}$, uniquely extended onto $\hat{\mathscr{G}}_{ph}$ as *ph*-uniformly continuous maps.

The above results are also valid for finite and denumerable infinite groups with the discrete topology; finite groups are already compact in this sense; denumerable infinite groups G must yet be made compact, where the topology of the compactification $\hat{\mathscr{G}}_{ph}$ onto $\mathscr{G}$ is the discrete topology!

In cases in which misunderstandings are unlikely, we will often write $\hat{\mathscr{G}}$ instead of $\hat{\mathscr{G}}_{ph}$.

Th. 1.1.1. *The set $\hat{\mathscr{G}}$ may be partitioned under (left and right) multiplication with the elements of $\mathscr{G}$ in invariant subsets $\mathscr{G}$ and $\hat{\mathscr{G}}\backslash\mathscr{G}$.*

PROOF. If $b \in \hat{\mathscr{G}}\backslash\mathscr{G}$ and $b_\nu \to b$ with $b_\nu \in \mathscr{G}$, then for $a \in \mathscr{G}$ it follows that $ab_\nu \to ab$. Suppose $ab = c \in \mathscr{G}$, then from $ab_\nu \to c$ it also follows that $a^{-1}(ab_\nu) \to a^{-1}c = b \in \mathscr{G}$.

The physical meaning of the uniform structure *ph* should also be evident in the representation of $\mathscr{G}$.

Since, by assumption, $\mathscr{G}$ is complete and $\mathscr{F}$ is denumerable (see III, §3), $\mathscr{G}$ cannot, in general, be represented by p-continuous r-automorphisms. Since $\mathscr{G}$ is separable, there exists a countable dense subgroup $\tilde{\mathscr{G}}$ in $\mathscr{G}$. We shall now assume that there exists a representation of such a subgroup $\tilde{\mathscr{G}}$ by means of p-continuous r-automorphisms.

It is reasonable to impose the following additional condition: If a is an element of $\tilde{\mathscr{G}}$ in the neighborhood of the unit element, then for all $w \in \mathscr{K}$ then, for all practical purposes, it is not possible to distinguish between the probabilities $\mu(w, \psi(f))$ and $\mu(w, \psi(af))$, that is, f and af, for all preparation procedures, yield the same probabilities if a is sufficiently close to e. In this way, for fixed $w \in \mathscr{K}$ and arbitrary f the above probabilities cannot be distinguished. We shall formulate this in the following mathematical axiom:

AG 1. To each $f \in \mathscr{F}$ and $\varepsilon > 0$ there exists a neighborhood U_ε of the unit element in $\tilde{\mathscr{G}}$, such that

$$|\mu(w, \psi(f)) - \mu(w, \psi(af))| < \varepsilon$$

for all $w \in \mathscr{K}$ and $a \in U_\varepsilon$. To each $w \in \mathscr{K}$ and $\delta > 0$ there exists a neighborhood U_δ of the unit element in $\tilde{\mathscr{G}}$, such that

$$|\mu(w, \psi(f)) - \mu(w, \psi(af))| < \delta$$

for all $f \in \mathscr{F}$ and $a \in U_\delta$.

If AG 1 holds, then the following theorems hold:

Th. 1.1.2. *For fixed $y \in \mathscr{D}$ the map defined by $a \longrightarrow ay$ of $\tilde{\mathscr{G}} \longrightarrow \mathscr{D}$ is uniformly continuous in the norm topology of $\mathscr{D}$. For fixed $x \in \mathscr{B}$ the map of $\tilde{\mathscr{G}} \longrightarrow \mathscr{B}$ defined by $a \longrightarrow a'x$ is uniformly continuous in the norm topology of $\mathscr{B}$.*

PROOF. Since $\mathscr{L} = \psi\mathscr{F}$ is norm-dense in $L \cap \mathscr{D}$, it is sufficient to prove the first assertion for $y \in \mathscr{L}$ as follows:

$$\begin{aligned}\|a_1\psi(f) - a_2\psi(f)\| &= \sup_{w \in \mathscr{K}} |\mu(w, a_1[\psi(f) - a_1^{-1}a_2\psi(f)])| \\ &= \sup_{w \in \mathscr{K}} |\mu(a_1'w_1, \psi(f) - \psi(a_1^{-1}a_2 f)| \\ &= \sup_{\tilde{w} \in \mathscr{K}} |\mu(\tilde{w}, \psi(f) - \psi(a_1^{-1}a_2 f)|.\end{aligned}$$

According to AG 1 we therefore obtain

$$\|a_1\psi(f) - a_2\psi(f)\| < \varepsilon$$

for $a_1^{-1}a_2 \in U_\varepsilon$.

Since $\mathscr{K}$ is norm-dense in K it suffices to prove the second assertion for $x \in \mathscr{K}$ ($\mathscr{L}$ is σ-dense in L):

$$\begin{aligned}\|a_1'w - a_2'w\| &= \sup_{g \in L} \mu(a_1'w - a_2'w, 2g - \mathbf{1}) \\ &= 2 \sup_{g \in \mathscr{L}} \mu(a_1'w - a_2'w, g) \\ &= 2 \sup_{f \in \mathscr{F}} \mu(w, \psi(a_1 f) - \psi(a_2, f)) \\ &= 2 \sup_{\tilde{f} \in \mathscr{F}} \mu(w, \psi(\tilde{f}) - \psi(a_2 a_1^{-1}\tilde{f})).\end{aligned}$$

According to AG 1 we therefore obtain $\|a_1'w - a_2'w\| < \delta$ for $a_2 a_1^{-1} \in U_\delta$.

Th. 1.1.3. *By continuous continuation of the map $\tilde{\mathscr{G}} \longrightarrow \mathscr{D}$ described in* Th. 1.1.2 *we may obtain a representation of $\mathscr{G}$ by means of $\mathscr{B}$-continuous effect automorphisms which maps $\mathscr{D}$ into itself. For $x \in \mathscr{B}$, $y \in \mathscr{D}$ the maps $a \longrightarrow \mathbf{R}$ defined by $a \longrightarrow \mu(x, ay)$ are uniformly continuous.*

PROOF. Since $\tilde{\mathscr{G}} \longrightarrow \mathscr{D}$ is uniformly continuous and $\mathscr{D}$ is complete with respect to the norm it follows that its continuous completion to each $y \in \mathscr{D}$ defines a map $\mathscr{G} \longrightarrow \mathscr{D}$. If we write this map as $a \longrightarrow ay$, then it follows that a is linear. Since $\mathscr{L}$ is norm-dense in $L \cap \mathscr{D}$, it follows that $a(L \cap \mathscr{D}) \subset L \cap \mathscr{D}$ and that a is norm-continuous. The map $y \longrightarrow ay$ is also $\mathscr{B}$-continuous. This result follows from Th. 1.1.2, since, for all $\tilde{a} \in \tilde{\mathscr{G}}$ it follows that $\mu(x, \tilde{a}y) = \mu(\tilde{a}'x, y)$ and that the mapping $\tilde{a} \longrightarrow \tilde{a}'x$ is uniformly continuous with respect to the norm topology in $\mathscr{B}$ and therefore can be extended to a map $\mathscr{G} \longrightarrow \mathscr{B}$ for which, to each $a \in \mathscr{G}$ a mapping a' of $\mathscr{B}$ into $\mathscr{B}$ is determined. From $\mu(\tilde{a}'x, y) = \mu(x, \tilde{a}y)$ it follows that, in the limit $\mu(a'x, y) = \mu(x, ay)$ for all $a \in \mathscr{G}$, from which we have proven the $\mathscr{B}$-continuity of a. Since a (as a map of $\mathscr{D} \cap L$ into itself) is $\mathscr{B}$-continuous, it can be extended onto the compact set L. Therefore a is a $\mathscr{B}$-continuous effect morphism. Since $\tilde{a}'\mathscr{K}$ is, for all $\tilde{a} \in \tilde{\mathscr{G}}$, equal to $\mathscr{K}$ (and therefore norm-dense in K) it follows that $a'\mathscr{K}$ is norm-dense in K, from which we conclude that, according to V, Th. 4.1.4, a is an effect automorphism.

Since the map $a \to \mu(x, ay)$ is uniformly continuous as a map $\tilde{\mathscr{G}} \to \mathbf{R}$, its extension on $\mathscr{G}$ is also uniformly continuous.

We must now make the uniform structure ph of physical imprecision evident in the representations. We have seen that the mapping $a \to \mu(w, \psi(af))$ of $\tilde{\mathscr{G}}$ into $\mathbf{R}$ is uniformly continuous. If, the elements of $\tilde{\mathscr{G}}$ are to be physically distinguishable according to ph, then it is necessary to impose the following strong requirement:

AG 2. For each $w \in \mathscr{K}$ and $f \in \mathscr{F}$ the mapping $\tilde{\mathscr{G}}_{ph} \to \mathbf{R}$ defined by $a \to \mu(w, \psi(af))$ is uniformly continuous.

From AG 2 we obtain, by continuous extension:

Th. 1.1.4. *The mapping of* $\mathscr{G}_{ph} \to \mathbf{R}$ *defined by* $a \to \mu(x, ay)$ *for* $x \in \mathscr{B}$, $y \in \mathscr{D}$ *is uniformly continuous.*

1.2 Some General Properties of a Representation of $\mathscr{G}$ in $\mathscr{A}$

In the previous section it has become evident that the representation of groups by means of $\mathscr{B}$-continuous effect morphisms plays an important role. If such a representation is generated by r-automorphisms then this "generation" will only play a role for the interpretation question (see, for example, VII). Since neither the sets $\mathscr{Q}$ and $\mathscr{F}$ nor the sets $\mathscr{K}$ and $\mathscr{L}$ are axiomatically fixed, except that, for $\mathscr{K}$ each denumerable set in K which is norm-dense in K and for $\mathscr{L}$ each denumerable σ-dense set in L may be used, we are led to concentrate our efforts on questions about the representation of a group on representations in $\mathscr{A}$. Unfortunately, the representation theory of groups by means of positive automorphisms of base-norm spaces has not yet been sufficiently developed for us to be able to present a comprehensive outline of even a part of a "general" representation theory. This is, in part, due to the fact that the relationships between special structures of base-norm spaces (for example, the special structures formulated by axioms AV 1.1–AV 4 in III, §4) and the structure of the automorphism groups are not well known (see, for example, the attempts in this direction presented in [20]). For this reason in this book we shall often refer to the special structure of these automorphism groups as groups of $\mathscr{B}$-continuous effect automorphisms such as those which were considered for $\mathscr{B} = \mathscr{B}(\mathscr{H}_1, \mathscr{H}_2, \ldots)$ in V. Now, in the remaining part of this section we shall consider a number of general and easily formulated (and which will later be seen as "physically" meaningful) properties.

D 1.2.1. A $g \in L$ is called a $\mathscr{G}$-invariant effect if $\mathscr{G}g = g$. An $e \in G$ is said to be a $\mathscr{G}$-invariant decision effect if $Ge = e$.

It is easy to verify that $\lambda\mathbf{1}$ is, for all $0 \leq \lambda \leq 1$, a $\mathscr{G}$-invariant effect, and that $\mathbf{0}$ and $\mathbf{1}$ are $\mathscr{G}$-invariant decision effects.

For $a \in \mathscr{G}$ it follows that from $ag = g$ and the uniqueness of the spectral representation of g that $ae(\lambda) = e(\lambda)$ holds for the spectral family $e(\lambda)$ of g. Therefore it suffices to only consider invariant decision effects.

D 1.2.2. A representation is said to be irreducible if $\mathbf{0}$ and $\mathbf{1}$ are the only invariant decision effects.

A representation is therefore irreducible only if the only invariant effects are given by $\lambda\mathbf{1}$ where $0 \leq \lambda \leq 1$.

As we have already mentioned, to each a we may consider the adjoint a'; in this way $\mathscr{G}'$ is defined as set of transformations in K and $\mathscr{B}$.

D 1.2.3. Two representations $\mathscr{G}\mathscr{B}'_1$ and $\mathscr{G}\mathscr{B}'_2$ are said to be equivalent if there exists a $\mathscr{B}$-continuous effect isomorphism T of $\mathscr{B}'_1$ onto $\mathscr{B}'_2$ for which $Ta = aT$ for all $a \in \mathscr{G}$.

By analogy with the considerations in §1.1 we shall assume that the $a \in \mathscr{G}$ transform the subspace $\mathscr{D}$ into itself, that is,

(D 1) $\mathscr{G}\mathscr{D} \subset \mathscr{D}$.

Since $\mathscr{D}$ has not otherwise been specified by means of the axioms (except that $\mathscr{D}$ is norm-separable and that $\mathscr{D} \cap L$ is $\sigma(\mathscr{B}', \mathscr{B})$-dense in L—see III, §3) we shall now proceed in the opposite way and seek to obtain conditions for $\mathscr{D}$ with the aid of the group representations. Therefore (D 1) is one of the conditions which $\mathscr{D}$ must satisfy.

According to III, §3 we will assume that $\mathscr{D}$ is norm-separable. Here we shall note that the following definitions and concepts will apply to every norm-closed subspace $\mathscr{D}$ of $\mathscr{B}'$ for which $\mathscr{D} \cap L$ is $\sigma(\mathscr{B}', \mathscr{B})$-dense in L and is $\mathscr{G}$-invariant.

Since $\mathscr{G}$ transforms the space $\mathscr{D}$ into itself, $\mathscr{G}'$ is—on all of $\mathscr{D}'$—defined as mixture automorphisms of $\bar{K}^\sigma$ ($\bar{K}^\sigma$ is the $\sigma(\mathscr{D}', \mathscr{D})$-closures of K in $\mathscr{D}'$).

In analogy to the considerations presented in §1.1 we shall, in addition, assume that the whole group $\mathscr{G}$ is represented in $\mathscr{A}$. From axioms AG 1 and AG 2 it follows that there are certain continuity properties of this representation; here we shall only use two of these (!): According to Th. 1.1.3 we obtain:

(D 2) For $x \in \mathscr{B}$, $y \in \mathscr{D}$ the maps $\mathscr{G} \to \mathbf{R}$ defined by $a \to \mu(x, ay)$ are continuous.

According to Th. 1.1.4 we obtain:

(D 3) For $x \in \mathscr{B}$, $y \in \mathscr{D}$ the maps $\mathscr{G}_{ph} \to \mathbf{R}$ defined by $a \to \mu(x, ay)$ are uniformly continuous.

It is clearly obvious that (D 3) $\Rightarrow$ (D 2). For a representation of $\mathscr{G}$ in $\mathscr{A}$ we shall not yet assume that a representation must be obtained by means of r-automorphisms in the way described in §1.1. We will only assume that $\mathscr{G}$ is complete and that either (D 1), (D 2), or (D 3) is satisfied. We will then find that, for a representation of $\mathscr{G}$ in $\mathscr{A}$ the properties which have been derived from AG 1 in §1.1 automatically hold.

Th. 1.2.1. *If, according to* (D 2), $\mu(x, ay)$ *is continuous at* $e \in \mathscr{G}$ *(e the unit element of* $\mathscr{G}$*) as a function on* $\mathscr{G}$ *for each pair* $(x, y) \in \mathscr{B} \times \mathscr{D}$*, then the mapping of* $\mathscr{G} \to \mathscr{B}$ *defined by* $a \to a'x$ *is norm-continuous.*

PROOF. Let $x \in \mathscr{B}$ be fixed. From $|\mu(a'x - x, y)| = |\mu(x, ay) - \mu(x, y)| < \varepsilon$ if a is in a suitable neighborhood of the unit element, then it follows that the map $a \to a'x$ is $\sigma(\mathscr{B}, \mathscr{D})$-continuous at e. Thus since

$$\mu(a'x - b'x, y) = \mu(b'^{-1}a'x - x, by)$$

it follows that the map $a \to a'x$ is $\sigma(\mathscr{B}, \mathscr{D})$-continuous in all of $\mathscr{G}$.

We now consider the set $\mathscr{G}'x$ as a subset of $\mathscr{B}$ with the topologies induced by the norm and $\sigma(\mathscr{B}, \mathscr{D})$. Since $\mathscr{B}$ is norm-separable there exists a denumerable subset $\{a_\nu\} \subset \mathscr{G}$, for which $\{a'_\nu x\}$ is norm-dense in $\mathscr{G}'x$. We define the spheres K_ν^ε in $\mathscr{G}'x$:

$$K_\nu^\varepsilon = \{x' \mid x' \in \mathscr{G}'x, \|x' - a'_\nu x\| \leq \varepsilon\}.$$

Let M_ν^ε denote the inverse image of $a \to a'x$, that is,

$$M_\nu^\varepsilon = \{a \mid a \in G \text{ where } a'x \in K_\nu^\varepsilon\}.$$

Since the $a'_\nu x$ are norm-dense in $\mathscr{G}'x$ we obtain $\bigcup_\nu K_\nu^\varepsilon = \mathscr{G}'x$ from which it follows that $\bigcup_\nu M_\nu^\varepsilon = G$.

The set $\mathscr{G}'x$ may be considered to be a subset of $\mathscr{D}'$; since the norm of $\mathscr{D}'$ on $\mathscr{B}$ agrees with that of $\mathscr{B}$ (see proof of V, Th. 4.1.6). K_ν^ε can be considered to be the intersection of spheres in $\mathscr{D}'$ with $\mathscr{G}'x$. Since the unit sphere of $\mathscr{D}'$ is $\sigma(\mathscr{D}', \mathscr{D})$-compact and therefore closed, K_ν^ε is $\sigma(\mathscr{D}', \mathscr{D})$-closed in $\mathscr{G}'x$, that is, K_ν^ε is $\sigma(\mathscr{B}, \mathscr{D})$-closed in $\mathscr{G}'x$. Since the map $a \to a'x$ is $\sigma(\mathscr{B}, \mathscr{D})$-continuous, the inverse image M_ν^ε is closed in $\mathscr{G}$. Since $\mathscr{G}$ is a Baire space (see AII, §3) it follows that from $\bigcup_\nu M_\nu^\varepsilon = \mathscr{G}$ that there exists a set $M_{\nu 0}^\varepsilon$ which contains an open subset of $\mathscr{G}$.

Therefore there exists a element $a \in M_{\nu 0}^\varepsilon$ and a neighborhood $U(a)$ of a for which $U(a) \subset M_{\nu 0}^\varepsilon$, that is, for all $b \in U(a)$ we obtain $a'x \in K_{\nu 0}^\varepsilon$ and $b'x \in K_{\nu 0}^\varepsilon$ and therefore $\|b'x - a'x\| < 2\varepsilon$. $U(a)a^{-1}$ is a neighborhood V of the unit element. Since the norm is preserved by mixture automorphisms, for all $c \in V$ we obtain $\|c'x - x\| < 2\varepsilon$ whereby we have proven the norm-continuity of $c \to c'x$ in the point $e \in G$. Thus we easily obtain the norm-continuity in $\mathscr{G}$.

Th. 1.2.2. *If* $a \to a'x$ *for all* $x \in \mathscr{B}$ *is norm-continuous for fixed* x*, then* $a \to ay$ *for all* $y \in \mathscr{B}'$ *is* $\sigma(\mathscr{B}', \mathscr{B})$*-continuous.*

The proof follows directly from

$$|\mu(x, ay - by)| = |\mu(a'x - b'x, y)| \leq \|a'x - b'x\| \|y\|.$$

Th. 1.2.2 says that the map $a \to \mu(x, ay)$ is continuous for each pair $(x, y) \in \mathscr{B} \times \mathscr{B}'$ on all of $\mathscr{G}$.

The following theorem is a corollary of Th. 1.2.1 and Th. 1.2.2:

Th. 1.2.3. *If $a \longrightarrow \mu(x, ay)$ is continuous for all $(x, y) \in \mathscr{B} \times \mathscr{D}$ it follows that it is continuous for all $(x, y) \in \mathscr{B} \times \mathscr{B}'$.*

In the same manner we may prove the following theorems:

Th. 1.2.4. *If $a \longrightarrow \mu(x, ay)$ is continuous for all $(x, y) \in \mathscr{B} \times \mathscr{D}$ it follows that $a \longrightarrow ay$ is norm-continuous for all $y \in \mathscr{D}$.*

Th. 1.2.5. *If $a \longrightarrow ay$ is norm-continuous for all $y \in \mathscr{D}$ it follows that $a \longrightarrow ax$ is $\sigma(\mathscr{D}', \mathscr{D})$ continuous for all $x \in \mathscr{D}'$.*

Th. 1.2.6. *If $a \longrightarrow \mu(x, ay)$ is continuous for all $(x, y) \in \mathscr{B} \times \mathscr{D}$ then it is continuous for all $(x, y) \in \mathscr{D}' \times \mathscr{D}$.*

In all these theorems only (D 1) and (D 2) are required. According to (D 3) $\mu(x, ay)$ is *ph*-uniformly continuous for all $(x, y) \in \mathscr{B} \times \mathscr{D}$. In general we should expect that $\mu(x, ay)$ is neither *ph*-continuous for all $(x, y) \in \mathscr{D}' \times \mathscr{D}$ nor for all $(x, y) \in \mathscr{B} \times \mathscr{B}'$.

Therefore there exists a complete symmetry with respect to the representation of groups between the spaces $\mathscr{B}$ and $\mathscr{D}$ and their corresponding extensions $\mathscr{B} \subset \mathscr{D}'$ and $\mathscr{D}$ or $\mathscr{B}$ and $\mathscr{B}' \supset \mathscr{D}$. Previously we have generally not made use of the dual pair $\mathscr{D}', \mathscr{D}$ together with $\mathscr{B} \subset \mathscr{D}'$ as mathematical tools for the representation of physical problems of quantum mechanics. It is, however, possible that there exist new methods for the mathematical treatment of many problems using the methods of C^*-algebras if $\mathscr{D}$ is a C^*-algebra.

We will now examine the possible consequences of (D 3). In order to provide a comprehensive formulation of the following processes we now introduce the following spaces Δ_n, Δ_{ph} and Δ:

D 1.2.4. Δ_n is the set of all $y \in \mathscr{B}'$ for which the map $a \longrightarrow ay$ of $\mathscr{G}$ into $\mathscr{B}'$ is norm-continuous. Δ_{ph} is the set of all $y \in \mathscr{B}'$ for which the maps $a \longrightarrow \mu(x, ay)$ are *ph*-uniformly continuous for all $x \in \mathscr{B}$. $\Delta = \Delta_{ph} \cap \Delta_n$.

According to (D 3) and Th. 1.2.5 it follows that $\mathscr{D} \subset \Delta \subset \mathscr{B}'$.

Th. 1.2.7. *Δ_n, Δ_{ph} and Δ are norm-closed subspaces of $\mathscr{B}'$ which are invariant with respect to $\mathscr{G}$.*

Proof. The fact that Δ_{ph} is a subspace of $\mathscr{B}'$ is clear. Let $y \in \Delta_{ph}$ and let $c \in \mathscr{G}$; then $\mu(x, acy)$ is, as a function of a, *ph*-uniformly continuous since both maps $a \longrightarrow ac \longrightarrow \mu(x, acy)$ are *ph*-uniformly continuous; therefore $cy \in \Delta_{ph}$.

If $y_\nu \longrightarrow y$ is a norm-convergent sequence for which $y_\nu \in \Delta_{ph}$ then

$$|\mu(x, ay - by)| \leq |\mu(x, a(y - y_\nu)| + |\mu(x, ay - by_\nu)| + |\mu(x, b(y_\nu - y)|$$

$$\leq 2\|x\| \, \|y - y_\nu\| + |\mu(x, ay_\nu - by_\nu)|$$

from which it follows that $\mu(x, ay)$ is *ph*-uniformly continuous.

For Δ_n the same result follows more easily from the continuity of the maps $a \to ac \to (ac)y$ and from

$$\|ay - by\| \leq \|a(y - y_\nu)\| + \|ay_\nu - by_\nu\| + \|b(y_\nu - y)\|$$
$$\leq 2\|y - y_\nu\| + \|ay_\nu - by_\nu\|.$$

By analogy to D 1.2.4 we define:

D 1.2.5. Let Ξ_n be the set of all $x \in \mathscr{D}'$ for which the mapping $a \to a'x$ of $\mathscr{G}$ into $\mathscr{D}'$ is norm-continuous. Let Ξ_{ph} denote the set of all $x \in \mathscr{D}'$ for which the maps $a \to \mu(x, ay) \to \mu(a'x, y)$ are *ph*-continuous for all $y \in \mathscr{D}$. Let $\Xi = \Xi_n \cap \Xi_{ph}$.

According to (D 3) and Th. 1.2.1 $\mathscr{B} \subset \Xi \subset \mathscr{D}'$. By analogy with Th. 1.2.7 it follows that:

Th. 1.2.8. *Ξ_n, Ξ_{ph} and Ξ are norm-closed subspaces of $\mathscr{D}'$ which are invariant relative to $\mathscr{G}'$.*

Since $\mathscr{D}$ has not previously been restricted by means of axioms, we may now reverse the sequence of ideas presented above as follows: We begin with a topologically complete group $\mathscr{G}$ and a representation for which the map $\mathscr{G} \to \mathscr{G}'x$ is norm-continuous for all $x \in \mathscr{B}$. The "physical" uniform structure *ph* on $\mathscr{G}$ and the space $\mathscr{D}$ have previously not been specified. We shall now define topologies on $\mathscr{G}$ by means of its representations as follows:

D 1.2.6. We shall call the initial topology on $\mathscr{G}$ for which the maps

$$\mathscr{G} \xrightarrow{x} \mathscr{B} \quad \text{defined by } a \to a'x$$

are continuous for all $x \in \mathscr{B}$ (with respect to the norm topology in $\mathscr{B}$) the ($\mathscr{B}$)-topology.

The original topology in $\mathscr{G}$ is therefore finer than the ($\mathscr{B}$)-topology.

If the representation of $\mathscr{G}$ is not true, then there exists a invariant subgroup $\mathscr{N}$ of $\mathscr{G}$ which, in the representation of $\mathscr{G}$ is mapped onto the identity. The representation of the factor group $\mathscr{G}/\mathscr{N}$ is then true. The fact that the elements of $\mathscr{N}$ behave like the identity with respect to quantum mechanics has the physical meaning that the transformations in $\mathscr{N}$ are possibly meaningful outside the domain of microsystems, but in the domain of microsystems are equivalent to the identity. Thus for microsystems only the group $\mathscr{G}/\mathscr{N}$ is physically meaningful. From this point on we shall assume that the representation of $\mathscr{G}$ is true (this assertion we may not change, because in certain "subdomains" of the domain of microsystems nontrue representations may occur! See VII, end of §2). From the above arguments it seems reasonable, on physical grounds, to identify the topology of the group $\mathscr{G}$ with the ($\mathscr{B}$)-topology. First, we must determine whether $\mathscr{G}$ together with the ($\mathscr{B}$)-topology is always a topological group.

Th. 1.2.9. *$\mathscr{G}$ together with the ($\mathscr{B}$)-topology is a topological group (because the representation of $\mathscr{G}$ was assumed to be true, the ($\mathscr{B}$)-topology is separating). $\mathscr{G}$ is separable and metrizable in the ($\mathscr{B}$)-topology.*

PROOF. Since a' preserves the norm, TG 1 follows from

$$\|a'b'x - x\| \leq \|a'b'x - a'x\| + \|a'x - x\|$$
$$= \|a'(b'x - x)\| + \|a'x - x\| = \|b'x - x\| + \|a'x - x\|.$$

TG 2 follows from

$$\|a'^{-1}x - x\| = \|a'(a'^{-1}x - x)\| = \|x - a'x\|.$$

TG 3 follows from the fact that $\|b'x_1 - x_1\| < \varepsilon$ is equivalent to

$$\|a'^{-1}b'a'x_2 - x_2\| < \varepsilon$$

for $x_2 = a'^{-1}x_1$ since

$$\|a'^{-1}b'a'x_2 - x_2\| = \|b'a'x_2 - a'x_2\| = \|b'x_1 - x_1\|.$$

Since $\mathscr{G}$ was assumed to be separable in the original topology, $\mathscr{G}$ is separable in the ($\mathscr{B}$)-topology.

Every subset $A \subset \mathscr{B}$ for which the closed subspace spanned by A is equal to $\mathscr{B}$ generates the same initial topology as does $\mathscr{B}$. This result follows easily for linear combinations of elements of A and can be proven using the inequalities:

$$\|a_1x - a_2x\| \leq \|a_1x - a_1\hat{x}\| + \|a_1\hat{x} - a_2\hat{x}\| + \|a_2\hat{x} - a_2x\|$$
$$= 2\|x - \hat{x}\| + \|a_1\hat{x} - a_2\hat{x}\|.$$

Since $\mathscr{B}$ is separable, there exists a denumerable set A and therefore a denumerable neighborhood basis of the unit element in G.

For a subset $\Lambda \subset \mathscr{B}'$ let the norm-closed subspace of $\mathscr{B}'$ spanned by Λ be denoted by $\mathscr{D}_\Lambda$.

D 1.2.7. Let $\Lambda \subset \mathscr{B}'$ be a subset of $\mathscr{B}'$ which satisfies the condition $\mathscr{G}\Lambda \subset \mathscr{D}_\Lambda$. We shall call the initial structure for which the maps

$$\mathscr{G} \xrightarrow{(1)y} \mathscr{B}' \quad \text{defined by } a \longrightarrow ay,$$
$$\mathscr{G} \xrightarrow{(2)y} \mathscr{B}' \quad \text{defined by } a \longrightarrow a^{-1}y$$

are uniformly continuous for all $y \in \Lambda$ with respect to the $\sigma(\mathscr{B}', \mathscr{B})$-topology in $\mathscr{B}'$ the Λ-uniform structure on $\mathscr{G}$. We shall call the topology determined by the Λ-uniform structure the Λ-topology on $\mathscr{G}$.

It is easy to see that, for all $y \in \mathscr{D}_\Lambda$, the maps $a \rightarrow ay$ and $a \rightarrow ay$ are uniformly continuous. Since $\|ay - a\hat{y}\| = \|y - \hat{y}\|$ it follows from $\mathscr{G}\Lambda \subset \mathscr{D}_\Lambda$ that $\mathscr{G}\mathscr{D}_\Lambda = \mathscr{D}_\Lambda$, that is, $\mathscr{D}_\Lambda$ is an invariant subspace of $\mathscr{B}'$. Thus it easily follows that $ay = y$ for all $y \in \mathscr{D}_\Lambda$ is equivalent to $ay = y$ for all $y \in \Lambda$. A representation will be said to be Λ-true if $ay = y$ for all $y \in \Lambda$, it follows that $a = e$.

Th. 1.2.10. *The maps of $\mathscr{G}$ onto itself given by $a \to a^{-1}$, $a \to ab$ and $a \to ba$ (for fixed b) are uniformly continuous with respect to the Λ-uniform structure.*

PROOF. Since the composite maps $a \to a^{-1} \to a^{-1}y$ and $a \to a^{-1} \to (a^{-1})^{-1}y = ay$, that is, since the maps $a \to a^{-1}y$ and $a \to ay$ are uniformly continuous $a \to a^{-1}$ is uniformly continuous. From $a \to ab \to aby$ and $a \to ab \to (ab)^{-1}y = b^{-1}a^{-1}y$ are (for fixed b) uniformly continuous, it follows that $a \to ab$ is also such. The uniform continuity of $a \to aby$ is obtained with $\hat{y} = by$ from $a \to aby = a\hat{y}$ since $\hat{y} \in \mathscr{D}_\Lambda$. The uniform continuity of $a \to b^{-1}a^{-1}y$ follows from

$$\mu(x, b^{-1}a^{-1}y) = \mu(b'^{-1}x, a^{-1}y) = \mu(\hat{x}, a^{-1}y),$$

where $\hat{x} = b'^{-1}x \in \mathscr{B}$ and the uniform continuity of $a \to a^{-1}y$.

In this way we obtain the uniform continuity of $a \to ba$ from $a \to ba \to bay$ and $a \to ba \to a^{-1}b^{-1}y$ and from $\hat{y} = b^{-1}y$ and $\hat{x} = b'x$.

Th. 1.2.11. *The $(\mathscr{B})$-topology on $\mathscr{G}$ is finer than the Λ-topology. $\mathscr{G}$ is therefore also separable in the Λ-topology.*

PROOF. Let $\mathscr{G}$, together with the $(\mathscr{B})$-topology (or Λ-topology) be denoted by $\mathscr{G}_\mathscr{B}$ (or $\mathscr{G}_\Lambda$, respectively). We must now show that the identity map $\mathscr{G}_\mathscr{B} \to \mathscr{G}_\Lambda$ is continuous. $\mathscr{G}_\mathscr{B} \to \mathscr{G}_\Lambda$ is continuous providing that the composite maps $\mathscr{G}_\mathscr{B} \longrightarrow \mathscr{G}_\Lambda \xrightarrow{(1)y} \mathscr{B}'$ and $\mathscr{G}_\mathscr{B} \longrightarrow \mathscr{G}_\Lambda \xrightarrow{(2)y} \mathscr{B}'$ are continuous. This follows, however, from

$$|\mu(x, a_1y - a_2y)| = |\mu(a_1'x - a_2'x, y)| \leq \|a_1'x - a_2'x\| \, \|y\|$$

and Th. 1.2.9.

Th. 1.2.12. *The set $\Delta_n \cap L$ is $\sigma(\mathscr{B}', \mathscr{B})$-dense in L and the set Δ_n is therefore $\sigma(\mathscr{B}', \mathscr{B})$-dense in $\mathscr{B}'$.*

PROOF. For the P_φ for which $\varphi \in \bigcup_\nu \mathscr{H}_\nu$ we find that, in the norm of $\mathscr{B}$ $\|a'P_\varphi - P_\varphi\| < \varepsilon$ whenever a is in a suitable neighborhood U (either in the original topology or in the $\mathscr{B}$-topology) of the unit element. Since $a'P_\varphi = P_\psi = a^{-1}P_\varphi$ (see V, Th. 5.13), denoting the norms in $\mathscr{B}'$ and $\mathscr{B}$ by $\|\cdots\|_{\mathscr{B}'}$ and $\|\cdots\|_\mathscr{B}$, respectively, from the relation

$$\|P_\psi - P_\varphi\|_{\mathscr{B}'} \leq 2\|P_\psi - P_\varphi\|_\mathscr{B}$$

we obtain $\|a^{-1}P_\varphi - P_\varphi\|_{\mathscr{B}'} < 2\varepsilon$ for $a \in U$. Therefore for $y = P_\nu \in L$ we find that the map $a \to ay$ is norm continuous (since $a \to a^{-1}$ is continuous). Therefore the map $a \to ay$ is norm continuous for all finite linear combinations of elements of the form P_φ. The set of all such finite linear combinations is $\sigma(\mathscr{B}', \mathscr{B})$-dense in L.

Since L is $\sigma(\mathscr{B}', \mathscr{B})$-separable there exists denumerably many $y_\nu \in \Delta_n \cap L$ which are $\sigma(\mathscr{B}', \mathscr{B})$-dense in L. Since $\mathscr{G}_\mathscr{B}$ is separable, there exist denumerably many b_μ (the unit element of $\mathscr{G}$ may be among the b_μ) which are dense in the $\mathscr{B}$-topology in $\mathscr{G}$. The set Λ of all $b_\mu y_\nu$ is then denumerable, and we find that $\Lambda \subset L$ and Λ is $\sigma(\mathscr{B}', \mathscr{B})$-dense in L. Here we realize that the set y_ν does not need to be $\sigma(\mathscr{B}', \mathscr{B})$-dense in L if only the set of the $b_\mu y_\nu$ is $\sigma(\mathscr{B}', \mathscr{B})$-dense in L.

Since $y_\nu \in \Delta_n$ we find that $\|by_\nu - b_\mu y_\nu\| < \varepsilon$ for a $b \in \mathscr{G}$ and for b_μ sufficiently close to b. Therefore $\mathscr{G}\Lambda$ is contained in the norm-closure of Λ

and therefore we find that $\mathscr{G}\Lambda \subset \mathscr{D}_\Lambda$. Λ therefore satisfies the conditions in D 1.2.7 and is denumerable, and is $\sigma(\mathscr{B}', \mathscr{B})$-dense in L. Thus we find that $\mathscr{D}_\Lambda$ is norm-separable.

We may therefore use the space $\mathscr{D}_\Lambda$ as the space $\mathscr{D}$ of the theory if the "physical" uniform structure *ph* is finer than the Λ-uniform structure. If the Λ-topology is equal to the original topology on the group $\mathscr{G}$, then we may identify the physical uniform structure with the Λ-uniform structure, that is, by the choice of the set Λ we fix the choice of the physical uniform structure *ph*. In this way we recognize the close relationship between the designation of the space $\mathscr{D}$ and the designation of the uniform structure *ph* of "physical imprecision" on $\mathscr{G}$.

In practice it is often easy to show that the Λ-topology is identical to the original topology on $\mathscr{G}$. In applications $\mathscr{G}$ is at most a locally compact group. If we could find a Λ-neighborhood of the identity which is compact in the original topology, then the original topology and the weaker Λ-topology will coincide in this neighborhood, and therefore will have the same neighborhood system of the identity. The proof that there is such a Λ-neighborhood is very simple for the case of Lie groups (see VII, §8). In the following we will always assume that there exists a Λ-set in order that the Λ-topology will coincide with the original topology. Then the $(\mathscr{B})$-topology also coincides with the original topology.

With this result we note that we have not yet solved the problem of the space $\mathscr{D}$. For example, we may set $\mathscr{D} = \mathscr{D}_\Lambda$, but we may yet find that it is possible to have different sets Λ and also different spaces $\mathscr{D}_\Lambda$ which result in the same Λ-topology on $\mathscr{G}$. If, for example, $\mathscr{G}$ is compact, then, for all possible Λ-sets the Λ-topologies will coincide with the original topology on $\mathscr{G}$ and Δ is equal to Δ_n. This is not so surprising: If we need only consider compact groups in physics, then we would always be able to choose finite-dimensional Hilbert spaces $\mathscr{H}_\nu$ (see VII, §3 and AV, §10) and the problem of the selection of a space $\mathscr{D}$ would be nonexistent.

After the construction (with the physically uniform structure *ph* being the Λ-uniform structure) we obtain $\Delta \supset \mathscr{D}_\Lambda$. The conditions under which $\Delta = \mathscr{D}_\Lambda$ has not yet been investigated.

The introduction of the Λ-uniform structure in D 1.1.7 now appears to be unsymmetric with respect to $\mathscr{B}$ and $\mathscr{D} = \mathscr{D}_\Lambda$. This is, however, not the case, because the Λ-uniform structure is also the initial structure for which the following maps $\mathscr{G} \to \mathbf{R}$ of the form

$$a \to \mu(x, ay) = \mu(a'x, y),$$

$$a \to \mu(x, a^{-1}y) = \mu(a'^{-1}x, y)$$

are uniformly continuous for all $x \in \mathscr{B}$ and $y \in \mathscr{D} = \mathscr{D}_\Lambda$, that is, for which all maps of the form

$$a \to a'x \quad \text{and} \quad a \to a'^{-1}x$$

are uniformly continuous with respect to the $\sigma(\mathscr{B}, \mathscr{D}_\Lambda)$-topology in $\mathscr{B}$.

Therefore the above considerations are symmetric between $\mathscr{D}$ and $\mathscr{B}$. In particular, the question arises whether and under what conditions is it possible that $\Xi = \mathscr{B}$ (where Ξ is defined by D 1.1.5).

1.3 Topologies on the Group $\mathscr{A}$

Since the elements of $\mathscr{A}$ are maps, there are innumerable possibilities for the introduction of a topology on $\mathscr{A}$. We shall select three possibilities—these have already been encountered in §1.2.

To each $T \in \mathscr{A}$ there is a corresponding map $\mathscr{B} \times \mathscr{D} \xrightarrow{T} \mathbf{R}$ defined by $T \to \mu(x, Ty)$. A separating uniform structure on $\mathscr{A}$ is defined by means of the uniform structure of normal convergence on $\mathscr{B} \times \mathscr{D}$ (that is, the initial structure for all the maps $\mathscr{A} \xrightarrow{x,y} \mathbf{R}$ defined by $T \to \mu(x, Ty)$); we shall let $\mathscr{A}_{(\mathscr{B}\mathscr{D})}$ denote $\mathscr{A}$ together with this uniform structure.

In the same way $\mu(x, Ty)$ determines a mapping $\mathscr{B} \times \mathscr{B}' \to \mathbf{R}$ which, by analogy, determines $\mathscr{A}_{(\mathscr{B},\mathscr{B}')}$.

A mapping $\mathscr{B} \to \mathscr{B}$ is defined by means of the adjoint map T' as follows $x \to T'x$. If we use the norm topology of $\mathscr{B}$ in the image set, then a separating uniform structure in $\mathscr{A}$ is defined by means of the normal convergence of this map. We shall let $\mathscr{A}_{(\mathscr{B})}$ denote $\mathscr{A}$ endowed with this uniform structure.

It is easy to see that $\mathscr{A}_{(\mathscr{B})}$ is finer than $\mathscr{A}_{(\mathscr{B},\mathscr{B}')}$ and $\mathscr{A}_{(\mathscr{B},\mathscr{B}')}$ is finer than $\mathscr{A}_{(\mathscr{B},\mathscr{D})}$.

We may now express (D 2) and (D 3) as follows: The representation maps $\mathscr{G} \to \mathscr{A}_{(\mathscr{B},\mathscr{D})}$ or $\mathscr{G}_{ph} \to \mathscr{A}_{(\mathscr{B},\mathscr{D})}$ are uniformly continuous.

From (D 3) it does not follow that the maps $\mathscr{G}_{ph} \to \mathscr{A}_{(\mathscr{B},\mathscr{D})}$ can be extended onto all of $\hat{\mathscr{G}}_{ph}$ since $\mathscr{A}_{(\mathscr{B},\mathscr{D})}$ is, in general, not complete.

From the uniform continuity of $\mathscr{G}_{ph} \to \mathscr{A}_{(\mathscr{B},\mathscr{D})}$ in a trivial manner it follows that the map $\mathscr{G} \to \mathscr{A}_{(\mathscr{B},\mathscr{D})}$ is simply continuous.

Th. 1.2.2 states that, from the continuity of $\mathscr{G} \to \mathscr{A}_{(\mathscr{B},\mathscr{D})}$ it follows that the map $\mathscr{G} \to \mathscr{A}_{(\mathscr{B})}$ is continuous. Th. 1.2.3 is then only a trivial corollary because the map $\mathscr{G} \to \mathscr{A}_{(\mathscr{B},\mathscr{B}')}$ is continuous.

From the proof of Th. 1.2.10 it follows that in the special case in which we choose $\mathscr{G} = \mathscr{A}_{(\mathscr{B})}$ we obtain the first part of the following theorem:

Th. 1.3.1. *$\mathscr{A}_{(\mathscr{B})}$ is a topological group and is metrizable. $\mathscr{A}_{(\mathscr{B})}$ is separable. $\mathscr{A}_{(\mathscr{B})}$ is complete.*

PROOF. In order to show that $\mathscr{A}_{(\mathscr{B})}$ is separable, choose a denumerable set $\{x_\nu\}$ which is norm-dense in $\mathscr{B}$, and, for each ν, choose a denumerable set $T_{\nu\lambda} \in \mathscr{A}$ for which $\{T'_{\nu\lambda}x_\nu\}$ is norm-dense in $\mathscr{A}x_\nu$. Then $\{T_{\nu\lambda}\}$ is a dense set in $\mathscr{A}_{(\mathscr{B})}$ which follows directly from

$$\begin{aligned}\|(T' - T'_{\nu\lambda})x\| &\leq \|(T' - T_{\nu\lambda})x_\nu\| + \|(T' - T'_{\nu\lambda})(x - x_\nu)\| \\ &\leq \|(T' - T'_{\nu\lambda})x_\nu\| + 2\|x - x_\nu\|\end{aligned}$$

(choose ν such that $\|x - x_\nu\| < \varepsilon/4$, and then choose λ such that $\|(T' - T'_{\nu\lambda})x_\nu\| < \varepsilon/2$).

The fact that $\mathscr{A}_{(\mathscr{B})}$ is complete follows from the general theorem that a sequence T_ν for which $T'_\nu x$ is a Cauchy sequence for each x converges towards a $T \in \mathscr{A}$ (from $T'_\nu \to T'$ where T' is a norm-continuous map of $\mathscr{B}$ into itself, it follows that if T'_ν is positive, then T' is positive and from $\|T'_\nu\| = 1$ it also follows that $\|T'\| = 1$).

If, in addition to the requirements that $\mathscr{G}$ be separable and metrizable, we also require that it be locally compact, then with Th. 1.3.1 the following important mathematical theorem may be proven:

If we endow $\mathscr{G}$ and $\mathscr{A}_{(\mathscr{B})}$ with the Borel structure generated by the open sets and if the map $\mathscr{G} \to \mathscr{A}_{(\mathscr{B})}$ is measurable with respect to this structure then the map $\mathscr{G} \to \mathscr{A}_{(\mathscr{B})}$ is also continuous (for proof see [10]). This theorem therefore shows that if $\mathscr{G}$ is locally compact then we may start with a much weaker requirement than that $\mathscr{G} \to \mathscr{A}_{(\mathscr{B})}$ is continuous. All groups which occur in quantum mechanics are locally compact.

In V we have seen that the group $\mathscr{A}$ is "physically too large." Only the subgroup $\mathscr{A}^{(\mathscr{D})}$ can be physically meaningful where $\mathscr{A}^{(\mathscr{D})}$ is the set of all T in $\mathscr{A}$ for which $T\mathscr{D} \subset \mathscr{D}$.

Therefore $\mathscr{A}^{(\mathscr{D})}_{(\mathscr{B})}$ is a separable metrizable topological group.

Th. 1.3.2. *The maps $T \xrightarrow{y} Ty$ for $y \in \mathscr{D}$ of $\mathscr{A}^{(\mathscr{D})}_{(\mathscr{B})}$ in $\mathscr{D}$ are norm-continuous.*

Proof. This result follows directly from Th. 1.2.5, in which we replace $\mathscr{G}$ by $\mathscr{A}^{(\mathscr{D})}_{(\mathscr{B})}$.

On $\mathscr{A}^{(\mathscr{D})}$ we may introduce the topology of normal convergence of the maps $\mathscr{A}^{(\mathscr{D})} \xrightarrow{y} \mathscr{D}$ (with the norm-topology in $\mathscr{D}$); this topology will be called the $(\mathscr{D})$-topology and $\mathscr{A}^{(\mathscr{D})}$ together with this topology will be denoted by $\mathscr{A}^{(\mathscr{D})}_{(\mathscr{D})}$. From Th. 1.3.2 it follows that on $\mathscr{A}^{(\mathscr{D})}$ the $(\mathscr{B})$-topology is finer than the $(\mathscr{D})$-topology.

The results which we have obtained for the space $\mathscr{B}$ may also be obtained for the space $\mathscr{D}$, and we obtain:

Th. 1.3.3. *The maps of $\mathscr{A}^{(\mathscr{D})}_{(\mathscr{D})}$ into $\mathscr{B}$ given by $T \to T'x$ for $x \in \mathscr{B}$ are norm-continuous.*

From Th. 1.3.1, Th. 1.3.2, and Th. 1.3.3 it easily follows that:

Th. 1.3.4. *On the topological group $\mathscr{A}^{(\mathscr{D})}$ the $(\mathscr{D})$-topology is equal to the $(\mathscr{B})$-topology.*

The representations of a group which are of interest to us are therefore the continuous representation maps $\mathscr{G} \to \mathscr{A}^{(\mathscr{D})}_{(\mathscr{B})}$ which are (according to (D 3)) uniformly continuous as maps $\mathscr{G}_{ph} \to \mathscr{A}^{(\mathscr{D})}_{(\mathscr{B},\mathscr{D})}$.

1.4 The Representation of $\mathscr{G}$ in Phase Space Γ

The properties of group representations in phase space Γ are seldom investigated in quantum mechanics. Since this topic is of importance in understanding the relationship between quantum mechanics and classical mechanics we shall present a brief description of the fundamentals of the phase space representations.

The $\sigma(\mathscr{D}', \mathscr{D})$-closure $\bar{K}^\sigma$ of K in $\mathscr{D}'$ is $\sigma(\mathscr{D}', \mathscr{D})$- compact. Therefore, the convex set $\bar{K}^\sigma$ is, according to the Krein–Milman theorem, generated by the set $\partial_e \bar{K}^\sigma$ of its extreme points.

D 1.4.1. We shall call the set $\partial_e \bar{K}^\sigma$ together with the uniform structure generated by the $\sigma(\mathscr{D}', \mathscr{D})$-topology the phase space (which we denote by Γ).

Γ is precompact as a subset of $\bar{K}^\sigma$. Since $\mathscr{D}'$ is separable and metrizable in the $\sigma(\mathscr{D}', \mathscr{D})$-topology, Γ is separable and metrizable, and its points describe in a physically meaningful way (see [1], §9) the "idealized" preparation possibilities for the systems in M.

Since the elements of $\mathscr{G}'$ can also be considered to be mixture automorphisms of $\bar{K}^\sigma$, Γ is $\mathscr{G}'$ invariant. According to Th. 1.2.6, for $a \in \mathscr{G}$, $\gamma \in \Gamma$ the map $a \to a'\gamma$ is continuous. In addition, we find that

Th. 1.4.1. *The mapping of $\mathscr{G} \times \Gamma \to \Gamma$ defined by $(a, \gamma) \to a'\gamma$ is continuous.*

PROOF. For fixed $a \in \mathscr{G}$ and fixed $\gamma \in \Gamma$ and for $y \in \mathscr{D}$ we obtain

$$\begin{aligned}|\mu(a'\gamma - \tilde{a}'\tilde{\gamma}, y)| &\leq |\mu(a'\gamma - a'\tilde{\gamma}, y)| + |\mu(a'\tilde{\gamma} - \tilde{a}'\tilde{\gamma}, y)| \\ &\leq |\mu(\gamma - \tilde{\gamma}, ay)| + |\mu(\tilde{\gamma}, ay - \tilde{a}y)| \\ &\leq |\mu(\gamma - \tilde{\gamma}, ay)| + \|ay - \tilde{a}y\|.\end{aligned}$$

From which it follows that, since $a \to ay$ is norm-continuous, that the map $\mathscr{G} \times \Gamma \to \Gamma$ is continuous.

D 1.4.2. We shall call the representation of $\mathscr{G}'$ on Γ by means of point transformations the associated phase space representation of $\mathscr{G}$ corresponding to the original representation of $\mathscr{G}$.

For quantum mechanics the structure of the associated phase space representation is surprisingly unfamiliar. In the case of "physical objects" (see III, §4.1 and [1], §12) we are able to determine the phase space Γ by means of a particular choice of the group $\mathscr{G}$ (Galileo group or a direct product of Galileo groups) and the uniform structure ph of physical imprecision, by requiring that $\mathscr{D} = \Delta$ (see [21]). Here for Γ we obtain the usual Γ-space of classical mechanics. The axioms in III, §4 and the specification of the group $\mathscr{G}$ permit us, therefore, to deduce the "usual" phase space of classical mechanics in the case of the description of physical objects.

An analogous description for quantum mechanics is, up to now, not commonly in use because the set $\Gamma = \partial_e \bar{K}^\sigma$ is, at present, mathematically not as accessible. For quantum mechanics there is no pressing necessity to investigate the set $\partial_e \bar{K}^\sigma$ as in the case for classical mechanics, because $\partial_e K$ is not only nonempty, but it also satisfies (in the norm-closure of B): $\overline{\text{co}}\ \partial_e K = K$. Since $\mathscr{B}$ is the dual Banach space corresponding to $\mathscr{L}_{cr}$ where the latter is the subspace of all compact operators of $\mathscr{B}'$ it follows that in quantum mechanics K has the property $\overline{\text{co}}\ \partial_e K = K$ (see AIV, §11). $\mathscr{L}_{cr}$ is norm-separable. It is not difficult to show that $\mathscr{L}_{cr} \subset \Delta_n$. In a pure formal way we may choose $\mathscr{D} = \mathscr{L}_{cr}$; then we would find that $\mathbf{1} \notin \mathscr{D}$; however the selection of $\mathscr{D}$ as the norm-closed subspace of $\mathscr{B}$ spanned by $\mathbf{1}$ and $\mathscr{L}_{cr}$ does not appear to be physically meaningful.

The above considerations show why $\partial_e K$, that is, the set of all P_φ, is used as a substitute for the phase space Γ. For this reason we must put up with the fact that for a decision scale observable A which has a continuous spectrum there cannot be an element of $\partial_e K$, that is, a P_φ for which $\mu(P_\varphi, (A - \alpha 1)^2) = 0$ (for α in the continuous spectrum). For "physical" decision observables A, that is, for $A \in \mathscr{D}$ there exists, for each α in the continuous spectrum an element $w \in \partial_e \bar{K}^\sigma$ for which $\mu(w, (A - \alpha\mathbf{1})^2) = 0$!

2 The $\mathscr{G}$-invariant Structure Corresponding to a Group Representation

As we have seen in V, §5, every $\mathscr{B}$-continuous effect automorphism is uniquely determined by a $\perp$-automorphism of G and every $\perp$-automorphism of G determines a $\mathscr{B}$-continuous effect automorphism. Thus the representation of $\mathscr{G}$ by means of $\mathscr{B}$-continuous effect isomorphisms is uniquely determined by means of the representation of $\mathscr{G}$ determined by $\perp$-automorphisms of G.

Of special importance are two subsets of G, first the set of $\mathscr{G}$-invariant decision effects (see D 1.2.1).

Th. 2.1. *The set of $\mathscr{G}$-invariant decision effects forms a complete orthocomplemented sublattice of G which is $\sigma(\mathscr{B}', \mathscr{B})$-closed in G.*

PROOF. Follows directly from the fact that each element $a \in \mathscr{G}$ defines a $\perp$-automorphism of G which is $\sigma(\mathscr{B}', \mathscr{B})$-continuous by means of the map $e \to ae$. $(e \in G)$.

D 2.1. Let $G(\mathscr{G})$ denote the set of $\mathscr{G}$-invariant decision effects, $L(\mathscr{G})$ denote the set of $\mathscr{G}$-invariant effects and $\mathscr{B}'(\mathscr{G})$ denote the set of $\mathscr{G}$-invariant elements of $\mathscr{B}'$.

Th. 2.2. *$L(\mathscr{G}) = \overline{\text{co}}\ G(\mathscr{G})$ where the closure is to be taken with respect to the $\sigma(\mathscr{B}', \mathscr{B})$-topology. $\mathscr{B}'(\mathscr{G})$ is the $\sigma(\mathscr{B}', \mathscr{B})$-closed subspace of $\mathscr{B}'$ which is spanned by $G(\mathscr{G})$.*

PROOF. If $A \subset \mathscr{B}'$ is a set of $\mathscr{G}$-invariant elements in $\mathscr{B}'$ then $\text{co}(A)$ and the $\sigma(\mathscr{B}', \mathscr{B})$-closed subspace spanned by A are also sets of $\mathscr{G}$-invariant elements because the elements of $\mathscr{A}$ are $\mathscr{B}$-continuous effect automorphisms. If g is a $\mathscr{G}$-invariant effect, then from the uniqueness of the spectral representation of g it follows that $g \in \overline{\text{co}}\, G(\mathscr{G})$. Thus for a $\mathscr{G}$-invariant $y \in \mathscr{B}'$ it follows that y lies in the $\sigma(\mathscr{B}', \mathscr{B})$-closed subspace spanned by $G(\mathscr{G})$.

Let $\mathscr{B}'(\mathscr{G})^{\perp}$ denote the set of all $x \in \mathscr{B}$ for which $\mu(x, y) = 0$ for all $y \in \mathscr{B}'(\mathscr{G})$. Then $\mathscr{B}/\mathscr{B}'(\mathscr{G})^{\perp}$ is a Banach space (see [33]).

D 2.2. Let $\mathscr{B}(\mathscr{G})$ be an abbreviation for $\mathscr{B}/\mathscr{B}'(\mathscr{G})^{\perp}$.

Th. 2.3. *$\mathscr{B}(\mathscr{G})$ is a base-norm space and $\mathscr{B}'(\mathscr{G})$ may be identified with the dual space of $\mathscr{B}(\mathscr{G})$ by means of the map*

$$(\tilde{x}, y) = \mu(x, y),$$

where $x \in \tilde{x} \in \mathscr{B}(\mathscr{G})$ and $y \in \mathscr{B}'(\mathscr{G})$ (here $(\tilde{x}, y)$ is the canonical bilinear form to $\mathscr{B}(\mathscr{G})$ and the dual space for $\mathscr{B}(\mathscr{G})$). $\mathscr{B}'(\mathscr{G})$ is an order unit space.

PROOF. Since $\mathscr{B}'(\mathscr{G})$ is $\sigma(\mathscr{B}', \mathscr{B})$-closed, $\mathscr{B}'(\mathscr{G})$ can be identified with the dual space of $\mathscr{B}(\mathscr{G})$ (see [33]). $\mathscr{B}(\mathscr{G})$ is a base norm space if $\mathscr{B}'(\mathscr{G})$ is an order unit space (see AIII, §6 and [33]). The unit sphere of $\mathscr{B}'(\mathscr{G})$ is given by $\mathscr{B}'(\mathscr{G}) \cap (2\mathrm{L} - \mathbf{1})$. Since $\mathbf{1} \in \mathscr{B}'(\mathscr{G})$ and $L \cap \mathscr{B}'(\mathscr{G}) = L(\mathscr{G})$ we obtain $\mathscr{B}'(\mathscr{G}) \cap (2L - 1) = 2L(\mathscr{G}) - \mathbf{1}$, that is, it is equal to the order interval $[-\mathbf{1}, \mathbf{1}]$ in $\mathscr{B}'(\mathscr{G})$.

D 2.3. Let $K(\mathscr{G})$ denote the basis of $\mathscr{B}(\mathscr{G})$.

Do $K(\mathscr{G})$, $L(\mathscr{G})$, $\mathscr{B}(\mathscr{G})$, $\mathscr{B}'(\mathscr{G})$ satisfy all the axioms and theorems which have been formulated in III, §3? We will not pursue this question for the general case. In the case of the special quantum mechanical structure of K and L, we shall return to the question of the structure of $K(\mathscr{G})$, $L(\mathscr{G})$ in VII, §2 and VIII, §1 for physically important groups $\mathscr{G}$.

3 Properties of Representations of $\mathscr{G}$ which are Dependent on the Special Structure of $\mathscr{A}_{(\mathscr{B})}$ in Quantum Mechanics

In this section we shall use the special structure of $\mathscr{A}$ which was described in V, and we present a outline of the properties of a representation of $\mathscr{G}$.

3.1 The Topological Structure of the Group $\mathscr{A}_{(\mathscr{B})}$

According to V (5.4) each element $T \in \mathscr{A}$ has the form

$$Ty = \sum_{\nu} U_{p(\nu)\nu} y_{\nu} U_{p(\nu)\nu}^{-1}, \tag{3.1.1}$$

where the $U_{p(\nu)\nu}$ are isomorphic or anti-isomorphic maps of $\mathscr{H}_{\nu}$ upon $\mathscr{H}_{p(\nu)}$, and each operator T of the form (3.1.1) is an element of $\mathscr{A}$. T uniquely determines the permutation P of the indices ν and the $U_{p(\nu)\nu}$ up to a phase factor $e^{i\alpha_\nu}$.

Therefore T uniquely determines the subset of those ν for which the $U_{p(\nu)\nu}$ is an anti-isomorphic map. We shall denote this subset of the indices by I.

D 3.1.1. Let $\mathscr{A}_{(\mathscr{B})}(p, I)$ denote the subset of all $T \in \mathscr{A}$ which determine the same permutation p and the same index subset I endowed with the topology induced by $\mathscr{A}_{(\mathscr{B})}$.

Th. 3.1.1. *The topology of $\mathscr{A}_{(\mathscr{B})}$ is identical to the initial topology generated by the maps $d_{\varphi_1\varphi_2}$:*

$$T \longrightarrow |\langle \varphi_2, U_{p(\nu)\nu}\varphi_1 \rangle|$$

(where $\varphi_1 \in \mathscr{H}_\nu$ and $\varphi_2 \in \mathscr{H}_{p(\nu)\nu}$) for all $\varphi_1, \varphi_2 \in \mathscr{H}$.

PROOF. First we shall show that $d_{\varphi_1\varphi_2}$ is a continuous mapping of $\mathscr{A}_{(\mathscr{B})}$ in $\mathbf{R}$.

The map $T \xrightarrow{x} T'x$ of $\mathscr{A}_{(\mathscr{B})}$ in $\mathscr{B}$ is continuous for each $x \in \mathscr{B}$. The map $x \longrightarrow \mu(x, y)$ of $\mathscr{B}$ in $\mathbf{R}$ is continuous for each $y \in \mathscr{B}'$. Therefore the map

$$T \longrightarrow \mu(T'x, y) = \mu(x, Ty)$$

of $\mathscr{A}_{(\mathscr{B})} \longrightarrow \mathbf{R}$ is continuous for each $(x, y) \in \mathscr{B} \times \mathscr{B}'$; in particular, the map

$$T \longrightarrow \mu(P_{\varphi_2}, TP_{\varphi_1}) = \operatorname{tr}(P_{\varphi_2}(TP_{\varphi_1}))$$

$$= \operatorname{tr}(P_{\varphi_2}P_{U_{p(\nu)\nu}\varphi_1}) = |\langle \varphi_2, U_{p(\nu)\nu}\varphi_1 \rangle|^2$$

is continuous, and therefore the map $d_{\varphi_1\varphi_2}$ is also continuous.

The initial topology corresponding to the maps $d_{\varphi_1\varphi_2}$ is therefore coarser than the $(\mathscr{B})$-topology. We will show that it is also finer, that is, for a sequence $T^{(n)}$ the relation $T^{(n)\prime}x \longrightarrow T'x$ is satisfied in the norm topology for every $x \in \mathscr{B}$ if the maps $d_{\varphi_1\varphi_2}$ are continuous.

Since each $x \in \mathscr{B}$ has the form $x = \sum_\nu \lambda_\nu P_{\varphi_\nu}$, where $\sum_\nu |\lambda_\nu| < \infty$, from

$$\|(T^{(n)\prime} - T')x\| \leq \sum_{\nu=1}^{N} |\lambda_\nu| \, \|(T^{(n)\prime} - T')P_{\varphi_\nu}\| + 2 \sum_{\nu=N+1}^{\infty} |\lambda_\nu|$$

it follows that $T^{(n)\prime}x \longrightarrow T'x$ for all $x \in \mathscr{B}$ if $T^{(n)\prime}P_\varphi \longrightarrow T'P_\varphi$ for all P_φ.

From V (5.7) for all $\varphi \in \mathscr{H}_{p(\nu)}$ it follows that

$$T'P_\varphi = P_\psi \quad \text{with} \quad \psi = U^{-1}_{p(\nu)\nu}\varphi. \tag{3.1.2}$$

Therefore, letting $P_{\psi_n} = T^{(n)\prime}P_\varphi$ we obtain (where $\|\ldots\|$ is the norm of $\mathscr{B}'$)

$$\|(T^{(n)\prime} - T')P_\varphi\| = \|P_{\psi_n} - P_\psi\|.$$

In the spectral representation of $P_\psi - P_\varphi$ there are only two (nondegenerate) eigenvalues $\alpha_1^{(n)}$ and $\alpha_2^{(n)}$; therefore, for the norm in $\mathscr{B}$ we obtain:

$$\|P_{\psi_n} - P_\psi\|_s = |\alpha_1^{(n)}| + |\alpha_2^{(n)}|.$$

Therefore $\|P_{\psi_n} - P_\psi\|_s \longrightarrow 0$ is therefore equivalent to $|\alpha_1^{(n)}| + |\alpha_2^{(n)}| \longrightarrow 0$ and therefore also to $(\alpha_1^{(n)})^2 + (\alpha_2^{(n)})^2 \longrightarrow 0$, that is, $\operatorname{tr}((P_{\psi_n} - P_\psi)^2) \longrightarrow 0$. From $\operatorname{tr}((P_{\psi_n} - P_\psi)^2) = 2 - 2|\langle \psi_n, \psi \rangle|^2$ it follows that $T^{(n)\prime}x \longrightarrow T'x$ for all $x \in \mathscr{B}$ if

$$|\langle \psi_n, \psi \rangle| \longrightarrow 1 \tag{3.1.3}$$

for all $\varphi \in \mathscr{H}$.

For $\varphi \in \mathcal{H}_{p(v)}$ we obtain

$$\begin{aligned}|\langle\psi_n, \psi\rangle| &= |\langle U^{(n)-1}_{p(v)v}\varphi, U^{-1}_{p(v)v}\varphi\rangle| \\ &= |\langle U_{p(v)v}\psi, U^{(n)}_{p(v)v}\psi\rangle|,\end{aligned}$$

where ψ is given in (3.1.2). If the sequence $T^{(n)}$ is convergent in the initial topology corresponding to $d_{\varphi_1\varphi_2}$ we therefore obtain:

$$|\langle\varphi_2, U^{(n)}_{p(v)v}\varphi_1\rangle| \to |\langle\varphi_2, U_{p(v)v}\varphi_1\rangle|,$$

where $\varphi_1 = \psi$ and $\varphi_2 = U_{p(v)v}\psi$, that is,

$$|\langle U_{p(v)v}\psi, U^{(n)}_{p(v)v}\psi\rangle| \to |\langle U_{p(v)v}\psi, U_{p(v)v}\psi\rangle| = 1,$$

where (3.1.3) is proven.

It is easy to show that the Borel structure corresponding to the initial toplogy is equal to the initial Borel structure, that is, according to Th. 3.2: The Borel structure of $\mathcal{A}_{(\mathcal{B})}$ is the initial Borel structure associated with the maps $d_{\varphi_1\varphi_2}$. On the basis of the theorem which was mentioned at the conclusion of Th. 1.3.1 we therefore obtain:

If, for a local compact group $\mathcal{G}$, the maps $a \to |\langle\varphi_1, U_{p(v)v}(a)\varphi_2\rangle|$ of $\mathcal{G}$ in $\mathbf{R}$ are measurable (where $U_{p(v)v}(a)$ is the map of $\mathcal{H}_v$ in $\mathcal{H}_{p(v)}$ described by (3.1.1) corresponding to a) then $\mathcal{G} \to \mathcal{A}_{(\mathcal{B})}$ is continuous.

In the following, for the most part, we shall only use the following fact which is a simple corollary of Th. 3.2: $\mathcal{G} \to \mathcal{A}_{(\mathcal{B})}$ is continuous if and only if, for each $\varphi_1, \varphi_2 \in \mathcal{H}$ the maps

$$a \to |\langle\varphi_1, U_{p(v)v}(a)\varphi_2\rangle|$$

are continuous.

Th. 3.1.2. *$\mathcal{A}_{\mathcal{B}}(p, I)$ are the connected components of $\mathcal{A}_{(\mathcal{B})}$. In particular, $\mathcal{A}_{(\mathcal{B})}(1, \varnothing)$ (where 1 is the unit permutation) is the component which is connected to the unit element of $\mathcal{A}_{(\mathcal{B})}$.*

PROOF. The fact that the set $\mathcal{A}_{(\mathcal{B})}(1, \varnothing)$ is connected to the unit element follows directly from the spectral representation

$$U = \int_0^{2\pi} e^{i\omega}\, dE(\omega)$$

of a unitary operator and that the function $\langle\varphi, U_t\varphi\rangle$ is continuous in t (where

$$U_t = \int_0^{2\pi} e^{i\omega t}\, dE(\omega)$$

and that $\langle\varphi, U_0\varphi\rangle = 1, \langle\varphi, U_1\varphi\rangle = \langle\varphi, U\varphi\rangle$).

Therefore, if $\mathcal{N}$ is the component which is connected to the unit element, then $\mathcal{A}_{(\mathcal{B})}(1, \varnothing) \subset \mathcal{N}$.

Since $\mathcal{N}$ is connected, it follows that, for each neighborhood $\mathcal{V}$ of the unit element of $\mathcal{A}_{(\mathcal{B})}$, that all elements of the subgroup $\mathcal{N}$ may be represented as products of finitely many elements of $\mathcal{V} \cap \mathcal{N}$ (see AV, §10).

Let $\varphi_1, \varphi_2, \ldots, \varphi_n$ be normed vectors for which $\varphi_\nu \in \mathscr{H}_\nu$. For $\mathscr{V}$ we choose the neighborhood determined by

$$1 - |\langle \varphi_\nu, U_{p(\nu)\nu}\varphi_\nu \rangle| < \varepsilon \quad \text{for } \nu = 1, 2, \ldots, n.$$

From this result it follows that $p(\nu) = \nu$ for $\nu = 1, 2, \ldots, n$, that is, for all elements in $\mathscr{N}$ the relation $p(\nu) = \nu$ for $\nu = 1, 2, \ldots$, n. Since n was arbitrary it follows that for all elements of $\mathscr{N}$ the permutation p is equal to the unit permutation 1.

We choose the neighborhood $\mathscr{V}$ which is determined by

$$1 - |\langle \psi_\nu, U_{\alpha\alpha}\psi_\nu \rangle| < \varepsilon \qquad (\nu = 1, 2, 3, 4),$$

where ψ_1, ψ_2 are unit vectors in $\mathscr{H}_\alpha$ and where $\psi_3 = (1/\sqrt{2})(\psi_1 + \psi_2)$, $\psi_4 = (1/\sqrt{2})(\psi_1 + i\psi_2)$. It is easy to show that $U_{\alpha\alpha}$ is unitary. Thus it follows that $U_{\alpha\alpha}$ must be unitary for all products of elements in $\mathscr{V}$, and also for all elements of $\mathscr{N}$. Since α was arbitrary, it therefore follows that, for all elements of $\mathscr{N}$, the set of indices $\mathscr{I} = \varnothing$. Therefore $\mathscr{N} = \mathscr{A}_{(\mathscr{B})}(1, \varnothing)$.

It is well known that $\mathscr{N} = \mathscr{A}_{(\mathscr{B})}(1, \varnothing)$ is an invariant subgroup of $\mathscr{A}_{(\mathscr{B})}$ (see AV, §10.1). It is easy to show that the cosets of $\mathscr{A}_{(\mathscr{B})}(1, \varnothing)$ are precisely the $\mathscr{A}_{(\mathscr{B})}(p, I)$. Thus the $\mathscr{A}_{(\mathscr{B})}(p, I)$ are the connected components of $\mathscr{A}_{(\mathscr{B})}$ (see AV, §10.1).

Under the following rule for multiplication:

$$(p_1, I_1)(p_2, I_2) = (p_1 p_2, p_2^{-1} I_1 \dotplus I_2)$$

(where $A \dotplus B$ is the symmetric difference $(A \backslash A \cap B) \cup (B \backslash A \cap B)$) the elements (p, I) form a group $\mathscr{F}$; it is easy to see that this group is isomorphic to the factor group $\mathscr{A}_{(\mathscr{B})}/\mathscr{A}_{(\mathscr{B})}(1, \varnothing)$.

3.2 The Topological Properties of a Representation of $\mathscr{G}$

Let $\mathscr{k}$ be the component of $\mathscr{G}$ which is connected to the unit element of $\mathscr{G}$. $\mathscr{k}$ is an invariant subgroup; therefore for the homomorphic map $\mathscr{G} \to \mathscr{A}_{(\mathscr{B})}$ we obtain $\mathscr{k} \to \mathscr{A}_{(\mathscr{B})}(1, \varnothing)$ (see AII, §4). Since $\mathscr{G}$ is separable, the factor group $\mathscr{G}/\mathscr{k}$ is at most denumerable and is a discrete group.

Let $\mathscr{J}$ denote the set of those elements in $\mathscr{G}$ which are mapped into $\mathscr{A}_{(\mathscr{B})}(1, \varnothing)$. Clearly $\mathscr{k} \sim \mathscr{J}$. It is easy to show that $\mathscr{J}$ is a subgroup of $\mathscr{G}$.

For $j \in \mathscr{J}$ and $a \in \mathscr{G}$ the product aja^{-1} (considered as an element of $\mathscr{A}_{(\mathscr{B})}$) is a map for which $p = 1$ and $I = \varnothing$, that is, $a\mathscr{J}a^{-1} \subset \mathscr{J}$ for all $a \in \mathscr{G}$. $\mathscr{J}$ is therefore an invariant subgroup in $\mathscr{G}$; then, according to the isomorphism theorem (AV, §4) we obtain:

$$\frac{\mathscr{G}/\mathscr{k}}{\mathscr{J}/\mathscr{k}} \cong \frac{\mathscr{G}}{\mathscr{J}}.$$

From the mapping $\mathscr{G} \to \mathscr{A}_{(\mathscr{B})}$ we obtain a homomorphic map $\mathscr{G}/\mathscr{J} \to \mathscr{A}_{(\mathscr{B})}/\mathscr{A}_{(\mathscr{B})}(1, \varnothing)$. An element of $\mathscr{G}/\mathscr{J}$ is the union of all those cosets in $\mathscr{G}/\mathscr{k}$ which are mapped by means of the homomorphic map $\mathscr{G} \to \mathscr{A}_{(\mathscr{B})}$ into one and the same $\mathscr{A}_{(\mathscr{B})}(p, I)$. An element of $\mathscr{G}/\mathscr{k}/\mathscr{J}/\mathscr{k}$ is precisely the set of those cosets of $\mathscr{G}/\mathscr{k}$ which are mapped by the homomorphic map $\mathscr{G} \longrightarrow \mathscr{A}_{(\mathscr{B})}$ into one and the same $\mathscr{A}_{(\mathscr{B})}(p, \mathscr{I})$. Therefore, on the basis of the isomorphism theorem a homomorphism $\mathscr{G}/\mathscr{k}/\mathscr{J}/\mathscr{k} \longrightarrow \mathscr{A}_{(\mathscr{B})}/\mathscr{A}_{(\mathscr{B})}(1, \varnothing)$ is defined by the map $\mathscr{G} \longrightarrow \mathscr{A}_{(\mathscr{B})}$, and, consequently, we may identify $\mathscr{A}_{(\mathscr{B})}/\mathscr{A}_{(\mathscr{B})}(1, \varnothing)$ with the group $\mathscr{F}$ which was given at the end of §3.1.

In nonrelativistic quantum mechanics we only consider those representations in which the group $\mathscr{G}$ is mapped onto portions of the form $\mathscr{A}_{(\mathscr{B})}(1, I)$, that is, for $p \neq 1$ there are no images of elements in portions of the form $\mathscr{A}_{(\mathscr{B})}(p, I)$. In nonrelativistic quantum mechanics it is also possible to use representations of groups in which there exist images of group elements in portions of the form $\mathscr{A}_{(\mathscr{B})}(p, I)$ where $p \neq 1$(!). Such representations are of little physical significance in nonrelativistic quantum mechanics. The situation is somewhat different in the case of relativistic quantum mechanics. Here, on physical grounds, it is meaningful, for example, to choose a transformation of a portion of $\mathscr{A}_{(\mathscr{B})}(p, \varnothing)$ where $p \neq 1$ as the homomorphic image of reflection (parity inversion).

In this book we shall only be concerned with nonrelativistic quantum mechanics for the following two reasons:

(1) By analogy with the case of the Galileo group (which is described in VII) we could consider the problem of the representation of the Poincaré group (since it has been solved, as is the case of the Galileo group); however. the interaction problem has been solved in nonrelativistic quantum mechanics by the consideration of combined systems (see VIII). The interaction problem has not yet been satisfactorily solved in relativistic quantum mechanics, that is, for "elementary particle theory." We shall again discuss this problem in VII and VIII.

(2) This book should provide an insight into the structure of a "closed" physical theory. For this purpose we must abstain from the consideration of fragments of other theories (not only that of relativistic quantum mechanics) although these fragments would fit into the context of II–VI. See, for example, the discussion with respect to the case of nuclear physics in VII, §2.

Since we shall only make use of the partitions $\mathscr{A}_{(\mathscr{B})}(1, \mathscr{I})$ of $\mathscr{A}_{(\mathscr{B})}$ it is therefore tedious to use the whole spaces $\mathscr{B}(\mathscr{H}_1, \mathscr{H}_2, \ldots)$ and $\mathscr{B}'(\mathscr{H}_1, \mathscr{H}_2, \ldots)$ for computational purposes. It suffices to consider for each $\mathscr{H}_\nu$ the spaces $\mathscr{B}(\mathscr{H}_\nu)$, $\mathscr{B}'(\mathscr{H}_\nu)$ separately, together with the group $\mathscr{A}_{(\mathscr{B})}$ of all $\mathscr{B}$-continuous effect isomorphisms of $L(\mathscr{H}_\nu)$. In more general circumstances, or in circumstances in which it is clear which system type (as characterized by an atom of the center—see IV, §8) we are concerned with, we shall ignore the index ν for $\mathscr{H}_\nu$ and, instead, consider $\mathscr{B}(\mathscr{H})$, $\mathscr{B}'(\mathscr{H})$ together with the subsets $K(\mathscr{H})$, $L(\mathscr{H})$. Then the group $\mathscr{A}_{(\mathscr{B})}$ will consist of two connected parts—$\mathscr{A}_{(\mathscr{B})}(u)$, which is connected to $\mathbf{1}$ and consists of effect isomorphisms described by unitary transformations and $\mathscr{A}_{(\mathscr{B})}(a)$ which is the set of effect isomorphisms described by anti-unitary transformations in $\mathscr{H}$. The subgroup $\mathscr{k}$ of $\mathscr{G}$

therefore permits a representation in $\mathscr{A}_{(\mathscr{B})}(u)$. Only elements of the cosets $a\mathscr{k}$ for which $a\mathscr{k} \neq \mathscr{k}$ can have images in $\mathscr{A}_{(\mathscr{B})}(a)$.

Readers who are not well versed in physics will have some difficulty accepting such a "reduced" description of physical experiments because, to a given preparation procedure $a \in \mathscr{Q}'$ it is possible to find a $\varphi(a) \in K(\mathscr{H}_1, \mathscr{H}_2, \ldots)$ which has nonzero components in more than one $K(\mathscr{H}_\nu)$, that is,

$$\varphi(a) = (W_1, W_2, \ldots),$$

where more than one of the W_ν are nonzero. Such a preparation apparatus (described by a) clearly does not produce only microsystems of a single type. However, mixtures of different system types in the ratios

$$\operatorname{tr}(W_1): \operatorname{tr}(W_2): \operatorname{tr}(W_3): \ldots$$

are "physically trivial" and are therefore uninteresting. Similarly, for those effects $\psi(b_0, b) = (F_1, F_2, \ldots)$ having more than one nonzero F_ν we find that the probabilities are computed according to the mixture formula

$$\mu(\varphi(a), \psi(b_0, b)) = \sum_\nu \operatorname{tr}(W_\nu F_\nu).$$

Here only the individual terms $\operatorname{tr}(W_\nu F_\nu)$ are of interest. Combinations of several system types are only of interest if there are "physically interesting" effect isomorphisms in $\mathscr{A}_{(\mathscr{B})}(p, \mathscr{I})$ for which $p \neq 1$. Such is the case only in relativistic quantum mechanics. For the present we shall only be interested in the simplified description of quantum mechanics in which only a single Hilbert space is used—considering different system types separately.

3.3 Unitary and Anti-unitary Representations Up to a Factor

Each element a of the group $\mathscr{G}$ (considered as an element of $\mathscr{A}_{(\mathscr{B})}$ by means of the representation in $\mathscr{A}_{(\mathscr{B})}$ (where $\mathscr{B} = \mathscr{B}(\mathscr{H})$!) corresponds to a unitary or anti-unitary transformation of $\mathscr{H}$ into itself as follows:

$$a \longrightarrow U(a). \tag{3.3.1}$$

$U(a)$ is determined by a up to a factor (which depends upon a). We may require that the unit element e of $\mathscr{G}$ corresponds to the unit operator $\mathbf{1}$ in $\mathscr{H}$:

$$e \longrightarrow \mathbf{1}. \tag{3.3.2}$$

From the representation properties of $\mathscr{G}$ in $\mathscr{A}_{(\mathscr{B})}$ it follows that $U(ab)$ must determine (according to (3.3.1)) the same transformation in $\mathscr{A}_{(\mathscr{B})}$ as does $U(a)U(b)$, that is, $U(ab)$ and $U(a)U(b)$ can differ only by a factor of magnitude 1:

$$U(ab) = \omega(a, b)U(a)U(b), \tag{3.3.3}$$

where $|\omega(a, b)| = 1$.

Conversely, if we have a representation (3.3.1) of a group $\mathscr{G}$ for which (3.3.2) and (3.3.3) hold, then to each $a \in \mathscr{G}$ there exists a unique element in $\mathscr{A}_{(\mathscr{B})}$ corresponding to $U(a)$ which we denote simply by a.

Let $\mathscr{U}(\mathscr{H})$ denote the group of unitary and anti-unitary transformations of $\mathscr{H}$.

A map of $\mathscr{G}$ into $\mathscr{U}(\mathscr{H})$ of the form (3.3.1) for which (3.3.2) and (3.3.3) are valid will be called a unitary (or anti-unitary) representation of $\mathscr{G}$ up to a factor. Each unitary (anti-unitary) representation of $\mathscr{G}$ up to a factor uniquely corresponds to a representation of $\mathscr{G}$ in $\mathscr{A}_{(\mathscr{B})}$. Each representation of $\mathscr{G}$ in $\mathscr{A}_{(\mathscr{B})}$ corresponds to a representation up to a factor, where two representations U_1, U_2 determine the same representation in $\mathscr{A}_{(\mathscr{B})}$ if

$$U_1(a) = e^{i\delta(a)}U_2(a), \tag{3.3.4}$$

where $\delta(a)$ is a real function.

For $a = e$ or $b = e$, from (3.3.3) and (3.3.2) it follows that

$$\omega(e, b) = \omega(a, e) = 1. \tag{3.3.5}$$

From (3.3.3) it follows that

$$\begin{aligned} U(a(bc)) &= \omega(a, bc)U(a)U(bc) \\ &= \omega(a, bc)\omega(b, c)U(a)U(b)U(c) \end{aligned}$$

and

$$\begin{aligned} U((ab)c) &= \omega(ab, c)U(ab)U(c) \\ &= \omega(ab, c)\omega(a, b)U(a)U(b)U(c) \end{aligned}$$

and we therefore obtain

$$\omega(a, bc)\omega(b, c) = \omega(ab, c)\omega(a, b). \tag{3.3.6}$$

The relations (3.3.5) and (3.3.6) are, in a sense, characteristic for the multipliers $\omega(a, b)$ because, to each "solution" of (3.3.5) and (3.3.6) there also exists a representation up to a factor (see [10]).

From (3.3.4) it follows that the multipliers ω_1 and ω_2 for two representations U_1, U_2 which correspond to the same representation in $\mathscr{A}_{(\mathscr{B})}$ satisfy the equation:

$$\omega_2(a, b) = \omega_1(a, b)e^{i[\delta(a) + \delta(b) - \delta(ab)]}. \tag{3.3.7}$$

Two multipliers ω_1 and ω_2 are said to be equivalent if there exists a real function δ on $\mathscr{G}$ such that (3.3.7) holds. Here we note that nothing more about $\omega(a, b)$ has been assumed—nothing about continuity, measurability, etc. has been assumed.

Therefore the problem remains to put forward a clever choice of special multiplier from a class of equivalent multipliers.

Thus the problem of finding the representation of a group $\mathscr{G}$ in $\mathscr{A}_{(\mathscr{B})}$ is equivalent to the problem of finding all unitary (or anti-unitary) representations up to a factor and the selection of a particularly "simple" multiplier from each equivalence class of multipliers.

In accord with Th. 3.1.1 we now propose the following continuity assumption for the representation map $U: \mathscr{G} \to U(\mathscr{H})$:

$$a \to |\langle \varphi, U(a)\psi \rangle| \tag{3.3.8}$$

is continuous for all $\varphi, \psi \in \mathscr{H}$.

The problem posed above can be solved for a certain type of group which contains all "physically relevant" groups. A more precise description of the solution of this representation problem will require a special book, and would result in the loss of continuity of the train of thought. Since monographs already exist on this topic (see, for example, [10]), in the appendix AV, §10 of this book we shall only present a brief summary for the case of a compact group in order that we may obtain a better understanding of the rotation groups.

In closing this section we shall now characterize the concepts introduced at the beginning of §1.2 in terms of the form of a representation up to a factor.

A $\mathscr{G}$-invariant effect (as defined in D 1.2.1) is therefore an element g which commutes with all $U(a)$, $a \in \mathscr{G}$, that is, $U(a)g = gU(a)$.

Therefore, according to D 1.1.2 a representation is irreducible if the only operators which commute with all the $U(a)$, $a \in \mathscr{G}$ are multiples of the **1**-operator. Otherwise, to each operator A which commutes with all $U(a)$ we would have $A^+U(a)^+ = U(a)^+A^+$, that is, $A^+U(a^{-1}) = U(a^{-1})A^+$; then all $U(a)$ would commute with the operator A and hence with the self-adjoint operator $A + A^+$. Then there would be a projection operator E ($\neq 0, \neq 1$) (for example, from the spectral family of $A + A^+$) which commutes with all $U(a)$.

If a representation is not irreducible, then there exists a projection operator E ($\neq 0, \neq 1$) which commutes with all $U(a)$; this is equivalent to the condition that there exists an invariant subspace of $\mathscr{H}$ (different from $\mathscr{H}$ and $\{0\}$), namely, the projection spaces belonging to E.

With the help of V (5.3) we may, in principle, rewrite condition D 1.2.3 for equivalent representations. We will do this for the case of two representations in $\mathscr{B}'(\mathscr{H})$ and $\mathscr{B}'(\tilde{\mathscr{H}})$ because these are the only significant ones in nonrelativistic quantum mechanics.

Suppose that two representations by means of $\mathscr{B}$-continuous effect isomorphisms are given in terms of representations up to a factor $U(a)$ in $\mathscr{H}$ and $\tilde{U}(a)$ in $\tilde{\mathscr{H}}$. They are equivalent if there exists an isomorphism (or anti-isomorphism) V of $\mathscr{H}$ onto $\tilde{\mathscr{H}}$ such that

$$U(a)yU(a)^{-1} = V^{-1}\tilde{U}(a)VyV^{-1}\tilde{U}(a)^{-1}V \tag{3.3.9}$$

holds for all $y \in \mathscr{B}'(\mathscr{H})$. From (3.3.9) we find that the above is equivalent to

$$U(a) = e^{i\delta(a)}V^{-1}\tilde{U}(a)V, \tag{3.3.10}$$

where $\delta(a)$ is a real function in $\mathscr{G}$. If $\mathscr{H} = \tilde{\mathscr{H}}$ then V is a unitary or anti-unitary operator in $\mathscr{H}$. By clever selection of the factors for a representation we may, according to (3.3.10) obtain $\delta(a) = 0$. Then (3.3.10) reduces to the "usual" form for the definition of equivalence of two unitary representations.

In particular, it follows from (3.3.10) that for $V = \mathbf{1}$, that two representations which differ only by factors of the form $e^{i\delta(a)}$ (and which, according to (3.3.7), correspond to the equivalent multipliers) are equivalent.

CHAPTER VII

The Galileo Group

In the investigation of microsystems an important role is played by a particular component of the physical–technical structure of those experiments which are composed of a preparation procedure and a registration method. Every experimental physicist is acquainted with this component, which underlies the problem of fixing the spatial and time relationships between the preparation and registration apparatuses. Earlier we have briefly described this question in the definition of C in II (4.3.1) and in III, §1. We must now introduce a corresponding mathematical structure into our mathematical formulation which describes the spatial and time relationships between the preparation and registration apparatuses.

1 The Galileo Group as a Set of Transformations of Registration Procedures Relative to Preparation Procedures

Underlying each element $b_0 \in \mathscr{R}_0$ there exists a whole technology of the construction of the apparatus to which the registration method b_0 belongs. Here it is not possible to discuss the technology involved. In this respect a large series of pre-theories (see [1], [2], III and [13]) is required for quantum mechanics. It is, however, essential that quantum mechanics itself is not used in the physical description of b_0. Here it is possible to raise objections to this assertion; in XVII we will examine such problems in more detail. In [2], XVI it will be obvious what we mean by the above assertions. In [13] this problem is treated in considerable detail.

Although we do not need to go into all of the details of the construction of b_0, we must, however, present a brief explanation of the special technical character associated with the pre-theory of space–time measurement.

The preparation and registration procedures always refer to a space–time reference system—an inertial system (or approximately inertial system) (see, for example, [2], II, VII, and IX). Every experimental physicist is aware of the importance of the spatial and temporal relationships of the registration apparatus relative to the preparation apparatus. Here it is often necessary to use the most modern measurement techniques. It is important that the technical specification of such a space–time reference system has nothing to do with quantum mechanics. The fact that this situation is often not sufficiently understood is, in part, the cause for many conceptual errors in quantum mechanics.

We shall now describe what we believe to be the correct meaning of space and time in quantum mechanics: Space and time coordinates refer only to the preparation and registration apparatuses and do not(!!) refer to the individual microsystems. In the formulation of quantum mechanics presented in this book we have not introduced the position of a microsystem as a basic concept because it is not clear how such a concept can be defined in terms of the pre-theories. The mapping principles do not permit us to use concepts which cannot be defined by the pre-theories (see [1] and [2], III). Instead of the "position of a microsystem" only the spatial relationship between the preparation and registration apparatuses is defined by means of the pre-theories, and, in this respect, is permitted by the mapping principles.

We shall not provide a precise mathematical picture of the relative placement of the registration apparatus relative to the laboratory spatial coordinate system. Here we shall be interested only in a particular aspect of this structure, which is described below. This description is a substitute for a more complicated description in terms of pre-theories (see [13]).

We shall assume that a registration method $b_0 \in \mathscr{R}_0$ is not only characterized by the "inner" structure of the registration apparatus but also by its position and motion relative to the space–time laboratory reference system. In this way we obtain a new structure in the domain of the registration methods and the corresponding registration procedures. Two registration methods b_{01} and b_{02} can only differ in their position and motion in the laboratory reference system if the corresponding apparatuses do not differ in their internal structure. We shall now introduce a mathematical description of such a situation between two registration methods.

How can two such registration methods differ? If we use the Newtonian space–time structure, they may differ only by a Galileo transformation. If the space–time structure of special relativity is used, they will differ by one of the transformations of the Poincaré group. For a discussion of the Galileo group see [2], VI, §1.2; for the Poincaré group see [2], IX, §4.3. Here the physical meaning of a group transformation is that b_{01} may be obtained from b_{02} by means of a spatial translation (in the reference system under consideration) or by a time translation, or that the apparatus for b_{01} moves with constant velocity relative to that of b_{02}.

A "time translation" for the time τ of b_{01} relative to b_{02} means that the apparatus corresponding to b_{02} is, placed into operation at a time interval τ later than the apparatus corresponding to b_{02}, for example, a voltage is turned on at a later time τ. Later we shall find that the problem of time displacement is not trivial with respect to the combination problem of a preparation and a registration apparatus described in II, §4.3.

We will consider only the mathematical formulation of the structure described above for the case of the Galileo group $\mathscr{G}$. The formulation of the analogous case for the Poincaré group is similar and trivial. $\mathscr{G}$ is a local compact, separable topological group. The elements of $\mathscr{G}$ can be given by the transformations (see [2], VI (1.2.1)) as follows:

$$x'_\nu = \sum_{\mu=1}^{3} \alpha_{\nu\mu} x_\mu + \delta_\nu t + \eta_\nu \qquad (\nu = 1, 2, 3),$$
$$t' = t + \gamma, \tag{1.1}$$

where $A = (\alpha_{\nu\mu})$ is the matrix of a spatial rotation, that is, $A' = A^{-1}$. We shall consider only transformations (1.1) which can be continuously transformed into the unit element, that is, those for which the determinant $|A| = 1$.

We may represent an element characterized by (1.1) as follows:

$$(A, \vec{\delta}, \vec{\eta}, \gamma). \tag{1.2}$$

We now make the following assertion: There exists at least one denumerable subgroup $\tilde{\mathscr{G}} \subset \mathscr{G}$ (we can choose $\tilde{\mathscr{G}}$ to be the set of all transformations (1.1), where $\alpha_{\mu\nu}, \delta_\nu, \eta_\nu, \gamma$ are rational numbers) which, for all $g \in \tilde{\mathscr{G}}$ there exists a map $\mathscr{R} \xrightarrow{g} \mathscr{R}$ (which we shall denote by g) for which $g(\mathscr{R}_0) \subset \mathscr{R}_0$. Its physical meaning (as we mentioned earlier) is that, for $b_0 \in \mathscr{R}_0$, the method gb_0 is obtained from b_0 by means of the Galileo transformation g and that for $b \in \mathscr{R}$, $b \subset b_0$ the procedure gb is precisely that which is obtained from b by means of the Galileo transformation g. The "meaning" described here is a mapping principle (in the sense of [1], §5 or [2], III, §4) for the relation mathematically described by $\mathscr{R} \xrightarrow{g} \mathscr{R}$. This short outline must suffice for the present. It is important, however, to note that the "physical meaning" of g and gb_0, gb is already determined by means of the pre-theories! The reason why we consider only a denumerable subgroup $\tilde{\mathscr{G}}$ is concerned with the "finiteness of physics" assumption (see [1], §9 or [2], III, §8) that $\mathscr{R}$ must be denumerable (see VI, §1.1).

We may express the fact that the inner structure of the registration apparatus remains unchanged by the application of the Galileo transformation by asserting that the mapping $\mathscr{R} \xrightarrow{g} \mathscr{R}$ is an r-automorphism. The fact that it is essential to consider only Galileo transformations is made clear by the fact that the acceleration of an apparatus can modify its inner structure. Therefore an apparatus b_{01} which is accelerated relative to an apparatus b_{02} cannot be characterized by means of a r-automorphism.

We now present a summary of the above considerations in axiomatic form:

For each $g \in \tilde{\mathscr{G}}$ there exists an r-automorphism $\mathscr{R} \xrightarrow{g} \mathscr{R}$ and we obtain a representation of the group $\mathscr{G}$ by means of r-automorphisms.

This axiom can be directly obtained as a theorem from the pre-theories—see, for example, [13].

(For the preparation and registration of *macro*systems this assertion about the possibility of representation of the group $\tilde{\mathscr{G}}$ by means of r-automorphisms is not correct for all time translations γ because we have chosen the time point $t = 0$ to be the time before which the preparation is complete and after which the registration begins (see III, §1, [2], XV, and [13]). In [13] it is shown that the time translation of the registration apparatus makes sense only for $\gamma \geq 0$.)

Our description of the application of Galilean transformations for registration procedures can also be directly carried out for preparation procedures. For such a transformation we shall write $a \longrightarrow ga$. The transformations of registration procedures and preparation procedures are not mutually independent. For a pair $a \in \mathscr{Q}'$, $b_0 \in \mathscr{R}'_0$ from a Galileo transformation g there arises an experiment (a, gb_0) if $(a, gb_0) \in C$. If, instead, we transform the preparation procedure by means of g^{-1}, then we obtain the pair $(g^{-1}a, b_0)$. Here (a, gb_0) and $(g^{-1}\mathrm{a}, \mathrm{b}_0)$ differ only in the fact that the complete experiment (a, gb_0) is transformed by g relative to $(g^{-1}a, b_0)$, while the "relative" position of the preparation and registration apparatus is the same in both cases. For this reason the following assertions are "almost trivial"

$$(a, gb_0) \in C \Leftrightarrow (g^{-1}a, b_0) \in C$$

and

$$\mu(a, g(b_0, b)) = \mu(g^{-1}a, (b_0, b)).$$

From V, Th. 3.4 it follows directly that $\mathscr{R} \xrightarrow{g} \mathscr{R}$ is a p-continuous r-automorphism and that $\mathscr{Q} \xrightarrow{g^{-1}} \mathscr{Q}$ is an r-continuous p-automorphism.

We may combine the above considerations by asserting that $\tilde{\mathscr{G}}$ may be represented by means of p-continuous r-automorphisms by means of the maps $\mathscr{R} \xrightarrow{g} \mathscr{R}$.

According to VI, §1.1 it follows that to each g there corresponds a unique $\mathscr{B}$-continuous effect automorphism which, in turn, corresponds to a representation $\tilde{\mathscr{G}} \longrightarrow \mathscr{A}$ of $\tilde{\mathscr{G}}$ into the group $\mathscr{A}$ of $\mathscr{B}$-continuous effect automorphisms (see VI, §1.2). In addition the elements of $\mathscr{A}$ which correspond to elements of $\tilde{\mathscr{G}}$ leave the subspace $\mathscr{D}$ of $\mathscr{B}'$ invariant.

For the representation of $\tilde{\mathscr{G}}$ by means of p-continuous r-automorphisms we shall require that AG 1 from VI, §1.1 holds. AG 1 is the mathematical expression for the condition that small errors in adjusting the registration apparatus in space and time cannot be detected by means of the probability distributions. In principle this is nothing other than the assumption that small errors are of a statistical character.

In this way we may, therefore, by VI, §1.1 and §1.2 consider the representation of the complete Galileo group $\mathscr{G}$ in $\mathscr{A}$ where $\mathscr{G} \rightarrow \mathscr{A}$ is continuous according to VI, §1.3. According to VI, §3.3. we may also consider separate representations of the Galileo group (up to a factor) for each of the Hilbert spaces $\mathscr{H}_v$ of different "system types" because the Galileo group (where we assume that the determinant of the rotation matrix is $+1$) is connected. For these representations we may impose the continuity condition VI, (3.3.8).

2 Irreducible Representations of the Galileo Group and Their Physical Meaning

The irreducible representations of the Galileo group play a fundamental physical role. As we have found at the end of the previous section, we can consider each system type and its corresponding Hilbert space $\mathscr{H}_v$ separately.

D 2.1. A system type (IV, D 8.1.1) is said to be "elementary" if its corresponding representation of the Galileo group (that is, its representation up to a factor in $\mathscr{H}_v$) is irreducible; otherwise, it is said to be "composite."

We may sometimes speak (less precisely) of $x \in p(e)$ (where e is an elementary system type) simply of elementary systems x of type e (for $p(e)$ see IV, §8.1).

This definition is mathematically clear. However, its physical interpretation may be misunderstood. For this reason we shall make a number of explanatory remarks.

Every theory refers to a certain "fundamental domain" where it is usuable (see, for example, [2], III, §2 and §4 or [1], §3 and §5). If we then have a more comprehensive theory, then the corresponding fundamental domain will probably be larger (see [2], III, §7 or [1], §8). If at a given point in the development of a theory there is a certain fundamental domain, the theory will describe certain real factual content—for example, atomic nuclei—as elementary systems. In a more comprehensive theory these systems need not necessarily be elementary. For example, if, in the fundamental domain, only low-energy processes are admitted, then the atom nuclei may be described as elementary systems (for example, in atom and molecular physics). If we extend the fundamental domain by admitting such processes as nuclear reactions, then we must use more comprehensive theories. In such theories the atomic nuclei (except for the neutron and proton) must be described as composite systems. We must be careful and avoid the mistakes made by considering the concepts in a physical theory to be absolute, instead of understanding that they are part of the description of a certain fundamental domain. In physics we often restrict the fundamental domain and consider simplified approximate theories for such restricted fundamental domains. In such a simplified and approximate theory a complete atom may be considered to be an "elementary" system. The fact that this situation does not result in a contradiction can be seen in the fact that in the treatment of composite systems we find that the "center of mass" of a composite system itself behaves as if it were an elementary system.

Since we may consider each system type separately—as we have already discussed in VI, §3—in the following we shall always consider only a single Hilbert space $\mathscr{H}$.

As we mentioned in VI, §3.3, we cannot present an exact derivation of the irreducible representations of the Galileo group here, because it would require an entire book to do so. The reader who is interested in this task is referred to [10]. The derivation presented there can be directly applied here because the continuity of the maps, VI (3.3.8), is the central assumption in [10] (if necessary, using the weaker requirement that the maps, VI (3.3.8), be only measurable, then continuity would follow for locally compact groups) for the derivation of the possible inequivalent representations. Here again we note that the assumptions made in §1 are completely sufficient to apply all the theorems presented in [10]. Every irreducible representation of the Galileo group in terms of $\mathscr{B}$-continuous effect isomorphisms can be given by the corresponding unitary representation up to a factor. We shall now give a brief summary of the structure of such representations.

The Galileo group (1.1) contains the abelian group of spatial translations $\vec{\eta}$ and the "proper" Galileo transformations $\vec{\delta}$ as subgroups. For a one-parameter group (for example, the translations in the 1-direction with the parameter η) the factors may be so chosen that we obtain a unitary representation as follows:

Let α denote the parameter of the group element, that is,

$$a(\alpha_1)a(\alpha_2) = a(\alpha_1 + \alpha_2)$$

from VI (3.3.5) and (3.3.6) where

$$\omega(a(\alpha_1), a(\alpha_2)) = e^{i\gamma(\alpha_1, \alpha_2)}$$

it follows that

$$\gamma(0, \alpha) = \gamma(\alpha, 0) = 0, \tag{2.1}$$

$$\gamma(\alpha_1, \alpha_2 + \alpha_3) + \gamma(\alpha_2, \alpha_3) = \gamma(\alpha_1 + \alpha_2, \alpha_3) + \gamma(\alpha_1, \alpha_2). \tag{2.2}$$

According to VI (3.3.7) the question arises whether there exists a $\delta(\alpha) = \delta(a(\alpha))$ for which $\omega_2(a, b) = 1$, that is,

$$\gamma(\alpha_1, \alpha_2) + \delta(\alpha_1) + \delta(\alpha_2) - \delta(\alpha_1 + \alpha_2) = 0. \tag{2.3}$$

On the basis of condition (2.1) it is always possible to find such a $\delta(\alpha)$. This can be proven without additional assumptions (see [10]). This can be easily seen if we assume that γ is twice differentiable, and we differentiate (2.2) first with respect to α_1 and then with respect to α_3.

For a family of unitary opertors $U(\alpha)$ with $U(0) = \mathbf{1}$ and $U(\alpha_1 + \alpha_2) = U(\alpha_1) + U(\alpha_2)$ it follows that (see [35]) there exists a spectral family $E(k)$ for which

$$U(\alpha) = \int_{-\infty}^{\infty} e^{ik\alpha}\, dE(k). \tag{2.4}$$

From (2.4) it follows that there exists a (not necessarily bounded) self-adjoint operator

$$K = \int_{-\infty}^{\infty} k\, dE(k), \tag{2.5}$$

where

$$U(\alpha) = e^{iK\alpha}. \tag{2.6}$$

The above procedure can also be carried out for the three parameter abelian group $(1, 0, \vec{\eta}, 0)$ if we only replace α by a three-dimensional vector $\vec{\eta}$. Then it follows that we may choose $U(1, 0, \vec{\eta}, 0)$ in such a way that we obtain a representation without factors, that is, the $U(1, 0, \vec{\eta}, 0)$ form an abelian group. From (2.6) we find that there exist self-adjoint operators K_1, K_2, K_3 (or $\vec{K}$) for which

$$U(1, 0, \vec{\eta}, 0) = e^{i\vec{K}\cdot\vec{\eta}}. \tag{2.7}$$

Since the $U(1, 0, \vec{\eta}, 0)$ form an abelian group, the K_ν must commute. $\vec{K}$ is not, however, uniquely determined by the choice of factors. It is easy to see that the $U(1, 0, \vec{\eta}, 0)$ are uniquely determined up to factors of the form $e^{i\vec{\kappa}\cdot\vec{\eta}}$ where $\vec{\kappa}$ is an arbitrary vector. Therefore $\vec{K}$ is determined only up to additive term $\vec{\kappa}\mathbf{1}$.

For the elements of the Galileo group we find that

$$(1, 0, A\vec{\eta}, 0) = (A, 0, 0, 0)(1, 0, \vec{\eta}, 0)(A^{-1}, 0, 0, 0). \tag{2.8}$$

For the representation we obtain

$$U(A, 0, 0, 0)U(1, 0, \vec{\eta}, 0)U(A, 0, 0, 0)^{-1} = \lambda(A, \vec{\eta})U(1, 0, A\vec{\eta}, 0). \tag{2.9}$$

The multipliers $\lambda(A, \vec{\eta})$ do not depend on the choice of factors for $U(A, 0, 0, 0)$. We may set $\lambda(A, \vec{\eta}) = 1$ by making a suitable choice of the factor $e^{i\vec{\kappa}\cdot\vec{\eta}}$ as follows: From (2.9) it easily follows (where $\lambda(A, \vec{\eta}) = e^{ig(A,\vec{\eta})}$, that

$$g(A, \vec{\eta}_1 + \vec{\eta}_2) = g(A, \vec{\eta}_1) + g(A, \vec{\eta}_2)$$

and

$$g(A_1 A_2, \vec{\eta}) = g(A_2, \vec{\eta}) + g(A_1, A_2\vec{\eta}).$$

From the first equation it follows that

$$g(A, \vec{\eta}) = \vec{h}(A)\cdot\vec{\eta}$$

and from the second equation we obtain

$$\vec{h}(A_1 A_2) = \vec{h}(A_2) + A_2^{-1}\vec{h}(A_1).$$

This equation fixes $\vec{h}(A)$ up to an arbitrary vector $\vec{h}(A_0)$ for an A_0 for which $\vec{h}(A_0) \neq 0$. The solution up to an arbitrary vector is given by

$$\vec{h}(A) = \vec{\kappa} - A^{-1}\vec{\kappa}.$$

Thus, from (2.9) it follows that the factors $e^{i\vec{\kappa}\cdot\vec{\eta}}$ of the $U(1, 0, \vec{\eta}, 0)$ can be chosen such that the $\lambda(A, \vec{\eta})$ are equal to 1.

From (2.9), using (2.7), we find that

$$U(A, 0, 0, 0)\vec{K}U(A, 0, 0, 0)^{-1} = A^{-1}\vec{K}. \tag{2.10}$$

By choice of $\lambda(A, \vec{\eta}) = 1$ in (2.9) the operator vector $\vec{K}$ is uniquely determined.

The same procedure can be applied to the subgroup of the proper Galileo transformation $(1, \vec{\delta}, 0, 0)$. Again the factors can be chosen such that this subgroup has a unitary representation:

$$U(1, \vec{\delta}, 0, 0) = e^{i\sum_{\nu=1}^{3} X_\nu\delta_\nu} = e^{i\vec{X}\cdot\vec{\delta}} \tag{2.11}$$

with mutually commuting self-adjoint operators X_ν which, by analogy with (2.10) satisfy the equation

$$U(A, 0, 0, 0)\vec{X}U(A, 0, 0, 0)^{-1} = A^{-1}\vec{X} \tag{2.12}$$

and $\vec{X}$ is uniquely determined.

Although all elements $(1, \vec{\delta}, \vec{\eta}, 0)$ form an abelian subgroup of the Galileo group, it is not necessary that the $U(1, \vec{\delta}, 0, 0)$ commute with the $U(1, 0, \vec{\eta}, 0)$ because from

$$(1, \vec{\delta}, 0, 0)(1, 0, \vec{\eta}, 0) = (1, 0, \vec{\eta}, 0)(1, \vec{\delta}, 0, 0)$$

it only follows that

$$e^{iX_\nu\delta_\nu}e^{iK_\mu\eta_\mu} = \lambda_{\nu\mu}(\delta_\nu, \eta_\mu)e^{iK_\mu\eta_\mu}e^{iX_\nu\delta_\nu}, \tag{2.13}$$

where $|\lambda_{\mu\nu}| = 1$. Since we have no more free choice of factors for $U(1, \vec{\delta}, 0, 0)$ and $U(1, 0, \vec{\eta}, 0)$ we must yet specify what coefficients $\lambda_{\mu\nu}$ can occur in the equation (2.13). We will now simplify the answer of this question by assuming that the coefficients $\lambda_{\mu\nu}$ are differentiable, although no additional assumptions are necessary (see [10]). If we multiply (2.13) on the left with $e^{-iK_\mu\eta_\mu}$, differentiate with respect to η_μ and then set $\eta_\mu = 0$ we obtain

$$-iK_\mu e^{iX_\nu\delta_\nu} + ie^{iX_\nu\delta_\nu}K_\mu = \left(\frac{\partial\lambda_{\nu\mu}}{\partial\eta_\mu}\right)_{\eta_\mu=0} e^{iX_\nu\delta_\nu}. \tag{2.14}$$

If we then multiply on the right with $e^{-iX_\nu\delta_\nu}$ then differentiate with respect to δ_ν and finally set $\delta_\nu = 0$ we then obtain

$$-X_\nu K_\mu + K_\mu X_\nu = \left(\frac{\partial^2\lambda_{\nu\mu}}{\partial\eta_\mu\partial\delta_\nu}\right)_{\eta_\mu=\delta_\nu=0} \mathbf{1}. \tag{2.15}$$

From (2.10) and (2.12) we obtain

$$\left(\frac{\partial^2\lambda_{\mu\nu}}{\partial\eta_\mu\partial\delta_\nu}\right)_{\eta_\mu=\delta_\nu=0} = im\delta_{\nu\mu},$$

where m must be real, because X_ν and K_μ are self-adjoint. Therefore we obtain (2.15)

$$K_\mu X_\nu - X_\nu K_\mu = im\delta_{\nu\mu}\mathbf{1}. \tag{2.16}$$

We may choose $m \geq 0$ because m and $-m$ lead to an equivalent representation of $\mathscr{G}$ in $\mathscr{A}$; since m transforms into $-m$ by means of a anti-unitary transformation V because i transforms into $-i$:

$$(VK_\mu V^{-1})(VX_\nu V^{-1}) - (VX_\nu V^{-1})(VK_\mu V^{-1}) = i(-m)\delta_{\nu\mu}\mathbf{1}.$$

We have to distinguish between two cases: $m = 0$ and $m \neq 0$. For $m \neq 0$ the K_ν do not commute with the X_ν. However, (2.16) is only an abbreviated notation for (2.13):

$$e^{i\vec{X}\cdot\vec{\delta}}e^{i\vec{K}\cdot\vec{\eta}} = e^{im\vec{\delta}\cdot\vec{\eta}}e^{i\vec{K}\cdot\vec{\eta}}e^{i\vec{X}\cdot\vec{\delta}}. \tag{2.17}$$

The result of these considerations (we again mention) is valid without any assumption about $\lambda_{\nu\mu}$ (see [10]). Different values of m lead to inequivalent representations, because factors in $U(1, \vec{\delta}, 0, 0)$ and $U(1, 0, \vec{\eta}, 0)$ cannot be varied more and the number $|m|$ remains unchanged under unitary or anti-unitary transformations.

The elements $(1, 0, 0, \gamma)$ form an abelian subgroup; therefore there exists a unitary representation

$$U(1, 0, 0, \gamma) = e^{iH\gamma}, \tag{2.20}$$

where H is a self-adjoint operator, which is uniquely determined up to additive term $\varepsilon\mathbf{1}$.

From $(A, 0, 0, 0)(1, 0, 0, \gamma)(A^{-1}, 0, 0, 0) = (1, 0, 0, \gamma)$ it follows that

$$U(A, 0, 0, 0)e^{iH\gamma}U(A, 0, 0, 0)U(A, 0, 0, 0)^{-1} = \alpha(A)e^{iH\gamma},$$

where $\alpha(A_1A_2) = \alpha(A_1)\alpha(A_2)$. Since the only one-dimensional representation of the rotation group is the identity (see §3) it follows that $\alpha = 1$, that is,

$$U(A, 0, 0, 0)HU(A, 0, 0, 0)^{-1} = H. \tag{2.21}$$

Equation (2.21) describes the rotation invariance of H. From

$$(1, 0, \vec{\eta}, 0)(1, 0, 0, \gamma)(1, 0, \vec{\eta}, 0)^{-1} = (1, 0, 0, \gamma)$$

it follows that

$$e^{iK_\nu\eta_\nu}e^{iH\gamma}e^{-iK_\nu\eta_\nu} = \beta_\nu(\eta_\nu, \gamma)e^{-iH\gamma}.$$

From the preceding results and from (2.10) it follows that $\beta_\nu = 1$, that is, H and the K_ν commute.

From

$$(1, 0, 0, \gamma)(1, \vec{\delta}, 0, 0)(1, 0, 0, \gamma)^{-1} = (1, \vec{\delta}, -\vec{\delta}\gamma, 0)$$

it follows that

$$e^{iH\gamma}e^{i\vec{X}\cdot\vec{\delta}}e^{-iH\gamma} = \rho(\gamma, \vec{\delta})e^{i\vec{X}\cdot\vec{\delta}}e^{-i\vec{K}\cdot\vec{\delta}\gamma}.$$

If we differentiate with respect to γ and set $\gamma = 0$ we obtain (for $\rho(0, \vec{\delta}) = 1$))

$$iHe^{i\vec{X}\cdot\vec{\delta}} - ie^{i\vec{X}\cdot\vec{\delta}}H = \left(\frac{\partial\rho}{\partial\gamma}\right)_{\gamma=0} e^{i\vec{X}\cdot\vec{\delta}} - ie^{i\vec{X}\cdot\vec{\delta}}\vec{K}\cdot\vec{\delta}.$$

Multiplying on the left with $e^{-i\vec{X}\cdot\vec{\delta}}$, differentiating with respect to $\vec{\delta}$ and setting $\vec{\delta} = 0$ we obtain

$$\vec{X}H - H\vec{X} = \left(\frac{\partial^2\rho}{\partial\gamma\partial\vec{\delta}}\right)_{\vec{\delta}=0,\,\gamma=0} \mathbf{1} - i\vec{K}. \tag{2.22}$$

According to (2.12) we must obtain

$$\left(\frac{\partial^2\rho}{\partial\gamma\partial\vec{\delta}}\right)_{\vec{\delta}=0,\,\gamma=0} = 0.$$

Then combining the results of (2.22) with (2.17) we obtain

$$H = \frac{1}{2m}\vec{K}^2 + H_i \tag{2.23}$$

whereby we find that the X_ν and K_μ commute with H_i.

From (2.17) it follows that the Hilbert space $\mathscr{H}$ can be represented in the following way, that is, $\mathscr{H}$ together with the operators $e^{i\vec{X}\cdot\vec{\delta}}$, $e^{i\vec{K}\cdot\vec{\eta}}$ is isomorphic to the following form:

$$\mathscr{H} = \mathscr{H}_b \times \mathscr{H}_i, \tag{2.24}$$

where $\mathscr{H}_b$ can be chosen as the space $\mathscr{L}^2(\mathbf{R}^3, dk_1\, dk_2\, dk_3)$. Then it is easy to obtain

$$\begin{aligned} e^{i\vec{K}\cdot\vec{\eta}}&: e^{i\vec{K}\cdot\vec{\eta}} \times \mathbf{1}, \\ e^{i\vec{X}\cdot\vec{\delta}}&: e^{i\vec{X}\cdot\vec{\delta}} \times \mathbf{1}. \end{aligned} \tag{2.25}$$

In the above equation $\vec{K}$ and $\vec{X}$ are operators only in $\mathscr{H}_b$; and for $\varphi(k) \in \mathscr{L}^2(\mathbf{R}^3, dk_1\, dk_2\, dk_3)$ we obtain

$$e^{i\vec{K}\cdot\vec{\eta}}\varphi(\vec{k}) = e^{i\vec{k}\cdot\vec{\eta}}\varphi(\vec{k}), \tag{2.26}$$

$$e^{i\vec{X}\cdot\vec{\delta}}\varphi(\vec{k}) = \varphi(\vec{k} + m\vec{\delta}). \tag{2.27}$$

Thus it follows that any operator which commutes with all the operators $e^{i\vec{X}\cdot\vec{\delta}}$, $e^{i\vec{K}\cdot\vec{\eta}}$ must have the form $\mathbf{1} \times A$. In particular, we may write (2.23) in the form

$$H = \frac{1}{2m}\vec{K}^2 \times \mathbf{1} + \mathbf{1} \times H_i. \tag{2.28}$$

In this way we obtain

$$\frac{1}{2m}\vec{K}^2\varphi(\vec{k}) = \frac{1}{2m}k^2\varphi(\vec{k}) \tag{2.29}$$

and

$$e^{i(1/2m)K^2\gamma}\varphi(\vec{k}) = e^{(i/2m)\vec{k}^2\gamma}\varphi(\vec{k}). \tag{2.30}$$

The proof of these relationships is given in [10]; certain aspects of the proof can be found in IX.

On the basis of (2.26) and (2.27) it follows that the space $\mathscr{H}_b$ is irreducible relative to the transformations $e^{i\vec{K}\cdot\vec{\eta}}$ and $e^{i\vec{X}\cdot\vec{\delta}}$.

We have not yet given an explicit description of the representation of the subgroup $\mathscr{D}_{\mathscr{G}}$ of spatial rotations. In the previously given relations such as (2.10), (2.12), and (2.21) the choice of factors in $U(A, 0, 0, 0)$ were arbitrary. Since $\mathscr{D}_{\mathscr{G}}$ is compact, each irreducible representation of $\mathscr{D}_{\mathscr{G}}$ is (up to a factor) finite-dimensional (see [10] and AV, §10), and each representation (up to a factor) in $\mathscr{H}$ can be reduced to the form

$$\mathscr{H} = \sum_n \oplus \mathscr{H}_n,$$

where the $\mathscr{H}_n$ are invariant with respect to the representation of $\mathscr{D}_{\mathscr{G}}$ and are irreducible subspaces of $\mathscr{H}$. Since $\mathscr{D}_{\mathscr{G}}$ is compact each representation is, up to a factor, equivalent to a normal unitary representation of the covering group $\mathscr{D}_{\mathscr{G}}^*$ of $\mathscr{D}_{\mathscr{G}}$ (see [10] and AV, §10.7).

For $A \in \mathscr{D}_{\mathscr{G}}$ we may define an operator $V(A)$ in $\mathscr{H}_b$ by

$$V(A)\varphi(\vec{k}) = \varphi(A^{-1}\vec{k}). \tag{2.31}$$

It is easy to see that $V(A)$ is unitary, and that $V(A)$ defines a representation of $\mathscr{D}_{\mathscr{G}}$. We then obtain

$$V(A)\vec{K}V(A)^{-1} = A^{-1}\vec{K} \tag{2.32}$$

and, from (2.27) we obtain

$$V(A)\vec{X}V(A)^{-1} = A^{-1}\vec{X}. \tag{2.33}$$

From (2.10) and (2.12) it follows that

$$R(A) \stackrel{\text{def}}{=} V(A)^{-1}U(A, 0, 0, 0) \tag{2.34}$$

commutes with all $e^{i\vec{X}\cdot\vec{\delta}}$ and $e^{i\vec{K}\cdot\vec{\eta}}$, that is, is of the form $R(A)$: $\mathbf{1} \times R(A)$ and that we may write

$$U(A, 0, 0, 0) = V(A) \times R(A). \tag{2.35}$$

Since the $U(A, 0, 0, 0)$ form a representation of $\mathscr{D}_{\mathscr{G}}$ up to a factor, this must also be true of $R(A)$. Then, from (2.21) and (2.28) it follows that (since $V(A)$ commutes with $\vec{K}^2$):

$$H_i R(A) = R(A)H_i, \tag{2.36}$$

that is, the H_i commutes with all $R(A)$.

Therefore we may obtain an irreducible representation of the Galileo group (for $m \neq 0$) only if the representation of $\mathscr{D}_{\mathscr{G}}^*$ by means of $R(A)$, that is, in $\mathscr{H}_i$ is irreducible. Thus it follows that $\mathscr{H}_i$ is finite dimensional and (from (2.36)) we find that we must have

$$H_i = \lambda\mathbf{1}, \tag{2.37}$$

that is, H is, according to (2.28) uniquely determined up to an additive constant, which, by choosing a suitable factor, can be set equal to 0.

The irreducible representation space $\mathscr{H}_i$ is called the "spin space" of the elementary system; for elementary systems we shall use the notation $\imath_s$ instead of $\mathscr{H}_i$, that is, where we use the symbol $\imath_s$ we shall be considering the spin space of an elementary system.

In §3 we shall consider the representations of $\mathscr{D}_{\mathscr{G}}^*$ in $\mathscr{H}_b$ by means of $V(A)$ and in $\imath_s$ by means of $R(A)$ separately, because these representations play a central role in the applications of quantum mechanics. We will see that each irreducible representation is uniquely characterized by a number $s = 0, \frac{1}{2}, 1, \frac{3}{2}, \ldots$—and that this number will be used as an index for $\imath_s$.

Thus we find that the Galileo group $\mathscr{G}$, as a group of transformations of registration procedures relative to preparation procedures leads, without any additional assumptions other than those introduced in §1 (!) to the following structure which is of central importance in quantum mechanics:

For the case $m \neq 0$ to each type of elementary systems there are two parameters m and s and the corresponding irreducible representations of the Galileo group. The necessarily infinite-dimensional Hilbert space $\mathscr{H}$ of such a type of elementary systems can be written in the form $\mathscr{H}_b \times \imath_s$ where $\mathscr{H}_b = \mathscr{L}^2(\mathbf{R}^3, dk_1\, dk_2\, dk_3)$ and the rules for the transformation of the Galileo group are given by (2.26), (2.27), (2.35), and (2.31) and by

$$e^{iH\gamma} = e^{(i/2m)\vec{K}^2\gamma} \times \mathbf{1}. \tag{2.38}$$

In §4 we will see that (2.16) is equivalent to the Heisenberg uncertainty relations.

The typical quantum mechanical structure obtained from the representation of the Galileo group by means of $\mathscr{B}$-continuous effect automorphisms is not a consequence (!) of the introduction (§1) of the structure of the Galileo transformation as transformations of registration procedures. Everything introduced in §1 also holds for classical systems. The distinction between classical systems and microsystems lies exclusively in axiom AV 4s in III, §3. If, on the contrary, we make the assumption that the systems under consideration are physical objects (see the remarks in III, §3 following AV 4s), then it follows for elementary systems that they are "mass points which move with constant velocity in a straight line between the preparation and registration apparatus" (for proof see [21]).

The fact that every elementary quantum mechanical system type uniquely determines two parameters m and s does not, of course, mean that every elementary system type is uniquely characterized by these two parameters. It is possible to give other "objective" properties in addition to m and s for elementary systems (for objective properties, see IV, §8.1). The fact that m and s are objective properties follows directly from the definition that they are parameters which correspond to atoms in the center Z of G.

D 2.2. The parameter m is called the mass of the elementary system type; the parameter s is called the spin of the elementary system type.

We shall later compare (see VIII, §6 and XVII, §6.2) the concept of "mass" described above with the "usual" concept. The meaning of the parameter s will be explained in §3 and §5.

The fact that in this formulation of quantum mechanics there is no "constant" $\hbar = h/2\pi$ (where h is Planck's constant) is not a defect. On the contrary, it merely expresses the fact that this formulation considers only the essential structure of quantum mechanics. This structure shows that the quantum mechanical laws are not invariant under transformations $m \to \lambda m$ where $\lambda > 0$. Because all previous classical theories exhibit this invariance, it seems advantageous to introduce a particular unit of mass in *classical* physics. For quantum mechanics it appears to be somewhat "artificial" to introduce a special unit of mass; if this was done, then it would be necessary to introduce a factor between the "natural" unit in quantum mechanics and the "artificial" unit in classical mechanics. The natural unit is $(\text{cm})^{-1}$ if the velocity of light is taken to be 1, that is, if the time is also measured in cm. Then we would have

$$1\,(\text{cm})^{-1} = \hbar\,(\text{gram}), \tag{2.39}$$

where gram is the usual mass unit. The conversion factor in (2.39) can be found if we measure, for example, the mass of a hydrogen atom in $(\text{cm})^{-1}$ and determine what the mass of a $(\text{cm})^3$ of water is compared to that of a hydrogen atom (see XVII, §6.2).

We have not yet discussed the case $m = 0$ in (2.16). In [10] the remaining possible irreducible representations of the Galileo group are given. Here we briefly mention these representations and we will make it experimentally evident that such systems are not found in nature, this evidence will be formulated in terms of the following axiom—that $m \neq 0$. (Light quanta cannot be described in terms of the Galileo group; here the representations of the Poincaré group must be used—see the remarks in §1 and at the end of §2.)

For $m = 0$ it follows from (2.16) that all K_ν and X_μ commute. With respect to the above derivation not much is changed; it is only necessary to set $m = 0$. This occurs only in the transition between (2.22) and (2.23), that is, in the determination of H in the equation (2.22);

$$\vec{X}H - H\vec{X} = -i\vec{K}. \tag{2.40}$$

Since $\vec{X}$ and H commute with $\vec{K}$ we may treat $\vec{K}$ in (2.40) like a number. From (2.40) by analogy with (2.17) and (2.16) it follows that

$$e^{i\vec{X}\cdot\vec{\delta}}e^{iH\gamma}e^{-i\vec{X}\cdot\vec{\delta}} = e^{i(H\gamma + \vec{K}\cdot\vec{\delta})}. \tag{2.41}$$

From (2.41) it follows that we may take (for an irreducible representation) the space of quadratic integrable functions $\varphi(k_0, \vec{k})$ with fixed $|\vec{k}| = r$ with integration measure $dk_0 r\, d\omega$ (where $d\omega$ is an element of solid angle or a surface element of the unit sphere) for $\mathscr{H}_b$ in (2.24). From (2.10) it follows that all directions are needed for $\vec{k}$. The equations for the representation operators are given by

$$e^{i\vec{K}\cdot\vec{\eta}}\varphi(k_0, \vec{k}) = e^{i\vec{k}\cdot\vec{\eta}}\varphi(k_0, \vec{k}),$$
$$e^{iH\gamma}\varphi(k_0, \vec{k}) = e^{ik_0\gamma}\varphi(k_0, \vec{k}),$$
$$e^{i\vec{X}\cdot\vec{\delta}}\varphi(k_0, \vec{k}) = \varphi(k_0 + \vec{k}\cdot\vec{\delta}, \vec{k}).$$

For $r = 0$, $\vec{X}$ commutes with H. For such an irreducible representation $e^{iH\gamma}$ reduces to a multiple of the unit operator.

The following experimental evidence shows that there are no physical systems which correspond to the representation for which $m = 0$.

For $r = 0$ it follows that all effects are invariant under time displacements (since $e^{iH\gamma}$ commutes with all $F \in L$). With the exception of a "vacuum" there is no physical system known (see below) for which such a "time invariance" is valid.

For $r \neq 0$ we have $|\vec{k}| = r$. Suppose we produce an ensemble W for which $\mathrm{tr}(WU(1, 0, \eta_\nu, 0)FU(1, 0, \eta_\nu, 0)^{-1})$ changes slowly with η_1, η_2 for all F. Then it follows that this expression will also vary weakly with η_3. This contradicts experience because experience has shown that it is possible to make ensembles which depend weakly on displacements η_1, η_2 but strongly on displacements η_3 in the third direction (in all scattering problems we seek to produce ensembles which weakly depend on η_1, η_2 in order that a "beam" of systems can be directed in the third direction—see XVI, §6.3).

Often the following argument is also introduced: For $m = 0$ there exists no decision observable for position in the sense discussed in §4. This objection is more than questionable because it is practically impossible to prove experimentally the existence of the decision observable for position (constructed in §4) because the latter is an idealization which is only obtained approximately (with difficulty) in terms of real constructable registration methods b_0 (in the sense of IV, §4). The nonexistence of a *decision* observable for position will therefore not immediately contradict all known experiments because it is conceivable that there yet exists a position observable (in the general sense, that is, a measure $\Sigma \xrightarrow{F} L$ with Σ as the ring of the "position domain" in the sense of §4) which describes what is measured (see [22]) where the case of a light quantum is a nice example. The assertion of the existence of a decision observable for position on arbitrary *a priori* grounds will absolutely contradict the concepts of physics as carried out here for the example of quantum mechanics and as described in general in [1]. Such *a priori* principles are not admissible in the development of an axiomatic basis for quantum mechanics. This, of course, does not mean that intuitive concepts such as the concept of "position" cannot be used in order to guess (or "discover") a $\mathscr{PT}$.

There exists a trivial irreducible representation of the Galileo group: the identity in a one-dimensional Hilbert space. The only other additional elementary system types are those whose corresponding Hilbert spaces are one-dimensional. There are no objective properties (IV, §8.1) which are experimentally known which permit us to distinguish such elementary system types whose corresponding Hilbert spaces are one-dimensional. We therefore impose the axiom that there exists only a single one-dimensional Hilbert space $\mathscr{H}_v$. For this one-dimensional Hilbert space we shall use the index 0: $\mathscr{H}_0$. The corresponding system type will be called the "vacuum." The corresponding objective property (as a subset of M) will be denoted by M_0. According to IV, §8.1 we therefore obtain

$$M_0 = \left[\bigcup_{\substack{a \in \mathscr{Q}' \\ \varphi(a) \in K_1(e_0)}} a \right] \cup \left[\bigcup_{\substack{b \in \mathscr{R} \\ \psi(b_0, b) \leq e_0}} b \right],$$

where e_0 is the atom of the center Z of G which projects onto $\mathscr{H}_0$.

The language used by the physicist for the situation $x \in M_0$ is that "no" microsystem is "present," and the set $M \backslash M_0$ is often called the set of "proper" microsystems. For the formulation presented in II it is conceptually important not to exclude the possibility that $x \in M_0$ because it is conceptually impossible to describe the set $M \backslash M_0$ before the introduction of the concept of a "system type." Therefore the set M in II is only an aid to mediate between the preparation and registration. Conceptually this would be clearer if we construct quantum mechanics without the aid of a set M of microsystems and instead use only mathematical structures which describe the preparation and registration apparatuses and the connection between them (see [3], [2], XVI, and [13]).

The discussions in §1 and §2 may also be carried out for the case of the Poincaré group. The same is true for parts of the discussion in sections §4–§7, but not for the considerations in VIII. The experiments of "elementary particle" physics lead us to suspect that there are actually no "elementary" systems in the sense of D 2.1 (with respect to the Poincaré group). The concept of an elementary particle (as introduced in §2) appears to have a meaningful application only in the realm of nonrelativistic physics of microsystems.

3 Irreducible Representations of the Rotation Group

Since the unitary representations (up to a factor) of the rotation group $\mathscr{D}_g$ are equivalent to the unitary representations of the covering group $\mathscr{D}_g^*$ (see [10] and AV, §10.7) we may restrict our consideration to the unitary representations of $\mathscr{D}_g^*$. Since $\mathscr{D}_g^*$ is compact, all irreducible representations are finite dimensional (see [10] and AV, §10.6). $\mathscr{H}$ may be a finite-dimensional Hilbert space and $U(A)$ a unitary representation of $\mathscr{D}_g^*$ in $\mathscr{H}$. For a rotation A of angle α about the 3-axis the equation $U(A) = U(\alpha)$ defines the representation of a one-parameter group satisfying $U(\alpha_1 + \alpha_2) = U(\alpha_1)U(\alpha_2)$ for which an infinitesimal rotation $\mathscr{I}_3$ is defined by

$$U(\alpha) = e^{\alpha \mathscr{I}_3}.$$

Since $\mathscr{H}$ is finite dimensional, $\mathscr{I}_3$ is defined in all of $\mathscr{H}$. Since U is unitary $i\mathscr{I}_3$ is a self-adjoint operator. In the same way infinitesimal rotations $\mathscr{I}_1$, $\mathscr{I}_2$ are defined for the axes 1 and 2. We therefore set

$$L_\nu = i\mathscr{I}_\nu \qquad (\nu = 1, 2, 3). \tag{3.1}$$

Thus the L_ν are self-adjoint operators.

From the representation property of the $U(A)$ it follows that (see AV, §10.5) the $\mathscr{I}_\nu$ in $\mathscr{H}$ satisfy the same commutation relations as the corresponding infinitesimal rotations in $\mathscr{D}_\mathscr{G}$ itself, that is,

$$\mathscr{I}_\nu\mathscr{I}_\mu - \mathscr{I}_\mu\mathscr{I}_\nu = \mathscr{I}_\rho \quad \text{where } \nu, \mu, \rho = \begin{cases} 1, 2, 3, \\ 2, 3, 1, \\ 3, 1, 2. \end{cases} \tag{3.2}$$

For the L_ν it follows that

$$L_\nu L_\mu - L_\mu L_\nu = iL_\rho \quad \text{where } \nu, \mu, \rho = \begin{cases} 1, 2, 3, \\ 2, 3, 1, \\ 3, 1, 2. \end{cases} \tag{3.3}$$

Since $\mathscr{H}$ is finite dimensional, the L_ν have (as is the case of all self-adjoint operators in $\mathscr{H}$) a discrete spectrum of eigenvalues and a complete orthonormal basis of eigenvectors. For that reason it is easy to carry out the following computations.

We replace L_1, L_2 by means of the operators

$$N = L_1 + iL_2 \quad \text{and} \quad N^+ = L_1 - iL_2 \tag{3.4}$$

and we define

$$\vec{L}^2 = L_1^2 + L_2^2 + L_3^2. \tag{3.5}$$

It follows that

$$L_3N - NL_3 = N, \qquad L_3N^+ - N^+L_3 = -N^+ \tag{3.6}$$

and

$$NN^+ = \vec{L}^2 + L_3 - L_3^2, \qquad N^+N = \vec{L}^2 - L_3 - L_3^2. \tag{3.7}$$

If v is an arbitrary eigenvector of L_3 in $\mathscr{H}$ which satisfies

$$L_3v = \mu v \tag{3.8}$$

then, from (3.6) it follows that:

$$L_3Nv = (NL_3 + N)v = (\mu + 1)Nv \tag{3.9}$$

and

$$L_3N^+v = (N^+L_3 - N^+)v = (\mu - 1)N^+v. \tag{3.10}$$

If $Nv \neq 0$ then Nv is an eigenvector of L_3 with eigenvalues $(\mu + 1)$; if $N_v^+ \neq 0$, then N_v^+ is an eigenvector of L_3 with eigenvalue $(\mu - 1)$. If we apply N repeatedly we obtain increasing eigenvalues of L_3 providing that we do not obtain the null vector. Since $\mathscr{H}$ is finite dimensional there exists an integer n such that

$$N^n v \neq 0, \qquad L_3 N^n v = (\mu + n)N^n v \quad \text{and} \quad N^{n+1} v = 0.$$

Let $N^n v$ be denoted by $u_j (j = \mu + n)$ we therefore obtain

$$L_3 u_j = j u_j \quad \text{and} \quad N u_j = 0. \tag{3.11}$$

From (3.11) it follows that $N^+ N u_j = 0$ and from (3.7) and (3.11) we obtain

$$\vec{L}^2 u_j = j(j+1) u_j. \tag{3.12}$$

Since we may assume that u_j is normalized, we may recursively define:

$$N^+ u_m = \tau_m u_{m-1}, \qquad m = j, j-1, \ldots, \tag{3.13}$$

where τ_m is so chosen that $\|u_m\| = 1$ for all m. Since we have required only that $\|u_m\| = 1$, we (arbitrarily) choose τ to be real and >0. The sequence of the u_m exists providing that $N^+ u_{m'} = 0$ for a value of m'. From (3.10) we find that for all u_m defined according to (3.13) that

$$L_3 u_m = m u_m. \tag{3.14}$$

From the commutation relations (3.3), and intuitively, from the fact that $\vec{L}^2$ is the square of the magnitude of a vector, it can easily be seen that $\vec{L}^2$ commutes with the rotations and, therefore, with all $\mathscr{I}_v$. Therefore $\vec{L}^2$ commutes also with N and N^+. From (3.12) and (3.13) it follows that

$$\vec{L}^2 u_m = j(j+1) u_m \tag{3.15}$$

for all m.

Since $\mathscr{H}$ is finite dimensional, for a particular value of m' (for which $u_{m'} \neq 0$) it follows that the relationship $N^+ u_{m'} = 0$ must hold. From (3.7) and (3.15) it follows that $m'(m'-1) = j(j+1)$. Since, according to the definition of the u_m, for m' the relation $m' \leq j$ holds, we therefore obtain $m' = -j$. Conversely, from

$$\begin{aligned} \|N^+ u_{-j}\|^2 &= \langle u_{-j}, N N^+ u_{-j} \rangle \\ &= \langle u_{-j}, (\vec{L}^2 - L_3 - L_3^2) u_{-j} \rangle = 0 \end{aligned}$$

it follows that $N^+ u_{-j} = 0$.

The sequence of the u_m runs as follows: $m = j, j-1, \ldots, -j$. Since the number $(2j+1)$ must be an integer, we find that j may only take on half-integer values—$0, \frac{1}{2}, 1, \frac{3}{2}, 2, \ldots$, etc.

We obtain the normalization condition from (3.13) as follows:

$$|\tau_m|^2 = \|N^+ u_m\|^2 = \langle u_m, N N^+ u_m \rangle$$

and, from (3.7), (3.14), and (3.15) (since we have chosen τ_m to be positive real) we obtain:

$$\tau_m = \sqrt{j(j+1) + m - m^2} = \sqrt{(j+m)(j-m+1)}. \tag{3.16}$$

Equation (3.13) then becomes

$$N^+ u_m = \sqrt{(j+m)(j-m+1)}\, u_{m-1}. \tag{3.17}$$

Since N is the adjoint operator to N^+ and since the u_m are orthonormal it follows that

$$N u_m = \sqrt{(j-m)(j+m+1)}\, u_{m+1}. \tag{3.18}$$

The subspace spanned by the u_m is therefore invariant under N, N^+, L_3 and therefore under $\mathscr{I}_1, \mathscr{I}_2, \mathscr{I}_3$. Since the set of operators given by

$$U(A) = e^{\sum_\nu \alpha_\nu \mathscr{I}_\nu} \tag{3.19}$$

contains the representative operators for all rotations A (see AV, §10.4) the subspace spanned by u_m is invariant under all the operators in $U(A)$. Since $\mathscr{H}$ is irreducible, the u_m span all of $\mathscr{H}$.

If, conversely, we abstractly construct a space $\mathscr{H}_j$ by specifying the $(2j+1)$ vectors u_m $(m = -j, -j+1, \ldots, j)$ as a complete orthonormal basis in $\mathscr{H}_j$ and define the operators L_3, N, N^+ by the equations (3.14), (3.17), and (3.18) and define the operators L_1, L_2 by (3.4) then, for the L_ν it follows that the commutation relations (3.3) hold. Then, for the $\mathscr{I}_\nu$ defined by (3.1) we find that the commutation rules (3.2) hold. Each $A \in \mathscr{D}_{\mathscr{G}}^*$ can be described by a rotation axis and a rotation angle, that is, by three parameters $\alpha_1, \alpha_2, \alpha_3$ (see AV, §10.7). Since the $\mathscr{I}_\nu$ satisfy the commutation relations (3.2) it follows that the $U(A)$ form a representation of the covering group. That this representation is irreducible follows from the construction because to every invariant subspace there exists an eigenvector of L_3 which must coincide with one of the u_m, from which we find that N and N^+ "generate" the entire space.

In the sequel we shall denote the above irreducible representation in $\mathscr{H}_j$ by $\mathscr{D}_j$ (in particular, for the simplification of the discussion in XI–XVI). $\mathscr{D}_j$ also represents a characterization for a class of equivalent representations, independent of which vector space by which it is realized.

The representations $\mathscr{D}_j$ may be obtained in a purely algebraic fashion, without the use of the theorems of Lie groups. The latter approach is often of great practical value. For this reason we shall explicitly derive the representation $\mathscr{D}_{1/2}$ of $\mathscr{D}_{\mathscr{G}}^*$ in $\mathscr{H}_{1/2}$.

We shall denote the basis vectors $u_{+1/2}, u_{-1/2}$ of $\mathscr{H}_{1/2}$ by u_+ and u_-. In $\mathscr{H}_{1/2}$ we find that, according to (3.14), (3.17), and (3.18), the infinitesimal rotations $\mathscr{I}_\nu$ satisfy the equations:

$$\begin{aligned}
\mathscr{I}_3 u_+ &= -\frac{i}{2} u_+, & \mathscr{I}_3 u_- &= \frac{i}{2} u_-,\\
\mathscr{I}_1 u_+ &= -\frac{i}{2} u_-, & \mathscr{I}_1 u_- &= -\frac{i}{2} u_+,\\
\mathscr{I}_2 u_+ &= \frac{1}{2} u_-, & \mathscr{I}_2 u_- &= -\frac{1}{2} u_+.
\end{aligned} \tag{3.20}$$

Thus from (3.2) it follows that the following equations hold for $\sigma_\nu = 2i\mathscr{I}_\nu$:

$$\sigma_\nu\sigma_\mu - \sigma_\mu\sigma_\nu = 2i\sigma_\lambda \quad \text{where } \nu, \mu, \lambda = \begin{cases} 1, 2, 3, \\ 2, 3, 1, \\ 3, 1, 2. \end{cases} \tag{3.21}$$

The σ_ν are self-adjoint and satisfy the equations:

$$\sigma_\nu\sigma_\mu + \sigma_\mu\sigma_\nu = \mathbf{0} \quad \text{if } \nu \neq \mu \quad \text{and} \quad \sigma_\nu^2 = \mathbf{1}. \tag{3.22}$$

If we again use the parameters $\alpha_1, \alpha_2, \alpha_3$ for the rotation A where $\alpha = \sqrt{\alpha_1^2 + \alpha_2^i + \alpha_3^2}$ represents the rotation angle and $w_\nu = \alpha_\nu/\alpha$ the components of the rotation axis, then from (3.19) it follows that

$$U(A) = e^{\sum_\nu \alpha_\nu \mathscr{I}_\nu} = e^{-i(\alpha/2)\sum_\nu w_\nu\sigma_\nu}. \tag{3.23}$$

From (3.22) it follows that

$$\left(\sum_\nu w_\nu\sigma_\nu\right)^2 = \mathbf{1} \tag{3.24}$$

and we therefore obtain

$$U(A) = e^{-i(\alpha/2)\sum_\nu w_\nu\sigma_\nu} = \mathbf{1}\cos\frac{\alpha}{2} - i\left(\sum_\nu w_\nu\sigma_\nu\right)\sin\frac{\alpha}{2}. \tag{3.25}$$

The element $A = [e, \mathscr{C}]$ of the fundamental group of $\mathscr{D}_{\mathscr{G}}^*$ corresponds to a continuous variation of the angle α from 0 to 2π (see AV, §10.7). From (3.25) it follows that

$$U([e, \mathscr{C}]) = -\mathbf{1}.$$

The operator $U(A) = \mathbf{1}$ corresponds to $\cos \alpha/2 = 1$, that is, $\alpha = 4n\pi$. All these values of α correspond to the unit element of the covering group. Therefore the representation in $\mathscr{H}_{1/2}$ is an isomorphic representation of $\mathscr{D}_{\mathscr{G}}^*$. We shall now use algebraic methods to obtain the above result.

First we shall show that the $U(A)$ contain all unitary operators in $\mathscr{H}_{1/2}$ which have determinant 1. From (3.25) and (3.20) it immediately follows that the matrix of $U(A)$ is given by:

$$\begin{pmatrix} \cos(\alpha/2) - iw_3\sin(\alpha/2) & -(iw_1 + w_2)\sin(\alpha/2) \\ (-iw_1 + w_2)\sin(\alpha/2) & \cos(\alpha/2) + iw_3\sin(\alpha/2) \end{pmatrix}. \tag{3.26}$$

This is a unitary matrix with determinant 1.

Conversely, let

$$\begin{pmatrix} a_{11} & a_{12} \\ a_{21} & a_{22} \end{pmatrix} \tag{3.27}$$

be a unitary matrix having determinant 1. Then we must also have:

$$\begin{gathered} |a_{11}|^2 + |a_{12}|^2 = 1, \qquad |a_{21}|^2 + |a_{22}|^2 = 1, \\ a_{11}\bar{a}_{21} + a_{12}\bar{a}_{22} = 0, \\ a_{11}a_{22} - a_{21}a_{12} = 1. \end{gathered} \tag{3.28}$$

From the third equation of (3.28) it follows that

$$a_{21} = -\lambda\bar{a}_{12}, \qquad a_{22} = \lambda\bar{a}_{11}$$

because, from the first equation, it is not possible that both a_{11} and a_{12} be zero. From the second equation of (3.28) it follows that

$$|\lambda|^2(|a_{12}|^2 + |a_{11}|^2) = |\lambda|^2 = 1.$$

From the fourth equation of (3.28) we finally obtain

$$\lambda|a_{11}|^2 + \lambda|a_{12}|^2 = \lambda = 1.$$

Therefore (3.27) has the form

$$\begin{pmatrix} a_{11} & a_{12} \\ -\bar{a}_{12} & \bar{a}_{11} \end{pmatrix} \quad \text{with } |a_{11}|^2 + |a_{12}|^2 = 1. \tag{3.29}$$

The matrix (3.29) is unitary with determinant 1. If, in (3.29) we then set $a_{11} = \beta_4 - i\beta_3, a_{12} = -(\beta_2 + i\beta_1)$ where the β_ν are real, we obtain

$$\begin{pmatrix} \beta_4 - i\beta_3 & -(\beta_2 + i\beta_1) \\ \beta_2 - i\beta_1 & \beta_4 + i\beta_3 \end{pmatrix} \quad \text{with } \beta_1^2 + \beta_2^2 + \beta_3^2 + \beta_4^2 = 1. \tag{3.30}$$

Otherwise the β_ν may be freely chosen. From the auxiliary condition in (3.30) we may introduce an angle α and set

$$\beta_4 = \cos(\alpha/2).$$

We introduce the w_ν by means of the equations

$$\beta_\nu = w_\nu \sin(\alpha/2) \quad \text{for } \nu = 1, 2, 3.$$

The auxiliary condition in (3.30) then reduces to

$$w_1^2 + w_2^2 + w_3^2 = 1.$$

Therefore the matrix in (3.30) takes on the form (3.26).

The group of two-dimensional unitary operators with determinant 1 is often called SU_2. SU_2 is isomorphic to $\mathscr{D}_{\mathscr{G}}^*$ by the correspondence (3.25). We shall now show that this is the case directly by algebraic methods:

For $\sigma_x = \sum_\nu x_\nu \sigma_\nu$ where the x_ν are real we find that σ_x is self-adjoint and that $\sigma_x^2 = (\sum_\nu x_\nu^2)\,\mathbf{1}$. If ρ is a self-adjoint operator in $\mathscr{H}_{1/2}$ and $\operatorname{tr}(\rho) = 0$ we find that $\rho = \sum_\nu a_\nu \sigma_\nu$ where the a_ν are real as follows: Since the four operators σ_ν, $\mathbf{1}$ form a complete linearly independent system of operators in $\mathscr{H}_{1/2}$ every operator ρ has the form $\rho = \sum_\nu a_\nu\sigma_\nu + a_0\mathbf{1}$; from $\operatorname{tr}(\rho) = 0$ it follows that $a_0 = 0$. From $\rho = \rho^+$ it follows that $\bar{a}_\nu = a_\nu$. Therefore, for a unitary operator U in $\mathscr{H}_{1/2}$ it follows that

$$U\sigma_\nu U^+ = \sum_\mu \sigma_\mu \alpha_{\mu\nu},$$

where $\alpha_{\mu\nu}$ are real. Thus it follows that

$$U\sigma_x U^+ = \sum_\nu x'_\nu \sigma_\nu \quad \text{with } x'_\nu = \sum_\mu \alpha_{\nu\mu} x_\mu.$$

From

$$\left(\sum_\nu x_\nu'^2\right)\mathbf{1} = (U\sigma_x U^+)^2 = U\sigma_x^2 U^+ = U\left(\sum_\nu x_\nu^2\right)\mathbf{1}U^+ = \left(\sum_\nu x_\nu^2\right)\mathbf{1}$$

it follows that $(\alpha_{\nu\mu})$ is the matrix of a three-dimensional rotation. It is easy to see that the correspondence $U \longrightarrow (\alpha_{\mu\nu})$ is a representation of SU_2 by means of three-dimensional rotations. That this correspondence is surjective on $\mathscr{D}_\mathscr{G}$ can be easily seen as follows:

For U defined according to equation (3.26), if we set $w_1 = w_2 = 0$ we then obtain for the matrix $(\alpha_{\nu\mu})$ a rotation about the 3-axis. For $w_2 = w_3 = 0$ we obtain a rotation about the 1-axis. Any other rotation can be obtained by multiplications of rotations about the 1-axis and the 3-axis (see a discussion of Euler angles, for example, [2], VI,§3.1).

Since SU_2 is isomorphic to $\mathscr{D}_\mathscr{G}^*$ we may therefore obtain all unitary representations of $\mathscr{D}_\mathscr{G}^*$ as unitary representations of SU_2. The irreducible unitary representations of SU_2 may also be constructed simply using algebraic methods.

With the help of u_+, u_- we may easily define a $(v + 1)$-dimensional vector space $\mathscr{T}_{v/2}$ which is generated by the basis vectors u_+^v, $u_+^{v-1}u_-$, $u_+^{v-2}u_-^2, \ldots, u_-^v$ the vectors of which are all homogeneous polynomials of the v-degree in the unknowns u_+, u_-. For $U \in SU_2$ there exists a representation of SU_2 by means of linear transformations in $\mathscr{T}_{v/2}$ generated by $V(u_+^\alpha u_-^\beta) = (Uu_+)^\alpha(Uu_-)^\beta$. It is a simple matter to define an inner product in $\mathscr{T}_{v/2}$ in such a way that the above representation is unitary. The definition of this inner product is suggested by the following considerations:

If $a_+u_+ + a_-u_-$ is a vector in $\mathscr{H}_{1/2}$, then under a unitary transformation in $\mathscr{H}_{1/2}$ the expression $\|a_+u_+ + a_-u_-\|^2 = \bar{a}_+a_+ + \bar{a}_-a_-$ will be invariant; the same result therefore holds for the expression

$$(\bar{a}_+a + \bar{a}_-a)^v = \sum_{r=0}^{v}\binom{v}{r}\bar{a}_+^{v-r}a_+^{v-r}\bar{a}_-^r a_-^r. \tag{3.31}$$

An arbitrary vector in $\mathscr{T}_{v/2}$ has the form

$$\sum_r c_r u_+^{v-r}u_-^r.$$

The coefficients c_r are transformed in the representation of SU_2 in the same way as the coefficients of

$$(a_+u_+ + a_-u_-)^v = \sum_{r=0}^{v}\binom{v}{r}a_+^{v-r}a_-^r u_+^{v-r}u_-^r, \tag{3.32}$$

that is, in the same way as

$$\binom{v}{r}a_+^{v-r}a_-^r. \tag{3.33}$$

From (3.31) it follows that the expression

$$\frac{1}{v!}\sum_{r=0}^{v} r!(v-r)!\bar{c}_r c_r \tag{3.34}$$

is an invariant for all such transformations. We shall now introduce the following set of basis vectors in $\mathscr{T}_{v/2}$

$$v_m = \frac{u_+^{v-r}u_-^r}{\sqrt{r!(v-r)!}} = \frac{u_+^{j+m}u_-^{j-m}}{\sqrt{(j+m)!(j-m)!}}, \tag{3.35}$$

where $j = v/2$, $m = j - r$ (that is, $m = -j, -j+1, \ldots, j$) and define an inner product $\langle v_m, v_{m'}\rangle = \delta_{mm'}$. Thus we find that the above representation of SU_2 in $\mathscr{T}_j$ is unitary.

We will now show that the above representation is identical with $\mathscr{D}_j$ which was derived from infinitesimal transformations $\mathscr{I}_\nu$ in $\mathscr{H}_j$. For this purpose we shall now compute the infinitesimal transformation $\mathscr{I}_\nu$ in $\mathscr{T}_j$. For the $\mathscr{I}_\nu$ in $\mathscr{H}_{1/2}$ we find that

$$\left[\frac{d}{d\alpha}U_\nu(\alpha)\right] = \mathscr{I}_\nu,$$

where $U_\nu(\alpha)$ is given by (3.26) for $w_\mu = 0$ and $\mu \neq \nu$. From (3.20) it follows that for the $\mathscr{I}_\nu$ in $\mathscr{T}_j$ we obtain

$$\begin{aligned}
\mathscr{I}_\nu v_m &= \left[\frac{d}{d\alpha}\frac{(U_\nu(\alpha)u_+)^{j+m}(U_\nu(\alpha)u_-)^{j-m}}{\sqrt{(j+m)!(j-m)!}}\right]_{\alpha=0} \\
&= \frac{(j+m)u_+^{j+m-1}u_-^{j-m}\mathscr{I}_\nu u_+}{\sqrt{(j+m)!(j-m)!}} \\
&\quad + \frac{(j-m)u_+^{j+m}u_-^{j-m-1}\mathscr{I}_\nu u_-}{\sqrt{(j+m)!(j-m)!}}.
\end{aligned}$$

Since the $\mathscr{I}_\nu$ in $\mathscr{H}_{1/2}$ are given in (3.20) we obtain

$$\begin{aligned}
\mathscr{I}_3 v_m &= -imv_m, \\
(\mathscr{I}_1 + i\mathscr{I}_2)v_m &= -iNv_m = -i\sqrt{(j-m)(j+m+1)}\,v_{m+1}, \\
(\mathscr{I}_1 - i\mathscr{I}_2)v_m &= -iN^+v_m = -i\sqrt{(j+m)(j-m+1)}\,v_{m-1},
\end{aligned} \tag{3.36}$$

that is, the relations (3.14), (3.17), and (3.18) are satisfied. Since SU_2 is isomorphic to $\mathscr{D}_{\mathscr{G}}^*$ the representations in $\mathscr{H}_j$ and $\mathscr{T}_j$ are identical.

By means of the algebraic construction of the representation $\mathscr{D}_j$ in $\mathscr{T}_j$ we may easily determine which values of j for which the representation is unitary and the values of j which correspond to a "multiple valued" representation of $\mathscr{D}_{\mathscr{G}}$. For this purpose we need only the representation of the element $-\mathbf{1}$ of SU_2 in $\mathscr{T}_j$. Since $\mathscr{T}_j$ consists of the homogeneous polynomials of $(2j)$th degree we will therefore find that $-\mathbf{1}$ will be represented by $(-1)^{2j}\mathbf{1}$. The integral values of j therefore result in unique representations of $\mathscr{D}_{\mathscr{G}}$; the half integer values lead to two valued representations of $\mathscr{D}_{\mathscr{G}}$.

All reducible representations up to a factor of $\mathscr{D}_{\mathscr{G}}$ in a Hilbert space $\mathscr{H}$ may (since $\mathscr{D}_{\mathscr{G}}^*$ is compact) be decomposed into irreducible representations which decompose $\mathscr{H}$ into a direct sum

$$\mathscr{H} = \sum_{\mu} \oplus \mathscr{H}_{\mu}, \tag{3.37}$$

where each subspace $\mathscr{H}_{\mu}$ is invariant and irreducible with respect to the representation. However, not any arbitrary irreducible representation (up to a factor) can occur in the $\mathscr{H}_{\mu}$. Since the representation in $\mathscr{H}$ has to be a representation up to a factor, the fundamental group of $\mathscr{D}_{\mathscr{G}}^*$ must be represented isomorphically in each of the $\mathscr{H}_{\mu}$ (see AV, §10.7). A representation up to a factor of $\mathscr{D}_{\mathscr{G}}$ in $\mathscr{H}$ contains either only representations with half integer j or only those with integer values of j.

It now remains to show that we have determined all of the irreducible representations of $\mathscr{D}_{\mathscr{G}}^*$, that is, of SU_2, that is, all representations (up to a factor) in terms of the representations $\mathscr{D}_j$ in the $\mathscr{T}_j$. This we shall show on the basis of the completeness of the characters of the representations in the $\mathscr{T}_j$ as class functions in SU_2 (see AV, §10.5). If $U \in SU_2$ then there exists a $V \in SU_2$ such that $W = VUV^{-1}$ has the form

$$Wu_+ = e^{i\alpha}u_+, \qquad Wu_- = e^{-i\alpha}u_-. \tag{3.38}$$

This follows from the fact that U must have two orthogonal eigenvectors v_1, v_2 with eigenvalues $e^{i\alpha}$, $e^{i\beta}$. V needs only be chosen as a transformation which transforms v_1 into u_+ and v_2 into u_-. If the determinant of V is not equal to 1, so that we may obtain this result by the multiplication of V with a factor such that a multiplication of V does not change VUV^{-1}. Since the determinant of W and that of U must be equal to 1 we must have $e^{i\beta} = e^{-i\alpha}$. It follows that two transformations from SU_2 which have the same eigenvalue belong to the same class of conjugate elements. We will run through the different classes when the parameter α runs through the values between 0 and π because the pair $e^{i\alpha}$, $e^{-i\alpha}$ of eigenvalues will run through all possible values. The character χ_j of the transformations W in $\mathscr{T}_j$ is determined by the equation

$$W(u_+^{j+m}u_-^{j-m}) = (Wu_+)^{j+m}(Wu_-)^{j-m}$$

it follows that

$$\chi_j(W) = e^{i2\alpha j} + e^{i2\alpha(j-1)} + \cdots + e^{-i2\alpha j}. \tag{3.39}$$

Therefore

$$\begin{aligned}
\chi_0(W) &= 1, \\
\tfrac{1}{2}[\chi_{1/2}(W)] &= \cos\alpha, \\
\tfrac{1}{2}[\chi_1(W) - \chi_0(W)] &= \cos 2\alpha, \\
\tfrac{1}{2}[\chi_{3/2}(W) - \chi_{1/2}(W)] &= \cos 3\alpha, \\
&\ldots\ldots\ldots\ldots
\end{aligned}$$

The functions $\cos(n\alpha)$ for $n = 0, 1, 2, \ldots$ in the interval $0 \leq \alpha \leq \pi$ form a complete function system, so that there exist no additional irreducible representations of SU_2.

As we have seen in §2, to an elementary system type there corresponds a spin space $\mathscr{r}_s$ in which an irreducible representation is given (up to a factor) of $\mathscr{D}_{\mathscr{G}}$. $\mathscr{r}_s$ must then be isomorphic to one of the $\mathscr{T}_j$. As an index s we use the same index as in the case of $\mathscr{T}_j$, that is, the spin s (see D 2.2) can take on the values $0, \frac{1}{2}, 1, \frac{3}{2}, \ldots$.

The representation (2.35) of the rotation group in $\mathscr{H}$ is not irreducible even if $R(A)$ is the operator which represents an irreducible representation $\mathscr{D}_s$ in $\mathscr{r}_s$. With this the problem arises of reducing the representation given by (2.35). For this purpose we will reduce the representation given by $V(A)$ in $\mathscr{H}_b$.

In order to derive some frequently used formulas we shall now consider the isomorphic map (see AIV, §13).

$$\varphi(\vec{k}) \rightarrow \psi(\vec{r}) = \frac{1}{(2\pi)^{3/2}} \int e^{i\vec{k}\cdot\vec{r}} \varphi(\vec{k})\, dk_1\, dk_2\, dk_3 \tag{3.40}$$

from the space $\mathscr{L}^2(\mathbf{R}^3, dk_1\, dk_2\, dk_3)$ to $\mathscr{L}^2(\mathbf{R}^3, dx_1\, dx_2\, dx_3)$. For simplicity we shall also denote this space by $\mathscr{H}_b$ since $\mathscr{L}^2(\mathbf{R}^3, dk_1\, dk_2\, dk_3)$ and $\mathscr{L}^2(\mathbf{R}^3, dx_1\, dx_2\, dx_3)$ can be considered to be different representations (see XI, §2) of "the same" Hilbert spaces $\mathscr{H}_b$.

It is easy to see that, according to (3.40)

$$V(A)\varphi(\vec{k}) = \varphi(A^{-1}\vec{k}) \rightarrow \psi(A^{-1}\vec{r}) = \tilde{V}(A)\psi(\vec{r}),$$

where $\tilde{V}(A)$ is the image of the operator $V(A)$ in $\mathscr{L}^2(\mathbf{R}^3, dx_1\, dx_2\, dx_3)$. Instead of $\tilde{V}(A)$ we shall write $V(A)$. The representation of $\mathscr{D}_{\mathscr{G}}$ in $\mathscr{H}_b$ which we must reduce is therefore given by

$$V(A)\psi(\vec{r}) = \psi(A^{-1}\vec{r}). \tag{3.41}$$

Since (3.41) is a unique representation of $\mathscr{D}_{\mathscr{G}}$, in the reduced representation we may only have integer values of j.

Consider a rotation A about the 3-axis of angle α, that is,

$$A = \begin{pmatrix} \cos\alpha & -\sin\alpha & 0 \\ \sin\alpha & \cos\alpha & 0 \\ 0 & 0 & 1 \end{pmatrix}$$

thus, for an infinitesimal rotation we obtain:

$$\mathscr{I}_3\psi(\vec{r}) = \left[\frac{d}{d\alpha}\psi(A^{-1}\vec{r})\right]_{\alpha=0} = \left(x_2\frac{\partial}{\partial x_1} - x_1\frac{\partial}{\partial x_2}\right)\psi(\vec{r}).$$

For the $L_\nu = i\mathscr{I}_\nu$ it follows that, in general:

$$L_\nu = x_\mu \frac{1}{i}\frac{\partial}{\partial x_\rho} - x_\rho \frac{1}{i}\frac{\partial}{\partial x_\mu}; \qquad \nu, \mu, \rho = \begin{cases} 1, 2, 3, \\ 2, 3, 1, \\ 3, 1, 2. \end{cases} \tag{3.42}$$

The space $\mathscr{H}_b = \mathscr{L}^2(\mathbf{R}^3, dx_1\, dx_2\, dx_3)$ may be represented in the following form by means of polar coordinates (see AIV, §14):

$$\mathscr{H}_b = \mathscr{H}_r \times \mathscr{H}_{\theta,\varphi}, \tag{3.43a}$$

where

$$\mathscr{H}_r = \mathscr{L}^2(\mathbf{R}_+, r^2\, dr), \qquad \mathscr{H}_{\theta,\varphi} = \mathscr{L}^2(\Omega, d\omega), \tag{3.43b}$$

where Ω is the surface of the unit sphere and $d\omega$ is the surface element (solid angle) $d\omega = \sin\theta\, d\theta\, d\varphi$. With respect to (3.43a) $V(A)$ takes on the form

$$V(A){:}\ 1 \times V(A). \tag{3.44}$$

We shall now reduce the representation $V(A)$ in $\mathscr{H}_{\theta,\varphi}$.

Similarly, the L_ν in (3.42) must take the form (3.44). By conversion to polar coordinates, for L_3 and $N^+ = L_1 - iL_2$ we obtain:

$$L_3 = \frac{1}{i}\frac{\partial}{\partial\varphi}, \tag{3.45}$$

$$N^+ = e^{-i\varphi}\left(-\frac{\partial}{\partial\theta} + i\frac{\cos\theta}{\sin\theta}\frac{\partial}{\partial\varphi}\right). \tag{3.46}$$

Let the components of a unit vector be given by

$$e_1 = \sin\theta\cos\varphi, \qquad e_2 = \sin\theta\sin\varphi, \qquad e_3 = \cos\theta.$$

Then, by the Weierstrass approximation theorem the set of all $e_1^{\alpha_1}e_2^{\alpha_2}e_3^{\alpha_3}$ (where the $\alpha_\nu \geq 0$ are integers) span the entire space $\mathscr{L}^2(\Omega, d\omega)$. The $e_1^{\alpha_1}e_2^{\alpha_2}e_3^{\alpha_3}$ with fixed sum $\alpha_1 + \alpha_2 + \alpha_3 = l$ span a finite-dimensional subspace $\mathscr{H}_{\theta,\varphi}^l$ of $\mathscr{H}_{\theta,\varphi}$.

Since the e_ν transform linearly under rotations, $\mathscr{H}_{\theta,\varphi}^l$ is obviously an invariant subspace under $V(A)$. We now seek the irreducible subspace in $\mathscr{H}_{\theta,\varphi}^l$ which contains the largest eigenvalue of L_3. According to (3.45) the eigenvectors of L_3 obviously have the form $e^{im\varphi}g(\theta)$. If, instead of e_1, e_2, e_3 we introduce the three functions $e = e_1 + ie_2 = e^{i\varphi}\sin\theta$, $\bar{e} = e_1 - ie_2 = e^{-i\varphi}\sin\theta$, $e_3 = \cos\theta$, then $\mathscr{H}_{\theta,\varphi}^l$ will be spanned by all $e^{\beta_1}\bar{e}^{\beta_2}e^{\beta_3}$ for which $\beta_1 + \beta_2 + \beta_3 = l$. The $e^{\beta_1}\bar{e}^{\beta_2}e^{\beta_3}$ are then precisely the eigenvectors of L_3 with eigenvalues $\beta_1 - \beta_2$. Therefore the largest eigenvalue of L_3 is obtained when $\beta_1 = l$ and $\beta_3 = \beta_3 = 0$. Its value is l. The eigenvalue l of L_3 is nondegenerate in $\mathscr{H}_{\theta,\varphi}^l$. The corresponding eigenfunction is

$$u_l = c_l e^{il\varphi}(\sin\theta)^l. \tag{3.47}$$

Since $N = L_1 + iL_2$ cannot take us out of $\mathscr{H}_{\theta,\varphi}^l$ we must have $Nu_l = 0$ and the following equation must be satisfied:

$$\vec{L}^2 u_l = l(l+1)u_l. \tag{3.48}$$

With the aid of N^+ we may, using (3.17), obtain the desired irreducible representation space of the form which is spanned by the u_m.

Since u_m lies in $\mathscr{H}^l_{\theta,\varphi}$ and $L_3 u_m = m u_m$, u_m must be a linear combination of vectors of the form $e^{\beta_1}\bar{e}^{\beta_2}e^{\beta_3}$ where $\beta_1 - \beta_2 = m, \beta_1 + \beta_2 + \beta_3 = l$:

$$u_m = c_m e^{im\varphi}(\sin\theta)^{-m} Q_m(\cos\theta), \tag{3.49}$$

where Q_m is a polynomial in $\cos\theta$.

With N^+ defined by (3.46) and from (3.17) we obtain

$$\begin{aligned} N^+ u_m &= c_m e^{i(m-1)\varphi}(\sin\theta)^{-m+1} Q'_m(\cos\theta) \\ &= \sqrt{(l+m)(l-m+1)}\, u_{m+1} \\ &= \sqrt{(l+m)(l-m+1)}\, c_{m-1} e^{i(m-1)\varphi}(\sin\theta)^{-m+1} Q_{m+1}(\cos\theta), \end{aligned}$$

where $Q'(\xi)$ is the derivative of $Q(\xi)$ with respect to ξ. Therefore the Q_m may be recursively defined by means of $Q'_m = Q_{m-1}$ and, for the normalization constant c_m we obtain the recursion formula:

$$c_m = \sqrt{(l-m+1)(l+m)}\, c_{m-1}. \tag{3.50}$$

From (3.47) it follows that $Q_l(\xi) = (1-\xi^2)^l$ and we therefore obtain:

$$Q_m(\xi) = \frac{d^{l-m}}{d\xi^{l-m}}(1-\xi^2)^l. \tag{3.51}$$

From (3.50) it follows that c_m takes on the value (with a yet to be determined normalization constant a):

$$c_m = a\sqrt{\frac{(l+m)!}{(l-m)!}}. \tag{3.52}$$

The factor a may be determined by means of the normalization condition

$$\int |u_0|^2 \sin\theta\, d\theta\, d\varphi = 1. \tag{3.53}$$

The integral (3.53) may be recursively calculated (see, for example, [2], XI, §5.3). We obtain

$$c_m = \frac{1}{\sqrt{2\pi}}\sqrt{\frac{2l+1}{2}\frac{(l+m)!}{(l+m)!}}\frac{1}{2^l l!}. \tag{3.54}$$

The functions u_m are generally known as "spherical harmonics"; the customary notation for them is $Y^l_m(\theta, \varphi)$. From (3.49), (3.53), and (3.54) we finally obtain

$$Y^l_m(\theta,\varphi) = \frac{1}{\sqrt{2\pi}}\sqrt{\frac{2l+1}{2}\frac{(l+m)!}{(l+m)!}}\frac{1}{2^l l!} e^{im\varphi}(\sin\theta)^{-m} Q^l_m(\cos\theta), \tag{3.55}$$

where

$$Q^l_m(\xi) = \frac{d^{l-m}}{d\xi^{l-m}}(1-\xi^2)^l.$$

The Y^l_m span an irreducible subspace $\mathscr{T}_l$ of $\mathscr{H}_{\theta,\varphi}$. We will now show that

$$\mathscr{H}_{\theta,\varphi} = \sum_{l=0}^{\infty} \oplus \mathscr{T}_l \tag{3.56}$$

that is, the Y^l_m completely span $\mathscr{H}_{\theta,\varphi}$. For this purpose it is sufficient to show that

$$\mathscr{H}^l_{\theta,\varphi} = \mathscr{T}_l \oplus \mathscr{T}_{l-2} \oplus \mathscr{T}_{l-4} \oplus \cdots \tag{3.57}$$

In fact, we may consider the $\mathscr{T}_{l-2}, \mathscr{T}_{l-4}, \ldots$ to be homogeneous polynomials of degree l; for example, obtained from multiplication of polynomials of $(l-2)$ degree by $1 = e^2 + e_2^2 + e_3^2$ (and similarly for $\mathscr{T}_{l-4}$). The right side of (3.57) is therefore a subspace of $\mathscr{H}^l_{\theta,\varphi}$. Since, for different l the $\mathscr{T}_l$ must be orthogonal, the dimension of the right-hand side must be equal to

$$\begin{aligned}&[2l+1] + [2(l-2)+1] + [2(l-4)+1] + \cdots \\ &\quad = [(l+1)+l] + [(l-1)+(l-2)] + [(l-3)+(l-4)] + \cdots\end{aligned}$$

The dimension $n(l)$ of $\mathscr{H}^l_{\theta,\varphi}$ is equal to the number of the $e_1^{\alpha_1}e_2^{\alpha_2}e_3^{\alpha_3}$ for which $\alpha_1 + \alpha_2 + \alpha_3 = l$. Thus it follows that $n(l+1) = n(l) + [n+l] + 1$. Therefore $n(l) = [l+1] + l + [l-1] + \cdots$. Thus we have proven (3.57) and (3.56).

Since (3.42) and (3.44) the reduction of the representation in $\mathscr{H}_b$ is very simple. Choose a complete orthonormal basis $\chi_\nu(r)$ in $\mathscr{H}_r$. The $\psi^l_{\nu\mu}(\vec{r}) = \chi_\nu(r)Y^l_m(\theta, \varphi)$ span (for fixed ν, l) an irreducible subspace $\mathscr{H}^l_{b\nu}$ of $\mathscr{H}_b$. From (3.56) we therefore obtain

$$\mathscr{H}_b = \sum_{\nu,l} \oplus \mathscr{H}^l_{b\nu}. \tag{3.58}$$

The reduction of the representation $U(A) = V(A) \times R(A)$ in $\mathscr{H}_b \times \imath_s$ ($\imath_s$ is an irreducible representation space) will be postponed until XI, §10. It is easy to see that this reduction is achieved if the representation $V(A) \times R(A)$ can be reduced in $\mathscr{T}_l \times \imath_s$. For a representation in a product space we write a $\times$-sign, for example, $\mathscr{D}_{j_1} \times \mathscr{D}_{j_2}$ for a product representation in the product space $\mathscr{T}_{j_1} \times \mathscr{T}_{j_2}$ where the representation operators have the form $V(A) \times R(A)$ and $\mathscr{T}_{j_1}$ is irreducible with respect to the $V(A)$ and $\mathscr{T}_{j_2}$ is with respect to $R(A)$.

4 Position and Momentum Observables

In §2 and §3 we encountered infinitesimal transformations. Using the latter we defined self-adjoint operators (not necessarily bounded) such as $\vec{K}$, $\vec{X}$ and H in §2 and $\vec{L}$ in §3. According to IV, D 2.5.6 to each such operator there corresponds a scale observable (which, according to D 2.5.6 is also a decision observable). These scale observables are often given names. The introduction of observables by means of infinitesimal transformations is common practice in quantum mechanics. In this way the problem of how these observables are to be measured is often ignored.

In an analogous procedure in classical mechanics the observables are functions in the Γ-space (p_ν, q_ν) of the system. Since the pre-theories of classical mechanics should define how position and momentum, and finally how p_ν, q_ν, are to be measured, we should be satisfied if we are able to specify the observables as functions in Γ-space; then their measurement will be described in terms of the pre-theories of classical mechanics.

The situation in quantum mechanics is totally different. Here the pre-theories make possible the description of registration methods and registration procedures. However, the correspondence $\mathscr{F} \xrightarrow{\psi} L$ is itself determined by quantum mechanics! Here the problem described in IV, §4 appears in clear focus. Here we must concede that the specification of self-adjoint operators and their corresponding scale observables is primarily of a "pure theoretical nature." According to the theory there should be "in principle" approximate measurement methods in the sense of IV, §4 for these observables. However, it appears that they cannot be obtained from previously defined theories!

The quantity X_ν (multiplied by m^{-1}) introduced in §2 is often called the "position observable at time $t = 0$." However, from the theory we can neither say how this may be measured or whether a particular measurement method b_0 measures these X_ν (even approximately). With respect to the latter point we may make an additional step in that direction in the following way:

In the above definition of $m^{-1}X_\nu$ as the "position at time $t = 0$" it is unclear not only what is meant by the expression "position" (unless we are willing to accept this expression as a mere name without any meaning) but also what the physical meaning of the expression "at time $t = 0$" should be. At present there are many different and varied conceptions and (apparent) interpretations about this problem in circulation. Here we shall only mention that the expression "at time $t = 0$" cannot mean (as we shall find in more precise terms in XVII) that the "measurement takes place at time $t = 0$." Such a "point in time" at which a measurement takes place does not exist.

On this basis other priorities will take precedence over the definition of a decision observable for "position at time t." Such a definition will necessarily refer in a precise manner to a laboratory fixed space–time reference system as defined in §1 for the physical interpretation of the Galileo group. Consider an apparatus b_0 for which the scale response refers to the spatial domain, that is, the scale of the measurement apparatus defines an isomorphism between $\mathscr{R}(b_0)$ and the "Boolean ring Σ of a region in three-dimensional space." Since we cannot expect that there is a real apparatus with a rigorous isomorphism between $\mathscr{R}(b_0)$ and Σ we proceed instead by making an idealization, that is, with an observable $\Sigma \xrightarrow{F} L$ where Σ is the "Boolean ring of the spatial region." How can we define Σ mathematically?

Let $\mathbf{A}$ be the σ-algebra of the Lebesgue measurable sets of $\mathbf{R}^3$—in this case the three-dimensional space defined by the laboratory fixed spatial reference system, that is, $\mathbf{R}^3$ is the set of coordinate tripels (x_1, x_2, x_3). Let $\mathbf{J}$ be the family of the sets of measure 0. $\Sigma = \mathbf{A}/\mathbf{J}$ is a complete Boolean ring, which we shall call the Boolean ring of the "spatial region." Each element $\sigma \in \Sigma$

therefore represents a possible response b of the *idealized* measurement apparatus b_0.

In the sense of IV, D 2.5.5 $\Sigma = \Sigma(x_1, x_2, x_3)$ where the x_ν are the measurable spatial coordinates in the laboratory reference system under consideration. These coordinates have nothing to do with quantum mechanics. They are defined by classical measurement techniques and procedures. $\Sigma = \Sigma(x_1, x_2, x_3) \xrightarrow{F} L$ is therefore, in the sense of IV, §2.5, an observable with the sufficient scales x_1, x_2, x_3.

We now seek to sharpen our assertion about the desired position observable. Let us set $\Sigma(x_1, x_2, x_3) \xrightarrow{F} G$, that is, let us consider a decision observable.

Since there is often much misunderstanding concerning the meaning of measurement, that is, of the registrations b which correspond to the $\sigma \in \Sigma(x_1, x_2, x_3)$, we again stress the fact that the registrations b on the apparatus under consideration do not have an immediate connection with the technical aspects of the measurement of the coordinates x_1, x_2, x_3! If, for example, a registration b corresponds to

$$\sigma = \{(x_1, x_2, x_3) | x_\nu^0 - \varepsilon < x_\nu < x_\nu^0 + \varepsilon\},$$

where ε is small, this means only that the apparatus b_0 records (for example, with the aid of a computer) the "measurement values" (x_1^0, x_2^0, x_3^0). It does not mean (!) that x_1^0, x_2^0, x_3^0 are the technically measured spatial coordinates of some macroscopic event or process associated with the registration b. The "responses" of the measurement apparatus have nothing to do with the technical aspects of measurement of spatial position. However, the usage of a measurement apparatus characterized by b_0 has much to do with the technical aspects of spatial measurement, as we have explained in §1 and which we shall now discuss.

The measurement of the "position of a microsystem at time t" (assuming that such an observable $\Sigma \xrightarrow{F} G$ exists which satisfies all the requirements which we will impose on it) by an apparatus is therefore different than, for example, the technical measurement of the position of a space ship at time t. The technical measurement of a space ship is explainable without the use of Newtonian mechanics. The construction of the desired position measurement apparatus b_0 for microsystems cannot be explained without quantum mechanics; the position of a microsystem is only indirectly measurable in quantum mechanics (in the sense of the discussions in [1], §10 or [2], III, §9). We shall now seek to define the position observable in this indirect way.

It must be possible to adjust the spatial position of the apparatus b_0 relative to the laboratory system in order that Galileo transformations have the meaning which was described in §1, for instance, that of a spatial translation $(1, 0, \vec{\eta}, 0)$. We shall now investigate the requirements which shall be imposed on $\Sigma(\mathrm{x}_1, \mathrm{x}_2, \mathrm{x}_3) \xrightarrow{E} G$ in order to relate indirectly the "measurement values" x_1, x_2, x_3 of b to the spatial coordinates. Intuitively we find that if we try to interpret the registration b for the apparatus b_0 as the "determination" of the position in σ then the apparatus b_0' which is obtained from b_0 by means of a translation $(1, 0, \vec{\eta}, 0)$ and the corresponding response b' must correspond to the determination that the position is in $\sigma' = \sigma + \vec{\eta}$ (where $\sigma + \vec{\eta}$ is, of course, the spatial domain for which σ is displaced by $\vec{\eta}$). According to §1 and §2 we obtain

$$\psi(b'_0, b') = U(1, 0, \vec{\eta}, 0)\psi(b_0, b)U(1, 0, \vec{\eta}, 0)^{-1}. \tag{4.1}$$

This equation should also be satisfied if $\psi(b_0, b)$ is replaced by $E(\sigma)$ and $\psi(b'_0, b')$ is replaced by $E(\sigma + \vec{\eta})$, since $E(\sigma)$ is the idealization of $\psi(b_0, b)$.

We now impose our first requirement upon $\Sigma(x_1, x_2, x_3) \xrightarrow{E} G$:

$$U(1, 0, \vec{\eta}, 0)E(\sigma)U(1, 0, \vec{\eta}, 0)^{-1} = E(\sigma + \vec{\eta}). \tag{4.2}$$

For rotations we may make a similar argument; we require that

$$U(A, 0, 0, 0)E(\sigma)U(A, 0, 0, 0)^{-1} = E(A\sigma), \tag{4.3}$$

where $A\sigma$ is the domain which is obtained by rotating σ by A.

The requirement (4.2) is "in principle" experimentally verifiable in the form (4.1) as follows: If we have built an apparatus and set it relative to the laboratory system, then there is a corresponding b_0. Then we easily obtain b'_0 (macroscopically) as the spatial translation of the apparatus b_0. Here (4.1) can be approximately controlled by probability measurements for different $a \in \mathcal{Q}'$. Thus it is clear that the requirements (4.2) and (4.3) refer to registrations and preparations—as we have described in II and in §1.

As we have found in IV, §2.5, the decision observable $\Sigma(x_1, x_2, x_3) \xrightarrow{E} G$ is uniquely determined by the scale-observables which correspond to the scales as follows:

$$Q_\nu = \int \lambda \, dE_\nu(\lambda). \tag{4.4}$$

Then, from IV, §2.5 we obtain

$$E_\nu(\lambda) = E(\sigma_\nu(\lambda)) \quad \text{where } \sigma_\nu(\lambda) = \{(x_1, x_2, x_3) \mid x_\nu \le \lambda\}. \tag{4.5}$$

The operators Q_ν introduced in (4.4) are self-adjoint (not bounded) and are not defined in all of $\mathscr{H}$. Nevertheless these operators uniquely determine the corresponding spectral families $E_\nu(\lambda)$ (see AIV, §10) and we therefore also obtain (as proven in IV, §2.5) the complete observable $\Sigma(x_1, x_2, x_3) \xrightarrow{E} G$. Later we shall return to the discussion of the use of infinite extended scales.

The requirements (4.2) and (4.3) for the observable $\Sigma(x_1, x_2, x_3) \xrightarrow{E} G$ are not sufficient to uniquely determine this observable. In addition, we must also take into account the use of the laboratory time scale t.

We shall now again consider the intuitive idea that b_0 represents a measurement apparatus for which the measurement result can be interpreted as the registration of "position at time t." What should this mean? In order to answer this question we shall now consider the original apparatus b_0 together with its response b and a second experiment b''_0 and response b'' where b''_0 and b_0 are identical except that b''_0 moves relative to b_0 with velocity $\vec{\delta}$ in such a manner that, with respect to the laboratory time scale the apparatus b_0'' takes the same spatial position as b_0 does at the time $\tilde{t}$. It is obvious that the apparatuses for b_0 and b_0'' cannot be applied to the same experiment. Two experimental series, one with b_0, the other with b_0'' must be carried out in order to measure the frequencies of the responses. Then b_0'' may therefore be obtained by the application of the following Galileo transformation to b_0;

$$\begin{aligned} x_\nu'' &= x_\nu + \delta_\nu(t - \tilde{t}), \\ t'' &= t. \end{aligned} \tag{4.6}$$

Equation (4.6) may be written as a product as follows:

$$(1, 0, 0, \tilde{t})(1, \vec{\delta}, 0, 0)(1, 0, 0, -\tilde{t}). \tag{4.7}$$

The form (4.7) permits us to use the operators described in §2.

If b_0 "registers the position at time $\tilde{t}$" by the response b then b_0'' should also be the same because both apparatuses will be in coincidence at time $\tilde{t}$. We will therefore require that $\psi(b_0, b) = \psi(b_0'', b'')$. We may transform $\psi(b_0, b)$ to its idealized version $E(\sigma)$ and therefore require that

$$\begin{aligned} &U(1, 0, 0, \tilde{t})U(1, \vec{\delta}, 0, 0)U(1, 0, 0, \tilde{t})^{-1}E(\sigma) \\ &\qquad \cdot U(1, 0, 0, \tilde{t})U(1, \vec{\delta}, 0, 0)^{-1}U(1, 0, 0, \tilde{t})^{-1} = E(\sigma), \end{aligned} \tag{4.8}$$

where we have used the form (4.7) of the Galileo transformation (4.6).

We shall now show that there exists exactly one decision observable $\Sigma(x_1, x_2, x_3) \xrightarrow{E} G$ which satisfies the conditions (4.2), (4.3), and (4.8). This observable will be called "position at time $\tilde{t}$" and we shall denote the corresponding scale observables (4.4) by $Q_\nu(\tilde{t})$. Then the real physical meaning of the time parameter $\tilde{t}$ is characterized by (4.8) because (4.8) can only be satisfied by a single $\tilde{t}$. The time $\tilde{t}$ has nothing to do with the time of occurrence of the macroscopic response b; in addition it has nothing to do with the notion of the "time of measurement." Such a "time of measurement" does not refer to an instant of time, but rather to a time interval in which the interaction between the microsystem and the apparatus takes place. Rather, the time $\tilde{t}$ is that for which the moving apparatus b_0'' comes into coincidence with the stationary apparatus b_0. Therefore, $\tilde{t}$ may be determined macroscopically and has nothing to do with the temporal evolution of the interaction of the microsystem with the measurement apparatus. $\tilde{t}$ is obtained from the spatial alignment of the two registration methods b_0 and b_0''. Since we are unable to measure the position of a microsystem in a technical sense, it is not meaningful to speak of a measurement "at time $\tilde{t}$." Since the x_1, x_2, x_3 and t are actually parameters of the underlying reference system and are only adjustable parameters for the registration apparatus relative to the reference system, they are *not* observables in the quantum mechanical sense, as these are defined and studied in IV.

We shall now prove the above assertion that the $Q_\nu(\tilde{t})$ are uniquely defined (for a precise mathematical formulation and proof see [10] and [22]). For this purpose we shall first consider the special case in which $\tilde{t} = 0$. Then from (4.8) we obtain

$$U(1, \vec{\delta}, 0, 0)E(\sigma)U(1, \vec{\delta}, 0, 0)^{-1} = E(\sigma), \tag{4.9}$$

that is, $U(1, \vec{\delta}, 0, 0)$ (of (2.11)) commutes with $E(\sigma)$.

From (4.2) it follows that, with (2.9)

$$e^{i\vec{K}\cdot\vec{\eta}}E(\sigma)e^{-i\vec{K}\cdot\vec{\eta}} = E(\sigma + \vec{\eta}). \tag{4.10}$$

Thus, for all $Q_v(0)$ from (4.4) it follows that

$$e^{i\vec{K}\cdot\vec{\eta}}\vec{Q}(0)e^{-i\vec{K}\cdot\vec{\eta}} = \vec{Q}(0) - \vec{\eta}1. \tag{4.11}$$

From (2.17) we obtain a similar equation

$$e^{i\vec{K}\cdot\vec{\eta}}\vec{X}e^{-i\vec{K}\cdot\vec{\eta}} = \vec{X} - m\vec{\eta}1. \tag{4.12}$$

Therefore $\vec{Y} = \vec{Q}(0) - m^{-1}\vec{X}$ is an operator which commutes with $U(1, \vec{\delta}, 0, 0)$ and with $U(1, 0, \vec{\eta}, 0)$. Therefore we obtain:

$$\vec{Y}: \mathbf{1} \times \vec{Y},$$

that is,

$$\vec{Q}(0) = m^{-1}\vec{X} \times \mathbf{1} + \mathbf{1} \times \vec{Y}. \tag{4.13}$$

From (4.3) it follows that, using (2.35):

$$R(A)\vec{Y}R(A)^{-1} = A^{-1}\vec{Y}. \tag{4.14}$$

Since the $Q_v(0)$ mutually commute, the Y_v must also commute. Since $\mathscr{h}_s$ is finite dimensional, the Y_v have a common system of eigenspaces which span all of $\mathscr{h}_s$. From (4.14) it follows that $R(A)Y_vR(A)^{-1}$ are linear combinations of the Y_v and therefore the common eigenspaces of Y_v are also eigenspaces of the $R(A)Y_vR(A)^{-1}$, that is, the $R(A)$ leave the eigenspaces of Y_v invariant. Since the representation of the rotation group in $\mathscr{h}_s$ is irreducible, there are no proper invariant subspaces. Therefore $Y_v = \lambda_v\mathbf{1}$. From (4.14) it follows that $\lambda_1 = \lambda_2 = \lambda_3 = 0$ and therefore, according to (4.13) we obtain:

$$\vec{Q}(0) = m^{-1}\vec{X}. \tag{4.15}$$

Conversely, if we define a decision observable according to IV, §2.5 $\Sigma(x_1, x_2, x_3) \xrightarrow{E} G$ by means of (4.15) then the so-defined observable will satisfy conditions (4.2), (4.3), and (4.8).

If $\Sigma(x_1, x_2, x_3) \xrightarrow{E} G$ is the position observable for time $t = 0$ and we define (using (2.38))

$$\begin{aligned}\tilde{E}(\sigma) &= U(1, 0, 0, \tilde{t})E(\sigma)U(1, 0, 0, \tilde{t})^{-1} \\ &= e^{iH\tilde{t}}E(\sigma)e^{-iHt}\end{aligned}$$

then, from (4.9) it follows that

$$\begin{aligned}&U(1, \vec{\delta}, 0, 0)U(1, 0, 0, \tilde{t})^{-1}E(\sigma)U(1, 0, 0, \tilde{t})U(1, \vec{\delta}, 0, 0)^{-1} \\ &\quad = U(1, 0, 0, \tilde{t})^{-1}\tilde{E}(\sigma)U(1, 0, 0, \tilde{t}),\end{aligned}$$

that is, (4.8) holds for $\tilde{E}(\sigma)$. It is easy to prove that $\tilde{E}(\sigma)$ also satisfies conditions (4.2) and (4.3). The scale observables

$$Q_\nu(t) = e^{iHt}Q_\nu(0)e^{-iHt} \tag{4.16}$$

therefore exactly satisfy the conditions for the desired observables "position at time t." Since the $Q_\nu(t)$ must have the form (4.16) it therefore follows that the observable defined by

$$e^{-iHt}Q_\nu(t)e^{iHt}$$

must satisfy the conditions (4.2), (4.3), and (4.9).

Thus, in this way we have uniquely defined the decision observable "position at time t" for elementary systems.

We have not yet constructed a registration method b_0 which will permit (in the sense of IV, §4) an approximate realization of the "position observable at time t." We have, however, described a type of experiment which may be used in order to determine whether a given registration method approximates the "position observable at time t."

A similar method can be used for the definition of a "momentum observable." We begin with a decision observable $\Sigma(p_1, p_2, p_3) \xrightarrow{E} G$ and impose the following requirements: The observable is invariant under spatial translations:

$$U(1, 0, \vec{\eta}, 0)E(\sigma)U(1, 0, \vec{\eta}, 0)^{-1} = E(\sigma). \tag{4.17}$$

Under rotations we require that

$$U(A, 0, 0, 0)E(\sigma)U(A, 0, 0, 0)^{-1} = E(A\sigma). \tag{4.18}$$

Under proper Galileo transformations we require that

$$U(1, \vec{\delta}, 0, 0)E(\sigma)U(1, \vec{\delta}, 0, 0)^{-1} = E(\sigma - m\vec{\delta}). \tag{4.19}$$

Here (4.19) corresponds to the intuitive idea that the motion of the registration apparatus with velocity $\vec{\delta}$ results in a change of momentum of the microsystems relative to the moving registration apparatus of magnitude $(-m\vec{\delta})$.

The observable $\Sigma(p_1, p_2, p_3) \xrightarrow{E} G$ can be characterized by three scale observables

$$P_\nu = \int \lambda \, dE_\nu(\lambda) \tag{4.20}$$

with respect to the scales p_1, p_2, p_3.

From (4.17) it follows that the P_ν commute with the $e^{i\vec{K}\cdot\vec{\eta}}$. From (2.17) and (4.19) it follows that the

$$Z_\nu = P_\nu + K_\nu$$

commute with the X_ν and the K_ν, that is, we must obtain

$$P_\nu = -K_\nu \times \mathbf{1} + \mathbf{1} \times Z_\nu.$$

Similarly, as in the case of (4.13), from (4.18) it follows that we must have $Z_\nu = 0$.

Therefore, a momentum observable is uniquely determined by the requirements (4.17), (4.18), and (4.19) which is characterized by the scale observables

$$P_\nu = -K_\nu \times \mathbf{1}. \qquad (4.21)$$

It is easy to see that $e^{iH\gamma}$ commutes with the P_ν, that is, a time translation of the idealized registration methods corresponding to the P_ν do not produce any change in the observables.

From (2.16), (4.15), and (4.21), we obtain the famous Heisenberg commutation relation for position and momentum

$$P_\mu Q_\nu - Q_\nu P_\mu = \frac{1}{i}\delta_{\mu\nu}\mathbf{1}. \qquad (4.22)$$

Since Planck's constant does not appear in the theory formulated here (as we have already discussed), it is clear that we have correctly formulated the fundamental structure of quantum mechanics. Equation (4.22) is an indirect consequence of the representations of the Galileo group and is therefore only a consequence of axiom AV 4s in III, §3. For a discussion of the corresponding Heisenberg uncertainty relations, see IV, §8.3.

These observables P_ν, Q_μ provide our first example for the use of "unbounded" self-adjoint operators. In quantum mechanics the meaning of unbounded operators, their domain of definition, and the precise formulation of the relation (4.22) are often treated as a mystery. Here we have found that there is, in principle, nothing unusual underlying the introduction of the unbounded operators P_ν, Q_μ described above. The conceptual structure of quantum mechanics has nothing to do with this occurrence of unbounded operators. It arises exclusively (and, for the most part, effortlessly) from a mathematical idealization which has no particular physical meaning (see, for example, [1], §9 and [2], III, §8). If, for example, we describe the laboratory reference system by Euclidean geometry, it is then practical to use the Euclidean rectangular coordinates x_ν as scales, although, in principle, it is not necessary; we could instead use other finite scales. Since the lack of finiteness for the scale x_ν has no physical meaning, it follows that in the real world arbitrarily large x_ν have no physical content (see, for example, [1], §9 and [2], IX, X).

Therefore unbounded self-adjoint operators for scale observables occur if we introduce unbounded scales. The scales for such observables are only a practical tool, as we have described in IV, §2.5.

There is, however, a second case where unbounded self-adjoint operators (and therefore also their spectra) have, in a natural way, a physical meaning. This situation arises if the self-adjoint operators occur as infinitesimal operators in the representation of a group which has a physical interpretation. We have encountered such self-adjoint operators of the form X_ν, K_ν, H, $\vec{L}$ in §2 and §3. Here the spectrum, that is, the scale values are of crucial importance for the structure of the corresponding group representations.

Both of the above viewpoints—scales as only a practical tool for the ordering of a Boolean ring of an observable (as described in IV, §2.5) and scales as a characteristic of an infinitesimal transformation—are often confused with each other. In particular, this occurs when the infinitesimal transformations are, because of their representations in terms of self-adjoint operators, called observables and are given special names.

Naturally it is permissible to relate some scale observables in this way with infinitesimal transformations; here the scales are, on the basis of their definition, no longer arbitrary. Often a sufficient distinction is not made between the case where, in an experiment, a registration method b_0 is used with a scale which corresponds to a theoretically defined observable, and the case in which the registrations which were carried out reflect the transformations and indirectly permit the conclusions concerning the spectrum of the infinitesimal transformation. In the applications presented in this book we shall be careful to point out which are preparations, which are registrations, and which are transformations.

In closing this section we shall now make a few remarks concerning the parallelism between the case in which the Galileo group is replaced by the Poincaré group. Obviously (4.8) cannot be applied to the relativistic case because it is not possible to bring two moving systems into "coincidence." Here it can be shown that the position decision observable for elementary systems with nonzero mass conditions analogous to (4.2) and (4.3) can be satisfied. These position observables are, however, not uniquely determined. For light quanta (systems of zero mass) there is no such position decision observable. This is, however, not an argument against the theory. If, in an experiment with light quanta something similar to a position is measured, this measurement does not correspond to a decision observable [22]. Here we have an example that shows that it is somewhat risky to *only* discuss such an observable concept which we have called a decision observable. Such a restricted concept of an observable (namely that of a decision observable) would not fit all essential experimental procedures.

5 Energy and Angular Momentum Observables

In this section we shall consider two observables for which the theory of measurement methods is less well known than is the case for the position and momentum observables described in §4. First, for purely formal reasons, we may call the observable H of infinitesimal time translations (2.20):

$$U(1, 0, 0, \gamma) = e^{iH\gamma} \tag{5.1}$$

the energy observable. This does not mean that (as we have already mentioned in §4) we are able to construct a b_0 for which $\psi(b_0, b)$ corresponds to an approximate measurement of H (in the sense of IV, §4).

If there exist elementary systems, then according to (2.38) H is a function of $\vec{K}$ and therefore also of $\vec{P}$:

$$H = \frac{1}{2m}\vec{P}^2. \tag{5.2}$$

H is therefore (in the sense of IV, D 2.5.5 and IV, D 2.5.6) a scale partial observable for the momentum observable, that is, H is automatically determined by the measurement of momentum. This is no longer the case for composite systems, as we shall find in VIII, §1.

For certain systems experimental physicists have suitable registration methods for the measurement of H. Often, however, H can only be experimentally determined indirectly—only by means of $U(1, 0, 0, \gamma)$. For this reason we must carefully use the notion of an energy observable H in applications.

The operators L_ν are defined in terms of $U(A, 0, 0, 0)$ by (3.1) in a similar manner as H is defined in terms of $U(1, 0, 0, \gamma)$. The scale observables L_ν are called the components of angular momentum. Again, we are not told how these observables may be measured. In [2], XI, §7.2 and [2], XII, §2.2 we show how angular momentum can be measured by means of the *Stern–Gerlach* experiment. In many applications, however, the role of the components of angular momentum as infinitesimal rotations is more important—a typical example is given by atomic spectra, where the latter is discussed in XI–XIV.

For elementary systems (2.35) holds, where $R(A)$ are the operators of irreducible representations D_s in $\imath_s$. In this way $U(A, 0, 0, 0)$ defines not only the total angular momentum (denoted by $\vec{J}$) but, in addition, $V(A)$ defines the orbital angular momentum (denoted by $\vec{L}$—as an operator in $\mathscr{H}_b$) and $R(A)$ defines the spin angular momentum (denoted by $\vec{S}$—as an operator in $\imath_s$). By differentiation of (2.35) we obtain

$$\vec{J} = \vec{L} \times \mathbf{1} + \mathbf{1} \times \vec{S}, \tag{5.3}$$

$\vec{L}$, as an operator in $\mathscr{H}_b = \mathscr{L}^2(\mathbf{R}^3, dx_1\, dx_2\, dx_3)$ is given by (3.42). $\vec{S}$, as an operator in $\imath_s$ is given by (3.14), (3.15), (3.17), and (3.18) with s instead of j. Here we say that the spin has the fixed value s, where by this we mean that, according to (3.15), in $\imath_s$ the relation $\vec{S}^2 = s(s+1)\mathbf{1}$ holds.

These mathematical formulas for $\vec{L}, \vec{S}, \vec{J}$ provide no instructions for the construction of measurement apparatuses for these observables. In the theory of atomic spectra the infinitesimal transformations characterized by the $\vec{L}, \vec{S}, \vec{J}$ play a very important role (see XI–XIV).

6 Time Observable?

In the literature of quantum mechanics the discussion about the so-called "time observable" has reached vast and overwhelming proportions. Most of this discussion rests upon a misunderstanding of the concept of an observable. In the "usual" interpretation of quantum mechanics—the one most frequently heard by the student—the observable concept is used as a fundamental concept (see, for example, [2], XI, §1.7 where we have pointed out the inadequacy of this interpretation). It is often necessary to go to great lengths in order to provide an intuitive justification of this concept of an observable, and the discussion of the difficulties associated with this concept are avoided in order to minimize the difficulties with this approach in order not to frighten the student excessively.

In this approach the observable concept is introduced as a "quantity measured by an observer" or as a "measurable quantity," etc. Clearly, in the laboratory there exist clocks by which we may "measure time"; therefore time should also be an observable. In order to counter all such erroneous interpretations of quantum mechanics we have laid the foundations of the interpretation of quantum mechanics in II and extended this interpretation by the structure introduced in VII, §1. Hence it clearly follows that measurements with "meter sticks" and "clocks" do not constitute measurements of quantum mechanical observables. For this reason we have developed the concept of an observable as a derived concept in IV (for a discussion of derived concepts see [1], §10). Meter sticks and clocks are used only for the purpose of adjusting and calibrating preparation and registration apparatuses.

Therefore in quantum mechanics the spatial coordinates x_1, x_2, x_3 and the time coordinate t are only parameters given by the laboratory reference system! As we have found in §4 the measurement of the position observable $Q_v(t)$ does not mean that at time t (clock time) the coordinates x_1, x_2, x_3 of a microsystem are measured by means of meter sticks because the microsystem as such is "not there" in the sense that it can be measured in this way. The microsystem may only be detected by producing a response in a registration apparatus.

The introduction of the position observable $Q_v(t)$ in §4 clearly shows that it is concerned only with the possibilities of (idealized) registrations. The claim that the coordinates x_1, x_2, x_3 are *defined* as the measured values of the $Q_v(t)$ is a misunderstanding of quantum mechanics. In fact it is just the opposite—the technical process by which the coordinates x_1, x_2, x_3 are defined in the laboratory system must be explained independently of quantum mechanics. After it is understood that an observable $\Sigma \xrightarrow{E} G$ (here Σ is the Boolean ring for the region of space under consideration) is determined by certain requirements, it is reasonable to also choose the previously defined x_1, x_2, x_3 as the scales for this observable. Therefore the x_1, x_2, x_3 are definite scales for the Boolean ring Σ of the region of space under consideration and are already determined by the pre-theories. After x_1, x_2, x_3 are defined, the quantum mechanics of the registration process comes into play as a map $\Sigma \xrightarrow{E} G$.

The next question which we would like to ask, and is meaningful in the context of quantum mechanics is whether there exists a decision observable $\Sigma(t) \xrightarrow{E} G$ for which $\Sigma(t)$ is the Boolean ring of the "time domain" which satisfies the following reasonable conditions (by analogy with (4.2)):

$$U(1, 0, 0, \gamma)E(\sigma)U(1, 0, 0, \gamma)^{-1} = E(\sigma + \gamma). \tag{6.1}$$

Using the spectral family $E(\lambda) = E(\sigma)$ for $\sigma = \{t \mid t \leq \lambda\}$ from (6.1) we obtain the relation:

$$U(1, 0, 0, \gamma)E(\lambda)U(1, 0, 0, \gamma)^{-1} = E(\lambda + \gamma), \tag{6.2}$$

where from

$$U(1, 0, 0, \gamma) = e^{iH\gamma}$$

and

$$e^{iT\alpha} = \int e^{i\lambda\alpha}\, dE(\lambda)$$

we obtain the relation

$$e^{iH\gamma}e^{iT\alpha}e^{-iH\gamma} = e^{i\gamma\alpha}e^{iT\alpha}$$

which we can also write in the form

$$e^{-iT\alpha}e^{iH\gamma}e^{iT\alpha} = e^{i\gamma\alpha}e^{iH\gamma}.$$

If we let $\tilde{E}(\omega)$ denote the spectral family of H it follows that

$$e^{-iT\alpha}\tilde{E}(\omega)e^{iT\alpha} = \tilde{E}(\omega + \alpha).$$

If (for $\omega_2 > \omega_1$) $\tilde{E}(\omega_2) - \tilde{E}(\omega_1) \neq \mathbf{0}$ then we also find that

$$\tilde{E}(\omega_2 + \alpha) - \tilde{E}(\omega_1 + \alpha) \neq \mathbf{0}.$$

Since α may be chosen arbitrarily, it follows that the spectrum of H varies between $-\infty$ to $+\infty$ in contradiction to (2.38).

Therefore a decision observable $\Sigma(t) \xrightarrow{E} G$ which satisfies (6.2) does not exist. For elementary systems this is purely a consequence of the axioms cited in III, §3 and the conditions imposed on Galileo transformations of the registration procedures in §1. It does, however, also hold for composite systems because the Hamiltonian operators of time translation

$$U(1, 0, 0, \gamma) = e^{iH\gamma}$$

are, in all cases, bounded from below (see VIII, §5). Why, however, should an observable which satisfies (6.2) exist? Is it only because it is "desirable"—even though such an observable is not realizable?

It is necessary to go beyond quantum mechanics to a more comprehensive theory which permits an apparatus which registers the "desired" observable if we succeed in constructing an apparatus which measures a "time observable" which satisfies (6.2). Clearly a registration apparatus which contradicts the Heisenberg uncertainty relations or contradicts the assertion of the nonexistence of an observable $\Sigma(t) \xrightarrow{E} G$ which satisfies (6.2) has not yet been constructed. Therefore we may consider the nonexistence of the observable $\Sigma(t) \xrightarrow{E} G$ satisfying (6.2) as a statement about the structure of the real world.

Clearly there exist apparatuses—for example, a particle counter—which registers the time upon the detection of a particle. Is such a counter a realization of a type of time observable?

This is indeed correct. Let b_0 denote such a particle counter (including its spatial orientation with respect to the preparation apparatus), then we apparently can register whether a response signal has occurred in the time interval between t_1 to t_2. Let b_n be the registration that the counter has not responded at all; then the various $b \subset b_0 \backslash b_n$ register the time domain into which the signal has occurred. If we consider the ideal case where the length of the signal can be ignored than it is reasonable to proceed from $\mathscr{R}(b_0)$ to the following Boolean ring $\Sigma'(t)$: In $\Sigma'(t)$ there is a particular $\sigma_n \in \Sigma'(t)$ which is an atom of $\Sigma'(t)$; the set $\{\sigma \mid \sigma < \varepsilon \dot{+} \sigma_n\}$ (ε is the unit element of $\Sigma'(t)$) forms a Boolean ring (with $\varepsilon \dot{+} \sigma_n$ as the unit element) which is isomorphic to the Boolean ring of the time domain which was denoted by $\Sigma(t)$. Therefore $\mathscr{R}(b_0)$ can be considered to be an approximation of $\Sigma'(t)$, where the $\psi(b_0, b)$ represent an approximation to an observable $\Sigma'(t) \xrightarrow{F} L$. Therefore $\Sigma'(t) \xrightarrow{F} L$ is obviously a type of "time observable."

Does this observable satisfy the following relationship

$$U(1, 0, 0, \gamma)F(\sigma)U(1, 0, 0, \gamma)^{-1} = F(\sigma + \gamma) \tag{6.3}$$

which is analogous to (6.1)? For real counters the $\psi(b_0, b)$ cannot, of course, exactly satisfy (6.3) because the counter can only be turned on for a finite time, and is therefore usable only for certain registrations b which are not exactly at the beginning or the end of the "on" cycle of the counter. For this reason we should find that (6.3) is satisfied by $\psi(b_0, b) = F(\sigma)$ if γ is sufficiently small. Therefore it is conceivable to require (6.3) for the idealization $\Sigma'(t) \xrightarrow{F} L$. The following additional idealization is, according to the previous discussions, not allowed: $\Sigma'(t) \xrightarrow{F} L$ cannot be a decision observable!

Apparently this is precisely the point where errors are often made. Since we are often only familiar with observables which are decision observables it is often thought that the signal for a counter is a "yes–no" response of a decision observable, that is, that the registration "the signal occurs in the interval t, to t_2" must correspond to a projection operator (in our notation, to a $\psi(b_0, b) \in G$). This is clearly an error arising from an inadequate interpretation of the mathematical framework of quantum mechanics.

In order to show that there exist observables $\Sigma'(t) \xrightarrow{F} L$ which satisfy (6.3) we shall now give an example. We note that (6.3) is equivalent to the following equation (which is analogous to (6.2))

$$U(1, 0, 0, \gamma)F(\lambda)U(1, 0, 0, \gamma)^{-1} = F(\lambda + \gamma) \tag{6.4}$$

so that we need only to exhibit a general spectral family which satisfies (6.4).

In $\mathscr{H}_b = \mathscr{L}^2(\mathbf{R}^3, dx_1\, dx_2\, dx_3)$ we set

$$F(\lambda)\psi = a\int_{-\infty}^{\lambda} e^{iHt}e^{-r^2/\rho^2}e^{-iHt}\psi(\vec{r})\, dt, \tag{6.5}$$

where e^{-r^2/ρ^2} is the operator consisting of multiplication by e^{-r^2/ρ^2}. Since e^{-r^2/ρ^2} is a positive operator, the same is true for $e^{iHt}e^{-r^2/\rho^2}e^{-iHt}$. Therefore the operator

$$\int_{\lambda_1}^{\lambda_2} e^{iHt}e^{-r^2/\rho^2}e^{-iHt}\, dt$$

is always a positive operator. If we show that

$$\int_{-\infty}^{\infty} e^{iHt} e^{-r^2/\rho^2} e^{-iHt}\, dt \tag{6.6}$$

is a bounded operator, then a in (6.5) can be chosen such that $F(\infty) \leq \mathbf{1}$. From $F(\sigma_n) = \mathbf{1} - F(\infty)$ we obtain an example for an observable $\Sigma'(t) \xrightarrow{F} L$ which satisfies (6.3) because from (6.5) it follows that

$$\begin{aligned} e^{iH\gamma} F(\lambda) e^{-iH\gamma} &= a \int_{-\infty}^{\lambda} e^{iH(t+\gamma)} e^{-r^2/\rho^2} e^{-iH(t+\gamma)}\, dt \\ &= a \int_{-\infty}^{\lambda+\gamma} e^{iH\tau} e^{-r^2/\rho^2} e^{-iH\tau}\, d\tau = F(\lambda + \gamma). \end{aligned}$$

In order to show that (6.6) is a bounded operator we transform from $\mathscr{L}^2(\mathbf{R}^3, dx_1\, dx_2\, dx_3)$ to $\mathscr{L}^2(\mathbf{R}^3, dk_1\, dk_2\, dk_3)$ using the inverse formula to (3.40)

$$\varphi(\vec{k}) = \frac{1}{(2\pi)^{3/2}} \int e^{-i\vec{k}\cdot\vec{r}} \psi(\vec{r})\, dx_1\, dx_2\, dx_3.$$

In this space the operator (6.6) takes on the form:

$$\varphi(\vec{k}) \longrightarrow \varphi'(\vec{k}) = \alpha_1 \int_{-\infty}^{\infty} dt \int e^{(i/2m)k^2 t} e^{-(\rho^2/4)|\vec{k}-\vec{k}'|^2} e^{-(i/2m)k'^2 t} \varphi(k')\, dk_1'\, dk_2'\, dk_3', \tag{6.7}$$

where the factor α_1 is not of any further interest. The positive operator (6.6) is bounded if

$$\begin{aligned} \langle \varphi(\vec{k}), \varphi'(\vec{k}) \rangle &= \int \overline{\varphi(\vec{k})} \varphi'(\vec{k})\, dk_1\, dk_2\, dk_3 \\ &< C \int |\varphi(\vec{k})|^2\, dk_1\, dk_2\, dk_3 \end{aligned}$$

is satisfied for a particular value of C. It is sufficient to show that this condition is satisfied for continuous $\varphi(\vec{k})$ because the latter are dense in $\mathscr{H}$. For continuous $\varphi(\vec{k})$ we may obtain the integration of t in (6.7) from the Fourier transform. If we introduce polar coordinates for $\vec{k}$ and $\vec{k}'$ we obtain

$$\langle \varphi(\vec{k}), \varphi'(\vec{k}) \rangle = \alpha_2 \int \overline{\varphi(k, \theta, \varphi)} e^{-(k^2\rho^2/4)|\vec{e}-\vec{e}'|^2} \varphi'(k, \theta', \varphi') k^3\, dk\, d\omega'\, d\omega, \tag{6.8}$$

where $d\omega$, $d\omega'$ are area elements for the unit sphere (or elements of solid angle) and $\vec{e}$, $\vec{e}'$ are unit vectors in the direction θ, φ or θ', φ'.

If we introduce the expansion

$$\varphi(k, \theta, \varphi) = \sum_{l,m} \chi_{l,m}(k) Y_m^l(\theta, \varphi)$$

into (6.8) we must compute the following integrals:

$$\int Y_m^l(\theta, \varphi) e^{-(k^2\rho^2/4)|\vec{e} - \vec{e}'|^2} Y_{m'}^{l'}(\theta', \varphi')\, d\omega\, d\omega'. \tag{6.9}$$

Since the operator defined by the integral kernel

$$e^{-(k^2\rho^2/4)|\vec{e} - \vec{e}'|^2}$$

as an operator on functions on the unit sphere, commutes with rotations, the Y_m^l must be eigenfunctions of this operator, where for fixed l all Y_m^l must correspond to the same eigenvalue:

$$\int e^{-(k^2\rho^2/4)|\vec{e} - \vec{e}'|^2} Y_{m'}^{l'}(\theta', \varphi')\, d\omega' = c_l(k) Y_{m'}^{l'}(\theta, \varphi). \tag{6.10}$$

We only need to estimate $c_l(k)$. For this purpose we shall set $m = 0$ in (6.10). Then, according to (3.55) Y_0^l is real. Let θ_m denote the location of the maximum of $Y_0^l(\theta)$ and let $Y_m^l(\theta_m) = d_l$. For $\vec{e} = \vec{e}_m$ in the direction θ_m from (6.10) it follows that

$$\begin{aligned} c_l(k) d_l &= \int e^{-(k^2\rho^2/4)|\vec{e}_m - \vec{e}'|^2} Y_0^l(\theta')\, d\omega' \\ &\leq d_l \int e^{-(k^2\rho^2/4)|\vec{e}_m - \vec{e}'|^2}\, d\omega'. \end{aligned} \tag{6.11}$$

Since the last integral is independent of the direction of $\vec{e}_m$ we may choose $\vec{e}_m$ to be the direction of the polar axis and by setting $\xi = \cos\theta$ we obtain

$$\begin{aligned} \int e^{-(k^2\rho^2/4)|\vec{e}_m - \vec{e}'|^2}\, d\omega' &= 2\pi \int_{-1}^{1} e^{-(k^2\rho^2/2)(1-\xi)}\, d\xi \\ &= \frac{4\pi}{k^2\rho^2}(1 - e^{-k^2\rho^2}). \end{aligned}$$

Thus, with the above result, and (6.11) we obtain, for $c_l(k)$:

$$0 \leq c_l(k) \leq \frac{4\pi}{k^2\rho^2}(1 - e^{-k^2\rho^2}). \tag{6.12}$$

Combining (6.8) with (6.9)-(6.12) we obtain

$$\langle \varphi(\vec{k}), \varphi'(\vec{k}) \rangle \leq \alpha_3 \sum_{l,m} \int_0^\infty |\chi_{l,m}(k)|^2 \frac{1 - e^{-k^2\rho^2}}{k}\, dk. \tag{6.13}$$

From

$$\frac{1 - e^{-k^2\rho^2}}{k\rho} \leq 1$$

we finally obtain

$$\langle\varphi(\vec{k}), \varphi'(\vec{k})\rangle \leq C \sum_{l,m} \int_0^\infty |\chi_{l,m}(k)|^2 k^2\, dk$$

$$= C \int |\varphi(\vec{k})|^2\, dk_1\, dk_2\, dk_3.$$

Thus we have proven that the operator (6.6) is bounded.

For a real counter we neither have $\psi(b_0, b) \in G$ nor does (6.3) hold for all σ and γ. Therefore no experimental evidence exists against the assertion of the theory that there is no observable $\Sigma(t) \xrightarrow{E} G$ which satisfies (6.1). The resistance to this fact is analogous to the resistance to the entropy theorem—because it contradicts cherished beliefs.

The existence of the observable "position at time t" introduced in §4 represents a type of position-time measurement. In contrast with the usual situation in classical mechanics, the observables "position at time t_1" and "position at time t_2" are, for $t_1 \neq t_2$, not commensurable (in the sense of IV, D 3.1 and D 3.2). This fact follows simply from the fact that $Q_\nu(t_1)$ and $Q_\nu(t_2)$ do not commute for $t_1 \neq t_2$ (see IV, Th. 3.2).

From (4.16), (4.21), and (2.41) it follows that

$$Q_\nu(t_2) = Q_\nu(t_1) + \frac{t_2 - t_1}{m} P_\nu \tag{6.14}$$

and therefore follows that

$$Q_\nu(t_2)Q_\nu(t_1) - Q_\nu(t_1)Q_\nu(t_2) = \frac{(t_2 - t_1)}{m}[P_\nu Q_\nu(t_1) - Q_\nu(t_1)P_\nu] = \frac{(t_2 - t_1)}{im}\mathbf{1}. \tag{6.15}$$

Therefore, there is no possible way to jointly measure the positions of a microsystem at two different times! This does not, however, mean that a microsystem cannot exist having the two pseudoproperties, "position in σ_1 at time t_1" and "position in σ_2 at time t_2" (IV, §8.3), for example, x was prepared having a position in σ_1 at time t_1 and was registered as having a position in σ_2 at time t_2.

7 Spatial Reflections (Parity Transformations)

Up to now we have considered the Galileo group as the group which is continuously connected to the unit element. This group can be given a well-defined meaning in terms of transformations of the registration apparatus. In practical applications, however, discontinuous transformations—such as the space reflection $r\colon x'_\nu = -x_\nu$—play an important role. If we expand the

Galileo group $\mathscr{G}$ by admitting transformations A for which the determinant $|A| = -1$, we obtain a new group $\mathscr{G}^{(r)}$ which can be decomposed into two disconnected components—$\mathscr{G}$ as an invariant subgroup of $\mathscr{G}^{(r)}$ and the coset $r\mathscr{G}$ where r is the spatial reflection transformation. If we choose, we may replace the reflection r by a transformation in which only one of the coordinates is reversed, that is, by

$$r_1\colon \begin{aligned} x_1' &= -x_1, \\ x_2' &= x_2, \\ x_3' &= x_3. \end{aligned} \tag{7.1}$$

We then obtain $r\mathscr{G} = r_1\mathscr{G}$.

We may obtain a physical interpretation of the entire group $\mathscr{G}^{(r)}$ if we give meaning to one of the transformations r or r_1. Here we are confronted by a problem which has often been neglected. While it is clear what it means to rotate or translate an apparatus, it is not clear what it means to subject an apparatus to a reflection r. The transformation (7.1) does not establish what should be done with an entire apparatus.

We can visualize (7.1) as reflecting the apparatus in a mirror located at the (2, 3) plane. Here we would see that mirror image of the apparatus. The mirror image is, however, not an actual apparatus. The fact that the production of an apparatus which is the mirror image of the original apparatus is not trivial can be seen from the example of a person—it may well be impossible.

However, the mirror image only establishes the spatial organization of the components of the apparatus—by means of the transformation (7.1). However, how do such things as electric charges, electric and magnetic fields, etc., change in the apparatus? The transformation (7.1) therefore does not establish how we should determine the corresponding transformations of registration apparatuses. The arbitrariness of the application of (7.1) to an apparatus is not sufficiently noted, and plays an important role for the so-called "elementary particle physics."

In nonrelativistic quantum mechanics the action of the reflection r (it is mathematically simpler to deal with r rather than with r_1) is defined as follows: for the apparatus b_0 a new apparatus is built in which the spatial organization of the components and the spatial placement are changed in the sense of the transformation r without changing the charges present. In spite of the objections implicit in the example of a person described above, we assume that we may build such a "reflected" apparatus. Axiomatically we require that for r there exists a p-continuous r-automorphism such that, together with the interpretation of elements of $\mathscr{G}$ in §1 we obtain a representation of $\mathscr{G}^{(r)}$ by means of p-continuous r-automorphisms.

From the representation of $\mathscr{G}^{(r)}$ by means of p-continuous r-automorphisms there arises a representation by means of $\mathscr{B}$-continuous effect automorphisms. Since $\mathscr{G}$ is not connected with $r\mathscr{G}$, according to VI, §3.2 we cannot exclude the possibility that the elements of $r\mathscr{G}$ transform one system type into another. Only experience will lead us to impose the requirement that the $\mathscr{B}$-continuous effect automorphism corresponding to r transforms an F of the form $(0, 0, \ldots, F_\nu, 0, \ldots)$ into an F of the same form $(0, 0, \ldots, F'_\nu, 0, \ldots)$. This is equivalent to the condition that the effect automorphism corresponding to r leave the "objective properties" (IV, §8.1) invariant. Thus, according to VI, §3.2 and §3.3 we may again restrict ourselves to a Hilbert space of a single elementary system type and consider the representation of $\mathscr{G}^{(r)}$ in $\mathscr{H}$ by means of unitary or anti-unitary operators up to a factor. According to IV, §3.2 we must be able to represent $\mathscr{G}$ by means of unitary transformations, that is, we may assume the previous results about the representation of $\mathscr{G}$; in principle the elements of $r\mathscr{G}$ may also be represented by means of anit-unitary operators, in particular, the same is true of r itself. Let $U(r)$ denote the operator representing r.

From $(A, \vec{\delta}, \vec{\eta}, \gamma)r = r(A, -\vec{\delta}, -\vec{\eta}, \gamma)$ it follows that

$$U(A, \vec{\delta}, \vec{\eta}, \gamma)U(r)U(A, -\vec{\delta}, -\vec{\eta}, \gamma)^{-1} = \lambda(A, \vec{\delta}, \vec{\eta}, \gamma)U(r). \tag{7.2}$$

For $\varphi(\vec{k}) \in \mathscr{H}_b = \mathscr{L}^2(\mathbf{R}^3, dk_1\, dk_2\, dk_3)$ we define a unitary operator in $\mathscr{H} = \mathscr{H}_b \times \imath$ by $R \times \mathbf{1}$, where

$$R\varphi(\vec{k}) = \varphi(-\vec{k}).$$

It follows then, by simple computation, that

$$U(A, \vec{\delta}, \vec{\eta}, \gamma)(R \times \mathbf{1}) = (R \times \mathbf{1})U(A, -\vec{\delta}, -\vec{\eta}, \gamma).$$

If we multiply (7.2) on the right by $R \times \mathbf{1}$ we obtain:

$$U(A, \vec{\delta}, \vec{\eta}, \gamma)U(r)(R \times \mathbf{1})U(A, \vec{\delta}, \vec{\eta}, \gamma)^{-1} = \lambda(A, \vec{\delta}, \vec{\eta}, \gamma)U(r)(R \times \mathbf{1}). \tag{7.3}$$

Thus it follows that the set of $\lambda(A, \vec{\delta}, \vec{\eta}, \gamma)$ form a one-dimensional unitary representation of the Galileo group and therefore must be equal to 1. Thus (7.3) states that $U(r)(R \times \mathbf{1})$ commutes with all $U(A, \vec{\delta}, \vec{\eta}, \gamma)$. It follows that for an elementary system $U(r)(R \times \mathbf{1})$ cannot be anti-unitary; therefore $U(r)$ is not anti-unitary; therefore $U(r)(R \times \mathbf{1})$ must be a multiple of the unit operator. Since a factor in $U(r)$ is arbitrary, we may therefore choose this factor such that

$$U(r) = R \times \mathbf{1}. \tag{7.4}$$

Here we stress the fact that (7.4) means that in the spin space $U(r)$ behaves like the unit operator.

We have given a physical meaning to the reflection as a transformation of registration procedures. In the applications of quantum mechanics we will use additional unitary symmetry transformations. Not all of these can be physically interpreted. This is true, for example, for many of the permutations used in VIII, §4 and XII–XV. The application of such symmetry transformations is legitimate if it illuminates the mathematical structure for a problem.

In addition to the spatial reflection r we shall often consider another type of reflection transformation—"motion reversal." We will introduce this transformation and its physical meaning in X, §4.

8 The Problem of the Space $\mathscr{D}$ for Elementary Systems

In this section we shall consider the problems of the space $\mathscr{D}$ which was previously discussed in VI only for the case of elementary systems. For composite systems we shall present a brief discussion in VIII, §7.

A single type of elementary systems is described by a single Hilbert space $\mathscr{H}$ and an irreducible representation of the Galileo group characterized by mass m and spin s.

The following attempt to introduce the space $\mathscr{D}$ is particularly fascinating for physics: For the Galileo group we shall define the uniform structure which characterizes physical imprecision and is denoted by ph in VI, §1.1. It is physically reasonable to do this in the following manner: In both the three-dimensional spaces $\vec{\eta}$ and $\vec{\delta}$ we introduce a uniform structure in an analogous way. We shall now write the formula only for $\vec{\eta}$.

The equations $\vec{e} = \vec{\eta}/|\vec{\eta}|$ and $\rho = \arctan(|\vec{\eta}|)$ define a bijective map $\vec{\eta} \leftrightarrow \rho\vec{e}$ of the infinite three-dimensional space onto the finite three-dimensional space of points within a sphere of radius $\pi/2$. We define the uniform structure ph in the space of $\vec{\eta}$ as the Euclidean uniform structure in this sphere. It is easy to verify that the uniform structure ph in the space of $\vec{\eta}$ also generates the Euclidean topology. It is easy to compactify the space of $\vec{\eta}$ with respect to ph—we need only add the surface to the interior of the sphere of radius $\pi/2$ to the space of $\rho\vec{e}$.

We proceed in the same way for $\vec{\delta}$ and for the time translation γ. Thus we obtain a uniform structure ph in $\mathscr{G}$. This expresses the fact that for large $|\vec{\eta}|$ that the physical distinguishability for a pair of group elements $(1, 0, \vec{\eta}_1, 0)$ and $(1, 0, \vec{\eta}_2, 0)$ is good only with respect to the direction $\vec{e}$; for increasing $|\vec{\eta}|$ physical distinguishability deteriorates. In this way we eliminate the idealized "infinity" from the transformations. A similar situation is found for the case of the transformations $(1, \vec{\delta}, 0, 0)$ and for the time translations $(1, 0, 0, \gamma)$.

For very rapid motion of the registration apparatus relative to the laboratory reference system, the more certain is the direction of motion, but the absolute magnitude of the velocity of motion becomes less certain. In a similar way the magnitude of the displacement becomes less certain for large time translations.

Therefore it is physically reasonable that the real registration procedures are such that the probabilities under displacement of the apparatus for large $\vec{\eta}$ do not depend strongly upon $|\vec{\eta}|$, similarly for large $\vec{\delta}$ and for large γ. We may therefore assert that $\mathscr{D}$ is identical to the space Δ of VI, D 1.2.4. To this end we need to prove that the space Δ is norm-separable, as we assumed. We shall not prove this here; see [19] for the proof.

For elementary systems the space Δ has not yet been sufficiently analyzed. In addition Δ' (the dual Banach space to Δ, in which $\mathscr{B}$ can be embedded) and the closure $\bar{K}^\sigma$ of K in Δ' in the $\sigma(\Delta', \Delta)$-topology is not well known (see [19]). Perhaps in the structure of $\partial_e\bar{K}^\sigma$ there is a path to a new mathematical method for the treatment of quantum mechanical problems.

Since Δ is norm-separable there are good physical reasons for the assertion that $\mathscr{D} = \Delta$. Since the structure of Δ is not known in detail, the reader should be aware of additional possibilities for the definition of $\mathscr{D}$.

For the above reason, we shall now proceed in the opposite direction (as shown in VI, §1.2): Using the subset Λ we choose $\mathscr{D} = \mathscr{D}_\Lambda$ and we choose the Λ-uniform structure (VI, D 1.27) in $\mathscr{G}$ as the *ph*-uniform structure. As a result we lose the physical intuition for *ph*, but we do realize the possibility to freely choose Λ within certain limitations.

A first but somewhat radical choice for Λ consists of the selection of $\Lambda = \mathscr{D}_\Lambda$ and we construct $\mathscr{D}$ from the position, momentum, and angular momentum observables. Consider the set of continuous functions χ in $\mathbf{R}^3$ with compact support. Then we construct the norm-closure algebra generated by the $\chi(\vec{P})$, $\chi(\vec{Q})$ and all the $U(A, 0, 0, 0)$. This algebra is norm-separable by construction. For $\mathscr{D}_\Lambda$ we choose the subspace of all self-adjoint operators from this algebra. Since the representation of the Galileo group is irreducible, $\mathscr{D}_\Lambda$ separates the elements of K and is therefore $\sigma(\mathscr{B}', \mathscr{B})$-dense in $\mathscr{B}'$.

It is easy to see that the Λ-uniform structure on $\mathscr{G}$ generates the same topology as does the original topology on $\mathscr{G}$. Thus we need only construct a Λ-neighborhood of the unit element of $\mathscr{G}$ which is compact in the original topology, that is, a subset in $\mathscr{G}$ which is bounded in $\vec{\eta}$, $\vec{\delta}$, γ. Thus, together with U as the representative transformation we need only consider

$$\operatorname{tr}(U^+ P_\varphi U \chi(\vec{P})), \operatorname{tr}(U^+ P_\psi U \chi(\vec{Q}))$$

for a φ for which $\langle \vec{p} \mid \varphi \rangle$ is concentrated in momentum space, a ψ for which $\langle \vec{r} \mid \psi \rangle$ is concentrated in position space,[1] and a χ which is concentrated in $\mathbf{R}^3$.

By means of the algebraic construction we have obtained the possibility for the selection of $\mathscr{D}$; however, we note that because of its algebraic nature, it does not have a clear intuitive physical interpretation. An essential aspect of the description of quantum mechanics presented in this book is that algebraic operations such as products of operators in Hilbert space do not have a clearly evident physical interpretation.

We note, however, that even for this choice of $\mathscr{D}$ the structure of $\mathscr{D}'$, $\bar{K}^\sigma$, $\partial_e\bar{K}^\sigma$ are not well understood (see [19]).

The following approach would be more satisfying. Consider the previously described set of continuous functions χ in $\mathbf{R}^3$ satisfying $0 \le \chi \le 1$ and having compact support. Then it is plausible that if we consider $\vec{Q}(0)$ to be the

1. For $\langle \vec{p} \mid \varphi \rangle$, $\langle \vec{r} \mid \psi \rangle$ see IX, §5 and §6.

"position operator at time $t = 0$" then we are able to register the effect $\chi(\vec{Q}(0))$. Then, by time displacement, we are able to measure the effects

$$e^{iHt}\chi(\vec{Q}(0))e^{-iHt} = \chi(\vec{Q}(t)) = \chi\left(\vec{Q}(0) + \frac{t}{m}\vec{P}\right).$$

Let us choose Λ to be the set of these effects. Clearly $\mathcal{G}\Lambda \subset \mathcal{D}_\Lambda$. Λ is norm-separable. The investigations in [23] indirectly show that Λ is $\sigma(\mathcal{B}', \mathcal{B})$-dense in $L(\mathcal{H})$; we may therefore choose $\mathcal{D} = \mathcal{D}_\Lambda$ (see VI, §1.2). We may therefore choose the Λ-uniform structure as the physical uniform structure on $\mathcal{G}$. It remains to show whether it generates the same topology on $\mathcal{G}$ as does the original topology.

Here we have pointed out several problems of a fundamental nature because we must describe the real measurement possibilities (that is, the actual ability to distinguish between different ensembles). Here we have not successfully obtained a unique "solution of these physical questions" by means of the formulation of axioms concerning $\mathcal{D}$ (see also the discussion of this problem in [19]). It is also important to state the open questions in order to give a better estimate of the current state of the theory.

9 The Problem of Differentiability

Differentiable functions, differentiable manifolds, etc. appear in many mathematical formulations $\mathcal{MT}$ of physical theories. Is differentiability an essential component of physics—that is, does differentiability represent an aspect of the structure of reality—or is it an artifact and a convenient mathematical idealization?

There has been much philosophical discussion about this question. If a physical theory is not based upon an axiomatic basis (see I, §1 or [1], §7.3) then there is little that can be said about this problem. Since we used an axiomatic basis for the development of quantum mechanics presented here, we are able to apply the methods described in [1], §10. However, the question whether differentiability is not merely a mathematical idealization still remains.

Without using the methods of [1], §10 we may explicitly determine how differentiability arises in the mathematical formulation $\mathcal{MT}$ of quantum mechanics. Obviously the structure introduced in II–VII, §1 does not have any axioms about differentiability. Note that the Galileo group is initially introduced as a topological group, where the group structure and the topology (more precisely, the uniform structure *ph*) reflect certain aspects of reality.

By selecting a particular parameterization of the Galileo group we obtain a differentiable manifold: $\mathcal{G}$ then becomes a Lie group. Pontrjagin [24] has shown that, for compact and finite-dimensional groups there always exist such parameters by which the group $\mathcal{G}$ can be made into a Lie group. This is also the case for many locally compact groups, in particular, for the Galileo group [25].

For these groups the structure "differentiable manifold" may be derived from the group structure and the topology. Therefore it is not necessary to introduce additional axioms. This fact is important for the following two reasons:

First, the structure "differentiable manifolds" represents a structure (clearly, in idealized form) which represents certain aspects of reality. Second, this structure represents nothing which is not already present in the topology (more accurately in the uniform structure *ph* of physical imprecision) together with the group structure.

Nevertheless it is, of course, correct that the differentiability structure represents an idealization about reality, the basis of which is nothing other than the idealization of a topological group. Let us consider the case of a translation group. For a given translation a it is always possible to find another translation b, the square of which generates the given translation, that is, $b^2 = a$. However, is this true for smaller and smaller translations? Consider, for example, translations of an apparatus in the laboratory, as we have done in §1. Obviously we cannot give an answer to this question at the present time. We can express our lack of knowledge in mathematical terms by the following idealization: There exist arbitrarily small translations; but in order to proceed away from idealizations it is necessary to introduce "uncertainty sets" in the neighborhood of the unit element of the group, that is, make $\mathscr{G}$ into a topological group (see VI, §1). The idealization of group elements which are "arbitrarily close" to the unit element leads to the differentiability structure for the Galileo group.

The above assertion that the differentiable structure represents "something" of the structure of reality may yet be made more precise as follows: It expresses in idealized form the fact that the group structure describes what happens in the neighborhood of the unit element idealized in the form of an infinitesimal transformation, that is, in terms of the Lie algebra which corresponds to the structure of the group. The well-known mathematical result that the Lie algebra determines the group locally is only a mathematical expression of this fact.

If we have recognized the physical meaning of the mathematical structure, we could hardly then quarrel about whether it is physically correct to use Hilbert space for the description of quantum mechanics, or whether it is more correct to use subspaces in which all finite products of the operators $\vec{K}$, $\vec{X}$, $\mathscr{I}_v$, H are defined. The answer to such a dispute is very simple: It is not a physical problem but a question about the method of computation, that is, concerning which mathematical methods are best suited for the solution of physical problems. Methods can be judged only by their usefulness. Here the use of Hilbert space is already a mathematical mode of description which permits us to avoid the structure of the "original" set K of ensembles, of the set L of effects, and of the form $\mu(w, g)$ for the probability function. Thus it is permissible to introduce new methods which facilitate the solution of practical problems such as those which will be described in IX. Such investigations have been undertaken extensively using the Gel'fand space triple as a tool. Here we shall refer readers to the literature [26] because we cannot describe all possible more or less practical methods especially when they do not result in any new physical structure. Unfortunately it is not always easy to determine from the literature whether we are dealing only with practical methods or with new physical structures.

A problem of physical meaning can be directly related to the differentiability problem, namely in the area of the problem of the space $\mathscr{D}$ described in §8.

In $\mathscr{B}(\mathscr{H})$ as well as in $\mathscr{B}'(\mathscr{H})$ we may accentuate the "differentiable" elements by asserting that $W \in K$ is "differentiable" if

$$i(W\vec{K} - \vec{K}W), \qquad i(W\vec{X} - \vec{X}W), \qquad i(W\vec{L} - \vec{L}W), \qquad i(WH - HW)$$

are elements of $\mathscr{B}(\mathscr{H})$; similarly, $F \in L$ is "differentiable" if

$$i(F\vec{K} - \vec{K}F), \qquad i(F\vec{X} - \vec{X}F), \qquad i(F\vec{L} - \vec{L}F), \qquad i(FH - HF)$$

are elements of $\mathscr{B}'(\mathscr{H})$.

While the subset of differentiable ensembles has only a practical meaning and does not have physical meaning, the set of differentiable effects can be given a physical meaning insofar as the space $\mathscr{D}$ contains this set or not or whether the norm-closed subsets of $\mathscr{B}'$ spanned by this set is the space $\mathscr{D}$.

CHAPTER VIII

Composite Systems

The real great achievement of quantum mechanics is not its successful treatment of elementary systems, the basis for which was presented in VII, but its successful description of composite systems. According to VI, D 2.1 the representation of the Galileo group in Hilbert space for composite systems is reducible. In addition, there exist decision effects $E \subset G(\mathscr{H})$ which are different from $\mathbf{0}$ and $\mathbf{1}$ which are left invariant under transformations of the Galileo group.

1 Registrations and Effects of the Inner Structure

We shall first consider the structure of those effects which are left invariant under the Galileo group as a whole, or are left invariant under subgroups of the Galileo group. We have already become familiar with some of the general properties of the set of these effects in VI, §2. In the case of the Galileo group and its subgroups we can yet say something more about the structure of these invariant effects. For this purpose we shall now consider some of the results from VII, §2. It is easy to verify that in the derivations up to VII (2.23) no use has been made of the fact that the representation is irreducible. The decisive next step in VII, §2 was that the Hilbert space $\mathscr{H}$ can be represented in the form (2.24) where the operators $\vec{K}$ and $\vec{X}$ obey the operator rules (2.25). (We have not proven these results in VII, §2; an indirect elementary proof will be provided in the discussion in IX, §5. For a proof which uses the group representation or the algebra generated by $e^{i\vec{K}\cdot\vec{\eta}}$ and $e^{i\vec{X}\cdot\vec{\delta}}$ see, for example, [10] and [28].

Thus, for composite systems we may use the form

$$\mathscr{H} = \mathscr{H}_b \times \mathscr{H}_i \tag{1.1}$$

of Hilbert space from VII (2.24) and use the operator rules for the operators $\vec{P} = -\vec{K}$ and $\vec{Q} = (1/M)\vec{X}$ in $\mathscr{H}_b$ from VII, §2 together with the commutation relation VII (2.17), where we replace m by M. For the definition of $\vec{Q}$ we have already made use of the requirement that, for composite systems, the parameter M which occurs in VII (2.17) is also nonzero, a result which is not contradicted by any experiment in the fundamental domain of quantum mechanics (that is, in atomic and molecular physics).

The observable $\vec{P}$ will be called the "total momentum" and the observable $\vec{Q}$ will be called the "position of the center of mass" for the composite system. These observables $\vec{P}$ and $\vec{Q}$ are, however, not uniquely determined by the conditions set down in VII, §4 (of course, $\vec{P}$ and $\vec{Q}$ satisfy all those requirements).

According to VII (2.28) the Hamiltonian operator H is given by

$$H = \frac{1}{2M}\vec{P}^2 \times \mathbf{1} + \mathbf{1} \times H_i. \tag{1.2}$$

The term $(1/2M)\vec{P}^2$ is called the "kinetic energy" observable of the system; H_i is called the Hamiltonian operator of the "inner structure" or the observable of "rest energy," that is, the energy of the microsystems if they were prepared in such a way that the kinetic energy is (approximately) zero.

According to VII (2.35), for rotations we obtain

$$U(A, 0, 0, 0) = V(A) \times R(A), \tag{1.3}$$

where, according to VII (2.31), for $\mathscr{H}_b = \mathscr{L}^2(\mathbf{R}^3, dk_1\, dk_2\, dk_3)$ we obtain

$$V(A)\varphi(\vec{k}) = \varphi(A^{-1}\vec{k}) \tag{1.4}$$

or, for $\mathscr{H}_b = \mathscr{L}^2(\mathbf{R}^3, dx_1\, dx_2\, dx_3)$ from VII (3.41) we obtain

$$V(A)\psi(\vec{r}) = \psi(A^{-1}\vec{r}). \tag{1.5}$$

The $R(A)$ generate a representation of $\mathscr{D}_{\mathscr{G}}^*$ in $\mathscr{H}_i$.

For composite systems it is no longer the case that these representations in $\mathscr{H}_i$ are irreducible. Thus we find that VII (2.37) no longer holds. Instead, we only have

$$R(A)H_i = H_iR(A) \tag{1.6}$$

which describes the rotation invariance of H_i.

Thus we have already identified the structures which can be deduced from the structure of the Galileo group (considered as a transformation group of registration procedures) which was introduced in VII, §1.

We will now introduce new terminology, describe its usage, and, in addition, obtain some consequences from (1.6).

An effect is invariant under translations $(1, 0, \vec{\eta}, 0)$ and proper Galileo transformations $(1, \vec{\delta}, 0, 0)$ if and only if it is of the form $\mathbf{1} \times F$. Such effects will be called "inner structure effects." F need not commute with either the $R(A)$ or with H_i. This effect does not depend on the location or the velocity of the registration apparatus corresponding to F but may depend on the orientation in space or on the time the apparatus is switched on.

It is easy to see that the position and momentum observables will be uniquely determined if the requirements for elementary systems are supplemented by the following additional requirement: position and momentum observables are coexistent with all inner structure effects.

In the sense of the terminology described above elementary systems also have—according to VII, §2—an "inner structure"—one which is very simple: spin, where the latter is described by the irreducible representation of $\mathscr{D}_{\mathscr{G}}^*$ in $\imath_s$; for H_i then VI (2.37) holds where λ is a constant having no particular physical significance.

By analogy with the expression "inner structure effect" we shall call an observable $\Sigma \xrightarrow{F} L$ an "inner structure observable" if the elements of the range of the measure is of the form $\mathbf{1} \times F$. An "inner structure scale observable" (according to IV, D 2.4.5 a scale observable is always a decision observable) is uniquely defined by a self-adjoint operator of the form $\mathbf{1} \times B$. For the angular momentum operator $\vec{J}$ (for the definition see VII, §5) which corresponds to the infinitesimal rotation $R(A)$ the quantity $\mathbf{1} \times \vec{J}$ is an inner structure observable. Since the representation of $\mathscr{D}_{\mathscr{G}}^*$ can be completely reduced in $\mathscr{H}_i$ we may write $\mathscr{H}_i$ in the following way as a direct sum as follows:

$$\mathscr{H}_i = \sum_j \oplus \mathscr{H}^j, \tag{1.7}$$

where $\mathscr{H}^j$ is the eigenspace of the operator $\vec{J}^2$ with eigenvalues $j(j+1)$. Since the $R(A)$ form a representation up to a factor of $\mathscr{D}_{\mathscr{G}}$ (see VII, §2) the number j in the sum must be either an integer or half integer.

The operators $\mathbf{1} \times F$ are all effects which are invariant under the action of the subgroup $(1, \vec{\delta}, \vec{\eta}, 0)$. The effects which are left invariant under the action of the subgroup $(A, \vec{\delta}, \vec{\eta}, 0)$ are therefore of the form $\mathbf{1} \times F$, where F leaves each of the subspaces $\mathscr{H}^j$ invariant. Let F^j denote the part of F which acts in $\mathscr{H}^j$.

Since $\mathscr{H}^j$ can itself be completely reduced with respect to irreducible representations, we may introduce a complete orthonormal basis $u_m^{(\nu)}$ where $m = -j, \ldots, j+1$ where the $u_m^{(\nu)}$ span an irreducible subspace $\imath^{(\nu)}$ and the $u_m^{(\nu)}$ transform as described in VII, §3. That means nothing other than that $\mathscr{H}^j$ can be written in the form

$$\mathscr{H}^j = \hbar^j \times \hbar^j, \tag{1.8}$$

where the restriction of the operators $R(A)$ on $\mathscr{H}^j$ have the form $R^j(A) \times \mathbf{1}$ with respect to (1.8) (see the general questions related to (1.7) and (1.8) concerning operators which commute with a completely reducible representation in AIV, §14). Since the F^j commute with all the $R^j(A) \times \mathbf{1}$, they must have the form $\mathbf{1} \times F^j$ with respect to (1.8). Thus all the effects (and therefore all scale observables) are known which are invariant with respect to the subgroup $\mathscr{G}_a$ of all $(A, \vec{\delta}, \vec{\eta}, 0)$, that is, the corresponding registration apparatuses may be arbitrarily translated, given velocities and arbitrary orientations in space. They are the set of all effects of the form $\mathbf{1} \times F$ (with respect to (1.1)) where F (with respect to (1.7)) transforms each subspace $\mathscr{H}^j$ into itself and in $\mathscr{H}^j$ takes on the form $\mathbf{1} \times F^j$ (with respect to (1.8)).

This is particularly true for the case of the rest energy observable H_i (see (1.6)): To each angular momentum eigenvalue j there is a part H_i^j of the operator H_i for rest energy where H_i^j operates in $\mathscr{H}^j$. For an effect which is invariant under the entire Galileo group the corresponding F^j must also commute with the H_i^j.

Without violating any of the previously introduced structures in the theory we may therefore begin with a series of arbitrary Hilbert spaces $\mathscr{k}_i^j$, define arbitrary self-adjoint operators H_i^j, and then construct the $\mathscr{H}^j$ using the spaces $\mathscr{k}^j$ constructed in VII, §3 for an irreducible representation D_j of $\mathscr{D}_{\mathscr{G}}^*$ according to (1.8). From the $\mathscr{H}^j$ (here we observe that only whole or half integer values of j may occur) we may construct $\mathscr{H}_i$ according to (1.7) and finally construct $\mathscr{H}$ according to (1.1).

In this manner we find that the previously introduced structures yield a theory which does not contradict experience, but does contain too much "arbitrariness" to be useful in clarifying the structure of atoms and molecules. In the sense of [1], §10.3 or of [2], III, §7–§9 the preceding theory is not g.G.-closed: not everything which is possible in the theory really occurs. This represents a challenge to strengthen the theory by means of additional structures and axioms (that is, in the sense of [1], §8 or [2], III, §7) in order to proceed to a standard extension ("Standarderweiterung" in [1], §8)

2 Composite Systems Consisting of Two Different Elementary Systems

In VII, §2 we have found that an elementary system can be characterized by the mass m, the spin s, and perhaps (in case there is more than one elementary system type having the same mass and same spin) other discrete quantities. In nonrelativistic quantum mechanics it is possible to characterize all elementary systems by the mass m, spin s, and electric charge; we shall discuss such a description below. In most cases the charge is uniquely determined by the mass and the spin. At present there is no satisfactory physical theory which accurately predicts the masses of elementary particles. In quantum mechanics the masses of the elementary systems are "determined by experiment."

What do we mean by "experimentally determine" in the context of a physical theory?

It means that we take the (finitely many!) results of experiments and after expressing them in mathematical form "adding" them (possibly as axioms) to the mathematical framework of the theory. A more precise description of this process is given in [1], §5 or in [2], III, §4. From these experimental results we then seek to deduce the value of the mass (as lying within a certain interval of the real line—the interval describing the so-called experimental errors). Since an analogous situation exists in the case of classical mechanics, there is some discussion of this topic in [1], §10.5. It is not necessary to present a detailed discussion in order to establish the fact that few parameters of a theory can be determined by proceeding "backwards" from experiment—such is the case for the mass m and the spin s for elementary systems. Certainly it is desirable to have a theory which theoretically determines these parameters, so that we may be able to determine whether the theory is or is not contradicted by experiment. If we then would have such a theory then we would have established a more comprehensive description of the structure of the real world. If such a theory is not at hand then we must be satisfied in experimentally determining which elementary systems have which values of mass and spin. Here we shall only consider a restricted but yet vastly large area of knowledge—that of atoms and molecules. Here it suffices to introduce, as elementary systems:

(1) The so-called electrons, $m = 4.13 \times 10^9\ \text{cm}^{-1}$, $s = \frac{1}{2}$ and charge e where e is the negative of the so-called elementary unit of charge; in the units used here e is dimensionless and has magnitude $\sim \sqrt{(1/137)}$.
(2) The different atomic nuclei with specific mass M (which is more than 1800 times that of the electron mass), is positively charged with charge given by Ze where Z is an integer—the so-called charge number of the nucleus. The spin of the different nuclei can be experimentally determined. There exist tables of nuclei labeled with the corresponding values of M, Z, and s.

Certainly there exists a theory of atomic nuclei which permits the values of M, Z, and s to be correctly assigned and which also permits the description of the structure of nuclei. The theory of atomic nuclei permits the use of the formulation of quantum mechanics presented here in which we consider protons and neutrons as elementary systems. However, the problem of constructing the Hamiltonian operator for the composite nucleus is much more difficult than is the case for atoms and molecules; the case for the latter may be handled by means of postulating axioms (see (5.8)). For this reason we shall not be concerned with the application of quantum mechanics for problems in nuclear physics. Since we seek to develop a more illuminating discussion of the fundamentals of a physical theory, we shall confine our interest to the fundamental domain of atomic and molecular physics, where we may treat the atomic nuclei as elementary systems. On the other hand it is important to note that physicists generally seek to extend a successful theory to new areas, that is, to break new ground. Such efforts are attempts to discover new structure laws obtained from using known theories combined with the aid of intuitive guesswork. We shall not discuss such attempts in this book; the interested reader is referred to the presentation in certain sections in [2].

For the desired application domain we shall assume that the elementary systems in the theory consist of electrons and nuclei, where the different types of nuclei are listed in tabular form. The electrons are completely (with respect to the above domain of application) characterized by their mass. The nuclei are uniquely characterized by their mass and charge. Electrons have spin $\frac{1}{2}$; the nuclear spin will not play a role in the applications presented in this book, since we shall consider those experiments in which the nuclear spin does not play a role, or its effect is very small. It is, however, not difficult to extend the theory developed here to take into account the effect of the spin of the nucleus (hyperfine structure of spectral lines).

We shall also identify the experiments which prove that the electron spin is $\frac{1}{2}$ in the development of applications (XI, §9 and §11).

We have presented the above overview of elementary systems in order to formulate additional structure axioms for composite systems. If the reader has studied physics for a few semesters he will encounter the well-known "fact" that all atoms consist of "atomic nuclei and electrons." What, however, should such a statement mean? It is a problem of the theory to subject such statements to scrutiny and to formulate them more precisely—by this we mean that axioms should be introduced in the mathematical framework which provide meaning for the short-hand statement that, for example, an atom consists of a nucleus and a number of electrons.

In order to clarify the fundamentals we shall begin by considering a composite system consisting of two elementary systems. Let $\mathscr{H}_1$ and $\mathscr{H}_2$ denote the Hilbert spaces for the elementary systems of different types (1) and (2). Let $\mathscr{H}_3$ denote the Hilbert space of the composite system. Previously there existed no relationship between the different Hilbert spaces $\mathscr{H}_\nu$ in $(\mathscr{H}_1, \mathscr{H}_2, \ldots)$ which were introduced in AQ (III, §5) or by the theorem in III, §3. We now wish to impose additional requirements which will establish connections between these $\mathscr{H}_\nu$. We shall therefore begin by examining the relationships between $\mathscr{H}_1$, $\mathscr{H}_2$ (as Hilbert spaces of two elementry systems), and $\mathscr{H}_3$ (as the Hilbert space of the composite system).

Each effect $F = (F_1, F_2, F_3, F_4, \ldots)$ has three components F_1, F_2, F_3 which refer to these three system types; similarly each ensemble $W = (W_1, W_2, W_3, W_4, \ldots)$ has three corresponding components W_1, W_2, W_3. We may consider W_1 as an element of $\mathscr{B}(\mathscr{H}_1)$, F_1 of $\mathscr{B}'(\mathscr{H}_1)$, and similarly for (2) and (3).

From experience we have found that it is not difficult to carry out experiments with pairs of systems, one each of types (1) and (2). Every scattering experiment of a system (1) onto a system (2) is of this form, and that pairs, that is, a pair of (1) and (2) must be prepared as a new system (3). We will consider the experimental situation of the scattering process in some detail in XVI but only after we have explained what we mean by a composite system (3) of the pair (1) and (2). For a mathematical basis it is certainly desirable to only use such structures which are related to the experimental situation (as is the case for scattering experiments) in order to derive other structures which describe the situation: (3) is composed of (1) and (2) from the original structure. However, this route has not as yet led to success.

We shall therefore proceed to extrapolate from a few experimentally motivated structures without being certain whether we will eventually encounter contradictions with experience.

If we prepare pairs of systems (1) and (2) where, for example, the partners (1) of each pair is on the moon and the other (2) is on the earth, then it would appear to be possible to register the effects of system (1) independently of registration of system (2). For certain preparations it also seems to make sense that there are effects F_3 for the pairs which correspond to one of the effects caused by the partner (1).

Physicists often prefer to paraphrase the expression "for a specified preparation" as follows: For systems "without interaction" certain effects F of a "pair" are singled out as those which are caused by the "partner" (1). Before we proceed to extrapolate this "structure" onto systems with interaction we shall first seek to mathematically describe this intuitive structure.

The obvious method is to introduce the following new structure $L_1 \xrightarrow{T_1} L_3$ with the following physical interpretation: $F_3 = T_1(F_1)$ is precisely the effect of the pairs (3) which are actuated by (1) and correspond to the effect F_1 of the elementary system (1). We shall also introduce a corresponding map $L_2 \xrightarrow{T_2} L_3$. We shall now introduce the following axioms about T_1 and T_2 in order to mathematically describe the experimental situation "pairs without interaction" on which we have earlier provided an intuitive description.

(α) T_1 and T_2 are $\mathscr{B}$-continuous injective effect morphisms. If the systems (1) and (2) are prepared widely separated then we may obviously measure "together" both the effects caused by (1) and (2). We therefore require that:

(β) Each effect in T_1L_1 is coexistent with each effect in T_2L_2.

Furthermore, it is reasonable to require that, for decision effects E_1 for (1), we cannot expect that there exist an $F_3 \in L_{30}K_{30}(T_1E_1)$ which is more sensitive than T_1E_1 only because the partner (2) is far removed from (1). We therefore require that:

(γ) $T_1G_1 \subset G_3$, $T_2G_2 \subset G_3$.

In order to express the fact that system (3) does not contain anything in addition to systems (1) and (2) we require that:

(δ) With exception to $\lambda\mathbf{1}$ there exist no effects which are coexistent with all $F \in T_1L_1 \cup T_2L_2$.

From these results it follows that (for proof see [29]) $\mathscr{H}_3$ can be written in the following form:

$$\mathscr{H}_3 = \mathscr{H}_1 \times \mathscr{H}_2. \tag{2.1}$$

and that the maps T_1, T_2 have the form

$$T_1(F_1) = F_1 \times \mathbf{1}, \qquad T_2(F_2) = \mathbf{1} \times F_2. \tag{2.2}$$

Here we note (see the proof in [29]) that conditions (α)–(δ) cannot be satisfied if the number field for the Hilbert spaces is either **R** or **Q**—it must be **C** (see III, §3 and §5).

Without going into the proof cited above, we may base our structure on (2.1), and define T_1, T_2 by means of (2.2). Then we may easily prove that the relations (α)–(δ) hold. The structure characterized by means of (α)–(δ) or by (2.1) and (2.2) is therefore only a representation of the physical situation of a "pair without interaction." In scattering theory (described in XVI) the partners "before" and "after" scattering have practically no interaction. Then there exists such a structure $T_1^{(i)}$, $T_2^{(i)}$ which describes the effects of the partner (1) or (2) before scattering and a similar structure $T_1^{(f)}$, $T_2^{(f)}$ which describes the effects of the partner (1) or (2) after the scattering. For scattering theory it is decisively important that $T_1^{(f)} \neq T_1^{(i)}$, $T_2^{(f)} \neq T_2^{(i)}$ a result which we shall examine more closely in XVI, §1 and §4.4.

For the nonrelativistic theory of interaction—as is applied in the case of quantum mechanics—it is essential that we extrapolate the structure T_1, T_2 for the case of interaction. As the scattering experiments have shown we cannot specify any "fixed" structure T_1, T_2 because in the case of interaction each effect which is triggered only by partner (1) cannot be coexistent with each effect which is triggered only by partner (2), as is required in (β). This is the case because the interaction of the partner (the measurement process on partner (1)) will influence, as a result of the interaction, the effect triggered later by partner (2). These intuitive ideas suggest the following attempt at extrapolation.

AZ. To each time point t of the laboratory time scale there exists maps T_{1t}, T_{2t} which satisfy (α)–(δ). For the Galileo transformation $U_\gamma = U(1, 0, 0, \gamma)$ in $\mathscr{H}_3$ and the corresponding Galileo transformations $U_\gamma^{(1)}$, $U_\gamma^{(2)}$ in $\mathscr{H}_1$ and $\mathscr{H}_2$ we require that

$$T_{i(t+\gamma)} U_\gamma^{(i)} F_i U_\gamma^{(i)+} = U_\gamma (T_{it} F_i) U_\gamma^+$$

for $i = 1, 2$.

The intuitive idea here is that the effects $T_{1t}F_1$ are such that they will be triggered by partner (1) if (1) is placed into interaction with a measurement apparatus for a very short time at time t. Here we assume that there exist measurement apparatuses having interaction processes which act "momentarily" between the microsystem being measured and the apparatus and that during the brief interval during which the interaction takes place there is no noticeable change caused by the interaction between the partners, that is, if $\Delta\tau$ is the duration of the interaction, then

$$T_{it} U_{\Delta t}^{(i)} F_i U_{\Delta t}^{(i)} \approx U_{\Delta t}(T_{it} F_i) U_{\Delta\tau}^+ .$$

These considerations show that in the case of elementary particle physics we cannot expect to satisfy such a condition. The long-range electromagnetic interaction of atomic and molecular structure and the small value of the elementary charge $e \approx \sqrt{(1/137)}$ combine to make the use of AZ possible.

The requirement imposed by (AZ)

$$T_{1(t+\gamma)}U_\gamma^{(1)}F_1U_\gamma^{(1)+} = U_\gamma(T_{1t}F_1)U_\gamma^+$$

is self-evident on the basis of the physical interpretation of the Galileo group presented in VII. If $T_{1t}F_1$ is an effect for which a measurement interaction takes place at time t then $U_\gamma(T_{1t}F_1)U_\gamma^+$ is the effect which is produced if the measurement apparatus is applied at a time γ later, that is, when the measurement interaction takes place at time $(t + \gamma)$ later; this is again an effect triggered by partner (1) which corresponds to the effect $T_{1(t+\gamma)}U_\gamma^{(1)}F_1U_\gamma^{(1)+}$ as if system (1) had been triggered at the displaced time. The requirement

$$T_{1(t+\gamma)}U_\gamma^{(1)}F_1U_\gamma^{(1)+} = U_\gamma(T_{1t}F_1)U_\gamma^+$$

is mathematically only a definition for T_{1t} if T_{10} (that is T_{1t} for $t = 0$) is known.

According to (2.1) and (2.2) there exists a product representation $\mathscr{H}_3 = \mathscr{H}_1 \times \mathscr{H}_2$ corresponding to the maps T_{10}, T_{20}; a similar situation also holds for T_{1t}, T_{2t}. Since the product representation of $\mathscr{H}_3$ obtained in this way changes with time, it is not well suited for practical problems. In X we shall become familiar with the reformulation of the time variability described by U_γ in the form of the Schrödinger picture and the interaction picture, where the product representation of $\mathscr{H}_3$ can be considered not to change with time.

Now that we have introduced the maps T_{1t}, T_{2t} we must warn the reader about a possible error concerning its physical interpretation: If b_0 is a registration procedure by which the effect F_1 can be registered by systems of type (1), that is, for $b \subset b_0$

$$\psi(b_0, b) = (F_1, F_2, F_3, \ldots)$$

it no longer follows (also in the case if $F_2 = 0$!) that there exists a time t for which $F_3 = T_{1t}F_1$. The registration procedure which registers the effect F_1 for the systems of type (1) need no longer register the effect $T_{1t}F_1$ for composite systems of type (3). If, however, in the special case in which b_0 is a registration method in the sense described above—that is, for all practical purposes it is only in interaction at time t—then for the case $F_2 = 0$ we would expect that $F_3 = T_{1t}F_1$.

We may extend the requirement AZ as follows: There exist registration methods b_0 and registration procedures b for which

$$\psi(b_0, b) = (F_1, 0, T_{1t}F_1, \ldots)$$

On the basis of (2.1) and (2.2) "at time $t = 0$" we may, in a formal sense(!), transfer the rules for Galileo transformations for $\mathscr{H}_1$, $\mathscr{H}_2$, onto $\mathscr{H}_3$—and even carry out different Galileo transformations in $\mathscr{H}_1$ and $\mathscr{H}_2$. That is, for the group $\mathscr{G} \times \mathscr{G}$ (see AV, §5) with elements $((A_1, \vec{\delta}_1, \vec{\eta}_1, \gamma_1), (A_2, \vec{\delta}_2, \vec{\eta}_2, \gamma_2))$ where we define

$$U[(A_1, \vec{\delta}_1, \vec{\eta}_1, \gamma_1), (A_2, \vec{\delta}_2, \vec{\eta}_2, \gamma_2)] = U^{(1)}(A_1, \vec{\delta}_1, \vec{\eta}_1, \gamma_1) \times U^{(2)}(A_2, \vec{\delta}_2, \vec{\eta}_2, \gamma_2), \tag{2.3}$$

where $F_1 \times 1$ transforms into

$$U[(A_1, \ldots), (A_2, \ldots)](F_1 \times \mathbf{1})U[(A_1, \ldots), (A_2, \ldots)]^+ = U^{(1)}(A_1, \ldots)F_1 U^{(1)}(A_1, \ldots)^+ \times \mathbf{1}. \tag{2.4}$$

A similar result holds for $1 \times F_2$. Here $F_1 \times 1$ transforms according to the first Galileo transformation $(A_1, \ldots)$ and $1 \times F_2$ according to the other transformation $(A_2, \ldots)$.

Is the formal expression (2.4) physically meaningful?

We shall now consider a pair (1), (2) of systems which do not interact (for example, (1) and (2) may be widely separated) then it appears to be possible to subject the registration methods for (1) and (2) to separate(!) Galileo transformations (which are not very distant from the unit element). However, for the case of interaction we cannot simply combine two different registration methods for systems (1) and (2) into a new registration method for the pair (3). Indeed $T_{10}F_1 = F_1 \times \mathbf{1}$ and $T_{20}F_2 = \mathbf{1} \times F_2$ are coexistent, that is, there exists a registration method b_0 and registration procedures $b_1 \subset b_0$, $b_2 \subset b_0$ such that

$$\psi(b_0, b_1) = (\cdot, \cdot, F_1 \times \mathbf{1}, \ldots),$$

$$\psi(b_0, b_2) = (\cdot, \cdot, \mathbf{1} \times F_2, \ldots).$$

However, we do not know how to construct the apparatus for b_0. At least we do not expect that parts of the apparatus for b_0 can be subjected to distinct Galileo transformations. Thus it appears to be more reasonable not to generally interpret the transformations in (2.3) as transformations of registration methods.

On the contrary it appears only reasonable to investigate the relationship between the Galileo transformations of the entire registration method and their representations in Hilbert space $\mathscr{H}_3$ for systems of type (3). From the extrapolation of systems (1) and (2) without interaction we obtain the following answer which appears to be reasonable: In (2.3) we require that $(A_2, \vec{\delta}_2, \vec{\eta}_2, \gamma_2) = (A_1, \vec{\delta}_1, \vec{\eta}_1, \gamma_1)$, that is, the representation of the Galileo group in $\mathscr{H}_3$ with respect to the product representation of $\mathscr{H}_3$ "at time $t = 0$" ($\mathscr{H}_3 = \mathscr{H}_1 \times \mathscr{H}_2$) is given as follows:

$$U(A, \vec{\delta}, \vec{\eta}, \gamma) = U^{(1)}(A, \vec{\delta}, \vec{\eta}, \gamma) \times U^{(2)}(A, \vec{\delta}, \vec{\eta}, \gamma). \tag{2.5}$$

Equation (2.5) cannot hold in general, that is, it leads to contradictions with experience (for example, for scattering processes—see XVI) since for time translations it describes systems "without interaction." The structure of the representation of the Galileo group for composite systems which was described in general in §1 may not therefore be extended by means of the severe condition (2.5). The fact that it is possible to solve the interaction problem for the case of nonrelativistic quantum mechanics by means of a small change of (2.5) has led to overwhelming success in the applications of quantum mechanics. However, the fact that an analogous solution is not possible for relativistic quantum mechanics suggests that such a similar closed theory for elementary particles does not exist.

The physically functional but not very elegant solution of the interaction problem for nonrelativistic quantum mechanics is obtained as follows: With respect to the product representation of $\mathscr{H}_3$ at time $t = 0$ (2.5) is required only for $\gamma = 0$, that is, for the subgroup of the Galileo group denoted by $\mathscr{G}_a$ in §1 we require that:

$$U(A, \vec{\delta}, \vec{\eta}, 0) = U^{(1)}(A, \vec{\delta}, \vec{\eta}, 0) \times U^{(2)}(A, \vec{\delta}, \vec{\eta}, 0). \tag{2.6}$$

Thus, for the infinitesimal transformations "at time $t = 0$" it follows that:

$$\begin{aligned} \vec{K} &= \vec{K}_1 \times \mathbf{1} + \mathbf{1} \times \vec{K}_2, \\ \vec{X} &= \vec{X}_1 \times \mathbf{1} + \mathbf{1} \times \vec{X}_2, \\ \vec{J} &= \vec{J}_1 \times \mathbf{1} + \mathbf{1} \times \vec{J}_2, \end{aligned} \tag{2.7}$$

where $\vec{J}, \vec{J}_1, \vec{J}_2$ are the angular momentum operators. From the first two equations and VII (2.16) we obtain the commutation relations

$$\begin{aligned} K_\mu X_\nu - X_\nu K_\mu &= iM\delta_{\nu\mu}\mathbf{1} \\ &= (K_{1\mu}X_{1\nu} - X_{1\nu}K_{1\mu}) \times \mathbf{1} + \mathbf{1} \times (K_{2\mu}X_{2\nu} - X_{2\nu}K_{2\mu}) \\ &= im_1\delta_{\nu\mu}\mathbf{1} \times \mathbf{1} + \mathbf{1} \times im_2\delta_{\mu\nu}\mathbf{1}. \end{aligned}$$

It follows that $M = m_1 + m_2$; the mass M of the composite system (3) is equal to the sum of the masses for the two systems (1) and (2).

For the position and momentum observables given by VII, §4, that is, for

$$\vec{P}_1 = -\vec{K}_1, \qquad \vec{P}_2 = -\vec{K}_2, \qquad \vec{Q}_1 = \frac{1}{m}\vec{X}_1, \qquad \vec{Q}_2 = \frac{1}{m_2}\vec{X}_2$$

from (2.7) it follows that the total momentum defined in §1 is given by

$$\vec{P} = \vec{P}_1 \times \mathbf{1} + \mathbf{1} \times \vec{P}_2 \tag{2.8}$$

and the position of the center of mass $\vec{Q} = \vec{X}/M$ is given by

$$\vec{Q} = \frac{m_1\vec{Q}_1 \times \mathbf{1} + m_2\mathbf{1} \times \vec{Q}_2}{m_1 + m_2}. \tag{2.9}$$

The term "center of mass" arises from the form of the right side of (2.9).

If (2.5) were correct, then it would follow that

$$H = \frac{1}{2m_1}\vec{P}_1^2 \times \mathbf{1} + \frac{1}{2m_2}\mathbf{1} \times \vec{P}_2^2. \tag{2.10}$$

In the case in which misunderstandings are unlikely we shall use the more familiar notation of the physicist and write (2.10) in the form:

$$H = \frac{1}{2m_1}\vec{P}_1^2 + \frac{1}{2m_2}\vec{P}_2^2. \tag{2.11}$$

If H in (1.2) has the form (2.11) we then say that it describes the situation "no interaction." We write (1.2) in the form

$$H = \frac{1}{2m_1}\vec{P}_1^2 + \frac{1}{2m_2}\vec{P}_2^2 + H_J, \tag{2.12}$$

where H_J is called the "interaction operator."

In (2.12) we may replace $\vec{P}_1$, $\vec{P}_2$ by $\vec{P}$ in (2.8) and by the following new operator:

$$\vec{P}_r = \frac{m_2}{m_1 + m_2}\vec{P}_1 - \frac{m_1}{m_1 + m_2}\vec{P}_2. \tag{2.13}$$

Here we call $\vec{P}_r$ the relative momentum. Then (2.12) transforms into

$$H = \frac{1}{2M}\vec{P}^2 + \frac{1}{2m}P_r^2 + H_J \quad \text{where } m = \frac{m_1 m_2}{m_1 + m_2}; \tag{2.14}$$

m is called the reduced mass. It is easy to verify that $\vec{P}_r$ commutes with $\vec{X}$ and with all of the $U(1, \vec{\delta}, \vec{\eta}, 0)$. Therefore, relative to $\mathscr{H}_3 = \mathscr{H}_1 \times \mathscr{H}_2$, that is, relative to (1.1) we obtain

$$\frac{1}{2m}\vec{P}_r^2 = \mathbf{1} \times \frac{1}{2m}\vec{P}_r^2. \tag{2.15}$$

Since (1.2) holds, H_J must have the form

$$H_J = \mathbf{1} \times H_J \tag{2.16}$$

relative to (1.1), from which it follows from (1.2) that

$$H_i = \frac{1}{2m}\vec{P}_r^2 + H_J. \tag{2.17}$$

(The notation of the physicist permits us to consider two different representations of the same Hilbert space as a product space without changing symbols or indices—namely, $\mathscr{H}_3 = \mathscr{H}_b \times \mathscr{H}_i = \mathscr{H}_1 \times \mathscr{H}_2$, and to write the corresponding operators accordingly. Here we must always take care to note which representation of a product space is used in formulas like (2.7) and (2.16)!)

In addition to the center of mass position, we introduce the relative position operator $\vec{Q}_r = \vec{Q}_1 - \vec{Q}_2$; it is easy to show that $\vec{Q}_r$ commutes with $\vec{Q}$ and $\vec{P}$, that is, it is of the form

$$\vec{Q}_r = \mathbf{1} \times \vec{Q}_r \tag{2.19}$$

relative to (1.1). On the other hand it follows that

$$P_{r\nu}Q_{r\mu} - Q_{r\mu}P_{r\nu} = \frac{1}{i}\delta_{\nu\mu}\mathbf{1}. \tag{2.20}$$

The "coordinate transformation" of $\vec{Q}_1, \vec{P}_1, \vec{Q}_2, \vec{P}_2$ to $\vec{Q}, \vec{P}, \vec{Q}_s, \vec{P}_s$ leads to the following new notation: For

$$\mathscr{H}_1 = \mathscr{H}_{1b} \times \imath_{1s}, \qquad \mathscr{H}_2 = \mathscr{H}_{2b} \times \imath_{2s}$$

we obtain

$$\mathscr{H}_1 \times \mathscr{H}_2 = (\mathscr{H}_{1b} \times \mathscr{H}_{2b}) \times \imath_{1s} \times \imath_{2s} \tag{2.21}$$

and

$$\mathscr{H}_{1b} \times \mathscr{H}_{2b} = \mathscr{H}_b \times \mathscr{H}_{rb}, \tag{2.22}$$

where $\vec{Q}, \vec{P}$ are operators in $\mathscr{H}_b$ and $\vec{Q}_r, \vec{P}_r$ are operators in $\mathscr{H}_{rb}$. From (2.22), (2.21), and (1.1) it follows that

$$\mathscr{H}_i = \mathscr{H}_{rb} \times \imath_{1s} \times \imath_{2s}. \tag{2.23}$$

If the subsystem (2) has no spin, or if we can neglect the spin of (2) (a more precise formulation of what is meant by being able to "neglect" spin is given in [2], XI, §7.5) then we can ignore the space $\imath_{2s}$ in (2.23) and we obtain

$$\mathscr{H}_i = \mathscr{H}_{rb} \times \imath_{1s}. \tag{2.24}$$

Not only the form (2.24) but also the algebra of the operators $\vec{P}_r, \vec{Q}_r$ and the rotations (as we shall later prove) are completely equivalent with that of the elementary system (1) with the exception of the Hamiltonian operator H_i described by (2.17) and the mass factor m (m is the reduced mass in (2.14)). It is now necessary to consider the above assertion about rotations.

From $U(A) = U^{(1)}(A) \times U^{(2)}(A)$ relative to (2.1) it follows that with $U^{(1)}(A) = V^{(1)}(A) \times R^{(1)}(A)$, $U^{(2)}(A) = V^{(2)}(A) \times R^{(2)}(A)$

$$U(A) = (V^{(1)}(A) \times V^{(2)}(A)) \times R^{(1)}(A) \times R^{(2)}(A)$$

relative to (2.21). Here $V^{(1)}(A) \times V^{(2)}(A)$ is an operator in $\mathscr{H}_{1b} \times \mathscr{H}_{2b}$. It is easy to show that from the change in the product representation according to (2.22) we obtain

$$V^{(1)}(A) \times V^{(2)}(A) = V(A) \times V_r(A), \tag{2.25}$$

where the left (right) side of (2.25) corresponds to the left (right) side of (2.22). The operators $V^{(1)}, V^{(2)}, V, V_r$ behave in a manner which we have generally described for an orbit space in VII, §3.

According to (2.23) in $\mathscr{H}_i$ we have the representation

$$R(A) = V_r(A) \times R^{(1)}(A) \times R^{(2)}(A), \tag{2.26}$$

where $R(A)$ is defined according to (1.3). For the angular momentum, from (2.7) together with VII (5.3) we obtain

$$\begin{aligned} \vec{J} &= \vec{J}_1 + \vec{J}_2 = \vec{L}_1 + \vec{L}_2 + \vec{S}_1 + \vec{S}_2 \\ &= \vec{L} + \vec{L}_r + \vec{S}_1 + \vec{S}_2, \end{aligned} \tag{2.27}$$

where $\vec{L}$ is the orbital angular momentum in $\mathscr{H}_b$, L_r in $\mathscr{H}_{rb}$. Since the $V(A)$ and $V_r(A)$ behave in the same way as described in VII, §3 for an orbit space, from VII (3.42) it follows that

$$L = \vec{Q} \times \vec{P}, \qquad \vec{L}_r = \vec{Q}_r \times \vec{P}_r. \tag{2.28}$$

Therefore, if we may neglect the spin s_2 of the elementary system (2) we then obtain a description in $\mathscr{H}_i$ which is equivalent to that of an elementary system of mass m, spin s_1 and having the Hamiltonian operator (2.17). Such a description is applicable to a system consisting of an electron as system (1) and an atomic nucleus as system (2) (see XI, §3).

The structure for the dynamics of composite systems is therefore determined when the operator H_J is explicitly given. The form of the Hamiltonian operator will be discussed later in this book.

In closing we emphasize the fact that the observables $\vec{Q}_1 \times \mathbf{1}$, $\mathbf{1} \times \vec{Q}_2$, $\vec{P}_1 \times \mathbf{1}$, $\mathbf{1} \times \vec{P}_2$ for the representation $\mathscr{H}_3 = \mathscr{H}_1 \times \mathscr{H}_2$ refer to the time $t = 0$. They therefore correspond to the operator

$$\vec{Q}_{(1)}(0) = \vec{Q}_1 \times \mathbf{1}$$

"position of the subsystem (1) at time $t = 0$"; the "position of the subsystem (1) at time t," using the interpretation of the time translation operator

$$U_\gamma = U(1, 0, 0, \gamma)$$

corresponds to the operator

$$Q_{(1)}(t) = U_t \vec{Q}_{(1)}(0) U_t^+ = U_t(\vec{Q}_1 \times \mathbf{1}) U_t^+$$

which does not take the form $A \times 1$ with respect to the product representation $\mathscr{H}_3 = \mathscr{H}_1 \times \mathscr{H}_2$ defined by T_{10}, T_{20}. It does, however, have this form with respect to the product representation corresponding to T_{1t}, T_{2t}! A similar situation is found for the momenta of the subsystems (1) and (2). The momenta of the subsystems (and the angular momentum) are no longer constant with time!

3 Composite Systems Consisting of Two Identical Elementary Systems

If the system types (1) and (2) are identical, then the intuitive reasoning which led to (2.1) and (2.2) is fundamentally incorrect because it is impossible to distinguish between the "effect of subsystem (1)" from that of "system (2)." The following intuitive approach to this problem has been fruitful:

Let us suppose that we have constructed a product space at time $t = 0$ from a Hilbert space $\mathscr{H}_1$ for a system of type (1) as follows:

$$\mathscr{H}_1^2 = \mathscr{H}_1 \times \mathscr{H}_1 \tag{3.1}$$

(see §4 for more details). Here, of course, it does not make sense to apply (2.2) in this case. Since both systems are identical, all "effects" are invariant under exchange of the two systems and, consequently, (3.1) cannot be the "correct" Hilbert space for the composite system. How can we formulate this invariance in mathematical terms? For this purpose we define an exchange operator as a linear operator in $\mathscr{H}_1^2$ as follows:

$$\mathbf{P}\varphi_\nu(1)\varphi_\mu(2) = \varphi_\mu(1)\varphi_\nu(2), \tag{3.2}$$

where φ_ν is a complete orthonormal basis in $\mathscr{H}_1$.

It is easy to verify (see §4) that the operator $\mathbf{P}$ defined in (3.2) is independent of the complete orthonormal basis used in (3.2).

In addition $\mathbf{P}$ is obviously unitary, and satisfies $\mathbf{P}^2 = \mathbf{1}$. Therefore we may partition $\mathscr{H}_1^2$ into two subspaces $\{\mathscr{H}_1^2\}_+$ and $\{\mathscr{H}_1^2\}_-$ where $\{\mathscr{H}_1^2\}_+$ is the eigenspace of P with eigenvalue $+1$ and $\{\mathscr{H}_1^2\}_-$ is the eigenspace of P with eigenvalue -1. An operator A in $\mathscr{H}_1^2$ is said to be symmetric if

$$\mathbf{P}A = A\mathbf{P}. \tag{3.3}$$

All effects of systems (3) therefore should be (intuitively speaking) symmetric operators, that is, $\{\mathscr{H}_1^2\}_+$ and $\{\mathscr{H}_1^2\}_-$ are invariant subspaces with respect to symmetric effects. Since the set of symmetric effects is a proper subset of the set $L(\mathscr{H}_1^2)$, $\mathscr{H}_1^2$ can therefore not be a Hilbert space for the system type (3) because $L(\mathscr{H}_3)$ is the set of effects for system (3). However, we may identify $\mathscr{H}_3$ with $\{\mathscr{H}_1^2\}_+$ or with $\{\mathscr{H}_1^2\}_-$ because it is clear that every self-adjoint operator which leaves $\{\mathscr{H}_1^2\}_+$ or $\{\mathscr{H}_1^2\}_-$ invariant is also symmetric.

Both of these intuitively motivated possibilities have proven to be successful. We shall now formulate the additional structure needed to describe systems of type (3) which are composed of pairs of identical systems of type (1).

Case (+):

$$\mathscr{H}_3 = \{H_1^2\}_+ \tag{3.4}$$

together with the map T_0 (that is T_t for $t = 0$) of $L_1 = L(\mathscr{H}_1)$ into $L_3 = L(\mathscr{H}_3)$ defined by:

$$T_0(F_1) = \tfrac{1}{2}(F_1 \times \mathbf{1} + \mathbf{1} \times F_1). \tag{3.5}$$

Case (−):

$$\mathscr{H}_3 = \{\mathscr{H}_1^2\}_- \tag{3.6}$$

together with the map T_0 of $L_1 = L(\mathscr{H}_1)$ into $L_3 = L(\mathscr{H}_3)$ defined by:

$$T_0(F_1) = \tfrac{1}{2}(F_1 \times \mathbf{1} + \mathbf{1} \times F_1). \tag{3.7}$$

Here (3.6) and (3.7) are the analogs of (2.2). $T_0(F_1)$ is the effect which is triggered by one of the subsystems of type (1) as if the other were not present.

In §2 we have already mentioned the fact that the physical interpretation of this requirement is problematical.

Which of the cases (+) or (−) are we to choose? That depends on the elementary system type (1) and—according to experience and certain results from quantum field theory—depends only on the spin of the system type (1). If the spin takes on an integer value, then case (+) is to be chosen; then we call such systems "Bose systems." If the spin takes on half-integer values then case (−) is to be chosen; here the systems are called "Fermi systems."

Many of the ideas in §2 are applicable to the cases (+) and (−). Clearly (2.3) is not applicable; nevertheless the formula (2.6) is applicable. Thus the formulas (2.7)–(2.12) remain applicable and we find that H and $\vec{P}$ must commute. For the application of the material following (2.13) we must be cautious.

If we define, in a formal manner, the following using (2.13) (here $m_1 = m_2$!):

$$\vec{P}_r = \tfrac{1}{2}(\vec{P}_1 - \vec{P}_2) \tag{3.8}$$

then $\vec{P}_r$ cannot be an observable of system (3) because it is not symmetric. Of course, (2.14) is applicable because P_r^2 is symmetric.

We must now consider the decomposition

$$\mathscr{H}_3 = \{\mathscr{H}_1^2\}_\pm = \mathscr{H}_b \times \mathscr{H}_i. \tag{3.9}$$

$\mathscr{H}_b$ is defined as in §2, but $\mathscr{H}_i$ is not of the form (2.19). What is the structure of $\mathscr{H}_i$ in (3.9)?

This we may discover most simply by analogy to (3.8) if we assume the formal definition of (2.18). Then, corresponding to (2.21) and (2.22) we obtain:

$$\mathscr{H}_1^2 = \mathscr{H}_{1b}^2 \times \imath_{1s}^2 = \mathscr{H}_b \times \mathscr{H}_{rb} \times \imath_{1s}^2. \tag{3.10}$$

How does the exchange operator $\mathbf{P}$ behave? To answer this question we shall define $\mathbf{P}_s$ in an obvious way as an exchange operator in $\imath_s^2$ and the parity operator R in $\mathscr{H}_{rb} = \mathscr{L}(\mathbf{R}^3, dx_1\, dx_2\, dx_3)$ (see VII, §7) as follows:

$$R\psi(x_1, x_2, x_3) = \psi(-x_1, -x_2, -x_3). \tag{3.11}$$

Then, according to the representation on the right side of (3.10) we obtain

$$\mathbf{P} = \mathbf{1} \times R \times \mathbf{P}_s. \tag{3.12}$$

Therefore $\mathscr{H}_i$ in $\mathscr{H}_3 = \mathscr{H}_b \times \mathscr{H}_i$ is the subspace of $\mathscr{H}_{rb} \times \imath_{1s}^2$ which corresponds to the eigenvalue 1 of $R \times \mathbf{P}_s$ for Bose systems and -1 for Fermi systems. The operators H_i and H_J must, of course, commute with $R \times \mathbf{P}_s$.

If $\mathscr{H}_{rb+}$ is the eigenspace of R with eigenvalue $+1$ and $\mathscr{H}_{rb-}$ that for eigenvalue -1, then we obtain

$$\mathscr{H}_i = \mathscr{H}_{rb+} \times \{\imath_{1s}^2\}_+ \oplus \mathscr{H}_{rb-} \times \{\imath_{1s}^2\}_- \tag{3.13}$$

for Bose systems; for Fermi systems we obtain

$$\mathscr{H}_i = \mathscr{H}_{rb-} \times \{\imath_{1s}^2\}_+ \oplus \mathscr{H}_{rb+} \times \{\imath_{1s}^2\}_-. \tag{3.14}$$

4 Composite Systems Consisting of Electrons and Atomic Nuclei

Now that we have introduced the characteristic structure for the concept of composite systems for the case of two systems we shall seek to apply this structure without additional discussion to the case of larger numbers of subsystems. For applications of quantum mechanics to atoms and molecules we need only consider the case in which the elementary subsystems are electrons and atomic nuclei. We shall now present the construction of the structure of a "composite system" in the form of a prescription since this form is the most transparent.

First Step: We consider the different elementary system types from which the system under consideration is composed. Let the system types be denoted by (1), (2), ..., (k). For each of these let n_ν denote the number of systems of type (ν) which appear in the composite system.

Second Step: For each system type (ν) we then construct, with the aid of the Hilbert spaces $\mathscr{H}_\nu$, the Hilbert space $\mathscr{H}_{n_\nu} = \{\mathscr{H}_\nu^{n_\nu}\}_+$ or $\mathscr{H}_{n_\nu} = \{\mathscr{H}_\nu^{n_\nu}\}_-$ depending on whether the spin of the system type (ν) is an integer or a half-integer.

Third Step: We then construct the Hilbert space of the composite system as follows:

$$\mathscr{H} = \mathscr{H}_{n_1} \times \mathscr{H}_{n_2} \times \cdots \times \mathscr{H}_{n_K}. \tag{4.1}$$

For the product representation (4.1) the corresponding maps (at time $t = 0$) of $L(\mathscr{H}_\nu)$ into $L(\mathscr{H})$ are given by:

$$T_\nu F_\nu = \mathbf{1} \times \mathbf{1} \times \cdots \times \tilde{F} \times \cdots \times \mathbf{1},$$

where $\tilde{F} \in L(\mathscr{H}_{n_\nu})$ is in the νth position. With respect to the $\mathscr{H}_{n_\nu}$ the symmetric operator $\tilde{F}$ in $\mathscr{H}_\nu^{n_\nu} = H_\nu \times H_\nu \times \cdots \times H_\nu$ is defined by

$$\tilde{F} = \frac{1}{n_\nu}[F_\nu \times 1 \times \cdots \times 1 + 1 \times F_\nu \times \cdots \times 1 + \cdots + 1 \times 1 \times \cdots \times F_\nu].$$

It is only necessary to make the second step more mathematically precise. In order not to keep track of multiple indices we shall now consider one elementary system type with Hilbert space $\mathscr{H}$ and the n-fold product space $\mathscr{H}^n$. This will be correctly defined in the following; here we shall write the n-fold product space of $\mathscr{H}$ with itself as follows:

We define n Hilbert spaces $\mathscr{H}^{(1)}, \mathscr{H}^{(2)}, \ldots, \mathscr{H}^{(n)}$; suppose that for each $\mathscr{H}^{(\alpha)}$ we are given an isomorphic map V_α of $\mathscr{H}$ onto $\mathscr{H}^{(\alpha)}$. Here we shall consider the pair of effects $F \in L(\mathscr{H})$ and $V_\alpha F V_\alpha^{-1} \in L(\mathscr{H}^{(\alpha)})$ to be identical (by definition) that is, on the basis of the mapping principles they are interpreted as being identical. Such isomorphisms are, for the most part, defined by selecting a complete orthonormal basis ϕ_ν in $\mathscr{H}$ and considering the images $V_\alpha \phi_\nu$ in $\mathscr{H}^{(\alpha)}$. The $V_\alpha \phi_\nu$ are usually denoted by $\phi_\nu(\alpha)$. Such isomorphisms will be used in XII–XV.

$\mathscr{H}^n$ is then defined by $\mathscr{H}^{(1)} \times \mathscr{H}^{(2)} \times \cdots \times \mathscr{H}^{(n)}$ together with the specification of the isomorphisms V_α. In such a product $\mathscr{H}^n$ we may define linear permutation operators which also play an important role in practical applications (XII–XV).

Using the above defined $\phi_\nu(\alpha)$

$$\Psi_{\nu_1,\nu_2,\ldots,\nu_n} = \phi_{\nu_1}(1), \phi_{\nu_2}(2), \ldots, \phi_{\nu_n}(n) \tag{4.2}$$

defines a complete orthonormal basis in $\mathscr{H}^n$. If **P** is a permutation among a set of n items (as defined in AV, §9) we define an operator in $\mathscr{H}^n$ by the equation

$$\mathbf{P}\Psi_{\nu_1,\nu_2,\ldots,\nu_n} = \Psi_{\mu_1,\mu_2,\ldots,\mu_n}, \tag{4.3}$$

where the $\mu_1, \mu_2, \ldots, \mu_n$ is obtained from the permutation of the "n indices" $\nu_1, \nu_2, \ldots, \nu_n$ (see AV, §9). The operator **P** is (according to (4.3)) a unitary operator, since it transforms a complete orthonormal basis into another (in this case it only changes the order of the basis vectors).

The definition (4.3) of the operator **P** does not depend on the particular choice of the basis ϕ_ν but only on the product representation. The independence of the definition (4.3) from the choice of a basis follows directly from the fact that each vector of the form $\chi_1(1)\chi_2(2)\cdots\chi_n(n)$ is transformed into a vector $\chi_{\alpha_1}(1)\chi_{\alpha_2}(2)\ldots\chi_{\alpha_n}(n)$ by **P** where **P** permutes the n symbols $1, 2, \ldots, n$ into $\alpha_1, \alpha_2, \ldots, \alpha_n$. This can be proved very simply as follows: For a vector

$$\chi = \sum_{\nu_1,\ldots,\nu_n} \Psi_{\nu_1,\ldots,\nu_n} a_{\nu_1,\ldots,\nu_n}$$

we obtain

$$\mathbf{P}\chi = \sum_{\nu_1,\ldots,\nu_n} \Psi_{\mu_1,\ldots,\mu_n} a_{\nu_1,\ldots,\nu_n}.$$

If η_ρ is another complete orthonormal basis in $\mathscr{H}$ for which

$$\eta_\rho = \sum_\nu \phi_\nu b_{\nu_\rho}$$

then we have

$$\chi_{\rho_1,\ldots,\rho_n} \overset{\text{def}}{=} \eta_{\rho_1}(1), \ldots, \eta_{\rho_n}(n) = \sum_{\nu_1,\ldots,\nu_n} \Psi_{\nu_1,\ldots,\nu_n} b_{\nu_1\rho_1} b_{\nu_2\rho_2}, \ldots, b_{\nu_n\rho_n}$$

and therefore obtain

$$\mathbf{P}\chi_{\rho_1,\ldots,\rho_n} = \sum_{\nu_1,\ldots,\nu_n} \Psi_{\mu_1,\ldots,\mu_n} b_{\nu_1\rho_1}, \ldots, b_{\nu_n\rho_n},$$

where the $\mu_1, \ldots, \mu_n$ are obtained using **P** on the $\nu_1, \ldots, \nu_n$. If we apply the permutation **P** to the $b_{\nu_1\rho_1}, \ldots, b_{\nu_n\rho_n}$ we do not change the value of the product; they appear in the sequence $b_{\mu_1\sigma_1}, \ldots, b_{\mu_n\sigma_n}$ where the $\sigma_1, \ldots, \sigma_n$ are obtained from the $\rho_1, \ldots, \rho_n$ by means of the permutation **P** so that we obtain

$$\mathbf{P}\chi_{\rho_1,\ldots,\rho_n} = \sum_{\mu_1,\ldots,\mu_n} \Psi_{\mu_1,\ldots,\mu_n} b_{\mu_1\sigma_1}, \ldots, b_{\mu_n\sigma_n} = \chi_{\sigma_1,\ldots,\sigma_n}.$$

The operators **P** form a unitary representation of the symmetric group $\mathbf{S}_n$ (see AV, §9).

In this case the general form of an operator A which commutes with all the operators **P** may easily be determined according to AIV, §14 in the following way: To each element $e_{\alpha\beta}^{(\nu)}$ of the group ring (AV, §7, §9, and §10.6) there exists a corresponding operator $E_{\alpha\beta}^{(\nu)}$ in $\mathscr{H}^n$. The operators $E_{\alpha\alpha}^{(\nu)}$ and $E^{(\nu)} = \sum_\alpha E_{\alpha\alpha}^{(\nu)}$ are projection operators. Since $e = \sum_{\nu,\alpha} e_{\alpha\alpha}^{(\nu)}$ we therefore obtain $\sum_{\nu,\alpha} E_{\alpha\alpha}^{(\nu)} = \sum_\nu E^{(\nu)} = \mathbf{1}$. From this partition of unity the Hilbert space $\mathscr{H}^n$ decomposes into subspaces $\imath^{(\nu)} = E^{(\nu)}(\mathscr{H}^n)$ and the $\imath^{(\nu)}$, in turn, decompose into subspaces $\imath_\alpha^{(\nu)} = E_{\alpha\alpha}^{(\nu)}(\mathscr{H}^n)$: $\mathscr{H}^n = \sum_\nu \oplus\, \imath^{(\nu)}$ and $\imath^{(\nu)} = \sum_\alpha \oplus\, \imath_\alpha^{(\nu)}$. The Hilbert spaces $\imath_\alpha^{(\nu)}$ (for fixed ν) are transformed into themselves by the operator A; we may represent $\imath^{(\nu)}$ as follows: $\imath^{(\nu)} = \jmath^{(\nu)} \times \ell^{(\nu)}$ where the operators **P** are represented in the finite-dimensional space $\ell^{(\nu)}$ by the νth irreducible representation. The operators A leave $r^{(\nu)}$ invariant and have the form $(A^{(\nu)} \times \mathbf{1})$ in $\imath^{(\nu)}$ where $\mathbf{1}$ is the unit operator in $\ell^{(\nu)}$ and $A^{(\nu)}$ is the operator in $\jmath^{(\nu)}$; the permutation operators have the form $(\mathbf{1} \times \mathbf{P}^{(\nu)})$ in $\imath^{(\nu)}$ where $\mathbf{P}^{(\nu)}$ denotes the operator for the νth irreducible representation in $\ell^{(\nu)}$ and $\mathbf{1}$ is the unit operator in $\jmath^{(\nu)}$. The operators $A^{(\nu)}$ in $\jmath^{(\nu)}$ can be completely arbitrary. Thus it follows that the general form of all operators which commute with all the **P** (here $\sum \oplus$ represents the fact that the operators A and P leave the spaces $\imath^{(\nu)}$ invariant) will be given by:

$$A = \sum_\nu \oplus\, (A^{(\nu)} \times \mathbf{1}), \qquad \mathbf{P} = \sum_\nu \oplus\, (\mathbf{1} \times \mathbf{P}^{(\nu)}). \tag{4.4}$$

In addition we find that unbounded operators which commute with all the **P** also have the form $A = \sum_\nu \oplus\, (A^{(\nu)} \times \mathbf{1})$ since the domain of definition of A is transformed by the permutation operators into itself. Therefore we may construct the domain of definition of $A^{(\nu)}$ by use of $E_{\alpha\alpha}^{(\nu)}$ on the domain of definition of A. The problem of the spectral decomposition of A is also solved if we know the spectral decomposition of finitely many $A^{(\nu)}$. Therefore to each irreducible representation (ν) of the permutation group there exists a spectrum which corresponds to A (providing that none of the $\imath^{(\nu)}$ is equal to zero, a situation which can occur for a finite-dimensional Hilbert space $\mathscr{H}$; see the presentation for the spin space $\imath_s$ and the Hilbert space $\imath_s^n$ in XIV, §7).

We shall now consider the possibility that one of the spaces $\jmath^{(\nu)}$ has been chosen as the Hilbert space $\mathscr{H}_n$ for n identical systems. From experience we find that there are only two possible cases—either $\mathbf{P}^{(\nu)}$ belongs to the symmetric (that is, the identity representation) or it belongs to the anti-symmetric representation of the permutation group. Since these representations are one-dimensional we may identify $\jmath^{(\nu)}$ with $\imath^{(\nu)}$, that is, $\imath^{(\nu)} = \{\mathscr{H}^n\}_+$ or $\imath^{(\nu)} = \{\mathscr{H}^n\}_-$. We shall now impose the following postulate:

For elementary systems with integer values of spin we choose $\mathscr{H}_n = \{\mathscr{H}^n\}_+$; for elementary systems with half-integer values of spin we choose $\mathscr{H}_n = \{\mathscr{H}^n\}_-$.

For the antisymmetric representation the operator $E^{(v)}$ has the form (see AV, §9):

$$\frac{1}{n!}\sum_{\mathbf{P}}(-1)^{\mathbf{P}}\mathbf{P}, \tag{4.5}$$

where $(-1)^{\mathbf{P}}$ is $+1$ or -1 depending whether $\mathbf{P}$ is either an even or an odd permutation. For the $\Psi_{v_1,\ldots,v_n}$ defined according to (4.2) we find that the

$$\Gamma_{v_1,\ldots,v_n} = \frac{1}{n!}\sum_{\mathbf{P}}(-1)^{\mathbf{P}}\mathbf{P}\Psi_{v_1,\ldots,v_n} \tag{4.6}$$

span the space $\{\mathscr{H}^n\}$. The $\Gamma_{v_1,\ldots,v_n}$ are equal to zero if two of the indices are equal. The $\Gamma_{v_1,\ldots,v_n}$ (except for sign) are identical to $\Gamma_{\mu_1,\ldots,\mu_n}$ if the $\mu_1,\ldots,\mu_n$ are obtained from the $v_1,\ldots,v_n$ by a permutation. For the inner product of two $\Gamma_{v_1,\ldots,v_n}$ it follows that (here $\mathbf{R} = \mathbf{P}^{-1}\mathbf{Q}$) we obtain

$$\begin{aligned}
&\langle\Gamma_{v_1,\ldots,v_n}, \Gamma_{\mu_1,\ldots,\mu_n}\rangle \\
&= \frac{1}{(n!)^2}\sum_{\mathbf{P},\mathbf{Q}}(-1)^{\mathbf{PQ}}\langle\mathbf{P}\phi_{v_1}(1),\ldots,\phi_{v_n}(n), \mathbf{Q}\phi_{\mu_1}(1),\ldots,\phi_{\mu_n}(n)\rangle \\
&= \frac{1}{(n!)^2}\sum_{\mathbf{P}}\sum_{\mathbf{Q}}(-1)^{\mathbf{R}}\langle\phi_{v_1}(1),\ldots,\phi_{v_n}(n), \mathbf{R}\phi_{\mu_1}(1),\ldots,\phi_{\mu_n}(n)\rangle \\
&= \frac{1}{n!}\sum_{\mathbf{R}}(-1)^{\mathbf{R}}\langle\phi_{v_1}(1),\ldots,\phi_{v_n}(n), \mathbf{R}\phi_{\mu_1}(1),\ldots,\phi_{\mu_n}(n)\rangle.
\end{aligned}$$

From the above remark it suffices to choose an index sequence $v_1,\ldots,v_n$ from all the index sequences obtained from permutation and a definite corresponding $\Gamma_{v_1,v_2,\ldots,v_n}$. If $\mu_1,\mu_2,\ldots,\mu_n$ is a "different" index sequence, then there must be a v_i which does not occur in $\mu_1,\mu_2,\ldots,\mu_n$; then $\Gamma_{v_1,v_2,\ldots,v_n}$ will be orthogonal to $\Gamma_{\mu_1,\mu_2,\ldots,\mu_n}$. If both index sequences are identical (that is, $v_i = \mu_i$) then, (since no pair of indices v_i are equal and therefore only the summand for which $\mathbf{R} = \mathbf{1}$ yields a nonzero term), we obtain

$$\|\Gamma_{v_1,\ldots,v_n}\|^2 = \frac{1}{n!}. \tag{4.7}$$

In this way the so-called Slater determinant

$$\begin{aligned}
\gamma_{v_1,\ldots,v_n} &= \frac{1}{\sqrt{n!}}\sum_{\mathbf{P}}(-1)^{\mathbf{P}}\mathbf{P}\phi_{v_1}(1),\ldots,\phi_{v_n}(n) \\
&= \frac{1}{\sqrt{n!}}\begin{vmatrix} \phi_{v_1}(1) & \phi_{v_1}(2) & \cdots & \phi_{v_1}(n) \\ \phi_{v_2}(1) & \phi_{v_2}(2) & \cdots & \phi_{v_2}(n) \\ \vdots & \vdots & & \vdots \\ \phi_{v_n}(1) & \phi_{v_n}(2) & \cdots & \phi_{v_n}(n) \end{vmatrix}
\end{aligned} \tag{4.8}$$

defines a complete orthonormal basis in $\{\mathscr{H}^n\}_-$, where for each $\gamma_{v_1,\ldots,v_n}$ only one index sequence is chosen from the index sequences obtained by permutation.

For the identity representation the operator $E^{(\nu)}$ has the simple form (AV, §9):

$$\frac{1}{n!}\sum_{\mathbf{P}} \mathbf{P}. \tag{4.9}$$

We obtain a basis for $\{\mathcal{H}^n\}_+$ from the $\Psi_{\nu_1,\ldots,\nu_n}$ as follows:

$$\Omega_{\nu_1,\ldots,\nu_n} = \frac{1}{n!}\sum_{\mathbf{P}} \mathbf{P}\Psi_{\nu_1,\ldots,\nu_n}. \tag{4.10}$$

The $\Omega_{\nu_1,\ldots,\nu_n}$ are never equal to zero; however, we obtain the same element when we permute the index sequence. Therefore, for the $\Omega_{\nu_1,\ldots,\nu_n}$ we must select only one index sequence from all the possible index sequences obtained by permutation, that is, according to (4.11) we do not pay attention to the sequence of the indices $\nu_1, \ldots, \nu_n$ and only consider index sequences to be different if different indices are present. It is easy to see that the $\Omega_{\nu_1,\ldots,\nu_n}$ which differ in the index sequences are orthogonal.

In order to obtain normalized vectors we compute $\|\Omega_{\nu_1,\ldots,\nu_n}\|^2$. In order to do this we must know how many indices are identical. Since we do not consider the order we shall assume

$$\nu_1, \nu_2, \ldots, \nu_n = \mu_1, \mu_2, \ldots, \mu_{n_1}, \ldots, \rho_1, \rho_2, \ldots, \rho_{n_2}, \ldots, \eta_1, \eta_2, \ldots, \eta_{n_\nu}, \tag{4.11}$$

where the $\mu_i, \rho_i, \ldots, \eta_i$ are identical and that $n_1 + n_2 + \cdots + n_\nu = n$. Therefore we obtain (here $\mathbf{R} = \mathbf{P}^{-1}\mathbf{Q}$):

$$\begin{aligned}
&\|\Omega_{\nu_1,\ldots,\nu_n}\|^2 \\
&= \frac{1}{n!}\sum_{\mathbf{P},\mathbf{Q}} \langle \mathbf{P}\phi_{\nu_1}(1), \ldots, \phi_{\nu_n}(n), \mathbf{Q}\phi_{\nu_1}(1), \ldots, \phi_{\nu_n}(n)\rangle \\
&= \frac{1}{(n!)^2}\sum_{\mathbf{P}}\sum_{\mathbf{R}} \langle \phi_{\nu_1}(1), \ldots, \phi_{\nu_n}(n), \mathbf{R}\phi_{\nu_1}(1), \ldots, \phi_{\nu_n}(n)\rangle \\
&= \frac{1}{n!}\sum_{\mathbf{R}} \langle \phi_{\nu_1}(1), \ldots, \phi_{\nu_n}(n), \mathbf{R}\phi_{\nu_1}(1), \ldots, \phi_{\nu_n}(n)\rangle.
\end{aligned}$$

The inner products in the last sum vanish unless

$$\mathbf{R}(\mu_1, \mu_2, \ldots, \eta_{n_\nu}) = (\mu_1, \mu_2, \ldots, \eta_{n_\nu})$$

if this is the case then they are equal to 1. In the sum there are as many terms equal to 1 as there are permutations which transform the series (4.11) into itself; there are exactly $n_1!, n_2!, \ldots, n_\nu!$ permutations. Therefore we obtain

$$\|\Omega_{\nu_1,\ldots,\nu_n}\|^2 = \frac{1}{n!} n_1!, n_2!, \ldots, n_\nu! \tag{4.12}$$

Therefore, for a complete orthonormal basis for $\{\mathcal{H}^n\}_+$ we obtain:

$$\omega_{\nu_1,\ldots,\nu_n} = \frac{1}{\sqrt{n!, n_1!, n_2!, \ldots, n_\nu!}}\sum_{\mathbf{P}} \mathbf{P}\phi_{\mu_1}(1), \phi_{\mu_2}(2), \ldots, \phi_{\mu_{n_1}}(n_1), \ldots, \phi_{\mu_{n_\nu}}(n) \tag{4.13}$$

We will not carry out the separation of the center of mass as we have done in §2 and §3. We shall do this later in our discussion of applications.

We shall also not consider the detailed analysis of the representation of the Galileo group since, by analogy to (2.6), the transformations are given as a product of $\sum_{\alpha=1}^{k} n_\alpha$ factors. The time translation operator in $\mathscr{H}$ from (4.1) is given by

$$U(1, 0, 0, \gamma) = e^{i\gamma H}, \tag{4.14}$$

where H must commute with translations, rotations, and permutation operators.

In closing we shall now formulate the following axiom:

Every system type is either elementary (either an electron or an atomic nucleus) or consists of a composite of a finite number of electrons and atomic nuclei. Here the number n_1 of electrons and the numbers $n_2, n_3, \ldots$ of atomic nuclei of identical type uniquely define the type of the composite system.

This axiom is, in part, a description of microsystems and, in part, reflects the boundary of the fundamental domain to which we shall apply the theory.

This axiom places no restriction on the numbers n_α and $\sum_\alpha n_\alpha$. There exists grave doubts, indeed, strong evidence from experience, that the formal theory for system types cannot be g.G.-closed for systems with very large values of n_α, that is, must be extended by a more comprehensive theory (see, for example, [2], XV and [13]). This means that eventually we must place into question predictions of the theory for large n_α concerning physical possibilities (see [1], §10). We may only obtain reference points for what is possible in nature for small values of n_α. The larger the value of n_α the larger the possible discrepancy between the theoretical possibilities and the actual possibilities. We introduced these critical remarks in order to urge the reader to critically examine all the structure inherent in the theory.

5 The Hamiltonian Operator

The representation of the Galileo group is known for the case of composite systems (in the sense of a generalization of (2.6)) if the time translation operator (that is the Hamiltonian) in (4.14) is known. H can therefore be given for every system type. According to VII (5.2) for elementary systems we have

$$H = \frac{1}{2m}\vec{P}^2. \tag{5.1}$$

What is H for the case of composite systems?

The first answer to this question is disappointing. More precisely there exists no H, that is, the description already presented in §4 can only represent an approximation of experience, because it is not difficult to give examples of experiments which contradict (4.14). The physicist can rapidly determine where the deviations from (4.14) arise: As long as the emission of light (that is, of electromagnetic radiation) plays a role, (4.14) can only be an approximation. For elementary systems (5.1) is in agreement with experience.

We may strive to obtain a better description than that given in §4 in two different ways.

The first attempt seeks to add the electromagnetic radiation to the microsystem, that is, the action carriers from the preparation to the registration systems. Then we must give up the picture of a composite system of the type described in §4, since, for example, the composite system "atom" can emit new systems—light quanta. This method, known under the designation "Quantum Electrodynamics" has led to fantastic success. It is, however, not clear today whether it yields a mathematically correct description. For the latter reason we shall not consider this theory here; we shall only consider mathematically correct theories. Therefore we shall now consider the second attempt, in which we must take into consideration inadequate physical concepts and mathematically unpleasant formulations.

In the second approach (4.14) appears as a "first" approximation which yields good results as long as the radiation of light does not play a significant role. For this first approximation we will give a Hamiltonian operator without giving any additional theoretical reasons. The first estimate for H was obtained from the correspondence principle which can be found in every quantum mechanics textbook (see, for example, [2], XI). The most general estimate is obtained as an approximation from quantum electrodynamics; in (5.8) we shall give such an estimate for H. In the domain of the theory presented here (5.8) will therefore be an axiom. The emission of radiation will therefore be described in a second "approximation" step in the following way.

We may consider the registration of the light emitted by an atom or a molecule as an effect triggered by the atom or molecule. This conception is correct in terms of the meaning of the formulation of quantum mechanics presented here and also of the meaning of quantum electrodynamics. In XVII we shall see how the effects for a system can be produced "with the help" of other systems. According to quantum electrodynamics we may indeed express an effect produced by a light quantum as an effect which is produced by the atom or molecule (by analogy with the considerations of XVII, §2). In XI, §1 we will outline a route by which we may guess an operator $F \in L$ which describes these effects. The form of this effect operator can be considered to be an axiom in the context of the theory presented here.

In the second approximation step the effect of the emission of a quantum of radiation is treated by introducing a correction to (4.14). According to (4.14) the frequency for an effect F and for its time displacement by γ is calculated as follows:

$$\operatorname{tr}(WF) \text{ and } \operatorname{tr}(WF_\gamma) = \operatorname{tr}(We^{iH\gamma}Fe^{-iH\gamma}). \tag{5.2}$$

The required correction of (5.2) may be simply determined by realizing that W is also somewhat dependent on γ, that is, instead of writing the right-hand side of (5.2) as above we write

$$\mathrm{tr}(W_\gamma F_\gamma) = \mathrm{tr}(W_\gamma e^{iH\gamma} F e^{-iH\gamma}) \tag{5.3}$$

and for W_γ we introduce a differential equation of the form:

$$\frac{dW_\gamma}{d\gamma} = \sum_{n,m} [A_{nm} W_\gamma A_{nm}^+ - \tfrac{1}{2} A_{nm}^+ A_{nm} W_\gamma - \tfrac{1}{2} W_\gamma A_{nm}^+ A_{nm}], \tag{5.4}$$

where the A_{nm} are operators in the Hilbert space of the composite systems. From (5.4) it follows that

$$\frac{d}{d\gamma}\,\mathrm{tr}(W_\gamma) = 0 \tag{5.5}$$

and

$$\frac{d}{d\gamma}\langle\varphi, W_\gamma\varphi\rangle = \sum_{n,m}[\|W_\gamma^{1/2} A_{nm}^+\varphi\|^2 - \tfrac{1}{2}\langle\varphi, A_{nm}^+ A_{nm} W_\gamma\varphi\rangle - \tfrac{1}{2}\langle W_\gamma\varphi, A_{nm}^+ A_{nm}\varphi\rangle]. \tag{5.6}$$

Since for $\gamma = 0$ the expression $\langle\varphi, W_\gamma\varphi\rangle \geq 0$, it could only be negative if there exists a value γ_0 of γ for which $\langle\varphi, W_{\gamma_0}\varphi\rangle = 0$ and $[(d/d\gamma)\langle\varphi, W_\gamma\varphi\rangle]_{\gamma_0} < 0$. From $\langle\varphi, W_{\gamma_0}\varphi\rangle = 0$ it follows that $W_{\gamma_0}\varphi = 0$ and therefore $[(d/d\gamma)\langle\varphi, W_\gamma\varphi\rangle]_{\gamma_0} > 0$.

The form (5.4) therefore guarantees that a mixture morphism $S_\gamma: K \to K$ (which depends on γ) is defined by $W_0 \to W_\gamma$; it may not, however, be a mixture automorphism! We may therefore rewrite (5.3) as follows:

$$\mathrm{tr}((S_\gamma W) e^{iH\gamma} F e^{-iH\gamma}) = \mathrm{tr}(W[S_\gamma'(e^{iH\gamma} F e^{-iH\gamma})]). \tag{5.7}$$

For the case of radiation emission we shall first give an explicit equation for (5.4) in XI (1.18).

There is a great similarity between (5.3) and (5.4) and the interaction picture given in X, §3 except that the quantity S_γ described above is not an automorphism.

Now that we have presented an overview of the second method which we shall use in X–XVI of this book we shall make a few remarks concerning approximations. The Hamiltonian operator given in (5.8) is so complex that it is essential that simplified approximations be used for practical applications of the theory. Physicists are so accustomed to this situation that they have no difficulty in accepting it. Often mathematicians will not accept an approximation until they have obtained formulas for the estimation of errors. Physicists are seldom concerned with such estimates because they can compare the approximate theory directly with experiment. Since every physical theory is not an exact picture of reality, every physicist must learn, in terms of experience, how good the theory is. A conceptually detailed discussion about the methods of approximation can be found in [1], §8.

The Hamiltonian operator which is well suited for systems consisting of n_1 electrons and n_2 atomic nuclei of the same type with charge Z and mass M is given by (here latin indices are summation indices for electrons, greek indices are summation indices over the nuclei)

$$H = H_0 + H_1 + \cdots + H_6,$$

where the individual terms are given by

$$\begin{aligned}
H_0 &= \frac{1}{2m}\sum_i (\vec{P}_i - e\vec{A}(\vec{r}_i, t))^2 + \frac{1}{2M}\sum_\nu (\vec{P}_\nu - Z|e|\vec{A}(\vec{r}_\nu, t))^2 \\
&\quad + \sum_{i<k}\frac{e^2}{r_{ik}} + \sum_{\nu<\mu}\frac{Z^2e^2}{r_{\nu\mu}} - \sum_{i\nu}\frac{Ze^2}{r_{i\nu}} + \sum_i e\varphi(\vec{r}_i, t) \\
&\quad + \sum_\nu Z|e|\varphi(\vec{r}_\nu, t); \\
H_1 &= -\frac{1}{8m^3}\sum_i (\vec{P}_i)^4; \\
H_2 &= -\frac{e^2}{2m^2}\sum_{i<k}\frac{1}{r_{ik}^3}[\vec{P}_i\cdot\vec{P}_k r_{ik}^2 + (\vec{P}_i\cdot\vec{r}_{ik})(\vec{P}_k\cdot\vec{r}_{ik})]; \\
H_3 &= \frac{e^2}{2m^2}\sum_{i<k}\frac{1}{r_{ik}^3}[(\vec{r}_{ik}\times\vec{P}_k)\cdot\vec{S}_i + (\vec{r}_{ik}\times\vec{P}_i)\cdot\vec{S}_k]; \\
H_4 &= \frac{e^2}{m^2}\sum_{i<k}\frac{1}{r_{ik}^5}[\vec{S}_i\cdot\vec{S}_k r_{ik}^2 - 3(\vec{S}_i\cdot\vec{r}_{ik})(\vec{S}_k\cdot\vec{r}_{ik})]; \\
H_5 &= -\frac{e}{m}\sum_i \vec{S}_i\cdot\vec{B}(\vec{r}_i, t) + \frac{Ze^2}{2m^2}\sum_{i,\nu}\frac{1}{r_{i\nu}^3}(\vec{r}_{i\nu}\times\vec{P}_i)\cdot\vec{S}_i; \\
H_6 &= \frac{\pi Ze^2}{2m^2}\sum_{i,\nu}\delta(\vec{r}_{i\nu}).
\end{aligned} \tag{5.8}$$

Here we have also taken into account "external fields" where the latter are described by means of a given vector potential $\vec{A}(\vec{r}, t)$ and a scalar potential $\varphi(\vec{r}, t)$. We will systematically treat the problem of external fields in §6; we have included them in (5.8) for completeness. In (5.8) we have denoted the position operators by $\vec{r}_i$ or $\vec{r}_\nu$ and $\vec{r}_{ik} = \vec{r}_i - \vec{r}_k$, $r_{ik} = |\vec{r}_{ik}|$, etc., the spin operators for the ith electron are denoted by $\vec{S}_1\cdot\ \vec{B} = \operatorname{curl}\vec{A}$ is the external magnetic field.

The form of (5.8) can hardly be obtained from the correspondence principle. It follows directly from quantum electrodynamics (where the latter unfortunately does not have a correct mathematical representation).

In (5.8) we have only considered a single type of atomic nucleus. It is trivial to extend (5.8) to the case of several types of nuclei with different charges Z.

In addition, it is important to note that, mathematically, H as defined by (5.8) is not well defined. Since H is an unbounded operator it is necessary to specify its domain of definition. We shall not do so here. We only note that it must be a subset of the well-defined domain of definition of the kinetic energy

$$\frac{1}{2m}\sum_i \vec{P}_i^2 + \frac{1}{2M}\sum_\mu \vec{P}_\mu^2.$$

Difficulties in the specification of the domain of definition of H arise from singularities of the "position functions" in (5.8). In order to find a physically meaningful definition of H consider the point charge of an electron or nucleus as the limit of a charged sphere of fixed charge as the radius of the sphere tends to 0. This picture is helpful in overcoming some of the difficulties associated with the possible multiple meaning of (5.8), especially with respect to the δ-function $\delta(\vec{r}_{i\nu})$. In this book we shall not deal with the mathematically complicated problem of the determination of the domain of definition of the operator H. In the problems presented in XII–XIV we shall occasionally make reference to this problem.

According to VII, §7 the parity operator r can be represented in the form

$$U(r) \times U(r) \times \cdots \times U(r), \tag{5.9}$$

where $U(r)$ is given by VII (7.4). In this way we obtain a representation of the complete Galileo group $\mathscr{G}^{(r)}$ since the Hamiltonian operator (5.8) (excluding the case of external fields) commutes with the transformations (5.9). This commutivity plays an important role in the investigation of atomic spectra presented in XI–XIV.

6 Microsystems in External Fields

A frequently used and, for many experiments, a very important approximation is that of the external fields. We shall briefly discuss this method and its application.

In many experiments the composite system consists of a microsystem (electron, atom, molecule) and a much larger macroscopic system. If a microsystem enters the structure of a macrosystem—such as a particle in a counter—then we cannot use the approximation of the external field mentioned above. If, however, the microsystem remains away from the atomic structure of the macrosystem then the interaction can be described to good approximation with the help of the external field. We also require that the microsystem does not produce such changes which alter the external field produced by the macrosystem.

From this first, yet somewhat provisional limitation of the application domain for the method of the external field, we find that we leave the domain of application of quantum mechanics with its different discrete (!) system types; since we have incorporated macrosystems which can be described objectively in terms of external fields, as we have outlined in connection with axiom AV 4 in III, §3. Nevertheless it is possible to retain this approximation in the "domain" of quantum mechanics (see [1], §12.3), precisely because in this approximation the macrosystems appear only in the form of given fields, and are not influenced by the microsystems themselves. The influence of macrosystems by microsystems, after all, is decisively important for the "measurement process," that is, for a more precise physical analysis of the registration of microsystems. We have only superficially described these registration processes in terms of the structure $\mathscr{R}_0$, $\mathscr{R}$ introduced in II, §4.2 (a precise description is given in [13]).

As exterior fields we consider only electromagnetic fields which may be described by a vector potential $\vec{A}(\vec{r}, t)$ and scalar potential $\varphi(\vec{r}, t)$. Here $\vec{r}$, t are the technically specified laboratory space time coordinates. Here $(\vec{A}, \varphi)$ are considered to be fields which were specified in terms of the experimental arrangement. The electromagnetic field is obtained from $(\vec{A}, \varphi)$ in the usual way

$$\vec{E} = -\frac{\partial \vec{A}}{\partial t} - \operatorname{grad} \varphi, \qquad \vec{B} = \operatorname{curl} \vec{A}. \tag{6.1}$$

In terms of quantum mechanics we now have a "composite" system consisting of a microsystem of type (1) and a field $(\vec{A}, \varphi)$. We may characterize such a system by the "index" (1) $(\vec{A}, \varphi)$. Since the type (1) system will be held fixed in the following considerations, it suffices therefore to characterize a system type of composite systems by using $(\vec{A}, \varphi)$ as an "index." The "discrete" index for system types is then replaced by the "continuous" index $(\vec{A}, \varphi)$. Two different function pairs $(\vec{A}(\vec{r}, t), \varphi(\vec{r}, t))$ will therefore be considered to be different indices. Corresponding to this, an effect procedure (b_0, b) does not determine a single effect $F \in L(\mathscr{H})$, where $\mathscr{H}$ is the Hilbert space of the microsystems (as a subsystem of the composite system consisting of the microsystem and the field $(\vec{A}, \varphi)$), but an operator which depends on $\vec{A}, \varphi$: $\psi(b_0, b) = F(\vec{A}, \varphi)$ where $(\vec{A}, \varphi)$ occurs instead of the discrete index in the formula:

$$\psi(b_0, b) = (F_1, F_2, \ldots, F_i, \ldots).$$

$F(\vec{A}, \varphi)$ is, as an operator in $\mathscr{H}$, a function of the field $(\vec{A}, \varphi)$.

The fact that an effect procedure (b_0, b) generally determines an effect operator $F(\vec{A}, \varphi) \in L(\mathscr{H})$ which depends upon a field $(\vec{A}, \varphi)$ is usually neglected in quantum mechanics. Such neglect can lead to difficulties in the interpretation of the Galileo group.

If we wish to be more general we must introduce a σ-ring Σ of measurable subsets in the space of the fields $(\vec{A}, \varphi)$ and each individual preparation procedure there will correspond to an entire σ-additive measure $\Sigma \xrightarrow{W} K(\mathscr{H})$ as an ensemble. For such a measure the probability will be computed by the formula

$$\int \operatorname{tr}(dW(\vec{A}, \varphi) F(\vec{A}, \varphi)), \tag{6.2}$$

where the integration over $(\vec{A}, \varphi)$ corresponds to the summation over the different system types in the formula $\sum_i \operatorname{tr}(W_i F_i)$ for the probabilities. Here, however, we shall return to the usual description, where we assume that the field $(\vec{A}, \varphi)$ is, for all practical purposes, uniquely specified by the experiment, that is, was uniquely prepared, that is, the dispersion of the field can be neglected. Then the measure W is nonzero only in the neighborhood of a single point in the space of $(\vec{A}, \varphi)$. Instead of (6.2)

$$\operatorname{tr}(WF(\vec{A}, \varphi)) \tag{6.3}$$

is the probability that the effect $F(\vec{A}, \varphi)$ will occur for the "specified field" $(\vec{A}, \varphi)$. For the same preparation procedure a for which $\varphi(a) = W$ the probability for the effect procedure (b_0, b) for which $\psi(b_0, b) = F(\vec{A}, \varphi)$ will depend on the field $(\vec{A}, \varphi)$. The condition $\varphi(a) = W$ is independent of the field is correct only if we prepare the microsystems in a space region outside the field.

After these introductory remarks we will begin with (6.3), $W = \varphi(a)$ and $F(\vec{A}, \varphi) = \psi(b_0, b)$ as the starting point for the description of the microsystems under consideration in an external field. In this way we obtain a description which can be carried out using the Hilbert space $\mathscr{H}$ of the microsystems.

The essential structure which must be added to (6.3) consists of the specification of the Galileo group as the transformation group of the registration methods relative to the preparation procedures and therefore also relative to the external field $(\vec{A}, \varphi)$ as was physically interpreted in VII, §1.

Here we observe that this representation cannot be the representation of the Galileo group in the Hilbert space for system (1) (that is, without external fields) because the effects $F(\vec{A}, \varphi)$ span a different Banach space than $\mathscr{B}(\mathscr{H})$ does, because the set of $F(\vec{A}, \varphi)$ is a set of maps from the space of the fields $(\vec{A}, \varphi)$ into $\mathscr{B}(\mathscr{H})$! Here we shall not proceed any further towards a more precise determination of this Banach space of maps; we would have to introduce a uniform structure into the space of $(\vec{A}, \varphi)$ and require that the maps be uniformly continuous. A representation of the Galileo group in the space of the functions $F(\vec{A}, \varphi)$ is not uniquely determined, even less so than in the space $\mathscr{B}(\mathscr{H})$ of composite microsystems.

For a certain range of applications the following method leads to the determination of the representation of the Galileo group, that is, leads to a useful theory. The method is copied from that presented in §2.

To each time t there exists a map $T_t^{(1)}$ of $L(\mathscr{H})$ into the set of effects $F(\vec{A}, \varphi)$ for which

$$T_t^{(1)}F = F_t, \qquad T_{t+\gamma}^{(1)}F = F_{t+\gamma} = V(1, 0, 0, \gamma)F_t, \tag{6.4}$$

where $V(D, \vec{\delta}, \vec{\eta}, \gamma)$ represents the Galileo transformation of the $F(\vec{A}, \varphi)$. According to the second requirement in (6.4) it is sufficient to specify $T_t^{(1)}$ for $t = 0$.

The physical interpretation is that $F_t = T_t^{(1)}F$ represents an effect for a registration apparatus which is only in interaction with the microsystems for a short time at time t. The structure introduced by (6.4) is probably physically useful only when it is possible to make such relatively "quick" measurements, where we must later specify what we mean by "quick."

By analogy with §2 we require that for system (2)—that is, the field $(\vec{A}, \varphi)$—we are able to register the field independently of the microsystem at each time t. We express this as follows: The special effects

$$F_t(\vec{A}, \varphi) = \mathbf{1}f_t(\vec{A}, \varphi), \tag{6.5}$$

where $f_t(\vec{A}, \varphi)$ are real functions of the field (and its first derivatives) at time t and are to be interpreted as such effects which depend "only" on the field at time t. This does not mean that the signal which corresponds to the effect (6.5) occurs at time t but that the registration apparatus is only influenced by the behavior of the field during a short time span around t.

Let us now consider (6.4) and (6.5) for the special case where $t = 0$:

$$F_0 = T_0^{(1)}F, \tag{6.6}$$

$$F_0(\vec{A}, \varphi) = \mathbf{1}f_0(\vec{A}, \varphi). \tag{6.7}$$

Here F_0 in (6.6) corresponds to the effect $F_1 \times \mathbf{1}$ from §2 and $\mathbf{1}f_0$ corresponds to the effect $\mathbf{1} \times F_2$ from §2. The effects $F_1 \times F_2$ from §2 here correspond to an effect of the form $F_0 f_0(\vec{A}, \varphi)$.

It is reasonable to transform (2.6) into the following form. Let $U_1(D, \vec{\delta}, \vec{\eta}, 0)$ denote the representative of the Galileo transformation for the microsystem (1) without an external field. How may we introduce the Galileo transformations of the effects $f_0(\vec{A}, \varphi)$, that is, of the "effects of the pure fields" (without microsystems)? This formulation must be based on the physical interpretation of these effects.

If, for example, the registration apparatus characterized by $f_0(\vec{A}, \varphi)$ is translated by $\vec{\eta}$ then it is apparently subjected to the field

$$\vec{A}'(\vec{r}, t) = \vec{A}(\vec{r} + \vec{\eta}, t), \qquad \varphi'(\vec{r}, t) = \varphi(\vec{r} + \vec{\eta}, t), \tag{6.8}$$

that is, it responds as if the field (6.8) is acting on the original apparatus. The translated registration apparatus then corresponds to an f_0' for which

$$f_0'(A, \varphi) = f_0(A', \varphi'), \tag{6.9}$$

where $\vec{A}', \varphi'$ were defined in (6.8). We now generally define a Galileo transformation of the field by

$$k(D, \vec{\delta}, \vec{\eta}, \gamma)(\vec{A}, \varphi) = (\vec{A}', \varphi'), \tag{6.10}$$

where

$$k(1, 0, \vec{\eta}, \gamma)(\vec{A}(\vec{r}, t), \varphi(\vec{r}, t)) = (\vec{A}(\vec{r} - \vec{\eta}, t - \gamma), \varphi(\vec{r} - \vec{\eta}, t - \gamma)), \tag{6.11}$$

$$k(D, 0, 0, 0)(\vec{A}(\vec{r}, t), \varphi(\vec{r}, t)) = (D\vec{A}(D^{-1}\vec{r}, t), \varphi(D^{-1}\vec{r}, t)). \tag{6.12}$$

The $k(1, \vec{\delta}, 0, 0)$ are to be defined in a similar way as an approximation of the proper Lorentz transformations of $\vec{A}$ and φ. We will not enter into a discussion of such approximations.

We now define the transformation of $f_0(\vec{A}, \varphi)$ which corresponds to the transformation $(D, \vec{\delta}, \vec{\eta}, 0)$ as a generalization of (6.9)

$$f_0'(\vec{A}, \varphi) = f_0(k^{-1}(D, \vec{\delta}, \vec{\eta}, 0)(\vec{A}, \varphi)). \tag{6.13}$$

We transform (2.6) for this case as follows:

$$\begin{aligned} &V(D, \vec{\delta}, \vec{\eta}, 0)F_0 f_0(\vec{A}, \varphi) \\ &\quad = U_1(D, \vec{\delta}, \vec{\eta}, 0)F_0 U_1^+(D, \vec{\delta}, \vec{\eta}, 0) f_0(k^{-1}(D, \vec{\delta}, \vec{\eta}, 0)(\vec{A}, \varphi)). \end{aligned} \tag{6.14}$$

According to (6.14) $V(D, \vec{\delta}, \vec{\eta}, 0)F_0 f_0$ takes on the same form $F'f'$ as does $F_0 f_0$: the $T_0^{(1)}F = F_0$ and the $\mathbf{1}f_0$ are separately transformed by the Galileo transformations without time translation. If we require that (6.14) holds for the case of a time translation, that would mean that the external field exerts no influence on the microsystem. By analogy with the case described in §2 we define

$$V(1, 0, 0, \tau)F_0 f_0(A, \varphi) = U_\tau F_0 U_\tau^+ f_0(k^{-1}(1, 0, 0, \tau)(\vec{A}, \varphi)), \tag{6.15}$$

where the unitary family of operators U_τ ($U_0 = \mathbf{1}$) depends on the field $\vec{A}$, φ. For U_τ we require that the following differential equation be satisfied:

$$\frac{dU_\tau}{d\tau} = iH_\tau U_\tau, \tag{6.16}$$

where H_τ is a time-dependent Hamiltonian operator.

We will often make use of the special effects $T_0^{(1)}F = F_0$ defined in (6.15) for which we write $V(1, 0, 0, t)F_0 = F_t$. From (6.15) we find that

$$F_t = U_t F_0 U_t^+. \tag{6.17}$$

From (6.16) and (6.17) we obtain the following differential equation for F_t:

$$\frac{dF_t}{dt} = i(H_t F_t - F_t H_t). \tag{6.18}$$

According to the above interpretation of T_t and (6.4) F_t is an effect which is triggered only by the microsystem "at time t." If we apply the registration method for the effect F_t at time γ later, then we would register the ffect $F_{t+\gamma}$.

We may now make the transition from the effects to a scale observable as follows: Let A_0 denote the self-adjoint operator which corresponds to the measurement of a scale observable for a microsystem "at time zero." Here "at time zero" means that the interaction for the registration method takes place, for all practical purposes, only at $t = 0$ (in the time scale of the laboratory). If, instead, we make the measurement "at time t" (that is, we displace the time relative to 0) we therefore measure the observable A_t which (according to (6.17)) satisfies the equation

$$A_t = U_t A_0 U_t^+ \tag{6.19}$$

and, according to (6.18) satisfies the differential equation

$$\frac{dA_t}{dt} = i(H_tA_t - A_tH_t). \tag{6.20}$$

Examples of such special observables are the position and momentum observables of microsystems: Whether the system of type (1) under consideration is itself an elementary or a composite system in the sense of §2–§4, in every case the position and momentum observables $\vec{Q}_i(0)$, $\vec{P}_i(0)$ for the individual components of system (1) are defined by the representation $U(1, \vec{\delta}, \vec{\eta}, 0)$. Here $\vec{Q}_i(0)$ and $\vec{P}_i(0)$ are these observables at time $t = 0$. By analogy with (6.19) we obtain

$$\vec{P}_i(t) = U_t\vec{P}_i(0)U_t^+, \qquad \vec{Q}_i(t) = U_t\vec{Q}_i(0)U_t^+. \tag{6.21}$$

From (6.21) it follows that the components of $\vec{P}_i(t)$, $\vec{Q}_i(t)$–$P_{i\nu}(t)$, $Q_{i\nu}(t)$ satisfy the same commutation relations as the $P_{i\nu}(0)$, $Q_{i\nu}(0)$:

$$P_{i\nu}(t)Q_{k\mu}(t) - Q_{k\mu}(t)P_{i\nu}(t) = \delta_{ik}\delta_{\mu\nu}\frac{1}{i}\mathbf{1},$$

$$Q_{i\nu}(t)Q_{k\mu}(t) - Q_{k\mu}(t)Q_{i\nu}(t) = 0,$$

$$P_{i\nu}(t)P_{k\mu}(t) - P_{k\mu}(t)P_{i\nu}(t) = 0. \tag{6.22}$$

The Heisenberg commutation relations are therefore satisfied "for all time," but not, of course, for $P_{i\nu}(t)$, $Q_{k\mu}(t')$, where $t \neq t'$.

In (6.15) we may consider generalized effects $F_0(\vec{A}, \varphi)$ which are not only of the form $F_0 f_0(\vec{A}, \varphi)$. We obtain

$$V(1, 0, 0, \tau)F_0(\vec{A}, \varphi) = U_\tau F_0(k^{-1}(1, 0, 0, \tau)(\vec{A}, \varphi))U_\tau^+. \tag{6.23}$$

Here $F_0(\vec{A}, \varphi)$ depends only on the fields $\vec{A}$, φ (and their first derivatives) at time $t = 0$. $F_0(k^{-1}(1, 0, 0, \tau)(\vec{A}, \varphi))$ depends only on the fields at time $t = \tau$. We shall now simplify the notation of (6.23) as follows:

$$F_t(\vec{A}, \varphi) = V(1, 0, 0, t)F_0(\vec{A}, \varphi). \tag{6.24}$$

$F_t(A, \varphi)$ is therefore an effect which is measured by a registration method which is only affected by the microsystem and the field at time t. The meaning of the left side of (6.24) is expressed by the right side of equation (6.23) as follows: F_t is the time dependent function $U_tF_0U_t^+$ which depends on the functions $\vec{A}$, φ (and their first derivative) at time t. By differentiation, from (6.23) we obtain

$$\frac{d}{dt}(F_t(A, \varphi)) = i[H_tF_t(\vec{A}, \varphi) - F_t(\vec{A}, \varphi)H_t]$$

$$+ U_t\frac{d}{dt}F_0(k^{-1}(1, 0, 0, t)(\vec{A}, \varphi))U_t^+. \tag{6.25}$$

The second term on the right-hand side of (6.25) is nothing other than the partial derivative of $F_t(A, \varphi)$ with respect to the t which appears in the fields

$\vec{A}, \varphi$. If we write this derivative in the form $(\partial/\partial t)F_t(A, \varphi)$ then (6.25) is transformed into

$$\frac{d}{dt}(F_t(\vec{A}, \varphi)) = i(H_t F_t - F_t H_t) + \frac{\partial F_t(\vec{A}, \varphi)}{\partial t}. \tag{6.26}$$

For a corresponding scale observable $B_t(\vec{A}, \varphi)$ (where the latter is a time-dependent function of the fields and their first derivatives at time t) it follows that

$$\frac{dB_t}{dt} = i(H_t B_t - B_t H_t) + \frac{\partial B_t}{\partial t}. \tag{6.27}$$

The frequently used simplified notation (6.27) can often lead to misunderstanding if it is not precisely explained. (6.27) is often referred to as "the quantum mechanical equation of motion" because it appears to be the "most general" form of the Galileo transformation for a simple time translation.

Certainly (6.27) is (in a formal sense) the most general form since we may derive the others as special cases (for example, for the case $\vec{A} = 0$, $\varphi = 0$ which corresponds to the case of no external field) from (6.27). We note, however, that physically the applications of (6.27) are limited to the case of the external field approximation. For the general case of measurement (6.27) is, on the contrary, not sufficiently general.

In (6.26) we considered effects which "respond only at time t." In fact, realistic effects are somewhat more complicated—they are somewhat complicated functionals of the entire field $\vec{A}, \varphi$. Since, however, they do not easily permit a formulation in terms of simple equations analogous to (6.26) and (6.27) we shall restrict our consideration to the effects and observables described by (6.26) and (6.27) and leave the problem of the measurement of such special effects to the experimental physicist.

We shall now define the Hamiltonian operator H_t in (6.16) more precisely.

In principle we have already given it in (5.8). Here we need only state which operators $\vec{P}_i, \vec{Q}_i = \vec{r}_i, \vec{P}_\nu, \vec{Q}_\nu = \vec{r}_\nu$ in (5.8) are to be used. If the external fields do not depend on time t—that is, we have constant external fields—we may choose the operators $\vec{P}_i, \ldots, \vec{Q}_\nu$ to be the operators defined at $t = 0$, that is, we choose $H_t = H_0$ to be constant with time. We may then choose the $\vec{P}_i, \ldots, \vec{Q}_\nu$ at any other time since, for $A_t = H_t$ in (6.20) we obtain

$$H_t = H(\vec{P}(t), Q(t)) = H(P(0), Q(0)).$$

For U_t it then follows that

$$U_t = e^{iHt}. \tag{6.28}$$

If the external fields $\vec{A}, \varphi$ are time dependent, then in (5.8) we must choose in H_t the $\vec{P}_i, \ldots, \vec{Q}_\nu$ and the spins $\vec{S}_i$ as the $\vec{P}_i(t), \ldots, \vec{Q}_\nu(t), \vec{S}_i(t)$. In this way we obtain a representation of the Galileo group for systems in external fields. Here H_t is itself an observable of the form B_t described in (6.26). We therefore find that:

$$\frac{dH_t}{dt} = i(H_t H_t - H_t H_t) + \frac{\partial H_t}{\partial t} = \frac{\partial H_t}{\partial t}. \tag{6.29}$$

Here we shall not proceed further into the problem of the domain of definition of H_t and therefore the existence of solutions for the differential equation (6.16). In physics it is generally assumed that a family of unitary transformations U_t exists and that the domain of definition of H_t is the same as the region where U_t is differentiable. These physical ideas do not, of course, lead to a solution of the mathematical problem for a more or less well defined H_t given by (6.16) and (5.8). Here we refer the reader to [11] and [36].

In closing, we shall now state the form of an infinitesimal translation for the special effects introduced in (6.17). With the total momentum $\vec{P}(t)$ of system (1) it follows that for an infinitesimal $\vec{\eta}$:

$$V(1, 0, \vec{\eta}, 0)F_t = F_t + i\vec{\eta}[\vec{P}(t)F_t - F_t P(t)]. \tag{6.30}$$

We have given this formula in order to warn about possible errors in the use of $\vec{P}(0)$ in (6.30) for all F_t. It is of decisive importance that $V(\ldots)$ does not give a representation of the Galileo group by means of transformations in $\mathscr{B}(\mathscr{H})$ but gives a representation by means of transformations in the Banach space of the mappings of the space of the fields into the space $\mathscr{B}(\mathscr{H})$!

The "external fields" provide a useful means by which the mass of charged systems can be measured, especially for the case of elementary systems. For example, for an electron the Hamiltonian operator (5.8) (for small values of the momentum and neglecting the spin term in H_5) is given by

$$H = \frac{1}{2m}(\vec{P} - e\vec{A})^2 + e\varphi. \tag{6.31}$$

In XVII, §6.2 we shall briefly outline how we may develop measurement methods by which the quantity e/m (in cm) may be measured by the deflection produced by the field. If we then succeed in measuring the elementary charge (by measuring e/M for macroscopic bodies and then measuring M with the aid of other forces in units of gm—Milikan experiment and other modern experiments) we then obtain Planck's constant as a conversion factor between the units gm and cm^{-1} (see VII (2.39)). Here we, of course, assume that m is previously known in units of cm^{-1} from atomic processes (from the structure of atomic spectra (according to XI–XIV) or more directly from the motion of electrons).

7 Criticism of the Description of Interaction in Quantum Mechanics and the Problem of the Space $\mathscr{D}$

Since the position and momentum observables for elementary systems could be introduced in the manner described in VII, §4 it is, in principle, possible to determine whether a registration method measures position or momentum for an elementary system on the basis of the interpretation of the preparation and registration procedures described in II and the interpretation of the Galileo group described in VII, §1. For the case of the position and momentum observables defined in §2 for the case of individual systems in a composite system the situation is problematical, since the subsystems undergo mutual interaction.

The introduction of these observables was only a consequence of the newly introduced structure (2.1) for the Hilbert space of a composite system. Since every Hilbert space $\mathscr{H}_1$ is isomorphic to any other Hilbert space of the same dimension we find that $K(\mathscr{H}_1)$ is isomorphic to $K(\mathscr{H}_2)$ and $L(\mathscr{H}_1)$ is isomorphic to $L(\mathscr{H}_2)$ including the bilinear form $\operatorname{tr}(w, g)$. A mathematical structure of the form (2.1) can be introduced for any Hilbert space. Such a structure has no physical significance as long as this structure is not connected with an additional physical interpretation. In §2 we have attempted to give such an interpretation by singling out the set of all effects of the form $F_1 \times F_2$ for this purpose. However, this procedure is physically questionable.

Is, however, such a procedure absolutely necessary? If $L(\mathscr{H}_1)$, $L(\mathscr{H}_2)$, and $L(\mathscr{H}_1 \times \mathscr{H}_2)$ are isomorphic, how may we correctly make such a distinction? It makes possible a heuristic procedure for the formulation of the Hamiltonian operator according to (5.8); as we shall see in XI–XIV the eigenvalue spectrum of this Hamiltonian operator is in excellent agreement with experience. Where else is the structure (2.1) useful? Is this structure really needed as part of the structure associated with the physics? Does the subset of effects $F_1 \times F_2$ have any real physical significance? Can we simply forget about (2.1) once we have determined the spectral family of H, that is, of e^{iHt}?

Certainly we have the fact that it is possible to register the subsystems once they are no longer in interaction. Scattering theory, described in XVI is based on this experimental fact. In this sense we can also say that in an asymptotic sense, for large times (or large distances) the introduction of (2.1) correctly describes the asymptotic structure. In particular, it is not clear whether it is physically meaningful to extend such an asymptotic structure to situations in which the subsystems are in close interaction. Indeed, it may be meaningless in those circumstances to speak of "subsystems" because the concept of a subsystem depends upon (2.1)! In elementary particle physics it is physically meaningless (except as a crude approximation) to speak of subsystems of a system with respect to high energy scattering experiments.

In "nonrelativistic" quantum mechanics described here there are additional experimental facts which are explainable in terms of the structure (2.1). For example, the angular momentum structure is, according to §1, quite open; however, on the basis of experiments with external fields (see, for example, the Zeeman effect in XIV, §6) it is in agreement with the structure (2.1) (and its extension to the case of more than two subsystems). The intensities of spectral lines, calculated in XIV, §5 is in agreement with the structure given by (2.1). In addition, the structure of molecules (see XV) and therefore (indirectly) the theoretical description of the most important facts in chemistry are a consequence of the structures which were introduced in §2–§6.

In spite of all these successes there remains a cluster of problems which have already been cited in §2. Is it necessary to impose the condition of "relatively short measurement interaction times" in order to obtain a basis for (2.1)? This does not appear to be the case, but cannot be proven. The considerations in XVII appear to mean that there is no particular time (even approximately) required for a measurement providing that the Hamiltonian operator does not explicitly depend upon the time. The case in which we have a time varying external field remains problematical, as we have found in the discussion in §6. If we exclude the case of external fields it appears that the structure (2.1) is only needed asymptotically and that the extrapolation of (2.1) is a useful tool for such physical problems as atomic spectra including light emission (XI–XV) and the most complicated scattering problems (not only those described in XVI).

Certainly it is possible that the structure (2.1) or better, the structure (2.3) indirectly determines a physical structure—namely the space $\mathscr{D}$.

By analogy with VII, §8 we may seek to obtain the space $\mathscr{D}$ with the aid of the direct product of two (or more) Galileo groups. Since the spaces $\mathscr{B}'(\mathscr{H}_1)$, $\mathscr{B}'(\mathscr{H}_2)$, and $\mathscr{B}'(\mathscr{H}_1 \times \mathscr{H}_2)$ are isomorphic, the additional structure of the subspaces $\mathscr{D}_1 \subset \mathscr{B}'(\mathscr{H}_1)$, $\mathscr{D}_2 \subset \mathscr{B}'(\mathscr{H}_2)$, and $\mathscr{D} \subset \mathscr{B}'(\mathscr{H}_1 \times \mathscr{H}_2)$ describe an essential difference between, for example, $\mathscr{D}_1 \subset \mathscr{B}'(\mathscr{H}_1)$ and $\mathscr{D} \subset \mathscr{B}'(\mathscr{H}_1 \times \mathscr{H}_2)$. The convex sets $K(\mathscr{H}_1)$ and $K(\mathscr{H}_1 \times \mathscr{H}_2)$ are certainly isomorphic, that is, there exists a mixture isomorphism $K(\mathscr{H}_1) \to K(\mathscr{H}_1 \times \mathscr{H}_2)$; however, it is possible that an isomorphism does not exist which is $\sigma(K(\mathscr{H}_1), \mathscr{D}_1)$–$\sigma(K(\mathscr{H}_1 \times \mathscr{H}_2), \mathscr{D})$-continuous!

In the case of classical mechanics the analogous fact has been known for a long time: The set K is the set of all measures which are totally continuous with respect to Lebesgue measure, that is, which can be described by a measurable density function $\rho(x)$ in the Γ-space. The set L is the set of all measurable functions f for which $0 \leq f \leq 1$. For the space $\mathscr{D}$ we may choose the space spanned by the 1-function and the set of all continuous f with compact support. For a single mass point the space Γ_1 is six-dimensional. For two mass points the composite space is twelve-dimensional. If K_1, L_1 correspond to Γ_1 then there certainly are mixture isomorphisms $K_1 \to K$ but none which are $\sigma(K_1, \mathscr{D}_1)$–$\sigma(K, \mathscr{D})$-continuous because the dual map must map the set $\mathscr{D}_1$ onto $\mathscr{D}$ which is impossible because of the different dimensions of Γ_1 and Γ since a $\sigma(K_1, \mathscr{D}_1)$–$\sigma(K, \mathscr{D})$-continuous isomorphism must lead to a topologically homeomorphic map of Γ_1 onto Γ (the extremal points of $\bar{K}^\sigma$ may be identified in the $\sigma(\mathscr{B}, \mathscr{D})$-topology with the topological space Γ!).

It is conceivable that we may be able to solve the problem of composite systems in quantum mechanics by proceeding in the opposite direction—where we supplement the assertions in §1 by introducing axioms about the space $\mathscr{D}$. Perhaps it is possible to carry out the introduction of the structure "composite system" in an improved manner with respect to the sense of an "axiomatic basis" in this way rather than the route we have used in §2–§6. Here, by the "sense of an axiomatic basis" we mean that new relations are introduced only when these relations can be physically interpreted on the basis of pre-theories (see [47]). Until such a clarification, it must be seen that the concept of a composite system is (as far as this concept exceeds the exposition of §1 is concerned) unfortunately one of the least well-explained concepts of quantum theory.

APPENDIX I

Summary of Lattice Theory

1 Definition of a Lattice

A set M is said to be a partially ordered set (poset) if there is a relation defined among pairs $(a, b) \in M \times M$ (which we denote by $\leq$) which satisfies the following axioms:

$$(1) \quad a \leq a \quad \textit{for all } a \in M,$$

$$(2) \quad a \leq b, b \leq c \Rightarrow a \leq c,$$

$$(3) \quad a \leq b, b \leq a \Rightarrow a = b.$$

M is said to be totally ordered if, for each pair $a, b \in M$ either $a \leq b$ or $b \leq a$.

If N is a subset of an ordered set M, then an element $a \in M$ is said to be an upper bound of N if $b \leq a$ for all $b \in N$. We say that the element $a \in M$ is the least upper bound for N if $a \leq c$ for every upper bound c of N. Similarly, a lower bound for N is an element $d \in M$ for which $d \leq b$ for all $b \in N$; we say that the element $c \in M$ is the greatest lower bound for N if $d \leq c$ for every lower bound d of N. If the greatest lower bound or least upper bounds of the set N exist, they are uniquely determined by the set N.

A lattice is a partially ordered set M for which every finite subset N has a least upper bound and a greatest lower bound. By induction, it is easy to show that M is a lattice if each pair of elements a, b has a least upper bound (denoted by $a \vee b$) and a greatest lower bound (denoted by $a \wedge b$).

Instead of $a \leq b$ we frequently write $b \geq a$. The least upper bound of N is denoted by $\bigvee_{a \in N} a$, the greatest lower bound is denoted by $\bigwedge_{a \in N} a$.

According to the definitions it is an easy matter to show that: $a = a \wedge a = a \vee a$, $a \vee b = b \vee a$, $(a \wedge b) \wedge c = a \wedge (b \wedge c)$, $(a \vee b) \vee c = a \vee (b \vee c)$, $a \vee (a \wedge b) = a$, $a \wedge (a \vee b) = a$, $a \leq b \Leftrightarrow a \vee b = b$, $a \leq b \Leftrightarrow a \wedge b = a$.

Th. 1.1. *Let M be a set, and let two binary operations $\vee$, $\wedge$, be defined on M which satisfy the following conditions:*

(1) $a \wedge b = b \wedge a, a \vee b = b \vee a;$
(2) $a \vee (b \vee c) = (a \vee b) \vee c,$
$a \wedge (b \wedge c) = (a \wedge b) \wedge c;$
(3) $a \vee (a \wedge b) = a, a \wedge (a \vee b) = a,$

then the relation $\leq$ defined by $a \leq b$ whenever $a \wedge b = a$ is a partial order and M is a lattice.

PROOF. From the first formula in (3) it follows that $a = a \vee (a \wedge (a \vee b))$, and from the second formula in (3) it follows that $a = a \vee a$; similarly it follows that $a \wedge a = a$. Therefore we obtain $a \leq a$.

From $a \leq b$ and $b \leq a$ it directly follows that $a = b$.

If $a \leq b$ and $b \leq c$, that is, if $a \wedge b = a$ and $b \wedge c = b$ then from (2): $a \wedge c = (a \wedge b) \wedge c = a \wedge (b \wedge c) = a \wedge b = a$, that is, $a \leq c$. Therefore $\leq$ is a partial order.

If $a \leq b$, then from $a \wedge b = a$ and from (3) it follows that $a \vee b = (a \wedge b) \vee b = b$. If $a \vee b = b$ then from (3) it follows that $a \wedge b = a \wedge (a \vee b) = a$, that is, $a \leq b$. Therefore $a \leq b$ is equivalent to $a \vee b = b$.

From (3) it follows directly that $a \wedge b \leq a$ and $a \wedge b \leq b$. If $c \leq a$ and $c \leq b$, then it follows that $c \wedge (a \wedge b) = (c \wedge a) \wedge b = c \wedge b = c$, that is, $c \leq a \wedge b$. Therefore $a \wedge b$ is the greatest lower bound of a, b. In a similar way it follows that $a \vee b$ is the least upper bound of a, b.

A lattice M is said to be complete if each subset N of M has a least upper bound and a greatest lower bound.

Th. 1.2. *If, for every subset N of a partially ordered set M there exists a greatest lower bound (least upper bound) then there exists a least upper bound (greatest lower bound), that is, M is a complete lattice.*

PROOF. Suppose for each subset N there exists a least upper bound. Let a denote the least upper bound of the set $R = \{b \mid b \leq c \text{ for all } c \in N\}$. Here we find that a is the greatest lower bound of the set N if we show that $a \in R$: From $c \in N$ it follows that $c \geq b$ for all $b \in R$ and we therefore obtain $c \geq a$.

In a similar way we may prove that every subset has a least upper bound.

The hypothesis in Th. 1.2 requires that the empty set has a least upper bound, that is, there exists a minimal element in M; corresponding to the existence of the greatest lower bound there exists a maximal element in M. We denote the minimal and maximal elements by 0 and 1 (or e) respectively.

D 1.1. If, for a given element $a \in M$ there exists an element $a' \in M$ for which $a \wedge a' = 0$ and $a \vee a' = 1$ then a' is called a complement of a. A lattice M is said to be complementary if each element $a \in M$ has a complement.

Here we note that a is also a complement of a'.

D 1.2. A bijective map $a \longrightarrow a^{\perp}$ of M into itself is said to be an orthocomplementation if the following conditions are satisfied:

(1) $a \leq b \Rightarrow b^{\perp} \leq a^{\perp}$,
(2) $(a^{\perp})^{\perp} = a$,
(3) $a \wedge a^{\perp} = 0$.

An orthocomplemented lattice is a lattice with unit and null element for which in addition, there exists an orthocomplementation structure. Often the orthocomplement of a will be denoted by a^* instead of by $a^{\perp}$.

Remark. For a given lattice there can be more than one orthocomplementation!

For a subset N we define the following subset

$$N^{\perp} = \{a \mid a = b^{\perp}, b \in N\}.$$

Th. 1.3. *In an orthocomplemented lattice $a \vee a^{\perp} = 1$, that is, $a^{\perp}$ is a complement of a. In addition $(a \vee b)^{\perp} = a^{\perp} \wedge b^{\perp}$ and $(a \wedge b)^{\perp} = a^{\perp} \vee b^{\perp}$.*

If the greatest upper bound $\bigwedge_{a \in N} a$ exists for the subset N, then $\bigvee_{b \in N^{\perp}} b$ exists, and $\bigvee_{b \in N^{\perp}} b = (\bigwedge_{a \in N} a)^{\perp}$. A similar result holds for the least upper bound.

PROOF. From D 1.2, (1) and (2) it follows that $b^{\perp} \leq a^{\perp} \Rightarrow a \leq b$. From this result it directly follows that the map $a \longrightarrow a^{\perp}$ maps least upper (greatest lower) bounds into greatest lower (least upper) bounds, and we have proven the second part of the theorem. In particular, $0^{\perp} = 1$. Thus from D 1.2, (3) it follows that $a \vee a^{\perp} = (a^{\perp} \wedge a)^{\perp} = 0^{\perp} = 1$.

D 1.3. We say that a is orthogonal to b (written $a \perp b$) if $a \leq b^{\perp}$.

From D 1.2, (1) and (2) it directly follows that $a \leq b^{\perp} \Rightarrow b \leq a^{\perp}$, that is, the relation $a \perp b$ is symmetric.

D 1.4. A lattice M is said to be distributive, if, for all $a, b, c \in M$ the following conditions are satisfied:

(1) $(a \vee b) \wedge c = (a \wedge c) \vee (b \wedge c)$,
(2) $(a \wedge b) \vee c = (a \vee c) \wedge (b \vee c)$.

In the following we shall abbreviate conditions (1) and (2) in D 1.4 by $D(a, b, c)$ and $D^*(a, b, c)$ respectively.

Th. 1.4. *A lattice M is distributive if one of the relations $D(a, b, c)$ or $D^*(a, b, c)$ is satisfied for all $a, b, c \in M$.*

PROOF. If $D(a, b, c)$ holds for all $a, b, c \in M$, then it follows that $(a \vee c) \wedge (b \vee c) = (a \wedge (b \vee c)) \vee (c \wedge (b \vee c)) = (a \wedge (b \vee c)) \vee c = (a \wedge b) \vee (a \wedge c) \vee c = (a \wedge b) \vee c$, that is, $D^*(a, b, c)$. In the same way we may show that if $D^*(a, b, c)$ holds for all $a, b, c \in M$ then $D(a, b, c)$ holds.

2 Orthomodularity

D 2.1. A pair of elements a, b is said to be a *modular* pair, if for all $c \leq b$ the relation $D(c, a, b)$ is satisfied, that is, if, for all $c < b$ the relation

$$(c \vee a) \wedge b = c \vee (a \wedge b)$$

is satisfied. We shall denote a modular pair a, b by $M(a, b)$. A lattice M is said to be modular if $M(a, b)$ is satisfied for all a, b.

We will now assume that M is an orthocomplemented lattice.

D 2.2. A pair a, b is said to be compatible (abbreviated $C(a, b)$) if

$$a = (a \wedge b) \vee (a \wedge b^{\perp})$$

is satisfied (see also IV, Th. 1.3.4(iv)).

It follows that $a \perp b \Rightarrow C(a, b)$; $C(a, b) \Rightarrow C(a, b^{\perp})$; $a \leq b \Rightarrow C(a, b)$, since $a \wedge b = a$ and $a \wedge b^{\perp} = a \wedge b \wedge b^{\perp} = 0$.

Th. 2.1. *The following statements are equivalent:*

(i) *For all pairs a, b satisfying $a \perp b$ $M(a, b)$ holds.*
(ii) *For all a we obtain $M(a, a^{\perp})$.*
(iii) *For all pairs a, b satisfying $a \leq b$ $C(b, a)$ holds, that is, $b = a \vee (b \wedge a^{\perp})$.*
(iv) *For all pairs a, b satisfying $a \leq b$*

$$a = b \wedge (a \vee b^{\perp}).$$

(v) *For all pairs a, b: $C(a, b) \Rightarrow C(b, a)$.*
(vi) *For each triple a, b, c satisfying $a \perp b$, $a \perp c$, the following condition holds:*

$$a \vee b = a \vee c \Rightarrow b = c.$$

(vii) *For each triple a, b, c satisfying $a \perp b$, $a \perp c$, the following condition holds:*

$$a \vee b = a \vee c \quad \text{and} \quad b \leq c \Rightarrow b = c.$$

(viii) *For each a the following condition holds:* $b \perp a$ *and* $a \vee b = 1 \Rightarrow b = a^{\perp}$, *that is, for each a there exists only one orthogonal complement.*

PROOF. (i) $\Rightarrow$ (ii) is clear, since $a \perp a^{\perp}$.

(ii) $\Rightarrow$ (iii). From $a < b$ it follows that $b^{\perp} \leq a^{\perp}$ and from $M(a, a^{\perp})$ it follows that $(b^{\perp} \vee a) \wedge a^{\perp} = b^{\perp} \vee (a \wedge a^{\perp}) = b^{\perp}$; therefore, by applying $\perp$ we finally obtain $b = a \vee (b \wedge a^{\perp})$.

(iii) $\Rightarrow$ (iv). If $a \leq b$ then $b^{\perp} \leq a^{\perp}$ and, according to (iii) $a^{\perp} = b^{\perp} \vee (a^{\perp} \wedge b)$, from which it follows that $a = b \wedge (a \vee b^{\perp})$. (iv) $\Rightarrow$ (iii) can be proven in the same way.

(iii) $\Rightarrow$ (v). From $C(a, b)$, that is, from $a = (a \wedge b) \vee (a \wedge b^{\perp})$ it follows that $a^{\perp} = (a^{\perp} \vee b^{\perp}) \wedge (a^{\perp} \vee b)$ and $b \wedge a^{\perp} = b \wedge (a^{\perp} \vee b)$. Hence it follows that $(b \wedge a) \vee (b \wedge a^{\perp}) = (b \wedge a) \vee (b \wedge (a^{\perp} \vee b^{\perp})) = (b \wedge a) \vee (b \wedge (a \wedge b)^{\perp})$; since $b \wedge a \leq b$, from (iii) we finally obtain $(b \wedge a) \vee (b \wedge (b \wedge a)^{\perp}) = b$.

(v) $\Rightarrow$ (vi). Assuming (v) holds, we obtain $C(a, b) \Rightarrow C(a, b^{\perp}) \Rightarrow C(b^{\perp}, a) \Rightarrow C(b^{\perp}, a^{\perp}) \Rightarrow C(a^{\perp}, b^{\perp})$. From $a \vee b = a \vee c$, $b \perp a$ and $c \perp a$ it follows that $a^{\perp} \wedge b^{\perp} = a^{\perp} \wedge c^{\perp}$. Since $a \perp b \Rightarrow C(a, b) \Rightarrow C(b^{\perp}, a^{\perp})$ it follows that $b^{\perp} = (b^{\perp} \wedge a) \vee (b^{\perp} \wedge a^{\perp}) = a \vee (b^{\perp} \wedge a^{\perp}) = a \vee (c^{\perp} \wedge a^{\perp})$. Similarly it follows that $c^{\perp} = a \vee (c^{\perp} \wedge a^{\perp})$ and therefore $b^{\perp} = c^{\perp}$, that is, $b = c$.

(vi) $\Rightarrow$ (vii) is trivial.

(vii) $\Rightarrow$ (viii). From $a \vee b = 1 = a \vee a^{\perp}$ and $b \perp a$, that is, $b \leq a^{\perp}$ we obtain the special case in which $b = a^{\perp}$.

(viii) $\Rightarrow$ (iii): We assume (by (iii)) $a \leq b$. We define $d = a \vee (b \wedge a^{\perp})$. Thus we obtain $d \leq b$ and therefore $d \wedge b^{\perp} = (d \wedge b) \wedge b^{\perp} = 0$. On the other hand, $b^{\perp} \vee d = b^{\perp} \vee a \vee (b \wedge a^{\perp}) = (b \wedge a^{\perp})^{\perp} \vee (b \wedge a^{\perp}) = 1$. From (viii) it follows that $d = (b^{\perp})^{\perp} = b$, that is, $b = a \vee (b \wedge a^{\perp})$.

(iii) $\Rightarrow$ (i). According to (i) we assume that $a \perp b$ and $c \leq b$. Since $a \perp b$ $M(a, b)$ is equivalent to $(c \vee a) \wedge b = c$. Since $c \leq b$ and $c \leq c \vee a$ we find that $(c \vee a) \wedge b \geq c$. According to (iii), for $g = ((c \vee a) \wedge b) \wedge c^{\perp}$ we obtain the relation $(c \vee a) \wedge b = c \vee g$. Since $(c \vee a) \wedge b \leq b$ we obtain $g \leq b \wedge c^{\perp}$, that is, $b^{\perp} \vee c \leq g^{\perp}$. Since $(c \vee a) \wedge b \leq c \vee a$ it follows that $g \leq c \vee a$ and, since $a \perp b$, that is, $a < b^{\perp}$ we obtain $g \leq c \vee b^{\perp} \leq g^{\perp}$—from which we conclude $g = 0$, that is, $c = (c \vee a) \wedge b$.

D 2.3. An orthocomplemented lattice is said to be orthomodular if it satisfies one of the conditions (i)–(viii) of Th. 2.1.

D 2.4. Let M be an orthocomplemented lattice. A real function $M \xrightarrow{m} [0, 1]$ is said to be a normed orthomeasure on M if

(1) $m(1) = 1$,
(2) $a \perp b \Rightarrow m(a \vee b) = m(a) + m(b)$.

Th. 2.2. *If there exists a set K of orthomeasures on M such that* $m(a) = m(b)$ *for all* $m \in K$ *implies the relation* $a = b$ *(that is K separates M) then M is orthomodular.*

PROOF. According to Th. 2.1(viii) let $a \vee b = a \vee a^{\perp} = 1$ with $b \perp a$. Thus, for all $m \in K$, it follows that $m(a) + m(b) = m(a) + m(a^{\perp})$, that is, $m(b) = m(a^{\perp})$ and we obtain $b = a^{\perp}$.

Th. 2.2 is directly applicable to the lattice G of decision effects (see III, D 6.2, III, D 6.3 and III, D 6.6) if we set $m(e) = \mathrm{tr}(we)$. G is therefore orthomodular.

3 Boolean Rings

D 3.1. A complemented distributive lattice is called a Boolean lattice or a Boolean ring (for the designation "Ring" also see D 3.2).

Th. 3.1. *In a Boolean ring M each element a has exactly one complement a'. The mapping $a \longrightarrow a'$ is an orthocomplementation.*

PROOF. Suppose, therefore, that $a \wedge x = a \wedge y = 0$ and $a \vee x = a \vee y = 1$. It then follows that $x = x \wedge 1 = x \wedge (a \vee y) = (x \wedge a) \vee (x \wedge y) = x \wedge y$. Similarly we obtain

$$y = x \wedge y \quad \text{and therefore} \quad x = y.$$

Since a is also a complement of a', the mapping $a \longrightarrow a'$ is bijective and $(a')' = a$. According to D 1.2 we need only prove $a \leq b \Rightarrow b' \leq a'$. From $a \leq b$ it follows that $a = a \wedge b$ and therefore $a \wedge b' = a \wedge b \wedge b' = 0$. Thus we obtain $b' = b' \wedge 1 = b' \wedge (a \vee a') = (b' \wedge a) \vee (b' \wedge a') = b' \wedge a'$, that is, $b' \leq a'$.

The fact that a Boolean ring is orthomodular is trivial, because it satisfies the distributive law. In a Boolean ring we may therefore use $a^\perp$ to denote the complement of a. Instead of $a^\perp$ we may often use a^*.

Th. 3.2. *An orthocomplemented orthomodular lattice is a Boolean ring if and only if every pair of elements is compatible* (*see* D 2.2).

PROOF. The fact that $C(a, b)$ holds for all pairs a, b in a Boolean ring follows from $a = a \wedge 1 = a \wedge (b \vee b^\perp)$ and the distributive law. In order to prove the converse, we shall now show that: $C(a, b) \Leftrightarrow a \wedge b = a \wedge (b \vee a^\perp)$. $C(a, b)$ means that $a = (a \wedge b) \vee (a \wedge b^\perp)$ from which it follows that $a^\perp = (a \wedge b)^\perp \wedge (a^\perp \vee b)$. From this it follows that $a \wedge (a^\perp \vee b) \wedge (a \wedge b)^\perp = 0$. Since $a \wedge b \leq a \wedge (b \vee a^\perp)$, from Th. 2.1(iii) it follows that $a \wedge (b \vee a^\perp) = (a \wedge b) \vee (a \wedge (b \vee a^\perp) \wedge (a \vee b)^\perp) = a \wedge b$. Conversely, from $a \wedge b = a \wedge (b \vee a^\perp)$ we obtain the following relationships:

$$(a \wedge b) \vee (a \wedge b^\perp) = (a \wedge (b \vee a^\perp)) \vee (a \wedge b^\perp) = (a \wedge b^\perp) \vee (a \wedge (a \wedge b^\perp)^\perp) = a$$

where the latter are obtained from $a \wedge b^\perp \leq a$ and Th. 2.1(iii).

Since $a \wedge c \leq a \vee b$, $a \wedge c \leq c$, $b \wedge c \leq a \vee b$, $b \wedge c \leq c$ we find that $(a \vee b) \wedge c \geq (a \wedge c) \vee (b \wedge c)$. According to Th. 2.1(iii) we therefore obtain $(a \vee b) \wedge c = (a \wedge c) \vee (b \wedge c) \vee [(a \vee b) \wedge c \wedge ((a \wedge c) \vee (b \wedge c))^\perp]$. We need only show that the expression z in the square brackets is equal to 0. Next it follows that $z = (a \vee b) \wedge c \wedge ((a \wedge c) \vee (b \wedge c))^\perp = (a \vee b) \wedge c \wedge (a^\perp \vee c^\perp) \wedge (b^\perp \vee c^\perp)$. From $C(c, a^\perp)$ we find $c \wedge a^\perp = c \wedge (a^\perp \vee c^\perp)$ and we therefore obtain $z = (a \vee b) \wedge c \wedge a^\perp \wedge (b^\perp \vee c^\perp)$. From $C(c, b^\perp)$ we obtain $z = (a \vee b) \wedge a^\perp \wedge c \wedge b^\perp = c \wedge (a \vee b) \wedge (a \vee b)^\perp = 0$.

Th. 3.3. *An orthocomplemented orthomodular lattice is a Boolean ring if and only if each element a has only one complement.*

PROOF. The first part of this theorem follows directly from Th. 3.1. According to Th. 3.2 we need only show that each pair of elements is compatible if each element has only one complement.

According to Th. 2.1(iii) we obtain

$$(a \wedge b) \vee (a \wedge (a \wedge b)^{\perp}) = a.$$

Let $c = (a \wedge (a \wedge b)^{\perp})$ we obtain $c \wedge b = 0$ and, from $(a \wedge b)^{\perp} \geq b^{\perp}$ we obtain

$$c \wedge b^{\perp} = a \wedge (b^{\perp} \wedge (a \wedge b)^{\perp}) = a \wedge b^{\perp}.$$

a and b are therefore compatible if $c \wedge b^{\perp} = c$, that is, if c and b are compatible.

For $d = b \vee c$ and $e = c \vee d^{\perp}$ we obtain $b \vee e = b \vee c \vee d^{\perp} = d \vee d^{\perp} = 1$, and, according to Th. 2.2(iii) it follows that $e = d^{\perp} \vee (e \wedge d)$. Since $c \perp d^{\perp}$, from Th. 2.1(vi) it follows that $c = e \wedge d$. From $b \leq d$ it follows that

$$b \wedge e = b \wedge d \wedge e = b \wedge c = 0.$$

Therefore e is a complement of b. Since it only has one complement, we find that $e = b^{\perp}$ and therefore $c < b^{\perp}$, that is, $c \wedge b^{\perp} = c$.

Th. 3.4. *If the least upper bound $\bigvee_{a \in N} a$ (greatest lower bound $\bigwedge_{a \in N} a$) exists for a subset N of a Boolean ring M, then for each $b \in M$ the least upper bound $\bigvee_{a \in N}(b \wedge a)$ (greatest lower bound $\bigwedge_{a \in N}(b \wedge a)$) exists, and satisfies the distributive law:*

$$b \wedge \left(\bigvee_{a \in N} a\right) = \bigvee_{a \in N} (b \wedge a); \qquad b \vee \left(\bigwedge_{a \in N} a\right) = \bigwedge_{a \in N} (b \vee a).$$

PROOF. We must show that $x = b \wedge (\bigvee_{a \in N} a)$ is the least upper bound of the set of all $b \wedge a$ for which $a \in N$. Since $b \wedge a \leq b$ and $b \wedge a \leq \bigvee_{d \in N} d$ for all $a \in N$ it follows that $x \geq b \wedge a$ for all $a \in N$. Let $g \geq b \wedge a$ for all $a \in N$. We need to show that $g \geq x$. Since $g \geq b \wedge a$ implies $g \wedge b \geq b \wedge a$, it suffices to show that, for $u \geq b \wedge a$ with $u \leq b$ it follows that $u \geq x$. Suppose that $v \geq b^{\perp} \wedge a$ for all $a \in N$ and $v \leq b^{\perp}$. Thus it follows that $u \vee v \geq (b \wedge a) \vee (b^{\perp} \wedge a) = a$ for all $a \in N$, that is, $u \vee v \geq \bigvee_{a \in N} a$.

From $b \wedge u = u$ and $b \wedge v = 0$ it follows that

$$u = b \wedge u = (b \wedge u) \vee (b \wedge v) = b \wedge (u \vee v) \geq b \wedge \bigvee_{a \in N} a = x.$$

From Th. 1.3, from

$$b \vee \bigwedge_{a \in N} a = \left(b^{\perp} \wedge \bigvee_{a \in N} a^{\perp}\right)^{\perp} = \left[\bigvee_{a \in N} (b^{\perp} \wedge a^{\perp})\right]^{\perp} = \bigwedge_{a \in N} (b \vee a)$$

we obtain the second part of the theorem.

The following relations are frequently defined in a Boolean ring:

D 3.2. $a \cdot b = a \wedge b, \qquad a \dotplus b = (a \wedge b^{\perp}) \vee (b \wedge a^{\perp}).$

It is easy to show that a Boolean ring together with the operations $\cdot$ and $\dotplus$ is a (commutative) ring (algebra) for which $a \cdot a = a$ and $a \dotplus a = 0$.

We may express the operation $\vee$ by means of $\cdot$ and $\dotplus$ as follows:

$$a \vee b = a \dotplus b \dotplus a \cdot b.$$

Conversely, let M be a commutative ring (with unit element) with the operations $\cdot$ and $\dotplus$ for which $a \cdot a = a$ and $a \dotplus a = 0$; if we define $a \wedge b = a \cdot b$ and $a \vee b = a \dotplus b \dotplus a \cdot b$, then from Th. 1.1(1), (2), (3) it follows that M is a lattice. It is easy to show that this lattice is distributive, and that $1 \dotplus a$ is the complement of a.

A distributive complemented lattice can therefore be well characterized as a commutative ring (with unit element) for which $a \cdot a = a$ and $a \dotplus a = 0$.

The following more general concept of a Boolean ring is often used:

D 3.3. A lattice is said to be relatively complemented if to an $a \geq b$ there exists a c such that $b \wedge c = 0$, $b \vee c = a$. c is called the relative complement of b with respect to a.

D 3.4. A distributive relatively complemented lattice is called a generalized Boolean ring.

In a generalized Boolean ring the relative complement is uniquely determined.

In a generalized Boolean ring we may obtain a commutative algebra by defining $a \cdot b = a \wedge b$ and $a \dotplus b = \tilde{a} \vee \tilde{b}$, where $\tilde{a}$ is the relative complement of $a \wedge b$ in a and $\tilde{b}$ is the relative complement of $a \wedge b$ in b. A generalized Boolean ring is a Boolean ring if and only if it has a unit element.

From two Boolean rings M_1, M_2 (and from a finite number of Boolean rings) it is possible to construct new Boolean rings in two different ways.

In the first way we are given a pair of arbitrary partially ordered sets M_1, M_2; on the product set $M = M_1 \times M_2$ we introduce the following partial order:

$$(a_1, b_1) \leq (a_2, b_2): \qquad a_1 \leq a_2, b_1 \leq b_2.$$

It is easy to see that M is a (complete) lattice whenever M_1 and M_2 are (complete) lattices. In particular, we obtain

$$(a_1, b_1) \wedge (a_2, b_2) = (a_1 \wedge a_2, b_1 \wedge b_2),$$

$$(a_1, b_1) \vee (a_2, b_2) = (a_1 \vee a_2, b_1 \vee b_2).$$

If M_1 and M_2 have 1 and 0 elements, then so does M:

$$1 = (1_1, 1_2), \qquad 0 = (0_1, 0_2).$$

If M_1 and M_2 are complemented, then so is M:

$$(a_1, a_2)' = (a_1', a_2').$$

If M_1 and M_2 are distributive, then so is M. Therefore, if M_1 and M_2 are Boolean rings, then so is M.

The second way to construct a Boolean ring M from M_1 and M_2 may be carried out most simply with the aid of the algebraic operations $\dotplus$ and $\cdot$. This construction is analogous to the technical "logical" switching circuits—from M_1 and M_2 we construct the "free" algebra (ring) which consists of all possible formal sums and products of elements in M_1 and M_2 by means of finitely many operations $\dotplus$ and $\cdot$. In this case M is not, in general, equal to the product set.

The elements of M can be represented by all formal finite sums of the form

$$\sum_i \dotplus a_1^{(i)} \cdot a_2^{(j)},$$

where $a_1^{(i)} \in M_1$, $a_2^{(i)} \in M_2$ and $a_1^{(i)} \cdot a_1^{(j)} = 0$, $a_2^{(i)} \cdot a_2^{(j)} = 0$ for $i \neq j$. Then, by means of the operations $\dotplus$ and $\cdot$ we may obtain new sums as follows:

$$\left(\sum_i \dotplus a_1^{(i)} \cdot a_2^{(i)}\right) \cdot \left(\sum_k \dotplus b_1^{(k)} \cdot b_2^{(k)}\right) = \sum_{i,k} \dotplus (a_1^{(i)} \cdot b_1^{(k)}) \cdot (a_2^{(i)} \cdot b_2^{(k)}).$$

For $(\sum_i \dotplus a_1^{(i)} \cdot a_2^{(i)}) \dotplus (\sum_k \dotplus b_1^{(k)} \cdot b_2^{(k)})$ we do not immediately obtain $a_1^{(i)} \cdot b_1^{(k)} = 0$, etc. We can, however, attain this result stepwise providing that, instead of $a_1^{(i)}, b_1^{(k)}$ we use new elements of M_1: $a_1^{(i)} \cdot b_1^{(k)}$, $a_1^{(i)} \dotplus a_1^{(i)} \cdot b_1^{(k)}$, $b_1^{(k)} \dotplus a_1^{(i)} \cdot b_1^{(k)}$.

D 3.5. A subset I of a lattice M is said to be an ideal if the following conditions are satisfied:

(1) $a \in I$ and $b \leq a \Rightarrow b \in I$,
(2) $a_1, a_2 \in I \Rightarrow a_1 \vee a_2 \in I$.

Th. 3.5. *If M is a generalized Boolean ring, then I is also algebraically an ideal of M, that is, the conditions*

$$a \in I \text{ and } b \in M \Rightarrow a \cdot b \in I,$$

$$a_1, a_2 \in I \Rightarrow a_1 \dotplus a_2 \in I.$$

Conversely, an algebraic ideal I is also an ideal in the sense of D 3.5.

PROOF. Let I be an ideal according to D 3.5. From $a \in I$ and $a \cdot b = a \wedge b \leq a$ it follows that $a \cdot b \in I$. From $a_1, a_2 \in I$ and $a_1 \dotplus a_2 \leq a_1 \vee a_2$ it follows that $a_1 \dotplus a_2 \in I$. Let I be an algebraic ideal. From $a \in I$ and $b \leq a$ it follows that $b = b \wedge a = b \cdot a \in I$; from $a_1, a_2 \in I$ it follows that $a_1 \vee a_2 = (a_1 \dotplus a_2) \dotplus a_1 \cdot a_2 \in I$.

From the fact that I is also an algebraic ideal, it follows that M/I is a Boolean ring.

An equivalence relation $b_1 \sim b_2$ is defined by $b_1 = b_2 \dotplus a$ where $a \in I$. Since $b_1 = b_2 \dotplus a \Leftrightarrow b_1 \dotplus b_2 = a$ we may define $b_1 \sim b_2$ by $b_1 \dotplus b_2 \in I$. The set of classes M/I can easily be seen to be a Boolean ring as follows: $c_1 \cdot c_2$ is equal to the class of $b_1 \cdot b_2$ for $b_1 \in c_1$, $b_2 \in c_2$.

$c_1 \dotplus c_2$ is equal to the class of the $b_1 \dotplus b_2$ for $b_1 \in c_1$, $b_2 \in c_2$.

4 Set Lattices

If X is a set and $M \subset \mathscr{P}(X)$ ($\mathscr{P}(X)$ is the power set of X) then M is a partially ordered set with respect to the set theoretical relation $\subset$ of inclusion. If M is a lattice, we then speak of a set lattice. Here it is important to note that set theoretical union $\cup$ and intersection $\cap$ will not necessarily correspond to the lattice least upper bound $\vee$ and greatest lower bound $\wedge$, respectively, that is, for $a, b \in M$ it is not necessarily true that $a \wedge b = a \cap b$ and $a \vee b = a \cup b$. We do find, however, that $a \wedge b \subset a \cap b$ and $a \vee b \supset a \cup b$.

Th. 4.1. *If M has a maximal element, and if, for every $N \subset M$ $\bigcap_{a \in N} a \in M$, then M is a complete set lattice and*

$$a \wedge b = a \cap b, \qquad a \vee b = \bigcap_{\substack{c \in M \\ a \cup b \subset c}} c.$$

Similarly, if M contains a minimal element and if, for every $N \subset M$, $\bigcup_{a \in N} a \in M$ then

$$a \vee b = a \cup b \quad \text{and} \quad a \wedge b = \bigcup_{\substack{c \in M \\ c \subset a \cap b}} c.$$

The proof of this theorem is a simple consequence of Th. 1.2.

If M is complemented, then in general the a' of $a \in M$ is not necessarily equal to the set complement $e \backslash a$ of a in the maximal element e of M. If M contains the empty set, then it follows that $a' \subset e \backslash a$.

Let M be a Boolean ring of sets, let $c \in M$ and let $c \subset e \backslash a$ where e is the maximal element of M. Then $c \wedge a = 0$. Using $a^\perp$ we find that $(c \vee a^\perp) \wedge a = (c \wedge a) \vee (a^\perp \wedge a) = 0$ and $(c \vee a^\perp) \vee a = c \vee e = e$, that is, $c \vee a^\perp$ is also a complement; by the uniqueness of the complement we obtain $c \vee a^\perp = a^\perp$, that is, $c \subset a^\perp$. If the empty set is an element of M then we must have $a^\perp \cap a = 0$ and therefore $a^\perp \subset e \backslash a$. Then the set $\{c \mid c \in M \text{ and } c \subset e \backslash a\}$ has a maximal element, namely $a^\perp$.

APPENDIX II

Remarks about Topological and Uniform Structures

Here we shall provide a brief summary of some of the concepts and principal results which are used in various places in this book. For the proof of these theorems, see [32].

1 Topological Spaces

A topological space consists of a set X together with a structure $\mathcal{O} \subset \mathcal{P}(X)$ which satisfies the following properties:

(1) $\mathcal{O}$ contains the intersection of any finite collections of elements of $\mathcal{O}$.
(2) $\mathcal{O}$ contains arbitrary unions of elements of $\mathcal{O}$.

$\mathcal{O}$ is called the set of open sets. We often say that $\mathcal{O}$ defines a topology on X. The complements of sets in $\mathcal{O}$ are called closed sets. The "interior" A^0 of a set A is defined as the union of all open subsets of A. A^0 is open. The closure $\bar{A}$ of $A \subset X$ is defined as the intersection of all closed sets which contain A. B is said to be dense in A if $B \subset A$ and $A \subset \bar{B}$. X is said to be separable if there exists a countable subset in X which is dense in X.

A filter in X is a subset $\mathcal{F} \subset \mathcal{P}(X)$ for which $\mathcal{F} \neq \varnothing$, $\varnothing \notin \mathcal{F}$; $A \in \mathcal{F}$ and $B \supset A \Rightarrow B \in \mathcal{F}$; $A \in \mathcal{F}$ and $B \in \mathcal{F} \Rightarrow A \cap B \in \mathcal{F}$.

We may also, in an equivalent manner, define a topology by means of a neighborhood structure, as follows: To each $x \in X$ there corresponds a filter $\mathcal{F}_x$—the so-called neighborhood filter of x. For the $\mathcal{F}_x$ we require that $x \in U$ for all $U \in \mathcal{F}_x$; to each $U \in \mathcal{F}_x$ there exists a $V \in \mathcal{F}_x$ such that for each $y \in V$ the relation $U \in \mathcal{F}_y$ is satisfied.

We may show the equivalence in the following way:

(1) Let X and $\mathcal{O}$ be given. We define $\mathcal{F}_x$ as the set of all A for which there exists a $B \in \mathcal{O}$ for which $x \in B \subset A$. The $\mathcal{F}_x$ then satisfy the requirements for a neighborhood structure.
(2) Let X be given together with a neighborhood structure. We define $\mathcal{O}$ as the set of all A for which $y \in A \Rightarrow A \in \mathcal{F}_y$; the set A is therefore a neighborhood of each of its points.

It is a simple matter to show that the neighborhood structure defined according to procedure (1) using X and $\mathcal{O}$ is identical to that defined by this $\mathcal{O}$ according to procedure (2). Similarly, if we begin with a neighborhood structure in X and define the set $\mathcal{O}$ of open sets according to procedure (2) and if we then define a neighborhood structure using $\mathcal{O}$ according to procedure (1), we then obtain the same structure we began with.

Let X and Y be topological spaces. A mapping $X \xrightarrow{f} Y$ is said to be continuous if, for every open set A of Y the set $f^{-1}(A) = \{x \mid f(x) \in A\}$ is open. This is equivalent to the condition that for every closed set $A, f^{-1}(A)$ is closed.

A filter $\mathcal{G}$ is said to be finer than a filter $\mathcal{F}$ if $\mathcal{G} \supset \mathcal{F}$. We say that a filter $\mathcal{G}$ in X converges to an $x \in X$ if $\mathcal{G}$ is finer than the neighborhood filter $\mathcal{F}_x$. If x_n is a sequence, then $\mathcal{G} \overset{\text{def}}{=} \{A \mid A$ contains all x_n except for a finite number$\}$ is a filter. We say that the sequence x_n converges towards x and write $x_n \to x$ if the corresponding filter $\mathcal{G}$ converges to x. It follows that $x_n \to x$ if and only if each $U \in \mathcal{F}_x$ contains all x_n (with the exception of a finite number). For a filter $\mathcal{G}$ in X $x \in X$ is called an accumulation point of $\mathcal{G}$ if $x \in \bar{A}$ for all $A \in \mathcal{G}$. x is called an accumulation point of the sequence x_n if it is an accumulation point of the corresponding filter $\mathcal{G}$; this is the case only if, to each $U \in \mathcal{F}_x$ there exists an infinity of elements x_{n_i} of the sequence for which $x_{n_i} \in U$.

We say that a topology $\mathcal{T}_2$ on X is finer than a topology $\mathcal{T}_1$ on X if $\mathcal{O}_2 \supset \mathcal{O}_1$. If $\mathcal{T}_\alpha$ if a family of topologies on X then $\mathcal{O} = \bigcap_\alpha \mathcal{O}_\alpha$ defines the finest topology $\mathcal{T}_U$ which is coarser than each of the $\mathcal{T}_\alpha$. The fact that the set $\mathcal{O} = \bigcap_\alpha \mathcal{O}_\alpha$ satisfies the axioms for open sets, has, as a consequence of AI, §4, the result that the topologies on X form a complete lattice since $\mathcal{P}(X)$ satisfies all the axioms for a set $\mathcal{O}$. Therefore, to each family $\mathcal{T}_\alpha$ there exists a coarsest topology $\mathcal{T}_0$ which is finer than all the $\mathcal{T}_\alpha$. $\mathcal{T}_U$ is called the greatest lower bound of all the $\mathcal{T}_\alpha$ and $\mathcal{T}_0$ is the least upper bound of the $\mathcal{T}_\alpha$.

Suppose we are given a set X together with a family X_λ of topological spaces and a set of maps $X \xrightarrow{f_\lambda} X_\lambda$. The coarsest topology in X for which all the maps f_λ are continuous is called the initial topology generated by the maps $X \xrightarrow{f_\lambda} X_\lambda$. If Y is a topological space, then the mapping $Y \xrightarrow{g} X$ (X with the initial topology) is continuous if and only if all composite maps $Y \xrightarrow{g} X \xrightarrow{f_\lambda} X_\lambda$ of Y into X_λ are continuous.

If A is a subset of the space X with the topology $\mathcal{T}$ we then denote the initial topology on A defined by the canonical injection of A in X as the induced topology on A or (more simply) the topology $\mathcal{T}$ on A. The open sets of this topology are precisely the intersection of open sets B of X with A.

Let X_λ be a family of topological spaces, and let X by the Cartesian product of the X_λ. We define the product topology on X to be the initial topology defined by the set of projections $X \to X_\lambda$.

A topological space X is said to be a Hausdorf space if to each different pair of points x, y there exist neighborhoods U_x and U_y of x and y such that $U_x \cap U_y = \emptyset$.

2 Uniform Spaces

A uniform structure on a set X is a subset $\mathscr{W}$ of $\mathscr{P}(X \times X)$ which satisfies the following axioms:

(1) $\mathscr{W}$ is a filter.

We define: $W^{-1} = \{(y, x) \mid (x, y) \in W\}$, $V \cdot W = \{(x, z) \mid$ there exists a $y \in X$ for which $(x, y) \in V, (y, z) \in W\}$ and $\Delta = \{(x, x) \mid x \in X\}$.

(2) $W \in \mathscr{W} \Rightarrow \Delta \subset W$.
(3) $W \in \mathscr{W} \Rightarrow W^{-1} \in \mathscr{W}$.
(4) $W \in \mathscr{W} \Rightarrow$ there exists a $V \in \mathscr{W}$ such that $V \cdot V \in W$.

The elements of $\mathscr{W}$ are called vicinities.

A subset Z of $\mathscr{P}(X \times X)$ is called a fundamental system of vicinities or the basis of a uniform structure if the filter generated by Z satisfies axioms (2)–(4).

A topological structure is defined by a uniform structure in the following way: Let $W \in \mathscr{W}$; it is easy to show that, for each x the sets $\mathscr{F}_x = \{y \mid (x, y) \in W\}$ form a filter for which $x \in U$ for all $U \in \mathscr{F}_x$, and that, to each $U \in \mathscr{F}_x$ there exists a $V \in \mathscr{F}_x$ such that $U \in \mathscr{F}_y$ for each $y \in V$. This topology is called the topology generated by $\mathscr{W}$. A topology is said to be uniformizable if there exists a uniform structure which generates the topology.

A uniform structure $\mathscr{W}$ is said to be separated if $\bigcap_{W \in \mathscr{W}} W = \Delta$. For the topology generated by $\mathscr{W}$ to be separated it is necessary and sufficient that $\mathscr{W}$ is separated.

A mapping $X \xrightarrow{f} Y$ between two uniform spaces X and Y is said to be uniformly continuous if, to each vicinity V of Y there exists a vicinity W of X such that $f(W) \subset V$ where $f(W)$ is defined by $f(W) = \{(f(x), f(y)) \mid (x, y) \in W\}$.

Let $\mathscr{W}_1$ and $\mathscr{W}_2$, $\mathscr{W}_1 \subset \mathscr{W}_2$ be two uniform structures; we then say that $\mathscr{W}_1$ is coarser than $\mathscr{W}_2$ and $\mathscr{W}_2$ is finer than $\mathscr{W}_1$. If $\mathscr{W}_\lambda$ is a collection of uniform structures on X, then $\mathscr{W} = \bigcap_\lambda \mathscr{W}_\lambda$ it is easy to show that $\mathscr{W}$ is also a uniform structure. If X is a set, X_λ are uniform spaces, and f_λ are maps $X \xrightarrow{f_\lambda} X_\lambda$, then there is a coarsest uniform structure for which all mappings f_λ are uniformly continuous. This is the initial uniform structure for the maps $X \xrightarrow{f_\lambda} X$. A mapping g of a uniform space Y into X with the initial uniform structure is uniformly continuous if and only if all the composite maps $Y \xrightarrow{g} X \xrightarrow{f_\lambda} X_\lambda$ are uniformly continuous. The topology corresponding to the initial uniform structure on X is precisely the initial topology for which all maps $X \xrightarrow{f_\lambda} X_\lambda$ are continuous.

The product uniform structure on the product set $X = \prod_\lambda X_\lambda$ is the initial uniform structure in which all projections $X \xrightarrow{f_\lambda} X_\lambda$ are uniformly continuous. The induced uniform structure on a subset A of a uniform space is the initial structure for which the canonical injection $A \to X$ is uniformly continuous.

A filter $\mathscr{F}$ in a uniform space is called a Cauchy filter if for each vicinity W there exists an element $F \in \mathscr{F}$ for which $F \times F \subset W$. X is said to be complete if every Cauchy filter converges to a point in X. For each uniform space X it is possible to construct a complete separating(!) uniform space $\hat{X}$.

If X is itself separating, then we may identify X with a dense subset of $\hat{X}$; $\hat{X}$ is then uniquely determined (up to an isomorphism). $\hat{X}$ is called the completion of X.

A subset A of a separating complete uniform space X is a complete space if and only if A is closed in X.

If X, Y are separating uniform spaces and Y is complete, then a uniformly continuous map $X \xrightarrow{f} Y$ has a unique extension $\hat{X} \xrightarrow{f} Y$.

A metric on the set X is a real valued function $X \times X \xrightarrow{d} \mathbf{R}$ which satisfies the following conditions: $d(x, y) \geq 0$; $d(x, y) = 0 \Leftrightarrow x = y$ and the so-called triangle inequality—$d(x, z) \leq d(x, y) + d(y, z)$. A metric space is a set X together with a metric d.

It is easy to show that a metric defines a basis for a uniform structure by the sets $W_\varepsilon = \{(x, y) \mid d(x, y) < \varepsilon\}$. This uniform structure is called the uniform structure generated by the metric. A uniform space X is said to be metrizable if there is a metric which generates its uniform structure. X is metrizable if and only if X is separating and there exists a denumerable basis for the uniform structure. A topological space is said to be metrizable if there exists a metric for which the topology is that generated by the uniform structure generated by the metric.

A sequence $\{x_n\}$ is called a Cauchy sequence if the corresponding filter $\mathscr{G}$ is a Cauchy filter, that is, if to each vicinity W there exists an integer N such that $(x_n, x_m) \in W$ for $n, m > N$. A sequence in a metric space is therefore a Cauchy sequence if and only if, to each $\varepsilon > 0$ there exists an N such that $d(x_n, x_m) < \varepsilon$ for $n, m > N$. A metric space X is complete if and only if every Cauchy sequence has a limit element in X.

In a separating topological space the following three conditions are equivalent:

(1) Every covering of X by open sets contains a finite subcovering.
(2) The intersection of a set of closed sets is nonempty if and only if every finite subset of these closed sets has nonempty intersection.
(3) Every filter has an accumulation point in X.

If X satisfies one of these properties, then X is said to be a compact space.

Every compact space is uniformizable, and the corresponding uniform structure is uniquely determined by the topology, and is the uniform structure of the neighborhood filter of Δ in the topological space $X \times X$. In this way X becomes a complete separating uniform space. A separating uniform space is said to be precompact if its completion is compact. X is precompact if and only if to each vicinity W there exists a finite number of points $x_\nu \in X$ such that $\bigcup_\nu W(x_\nu) = X$ where $W(x_\nu) = \{x \mid (x, x_\nu) \in W\}$.

Every closed subset A of a compact space X is compact. A subset A of a topological space X (X is not necessarily compact) is said to be relatively compact if its closure is compact. The product $X = \prod_\lambda X_\lambda$ of compact spaces is compact (Tychonov's theorem). If $f: X \xrightarrow{f} Y$ is a continuous map of a compact space X into a separating topological space, then $f(X)$ is a compact subset of Y. For the special case in which f is bijective, X and Y are homeomorphic. Thus, with the aid of the identity mapping $X \to X$ we find that if X is compact with respect to the topology $\mathscr{T}_1$ and $\mathscr{T}_2$ is a coarser, separating topology on X then $\mathscr{T}_2$ is identical to $\mathscr{T}_1$.

A subset A of a precompact space is precompact; the product $X = \prod_\lambda X_\lambda$ of precompact spaces X_λ is precompact. X with the initial uniform structures generated by the maps $X \xrightarrow{f_\lambda} X_\lambda$ where the X_λ are compact is precompact. If the family of the f_λ is denumerable, then X is also metrizable.

A compact and metrizable space is separable.

A separating topological space X is said to be locally compact if each point x has a compact neighborhood.

3 Baire Spaces

A subset A of a topological space X is said to be nowhere dense in X if the interior of the closure $\bar{A}$ of A is empty. A is said to be meager in X if A is the union of a denumerable set of nowhere dense subsets. A topological space is said to be a Baire space if every open set is not meager.

If a Baire space X is the union of denumerable many closed sets A_ν, then at least one of the A_ν must contain an open set, otherwise, all the A_ν would be nowhere dense and X itself would be meager, although X is open.

Every locally compact and every complete metrizable space is a Baire space.

4 Connectedness

A topological space X is said to be connected if it is not the union of two open nonempty disjoint sets. This is equivalent to the condition that X is not the union of two closed nonempty disjoint sets.

If $X = X_1 \cup X_2$, where X_1, X_2 are open and $X_1 \cap X_2 = \varnothing$ then X_1, X_2 are also closed.

A subset $A \subset X$ is said to be connected if A together with the topology induced by X is a connected space.

If A is a connected subset of X, f is a continuous mapping $X \to Y$, then $f(A)$ is a connected subset of Y.

If A is a family of connected subsets of X and if $\bigcap_\lambda A_\lambda \neq \varnothing$ then $\bigcup_\lambda A_\lambda$ is a connected subset. If x is an element of X, then the union of all such connected A which contain x is a connected set, and is the largest connected set containing x. This set is called the connected component of x, and is a closed set.

Two connected components (of x and y) are either identical or disjoint, that is, the connected components partition X into equivalence classes.

A topological space is said to be locally connected if each point has a fundamental system of connected neighborhoods, that is, if to each neighborhood U of a point there exists a connected neighborhood V of x such that $V \subset U$.

A space is said to be linearly connected if, for each pair of points x, y there exists a continuous path from x to y, that is, a continuous map $[0, 1] \xrightarrow{f} X$ such that $f(0) = x$ and $f(1) = y$.

APPENDIX III

Banach Spaces

We cannot develop the theory of Banach spaces here. Instead, we shall only briefly present a summary of important results without proof in order that readers who are not familiar with these results will be able to find them in other books, for example, in [33].

1 Linear Vector Spaces

A linear vector space X over the field K is an additive abelian group (that is, to each pair $x_1, x_2, \in X$ there corresponds a $x_1 + x_2 \in X$; the following axioms are also satisfied:

$$x_1 + (x_2 + x_3) = (x_1 + x_2) + x_3,$$

$$x_1 + x_2 = x_2 + x_1,$$

there exists an element 0 such that $0 + x = x$ for all $x \in X$, to each $x \in X$ there exists a $y \in X$ for which $x + y = 0$; we write $y = -x$ and $x_1 + (-x_2) = x_1 - x_2$). In addition the elements of the field K define maps of X into X which satisfy the following axioms (here let α, $\beta \in K$ and let e denote the unit element of K):

$$ex = x, \qquad \alpha(\beta x) = (\alpha\beta)x,$$

$$\alpha(x_1 + x_2) = \alpha x_1 + \alpha x_2, \qquad (\alpha + \beta)x = \alpha x + \beta x.$$

We shall only consider the two following cases—$K = \mathbf{R}$ (where $\mathbf{R}$ is the field of real numbers) and $K = \mathbf{C}$ (where $\mathbf{C}$ is the field of complex numbers).

We shall assume that the reader is familiar with the simple computation rules which follow directly from these axioms.

2 Normed Vector Spaces and Banach Spaces

In a linear vector space (over the field $K = \mathbf{R}$ or $\mathbf{C}$) the norm is a real function $\|x\|$ over X which satisfies the following properties:

(1) $\|x\| \geq 0$ and $\|x\| = 0$ only for $x = 0$.
(2) $\|x + y\| \leq \|x\| + \|y\|$.
(3) $\|\alpha x\| = |\alpha| \, \|x\|$ where $|\alpha|$ is the absolute magnitude of α.

From (3) it follows that $\|-x\| = \|x\|$. From (2) it follows that $\|x - y\| \geq | \|x\| - \|y\| |$. If a norm is defined on X, X is called a normed space. $d(x, y) = \|x - y\|$ defines a metric (see AII, §2) and therefore a uniform structure and a topology. The operations $X \times X \xrightarrow{x+y} X$ and $K \times X \xrightarrow{\alpha x} X$ are uniformly continuous.

The notion of a Cauchy sequence has already been defined in AII, §2. A normed space X is complete if for every Cauchy sequence x_n there exists a limit point $x \in X$, that is, $x_n \to x$. A complete normed space is called a Banach space. Every normed space may be completed to form a Banach space.

The set of all $x \in B$ such that $\|x\| \leq 1$ is called the unit sphere $B_{|1|}$ of B and is closed in the norm topology.

3 The Dual Space for a Banach Space

A mapping $X \xrightarrow{l} K$ for which $l(x_1 + x_2) = l(x_1) + l(x_2)$ and $l(\alpha x) = \alpha l(x)$ (where K is the field of scalars for the vector space X) is called a linear form or a linear functional.

A linear form is continuous over a normed vector space if it is continuous for $x = 0$. This is equivalent to the condition that l is bounded, that is, there exists a real number c such that

$$|l(x)| \leq c\|x\|.$$

A bounded linear form is clearly uniformly continuous, and therefore has a continuous extension from the normed vector space onto its completion. Let X' denote the set of continuous linear forms; we find that $X' = \hat{X}'$. X' is the dual space for X.

If we introduce the norm in X' as follows:

$$\|l\| = \sup_{x \in X} \frac{|l(x)|}{\|x\|} = \sup_{\|x\| \leq 1} |l(x)| = \sup_{\|x\| \leq 1} l(x)$$

(the last equality is only valid if $K = \mathbf{R}$), then X' is a Banach space because it is easy to verify the fact that a Cauchy sequence l_n is a convergent sequence in x and that $l(x) = \lim_{n\to\infty} l_n(x)$ is an element of X' with $\|l_n - l\| \to 0$.

The fact that X' separates X, that is, from $l(x_1) = l(x_2)$ for all $l \in X'$ it follows that $x_1 = x_2$ follows directly from the Hahn–Banach theorem. Since $|l(x)| \leq \|l\|\,\|x\|$ it follows that

$$\|x\| > \sup_{l\in X'} \frac{|l(x)|}{\|l\|}.$$

If $\|x\| \leq 1$ then, according to the Hahn–Banach theorem there exists an l satisfying $|l(x)| > |l(y)|$ for $\|y\| \leq 1$ and we therefore find that $|l(x)| > \|l\|$. Thus we obtain

$$\|x\| > 1 \Rightarrow \sup_{l\in X'} \frac{|l(x)|}{\|l\|} > 1.$$

Thus for $x = x'/\|x'\|\,(1+\varepsilon)$, for arbitrary $\varepsilon > 0$ it follows that

$$(1+\varepsilon) \sup_{l\in X'} \frac{l(x')}{\|l\|} > \|x'\|$$

and we therefore obtain

$$\|x\| = \sup_{l\in X'} \frac{l(x)}{\|l\|}.$$

A bilinear form (the canonical bilinear form of the dual pair X, X') is defined on $X \times X'$ by $\langle x, l\rangle = l(x)$. For $x \in X$, $y \in X'\langle x, y\rangle$ is, for fixed x, a norm-continuous linear form over X'. X'', the set of all norm-continuous linear forms over X' is, in general, larger than X.

4 Weak Topologies

Let X be a Banach space, X' the dual Banach space for X and let $\langle x, y\rangle$ be the canonical bilinear form for the pair X, X'.

$\sigma(X', X)$ is the initial topology in X' for the maps $X' \xrightarrow{\langle x,y\rangle} \mathbf{R}$ (or $\mathbf{C}$) which are continuous for all $x \in X$. Similarly $\sigma(X, X')$ is the initial topology in X for which the maps $X \xrightarrow{\langle x,y\rangle} \mathbf{R}$ (or $\mathbf{C}$) are continuous for all $y \in X'$. These $\sigma(\ldots)$ topologies are often called the weak topologies corresponding to the dual pair X, X' because they are weaker than the norm topologies.

The set of all continuous linear forms over X in the $\sigma(X, X')$-topology is given by the $y \in X'$. Similarly, the set of all continuous linear forms over X' in the $\sigma(X', X)$-topology is given by the $x \in X$. The sets of continuous linear forms over X in both the norm topology and the $\sigma(X, X')$-topology are therefore identical.

Uniform structures are defined by means of the weak topology; for example, the uniform structure defined by $\sigma(X, X')$ with vicinities

$$\{(x, y) \mid x - y \in U \text{ where } U \text{ is a neighborhood of } 0\}.$$

Thus every weakly continuous linear form is also uniformly continuous.

Let X be a linear vector space over the field of real numbers. A subset K of X is called a convex set, if, for $x_1, x_2 \in K$, $\lambda x_1 + (1 - \lambda)x_2 \in K$ for $0 \leq \lambda \leq 1$. For $A \subset X$ let co A denote the smallest convex set in X which contains A. The unit sphere of a Banach space is convex.

Let X be a Banach space; let $\overline{\text{co}}\, A$ and $\overline{\text{co}}^\sigma A$ denote the smallest convex set closed in the norm-topology and the $\sigma(X, X')$-topologies, respectively, which contain A. In X we find that $\overline{\text{co}}^\sigma A = \overline{\text{co}}\, A$. This is not the case in X'! However, the unit sphere $X'_{|1|}$ of X' is not only $\sigma(X', X)$-closed but also $\sigma(X', X)$-compact. This result follows from the fact that $X'_{|1|} \xrightarrow{\langle x, y\rangle} [-1, 1] \subset \mathbf{R}$, that is, the images under these maps are relatively compact (see AII, §2) and that $X'_{|1|}$ is the polar set to $X_{|1|}$ (that is, the set of all $y \in X'$ for which $\langle x, y\rangle \leq 1$ for all $x \in X_{|1|}$) which is $\sigma(X', X)$-closed. The convex set generated by one (or finitely many) compact sets is compact, and therefore is closed!

An extreme point of a convex set A is an $x \in A$ for which the relation $x = \lambda x + (1 - \lambda)x_2$, $x_1, x_2 \in A$, $0 < \lambda < 1$ is satisfied only for $x_1 = x_2 = x$. We shall denote the set of extreme points of A by $\partial_e A$. If A is compact, then $A = \overline{\text{co}}(\partial_e A)$ according to the Krein–Millman theorem. Therefore we obtain $X'_{|1|} = \overline{\text{co}}^\sigma(\partial_e X'_{|1|})$.

A linear form over X' is $\sigma(X', X)$-continuous if and only if it is $\sigma(X', X)$-continuous over the unit sphere. This corresponds to the fact that a linear subspace of X' is $\sigma(X', X)$-closed when its intersection with the unit sphere is $\sigma(X', X)$-closed.

The topologies $\sigma(X', X)$ and $\sigma(X', X_{|1|})$ are identical on X', where $\sigma(X', X_{|1|})$ is the initial topology for which the maps $X' \xrightarrow{\langle x, y\rangle} \mathbf{R}$ are continuous for all $x \in X_{|1|}$. If A is a subset of X for which $\overline{\text{co}}(A \cup -A) = X_{|1|}$ then the topologies $\sigma(X', X)$ and $\sigma(X', A)$ are also identical on $X'_{|1|}$. This result follows directly from the fact that both the $\sigma(X', X_{|1|})$- and $\sigma(X', A)$-topologies are weaker than the $\sigma(X', X)$-topology and are also separating, and must therefore coincide on a $\sigma(X', X)$-compact set. If X is norm-separable we can choose A to be denumerable. We define a norm in X' as follows:

$$\|y\|_A \stackrel{\text{def}}{=} \sum_{x \in A} \lambda_x |\langle x, y\rangle|,$$

where the $\lambda_x > 0$ and $\sum_{x \in A} \lambda_x < \infty$. Again it is easy to verfiy that the topology determined by the norm $\|y\|_A$ coincides with $\sigma(X', X)$ on $X'_{|1|}$.

If X is norm-separable then the $\sigma(X', X)$-topology on $X'_{|1|}$ is compact and metrizable, and therefore, according to AII, §2, $X'_{|1|}$ is separable in the $\sigma(X', X)$-topology. Thus X' is also separable in the $\sigma(X', X)$-topology.

5 Linear Maps of Banach Spaces

If X_1 and X_2 are linear vector spaces, then a map T from $X_1 \xrightarrow{T} X_2$ is said to be linear if $T(x + y) = T(x) + T(y)$ and $T(\alpha x) = \alpha T(x)$. If X_1 and X_2 are Banach spaces, then T is continuous with respect to the norm topology only if there exists a real number C such that

$$\|Tx\| < C\|x\| \quad \text{for all } x.$$

Then T is also said to be bounded.

If T is bounded, then $\langle Tx, y\rangle = \langle x, y'\rangle$ is a norm-continuous linear form over X, that is, it defines a $y' \in X'$. It is easy to verify that $y \longrightarrow y'$ defines a linear and bounded map $X' \xrightarrow{T'} X'$ which we call the dual map to T or the adjoint map. T' is continuous with respect to the $\sigma(X'_2, X_2)$–$\sigma(X'_1, X_1)$-topologies. The following important relation is satisfied: $\langle Tx, y\rangle = \langle x, T'y\rangle$.

A linear map $X'_2 \xrightarrow{S} X'_1$ is σ-continuous only if there exists a bounded linear map $X_1 \xrightarrow{T} X_2$ for which $S = T'$. S is σ-continuous in X'_2 only if it is σ-continuous on the unit sphere of X'_2.

6 Ordered Vector Spaces

A linear vector space X over the field of real numbers $\mathbf{R}$ is said to be ordered if it is a partially ordered set in the sense of AI, §1 and if the following conditions are satisfied:

$$x_1 > x_2 \Rightarrow x_1 + x > x_2 + x \quad \text{for all } x \in X,$$

$$x > 0, \alpha \geq 0 \Rightarrow \alpha x > 0.$$

A convex cone C is defined as a convex set which has the property that if $x \in C$ then so does λx for all $\lambda \geq 0$. The cone C is said to be proper if $C \cap (-C) = \{0\}$.

It is easy to see that the specification of an order structure in X is equivalent to the specification of a proper cone C as follows:

$$x > 0 \quad \text{if and only if } x \in C.$$

This cone C is called the positive cone; the positive cone of X is often denoted by X_+.

A convex subset K of C is said to be a basis for the cone C if to each $x \in C$, $x \neq 0$, there exists exactly one number $\lambda(x)$ such that $\lambda(x)x \in K$. The set K defined by $\check{K} \overset{\text{def}}{=} \bigcup_{0 \leq \lambda \leq 1} \lambda K$ is equal to the set $\{x \mid x \in C \text{ and } x < w \in K\}$ and is called the truncated cone generated by the base K.

An ordered Banach space X is said to be base-normed if there exists a basis K for the cone X_+ for which K is norm-closed and $X_{|1|} = \overline{\text{co}}\,(K \cup (-K))$. It is a simple matter to show that X_+ will also be norm-closed. In addition, it can be shown (see [33]) that X is generating, that is, $X = X_+ - X_+$.

An affine functional on the convex set K is a map $K \xrightarrow{f} \mathbf{R}$ for which $w_1, w_2 \in K$, $0 \leq \lambda \leq 1 \Rightarrow f(\lambda w_1 + (1 - \lambda)w_2) = \lambda f(w_1) + (1 - \lambda)f(w_2)$. It is easy to show that each affine functional on a basis K of a base-norm Banach space may be uniquely extended to all of X as a linear functional because $X = X_+ - X_+$ and $X_+ = \bigcup_{\lambda \geq 0} \lambda K$. There is a 1:1 correspondence between the linear functionals over X and the affine functionals over K.

From $X_{|1|} = \overline{\text{co}}(K \cup -K)$ it directly follows that $\|w\| = 1$ for all $w \in K$ and that $\|x\|^{-1}x \in K$ for all $x \in X_+$.

Since X_+ is generating, each $x \in X$ may be expressed in the form $x = \alpha w_1 - \beta w_2$, where $w_1, w_2 \in K$ and $\alpha, \beta > 0$. Thus it follows that $\|x\| \leq \alpha + \beta$. To each $\varepsilon > 0$ we may choose w_1, w_2, α, β such that $\|x\| \geq \alpha + \beta - \varepsilon$ (see [33]). We say that X has the minimal decomposition property if w_1, w_2, α, β may be chosen so that $\|x\| = \alpha + \beta$. All the examples of base-norm spaces in this book satisfy the minimal decomposition property.

If $x_n \in X_+$ is a bounded increasing (or decreasing) sequence, then there exists an $x \in X_+$ for which $x_n \to x$. This follows directly from the fact that for $n > m$ $x_n - x_m \in X_+$ and therefore there exists a $w \in K$ such that $x_n = x_m + \lambda w$, where $\lambda \geq 0$ and with $x_m = \|x_m\| w_m$ $(w_m \in K)$ it follows that

$$x_n = (\|x_m\| + \lambda)\left[\frac{\|x_m\|}{\|x_m\| + \lambda} w_m + \frac{\lambda}{\|x_m\| + \lambda} w\right]$$

and we therefore obtain $\|x_n\| = \|x_m\| + \lambda = \|x_m\| + \|x_n - x_m\|$.

For the norm in the dual Banach space X' we obtain

$$\begin{aligned} \|y\| &= \sup\{|\langle x, y\rangle| \,|\, x \in X_{|1|} = \overline{\text{co}}(K \cup -K)\} \\ &= \sup\{|\langle x, y\rangle| \,|\, x \in K\}. \end{aligned}$$

X_+ determines a polar cone in X' as follows:

$$\begin{aligned} X'_+ &= \{y \,|\, \langle x, y\rangle \geq 0 \text{ for all } x \in X_+\} \\ &= \{y \,|\, \langle x, y\rangle \geq 0 \text{ for all } x \in K\}. \end{aligned}$$

X'_+ is not only $\sigma(X', X)$-closed, but is also $\sigma(X', X)$-complete (see [33]) because all positive linear functionals over X are norm-continuous. X'_+ determines an order for X' because from $y \in X' \cap -X'$ it easily follows that $y = 0$ from $X = X_+ - X_+$.

If $l(x)$ is a bounded linear functional on K, that is, if $l(x) < c$ for all $x \in K$ then from $x = aw_1 - \beta w_2$ (where $w_1, w_2 \in K$) and $\|x\| > \alpha + \beta - \varepsilon$ (see above) that $|l(x)| \leq \alpha l(w_1) + \beta l(w_2) \leq (\alpha + \beta)c < c\|x\| + \varepsilon c$ for arbitrary $\varepsilon > 0$ and therefore $|l(x)| \leq c\|x\|$. Every linear functional which is bounded on K (and hence each bounded affine functional on K) is therefore an element of X'.

$l(w) = 1$ for all $w \in K$ defines an element of X' which we shall denote by $\mathbf{1}$. The unit sphere of X' is therefore equal to

$$\begin{aligned} X'_{|1|} &= \{y \,|\, |\langle w, y\rangle| \leq 1 \text{ for all } w \in K\} \\ &= \{y \,|\, -1 \leq \langle w, y\rangle \leq 1 \text{ for all } w \in K\} \\ &= \{y \,|\, -\mathbf{1} \leq y \leq \mathbf{1}\}. \end{aligned}$$

We shall denote the set of all y for which $y_1 < y < y_2$ by $[y_1, y_2]$ and call it the order interval generated by y_1 and y_2. Therefore we obtain $X'_{|1|} = [-\mathbf{1}, \mathbf{1}]$. Because of this property we shall call X' an order unit space.

From $X'_{|1|} = [-\mathbf{1}, \mathbf{1}]$ it easily follows that X' is generating, that is, $X' = X'_+ - X'_+$.

If $y_n \in X'_+$ is a decreasing sequence, then there exists a $y \in X'_+$ to which the sequence converges in the $\sigma(X', X)$-topology, a result which follows directly from the fact that every set of the form $[\mathbf{0}, \alpha\mathbf{1}]$ is compact in the $\sigma(X', X)$-topology.

In the same manner in which every positive linear function over X is norm continuous, every positive linear map T that is $Tx > 0$ for $x > 0$) of a base norm space X_1 into a base norm space X_2 is norm-continuous (see [33]). Therefore, according to §5 the adjoint map $X'_2 \xrightarrow{T'} X'_1$ exists. T' is therefore also positive.

The set of norm continuous maps $X_1 \xrightarrow{T} X_2$ together with the norm

$$\|T\| = \sup\{\|Tx\| \mid \|x\| \leq 1\}$$

form a Banach space Y because a Cauchy sequence is also uniformly convergent. Y becomes an ordered vector space by means of the cone

$$Y_+ = \{T \mid T \text{ is positive}\}.$$

APPENDIX IV

Operators in Hilbert Space

Since the mathematics of Hilbert space is an essential tool in quantum mechanics, we shall briefly outline the proofs of important theorems. In particular, we shall provide a few examples of the application of a number of general theorems in AIII.

1 The Hilbert Space Structure Type

A Hilbert space is:

(I) A linear vector space $\mathscr{H}$ over a field K (as defined in AIII, §1). Here we shall only consider the case in which $K = \mathbf{C}$, the field of complex numbers.

(II) There is a map, the so-called inner product, defined on $\mathscr{H} \times \mathscr{H} \to \mathbf{C}$ which is denoted by $\langle x, y\rangle$ and satisfies the following axioms:

$$\text{for } \alpha \in \mathrm{C} \qquad \langle x, \alpha y\rangle = \alpha\langle x, y\rangle,$$

$$\langle x, y_1 + y_2\rangle = \langle x, y_1\rangle + \langle x, y_2\rangle,$$

$$\langle x, y\rangle = \overline{\langle y, x\rangle}; \langle x, x\rangle \geq 0, = 0 \text{ only if } x = 0.$$

From $\langle x, y\rangle = \overline{\langle y, x\rangle}$ it follows that $\langle x, x\rangle$ is real. From the axioms it easily follows that $\langle \alpha x, y\rangle = \bar{\alpha}\langle x, y\rangle$, $\langle x_1 + x_2, y\rangle = \langle x_1, y\rangle + \langle x_2, y\rangle, \langle x, 0\rangle = \langle 0, x\rangle = 0$.

Two vectors $x, y \in \mathscr{H}$ are said to be orthogonal if $\langle x, y\rangle = 0$. If $x \neq 0$ and if we define a vector z by

$$y = \frac{\langle x, y\rangle}{\langle x, x\rangle} x + z$$

then z satisfies $\langle z, x\rangle = 0$ and we obtain

$$\langle y, y\rangle = \frac{|\langle x, y\rangle|^2}{\langle x, x\rangle} + \langle z, z\rangle$$

from which we obtain the Schwarz inequality:

$$\langle x, x\rangle\langle y, y\rangle \geq |\langle x, y\rangle|^2$$

which is also valid for the case $x = 0$. Here the equality is satisfied if and only if $z = 0$, that is, $y = \lambda x$.

If we define $\|x\| = \langle x, x\rangle^{1/2}$, then, from the Schwarz inequality $|\langle x, y\rangle| \leq \|x\| \|y\|$; from $\|x - y\|^2 = \|x\|^2 + \|y\|^2 - \langle x, y\rangle - \langle y, x\rangle$ we obtain the triangle inequality $\|x + y\| \leq \|x\| + \|y\|$ and $\|x - y\| \geq |\|x\| - \|y\||$. Therefore $\|\ldots\|$ satisfies conditions (1)–(3) for a norm from AIII, §2. Therefore $\mathscr{H}$ is a normed space with norm $\|\ldots\|$. The convergence $x_n \to x$ of a sequence is defined in the sense of the norm, that is, $\|x_n - x\| \to 0$. With the help of the Schwarz inequality it easily follows that the inner product $\mathscr{H} \times \mathscr{H} \to \mathbf{C}$ is a continuous map, that is, from $x_n \to x$ and $y_m \to y$ we obtain the relation $\langle x_n, y_m\rangle \to \langle x, y\rangle$.

A pre-Hilbert space is defined as a set $\mathscr{H}$ which satisfies all the above axioms except $\langle x, x\rangle = 0 \Rightarrow x = 0$. From $\langle x, x\rangle = 0$ it follows directly that $\langle \alpha x, \alpha x\rangle = 0$. With $\langle x, y\rangle = |\langle x, y\rangle| e^{i\delta}$ it follows that

$$0 \leq \|x - e^{-i\delta} y\|^2 = \|x\|^2 + \|y\|^2 - 2|\langle x, y\rangle| \tag{1.1}$$

and from $\langle x, x\rangle = 0$ and $\langle y, y\rangle = 0$ we obtain $\langle x, y\rangle = 0$ and therefore $\|x + y\|^2 = 0$. Therefore the set $\mathscr{T}_0 = \{x \,|\, \langle x, x\rangle = \|x\| = 0\}$ is a subspace of $\mathscr{H}$. From (1.1) it follows that $\|y\| = 0$ and with λx ($\lambda > 0$) instead of x:

$$0 \leq \lambda^2 \|x\|^2 - 2\lambda|\langle x, y\rangle|$$

for all $\lambda > 0$ and therefore $\langle x, y\rangle = 0$ for $x \in \mathscr{H}$ and $y \in \mathscr{T}_0$. Therefore it easily follows that $x_1 \sim x_2$ if $\|x_1 - x_2\| = 0$ defines an equivalence relation, and the value of $\langle x, y\rangle$ depends only on the equivalence classes to which x and y belong. In this way $\mathscr{H}/\mathscr{T}_0$ is a linear vector space over $\mathbf{C}$ which satisfies II.

We now present the third axiom for a Hilbert space:

(III) $\mathscr{H}$ is a Banach space, that is, it is complete with respect to the norm.

Every Cauchy sequence in $\mathscr{H}$ therefore has a limit element. Each noncomplete space $\mathscr{H}$ which satisfies (I) and (II) can be completed (see AII, §2).

We shall only consider Hilbert spaces which satisfy the axiom:

(IV) $\mathscr{H}$ is separable.

(For the notion of separability, see AII, §1.)

A subset G of $\mathscr{H}$ is called a linear basis (or Hamel basis) if the span of G is dense in $\mathscr{H}$. From (IV) it follows that there exist denumerable linear bases. If there exists a countable or a finite linear basis G then all finite sums $\sum_\nu \alpha_\nu x_\nu$ where $x_\nu \in G$ and the α_ν rational complex numbers form a denumerable subset which is dense in $\mathscr{H}$, that is, IV is satisfied.

The smallest cardinality of a linear basis of $\mathscr{H}$ is called the linear dimension of $\mathscr{H}$ which, according to (IV) can be only either finite or denumerable.

We shall now give two important examples of Hilbert spaces (to prove that these examples satisfy axioms (I)–(IV) see [34]).

Let $\mathscr{H}$ be the set of all complex number sequences $x = (\alpha_1, a_2, \ldots)$ for which $\sum_\nu |\alpha_\nu|^2 < \infty$. For $x = (\alpha_1, \alpha_2, \ldots)$, $y = (\beta_1, \beta_2, \ldots)$, $x + y = (\alpha_1 + \beta_1, \alpha_2 + \beta_2, \ldots)$, $\alpha x = (\alpha\alpha_1, \alpha\alpha_2, \ldots)$ and $\langle x, y\rangle = \sum_\nu \bar{\alpha}_\nu \beta_\nu$, where the convergence of $\sum_\nu \bar{\alpha}_\nu \beta_\nu$ is easily proven with the aid of the Schwartz inequality

$$\left|\sum_{\nu=n}^{m} \bar{\alpha}_\nu \beta_\nu\right|^2 \le \left(\sum_{\nu=n}^{m} |\alpha_\nu|^2\right)\left(\sum_{\nu=n}^{m} |\beta_\nu|^2\right).$$

For the second example we consider a σ-ring $\mathscr{A}$ of subsets of a set M, that is, $\mathscr{A}$ is a Boolean ring with respect to the intersection, the union, complements and the union of countably many elements of $\mathscr{A}$ is again an element of $\mathscr{A}$ (for such an example consider the σ-ring in IV, §2.5). Suppose $M \in \mathscr{A}$.

On $\mathscr{A}$ let a σ-additive real measure μ be defined for which $\mu(\eta) \ge 0$, and where we permit $\mu(\eta) = \infty$ for some η. Let $\mu(\varnothing) = 0$. For a sequence $\eta_i \in \mathscr{A}$ satisfying $\eta_i \cap \eta_j = \varnothing$ for $i \ne j$ we therefore obtain

$$\mu\left(\bigcup_i \eta_i\right) = \sum_i \mu(\eta_i).$$

Thus it follows that $\mu(\eta) \ge \mu(\sigma)$ for $\eta \supset \sigma$ since $\eta = \sigma \cup (\eta \wedge \sigma^*)$ where σ^* is the complement of σ.

If $\mu(M) = +\infty$ then there may exist a sequence η_i such that $\eta_{i+1} \supset \eta_i$, $\mu(\eta_i)$ is finite and $M = \bigcup_i \eta_i$. In addition this sequence may be chosen such that every $\mathscr{A}_i \overset{\text{def}}{=} \{\sigma \mid \sigma \in \mathscr{A}, \sigma \subset \eta_i\}$ is separable with respect to the metric (this metric is described in IV, §1.4)

$$d(\sigma_1, \sigma_2) = \mu(\sigma_1 \dotplus \sigma_2) = \mu(\sigma_1 \cap \sigma_2^*) + \mu(\sigma_2 \cap \sigma_1^*).$$

A complex function $M \xrightarrow{f} \mathbf{C}$ is said to be quadratically integrable if it is measurable and

$$\int_M |f(x)|^2 \, d\mu(x) < \infty.$$

Two functions f_1, f_2 are said to be equivalent if the set $\{x \mid f_1(x) \neq f_2(x)\}$ is of μ-measure zero, or, equivalently

$$\int_M |f_1(x) - f_2(x)|^2 \, d\mu(x) = 0.$$

We define the Hilbert space $\mathscr{H}$ as the set of all classes of equivalent functions. Since $f(x) = f_1(x) + f_2(x)$, $\alpha f(x)$, and

$$\int_M \overline{f_1(x)} f_2(x) \, d\mu(x) \tag{1.2}$$

depend only on the classes, the operations $f_1(x) + f_2(x)$, $\alpha f(x)$ make $\mathscr{H}$ into a complex vector space, and an inner product $\langle f_1, f_2 \rangle$ is defined by (1.2).

It can be proven (see [34]) that, under the above assumptions about $(M, \mathscr{A}, \mu)\mathscr{H}$ is a separable Hilbert space. We shall denote this Hilbert space by $\mathscr{L}^2(M, d\mu)$.

An example for $(M, \mathscr{A}, \mu)$ is obtained by choosing M to be the set of **R** of real numbers, $\mathscr{A}$ the set of Lebesgue measurable sets and μ as Lebesgue measure. For the case of Lebesgue measure it is customary to replace $d\mu(x)$ in the above integral by the simpler notation dx.

2 Orthogonal Systems and Closed Subspaces

A sequence of vectors x_ν for which $\langle x_\nu, x_\mu \rangle = \delta_{\nu\mu} = 1$ for $\nu = \mu$, 0 for $\nu \neq \mu$ is called a normed orthogonal (orthonormal) system. The elements of an orthonormal system are linearly independent because if $\sum_\nu \lambda_\nu x_\nu = 0$, then by taking an inner product with x_μ we obtain $\lambda_\mu = \langle x, \sum_{\nu=1}^n \lambda_\nu x_\nu \rangle = 0$. For each $x \in \mathscr{H}$ a vector p is defined by $x = \sum_{\nu=1}^N \langle x_\nu, x \rangle x_\nu + p$ which satisfies $\langle x_\nu, p \rangle = 0$ for all $\nu \leq N$. Thus it follows that $\|x\|^2 = \sum_{\nu=1}^N |\langle x_\nu, x \rangle|^2 + \|p\|^2$ and we obtain Bessel's inequality $\sum_{\nu=1}^N |\langle x_\nu, x \rangle|^2 \leq \|x\|^2$ and we therefore obtain $\sum_{\nu=1}^\infty |\langle x_\nu, x \rangle|^2 < \infty$ from which we conclude that $\langle x_\nu, x \rangle \to 0$. Since $\|\sum_{\nu=n}^m x_\nu \langle x_\nu, x \rangle\|^2 = \sum_{\nu=n}^m |\langle x_\nu, x \rangle|^2$ the sum $\sum_{\nu=1}^\infty x_\nu \langle x_\nu, x \rangle$ converges in norm, and we therefore obtain

$$\sum_{\nu=1}^\infty |\langle x_\nu, x \rangle|^2 \leq \|x\|^2.$$

We shall now show that the expression $\|x - \sum_{\nu=1}^N \alpha_\nu x_\nu\|$ takes on its minimum value when $\alpha_\nu = \langle x_\nu, x \rangle$ because, for p defined above and $q = x - \sum_{\nu=1}^N \alpha_\nu x_\nu$ it follows that $q = \sum_{\nu=1}^N (\langle x_\nu, x \rangle - \alpha_\nu) x_\nu + p$ and we therefore obtain

$$\|q\|^2 = \sum_{\nu=1}^N |\langle x_\nu, x \rangle - \alpha_\nu|^2 + \|p\|^2.$$

An orthonormal system is said to be an orthonormal basis if it is a basis for $\mathscr{H}$. For an orthonormal basis we therefore obtain

$$x = \sum_\nu x_\nu \langle x_\nu, x \rangle$$

since the $\sum_\nu \alpha_\nu x_\nu$ are dense in $\mathscr{H}$ and, according to a previous result $\|p\| \leq \|q\|$.

The cardinality of an orthonormal basis is equal to the dimension of $\mathscr{H}$; thus it immediately follows that the cardinality cannot be less than the dimension of $\mathscr{H}$.

Next we show that the cardinality of an orthonormal basis cannot be greater than denumerable. According to (IV) there exists a denumerable set $\{y_\nu\}$ which is dense in $\mathscr{H}$. Therefore, to each x of an orthonormal basis there exists a $y_{\nu(x)}$ such that $\|y_{\nu(x)} - x\| < \frac{1}{4}$. To each pair of different x_1, x_2 of an orthonormal basis $y_{\nu(x_1)}$ and $y_{\nu(x_2)}$ must be different because if $y_{\nu(x_1)} = y_{\nu(x_2)}$ then it follows that

$$\|x_1 - x_2\| \leq \|x_1 - y_{\nu(x_1)}\| + \|x_2 - y_{\nu(x_1)}\| < \tfrac{1}{2}$$

in contradiction to $\|x_1 - x_2\|^2 = \|x_1\|^2 + \|x_2\|^2 = 2$. Therefore every orthonormal basis is at most countable. Thus, in the case in which the dimension of $\mathscr{H}$ is denumerable, the theorem is proven.

Let the dimension of $\mathscr{H}$ equal n (finite), then a basis consists of finitely many $y_1, \ldots, y_n$. Thus it follows that each $x \in \mathscr{H}$ can be written in the form $x = \sum_{\nu=1}^{n} \alpha_\nu y_\nu$ and thus there cannot be more than n linearly independent vectors in $\mathscr{H}$. Thus an orthonormal basis can have at most n elements.

We will now show that if M is an arbitrary denumerable subset of $\mathscr{H}$ then it is possible to construct an orthonormal set of vectors which has the same linear span as does M. For $y_\nu \in M$ set $x_1 = y_1/\|y_1\|$, $x_2 = p_2/\|p_2\|$ where $p_2 = y_2 - \langle x_1, y\rangle x_1$ providing that $p \neq 0$ (if $p = 0$ then y_2 and x_1 are linearly dependent, and we can simply eliminate y_2 from M and renumber the elements y_n). We will now assume that M is a linearly independent set. Recursively, we may set

$$x_n = p_n/\|p_n\| \quad \text{where } p_n = y_n - \sum_{\nu=1}^{n-1} \langle x_\nu, y\rangle x_\nu;$$

this procedure is known as the Schmidt orthogonalization procedure. It is easy to verify that the set of x_ν form an orthonormal basis which has the same linear span as does M.

$\mathscr{T}$ is called a closed subspace if $\mathscr{T}$ is a subspace and is closed in norm. It follows that $\mathscr{T}$ is complete. If $\mathscr{S}$ is a subspace, then the closure of $\mathscr{S}$ in $\mathscr{H}$ is a closed subspace. It follows directly that the intersection of arbitrary many closed subspaces of $\mathscr{H}$ is a closed subspace. According to AI, Th. 4.1 the closed subspaces of $\mathscr{H}$ form a complete lattice where $\mathscr{T}_1 \wedge \mathscr{T}_2 = \mathscr{T}_1 \cap \mathscr{T}_2$ and $\mathscr{T}_1 \vee \mathscr{T}_2$ is equal to the intersection of all closed subspaces $\mathscr{T}$ for which $\mathscr{T} \supset \mathscr{T}_1$ and $\mathscr{T} \supset \mathscr{T}_2$.

If M is a subset of $\mathscr{H}$ let (M) denote the subspace generated by M and $[M]$ be the closed subspace generated by M. Therefore we find that $[M]$ is the intersection of all closed subspaces $\mathscr{T}$ for which $\mathscr{T} \supset M$ and $[M]$ is therefore equal to the closure of (M).

If p is orthogonal to all elements of M, it directly follows that it is orthogonal to all elements of (M) , and from the continuity of the inner product, is also orthogonal to all elements in the closure of (M), that is, of $[M]$. In this way it follows that all elements p which are orthogonal to M form a closed subspace which we denote by $M^\perp$. Therefore we find that $M^\perp = (M)^\perp = [M]^\perp$. In addition it follows that $[M] = (M^\perp)^\perp$. Later we shall find that $[M] = (M^\perp)^\perp$.

First we shall show that if $\mathcal{T}$ is a closed subspace, then each $x \in \mathcal{H}$ may be uniquely represented in the form $x = q + p$ where $q \in \mathcal{T}$ and $p \in \mathcal{T}^\perp$. Since the uniqueness is trivial, we need only demonstrate the representation. For $x \in \mathcal{T}$ we obtain $q = x$ and $p = 0$. We now consider the case in which $x \notin \mathcal{T}$. Then $\operatorname{Min}_{y \in \mathcal{T}} \|x - y\| = \mu \neq 0$. Therefore there exists a sequence $y_\nu \in \mathcal{T}$ for which $\|x - y_\nu\| \longrightarrow \mu$. From

$$\|y_\nu - y_\mu\|^2 = 2\|y_\nu - x\|^2 + 2\|y_\mu - x\|^2 - \|y_\nu + y_\mu - 2x\|^2$$

it follows that $\frac{1}{2}(y_\nu + y_\mu) \in \mathcal{T}$ and we therefore obtain $\|\frac{1}{2}(y_\nu + y_\mu) - x\| \geq \mu$ from which we conclude that

$$\|y_\nu - y_\mu\| \leq 2\|y_\nu - x\|^2 + 2\|y_\mu - x\|^2 - 4\mu.$$

From this it follows that the y_ν form a Cauchy sequence and that $y_\nu - q \in \mathcal{T}$, from which it follows that $\|x - y_\nu\| \longrightarrow \|x - q\|$ and finally $\|x - q\| = \mu$. For $p = x - q$ it is only necessary to show that $p \in \mathcal{T}^\perp$, that is, $\langle h, p\rangle = 0$ for all $h \in \mathcal{T}$.

For $p = h\langle h, p\rangle/\|h\|^2 + r$ we find that $\|r\| \leq \|p\| = \mu$. For $p = x - q$ it follows that $r = x - (q + h\langle h, p\rangle/\|h\|^2)$. Since $q + h\langle h, p\rangle/\|h\|^2 \in \mathcal{T}$ we must have $\|r\| \geq \mu$. Therefore $\|r\| = \mu = \|p\|$ and since $\|p\|^2 = \|r\|^2 + |\langle h, p\rangle|^2/\|h\|^4$, we finally obtain $\langle h, p\rangle = 0$.

If $x \in (M^\perp)^\perp = ([M]^\perp)^\perp$, then from $x = p + q$, $q \in [M]$ and $p \in [M]^\perp$ it follows that $\langle x, p\rangle = 0$, so that $\langle x, p\rangle = \langle q + p, p\rangle = \|p\|^2 = 0$ from which it follows that $x \in [M]$, and we have shown that $[M] = (M^\perp)^\perp$.

Let G be a basis in $\mathcal{H}$ and let $\mathcal{T}$ be a closed subspace in $\mathcal{H}$; then each element x in G can be written in the form $x = p + q$ where $q \in \mathcal{T}$ and $p \in \mathcal{T}^\perp$. Let the set of q obtained in this way be denoted by $G_{\mathcal{T}}$, similarly let the set of p be denoted by $G_{\mathcal{T}^\perp}$. It easily follows that $G_{\mathcal{T}}$ is a basis for $\mathcal{T}$ and $G_{\mathcal{T}^\perp}$ is a basis for $\mathcal{T}^\perp$, and that $G_{\mathcal{T}} \cup G_{\mathcal{T}^\perp}$ is a basis for $\mathcal{H}$. With the help of the Schmidt orthogonalization procedure it is easy to show that it is possible to select an orthonormal basis such that the elements x_ν of which are either elements of $\mathcal{T}$ or of $\mathcal{T}^\perp$.

We say that $\mathcal{T}_1$ and $\mathcal{T}_2$ are orthogonal (written $\mathcal{T}_1 \perp \mathcal{T}_2$) if $\mathcal{T}_1 \subset \mathcal{T}_2^\perp$ and therefore $\mathcal{T}_2 \subset \mathcal{T}_1^\perp$ holds. If $\mathcal{T}_1 \perp \mathcal{T}_2$ then the set of all $x + y$, where $x \in \mathcal{T}_1$ and $y \in \mathcal{T}_2$ is a closed subspace. Here we need only prove that the subspace is closed.

From

$$\|x_n + y_n - (x_n + y_n)\|^2 = \|x_n - x_m\|^2 + \|y_n - y_m\|^2$$

it follows directly that the x_n and the y_n form a Cauchy sequence if the $x_n + y_n$ form a Cauchy sequence.

For $\mathcal{T}_1 \perp \mathcal{T}_2$ we therefore obtain $\mathcal{T}_1 \vee \mathcal{T}_2 = \mathcal{T}_1 + \mathcal{T}_2$. If $\mathcal{T}_1 \perp \mathcal{T}_2$ we then write $\mathcal{T}_1 \oplus \mathcal{T}_2$ intsead of $\mathcal{T}_1 + \mathcal{T}_2$. From AI, D1.2 it follows that the operation $\perp$ in the lattice of closed subspaces of $\mathcal{H}$ is an orthocomplementation.

3 The Banach Space of Bounded Operators

A linear map $\mathcal{H}_1 \xrightarrow{A} \mathcal{H}_2$ is called a linear operator, or more simply an operator and satisfies $A(\alpha x) = \alpha Ax$ and $A(x_1 + x_2) = Ax_1 + Ax_2$. A is continuous if and only if there exists a number C for which $\|Ax\| \leq C\|x\|$. A continuous operator is therefore also called a bounded operator (see also AIII, §5).

The values $\langle x, Ax\rangle$ for all $x \in \mathcal{H}$ uniquely determine an operator A. This fact is a direct consequence of the following simple identity:

$$\begin{aligned} 4\langle x, Ay\rangle &= \langle x + y, A(x + y)\rangle - \langle x - y, A(x - y)\rangle \\ &\quad - i\langle x + iy, A(x + iy)\rangle + i\langle x - iy, A(x - iy)\rangle. \end{aligned}$$

A is uniquely determined by the matrix $a_{\nu\mu} = \langle x_\nu, Ax_\mu\rangle$ with respect to a complete orthonormal basis $\{x_\nu\}$ since $Ax_\mu = \sum_\nu x_\nu\langle x_\nu, Ax_\mu\rangle$.

An operator $A = A_1 + A_2$ is defined by $Ax = A_1x + A_2x$ and is bounded if both A_1 and A_2 are bounded. We find that the operator $A = A_1A_2$, defined $Ax = A_1(A_2x)$ is bounded if A_1 and A_2 are bounded. For $\alpha \in \mathbf{C}$ αA is defined by $(\alpha A)x = \alpha(Ax)$. Let $\mathcal{L}(\mathcal{H})$ denote the set of bounded operators of $\mathcal{H}$. $\mathcal{L}(\mathcal{H})$ is therefore a vector space over $\mathbf{C}$ and is also an algebra. The unit element is the operator $\mathbf{1}x = x$, the null element is the operator $\mathbf{0}x = 0$.

It is easy to see that $\|A\| = \sup_{\|x\| \leq 1} \|Ax\|$ defines a norm in $\mathcal{L}(\mathcal{H})$. The fact that $\mathcal{L}(\mathcal{H})$ is a Banach space follows directly from general theorems; however, we shall show that this is the case below.

In addition to the norm-topology in $\mathcal{L}(\mathcal{H})$ we may also introduce the pointwise topology as follows: We say that a sequence $A_n \in \mathcal{L}(\mathcal{H})$ converges pointwise if, for each $x \in \mathcal{H}$ the A_nx form a Cauchy sequence. From $\|A_nx - A_mx\| \leq \|A_n - A_m\|\,\|x\|$ it directly follows that every Cauchy sequence in the norm topology also converges pointwise.

If A_n is a pointwise convergent sequence, then a linear operator is defined by $A_nx \to y$ where $y = Ax$. A is then also bounded, that is, $A \in \mathcal{L}(\mathcal{H})$.

PROOF. Now we shall show that the A_nx are uniformly bounded, that is, there exists a D such that $\|A_nx\| \leq D\|x\|$ for all n. For this purpose it is sufficient to show that there exists a y and a sphere $K_\delta(y) = \{x \mid \|x - y\| \leq \delta > 0\}$ such that $\|A_nx'\| \leq \alpha$ for all $x' \in K_\delta(y)$, because for arbitrary x

$$\begin{aligned} \|A_nx\| &= \frac{2\|x\|}{\delta}\left\|A_n\frac{\delta}{2}\frac{x}{\|x\|}\right\| = \frac{2\|x\|}{\delta}\left\|A_n\left(\frac{\delta}{2}\frac{x}{\|x\|} + y\right) - A_ny\right\| \\ &\leq \frac{2\|x\|}{\delta}\left[\left\|A_n\left(\frac{\delta}{2}\frac{x}{\|x\|} + y\right)\right\| + \|A_ny\|\right] \leq \frac{4\alpha}{\delta}\|x\|. \end{aligned}$$

If such a sphere $K_\delta(y)$ does not exist, then we may stepwise construct the following sequence: first, find a y_1 and an n_1 such that $\|A_{n_1}y_1\| \geq 2$. From the continuity of A_{n_1} we can find a sphere $K_{\rho_1}(y_1)$ for which $\rho_1 < 1$ and $\|A_{n_1}x\| \geq 1$ for $x \in K_{\rho_1}(y_1)$. Then we may find in the interior of $K_{\rho_1}(y_1)$ a y_2 and n_2 which satisfy $\|A_{n_2}y_2\| \geq 3$ together with a sphere $K_{\rho_2}(y_2) \subset K_{\rho_1}(y_1)$ where $\rho_2 < \frac{1}{2}$ and $\|A_{n_2}x\| \geq 2$ for $x \in K_{\rho_2}(y_2)$. In this way we obtain a sequence of spheres $K_{\rho_\nu}(y_\nu) \subset K_{\rho_{\nu-1}}(y_{\nu-1})$ for which $\rho_\nu < 1/\nu$ and $\|A_{n_\nu}\| \geq \nu$ for $x \in K_{\rho_\nu}(y_\nu)$. Thus for $\mu > \nu$ we obtain $\|y_\mu - y_\nu\| \leq 2\rho_\nu$ and therefore there exists a y for which $y_\nu \longrightarrow y$ and $y \in K_{\rho_\nu}(y_\nu)$ for all ν. Here $\|A_{n_\nu}y\| \geq \nu$ in contradiction to the fact that $A_n y$ is convergent.

From $\|A_n x\| < D\|x\|$ it follows that, in the limit, $\|Ax\| < D\|x\|$. In this way we may prove that, for a sequence of bounded linear forms $\mathscr{H} \xrightarrow{l_n} \mathbf{C}$ which converge pointwise, $\|l_n(x)\| < D\|x\|$ for all n and hence defines a bounded linear form l by $l_n(x) \rightarrow l(x)$.

For a pointwise convergent sequence $A_n x \longrightarrow Ax$ we write $A_n \longrightarrow A$, for norm-convergence we write $A_n \xrightarrow{\|\cdots\|} A$. If A_n is a norm-Cauchy sequence, then, for all x, $A_n x$ is a Cauchy sequence. Therefore, there exists an A for which $A_n \longrightarrow A$. We now show that $A_n \xrightarrow{\|\cdots\|} A$ as follows:

Let $A'_n = A - A_n$; A'_n is also a norm-Cauchy sequence. If we choose N such that, for $n, m > N$ the relationship $\|A'_n - A'_m\| < \varepsilon$ holds, then for $\|x\| \leq 1$ we obtain:

$$\begin{aligned}\|A'_n x\| &\leq \|(A'_n - A'_m)x\| + \|A'_m x\| \\ &\leq \|A'_n - A'_m\| + \|A'_m x\| \leq \varepsilon + \|A'_m x\|.\end{aligned}$$

For fixed x we obtain $\|A'_m x\| \longrightarrow 0$, and therefore it follows that $\|A'_n x\| \leq \varepsilon$ for $n > N$ and arbitrary x. Therefore $\|A'_n\| = \sup_{\|x\| \leq 1} \|A'_n x\| \leq \varepsilon$, that is, $\|A'_n\| = \|A_n - A\| \longrightarrow 0$.

We have shown that $\mathscr{L}(\mathscr{H})$ is a Banach space. It is easy to show that $\|AB\| \leq \|A\|\,\|B\|$. If this relation holds in an algebra which is also a normed space, then it is called a normed algebra; if it is also complete, it is called a Banach algebra.

From $A_n \xrightarrow{\|\cdots\|} A$, $B_m \xrightarrow{\|\cdots\|} B$ it easily follows that $A_n B_m \xrightarrow{\|\cdots\|} AB$. From $A_n \longrightarrow A$, $B_m \longrightarrow B$ it follows that $A_n B_m \longrightarrow AB$ which follows directly from

$$\begin{aligned}\|A_n B_m x - ABx\| &\leq \|A_n(B_m - B)x\| + \|(A_n - A)Bx\| \\ &\leq D\|(B_m - B)x\| + \|(A_n - A)Bx\| \longrightarrow 0.\end{aligned}$$

4 Bounded Linear Forms

As in the case of a Banach space (see AIII, §3) we may also investigate bounded linear forms for $\mathscr{H}$. We shall now show that if $\mathscr{H} \xrightarrow{l} \mathbf{C}$ is a bounded linear form, then there exists a $y \in \mathscr{H}$ for which $l(x) = \langle y, x \rangle$.

It is easy to see that the set $\mathscr{T}_0 = \{h \mid l(h) = 0\}$ is a closed subspace of $\mathscr{H}$. If $\mathscr{T}_0 = \mathscr{H}$ it follows that $l(x) = 0 = \langle 0, x \rangle$. If $\mathscr{T}_0 \neq \mathscr{H}$, then there exists, according to §2, a y which is orthogonal to $\mathscr{T}_0$; therefore $l(y) \neq 0$. For $p = x - (l(x)/l(y))y$ it follows that $l(p) = 0$, that is, $p \in \mathscr{T}_0$ and therefore

$\langle p, y\rangle = 0$. From $x = (l(x)/l(y))y + p$ it follows that, by taking the inner product with $h = (l(y)/\|y\|^2)y$ we obtain $\langle h, x\rangle = l(x)$. From the Schwarz inequality it easily follows that $\langle h, x\rangle$ is a bounded linear form for all $h \in \mathscr{H}$. Since $\langle h_1, x\rangle = \langle h_2, x\rangle$ for all x implies that $h_1 = h_2$, it follows that the map $l \to h$ with $l(x) = \langle h, x\rangle$ is a bijective map $\mathscr{H}' \to \mathscr{H}$ where $\mathscr{H}'$ is, in the sense of AIII, §3, the dual Banach space to $\mathscr{H}$. For this correspondence we obtain $l_1 + l_2 \to h_1 + h_2$ and $\alpha l \to \bar{\alpha} h$. From $\sup_{\|x\| \le 1} \langle h, x\rangle = \|h\|$ it follows that the norm defined on $\mathscr{H}'$ corresponds, according to this bijective mapping, to the norm in $\mathscr{H}$.

A convergent sequence in $\mathscr{H}'$ (in the sense of the topology $\sigma(\mathscr{H}', \mathscr{H})$ in AIII, §4) corresponds to a sequence y_n in $\mathscr{H}$ for which $\langle y_n, x\rangle$ converges pointwise. Such a sequence is called a weakly convergent sequence.

According to §3, to each pointwise convergent sequence l_n of linear forms there exists a bounded linear form l which satisfies $l_n(x) \to l(x)$ for all x as a limit and there exists a C such that $\|l_n(x)\| \le C\|x\|$. Therefore, for each weakly convergent sequence y^n in $\mathscr{H}$ there exists a C such that $\|y_n\| < C$ and a limit element y towards which the y_n converges (we denote the weak convergence of y by $y_n \rightharpoonup y$). $\mathscr{H}$ is therefore sequence-complete.

In §2 we saw that the relation $\langle x_\nu, x\rangle \to 0$ is satisfied for the elements x_ν of an orthonormal basis. If a sequence x_ν satisfies the relations $\langle x_\nu, x_\mu\rangle = 0$ for $\mu \neq \nu$ and there exists a C such that $\|x_\nu\| < C$ for all ν, then it follows that $x_\nu \rightharpoonup 0$.

From the general theorems in AIII, §4 it follows that every bounded set in $\mathscr{H}$ is weakly relative compact. This result can also be easily shown directly: If $\mathscr{M}$ is a bounded set (that is, $\|y\| < C$ for all $y \in \mathscr{M}$), then the set of $\langle y, x\rangle$ for fixed x is bounded for $y \in \mathscr{M}$ since $|\langle y, x\rangle| \le \|y\|\,\|x\| \le C\|x\|$. Let $\mathscr{G}$ be a denumerable dense subset of $\mathscr{H}$ and let x_ν ($\nu = 1, 2, \ldots$) denote the elements of $\mathscr{G}$. In the usual way we may choose a sequence $y_\mu^{(1)} \in \mathscr{M}$ for which $\langle y_\mu^{(1)}, x_1\rangle$ converges; from the sequence of the $y_\mu^{(1)}$ we may choose a subsequence $y_\mu^{(2)}$ for which $\langle y_\mu^{(2)}, x_2\rangle$ converges, etc. For the diagonal sequence $y_\mu^{(\mu)}$ $\langle y_\mu^{(\mu)}, x_\nu\rangle$ converges for all fixed x_ν for all $x_\nu \in \mathscr{G}$. We will show that this situation holds for all $x \in \mathscr{H}$. This follows from

$$\begin{aligned}|\langle y_\mu^{(\mu)} - y_\rho^{(\rho)}, x\rangle| &\le |\langle y_\mu^{(\mu)} - y_\rho^{(\rho)}, x_\nu\rangle| + |\langle y_\mu^{(\mu)} - y_\rho^{(\rho)}, x - x_\nu\rangle|\\ &\le |\langle y_\mu^{(\mu)} - y_\rho^{(\rho)}, x_\nu\rangle| + 2C\|x - x_\nu\|,\end{aligned}$$

if we first choose x_ν such that $2C\|x - x_\nu\| < \varepsilon$ and then, for fixed x_ν, choose N such that for $\mu, \rho > N$ we obtain $|\langle y_\mu^{(\mu)} - y_\rho^{(\rho)}, x_\nu\rangle| < \varepsilon$.

If A is a bounded operator, then $\langle y, Ax\rangle$ is, for fixed y a bounded linear form over x; therefore there exists a $y' \in \mathscr{H}$ for which $\langle y, Ax\rangle = (y', x\rangle$. Since y' is uniquely determined, $y \to y'$ defines a map $\mathscr{H} \to \mathscr{H}$; it is easy to verify the fact that the map is linear. We denote this operator defined by this map by A^+. Thus A^+ is determined by $\langle y, Ax\rangle = \langle A^+ y, x\rangle$.

Since

$$\sup_{\|y\| < 1} \|A^+ y\| = \sup_{\|y\| \le 1} \sup_{\|x\| \le 1} |\langle A^+ y, x\rangle|$$

it easily follows that A^+ is bounded, and that $\|A^+\| = \|A\|$. It is also easy to show that $(\alpha A)^+ = \bar{\alpha} A^+$, $(A + B)^+ = A^+ + B^+$, $(AB)^+ = B^+A^+$ and $(A^+)^+ = A$.

We call A^+ the adjoint operator corresponding to A. An operator A is said to be self-adjoint (or Hermitian) if $A^+ = A$.

An operator A is said to be compact (or completely continuous) if A maps the unit sphere (and therefore every bounded set) on to a relatively compact set (with respect to the norm). We shall now show that the above condition is equivalent to the condition that for every sequence $y_\nu \rightharpoonup 0$ the relation $Ay_\nu \to 0$ is satisfied, that is, $\|Ay_\nu\| \to 0$.

If A is compact, then for each denumerable set $\{y_\nu\}$ for which $\|y_\nu\| < 1$ the set Ay_ν has an accumulation point in the norm-topology. Since, for a weakly convergent sequence $y_\nu \rightharpoonup 0$ $\langle y_\nu, x\rangle$ is uniformly bounded, there exists a number C such that $\|y_\nu\| < C$. If we consider the sequence $y_\nu C^{-1}$ instead of y_ν, we may then assume that the y_ν are elements of the unit sphere, and that the Ay_ν must therefore have an accumulation point y in the norm topology. Therefore there exists a subsequence y_{ν_i} such that $Ay_{\nu_i} \to y$. Thus, for arbitrary $x \in \mathscr{H}$ we obtain $\langle Ay_{\nu_i}, x\rangle \to \langle y, x\rangle$, and since $y_{\nu_i} \rightharpoonup 0$ we also obtain $\langle Ay_{\nu_i}, x\rangle \to \langle y_{\nu_i}, Ax\rangle \to 0$; therefore $y = 0$. Thus the sequence Ay must converge: $Ay_\nu \to 0$.

Conversely, we assume that for $y_\nu \rightharpoonup 0$ it follows that $Ay_\nu \to 0$. Let z_ν be an arbitrary denumerable subset of the unit sphere. A will be compact if the Az_ν have an accumulation point in the norm topology. Since the unit sphere is weakly compact, there exists a subsequence z_{ν_i} which is weakly convergent: $z_{\nu_i} \rightharpoonup z$, where $\|z\| \leq 1$; for $y_{\nu_i} = \frac{1}{2}(z_{\nu_i} - z)$ we obtain $\|y_{\nu_i}\| \leq 1$ and $y_{\nu_i} \rightharpoonup 0$ from which we obtain $Ay_{\nu_i} \to 0$, that is, $Az_{\nu_i} \to Az$. The Az_ν therefore have an accumulation point (in norm).

Let $\mathscr{L}_c(\mathscr{H})$ denote the set of all compact operators in $\mathscr{L}(\mathscr{H})$. It is easy to verify that $\mathscr{L}_c(\mathscr{H})$ is a linear subspace of $\mathscr{L}(\mathscr{H})$. $\mathscr{L}_c(\mathscr{H})$ is, however, closed with respect to the norm topology and is itself a Banach space.

PROOF. Let $A_n \in \mathscr{L}_c(\mathscr{H})$ and let $A_n \xrightarrow{\|\cdots\|} A$. Let $y_\nu \rightharpoonup 0$, then $\|A_n y_\nu\| \underset{\nu}{\to} 0$. From

$$\|Ay_\nu\| \leq \|(A - A_n)y_\nu\| + \|A_n y_\nu\| \leq \|A - A_n\| \, \|y_\nu\| + \|A_n y_\nu\|$$

it follows that, since $\|y_\nu\| < C$ for a suitable value of C,

$$\|Ay_\nu\| \leq \|A - A_n\| C + \|A_n y_\nu\|$$

from which we conclude that $\|Ay_\nu\| \to 0$.

5 The Banach Space $\mathscr{L}_r(\mathscr{H})$

The set of self-adjoint bounded operators is evidently a linear vector space over the field of real numbers $\mathbf{R}$. We may also consider $\mathscr{L}(\mathscr{H})$ to be a vector space over $\mathbf{R}$, and the set of self-adjoint operators form a subspace of $\mathscr{L}(\mathscr{H})$ which we shall denote by $\mathscr{L}_r(\mathscr{H})$. If $\mathscr{L}_r(\mathscr{H})$ is norm-closed in $\mathscr{L}(\mathscr{H})$ then

$\mathscr{L}_r(\mathscr{H})$ will be a Banach space. This will be the case if the limit (in norm) of a sequence of self-adjoint operators is a self-adjoint operator.

Since $\|A^+\| = \|A\|$, it follows from $A_n \xrightarrow{\|\cdots\|} A$ that $A_n^+ \xrightarrow{\|\cdots\|} A^+$ ($A_n^+ \to A^+$ does not necessarily follow from $A_n \to A$!). $A_n^+ = A^+$ implies $A^+ = A$. Therefore $\mathscr{L}_r(\mathscr{H})$ is a Banach space. $\mathscr{L}_r(\mathscr{H})$ is, however, closed with respect to pointwise convergence, because from $A_n \to A$ and from $\langle x, A_n y\rangle = \langle A_n x, y\rangle$ it follows that $\langle x, Ay\rangle = \langle Ax, y\rangle$. Since $\mathscr{L}_c(\mathscr{H})$ is a Banach subspace of $\mathscr{L}(\mathscr{H})$, $\mathscr{L}_{cr}(\mathscr{H}) = \mathscr{L}_c(\mathscr{H}) \cap \mathscr{L}_r(\mathscr{H})$ is a Banach space.

For $A \in \mathscr{L}_r(\mathscr{H})$ we find that $\langle x, Ax\rangle = \langle Ax, x\rangle = \overline{\langle x, Ax\rangle}$ and we find that $\langle x, Ax\rangle$ is real.

We may introduce a partial order in $\mathscr{L}_r(\mathscr{H})$ as follows: $A \geq B$ if $\langle x, Ax\rangle \geq \langle x, Bx\rangle$ for all $x \in \mathscr{H}$. Here we note that $\mathscr{L}_r(\mathscr{H})$ is, in the sense of AIII, §6 an ordered vector space with positive cone

$$\mathscr{L}_{r_+}(\mathscr{H}) = \{A \mid A \in \mathscr{L}_r(\mathscr{H}) \text{ and } A > 0\}.$$

It is easy to show that $\mathscr{L}_{r_+}(\mathscr{H})$ is closed in the norm topology.

We will now show that $\mathscr{L}_r(\mathscr{H})$ is an order unit space (see AIII, §6), that is, the unit sphere of $\mathscr{L}_r(\mathscr{H})$ is the order interval

$$[-\mathbf{1}, \mathbf{1}] = \{A \mid -\mathbf{1} < A < \mathbf{1}\}.$$

From $\|Ax\| < C\|x\|$ it directly follows that $|\langle x, Ax\rangle| \leq \|x\|\,\|Ax\| < C\|x\|^2 = C\langle x, x\rangle$, that is, $-C\mathbf{1} \leq A \leq C\mathbf{1}$. Conversely, if $\mu_1\mathbf{1} < A < \mu_2\mathbf{1}$, then for $Ax \neq 0$ (the case $Ax = 0$ is trivial), setting $y = \|Ax\|^{-\frac{1}{2}}\|x\|Ax = \alpha Ax$, we obtain

$$\begin{aligned}
\|Ax\|^2 &= \left\langle \frac{1}{\alpha}Ax, y\right\rangle = \tfrac{1}{4}\left\langle A\left(\frac{1}{\alpha}x + y\right), \frac{1}{\alpha}x + y\right\rangle \\
&\quad - \tfrac{1}{4}\left\langle A\left(\frac{1}{\alpha}x - y\right), \frac{1}{\alpha}x - y\right\rangle \leq \tfrac{1}{4}\mu\left\|\frac{1}{\alpha}x + y\right\|^2 + \tfrac{1}{4}\mu\left\|\frac{1}{\alpha}x - y\right\|^2 \\
&= \tfrac{1}{2}\mu\left(\left\|\frac{1}{\alpha}x\right\|^2 + \|y\|^2\right) = \mu\|x\|\,\|Ax\|,
\end{aligned}$$

where $\mu = \operatorname{Max}\{|\mu_1|, |\mu_2|\}$. Therefore we obtain $\|Ax\| < \mu\|x\|$. For $C = 1$, $\mu = -1$, $\mu = 1$ we obtain our assertion about the unit sphere of $\mathscr{L}_r(\mathscr{H})$.

If A and B are commuting self-adjoint operators, then $AB = BA$ is a self-adjoint operator. In the special case in which $B > 0$ then $A^2B = ABA = BA^2 > 0$ since $\langle x, ABAx\rangle = \langle (Ax), B(Ax)\rangle$. We will now show that, from $A > 0$, $B > 0$ and $AB = BA$ that $AB > 0$.

Since A is bounded, there exists a number c such that $\|Ax\| \leq c\|x\|$. For $A_1 = c^{-1}A$ we have previously found that $0 < A_1 < \mathbf{1}$. Recursively we define $A_{n+1} = A_n - A_n^2$; we shall now show that $0 < A_n < \mathbf{1}$ by induction:

$$A_{n+1} = A_n^2(\mathbf{1} - A_n) + A_n(\mathbf{1} - A_n^2) > 0$$

and

$$\mathbf{1} - A_{n+1} = (\mathbf{1} - A_n) + A_n^2 > 0$$

providing $0 < A_n < \mathbf{1}$. From $A_1 = \sum_{\nu=1}^{n} A_\nu^2 + A_{n+1}$ it follows that, since $A_{n+1} > 0$, $\sum_{\nu=1}^{n} \|A_\nu x\|^2 \leq \langle x, A_1 x\rangle$ and we find that $\|A_\nu x\| \to 0$, that is, $A_\nu \to 0$. Thus we find that $A_1 = \sum_{\nu=1}^{\infty} A_\nu^2$. Therefore A_ν is self-adjoint and commutes with B, as can easily be proven by induction. Thus $AB = cA_1 B = c\sum_{\nu=1}^{\infty} A_\nu^2 B > 0$.

From the preceding results we obtain the following important convergence properties of monotone sequences of commuting operators $A_n \in \mathscr{L}_r(\mathscr{H})$. Suppose that A_n is monotonically decreasing and suppose that $A_n > 0$ for all n. According to the above theorem $(A_m - A_n)A_m > 0$ and $A_n(A_m - A_n) > 0$ for $m > n$, that is,

$$\langle x, A_m^2 x\rangle \geq \langle x, A_m A_n x\rangle \geq \langle x, A_n^2 x\rangle.$$

The sequence of the $\langle x, A_m A_n x\rangle$ must therefore converge to the same value as the monotonically decreasing sequence $\langle x, A_n^2 x\rangle = \|A_n x\|^2$ so that $\|(A_m - A_n)x\|^2 = \langle x, A_m^2 x\rangle + \langle x, A_n^2, x\rangle - 2\langle x, A_m A_n x\rangle \to 0$. The $A_n x$ therefore form a Cauchy sequence, and there exists an A such that $A_n \to A$. From $\langle x, A_n x\rangle \geq 0$ it follows that $\langle x, Ax\rangle \geq 0$, that is, $A > 0$.

If B_n is a monotonically decreasing (or increasing) sequence of commuting self-adjoint operators and there exists an operator B' which commutes with all B_n and satisfies $B_n > B'$ (or $B_n < B'$), then by considering the sequence $A_n = B_m - B'$ (or $A_n = B' - B_n$) we obtain the result that there exists a B such that $B_n \to B$ and $B > B'$ (or $B < B'$).

6 Projection Operators

Let $\mathscr{T}$ be a closed subspace of $\mathscr{H}$. Since each $x \in \mathscr{H}$ has a unique partition $x = q + p$, where $q \in \mathscr{T}$, $p \in \mathscr{T}^\perp$ (see §2), the relation $q = Px$ defines a linear operator P, which we call a projection operator on $\mathscr{T}$. Since $\|x\|^2 = \|q\|^2 + \|p\|^2$ and since $x = q$ for $x \in \mathscr{T}$ we find that $\|P\| = 1$ (providing that P is not equal to 0, that is, $\mathscr{T} = \{0\}$). From $q \in \mathscr{T}$ it follows that $P^2 x = Pq = Px$, that is, $P^2 = P$. For a corresponding partition of y, $y = r + s$, $r \in \mathscr{T}$, $s \in \mathscr{T}^\perp$ it follows that

$$\langle y, Px\rangle = \langle r + s, q\rangle = \langle r, q\rangle = \langle r, q + p\rangle = \langle Py, x\rangle,$$

that is, $P \in \mathscr{L}_r(\mathscr{H})$.

Conversely, if $P \in \mathscr{L}_r(\mathscr{H})$, and $P^2 = P$, then there exists a closed subspace $\mathscr{T}$ upon which P projects. Thus the set $\mathscr{T} = P\mathscr{H}$ is a closed subspace. The fact that P projects upon $\mathscr{T}$ follows directly from the identity $x = Px + (\mathbf{1} - P)x$; it is easy to show that $(\mathbf{1} - P)x \in \mathscr{T}^\perp$. Thus we find that $\mathbf{1} - P$ is a projection operator upon $\mathscr{T}^\perp$. We therefore write $\mathbf{1} - P = P^\perp$.

Thus we find that $P \to P\mathscr{H}$ is a bijection of the set of projection operators on the set of all closed subspaces of $\mathscr{H}$.

For the special case in which $\mathscr{T}$ is the one-dimensional subspace spanned by y (with $\|y\| = 1$) we shall often denote the projection operator on $\mathscr{T}$ by P_y. We obtain $P_y x = y\langle y, x\rangle$.

If, for a pair of closed subspaces $\mathcal{T}_1 \supset \mathcal{T}_2$ then from $x = q_1 + p_1 = q_2 + p_2$ where $q_1 \in \mathcal{T}_1$, $p_1 \in \mathcal{T}_1^\perp$, $q_2 \in \mathcal{T}_2$ and $p_2 \in \mathcal{T}_2^\perp$ it follows that $p_1 \in \mathcal{T}_2^\perp$. For the partition $q_1 = r_2 + s_2$ where $r_2 \in \mathcal{T}_2$ and $s_2 \in \mathcal{T}_2^\perp$ it follows that $x = r_2 + (s_2 + p_1)$, where $r_2 \in \mathcal{T}_2$ and $s_2 + p_1 \in \mathcal{T}_2^\perp$. Therefore $q_2 = r_2$ and $p_2 = s_2 + p_1$. If P_1 is the projector onto $\mathcal{T}_1$ and P_2 is the projector onto $\mathcal{T}_2$ we then obtain $P_1x = P_2x + (\mathbf{1} - P_2)P_1x$, $P_2x = P_2P_1x$. Thus it follows that $\|P_1x\|^2 \geq \|P_2x\|^2$ and $P_2 = P_2P_1$. Since $P_1^2 = P_1$ we find that $\|P_1x\|^2 \geq \|P_2x\|^2$ is equivalent to $P_1 > P_2$. Since P_2 is self-adjoint, from $P_2 = P_2P_1$ it follows that $P_2 = P_2^+ = P_1P_2$, that is, P_1, P_2 commute.

If, conversely the projection operators P_1, P_2 satisfy $P_1 > P_2$ then it follows that

$$\begin{aligned}\|(P_2 - P_1P_2)x\|^2 &= \langle P_2x - P_1P_2x, P_2x - P_1P_2x\rangle \\ &= \langle P_2x, P_2x\rangle + \langle P_1P_2x, P_1P_2x\rangle \\ &\quad - \langle P_1P_2x, P_2x\rangle - \langle P_2x, P_1P_2x\rangle \\ &= \|P_2x\|^2 - \|P_1P_2x\|^2\end{aligned}$$

and since $\|P_1y\|^2 \geq \|P_2y\|^2$ we obtain

$$\|(P_2 - P_1P_2)x\|^2 = \|P_2P_2x\|^2 - \|P_1P_2x\|^2 \leq 0,$$

that is, $P_2 - P_1P_2 = 0$. Thus, as above, it follows that $P_2 = P_2P_1 = P_1P_2$.

For $x \in P_2\mathcal{H}$, from $P_2 = P_1P_2$ it follows that $x = P_1x$ and therefore $x \in P_1\mathcal{H}$, that is, $P_1\mathcal{H} > P_2\mathcal{H}$. The above bijection between the projection operators and the corresponding projection spaces is therefore an order isomorphism.

From $P_2 = P_1P_2$ it follows that $(P_1 - P_2)^2 = P_1 - P_2$ and therefore $P_1 - P_2 > 0$, that is, $P_1 > P_2$. If $(P_1 - P_2)^2 = P_1 - P_2$ and therefore $P_1 > P_2$ then it also follows that $P_2 = P_1P_2 = P_2P_1$.

The following conditions are therefore equivalent:

$$P_2 = P_1P_2, \quad P_2 = P_2P_1, \quad (P_1 - P_2)^2 = P_1 - P_2, \quad P_1 > P_2, \quad P_1\mathcal{H} \supset P_2\mathcal{H}.$$

For two projection operators P and Q we denote the projector onto $(P\mathcal{H}) \cap (Q\mathcal{H})$ by $P \wedge Q$ and the projector onto $(P\mathcal{H}) \vee (Q\mathcal{H})$ by $P \vee Q$. Because of the above order isomorphism, the set of projection operators is, according to §2, an orthocomplemented lattice.

We will now show that $PQ = QP$ is equivalent to $P \wedge Q = PQ$. From $PQ = QP$ it follows that $(PQ)^2 = PQ$ and $(QP)^2 = QP$. Thus it follows that $PQ\mathcal{H} \subset P\mathcal{H}$ and $PQ\mathcal{H} = QP\mathcal{H} \subset Q\mathcal{H}$ and we obtain $PQ\mathcal{H} \subset P\mathcal{H} \wedge Q\mathcal{H}$. If $x \in P\mathcal{H} \wedge Q\mathcal{H}$ then it follows that $x = Px = Qx$ and that $x = PQx \in PQ\mathcal{H}$.

Conversely, if $PQ = P \wedge Q$, then PQ is self-adjoint, that is, $(PQ)^+ = QP = PQ$.

In general PQ is not self-adjoint and is not a projection.

For the special case in which $PQ = 0$ it follows that $PQ = QP$ and $Q = (\mathbf{1} - P)Q$ and therefore $Q\mathcal{H} \subset (P\mathcal{H})^\perp$, that is, $Q\mathcal{H}$ is orthogonal to

$P\mathscr{H}$. It follows that $P + Q$ is the projection operator onto $P\mathscr{H} \oplus Q\mathscr{H} = P\mathscr{H} \vee Q\mathscr{H}$. If $P + Q < \mathbf{1}$, then it follows that $P < \mathbf{1} - Q$ and that $P = P(\mathbf{1} - Q)$, that is, $PQ = 0$, $P\mathscr{H} \perp Q\mathscr{H}$ and $P + Q$ is a projection onto $P\mathscr{H} \oplus Q\mathscr{H}$.

If P_n is a decreasing (or increasing) sequence of projection operators, then from §5 it follows that P_n converges $P_n \to P$, since $P_n > 0$ ($P_n < \mathbf{1}$). From $P_n^2 = P_n$ it follows that $P^2 = P$, that is, P is also a projection operator. It is easy to show that $P\mathscr{H} = \bigcap_n (P_n\mathscr{H})$ (or $P\mathscr{H} = \bigvee_n (P_n\mathscr{H})$).

If P_n is a sequence of projection operators, then $\sum_{n=1}^{N} P_n$ is an increasing sequence. If $\sum_{n=1}^{N} P_n < \mathbf{1}$ for all N then $\sum_{n=1}^{\infty} P_n$ exists and is $< \mathbf{1}$. It is easy to show that the condition $\sum_{n=1}^{\infty} P_n < \mathbf{1}$ is equivalent to the condition that the P_n are pairwise orthogonal, that is, $P_n P_m = 0$ for $n \neq m$. Thus we find that $\sum_{n=1}^{\infty} P_n$ is a projection operator on $\bigvee_n (P_n\mathscr{H}) = \sum_n \oplus P_n\mathscr{H}$.

If P is a projection operator and $A \in \mathscr{L}_r(\mathscr{H})$, then $AP = PA$ is equivalent to $A = PAP + (\mathbf{1} - P)A(\mathbf{1} - P)$, where the latter can be proven easily with the aid of the identity

$$A = PAP + (\mathbf{1} - P)AP + PA(\mathbf{1} - P) + (\mathbf{1} - P)A(\mathbf{1} - P).$$

If $A = (PA)$ (or AP) then since $A \in \mathscr{L}_r(\mathscr{H})$ it follows that $A = (PA)^+ = AP$ (or $A = PA$). A therefore commutes with P and we obtain $A = PA = P^2A = PAP$. If $PA = 0$, then it follows that $PA = AP = 0$ and we obtain $(\mathbf{1} - P)A = A(\mathbf{1} - P) = (\mathbf{1} - P)A(\mathbf{1} - P) = A$.

If $0 < A < P$, then for $x \in (\mathbf{1} - P)\mathscr{H}$ it easily follows that $\langle x, Ax \rangle = 0$. Since $A > 0$, according to §5 we may write A in the form $A = c \sum_\nu A_\nu^2$. From $\langle x, Ax \rangle = 0$ it easily follows that $A_\nu x = 0$ and $Ax = 0$, that is, $A(\mathbf{1} - P) = 0$. Thus we therefore obtain

$$A = AP = PA = PAP.$$

7 Isometric and Unitary Operators

A linear operator V is said to be isometric if $\|Vx\| = \|x\|$ for all $x \in \mathscr{H}$. Thus we find that $V \in \mathscr{L}(\mathscr{H})$ and $\|V\| = 1$. From $\langle Vx, Vx \rangle = \langle x, V^+Vx \rangle = \langle x, x \rangle$ it follows that $V^+V = \mathbf{1}$ (see §3) and from $V^+V = \mathbf{1}$ we obtain an isometry.

For the operator $P = VV^+$ it follows that $P^+ = P$ and $P^2 = P$, that is, VV^+ is a projection operator. If the Hilbert space $\mathscr{H}$ is finite dimensional, then we must have $V^+ = V^{-1}$ and therefore $P = \mathbf{1}$. For infinite-dimensional $\mathscr{H}$ it is possible that $P \neq \mathbf{1}$.

$P = \mathbf{1}$ is equivalent to the statement that V^+ is also an isometry. An operator U for which both U and U^+ are isometric is said to be unitary. Therefore U is unitary if and only if $UU^+ = U^+U = \mathbf{1}$.

If V is isometric, then $V\mathscr{H}$ is a closed subspace of $\mathscr{H}$ because, from the continuity of V and V^+ it follows from $Vx_n \to y$ that $V^+Vx_n \to V^+y$, that is, $x_n \to V^+y = y'$ and that $Vx_n \to Vy' \in V\mathscr{H}$. From $P = VV^+$ it follows easily that $P\mathscr{H} \subset V\mathscr{H}$. From $\mathscr{H} \supset V\mathscr{H}$ it follows that $P\mathscr{H} \supset PV\mathscr{H} = VV^+V\mathscr{H} = V\mathscr{H}$. Therefore we obtain $P\mathscr{H} = V\mathscr{H}$. P is therefore a projection onto the image of V. An isometric operator is therefore unitary if and only if $V\mathscr{H} = \mathscr{H}$.

If the isometric operator V has a right-inverse, that is, there exists a V' such that $VV' = \mathbf{1}$, then it follows that $V^+VV' = V^+$ and $V' = V^+$, that is, $VV^+ = \mathbf{1}$. The statement that an isometric operator V is unitary is equivalent to the statement that V has a right inverse.

8 Spectral Representation of Self-adjoint and Unitary Operators

We shall now demonstrate the following theorem:

Let $A \in \mathscr{L}_{r_+}(\mathscr{H})$, then there exists a unique $B \in \mathscr{L}_{r_+}(\mathscr{H})$ such that $B^2 = A$; all operators which commute with A also commute with B. (B is called the positive root of A, $B = A^{1/2}$.)

PROOF. From $B_1 \in \mathscr{L}_+(\mathscr{H})$ and $B_1^2 = A_1 = \|A\|^{-1}A$, it follows that for $B = \|A\|^{1/2}B_1$, $B^2 = \|A\|B_1^2 = A$ and $B \in \mathscr{L}_{r_+}(\mathscr{H})$; from $B^2 = A$ and $B \in \mathscr{L}_{r_+}(\mathscr{H})$ and for $B_1 = \|A\|^{-1/2}$ it follows that $B_1^2 = A_1$ and $B_1 \in \mathscr{L}_{r+}(\mathscr{H})$. Since $\|A\|^{-1}A < \mathbf{1}$ it suffices to prove the theorem for $A < \mathbf{1}$. For this purpose we define a sequence B_n as follows: $B_0 = \mathbf{0}$, $B_{n+1} = B_n + \frac{1}{2}(A - B_n^2)$. Thus it follows that all operators which commute with A also commute with all B_n. B_n is an increasing sequence which satisfies $\mathbf{0} < B_n < \mathbf{1}$, because from $\mathbf{1} - B_{n+1} = \frac{1}{2}(\mathbf{1} - B_n)^2 + \frac{1}{2}(\mathbf{1} - A)$ it follows that $\mathbf{1} - B_{n+1} > 0$; by the induction hypothesis it follows from $B_{n+1} - B_n = \frac{1}{2}(B_n - B_{n-1})[(\mathbf{1} - B_{n-1}) + (\mathbf{1} - B_n)]$ that the relation $B_{n+1} - B_n > \mathbf{0}$ holds. The sequence B_n converges; therefore we obtain $B_n \to B$ where $\mathbf{0} < B < \mathbf{1}$. From $B_{n+1} = B + \frac{1}{2}(A - B_n^2)$ it follows that, in the limit, $B^2 = A$. Since the B_n commute with all operators which commute with A, the same holds for B.

In order to prove uniqueness, we assume that $C > \mathbf{0}$ and that $C^2 = A$. From $AC = C^2C = CC^2 = CA$, C commutes with A, and therefore also commutes with B. Earlier we have shown that there exist positive roots $B^{1/2}$ and $C^{1/2}$. For $x \in \mathscr{H}$ and $y = (B - C)x$ it follows that

$$\|B^{1/2}y\|^2 + \|C^{1/2}y\|^2 = \langle y, By\rangle + \langle y, Cy\rangle = \langle y, (B + C)y\rangle$$
$$= \langle y, (B + C)(B - C)x\rangle = \langle y, (B^2 - C^2)x\rangle = 0.$$

Therefore we obtain $B^{1/2}y = C^{1/2}y = 0$, from which we conclude that $By = 0$ and $Cy = 0$. Thus we obtain $\|(B - C)x\|^2 = \langle x, (B - C)^2x\rangle = \langle x, (B - C)y\rangle = 0$, that is, $(B - C)x = 0$. Since x was arbitrary, we have proven $B = C$.

Now let $B(\mu) > \mathbf{0}$; let $B(\mu)^2 = (A - \mu\mathbf{1})^2$. $B(\mu)$ is therefore uniquely determined and commutes with all operators which commute with A. Let $A(\mu) = A - \mu\mathbf{1}$, $A(\mu)_+ = \frac{1}{2}(B(\mu) + A(\mu))$ and $A(\mu)_- = \frac{1}{2}(B(\mu) - A(\mu))$. Therefore $A(\mu) = A(\mu)_+ - A(\mu)_-$ and $B(\mu) = A(\mu)_+ + A(\mu)_-$. Since $B(\mu)$ and $A(\mu)$ commute, we obtain

$$A(\mu)_+A(\mu)_- = A(\mu)_-A(\mu)_+ = \tfrac{1}{4}(B(\mu)^2 - A(\mu)^2) = \mathbf{0}.$$

Let $\mathcal{T}_\mu$ denote the set $\{x \mid x \in \mathcal{H} \text{ and } A(\mu)_+ x = 0\}$. From the continuity of $A(\mu)_+$ it follows that $\mathcal{T}_\mu$ is a closed subspace of $\mathcal{H}$. Let $E(\mu)$ denote the projector onto $\mathcal{T}_\mu$. We therefore obtain $A(\mu)_+ E(\mu) = \mathbf{0}$. Since $A(\mu)_+ A(\mu)_- = \mathbf{0}$ we therefore obtain $A(\mu)_- \mathcal{H} \subset \mathcal{T}_\mu$, that is, $E(\mu)A(\mu)_- = A(\mu)_-$.

If $C \in \mathcal{L}_r(\mathcal{H})$ commutes with A, then it also commutes with $B(\mu)$, $A(\mu)_+$ and $A(\mu)_-$. For $y \in \mathcal{T}_\mu$ it then follows that $A(\mu)_+ Cy = CA(\mu)_+ y = 0$, that is, $Cy \in \mathcal{T}_\mu$. Thus it follows that $CE(\mu) = E(\mu)CE(\mu)$. Computing the adjoint operator we obtain $E(\mu)C = CE(\mu)$.

Let $C = A(\mu)_-$; from $E(\mu)A(\mu)_- = A(\mu)_-$ it follows that $A(\mu)_- E(\mu) = A(\mu)_-$. Similarly, from $A(\mu)_+ E(\mu) = \mathbf{0}$ we obtain $E(\mu)A(\mu)_+ = \mathbf{0}$. Thus, from $\mathbf{0} < E(\mu) < \mathbf{1}$ and $B(\mu) > \mathbf{0}$ it follows that

$$
\begin{aligned}
&\mathbf{0} < E(\mu)B(\mu) = B(\mu)E(\mu) = E(\mu)[A(\mu)_+ + A(\mu)_-] = A(\mu)_-, \\
&\mathbf{0} < [\mathbf{1} - E(\mu)]B(\mu) = A(\mu)_+,
\end{aligned}
\tag{8.1}
$$

$$
E(\mu)A(\mu) = -A(\mu)_-, \quad (\mathbf{1} - E(\mu))A(\mu) = A(\mu)_+.
$$

For $\lambda \leq \mu$ we obtain $A(\lambda) - A(\mu) > \mathbf{0}$; thus, from $A(\lambda)_- > \mathbf{0}$ we obtain $A(\lambda)_+ - A(\mu)_+ + A(\mu)_- > \mathbf{0}$. By multiplication with $A(\mu)_+$ we obtain

$$
A(\mu)_+[A(\lambda)_+ - A(\mu)_+ + A(\mu)_-] > \mathbf{0},
$$

that is, since $A(\mu)_+ A(\mu)_- = 0$ we obtain $A(\mu)_+ A(\lambda)_+ > (A(\mu)_+)^2$, and we find that $\langle x, A(\mu)_+ A(\lambda)_+ x\rangle \geq \|A(\mu)_+ x\|^2$ for all x. From $A(\lambda)_+ x = 0$ it therefore follows that $A(\mu)_+ x = 0$, that is, $\mathcal{T}_\lambda \subset \mathcal{T}_\mu$ which is equivalent to $E(\lambda) < E(\mu)$ and to $E(\mu)E(\lambda) = E(\lambda)$.

The fact that, for A, there exists two constants α, β for which $\alpha\mathbf{1} < A < \beta\mathbf{1}$ implies that $A(\lambda) > \mathbf{0}$ for $\lambda \leq \alpha$ and therefore $B(\lambda) = A(\lambda)$ for $\lambda \leq \alpha$ because the positive root of $A(\lambda)^2$ is unique; therefore we obtain $A(\lambda)_+ = A(\lambda)$. Since $\langle x, A(\lambda)x\rangle = \langle x, A(\lambda)_+ x\rangle \geq (\alpha - \lambda)\|x\|^2$ it follows that for $\lambda < \alpha$ $A(\lambda)_+ x = 0$ only for $x = 0$ and we therefore obtain $E(\lambda) = \mathbf{0}$ for $\lambda < \alpha$. For $\lambda \geq \beta$ it follows $-A(\lambda) > \mathbf{0}$ and therefore $B(\lambda) = -A(\lambda)$, that is, $A(\lambda)_+ = \mathbf{0}$, from which we conclude that $E(\lambda) = \mathbf{1}$ for $\lambda \geq \beta$.

Since for $\lambda < \mu$, $E(\lambda) < E(\mu)$, $E(\mu) - E(\lambda)$ is a projection operator onto the space $\mathcal{T}_\lambda^+ \cap \mathcal{T}_\mu$. Let $E(\mathcal{I})$ denote $E(\mu) - E(\lambda)$; from the relation (8.1) we obtain:

$$
\begin{aligned}
\mathbf{0} < A(\lambda)_+ E(\mathcal{I}) &= A(\lambda)(\mathbf{1} - E(\lambda))E(\mathcal{I}) = A(\lambda)E(\mathcal{I}) \\
&= (A - \lambda\mathbf{1})E(\mathcal{I}), \\
\mathbf{0} < A(\mu)_- E(\mathcal{I}) &= -A(\mu)E(\mu)E(\mathcal{I}) = -A(\mu)E(\mathcal{I}) = (\mu\mathbf{1} - A)E(\mathcal{I}).
\end{aligned}
$$

Thus it follows that

$$
\lambda E(\mathcal{I}) < AE(\mathcal{I}) < \mu E(\mathcal{I}). \tag{8.2}
$$

We shall denote the limit of $E(\mu)$ as $\mu \to \lambda$ by $E(\lambda^+)$. Let $Q(\lambda) = E(\lambda^+) - E(\lambda)$. We obtain

$$
AQ(\lambda) = \lambda Q(\lambda) \quad \text{that is, } A(\lambda)Q(\lambda) = \mathbf{0}.
$$

In addition it follows that

$$A(\lambda)_+ Q(\lambda) = (\mathbf{1} - E(\lambda))A(\lambda)Q(\lambda) = \mathbf{0}$$

and we therefore find that $Q(\lambda)x \in \mathscr{T}_\lambda$ for all x, that is, $E(\lambda)Q(\lambda) = Q(\lambda)$. From $E(\lambda)E(\mathscr{I}) = 0$ it follows that, in the limit $\mu \to \lambda$ $E(\lambda)Q(\lambda) = 0$—thus we find that $Q(\lambda) = 0$ and therefore we obtain $E(\lambda_+) = E(\lambda)$, that is, $E(\lambda)$ is continuous from above.

Thus, for $P(\mu) = E(\mu) - E(\mu_-)$, for $\lambda \to \mu$, from (8.2) we obtain

$$AP(\mu) = \mu P(\mu). \tag{8.3}$$

If we partition the real interval $[\alpha - \delta, \beta]$ (for α, β used above, and arbitrary small $\delta > 0$) into subintervals $\mathscr{I}_k = (\lambda_k, \lambda_{k+1}]$, where $E(\mathscr{I}_k) = E(\lambda_{k+1}) - E(\lambda_k)$, from (8.2) we obtain

$$\sum_k \lambda_k E(\mathscr{I}_k) < \sum_k AE(\mathscr{I}_k) < \sum_k \lambda_{k+1} E(\mathscr{I}_k).$$

If the maximal length of the intervals $\mathscr{I}_k$ is equal to ε, then from $\sum_k E(\mathscr{I}_k) = E(\beta) - E(\alpha - \delta) = \mathbf{1}$ it follows that

$$\mathbf{0} < \sum_k (\lambda_{k+1} - \lambda_k)E(\mathscr{I}_k) < \varepsilon\mathbf{1}$$

and therefore

$$\mathbf{0} < \sum_k \lambda_{k+1} E(\mathscr{I}_k) - A < \varepsilon\mathbf{1},$$

$$\mathbf{0} < A - \sum_k \lambda_k E(\mathscr{I}_k) < \varepsilon\mathbf{1}$$

from which it follows that

$$\left\| A - \sum_k \lambda_{k+1} E(\mathscr{I}_k) \right\| \le \varepsilon \quad \text{and} \quad \left\| A - \sum_k \lambda_k E(\mathscr{I}_k) \right\| \le \varepsilon.$$

Thus $\sum_k \lambda_k E(\mathscr{I}_k)$ and $\sum_k \lambda_{k+1} E(\mathscr{I}_k)$ converge, as $\varepsilon \to 0$, in norm to A. The limit of the sum is written as the integral

$$A = \int_{\alpha-}^{\beta} \lambda \, dE(\lambda). \tag{8.4}$$

(8.4) is called the spectral representation of A.

From (8.4) for a polynomial p it follows that

$$p(A) = \int_{\alpha-}^{\beta} p(\lambda) \, dE(\lambda).$$

In addition, for a projection valued measurable function $f(\lambda)$ which is measurable relative to the projection valued measure defined by $E(\lambda)$ (see IV, §2.5) we may define the function $f(A)$ as follows

$$f(A) = \int_{\alpha-}^{\beta} f(\lambda) \, dE(\lambda).$$

In particular, for

$$\eta(\lambda) = \begin{cases} 1 & \text{for } \lambda_1 < \lambda \leq \lambda_2, \\ 0 & \text{otherwise,} \end{cases}$$

it follows that $\eta(A) = E(\lambda_2) - E(\lambda_1)$.

From (8.4) it easily follows that the solution of the eigenvalue problem $Ax = \mu x$ (or $(A - \mu\mathbf{1})x = 0$) is equivalent to

$$\|(A - \mu\mathbf{1})x\|^2 = \int_{\alpha-}^{\beta} (\lambda - \mu)^2 \, d\|E(\lambda)x\|^2 = 0$$

and we therefore obtain $x \in P(\mu)\mathscr{H}$ ($P(\mu)$ is defined in (8.3)). The eigenvalues are therefore the values at which $E(\mu)$ is discontinuous.

The uniqueness of the spectral family follows from the representation (8.4). Let $E'(\lambda)$ be a second spectral family which has projection operators which are increasing and continuous from above and satisfy (8.4). Then it follows that

$$A(\mu)_+ = \tfrac{1}{2}\int [\lambda - \mu + |\lambda - \mu|] \, dE'(\lambda).$$

The space $\mathscr{T}_\mu$ is therefore the set of all x for which

$$\int [\lambda - \mu + |\lambda + \mu|] \, dE'(\lambda)x = 0. \tag{8.5}$$

From which is follows that

$$E'(\mu + \varepsilon)x = x$$

for each $\varepsilon > 0$. Since $E'(\lambda)$ is continuous from above, we therefore obtain $E'(\mu)x = x$ for all $x \in \mathscr{T}_\mu$, that is, for all $x = E(\mu)y$, where $y \in \mathscr{H}$. Therefore $E'(\mu)E(\mu) = E(\mu)$, that is, $E'(\mu) > E(\mu)$. Conversely, if $E'(\mu)x = x$, then (8.5) is satisfied, that is, $A(\mu)_+x = 0$ and we obtain $x \in \mathscr{T}_\mu$. Therefore $E'(\mu)\mathscr{H} \subset \mathscr{T}_\mu$, that is, $E'(\mu) < E(\mu)$. Thus we have proven that $E'(\mu) = E(\mu)$.

If U is a unitary operator, then $A = \frac{1}{2}(U + U^+)$ and $B = (1/2i)(U - U^+)$ are two commuting self-adjoint operators which satisfy $\|A\| \leq 1$, $\|B\| \leq 1$. From $UU^+ = \mathbf{1}$ it follows that $A^2 + B^2 = \mathbf{1}$, that is, $B^2 = \mathbf{1} - A^2$. For A there exists a spectral representation

$$A = \int_{-1-}^{1} \mu \, dE(\mu).$$

Thus we obtain $B^2 = \int_{-1-}^{1} (1 - \mu^2) \, dE(\mu)$.

For a partition of B into positive and negative parts $B = B_+ - B_-$ we obtain $B^2 = B_+^2 + B_-^2$. Let F denote the projection operator onto the subspace of all x for which $B_-x = 0$. Then $B_+ = BF$ and $B_- = B(\mathbf{1} - F)$ and we obtain $B_+^2 = B^2F$ and $B_-^2 = B^2(\mathbf{1} - F)$, from which we obtain

$$B_+ = \int_{-1-}^{1} \sqrt{1 - \mu^2} \, dE(\mu)F, \qquad B_- = \int_{-1-}^{1} \sqrt{1 - \mu^2} \, dE(\mu)(\mathbf{1} - F).$$

For $\mu \pm i\sqrt{1-\mu^2} = e^{i\varphi}$ and

$$G(\varphi) = \begin{cases} (1 - E(\mu))F & \text{for } 0 \le \varphi < \pi, \\ E(\mu)(1 - F) + F & \text{for } \pi \le \varphi \le 2\pi \end{cases}$$

we obtain

$$U = \int_0^{2\pi} e^{i\varphi}\, dG(\varphi).$$

For $D(\varphi) = G(\varphi_+)$ we find that $D(\varphi)$ is continuous from above, and that

$$U = \int_{0-}^{2\pi} e^{i\varphi}\, dD(\varphi). \tag{8.6}$$

9 The Spectrum of Compact Self-adjoint Operators

If A is self-adjoint and compact (see §4) then A can only have a discrete spectrum, that is, the spectral family of A cannot be continuously increasing. In addition, the eigenspaces corresponding to nonzero eigenvalues, that is, all $P(\mu)\mathscr{H}$ which satisfy (8.3) for nonzero values of μ can only have finite dimension, and the sequence of eigenvalues must converge towards 0.

PROOF. If $E(\lambda)$ were continuously increasing, or if there exists an infinite dimensional eigenspace corresponding to a nonzero eigenvalue, then there would exist an interval λ to $\lambda + \varepsilon$ for which $[E(\lambda + \varepsilon) - E(\lambda)]$ was infinite dimensional and $\lambda + \varepsilon < 0$ or $\lambda > 0$. Therefore there would be an infinite sequence $x_\nu \in [E(\lambda + \varepsilon) - E(\lambda)]\mathscr{H}$ for which $\langle x_\nu, x_\mu\rangle = \delta_{\nu\mu}$ so that $x_\nu \rightharpoonup 0$ (see §4). Since A is compact, we must have $Ax_\nu \to 0$. From (8.4), it would then follow that

$$\|Ax_\nu\|^2 = \int_\lambda^{\lambda+\varepsilon} \mu^2\, d\|E(\mu)x_\nu\|^2 \ge \min\{\lambda^2, (\lambda + \varepsilon)^2\} \ne 0$$

in contradiction to $\|Ax_\nu\| \to 0$.

Therefore we obtain, for the eigenvalues μ of A,

$$A = \sum_\mu \mu P(\mu), \tag{9.1}$$

where 0 is the only accumulation point for the eigenvalues μ and $P(\mu)$ is finite dimensional for $\mu \ne 0$. Therefore we may choose a complete orthonormal basis x_ν, for which (9.1) is transformed into

$$A = \sum_\nu \mu_\nu P_{x_\nu} \quad \text{where } \mu_\nu \to 0. \tag{9.2}$$

We shall now show that if (9.2) holds, then A is compact. According to §4 it is sufficient to consider a sequence $y_n \rightharpoonup 0$ with $\|y_n\| \le 1$. From (9.2) it follows that (here we set $M_N = \operatorname{Max}\{|\mu_\nu| \mid \nu > N\}$:

$$\|Ay_n\| = \left\|\sum_\nu \mu_\nu x_\nu \langle x_\nu, y_n\rangle\right\| \le \left\|\sum_{\nu=1}^N \mu_\nu x_\nu \langle x_\nu, y_n\rangle\right\|$$

$$+ \left\|\sum_{\nu=N+1}^\infty \mu_\nu x_\nu \langle x_\nu, y_n\rangle\right\| \le \left(\sum_{\nu=1}^N \mu_\nu^2 |\langle x_\nu, y_n\rangle|\right)^{1/2} + M_N.$$

Now we choose N such that $M_N < \varepsilon$ and then let $n \to \infty$, and we obtain $\langle x_\nu, y_n \rangle \to 0$. Therefore we obtain $\|Ay_n\| \to 0$.

From (9.2) it follows that $\mathscr{L}_{cr}(\mathscr{H})$ is the norm-closed subspace of $\mathscr{L}_r(\mathscr{H})$ generated by the set of all P_x. A projection operator P is an element of $\mathscr{L}_{cr}(\mathscr{H})$ if and only if $P\mathscr{H}$ is finite dimensional.

10 Spectral Representation of Unbounded Self-adjoint Operators

In general, an unbounded linear operator A in $\mathscr{H}$ is not defined in all of $\mathscr{H}$. Let $\mathscr{D}_A$ denote the domain of definition of A, that is, A is defined as the map $\mathscr{D}_A \to \mathscr{H}$. From linearity, we may assume that $\mathscr{D}_A$ is a subspace (but not necessarily a closed subspace). Let $\mathscr{W}_A = A\mathscr{D}_A$ denote the range of A. $\mathscr{W}_A$ is also a subspace of $\mathscr{H}$.

If $E(\lambda)$ is a spectral family of projection operators, that is, $E(\lambda_1) \geq E(\lambda_2)$ for $\lambda_1 > \lambda_2$, $E(\lambda_+) = E(\lambda)$ and $E(\lambda) \to \mathbf{0}$ for $\lambda \to -\infty$, $E(\lambda) \to \mathbf{1}$ for $\lambda \to \infty$, then

$$A_{\alpha\beta} = \int_\alpha^\beta \lambda \, dE(\lambda)$$

defines an operator $A_{\alpha\beta}$ for which $\|A_{\alpha\beta}\| \leq \max(|\alpha|, |\beta|)$. For an $x \in \mathscr{H}$ and for $\alpha \to -\infty$ and $\beta \to \infty$ $A_{\alpha\beta}x$ is convergent if there exists a c for which $\|A_{\alpha\beta}x\| < c$ for all α, β. In particular, for all x for which

$$\int_{-\infty}^{\infty} \lambda^2 \, d\|E(\lambda)x\|^2 < \infty \tag{10.1}$$

there exists an operator A defined by

$$Ax = \int_{-\infty}^{\infty} \lambda \, dE(\lambda)x. \tag{10.2}$$

The operator A therefore has a natural definition domain consisting of the set of vectors which satisfy (10.1). If there does not exist a λ for which either $E(\lambda) = \mathbf{1}$ or $\mathbf{0}$, then it is easy to show that $\mathscr{D}_A \neq \mathscr{H}$. $\mathscr{D}_A$ is, however, dense in $\mathscr{H}$!

For a map $\mathscr{D}_A \xrightarrow{A} \mathscr{H}$ we may consider the graphs—as subsets of the topological product space $\mathscr{H} \times \mathscr{H}$, that is, the set of all pairs (x, Ax) for which $x \in \mathscr{D}_A$. Here it is particularly advantageous to consider the topological product of $\mathscr{H}$ with itself to be a Hilbert space $\mathscr{H} \oplus \mathscr{H} = \mathscr{H}_2$ where $(x_1, y_1) + (x_2, y_2) = (x_1 + x_2, y_1 + y_2)$, $\alpha(x, y) = (\alpha x, \alpha y)$ and $\langle (x_1, y_1), (x_2, y_2) \rangle = \langle x_1, x_2 \rangle + \langle y_1, y_2 \rangle$. For $x \in \mathscr{D}_A$ the graph $\mathscr{G}_A$ of A is the set of all vectors of the form $(x, Ax) \in \mathscr{H}_2$. $\mathscr{G}_A$ is a subspace of $\mathscr{H}_2$ if and only if A is linear. Conversely, each subspace $\mathscr{G}$ of $\mathscr{H}_2$ for which there is no element of the form $(0, y)$ with $y \neq 0$ defines a linear operator A for which $\mathscr{D}_A$ is the first component of the elements of $\mathscr{G}$.

The operator A is said to be closed if $\mathscr{G}_A$ is a closed subspace. This is equivalent to the condition that for $x_n \in \mathscr{D}_A$, $x_n \longrightarrow x$ and $Ax_n \longrightarrow y$ it follows that $(x, y) \in \mathscr{G}_A$, that is, $x \in \mathscr{D}_A$ and $Ax = y$.

For an operator A we define A^+ as follows: $\mathscr{D}_{A^+}$ is the set of all x for which there exists an x' for which $\langle x, Ay\rangle = \langle x', y\rangle$ for all $y \in \mathscr{D}_A$. $A^+x = x'$ is therefore defined only if $\langle z, y\rangle = 0$ for all $y \in \mathscr{D}_A$ it follows that $z = 0$, that is, if $\mathscr{D}_A$ is dense in $\mathscr{H}$.

If we define a unitary operator U in $\mathscr{H}_2$ by $U(x, y) = (y, -x)$, we find that $U^2 = -1$. The graph of A^+ is precisely $\mathscr{G}_{A^+} = (U\mathscr{G}_A)^\perp$. A^+ is therefore defined only if $(U\mathscr{G}_A)^\perp$ is a graph, that is, if it contains no element $(0, y)$ for which $y \neq 0$, that is, if $\mathscr{D}_A$ is dense in $\mathscr{H}$. Since $\mathscr{G}_{A^+} = (U\mathscr{G}_A)^\perp$ is a closed subspace, A^+ is closed.

We shall now show that if A is itself closed, then $\mathscr{D}_{A^+}$ is also dense in $\mathscr{H}$. If $\mathscr{D}_{A^+}$ is not dense in $\mathscr{H}$ then there exists a $y \in \mathscr{H}$ for which $y \perp \mathscr{D}_{A^+}$. We therefore obtain $(0, y) \perp U\mathscr{G}_{A^+}$, that is, $(0, y) \in (U\mathscr{G}_{A^+})^\perp = [U(U\mathscr{G}_A)^\perp]^\perp = [(U^2\mathscr{G}_A)^\perp]^\perp = U^2\mathscr{G}_A = \mathscr{G}_A$ in contradiction to the assumption that $\mathscr{G}_A$ is a graph.

If $\mathscr{D}_{A^+}$ is dense in $\mathscr{H}$ (but A is not necessarily closed!) then A^{++} exists, and it follows that

$$\mathscr{G}_{A^{++}} = (U\mathscr{G}_{A^+})^\perp = [U(U\mathscr{G}_A)^\perp]^\perp = [\mathscr{G}_A^\perp]^\perp = [\mathscr{G}_A],$$

that is, $\mathscr{G}_{A^{++}}$ is the closed subspace generated by $\mathscr{G}_A$.

We say that B is an extension of A (written $B \supset A$) if $\mathscr{G}_B \supset \mathscr{G}_A$. A^{++} is therefore, if it exists (that is, if $\mathscr{D}_{A^+}$ is dense in $\mathscr{H}$), an extension of A: $A^{++} \supset A$. A^{++} is closed. If $B \supset A$, and if B is closed, then $\mathscr{G}_B \supset \mathscr{G}_A$ and we therefore obtain $\mathscr{G}_B \supset [\mathscr{G}_A]$, that is, $B \supset A^{++}$.

We define $A + B$ by $\mathscr{D}_{A+B} = \mathscr{D}_A \cap \mathscr{D}_B$ and $(A + B)x = Ax + Bx$. We define AB by means of $\mathscr{D}_{AB}$ as the set of all $x \in \mathscr{D}_B$ for which $Bx \in \mathscr{D}_A$ and $(AB)x = A(Bx)$. If $Ax = 0$ only for $x = 0$, then we can define A^{-1} by $\mathscr{D}_{A^{-1}} = \mathscr{W}_A$ and $A^{-1}(Ax) = x$. It is easy to show that $\mathbf{0}A \subset \mathbf{0}$, $A(BC) = (AB)C$, $(A + B)C = AC + BC$, $A(B + C) \supset AB + AC$, $(AB)^{-1} = B^{-1}A^{-1}$ if A^{-1}, B^{-1} exist; $(A + B)^+ \supset A^+ + B^+$; $(AB)^+ \supset B^+A^+$.

A is said to be self-adjoint if $A^+ = A$. Clearly $\mathscr{D}_A$ must be dense in $\mathscr{H}$. A is said to be symmetric if $\mathscr{D}_A$ is dense in $\mathscr{H}$ and $\langle x, Ay\rangle = \langle Ax, y\rangle$ for all $x, y \in \mathscr{D}_A$. If A is symmetric, then $A^+ \supset A$. If A is self-adjoint, then it is also symmetric. From $A^+ \supset A$ it follows that for a symmetric A $\mathscr{D}_{A^+}$ is dense in $\mathscr{H}$ and that A^{++} exists and satisfies $A^{++} \subset A^+ = (A^+)^{++} = (A^{++})^+$—therefore A^{++} is also symmetric. A symmetric A is said to be maximal, if A has no symmetric extension. A maximal A must therefore satisfy $A^{++} = A$. A self-adjoint operator is maximal since $B \supset A$ implies $B^+ \subset A^+ = A$ and $B \subset B^+ \subset A$, that is, $B = A$. If $A^{++} = A^+$ then A^{++}, the closure of A, is self-adjoint. Then we say that A is essentially self-adjoint. It is easy to show that the operator A defined by (10.2) is self-adjoint. If we formulate quantum mechanics in the manner presented in this book, then unbounded operators occur in the representation of Lie groups, namely, as infinitesimal transformations (see VII and VIII). As such, they are defined on the basis of representation theory (see VII, VIII and [10]) in the form (10.2). However, the introduction of unbounded scale observables (for example: the position and momentum observables introduced in VII) occurs on account of the fact that first a decision observable $\Sigma \longrightarrow G$ was defined and then a spectral family is defined. Then the operator A is introduced according to (10.2), so to speak, as a "condensation" of the spectral family. For the formulation of quantum mechanics presented here it is sufficient to consider only (10.2). For applications it is important to note that the so-called Hamiltonian operator H, that is, the operator for the infinitesimal time translation is only determined by the Galileo group for elementary systems (see VII, §2). For composite systems, on the other hand, it is necessary to discover the Hamiltonian H. In this way we discover an operator which is not always well defined, as, for example, we find in VIII (5.8), that is, $\mathscr{D}_H$ is not fully known, and even less for the spectral family $E(\lambda)$ which is representable in terms of a time displacement by the unitary operator

$$U(t) = \int e^{i\lambda t}\, dE(\lambda) = e^{iHt}.$$

Often we must seek to find an extension for an operator H which was only given as a symmetric operator. The "correct" extension can only be obtained on the basis of "physical" considerations. The extension must be a self-adjoint operator (or at least essentially self-adjoint); only in this way is it possible to uniquely define the spectral family and the operator e^{iHt} as a unitary operator. The problem of the definition of H is especially important for the term scheme of atoms (see XI–XV) and for scattering theory (see XVI). In the following we shall consider the circumstances under which a symmetric operator can be extended to a self-adjoint operator. According to the previous considerations it therefore follows that the maximal symmetric operators which are not self-adjoint are "physically useless."

We now proceed from a symmetric operator A and seek to define $(A - i\mathbf{1})(A + i\mathbf{1})^{-1}$. This form is motivated by the Cayley transform $w = (z - i)/(z + i)$ which maps the real z-axis onto the unit circles in the w plane.

For $x \in \mathscr{D}_A$ we obtain

$$\|(A \pm i\mathbf{1})x\|^2 = \|Ax\|^2 + \|x\|^2 > \|x\|^2. \tag{10.3}$$

Thus we find that $(A + i\mathbf{1})x = 0$ only for $x = 0$, that is, $(A + i\mathbf{1})^{-1}$ exists where $\mathscr{D}_{(A+i\mathbf{1})^{-1}} = (A + i\mathbf{1})\mathscr{D}_A$. Therefore the operator

$$U = (A - i\mathbf{1})(A + i\mathbf{1})^{-1} \tag{10.4}$$

is well defined with $\mathscr{D}_U = (A + i\mathbf{1})\mathscr{D}_A$. It directly follows that $\mathscr{W}_U = (A - i\mathbf{1})\mathscr{D}_A$. Each $y \in \mathscr{D}_U$, therefore, is of the form $y = (A + i\mathbf{1})x$ where $x \in \mathscr{D}_A$ and, for $y' = Uy$ it follows that $y' = (A - i\mathbf{1})x$. Thus, from (10.3) it follows that $\|Uy\| = \|y\|$—that U is an isometric operator on $\mathscr{D}_U$ (if $\mathscr{D}_U = \mathscr{H}$ then U is, in the sense of §7, an isometric operator).

From $Uy = (A - i\mathbf{1})x$ and $y = (A + i\mathbf{1})x$ it follows that $x = (\mathbf{1} - U)y/2i$ and that $Ax = (\mathbf{1} + U)y/2$. $(\mathbf{1} - U)^{-1}$ exists, because from $(\mathbf{1} - U)y = 0$ it follows that $x = 0$ and that $y = (A + i\mathbf{1})0 = 0$. Thus we find that

$$A = i(\mathbf{1} + U)(\mathbf{1} - U)^{-1} \tag{10.5}$$

since $\mathscr{D}_A = (\mathbf{1} - U)\mathscr{D}_U$.

Conversely, suppose that U is an isometric operator in a subspace $\mathscr{D}_U$. Then it follows that $\langle x, y\rangle = \langle Ux, Uy\rangle$ for all $x, y \in \mathscr{D}_U$. Since U is isometric, U^{-1} is defined in $\mathscr{W}_U$. From $(\mathbf{1} - U)y = 0$ it follows that for $z \in (\mathbf{1} - U^{-1})\mathscr{W}_U = (\mathbf{1} - U^{-1})U\mathscr{D}_U = (U - \mathbf{1})\mathscr{D}_U = (\mathbf{1} - U)\mathscr{D}_U$, that is, for $z = (\mathbf{1} - U^{-1})x$ and $x \in \mathscr{W}_U$ we obtain

$$\langle z, y\rangle = \langle x, y\rangle - \langle U^{-1}x, y\rangle = \langle x, y\rangle - \langle x, Uy\rangle = \langle x, (\mathbf{1} - U)y\rangle = 0.$$

If $(\mathbf{1} - U)\mathscr{D}_U$ is dense in $\mathscr{H}$, then from $\langle z, y\rangle = 0$ for all $z \in \mathscr{H}$ it follows that $y = 0$.

If U is an isometric operator in $\mathscr{D}_U$ and $(\mathbf{1} - U)\mathscr{D}_U$ is dense in $\mathscr{H}$, then from (10.5) an operator A is defined for which $\mathscr{D}_A = (\mathbf{1} - U)\mathscr{D}_U$. Therefore $\mathscr{D}_A$ is dense in $\mathscr{H}$ and A^+ is defined. For $x, y \in \mathscr{D}_A$, that is, $x = (\mathbf{1} - U)v$, $y = (\mathbf{1} - U)w$, where $w, v \in \mathscr{D}_U$, it follows that

$$\begin{aligned}\langle x, Ay\rangle &= \langle (\mathbf{1} - U)v, i(\mathbf{1} + U)w\rangle = i\langle v, w\rangle - i\langle Uv, Uw\rangle \\ &\quad + i\langle v, Uw\rangle - i\langle Uv, w\rangle = i\langle v, Uw\rangle - i\langle Uv, w\rangle\end{aligned}$$

and

$$\begin{aligned}\langle Ax, y\rangle &= \langle i(\mathbf{1} + U)v, (\mathbf{1} - U)w\rangle = -i\langle v, w\rangle + i\langle Uv, Uw\rangle \\ &\quad + i\langle v, Uw\rangle - i\langle Uv, w\rangle = i\langle v, Uw\rangle - i\langle Uv, w\rangle,\end{aligned}$$

that is, $\langle x, Ay\rangle = \langle Ax, y\rangle$—$A$ is therefore symmetric.

For an operator A defined as above by U, it follows that (10.4) holds. Symmetric operators A and isometric operators U for which $(\mathbf{1} - U)\mathscr{D}_U$ is dense in $\mathscr{H}$ are uniquely related by (10.4) and (10.5) and $\mathscr{D}_U = (A + i\mathbf{1})\mathscr{D}_A$ and $\mathscr{D}_A = (\mathbf{1} - U)\mathscr{D}_U$. Each proper symmetric extension of A leads to a proper extension of U and vice versa. In this way the problem of symmetric extensions of A are directly related to that of isometric extensions of U.

From (10.3) it follows that for $z_n = (A + i\mathbf{1})x_n$ the sequence z_n is convergent only if the sequences Ax_n and x_n are convergent. If A is closed, then $\mathscr{D}_U$ is closed (and since U is continuous, U is also closed). Therefore $\mathscr{D}_U$ and $\mathscr{W}_U$ are closed subspaces of $\mathscr{H}$. Conversely, if $\mathscr{D}_U$ is closed (and therefore $\mathscr{W}_U$ is also closed) then it follows that A is closed. Since A^{++} is the smallest closed extension of A, A^{++} corresponds to that extension of U for which the domain of definition is the closure of $\mathscr{D}_U$; this extension of U is uniquely determined. For this reason we need only consider *closed* symmetric operators A. $\mathscr{D}_U$ and $\mathscr{W}_U$ are therefore closed subspaces; $\mathscr{D}_U^\perp$ and $\mathscr{W}_U^\perp$ are called the defect spaces of A. For $x \in \mathscr{D}_U^\perp$ and $y \in \mathscr{D}_A$ it follows that $\langle x, (A + i\mathbf{1})y\rangle = 0$, that is, $\langle x, Ay\rangle = \langle ix, y\rangle$. In this way $\mathscr{D}_U^\perp \subset \mathscr{D}_{A^+}$ and $A^+x = ix$ for $x \in \mathscr{D}_U^\perp$. In the same way, for $z \in \mathscr{W}_U^\perp$ we obtain $A^+z = -iz$. If, conversely, $A^+x = ix$, then, for all $y \in \mathscr{D}_A$ it follows that $\langle x, (A + i\mathbf{1})y\rangle = 0$, that is, $x \in \mathscr{D}_U^\perp$. Since $A = A^+$ for a self-adjoint operator, $\langle x, Ax\rangle$ is real; so that we conclude that $\mathscr{D}_U^\perp = \{0\}$, $\mathscr{W}_U^\perp = \{0\}$, that is, $\mathscr{D}_U = \mathscr{W}_U = \mathscr{H}$. According to §7 this is equivalent to the condition that U is unitary.

We now assume the converse—that is U is unitary, that is if $\mathscr{D}_U = \mathscr{W}_U = \mathscr{H}$ then A is self-adjoint. In order to show this we prove the following for a closed and symmetric A—$\mathscr{D}_{A^+} = \mathscr{D}_A + \mathscr{D}_U^\perp + \mathscr{W}_U^\perp$. For $x \in \mathscr{D}_A$ we partition $(A^+ + i1)x$ into components in $\mathscr{D}_U$ and in $\mathscr{D}_U^\perp$, that is, $(A^+ + i1)x = (A + i1)z + y$ where $z \in \mathscr{D}_A$ and $y \in \mathscr{D}_U^\perp$. Since $Az = A^+z$ and $A^+y = iy$ it follows that $(A^+ + i1)x = (A^+ + i1)z + (A^+ + i1)y'$ where $y' = (1/2i)y$. Therefore $A^+(x - z - y') = -i(x - z - iy')$, that is $x - z - y' \in \mathscr{W}_U^\perp$. Therefore $x = z + y' + r$, where $z \in \mathscr{D}_A$, $y' \in \mathscr{D}_U^\perp$, and $r \in \mathscr{W}_U^\perp$. Thus we have proven $\mathscr{D}_{A^+} = \mathscr{D}_A + \mathscr{D}_U^\perp + \mathscr{W}_U^\perp$.

If $\mathscr{D}_U^\perp + \mathscr{W}_U^\perp \neq \{0\}$, then $\mathscr{D}_A$ is a proper subset of $\mathscr{D}_{A^+}$ because for $y \in \mathscr{D}_U^\perp$, $r \in \mathscr{W}_U^\perp$; from the assumption that $y + r \in \mathscr{D}_A$ it follows that $(A + i1)(y + r) = (A^+ + i1)(y + r) = 2iy$ and $(A - i1)(y + r) = (A^+ - i1)(y + r) = -2ir$. Since $\mathscr{D}_U = (A + i1)\mathscr{D}_A$ we obtain $y \in \mathscr{D}_U$ and, since $y \in \mathscr{D}_U^\perp$: $y = 0$. From $(A - i1)\mathscr{D}_A = \mathscr{W}_U$ it also follows that $r = 0$. Therefore the partition $x = z + y + r$, where $x \in \mathscr{D}_{A^+}$ uniquely defines $z \in \mathscr{D}_A$, $y \in \mathscr{D}_U^\perp$, $r \in \mathscr{W}_U^\perp$ and we find that $\mathscr{D}_A + \mathscr{D}_U^\perp + \mathscr{W}_U^\perp$ is a direct sum of subspaces.

If U is unitary, then $A^+ = A$.

Since there is a 1:1 correspondence between the extensions of the isometric operators $\mathscr{D}_U \xrightarrow{U} \mathscr{W}_U$ and the extensions of A, we need only investigate the possibilities of finding isometric extensions of U. From the isometry it easily follows that, for an extension V of U that V maps $\mathscr{D}_V \cap \mathscr{D}_U^\perp$ isometrically onto $\mathscr{W}_V \cap \mathscr{W}_U^\perp$, that is, for $\mathscr{T} = \mathscr{D}_V \cap \mathscr{D}_U^\perp$ and $\mathscr{S} = \mathscr{W}_V \cap \mathscr{W}_U^\perp$ the mapping $\mathscr{T} \xrightarrow{V} \mathscr{S}$ is an isomorphism (in the sense of §13). Therefore $\mathscr{T}$ and $\mathscr{S}$ have the same dimension. Conversely, if $\mathscr{T}$ and $\mathscr{S}$ are subspaces having the same dimension for which $\mathscr{T} \perp \mathscr{D}_U$, $\mathscr{S} \perp \mathscr{W}_U$, then we obtain an isometric extension V of U with the aid of one of infinitely many isomorphic maps $\mathscr{T} \xrightarrow{\tilde{V}} \mathscr{S}$ by $V(x + y) = Ux + \tilde{V}y$ for $x \in \mathscr{D}_U$, $y \in \mathscr{T}$ which satisfies $\mathscr{D}_V = \mathscr{D}_U \oplus \mathscr{T}$, $\mathscr{W}_V = \mathscr{W}_U \oplus \mathscr{S}$.

Therefore, A has a self-adjoint extension if and only if $\mathscr{D}_U^\perp$ and $\mathscr{W}_U^\perp$ have the same dimension. If A is not self-adjoint, then there exist infinitely many self-adjoint extensions of A. Therefore, if we wish to discover a Hamiltonian operator H (for example, in the form VIII (5.8)), it is not sufficient to show that H is symmetric in a certain domain of definition $\mathscr{D}_H$. Indeed, if the defect spaces $\mathscr{D}_U^\perp$, $\mathscr{W}_U^\perp$ do not have the same dimension, the operator cannot be used as a Hamiltonian operator. If, however, the dimension of $\mathscr{D}_U^\perp$ and $\mathscr{W}_U^\perp$ are the same, then we are still not finished, as long as $\mathscr{D}_U \neq \mathscr{H}$. If $\mathscr{D}_U^\perp = \{0\} = \mathscr{W}_U^\perp$, then H^{++} is a self-adjoint operator, that is, H is essentially self-adjoint. Such an H is as good as a self-adjoint H because there is a uniquely defined closure of H which is self-adjoint. The search procedure for H can only be carried out using "physical considerations" because, in the case H is not essentially self-adjoint, there are infinitely many possible extensions. If H is essentially self-adjoint, then we can use the operator H^{++} instead of the operator H.

The condition that A is self-adjoint is equivalent to the condition that U is unitary. If U is unitary then, according to §8 there exists a spectral family $D(\varphi)$ which satisfies (8.6). From (10.5) it follows that

$$A = \int_0^{2\pi} i\,\frac{1 + e^{i\varphi}}{1 - e^{-i\varphi}}\, dD(\varphi) = \int_0^{2\pi} \left(-\cot\frac{\varphi}{2}\right) dD(\varphi)$$

because from $\mathscr{D}_A = (\mathbf{1} - U)\mathscr{H}$ for $x = (\mathbf{1} - U)y$ (with arbitrary y) we obtain $D(\varphi)x = \int_0^{\varphi} (1 - e^{i\varphi'})\, dD(\varphi')y$ and we obtain

$$\|D(\varphi)x\|^2 = \int_0^{\varphi} |1 - e^{i\varphi'}|^2\, d\|D(\varphi')y\|^2 = \int_0^{\varphi} 4\sin^2\frac{\varphi'}{2}\, d\|D(\varphi')y\|^2.$$

From which it follows that

$$\int_0^{2\pi} \cot^2\frac{\varphi}{2}\, d\|D(\varphi)x\|^2 = \int_0^{2\pi} 4\cot^2\frac{\varphi}{2}\sin^2\frac{\varphi}{2}\, d\|D(\varphi)y\|^2$$
$$= \int_0^{2\pi} 4\cos^2\frac{\varphi}{2}\, d\|D(\varphi)y\|^2 < \infty,$$

that is, $\int_0^{2\pi} (-\cot(\varphi/2))\, dD(\varphi)x$ exists. For $\lambda = \cot(\varphi/2)$ and $E(\lambda) = D(-2\,\mathrm{arccot}\,\lambda)$ it follows that

$$A = \int_{-\infty}^{\infty} \lambda\, dE(\lambda), \tag{10.6}$$

where the integral is defined as an operator in $\mathscr{D}_A = (\mathbf{1} - U)\mathscr{H}$. Since $i\mathbf{1} + \int_{-\infty}^{\infty} \lambda\, dE(\lambda)$ maps the subspace $(\mathbf{1} - U)\mathscr{H}$ surjectively onto $\mathscr{H}$, (10.6) cannot converge for other vectors other than those of $\mathscr{D}_A = (\mathbf{1} - U)\mathscr{H}$.

The uniqueness of the spectral family follows in a similar fashion as that described in §8.

11 The Trace as a Bilinear Form

Let $A \in \mathscr{L}_{r+}(\mathscr{H})$ and let x_ν be a complete orthonormal basis. The sum

$$\sum_\nu \langle x_\nu, Ax_\nu\rangle \tag{11.1}$$

defines a real number ≥ 0 or $+\infty$. For another complete orthonormal basis y_μ we obtain

$$\langle y_\mu, Ay_\mu\rangle = \|A^{1/2}y_\mu\|^2 = \sum_\nu |\langle x_\nu, A^{1/2}y_\mu\rangle|^2 = \sum_\nu |\langle y_\mu, A^{1/2}x_\nu\rangle|^2$$

where $A^{1/2}$ is the positive square root of A. Thus it follows that

$$\sum_{\mu=1}^{N} \langle y_\mu, Ay_\mu\rangle = \sum_{\nu=1}^{\infty} \sum_{\mu=1}^{N} |\langle y_\mu, A^{1/2}x_\nu\rangle|^2 \leq \sum_{\nu=1}^{\infty} \sum_{\mu=1}^{\infty} |\langle y_\mu, A^{1/2}x_\nu\rangle|^2$$
$$= \sum_{\nu=1}^{\infty} \|A^{1/2}x_\nu\|^2 = \sum_{\nu=1}^{\infty} \langle x_\nu, Ax_\nu\rangle,$$

and, if we exchange the x_ν and y_μ we obtain

$$\sum_{\mu=1}^{\infty} \langle y_\mu, Ay_\mu \rangle = \sum_{\nu=1}^{\infty} \langle x_\nu, Ax_\nu \rangle.$$

We find that (11.1) is independent of the choice of x_ν; we call this invariant the trace of A, and denote it by $\operatorname{tr}(A)$; for $A \in \mathscr{L}_{r+}(\mathscr{H})$ $\operatorname{tr}(A)$ is a real positive number or $+\infty$.

For an arbitrary $A \in \mathscr{L}_r(\mathscr{H})$ we define

$$\|A\|_s = \operatorname{tr}(\sqrt{A^2}) = \operatorname{tr}(A_+) + \operatorname{tr}(A_-), \tag{11.2}$$

where $\sqrt{A^2}$ is the positive root and A_+, A_- are the positive and negative parts of A.

We shall denote by $\mathscr{B}(\mathscr{H})$ the subset of all $A \in \mathscr{L}_r(\mathscr{H})$ for which $\|A\|_s < +\infty$. The operators in $\mathscr{B}(\mathscr{H})$ are called "operators of trace class."

For $A \in \mathscr{B}(\mathscr{H})$ both $\operatorname{tr}(A_+)$ and $\operatorname{tr}(A_-)$ are finite. Therefore $\operatorname{tr}(A)$ exists because $\operatorname{tr}(A) = \sum_\nu \langle x_\nu, Ax_\nu \rangle = \operatorname{tr}(A_+) - \operatorname{tr}(A_-)$. We will now prove that (11.2) is a norm in $\mathscr{B}(\mathscr{H})$ and that $\mathscr{B}(\mathscr{H})$ is a base norm space (see AIII, §6) and a Banach space.

We will now show that

$$\|A\|_s = \sup_{E_1, E_2} \{\operatorname{tr}(E_1 A E_1) - \operatorname{tr}(E_2 A E_2)\}, \tag{11.3}$$

where the supremum over all projection operators E_1, E_2 with finite dimension of $E_1\mathscr{H}, E_2\mathscr{H}$, so $\|A\|_s$ is a norm. It easily follows that for $A \in \mathscr{L}_r(\mathscr{H})$ and finite-dimensional $E\mathscr{H}$ the expressions $\operatorname{tr}(EA_+E)$, $\operatorname{tr}(EA_-E)$ and $\operatorname{tr}(EAE)$ exist and are finite. In particular, for $E = P_x$ we obtain

$$\operatorname{tr}(P_x A P_x) = \operatorname{tr}(P_x A) = \operatorname{tr}(AP_x) = \langle x, Ax \rangle.$$

For $A \in \mathscr{L}_{r+}(\mathscr{H})$ we obtain $EAE \in \mathscr{L}_{r+}(\mathscr{H})$. For finite or infinite dimensional $E\mathscr{H}$ it easily follows that $\operatorname{tr}(EAE) \le \operatorname{tr}(A)$. Let $E(\lambda)$ denote the spectral family of $A \in \mathscr{L}_r(\mathscr{H})$; then $A_- = -E(0)AE(0)$ and $A_+ = (\mathbf{1} - E(0))A(\mathbf{1} - E(0))$. From (11.2) it follows that

$$\|A\|_s = \operatorname{tr}((\mathbf{1} - E(0))A(\mathbf{1} - E(0)) - \operatorname{tr}(E(0)AE(0)).$$

Let x_ν be a complete orthonormal basis for $(\mathbf{1} - E(0))\mathscr{H} = (E(0)\mathscr{H})^\perp$ and let y_μ be a complete orthonormal basis for $E(0)\mathscr{H}$—it therefore follows that

$$\|A\|_s = \sum_\nu \langle x_\nu, Ax_\nu \rangle - \sum_\mu \langle y_\mu, Ay_\mu \rangle.$$

Let E_{1N}, E_{2M} be the projections onto the space spanned by $x_1, \ldots, x_N$ and $y_1, \ldots, y_M$, respectively. We therefore find that

$$\|A\|_s = \lim_{\substack{N \to \infty \\ M \to \infty}} [\operatorname{tr}(E_{1N} A E_{1N}) - \operatorname{tr}(E_{2M} A E_{2M})].$$

Therefore, using the sup ... from (11.3) we obtain

$$\sup_{E_1, E_2} \{\operatorname{tr}(E_1 A E_1) - \operatorname{tr}(E_2 A E_2)\} \ge \|A\|_s.$$

Since $0 \leq \operatorname{tr}(EA_+E) \leq \operatorname{tr}(A_+)$ and $0 \leq \operatorname{tr}(EA_-E) < \operatorname{tr}(A_-)$ it follows that

$$\operatorname{tr}(EAE) = \operatorname{tr}(EA_+E) - \operatorname{tr}(EA_-E) < \operatorname{tr}(A_+),$$

$$-\operatorname{tr}(EAE) = -\operatorname{tr}(EA_+E) + \operatorname{tr}(EA_-E) \leq \operatorname{tr}(A_-)$$

and we obtain $\operatorname{tr}(E_1AE_1) - \operatorname{tr}(E_2AE_2) \leq \operatorname{tr}(A_+) + \operatorname{tr}(A_-)$ from which it follows that $\sup_{E_1,E_2} \operatorname{tr}(E_1AE_1) - \operatorname{tr}(E_2AE_2) \leq \|A\|_s$. Therefore we have proven (11.3). $\mathscr{B}(\mathscr{H})$ is therefore a normed space with norm $\|\cdots\|_s$. $\mathscr{B}(\mathscr{H})$ is also a Banach space.

From $\|A\| = \sup_{\|x\| \leq 1} \langle x, Ax \rangle$ it is easy to show that $\|A\|_s > \|A\|$. Thus it follows that a Cauchy sequence $A_\nu \in \mathscr{B}(\mathscr{H})$ with respect to the norm $\|\cdots\|_s$ is also a Cauchy sequence with respect to the norm $\|\cdots\|$. Therefore there exists an $A \in \mathscr{L}_r(\mathscr{H})$ such that $A_\nu \xrightarrow{\|\cdots\|} A$. We will now show that $\|A - A_\nu\|_s \to 0$ (from this result and $\|A\|_s \leq \|A - A_\nu\| + \|A_\nu\|_s$ it follows that $\|A\|_s < \infty$).

We now assume (11.3) and consider the case in which $E\mathscr{H}$ has dimension N:

$$|\operatorname{tr}(E(A - A_\nu)E)| \leq |\operatorname{tr}(E(A - A_\rho)E)| + |\operatorname{tr}(E(A_\rho - A_\nu)E)|$$
$$\leq \|A - A_\rho\| N + \|A_\rho - A_\nu\|_s.$$

For a given $\varepsilon > 0$ choose M such that $\|A_\rho - A_\nu\|_s < \varepsilon$ for $\rho, \nu > M$. Therefore it follows that

$$|\operatorname{tr}(E(A - A_\nu)E)| \leq \varepsilon + N\|A - A_\rho\| \tag{11.4}$$

for all $\rho, \nu > M$ and all E (dimension of $E\mathscr{H} < N$), where M does not depend on N!

Thus, for two projections E_1, E_2 for fixed $\nu > M$ we obtain

$$\operatorname{tr}(E_1(A - A_\nu)E_1) - \operatorname{tr}(E_2(A - A_\nu)E_2) < 2\varepsilon + (N_1 + N_2)\|A - A_\rho\|,$$

where N_1, N_2 are the dimensions of $E_1\mathscr{H}$ and $E_2\mathscr{H}$. For $\rho \to \infty$ it follows that

$$\operatorname{tr}(E_1(A - A_\nu)E_1) - \operatorname{tr}(E_2(A - A_\nu)E_2) \leq 2\varepsilon$$

and we obtain $\|A - A_\nu\|_s \leq 2\varepsilon$ for all $\nu > M$. Therefore $\|A - A_\nu\|_s$ is finite, and $\|A - A_\nu\| \to 0$. Therefore $\mathscr{B}(\mathscr{H})$ is a Banach space.

Thus for $A \in \mathscr{B}(\mathscr{H})$ it is necessary and sufficient that A has the form (9.2) (that is, it has only a discrete spectrum) and $\sum_\nu |\mu_\nu| < \infty$ (that is, A must also be compact).

PROOF. From (9.2) it follows that $\|A\|_s = \sum_\nu |\mu_\nu|$. If $A \in \mathscr{B}(\mathscr{H})$ and if, as in §9, for the spectral representation, there exists an interval $\lambda, \lambda + \varepsilon$ for which $E(\lambda + \varepsilon) - E(\lambda)$ is infinite dimensional, then for a complete orthonormal basis x_ν from $[E(\lambda + \varepsilon) - E(\lambda)]\mathscr{H}$ it follows that

$$\|A\|_s \geq \left|\sum_\nu \langle x_\nu, Ax_\nu \rangle\right| = \infty.$$

Therefore A must have the form (9.2).

$\mathscr{B}(\mathscr{H})$ is an ordered vector space with positive cone $\mathscr{B}_+(\mathscr{H})$ of all $A > 0$. The set

$$K = \{W \mid W > 0, \operatorname{tr}(w) = 1\} \tag{11.5}$$

is a base for this cone (see AIII, §6) because from $A > 0$ it follows that $A\|A\|_s^{-1} = A(\text{tr}(A))^{-1} \in K$. For an $A \in \mathscr{B}(\mathscr{H})$ it follows that $A = \|A_+\|_s W_1 - \|A_-\|_s W_2$ where $W_1 = A_+\|A_+\|_s^{-1}$ and $W_2 = A_-\|A_-\|_s^{-1}$. A therefore has the minimal decomposition property (see AIII, §6). Thus it follows directly that the unit sphere of $\mathscr{B}(\mathscr{H})$ is equal to $\text{co}(K \cup -K)$ and that $\mathscr{B}(\mathscr{H})$ is a base norm space. We shall now show that $\mathscr{B}(\mathscr{H})$ is norm-separable. It suffices to show that K is norm-separable. From $W = \sum_\nu w_\nu P_{x_\nu}$ where $w_\nu \geq 0, \sum_\nu w_\nu = 1$ for $W \in K$, it suffices to show that the set of all P_x in K is norm-separable. Since $P_{x_1} - P_{x_2}$ is nonzero only in a two-dimensional space generated by x_1 and x_2, it easily follows that $\|P_{x_1} - P_{x_2}\|_s \leq 2\|P_{x_1} - P_{x_2}\|$ where $\|P_{x_1} - P_{x_2}\|$ is the operator norm in $\mathscr{L}_r(\mathscr{H})$. Since $\|P_{x_1} - P_{x_2}\| \leq 2\|x_1 - x_2\|$ it therefore follows that $\|P_{x_1} - P_{x_2}\|_s \leq 4\|x_1 - x_2\|$. Therefore, the set of P_x is norm-separable as a subset of $\mathscr{B}(\mathscr{H})$ if the set of x for which $\|x\| = 1$ is separable as a subset of $\mathscr{H}$. This is indeed the case, because $\mathscr{H}$ is separable.

A central and most important result is that, for $A \in \mathscr{B}(\mathscr{H})$, $B \in \mathscr{L}_r(\mathscr{H})$, the expression $\text{tr}(AB)$ is a continuous bilinear form on $\mathscr{B}(\mathscr{H}) \times \mathscr{L}_r(\mathscr{H})$ and that each continuous linear form over $\mathscr{B}(\mathscr{H})$ is of the form $\text{tr}(AB)$ with suitably chosen $B \in \mathscr{L}_r(\mathscr{H})$. Therefore $\mathscr{L}_r(\mathscr{H})$ may be identified with the dual Banach space $\mathscr{B}'(\mathscr{H})$ for $\mathscr{B}(\mathscr{H})$, and we find that $|\text{tr}(AB)| \leq \|A\|_s\|B\|$.

We now prove this result in several steps. First, we show that

$$\sum_\nu \langle x_\nu, ABx_\nu\rangle = \sum_\nu \langle x_\nu, BAx_\nu\rangle = \text{tr}(AB) = \text{tr}(BA)$$

is convergent and is independent of the choice x_ν of a complete orthonormal basis. It suffices to show that this result is satisfied for A_+ and A_- where $A = A_+ - A_-$, that is, for an $A > 0$.

Suppose $|\text{tr}(A_+B)| \leq \|A_+\|_s\|B\|$ and $|\text{tr}(A_-B)| < \|A_-\|_s\|B\|$ then it directly follows that $|\text{tr}(AB)| \leq \|A\|_s\|B\|$.

We shall now show that

$$\sum_\nu \langle x_\nu, A^{1/2}BA^{1/2}x_\nu\rangle = \text{tr}(A^{1/2}BA^{1/2}) \tag{11.6}$$

is convergent. This result follows directly from

$$\begin{aligned}\sum_{\nu=1}^N |\langle x_\nu, A^{1/2}BA^{1/2}x_\nu\rangle| &= \sum_{\nu=1}^N |\langle A^{1/2}x_\nu, BA^{1/2}x_\nu\rangle| \\ &\leq \|B\| \sum_{\nu=1}^N \|A^{1/2}x_\nu\|^2 \\ &\leq \|B\|\,\text{tr}(A) = \|B\|\,\|A\|_s.\end{aligned}$$

Substituting B_+ for B, we find that $\text{tr}(A^{1/2}B_+A^{1/2}) < \infty$, that is, $A^{1/2}B_+A^{1/2} \in \mathscr{B}_+(\mathscr{H})$. Thus we find that $A^{1/2}B_-A^{1/2} \in \mathscr{B}_+(\mathscr{H})$ and we obtain $A^{1/2}BA^{1/2} \in \mathscr{B}(\mathscr{H})$, and we have proven (11.6).

We will now show that

$$\sum_\nu \langle x_\nu, A^{1/2}BA^{1/2}x_\nu\rangle = \sum_\nu \langle x_\nu, BAx_\nu\rangle. \tag{11.7}$$

If this is the case, the right-hand side is independent of the orthonormal basis because the left-hand side is according to (11.6). That is, the existence of $\operatorname{tr}(BA)$ follows from (11.7). Since $A^{1/2}BA^{1/2}$ is self-adjoint, the left-hand side of (11.7) is real, and therefore so is the right-hand side. Since $\langle x_\nu, BAx_\nu\rangle = \langle x_\nu, ABx_\nu\rangle$, from the existence of $\operatorname{tr}(BA)$ the existence of $\operatorname{tr}(AB)$ is guaranteed and $\operatorname{tr}(AB) = \operatorname{tr}(BA)$. From the previous results we conclude that $|\operatorname{tr}(BA)| \leq \|A\|_s \|B\|$.

In order to prove (11.7) we need only show that, for $N \longrightarrow \infty$,

$$\sum_\nu \sum_{\mu=1}^{N} \langle A^{1/2}x_\nu, x_\mu\rangle\langle x_\mu, BA^{1/2}x_\nu\rangle \to \sum_\nu \left[\sum_{\mu=1}^{\infty} \langle A^{1/2}x_\nu, x_\mu\rangle\langle x_\mu, BA^{1/2}x_\nu\rangle\right]$$
$$= \sum_\nu \langle A^{1/2}x_\nu, BA^{1/2}x_\nu\rangle = \operatorname{tr}(A^{1/2}BA^{1/2}). \tag{11.8}$$

Then since

$$\sum_{\mu=1}^{N} \sum_\nu \langle x_\mu, BA^{1/2}x_\nu\rangle\langle x_\nu, A^{1/2}x_\mu\rangle = \sum_{\mu=1}^{N} \langle x_\mu, BA^{1/2}A^{1/2}x_\mu\rangle$$
$$\to \sum_{\mu=1}^{\infty} \langle x_\mu, BAx_\mu\rangle$$

and we have proven (11.7).

In order to prove (11.8) we introduce a projection E_N for the subspace spanned by $x_1, \ldots, x_N$. Clearly $E_N \to \mathbf{1}$. Thus (11.8) becomes

$$\sum_\nu \langle A^{1/2}x_\nu, E_N BA^{1/2}x_\nu\rangle \underset{N}{\to} \sum_\nu \langle A^{1/2}x_\nu, BA^{1/2}x_\nu\rangle.$$

(11.8) will be proven if we can show that

$$\sum_\nu \langle A^{1/2}x_\nu, (\mathbf{1} - E_N)BA^{1/2}x_\nu\rangle \underset{N}{\to} 0. \tag{11.9}$$

From

$$|\langle A^{1/2}x_\nu, (\mathbf{1} - E_N)BA^{1/2}x_\nu\rangle| = |\langle(\mathbf{1} - E_N)A^{1/2}x_\nu, (\mathbf{1} - E_N)BA^{1/2}x_\nu\rangle|$$
$$\leq \|(\mathbf{1} - E_N)A^{1/2}x_\nu\| \, \|B\| \, \|A^{1/2}x_\nu\|$$

we obtain

$$\left|\sum_\nu \langle A^{1/2}x_\nu, (\mathbf{1} - E_N)BA^{1/2}x_\nu\rangle\right|$$
$$\leq \|B\| \sum_\nu \|(\mathbf{1} - E_N)A^{1/2}x_\nu\| \, \|A^{1/2}x_\nu\|$$
$$\leq \|B\| \sum_{\nu=1}^{M} \|(\mathbf{1} - E_N)A^{1/2}x_\nu\| \, \|A^{1/2}x_\nu\|$$
$$+ \|B\| \sum_{\nu=M+1}^{\infty} \|A^{1/2}x_\nu\|^2.$$

Since $\sum_\nu \|A^{1/2}x_\nu\| = \mathrm{tr}(A)$ exists, it is possible to find an M such that $\|B\| \sum_{\nu=M+1}^{\infty} \|A^{1/2}x_\nu\|^2 < \varepsilon$. Fix M and choose N so large that $\|B\| \sum_{\nu=1}^{M} \|(1 - E_N)A^{1/2}x_\nu\| \|A^{1/2}x_\nu\| < \varepsilon$, from which (11.9) is proven.

Now we must show the converse, that each linear form on $\mathscr{B}(\mathscr{H})$ is of the form $l(A) = \mathrm{tr}(AB)$ for a suitably chosen $B \in \mathscr{L}_r(\mathscr{H})$.

Next we shall show that we may easily extend $l(A)$ onto the set of "all" operators of trace class. We say that $A \in \mathscr{L}(\mathscr{H})$ is an operator of trace class if $A_1 = \frac{1}{2}(A + A^+)$ and $A_2 = (1/2i)(A - A^+)$ are elements of $\mathscr{B}(\mathscr{H})$. We extend as follows:

$$\tilde{l}(A) = l(A_1) + il(A_2).$$

It is a simple matter to show that $\tilde{l}$ is a linear form. For $\tilde{l}$ it follows from $|l(A)| \leq c\|A\|_s$ for $A \in \mathscr{B}(\mathscr{H})$ that, in general, $|\tilde{l}(A)| \leq |l(A_1)| + |l(A_2)| \leq c(\|A_1\|_s + \|A_2\|_s)$.

We now define an operator which we denote by $x\langle y|$ as follows: $z \to x\langle y, z\rangle$. It follows that $(x\langle y|)^+ = y\langle x|$. Thus we obtain

$$\tilde{l}(x\langle y|) = l(\tfrac{1}{2}(x\langle y| + y\langle x|)) + il\left(\frac{1}{2i}(x\langle y| - iy\langle x|)\right).$$

The operator $\frac{1}{2}(x\langle y| + y\langle x|)$ acts only on the two-dimensional subspace $\mathscr{T}$ spanned by x and y, that is, on $\mathscr{T}^\perp$ this operator is a null operator. Thus it follows that $\|\frac{1}{2}(x\langle y| + y\langle x|)\|_s < 2\|\frac{1}{2}(x\langle y| + y\langle x|)\|$. We may easily estimate the operator norm in $\mathscr{L}(\mathscr{H})$ by $\|\frac{1}{2}(x\langle y| + y\langle x|)\| \leq \|x\| \|y\|$. In this way it follows that $\|\frac{1}{2}(x\langle y| - y\langle x|)\|_s \leq 2\|x\| \|y\|$. Similarly it follows that $\|(1/2i)(x\langle y| - iy\langle x|)\|_s \leq 2\|x\| \|y\|$. Thus we obtain $|\tilde{l}(x\langle y|)| \leq 4c\|x\| \|y\|$. $\tilde{l}(x\langle y|)$ is, therefore, for fixed y a bounded linear form on $\mathscr{H}$. According to §4 there exists a $z \in \mathscr{H}$ for which $\tilde{l}(x\langle y|) = \langle z, x\rangle$. From $|\tilde{l}(x\langle y|)| \leq 4c\|x\| \|y\|$ it follows that $|\langle z, x\rangle| \leq 4c\|x\| \|y\|$ and we obtain $\langle z, z\rangle \leq 4c\|z\| \|y\|$, that is, $\|z\| \leq 4c\|y\|$.

A linear operator B is defined by $y \to z$ which, from $\|z\| \leq 4c\|y\|$ it follows that the relation $\|B\| \leq 4c$ is satisfied, that is, $B \in \mathscr{L}(\mathscr{H})$. Therefore we obtain

$$\tilde{l}(x\langle y|) = \langle By, x\rangle = \langle y, B^+x\rangle.$$

Since $x\langle x| = P_x$ is self-adjoint, $\tilde{l}(x\langle x|) = l(x\langle x|) = \langle Bx, x\rangle$ is real, that is, $\langle Bx, x\rangle = \langle x, Bx\rangle$. With the aid of the identity

$$\langle By, x\rangle = \tfrac{1}{4}\langle B(x + y), x + y\rangle - \tfrac{1}{4}\langle B(x - y), x - y\rangle$$
$$- \frac{1}{4i}\langle B(x + iy), x + iy\rangle + \frac{1}{4i}\langle B(x - iy), x - iy\rangle$$

it easily follows that $\langle By, x\rangle = \langle y, Bx\rangle$ and $B = B^+$, that is, $B \in \mathscr{L}_r(\mathscr{H})$.

Since l is continuous over $\mathscr{B}(\mathscr{H})$ and each $A \in \mathscr{B}(\mathscr{H})$ is of the form $A = \sum_\nu \mu_\nu P_{x_\nu}$, where the $\sum_\nu$ converge in the trace norm. from $l(P_{x_\nu}) = \langle x_\nu, Bx_\nu\rangle$ it follows that $l(A) = \sum_\nu \mu_\nu\langle x_\nu, Bx_\nu\rangle = \sum_\nu \langle Ax_\nu, Bx_\nu\rangle = \mathrm{tr}(AB)$. In $\mathscr{B}'(\mathscr{H})$ the norm is defined by $\|l\| = \sup_{\|A\|_s \leq 1} |l(A)|$. For $l(A) = \mathrm{tr}(AB)$ it is simple to show that $\|l\| = \|B\|$. Thus we have proven that $\mathscr{B}'(\mathscr{H}) = \mathscr{L}_r(\mathscr{H})$.

Since $\mathscr{B}(\mathscr{H})$ is norm separable, $\mathscr{B}'(\mathscr{H})$ is separable in the $\sigma(\mathscr{B}'(\mathscr{H}), \mathscr{B}(\mathscr{H}))$-topology (see AIII, §4).

In closing our investigation of $\mathscr{B}'(\mathscr{H})$ we shall now show that every affine positive functional $K \xrightarrow{l} \mathbf{R}_+$ is of the form $l(w) = \operatorname{tr}(WB)$ for some $B \in \mathscr{B}'_+(\mathscr{H}) = \mathscr{L}_{r+}(\mathscr{H})$.

We say that l is an affine functional over K if $W = \lambda W_1 + (1 - \lambda)W_2$, $0 \leq \lambda \leq 1$ implies that $l(W) = \lambda l(W_1) + (1 - \lambda)l(W_2)$. It is easy to verify that l has a unique extension on all of $\mathscr{B}(\mathscr{H})$ because $\mathscr{B}(\mathscr{H})$ is spanned by K. If l is positive on K, then l is positive on $\mathscr{B}_+(\mathscr{H})$. If a linear functional l satisfies $l: \mathscr{B}_+(\mathscr{H}) \to \mathbf{R}_+$, then it is said to be positive. For $\mathscr{B}(\mathscr{H})$ we may easily prove the theorem mentioned in AIII, §6 that a positive linear functional is continuous.

Let $W_1 = A_+\|A_+\|_s^{-1} \in K$, $W_2 = A_-\|A_-\|_s \in K$; we obtain $A = \|A_+\| W_1 - \|A_-\| W_2$. From this result and from $\|W\|_s = 1$ for $W \in K$ it is easy to show that l is continuous only if there exists a number α for which $|l(W)| \leq \alpha$ for all $W \in K$. For a positive $l(W)$ we need therefore only show that $l(W) \leq \alpha$ for some α.

Suppose such an α does not exist. Then there exists a sequence $W_n \in K$ for which $l(W_n) > 2^n$. Clearly $W = \sum_{n=1}^{\infty} 2^{-n} W_n$ is an element of K. For $V_N = \sum_{\nu=1}^{N} 2^{-n} W_n$ it follows that $l(V_N) = N$. Since $W - V_N = \sum_{n=N+1}^{\infty} 2^{-n} W > 0$ it follows that $l(W) > l(V_N) = N$, which, as $N \to \infty$ leads to a contradiction to the fact that $l(W)$ is defined.

Therefore every positive affine functional l on K is of the form $l(W) = \operatorname{tr}(WB)$ where $B \in \mathscr{L}_r(\mathscr{H})$. Since l is positive, it immediately follows that B is positive.

Let us consider the Banach subspace $\mathscr{L}_{cr}(\mathscr{H})$ (see §5) of $\mathscr{B}'(\mathscr{H}) = \mathscr{L}_r(\mathscr{H})$; then, for each $A \in \mathscr{B}(\mathscr{H})$ $\operatorname{tr}(AB)$ defines a bounded linear form over $\mathscr{L}_{cr}(\mathscr{H})$. Since $A_1 = A_2$ follows simply from $\operatorname{tr}(A_1B) = \operatorname{tr}(A_2B)$ for all $B \in \mathscr{L}_{cr}(\mathscr{H})$ (all P_x are elements of $\mathscr{L}_{cr}(\mathscr{H})$), $\mathscr{B}(\mathscr{H})$ is therefore a subspace of the Banach space which is dual to $\mathscr{L}_{cr}(\mathscr{H})$. We will now show that $\mathscr{B}(\mathscr{H})$ is equal to the Banach space which is dual to $\mathscr{L}_{cr}(\mathscr{H})$.

Let l be a bounded linear form over $\mathscr{L}_{cr}(\mathscr{H})$, therefore, since $\mathscr{B}(\mathscr{H}) \subset \mathscr{L}_{cr}(\mathscr{H})$ (as a set of operators) l is a linear form over $\mathscr{B}(\mathscr{H})$. Since $l(B) \leq c\|B\|$ and $\|B\|_s > \|B\|$ l is also bounded as a linear form over $\mathscr{B}(\mathscr{H})$, that is, there exists an $A \in \mathscr{L}_r(\mathscr{H})$ such that $l(B) = \operatorname{tr}(AB)$ for all $B \in \mathscr{B}(\mathscr{H}) \subset \mathscr{L}_{cr}(\mathscr{H})$; in particular, $l(P_x) = \langle x, Ax \rangle$ for all $P_x \in \mathscr{L}_{cr}(\mathscr{H})$. Since each $B \in \mathscr{L}_{cr}(\mathscr{H})$ is of the form $B = \sum_\nu \mu_\nu P_{x_\nu}$, where $|\mu_\nu| \to 0$ we find that $\|B - \sum_{\nu=1}^{N} P_{x_\nu}\| \to 0$ as $N \to \infty$ and we therefore obtain $l(B) = \sum_{\nu=1}^{\infty} \mu_\nu (P_{x_\nu}) = \sum_{\nu=1}^{\infty} \mu_\nu \langle x_\nu, Ax_\nu \rangle$. Since $\|B\| = \max_\nu\{|\mu_\nu|\}$ we therefore obtain $|\sum_{\nu=1}^{\infty} \mu_\nu \langle x_\nu, Ax_\nu \rangle| < c\|B\| = c \max_\nu\{|\mu_\nu|\}$ for any complete orthonormal basis and for all sequences μ_ν for which $\mu_\nu \to 0$. By appropriate choice of the $\mu_\nu \geq 0$ and x_ν we obtain, for x_ν as a complete orthonormal basis for the space $(\mathbf{1} - E(0))\mathscr{H}$ where $E(\lambda)$ is the spectral family of A and $\mu_\nu > \mu_{\nu+1}$ we obtain

$$\sum_\nu \mu_\nu \langle x_\nu, Ax_\nu \rangle = \sum_\nu \mu_\nu \langle x_\nu, Ax_\nu \rangle < c\mu_1.$$

By choosing $\mu_1 = \mu_2 = \cdots = \mu_N = 1$ and $\mu_\nu = 0$ for $\nu > N$ we obtain $\sum_{\nu=1}^N \langle x_\nu, Ax_\nu\rangle \le c$ for all N and we therefore obtain $\operatorname{tr}(A_+) < \infty$. Similarly it follows that $\operatorname{tr}(A_-) < \infty$ and we find that $A \in \mathscr{B}(\mathscr{H})$.

It is easy to show that the norm in the Banach space which is dual to $\mathscr{L}_{cr}(\mathscr{H})$ is identical to the norm $\|\cdots\|_s$ for $\mathscr{B}(\mathscr{H})$.

From the general theorems in AIII, §4 it follows that the unit sphere $\operatorname{co}(K \vee -K)$ is compact in the $\sigma(\mathscr{B}(\mathscr{H}), \mathscr{L}_{cr}(\mathscr{H}))$-topology. Since $\mathscr{B}_+(\mathscr{H})$ is also $\sigma(\mathscr{B}(\mathscr{H}), \mathscr{L}_{cr}(\mathscr{H}))$-closed, the intersection of $\mathscr{B}_+(\mathscr{H})$ with $\operatorname{co}(K \vee -K)$, that is, the set $\check{K} = \bigcup_{\nu \le \lambda \le 1} \lambda K$ (see also AIII, §6) is compact in the $\sigma(\mathscr{B}(\mathscr{H}), \mathscr{L}_{cr}(\mathscr{H}))$-topology.

12 Gleason's Theorem

If E_n is a collection of pairwise orthogonal projection operators then, according to §6 there exists a projection operator E for which

$$E = \sum_n E_n. \tag{12.1}$$

For $W \in K$, according to §11 we find that

$$W = \sum_\nu w_\nu P_{x_\nu},$$

where $w_\nu \ge 0$ and $\sum_\nu w_\nu = 1$. For $E_N = \sum_{\nu=N+1}^\infty E_\nu$ it follows that

$$\begin{aligned} 0 \le \operatorname{tr}(WE) - \sum_{n=1}^N \operatorname{tr}(WE_n) &= \operatorname{tr}\left(W\left(E - \sum_{n=1}^N E_n\right)\right) \\ &= \sum_{\nu=1}^M w_\nu\langle x_\nu, (E - E_N)x_\nu\rangle + \sum_{\nu=M+1}^\infty w_\nu\langle x_\nu, (E - E_N)x_\nu\rangle \\ &\le \sum_{\nu=1}^M w_\nu\|(E - E_N)x_\nu\|^2 + \sum_{\nu=M+1}^\infty w_\nu. \end{aligned}$$

By choosing M sufficiently large, and for fixed M, let N become infinite, it follows that

$$\operatorname{tr}(WE) = \sum_{n=1}^\infty \operatorname{tr}(WE_n). \tag{12.2}$$

Let G denote the set of projection operators. For $W \in K$ there exists a positive real function $G \xrightarrow{\mu} \mathbf{R}_+$ given by $\mu(E) = \operatorname{tr}(WE)$ for which $\mu(\mathbf{1}) = 1$ and for pairwise orthogonal E_n we obtain $\mu(\sum_n E_n) = \sum_n \mu(E_n)$. Gleason's theorem says that the converse holds—from $G \xrightarrow{\mu} \mathbf{R}_+$, $\mu(\mathbf{1}) = 1$ and $\mu(\sum_n E_n) = \sum_n \mu(E_n)$ then it follows that $\mu(E) = \operatorname{tr}(WE)$ for some $W \in K$.

From $\mu(\sum_n E_n) = \sum_n \mu(E_n)$ it follows that for an orthonormal basis $\{x_\nu\}$ that $\mu(\sum_\nu P_{x_\nu}) = \sum_\nu \mu(P_{x_\nu})$. Conversely, if the last equation holds for all orthonormal vectors, then the first is satisfied for any set of pairwise orthogonal E_n. Since $\mu(\mathbf{1}) = 1$, it follows that, for any complete orthonormal basis that $\sum_\nu \mu(P_{x_\nu}) = 1$. For a subspace $\mathcal{T} = E\mathcal{H}$ and for an orthonormal basis in $\mathcal{T}$ it follows that $\sum_\nu \mu(P_{x_\nu}) = \mu(E) \leq 1$.

Let $A(G)$ denote the set of all P_x ($A(G)$ is the set of all atoms of the lattice G; see V, §5). Suppose we are given a real function $A(G) \xrightarrow{\mu} [0, 1]$ where $\sum_\nu \mu(P_{x_\nu}) = 1$ for all complete orthonormal bases $\{x_\nu\}$. μ can be extended in one and only one way to all of G such that $\mu(\sum_n E_n) = \sum_n \mu(E_n)$ holds.

We must therefore prove that $\sum_\nu \mu(P_{x_\nu})$ yields the same value for each complete orthonormal basis in $E\mathcal{H}$; in this way we may define $\mu(E)$. This, however, follows from the fact if $\{x_\nu\}$ and $\{y_\nu\}$ are two complete orthonormal bases from $E\mathcal{H}$ and if $\{z_\rho\}$ is a complete orthonormal basis for $E^\perp\mathcal{H}$ that the $\{x_\nu, z_\rho\}$ and $\{y_\nu, z_\rho\}$ form a complete orthonormal basis for $\mathcal{H}$ for which $\sum_\nu \mu(P_{x_\nu}) + \sum_\rho \mu(P_{z_\rho}) = \sum_\nu \mu(P_{y_\nu}) + \sum_\rho \mu(P_{z_\rho}) = 1$. It is easy to verify that such an extension of μ onto G satisfies the condition $\mu(\sum_n E_n) = \sum_n \mu(E_n)$.

It therefore suffices to consider maps $A(G) \xrightarrow{\mu} [0, 1]$ for which $\sum_\nu \mu(P_{x_\nu}) = 1$ for each complete orthonormal basis $\{x_\nu\}$in $\mathcal{H}$.

Let $\mathcal{S}$ denote the surface of the unit sphere in $\mathcal{H}$, that is, $\mathcal{S} = \{x \mid x \in \mathcal{H}$ and $\|x\| = 1\}$. A function $\mathcal{S} \xrightarrow{m} \mathbf{R}$ is called a frame function of weight c if, for every complete orthonormal basis $\{x_\nu\}$ the relation $\sum_\nu m(x_\nu) = c$. Since it is possible to replace one of the x_ν, for example, $x_{\nu 0}$ by $e^{i\alpha}x_{\nu 0}$ it follows that $m(e^{i\alpha}x) = m(x)$; in particular $m(-x) = m(x)$. For a frame function m we have not assumed that $m(x)$ is positive! If $m(x) \geq 0$ and $c = 1$, then the function $A(G) \xrightarrow{\mu} [0, 1]$ satisfies $\sum_\nu \mu(P_{x_\nu}) = 1$ for all complete orthonormal bases $\{x_\nu\}$ of $\mathcal{H}$ and defines a function $G \xrightarrow{\mu} [0, 1]$ for which $\mu(\sum_n E_n) = \sum_n \mu(E_n)$.

The restriction of a positive frame function of weight 1 for $\mathcal{H}$ to the surface of the unit sphere $\mathcal{S}_{\mathcal{T}}$ of a subspace $\mathcal{T}$ is a frame function for $\mathcal{T}$ of weight $c = \sum_\nu m(y_\nu)$, where $\{y_\nu\}$ is a complete orthonormal basis for $\mathcal{T}$.

A frame function m is said to be regular if $m(x) = \langle x, Ax \rangle$ for some $A \in \mathcal{B}(\mathcal{H})$. The weight of $\langle x, Ax \rangle$ is equal to $\mathrm{tr}(A)$. For a regular positive frame function of weight 1 it follows that there exists a unique measure $G \xrightarrow{\mu} [0, 1]$ for which $\mu(E) = \mathrm{tr}(WE)$ where $W \in K$.

It suffices to prove that every positive frame function is regular.

We shall carry out the proof in several steps using Gleason's approach. First we shall show that:

Every continuous (and therefore not necessarily positive) frame function in Euclidean space $\mathbf{R}^3$ is regular. ($\mathbf{R}^3$ is the real Hilbert space of three dimensions).

PROOF. Let $\mathcal{S}$ denote the surface of the unit sphere of $\mathbf{R}^3$; let $C(\mathcal{S})$ denote the set of all continuous functions $\mathcal{S} \xrightarrow{f} \mathbf{R}$. Since $\mathcal{S}$ is compact, $\|f\| = \sup_{x \in \mathcal{S}} |f(x)| < \infty$. With the norm $\|f\|$ $C(\mathcal{S})$ is a Banach space. Let $C_+(\mathcal{S})$ denote the set of all positive functions (with $C_+(\mathcal{S})$, $C(\mathcal{S})$ becomes an ordered vector space—see AIII, §6). Let F be the set of continuous frame functions; clearly $F \subset C(\mathcal{S})$; F is a subspace of $C(\mathcal{S})$. It is easy to verify that F is closed, that is, F is a Banach space.

Let $\mathcal{G}$ denote the group of rotations in $\mathbf{R}^3$. Let $\Delta \in \mathcal{G}$; we obtain a representation of $\mathcal{G}$ in $C(\mathcal{S})$ by $U_\Delta f(x) = f(\Delta^{-1}x)$. Clearly F is an invariant subspace of $C(\mathcal{S})$ under this representation, that is, F is itself a representation space for $\mathscr{G}$. According to VII, §3 we may represent F in terms of the subspaces $\mathscr{T}_l$ (where $\mathscr{T}_l$ is defined in VII, §3), that is, $F = \sum_l +(F \cap \mathscr{T}_l)$. For which value of l is $F \cap \mathscr{T}_l \neq \{0\}$?

For $f \in \mathscr{T}_1$ is $f(-x) = -f(x)$? Since for $m \in F$ we must have $m(-x) = m(x)$ we find that $F \cap \mathscr{T}_1 = \{0\}$. According to VII, §3 $\mathscr{T}_0 + \mathscr{T}_2$ is the set of all homogeneous polynomials of second degree, and $F \cap (\mathscr{T}_0 + \mathscr{T}_2)$ is therefore the set of all real homogeneous polynomials of second degree. These have the form $\langle x, Ax\rangle$ with (real) symmetric A.

Now we need only show that $\mathscr{T}_l \cap F = \{0\}$ for $l \geq 3$. If $\mathscr{T}_l \cap F \neq \{0\}$, then $\mathscr{T}_l \cap F$ would be equal to the set of all real functions in $\mathscr{T}_l$. According to VII (3.55) all functions of the form

$$\cos m\varphi(\sin\theta)^{-m}Q^l_m(\cos\theta), \qquad \sin\varphi(\sin\theta)^{-m}Q^l_m(\cos\theta),$$

where $m = -l, \ldots, l$ must belong to $\mathscr{T}_l \cap F$. The restriction of these functions onto the subspace determined by $\theta = \pi/2$ will also be frame functions. For $\theta = \pi/2$, that is, for $\sin\theta = 1$ and $\cos\theta = 0$, all functions of the form

$$\cos m\varphi \quad \text{and} \quad \sin m\varphi \quad \text{for } m = -l, \ldots, l$$

must be frame functions. Therefore, for $\cos m\varphi$ we must have

$$\cos m\varphi + \cos m(\varphi + \pi/2) = \text{const.}$$

This is only possible for $m = 0$ or for $1 + \cos m\pi/2 = 0$, which cannot be the case for all $m = -l, \ldots, l$ for $l \geq 3$.

We shall now show that every positive frame function on the surface $\mathscr{S}$ of the unit sphere in $\mathbf{R}^3$ is continuous.

The proof can be carried out in a simple way. Let us consider the polar coordinates introduced above. Let N denote the closed subset $0 \leq \theta \leq \pi/2$ of $\mathscr{S}$. A frame function is, since $m(x) = m(-x)$, uniquely determined by its values on N. The set of points $\theta =$ constant is called a circle (parallel) of latitude. Through each point $x \in N$ (with the exception of the pole $\theta = 0$) there exists a uniquely determined great circle which, at x, has the same tangent as the latitude circle passing through x. Let $H(x)$ denote the set of all points on this great circle. Clearly $x \in H(x)$.

Next we shall show that for $z \in N$ (z is not the pole) the set

$$X(z) = \{x \mid x \in N, \text{ there exists a } y \text{ such that } y \in H(x) \text{ and } z \in H(y)\}$$

has a nonempty interior.

Proof. It suffices to choose the point z with rectangular coordinates $(\sin\theta, 0, \cos\theta)$ for which $0 < \theta < \pi/2$. The set L of all y for which $z \in H(y)$ is precisely the set of all (ξ, η, ζ) for which $\psi \overset{\text{def}}{=} (\xi^2 + \eta^2)\cos\theta - \xi\zeta\sin\theta = 0$. If x is a point for which the quadratic form ψ is negative, then $L \cap H(x) \neq \varnothing$ because if $\zeta = 0$ $\psi > 0$. Thus $X(z)$ contains the open set of all points for which $\psi < 0$. This is nonempty because, for $\xi = \sin\theta', \eta = 0, \zeta = \cos\theta'$ where $0 < \theta' < \theta$: $\psi < 0$.

Let $\operatorname{osc}(f, X) = \sup\{f(x) \mid x \in X\} - \inf\{f(x) \mid x \in X\}$. Let f be a frame function on the unit sphere $\mathscr{S}$ in $\mathbf{R}^3$. There may be a point x for which there exists a neighborhood U for which $\operatorname{osc}(f, U) = \alpha$. Thus each point y of the great circle which has x as a pole (that is, in all directions y which are orthogonal to the x direction) has a neighborhood V for which $\operatorname{osc}(f, V) \leq 2\alpha$.

PROOF. We introduce polar coordinates for which x is the "north pole." There exists an $\varepsilon > 0$ such that all points for which $\theta < \varepsilon$ belong to U. Let x_0 be a point for which $\theta = \pi/2$ and $\varphi = \varphi_0$. Let y_0 be the point for which $\theta = (\pi + \varepsilon)/2$ and $\varphi = \varphi_0$. The point y with $\theta = \varepsilon/2$, $\varphi = \varphi_0$ is orthogonal to y_0; similarly x is orthogonal to x_0. The points x, y, x_0, y_0 lie on a great circle, x and y lie in U. If, instead of x_0 we choose another point z from a sufficiently small neighborhood V of x_0, then the following two points z', y' lie in U: z', y' on the great circle passing through x and y_0 and orthogonal to z and y_0.

For $z_1, z_2 \in V$ let z'_1, y'_1, z'_2, y'_2 denote the points chosen in the above manner. For the frame function f it therefore follows that

$$f(y_0) + f(y'_i) = f(z_i) + f(z'_i) \qquad (i = 1, 2).$$

By subtraction, we obtain

$$|f(z_1) - f(z_2)| = |f(y'_1) - f(y'_2) + f(z'_2) - f(z'_1)| \leq 2\alpha$$

since y', z' lie in U. Therefore we find that $\operatorname{osc}(f, V) < 2\alpha$.

We now prove the following lemma:

If, for a frame function f on the unit sphere $\mathscr{S}$ in $\mathbf{R}^3$ there exists an open set U for which $\operatorname{osc}(f, U) = \alpha$ *then, to each point $y \in \mathscr{S}$ there exists a neighborhood W for which* $\operatorname{osc}(f, W) < 4\alpha$.

PROOF. For $x \in U$ and $y \in \mathscr{S}$ there exists a point z which is orthogonal to x and y. z therefore lies on the great circle with x as the pole, and there exists a neighborhood V of z such that $\operatorname{osc}(f, V) \leq 2\alpha$. y lies on the great circle with z as the pole, so there exists a neighborhood W of y for which $\operatorname{osc}(f, W) \leq 2 \operatorname{osc}(f, V) \leq 4\alpha$.

We are now in a position to prove that every positive frame function on the unit sphere $\mathscr{S}$ in $\mathbf{R}^3$ is continuous.

Let $g \geq 0$ be a frame function. Then the function f defined by $f(x) = g(x) - \inf_y g(y)$ is also a frame function. Clearly $f(x) \geq 0$ and $\inf_x f(x) = 0$. Therefore, to each $\varepsilon > 0$ there exists a point y for which $f(y) \leq \varepsilon$. Let us choose a polar coordinate system with y as the north pole. To each x there exists a point x' which has the same coordinate θ as does x but for which φ differs by $\pi/2$. Clearly $h(x) = f(x) + f(x')$ is a frame function for which $h \geq 0$. If $f(x)$ is of weight c, then $h(x)$ is of weight $2c$.

For the special case in which x has the coordinates $\theta = \pi/2$ and $\varphi = \varphi_0$ the vectors y, x, x' are an orthonormal basis for $\mathbf{R}^{3\prime}$, and we find that $h(y) + h(x) + h(x') = 2c$. Since $f(-x) = f(x)$ we find that $h(x') = f(x) + f(x')$, and we find that $h(x) = c - h(y)/2 = c - f(y)$. Therefore we find that h is constant on the great circle with y as the pole.

For an arbitrary point $x \in N\backslash\{y\}$ $H(x)$ is well defined. $H(x)$ intersects the great circle with pole y in the point z which is orthogonal to x (if x is on the great circle with pole y, then such a z also exists). Therefore we find that $2c > g(x) + g(z) = g(x) + c - f(y)$ and we obtain

$$g(x) \leq c + f(y) \leq c + \varepsilon \quad \text{for all } x \in N\backslash\{y\}.$$

Let $u \in H(x) \cap N$ and $u' \in H(x) \cap N$ where u' is orthogonal to u. It follows that $g(x) + c - f(y) = g(x) + g(z) = g(u) + g(u') \leq g(u) + c + \varepsilon$ and thus we obtain $g(x) \leq g(u) + 2\varepsilon$ for all $x \in N\backslash\{y\}$ and $u \in H(x)$.

Let $\beta = \inf\{g(x) \mid x \in N\backslash\{y\}\}$; let $z \in N\backslash\{y\}$ such that $g(z) \leq \beta + \varepsilon$. For $x \in X(z)$ (where $X(z)$ was defined in the first lemma) there exists a $v \in H(x)$ and $z \in H(v)$. According to the above inequality we obtain

$$g(x) \leq g(v) + 2\varepsilon \quad \text{and} \quad g(v) \leq g(z) + 2\varepsilon$$

and

$$\beta \leq g(x) \leq g(z) + 4\varepsilon \leq \beta + 5\varepsilon.$$

The set $X(z)$ has a nonempty interior, that is, there exists an open set U (as a subset of $X(z)$) for which $\beta \leq g(x) \leq \beta + 5\varepsilon$ for all $x \in U$; thus we obtain $\operatorname{osc}(g, U) \leq 5\varepsilon$. According to the previous lemma for the point y there exists a neighborhood V for which $\operatorname{osc}(g, V) \leq 20\varepsilon$. Since $g(y) = 2f(y) \leq 2\varepsilon$ we obtain $\sup\{g(x) \mid x \in V\} = \operatorname{osc}(g, V) + \inf\{g(x) \mid x \in V\} \leq 20\varepsilon + 2\varepsilon$. Since $0 \leq f \leq g$ we therefore obtain $\operatorname{osc}(f, V) \leq 22\varepsilon$. Thus, according to the previous lemma each point $x \in \mathscr{S}$ has a neighborhood W for which $\operatorname{osc}(f, W) < 88\varepsilon$. Since ε can be chosen arbitrarily small, f is therefore continuous.

We have proven the following: Every positive frame function on the unit sphere $\mathscr{S}$ in $\mathbf{R}^3$ is regular. We shall now extend this result to complex Hilbert spaces and to higher dimensions.

A subset $\mathscr{R}$ of a Hilbert space $\mathscr{H}$ is called a real subspace if for each x, $y \in \mathscr{R}$ $x + y \in \mathscr{R}$ and for each $\alpha \in \mathbf{R}$ $\alpha x \in \mathscr{R}$. Note that $\mathscr{R}$ need not be a subspace of $\mathscr{H}$ if $\mathscr{H}$ is a vector space over $\mathbf{C}$. $\mathscr{R}$ is said to be completely real if the inner product on $\mathscr{R} \times \mathscr{R}$ only takes on real values. If $X \subset \mathscr{H}$ is a set for which the inner product on $X \times X$ takes on only real values, then the real linear span of X is a complete real subspace. The real linear span of an orthonormal basis is therefore a complete real subspace. If the complex subspace generated by a complete real subspace $\mathscr{R}$ is dense in $\mathscr{H}$, then a complete orthonormal basis of $\mathscr{R}$ is also a complete orthonormal basis of $\mathscr{H}$. Then the restriction of every frame function of $\mathscr{H}$ onto $\mathscr{R}$ is a frame function for $\mathscr{R}$.

We now prove that a frame function f on a complex two-dimensional Hilbert space for which $f \geq 0$ which is regular on every complete real subspace is regular in the entire Hilbert space.

PROOF. Let c denote the weight of f and let $d = \sup\{f(x) \mid \|x\| = 1\}$ where $0 \leq d \leq c$ since $f \geq 0$. We will now show that f takes on the value d.

There exists a sequence x_n for which $\|x_n\| = 1$ and $f(x_n) \to d$. Since the surface of the unit sphere is compact, we may assume that the sequence x_n converges: $x_n \to y$. For $\lambda_n = \langle x_n, y\rangle / |\langle x_n, y\rangle|$ we find that $|\lambda_n| = 1$ and $\lambda_n \to 1$. Thus it follows that

$\lambda_n x_n \to y$ and $f(\lambda_n, x_n) = f(x_n)$. Since $\langle y, \lambda_n x_n \rangle$ is real, for each n both vectors y, $\lambda_n x_n$ lie in a complete real subspace $\mathscr{R}_n$. If $x, y \in \mathscr{R}_n$ satisfy $\|x\| = \|y\| = 1$, then, for a symmetric A we obtain

$$|f(x) - f(y)| = |\langle y, Ay \rangle - \langle x, Ax \rangle| = |\langle x - y, A(x + y) \rangle|$$
$$\leq \|x - y\| \, \|x + y\| \, \|A\| < 2\|A\| \, \|x - y\| \leq 2c\|x - y\|.$$

From the above inequality we obtain the special case

$$|f(y) - d| \leq |f(y) - f(\lambda_n x_n)| + |f(x_n) - d|$$
$$\leq 2c\|y - \lambda_n x_n| + |f(x_n) - d|$$

from which we conclude that $f(y) = d$.

We now define the following function on all of $\mathscr{H}$:

$$F(x) = \begin{cases} \|x\|^2 f\left(\dfrac{x}{\|x\|}\right) & \text{for } x \neq 0, \\ 0 & \text{for } x = 0. \end{cases}$$

Note that $F(y) = f(y) = d$. If z is orthogonal to y and if $\|z\| = 1$ then it follows that $F(z) = f(z) = c - f(y) = c - d$. $F(x)$ is a real quadratic form on the complete real subspace of $\mathscr{H}$ generated by y and z which takes on its minimum d at y. Therefore, for real α, β we find that

$$F(\alpha y + \beta z) = \alpha^2 F(y) + \beta^2 F(z) = \alpha^2 d + \beta^2 (c - d).$$

For complex nonzero λ, μ it follows that

$$F(\lambda y + \mu z) = F\left(\frac{|\lambda|}{\lambda}(\lambda y + \mu z)\right) = F(|\lambda| z + |\mu| z'),$$

where $z' = (\mu|\lambda|/|\mu|\lambda)z$. Clearly z' is orthogonal to y and $\|z'\| = 1$. Therefore we obtain

$$F(\lambda y + \mu z) = |\lambda|^2 d + |\mu|^2 (c - d).$$

It is easy to show that the above formula is correct for $\lambda = 0$ or $\mu = 0$. Therefore $F(x) = \langle x, Tx \rangle$ where T is the diagonal matrix

$$\begin{pmatrix} d & 0 \\ 0 & c - d \end{pmatrix}$$

with respect to the complete orthonormal basis $\{y, z\}$. Therefore f is regular.

The extension of the preceding result to arbitrary dimension follows from the following theorem:

If f is a frame function in $\mathscr{H}$ for which $f \geq 0$ and f is regular on each two dimensional subspace, then f is regular on the entire $\mathscr{H}$.

PROOF. Again we define

$$F(x) = \begin{cases} \|x\|^2 f\left(\dfrac{x}{\|x\|}\right) & \text{for } x \neq 0, \\ 0 & \text{for } x = 0. \end{cases}$$

On each two-dimensional subspace $\mathscr{T}$ $F(x) = \langle x, Ax \rangle$ where A is a self-adjoint operator which is defined in $\mathscr{T}$. We define a function $\mathscr{H} \times \mathscr{H} \xrightarrow{G} \mathbf{C}$ as follows:

$$G(x, y) = \langle x, A_{\mathscr{T}} y \rangle,$$

where $\mathscr{T}$ is the subspace generated by x and y.

Now we must show that this definition is meaningful. If x and y are linearly independent, then $\mathscr{T}$ is uniquely determined by x and y and therefore $G(x, y)$ is uniquely defined. If $y = \lambda x$ then for $\mathscr{T}$ we may select an arbitrary two-dimensional subspace which contains x. It then follows that $G(x, \lambda x) = \lambda \langle x, A_{\mathscr{T}} x \rangle = \lambda F(x)$, which is independent of the choice of $\mathscr{T}$. Since $A_{\mathscr{T}}$ is self-adjoint in $\mathscr{T}$ we find that G satisfies the following relationships for all $x, y \in \mathscr{H}$ and $\alpha \in \mathbf{C}$:

$$\begin{aligned} G(x, \alpha y) &= \alpha(G(x, y), \\ G(y, x) &= \overline{G(x, y)}, \\ 4 \operatorname{Re} G(x, y) &= F(x + y) - F(x - y), \\ 2F(x) + 2F(y) &= F(x + y) + F(x - y). \end{aligned}$$

Thus it follows that

$$G(x, y) + G(x, z) = \operatorname{Re} G(x, y) + \operatorname{Re} G(x, z) + \operatorname{Im} G(x, y) + \operatorname{Im} G(x, z).$$

Since

$$\operatorname{Im} G(x, y) = i \operatorname{Re}(-iG(x, y)) = i \operatorname{Re}(G(x, -iy))$$

it follows that

$$G(x, y) + G(x, z) = \operatorname{Re} G(x, y) + \operatorname{Re} G(x, z) + i \operatorname{Re} G(x, -iy) + i \operatorname{Re} G(x, -iz).$$

From the above relationships it follows that

$$\begin{aligned} 8 \operatorname{Re} G(x, y) + 8 \operatorname{Re} G(x, z) &= 2F(x + y) - 2F(x - y) + 2F(x + z) - 2F(x - z) \\ &= F(2x + y + z) - F(2x - y - z) \\ &= 4 \operatorname{Re} G(2x, y + z) = 8 \operatorname{Re} G(x, y + z). \end{aligned}$$

Therefore we obtain

$$\begin{aligned} G(x, y) + G(x, z) &= \operatorname{Re} G(x, y + z) + i \operatorname{Re} G(x, i(y + z)) \\ &= G(x, y + z). \end{aligned}$$

In addition it follows that

$$\begin{aligned} |G(x, y)| &\leq |\operatorname{Re} G(x, y)| + |\operatorname{Im} G(x, y)| = |\operatorname{Re} G(x, y)| + |\operatorname{Re} G(x - iy)| \\ &\leq \tfrac{1}{4}F(x + y) + \tfrac{1}{4}F(x - y) + \tfrac{1}{4}F(x + iy) + \tfrac{1}{4}F(x - iy). \end{aligned}$$

From the definition of F, and from $0 \leq f \leq c$, where c is the weight of f, it follows that:

$$\begin{aligned} |G(x, y)| &\leq \tfrac{1}{4}C[\|x + y\|^2 + \|x - y\|^2 + \|x + iy\|^2 + \|x - iy\|^2] \\ &\leq C(\|x\|^2 + \|y\|^2). \end{aligned}$$

For unit vectors x, y we therefore obtain $|G(x, y)| \leq 2c$.

G is therefore a bounded linear form; thus, according to §4 there exists a bounded operator A such that $G(x, y) = \langle x, Ay \rangle$. Since $G(x, y) = \overline{G(y, x)}$ we conclude that $A = A^+$, that is, A is self-adjoint; from $G(x, x) \geq 0$ we conclude that $A > 0$.

We now prove the following theorem:

Every frame function f on a Hilbert space $\mathscr{H}$ for which $f \geq 0$ is regular when the dimension of $\mathscr{H} \geq 3$.

PROOF. f is a frame function on each complete real subspace. Each complete real two-dimensional subspace is also a subspace of a three-dimensional complete real subspace. In each complete real three-dimensional subspace (that is, in $\mathbf{R}^3$) f is regular, and is therefore regular on every two-dimensional subspace. Thus, from the last two theorems, it follows that f is regular on each two-dimensional subspace, and therefore in all of $\mathscr{H}$.

13 Isomorphisms and Anti-isomorphisms

Let $\mathscr{H}_1, \mathscr{H}_2$ be two Hilbert spaces, where we permit $\mathscr{H}_1 = \mathscr{H}_2$. A map $H_1 \xrightarrow{A} H_2$ is said to be linear if $A(x + y) = Ax + Ay$ and $A(\alpha x) = \alpha A(x)$. A is said to be antilinear if we substitute $A(\alpha x) = \bar{\alpha} A(x)$ for the previous equation. A is continuous if and only if it is bounded, that is, if there exists a number c for which $\|Ax\| < c\|x\|$.

$\mathscr{H}_1 \xrightarrow{U} \mathscr{H}_2$ is said to be an isomorphism if U is linear and bijective and $\|Ux\| = \|x\|$. If $\mathscr{H}_1 = \mathscr{H}_2$ then, according to §7 U is an isomorphism if it is unitary.

If for $\mathscr{H}_1 \xrightarrow{V} \mathscr{H}_2$ only $\|Vx\| = \|x\|$ then, from the results of §7 it follows that V is an isomorphism on a closed subspace of $\mathscr{H}$.

$\mathscr{H}_1 \xrightarrow{U} \mathscr{H}_2$ is said to be an anti-isomorphism if U is antilinear and bijective, and $\|Ux\| = \|x\|$. For the special case in which $\mathscr{H}_1 = \mathscr{H}_2$ then we say that U is anti-unitary. For a complete orthonormal basis $\{x_\nu\}$ for $\mathscr{H}$ a special anti-unitary operator is defined by $U_s x_\nu = x_\nu$; it follows that $U_s(\sum_\nu \alpha_\nu x_\nu) = \sum_\nu \bar{\alpha}_\nu x_\nu$. If $\mathscr{H}_1 \xrightarrow{U} \mathscr{H}_2$ is an anti-isomorphism, then it follows that if U_1 is an anti-unitary operator in $\mathscr{H}_1$ then $\tilde{U} = UU_1$ is an isomorphism $\mathscr{H}_1 \xrightarrow{\tilde{U}} \mathscr{H}_2$. Both U_1 and U_1^{-1} are anti-unitary. Therefore every anti-isomorphism U can be represented as a product $\tilde{U}U_1^{-1}$ of an isomorphism and an arbitrary anti-unitary map.

If U_2 is an anti-unitary map in $\mathscr{H}_2$ it follows that $\hat{U} = U_2 U$ is an isomorphism and that U may be represented in the form $U_2^{-1}\hat{U}$.

As expected, the product of two antilinear maps $\mathscr{H}_1 \xrightarrow{A} \mathscr{H}_2 \xrightarrow{B} \mathscr{H}_3$ is a linear map $\mathscr{H}_1 \xrightarrow{BA} \mathscr{H}_3$.

If U is a unitary map, then for $A \in \mathscr{B}(\mathscr{H})$ and $B \in \mathscr{B}'(\mathscr{H})$ it follows that

$$\operatorname{tr}(AU^+BU) = \sum_\nu \langle x_\nu, AU^+BUx_\nu\rangle = \sum_\nu \langle Ux_\nu, UAU^+BUx_\nu\rangle$$

$$= \sum_\nu \langle y_\nu, UAU^+By_\nu\rangle,$$

where x_ν is a complete orthonormal basis and $y_\nu = Ux_\nu$. Since U is unitary, the set y_ν is a complete orthonormal basis, and we find that

$$\mathrm{tr}(AU^+BU) = \mathrm{tr}(UAU^+B).$$

An analogous proof is applicable to the case in which U is anti-unitary. From $\|Ux\| = \langle Ux, Ux\rangle = \|x\| = \langle x, x\rangle$ it follows that if we replace x by $x + y$ and $x + iy$, then we obtain

$$\langle x, y\rangle = \langle Uy, Ux\rangle = \overline{\langle Ux, Uy\rangle}$$

from which we obtain:

$$\begin{aligned}\mathrm{tr}(AU^{-1}BU) &= \sum_\nu \langle x_\nu, AU^{-1}BUx_\nu\rangle \\ &= \sum_\nu \overline{\langle Ux_\nu, UAU^{-1}BUx_\nu\rangle} = \sum_\nu \overline{\langle y_\nu, UAU^{-1}By_\nu\rangle} \\ &= \overline{\mathrm{tr}(UAU^{-1}B)}.\end{aligned}$$

Since $A \in \mathscr{B}(\mathscr{H})$, for $A' = UAU^{-1}$ it follows that

$$\begin{aligned}\langle x, A'y\rangle &= \langle x, UAU^{-1}y\rangle = \langle AU^{-1}y, U^{-1}x\rangle \\ &= \langle U^{-1}y, AU^{-1}x\rangle = \langle UAU^{-1}x, y\rangle \\ &= \langle A'x, y\rangle\end{aligned}$$

and we find that $A' \in \mathscr{B}(\mathscr{H})$. Therefore $\mathrm{tr}(A'B)$ is real, and we find that $\mathrm{tr}(AU^{-1}BU) = \mathrm{tr}(UAU^{-1}B)$ holds for anti-unitary U.

14 Products of Hilbert Spaces

Let $\mathscr{H}_1$ and $\mathscr{H}_2$ be two Hilbert spaces. We construct the so-called product space $\mathscr{H} = \mathscr{H}_1 \times \mathscr{H}_2$ (where $\mathscr{H}$ is not the Cartesian product of $\mathscr{H}_1$ and $\mathscr{H}_2$—a fact that is not reflected in the usual notation) in the following way:

Let $x \in H_1$, $y \in H_2$; $f(x, y)$ is said to be an anti-bilinear form if $f(x_1 + x_2, y) = f(x_1, y) + f(x_2, y)$, $f(x, y_1 + y_2) = f(x, y_1) + f(x, y_2)$, $f(\alpha x, y) = \bar{\alpha} f(x, y)$ and $f(x, \alpha y) = \bar{\alpha} f(x, y)$.

An example of an anti-bilinear form is given by $f(x, y) = \langle x, z_1\rangle\langle y, z_2\rangle$ where $z_1 \in \mathscr{H}_1$, $z_2 \in \mathscr{H}_2$. This anti-bilinear form is often denoted by $z_1 z_2$.

An anti-bilinear form is continuous if and only if there exists a c for which $|f(x, y)| \le c\|x\|\,\|y\|$. $f = z_1 z_2$ is continuous with $c = \|z_1\|\,\|z_2\|$.

The set of all continuous bilinear forms is a linear vector space over $\mathbf{C}$. From $|f(x, y)| < c\|x\|\,\|y\|$ from §4 it follows that $f(x, y) = \langle x, z\rangle$ for some $z \in \mathscr{H}_1$. An antilinear map $\mathscr{H}_2 \xrightarrow{A} \mathscr{H}_1$ which satisfies $\|Ay\| < c\|y\|$ is defined by $y \to z$. Therefore $f(x, y) = \langle x, Ay\rangle$. For $f = z_1 z_2$ we obtain $Ay = z_1\langle y, z_2\rangle$. Since A is bounded, there exists, according to §6 a bounded antilinear operator $\mathscr{H}_1 \xrightarrow{A'} \mathscr{H}_2$ for which $\langle x, Ay\rangle = \langle y, A'x\rangle$. If A is a bounded antilinear operator then $\langle x, Ay\rangle$ is a bounded antilinear form. Therefore there is a bijective map between bounded antilinear forms and bounded antilinear operators $\mathscr{H}_2 \xrightarrow{A} \mathscr{H}_1$.

From $\langle x, Ay\rangle = \langle y, A'x\rangle$ it follows that, for $y = A'z$, the operator AA' defined by $\langle x, AA'z\rangle$ is bounded and linear in H_1; the same holds for $A'A$ in H_2. From $\langle x, AA'z\rangle = \langle A'z, A'x\rangle = \langle AA'x, z\rangle$ it follows that AA' is self-adjoint. For $z = x$ it follows that $\langle x, AA'x\rangle = \langle A'x, A'x\rangle = \|A'x\|^2$ and $\langle y, A'Ay\rangle = \langle Ay, Ay\rangle = \|Ay\|^2$. Therefore $AA' \in \mathscr{L}_{r+}(\mathscr{H}_1)$ and $A'A \in \mathscr{L}_{r+}(\mathscr{H}_2)$. For a complete orthonormal basis $\{x_\nu\}$ for $\mathscr{H}_1$ we obtain $\operatorname{tr}(AA') = \sum_\nu \langle x_\nu, AA'x_\nu\rangle = \sum_\nu \|A'x_\nu\|^2$. For a complete orthonormal basis $\{y_\nu\}$ for $\mathscr{H}_2$ we may define an antilinear operator V (anti-isometry; see §13) by $x_\nu = Vy_\nu$.

Thus $A'V \in \mathscr{L}_r(\mathscr{H}_2)$ and $\operatorname{tr}(AA') = \sum_\nu \|A'Vy_\nu\|^2 = \operatorname{tr}((A'V)^+A'V)$. It is easy to show that $(A'V)^+ = V'A$. Therefore $\operatorname{tr}(AA') = \operatorname{tr}((A'V)^+(A'V)) = \operatorname{tr}((A'V)(A'V)^+) = \operatorname{tr}((V'A)^+V'A) = \sum_\nu \|V'Ay_\nu\|^2$. Since $\|Vy\| = \|y\|$ we finally obtain $\operatorname{tr}(AA') = \sum_\nu \|Ay_\nu\|^2 = \operatorname{tr}(A'A)$.

We shall now introduce an inner product on the set of all those anti-bilinear forms f for which $\operatorname{tr}(AA') = \operatorname{tr}(A'A) < \infty$ as follows: Let $f_1 \leftrightarrow A_1$, $f_2 \leftrightarrow A_2$, we define $\langle f_1, f_2\rangle = \operatorname{tr}(A_2'A_1)$; in this way the above set becomes a Hilbert space—the so-called product space $\mathscr{H} = \mathscr{H}_1 \times \mathscr{H}_2$.

For the special case in which $f = z_1z_2$ we obtain $Ay = z_1\langle y, z_2\rangle$, $A'x = z_2\langle x, z_1\rangle$ and we therefore obtain $A'Ay = z_2\|z_1\|^2\langle z_2, y\rangle$ and $\operatorname{tr}(A'A) = \|z_2\|^2\|z_1\|^2$. For $f_1 = z_1^{(1)}z_2^{(1)}$ and for $f_2 = z_1^{(2)}z_2^{(2)}$ it follows that

$$A_2'A_1y = z_2^{(2)}\langle z_1^{(1)}, z_2^{(2)}\rangle\langle z_2^{(1)}, y\rangle$$

from which we find that

$$\langle f_1, f_2\rangle = \langle z_1^{(1)}, z_1^{(2)}\rangle\langle z_2^{(1)}, z_2^{(2)}\rangle.$$

Let $\{x_\nu\}$ and $\{y_\nu\}$ be complete orthonormal bases for $\mathscr{H}_1$ and $\mathscr{H}_2$, respectively. From $f(x, y) = \langle x, Ay\rangle$ it follows that $f = \sum_{\nu,\mu} a_{\nu\mu}x_\nu y_\mu$; similarly, for $g(x, y) = \langle x, By\rangle$ it follows that $g = \sum_{\mu\nu} b_{\mu\nu}x_\mu y_\nu$. We therefore obtain $\langle g, f\rangle = \sum_{\nu\mu} \bar{b}_{\nu\mu}a_{\nu\mu}$. Therefore the set of $x_\nu y_\mu$ is a complete orthonormal basis in $\mathscr{H} = \mathscr{H}_1 \times \mathscr{H}_2$.

We will now show that for a given $f \in \mathscr{H}_1 \times \mathscr{H}_2$ we can choose a complete orthonormal basis $x_\nu y_\mu$ such that

$$f = \sum_\nu \lambda_\nu x_\nu y_\nu \quad \text{where } \lambda_\nu \geq 0. \tag{14.1}$$

Since $A'A \in \mathscr{B}(\mathscr{H}_2)$ $A'A$ can be expressed in the form $A'A = \sum_\nu \mu_\nu P_{y_\nu}$. For $Ay_\nu = z_\nu$ it follows that

$$AA'z_\nu = AA'Ay_\nu = \mu_\nu Ay_\nu = \mu_\nu z_\nu.$$

If $z_\nu \neq 0$, then $x_\nu = z_\nu/\|z_\nu\|$ is a normed eigenvector of $A'A$ with eigenvalue μ_ν. Clearly $\|z_\nu\|^2 = \langle Ay_\nu, Ay_\nu\rangle = \langle y_\nu, A'Ay_\nu\rangle = \mu_\nu$. Since $A'A$ is positive,

$\mu_\nu \geq 0$. From $f(x, y) = \langle x, Ay\rangle$ it follows that for $y = \sum_\nu y_\nu\langle y_\nu, y\rangle$, we can write

$$f(x, y) = \sum_\nu \langle x, z_\nu\rangle\langle y, z_\nu\rangle = \sum_\nu \sqrt{\mu_\nu}\langle x, x_\nu\rangle\langle y, y_\nu\rangle$$

from which we conclude that f has the form (14.1).

A bounded linear operator B in $\mathscr{H} = \mathscr{H}_1 \times \mathscr{H}_2$ is completely defined by the special values Bz_1z_2 where $z_1 \in \mathscr{H}_1$, $z_2 \in \mathscr{H}_2$. For each $A_1 \in \mathscr{L}(\mathscr{H}_1)$ and $A_2 \in \mathscr{L}(\mathscr{H}_2)$ an operator $A_1 \times A_2 \in \mathscr{L}(\mathscr{H}_1 \times \mathscr{H}_2)$ is defined by $(A_1 \times A_2)z_1z_2 = (A_1z_1)(A_2z_2)$ and satisfies $\|A_1 \times A_2\| = \|A_1\|\,\|A_2\|$. It is a simple matter to show that $(A_1 \times A_2)(B_1 \times B_2) = (A_1B_1) \times (A_2B_2)$. As a special case we obtain $(A_1 \times \mathbf{1})(\mathbf{1} \times B_2) = (A_1 \times B_2) = (\mathbf{1} \times B_2)(A_1 \times \mathbf{1})$. Thus we find that $A \times \mathbf{1}$ and $\mathbf{1} \times B$ commute. Under certain circumstances there is a partial converse.

Let $M \subset \mathscr{L}(\mathscr{H})$ and suppose that M has the property that if $A \in M$ then $A^+ \in M$. If $\mathscr{T}$ is a subspace which is invariant under M (that is $M\mathscr{T} \subset \mathscr{T}$) then so is $\mathscr{T}^\perp$ since if $u \in \mathscr{T}^\perp$ for $A \in M$ and $x \in \mathscr{T}$ it follows that $\langle x, Au\rangle = \langle A^+x, u\rangle = 0$ since $A^+x \in \mathscr{T}$. Let M be a set and suppose we are given two maps $M \xrightarrow{j_1} \mathscr{L}(\mathscr{H}_1)$ and $M \xrightarrow{j_2} \mathscr{L}(\mathscr{H}_2)$. A bounded linear map $\mathscr{H}_1 \xrightarrow{B} \mathscr{H}_2$ is called a homomorphism with respect to the operator domain M if $B(j_1x) = (j_2x)B$ for all $x \in M$. Often the functions j_1 and j_2 are ignored, and we consider the elements of M as operators in $\mathscr{H}_1$ and $\mathscr{H}_2$, where we observe that two different elements of M can correspond to the same operator in $\mathscr{H}_1$ and $\mathscr{H}_2$. In this sense the elements of a subset $M \subset \mathscr{L}(\mathscr{H})$ can be considered to be operators in an invariant subspace $\mathscr{T}$ of $\mathscr{H}$. In order to emphasize the fact that the same operator domain M is under consideration for both $\mathscr{H}_1$ and $\mathscr{H}_2$ we shall often write $(M)\mathscr{H}_1$ and $(M)\mathscr{H}_2$.

Two Hilbert spaces $(M)\mathscr{T}_1$ and $(M)\mathscr{T}_2$ (for example, two invariant subspaces of $\mathscr{H}$ for which $M \subset \mathscr{L}(\mathscr{H})$) are said to be isomorphic if there exists an isomorphism $\mathscr{T}_1 \xrightarrow{U} \mathscr{T}_2$ such that $Ax_1 = U^{-1}AUx_1$ for all $A \in M$, that is, if $UA = AU$ holds (as maps of $\mathscr{T}_1$ into $\mathscr{T}_2$) for all $A \in M$.

A space $(M)\mathscr{T}$ is said to be irreducible if it contains no invariant subspaces (other than $\{0\}$ and $\mathscr{T}$). If $\mathscr{T}_1$ and $\mathscr{T}_2$ are isomorphic and if $\mathscr{T}_1$ is irreducible, then $\mathscr{T}_2$ is also irreducible. If $\mathscr{T}_1, \mathscr{T}_2$ are irreducible, and $\mathscr{T}_1 \xrightarrow{U_1} \mathscr{T}_2$, $\mathscr{T}_1 \xrightarrow{U_2} \mathscr{T}_2$ are two isomorphic maps, then for $U_2^{-1}U_1$ from $U_1A = AU_1$, and $U_2A = AU_2$ it follows that $U_2^{-1}U_1A = U_2^{-1}AU_1 = AU_2^{-1}U_1$, that is, the unitary operator $V = U_2^{-1}U_1$ commutes in $\mathscr{T}_1$ with all operators of M (as operators in $\mathscr{T}_1$). Therefore the spectral family of V commutes with all operators in M. If $V \neq e^{i\alpha}\mathbf{1}$ then there exists an element $E(\varphi)$ of the spectral family for which $\mathbf{0} \neq E(\varphi) \neq \mathbf{1}$. From $AE(\varphi) = E(\varphi)A$ it follows that $E(\varphi)\mathscr{T}_1$ is an invariant subspace of $\mathscr{T}_1$ in contradiction to the irreducibility of $\mathscr{T}_1$. Therefore $U_2^{-1}U_1 = e^{i\alpha}\mathbf{1}$, that is, $U_1 = e^{i\alpha}U_2$. Two isomorphic maps can therefore differ only by a "phase factor."

We will now show that if $\mathscr{T}$ is an irreducible space and if $B \in \mathscr{L}(\mathscr{T})$ commutes with all operators in M (here we then say that B commutes with M) then $B = \tau\mathbf{1}$. From $A \in M$, $BA = AB$ it follows that $A^+B^+ = B^+A^+$, since if $A \in M$ then $A^+ \in M$ and B^+ commutes with M. Therefore $B + B^+$ and $i(B - B^+)$ commute with M. If a self-adjoint operator D commutes with M, then M commutes with its spectral family. If a projection operator P commutes with M, then $P\mathcal{T}$ is an invariant subspace. Since $\mathcal{T}$ is irreducible, the elements of the spectral family can only be equal to $\mathbf{0}$ or $\mathbf{1}$; therefore $D = \lambda\mathbf{1}$. Therefore $B + B^+ = \tau_1\mathbf{1}$ and $i(B - B^+) = \tau_2\mathbf{1}$ and therefore $B = \tau\mathbf{1}$.

If the irreducible spaces $(M)\mathcal{T}_1$ and $(M)\mathcal{T}_2$ are not isomorphic, then it follows that for a bounded linear map $\mathcal{T}_1 \xrightarrow{B} \mathcal{T}_2$ which satisfies $AB = BA$ for all $A \in M$, that is, for a homomorphism B, $B = 0$. Conversely, $\mathcal{T}_1$ and $\mathcal{T}_2$ are isomorphic if there exists a homomorphism $B \neq 0$.

PROOF. $l(y) = \langle Bx, By\rangle$ defines a bounded linear form in $\mathcal{T}_1$, from which, according to §4, there exists a $z \in \mathcal{T}_1$ for which $l(y) = \langle z, y\rangle$. An operator $D \in \mathcal{L}(\mathcal{T}_1)$ is defined which satisfies $l(y) = \langle Dx, y\rangle$. From $\langle Bx, By\rangle = \overline{\langle By, Bx\rangle}$ it follows that $\langle Dx, y\rangle = \overline{\langle Dy, x\rangle} = \langle x, Dy\rangle$, that is, $D^+ = D$ and $\langle Bx, Bx\rangle \geq 0$ and therefore $D > \mathbf{0}$. Therefore $\langle Bx, By\rangle = \langle x, Dy\rangle$. From $AB = BA$ for all $A \in M$ it follows that for $A^+ \in M$ $\langle x, ADy\rangle = \langle A^+x, Dy\rangle = \langle BA^+x, By\rangle = \langle A^+Bx, By\rangle = \langle Bx, ABy\rangle = \langle Bx, BAy\rangle = \langle x, DAy\rangle$ and therefore we obtain $AD = DA$. Therefore $D = \lambda\mathbf{1}$ and from $D = D^+ > \mathbf{0}$ we obtain $\lambda \geq 0$.

If $\lambda \neq 0$ then if $U = \lambda^{1/2}B$ it follows $\langle Ux, Uy\rangle = \langle x, y\rangle$. U is therefore an isometric map of $\mathcal{T}_1$ into $\mathcal{T}_2$. Thus U is an isomorphic map of $\mathcal{T}_1$ onto $U\mathcal{T}_1 \subset \mathcal{T}_2$. $U\mathcal{T}_1$ is, on account of the commutivity of U with M, an invariant subspace. Since $\mathcal{T}_2$ is irreducible, we must have $U\mathcal{T}_1 = \mathcal{T}_2$ and therefore $(M)\mathcal{T}_1$ and $(M)\mathcal{T}_2$ are isomorphic, in contradiction to the fact that we assummed that they were not isomorphic. Therefore $\lambda = 0$, that is, $\|Bx\|^2 = 0$ and therefore $B = \mathbf{0}$.

From the same proof it also follows if $(M)\mathcal{T}_1$ is irreducible, but $(M)\mathcal{T}_2$ is not necessarily irreducible, then $B = \lambda U$ where U is an isomorphic map of $\mathcal{T}_1$ onto the invariant subspace $U\mathcal{T}_1$ of $\mathcal{T}_2$. Since $\mathcal{T}_1$ is irreducible, then $U\mathcal{T}_1$ is irreducible.

$(M)\mathcal{H}$ is said to be completely reducible, if, in every invariant subspace of $\mathcal{H}$ there exists an irreducible (invariant) subspace. The set $\{R | R$ is a set of pairwise orthogonal irreducible invariant subspaces$\}$ satisfies the conditions of Zorn's lemma (with respect to the partial order of inclusion). Thus it follows that there exists a maximal element R_m and that the elements $\mathcal{T}_\nu$ of R_m satisfy the relation $\mathcal{H} = \sum_\nu \oplus \mathcal{T}_\nu$. Let P_ν be the projectors on $\mathcal{T}_\nu$. Then $\sum_\nu P_\nu = \mathbf{1}$ and, for each $A \in M$ $A = \sum_\nu P_\nu A P_\nu$. We shall now investigate the structure of a $B \in \mathcal{L}(\mathcal{H})$ which commutes with M. It is easy to show that $AB = BA$ is equivalent to $P_\nu A P_\nu P_\nu B P_\mu = P_\nu B P_\mu P_\mu A P_\mu$ for all pairs ν, μ. If $\mathcal{T}_\nu, \mathcal{T}_\mu$ are not isomorphic spaces, then, as above, it follows that $P_\mu B P_\nu = 0$. If $\mathcal{T}_\nu$ and $\mathcal{T}_\mu$ are isomorphic, then there exists an isomorphic map $\mathcal{T}_\mu \xrightarrow{U_{\nu\mu}} \mathcal{T}_\nu$ from which it therefore follows that $P_\nu A P_\nu U_{\nu\mu} = U_{\nu\mu} P_\mu A P_\mu$. $F = U_{\nu\mu}^{-1} P_\nu B P_\mu$ therefore commutes with $P_\mu A P_\mu$ and therefore, as an operator in $\mathcal{T}_\mu$ is a multiple of the unit operator, that is, $F = \tau_{\nu\mu} P_\mu$, we find that $P_\nu B P_\mu = \tau_{\nu\mu} U_{\nu\mu} P_\mu$. Thus it follows that

$$B = \sum_{\nu,\mu} P_\nu B P_\mu = \sum_{\nu,\mu}{}' \tau_{\nu\mu} U_{\nu\mu} P_\mu, \tag{14.2}$$

where $\sum'$ is only taken over such pairs ν, μ for which $\mathcal{T}_\nu$ is isomorphic to $\mathcal{T}_\mu$.

The set of $\mathcal{T}_\nu$ can be divided into classes, whereby two belong to the same class when they are isomorphic. Let $\mathcal{S}_\alpha = \sum_\nu^{(\alpha)} \oplus \mathcal{T}_\nu$ where the sum is taken over the $\mathcal{T}_\nu$ belonging to the same class labeled with the index α. We find that $\mathcal{H} = \sum_\alpha \oplus \mathcal{S}_\alpha$. We will now show that the $\mathcal{S}_\alpha$ are uniquely determined.

If, in addition, $\mathcal{H} = \sum_\mu \oplus \mathcal{T}'_\mu$ and $\mathcal{S}'_\alpha = \sum_\mu^{(\alpha)} \oplus \mathcal{T}'_\mu$, where the index α means that the $\mathcal{S}'_\alpha$ contains precisely those $\mathcal{T}'_\mu$ for which the $\mathcal{T}_\nu$ from $\mathcal{S}_\alpha$ are isomorphic, the projection P'_ρ onto a $\mathcal{T}'_\rho$ in $\mathcal{S}'_\alpha$ therefore commutes with M. $P'_\rho P_\mu$ is therefore a mapping $\mathcal{T}_\mu \longrightarrow \mathcal{T}'_\rho$ which commutes with M, and is, therefore equal to zero if $\mathcal{T}_\mu$ and $\mathcal{T}'_\rho$ are not isomorphic. Therefore, it follows that for P'_ρ instead of B in (14.2)

$$P'_\rho = \sum_{\nu,\mu}^{(\alpha)} \tau_{\nu\mu} U_{\nu\mu} P_\mu,$$

where we only sum those pairs ν, μ for which $\mathcal{T}_\nu, \mathcal{T}_\mu \subset \mathcal{S}_\alpha$. Thus it follows that $\mathcal{T}'_\rho \subset \mathcal{S}_\alpha$, that is, $\mathcal{S}'_\alpha \subset \mathcal{S}_\alpha$. Thus we also obtain $\mathcal{S}_\alpha \subset \mathcal{S}'_\alpha$, that is, $\mathcal{S}_\alpha = \mathcal{S}'_\alpha$.

We shall now consider one of the $\mathcal{S}_\alpha = \sum_\nu^{(\alpha)} \oplus \mathcal{T}_\nu$; let $\mathcal{T}_{\nu_0}$ be one of the $\mathcal{T}_\nu$. Let $\mathcal{T}_{\nu_0} \xrightarrow{U_\nu} \mathcal{T}_\nu$ be isomorphisms. Then $U_{\nu\mu} = U_\nu U_\mu^{-1}$ is an isomorphism $\mathcal{T}_\mu \longrightarrow \mathcal{T}_\nu$. Let Q_α be the projection onto $\mathcal{S}_\alpha$; from (14.2) we obtain

$$Q_\alpha B Q_\alpha = \sum_{\nu\mu}^{(\alpha)} \tau_{\nu\mu} U_\nu U_\mu^{-1} P_\mu. \tag{14.3}$$

Let $x_{\mu\nu_0}$ be a complete orthonormal basis in $\mathcal{T}_{\nu_0}$, then the $x_{\mu\rho} = U_\rho x_{\mu\nu_0}$ (for fixed ρ) is a complete orthonormal basis for $\mathcal{T}_\rho$. From (14.3) it follows that

$$Q_\alpha B Q_\alpha x_{\mu\rho} = \sum_{\nu\rho}^{(\alpha)} \tau_{\nu\rho} x_{\mu\nu}. \tag{14.4}$$

For $A \in M$ from $Q_\alpha A Q_\alpha = \sum_\nu^{(\alpha)} P_\nu A P_\nu$ and from $P_\nu A P_\nu U_\nu = U_\nu P_{\nu_0} A P_{\nu_0}$ (see above, setting $U_{\nu\nu_0} = U_\nu$) it follows that $Q_\alpha A Q_\alpha = \sum_\nu U_\nu P_{\nu_0} A P_{\nu_0} U_\nu^{-1}$. Thus, for $A x_{\mu\nu_0} = \sum_\nu x_{\nu\nu_0} a_{\nu\mu}$ it follows that

$$A x_{\mu\rho} = \sum_\nu^{(\alpha)} x_{\nu\rho} a_{\nu\mu}. \tag{14.5}$$

We now introduce two Hilbert spaces $\mathcal{H}_1(\alpha)$ which has the dimension of $\mathcal{T}_\nu \subset \mathcal{S}_\alpha$ and $\mathcal{H}_2^{(\alpha)}$ which has the dimension equal to the number of the $\mathcal{T}_\nu \subset \mathcal{S}_\alpha$. Let z_ν, u_ρ be complete orthonormal bases for $\mathcal{H}_1^{(\alpha)}$ and $\mathcal{H}_2^{(\alpha)}$, respectively; to each operator $A \in M$ we assign an operator $A^{(\alpha)}$ as follows

$$A^{(\alpha)} z_\mu = \sum_\nu z_\nu a_{\nu\mu}. \tag{14.6}$$

Then $x_{\nu\rho} \longrightarrow z_\nu u_\rho$ defines an isomorphism $\mathcal{S}_\alpha \longrightarrow \mathcal{H}_1^{(\alpha)} \times \mathcal{H}_2^{(\alpha)}$ if we assign the operator $Q_\alpha A Q_\alpha$, that is, if, for the operator A in $\mathcal{S}_\alpha$ we assign the operator $A^{(\alpha)} \times \mathbf{1}$ where $A^{(\alpha)}$ is defined by (14.6). Here we often use the simpler notation $\mathcal{S}_\alpha = \mathcal{H}_1^{(\alpha)} \times \mathcal{H}_2^{(\alpha)}$ and for $A \in M$: $Q_\alpha A Q_\alpha = A^{(\alpha)} \times \mathbf{1}$. From (14.4) it follows that $Q_\alpha B Q_\alpha = \mathbf{1} \times B^{(\alpha)}$ where $B^{(\alpha)}$ is an operator in $\mathcal{H}_2^{(\alpha)}$.

We may therefore write $\mathscr{H}$ in the form

$$\mathscr{H} = \sum_{\alpha} \oplus \mathscr{H}_1^{(\alpha)} \times \mathscr{H}_2^{(\alpha)}, \tag{14.7}$$

where the $A \in M$ are given by

$$A = \sum_{\alpha} Q_\alpha (A^{(\alpha)} \times \mathbf{1}) Q_\alpha . \tag{14.8}$$

All operators which commute with M therefore have the form

$$B = \sum_{\alpha} Q_\alpha (\mathbf{1} \times B^{(\alpha)}) Q_\alpha , \tag{14.9}$$

where the $B^{(\alpha)} \in \mathscr{L}(\mathscr{H}_2^{(\alpha)})$ are arbitrary. The $\mathscr{H}_1^{(\alpha)}$ are irreducible with respect to the operators $A^{(\alpha)}$ and $\mathscr{H}_1^{(\alpha)}, \mathscr{H}_2^{(\beta)}$ are, for $\alpha \neq \beta$, not isomorphic with respect to $A^{(\alpha)}$ and $A^{(\beta)}$.

If B is a homomorphism of $(M)\mathscr{H}$ into $(M)\mathscr{T}$ and if $(M)\mathscr{H}$ is completely reducible, then, as we have shown above, a self-adjoint operator D in $\mathscr{H}$ which commutes with all operators in M is defined by $\langle x, Dx \rangle = \langle Bx, Bx \rangle$. According to (14.9)

$$D = \sum_{\alpha} Q_\alpha (1 \times D^{(\alpha)}) Q_\alpha .$$

From $\langle Bx, Bx \rangle = \langle x, Dx \rangle$ it also follows that $\langle By, Bx \rangle = \langle y, Dx \rangle$. For $x_1 \in \mathscr{H}_1^{(\alpha)}, x_2 \in \mathscr{H}_2^{(\alpha)}, y_1 \in \mathscr{H}_1^{(\beta)}, y_2 \in \mathscr{H}_2^{(\beta)}$ it therefore follows that

$$\langle By_1 y_2, Bx_1 x_2 \rangle = \delta_{\alpha\beta} \langle y_1, x_1 \rangle \langle y_2, D^{(\alpha)} x_2 \rangle . \tag{14.10}$$

The subspaces $Q_q \mathscr{H}$ will therefore, for different α, be mapped onto orthogonal subspaces (or $\{0\}$). Therefore (here $\bar{A}$ is the closure of A) we obtain

$$\overline{B\mathscr{H}} = \sum_{\alpha}{}' \oplus \overline{BQ_\alpha \mathscr{H}}, \tag{14.11}$$

where the prime indicates that the sum is to take place only over the $BQ_\alpha \mathscr{H} \neq \{0\}$. It is therefore sufficient to examine the structure of the individual $\overline{BQ_\alpha \mathscr{H}}$.

Let $\mathscr{H}_{2s}^{(\alpha)}$ denote the subspace of $\mathscr{H}_2^{(\alpha)}$ which is the eigenspace of $D^{(\alpha)}$ for which the eigenvalue is 0. From (14.10), for $\beta = \alpha$, it follows that

$$Bx_1 x_2 = U^{(\alpha)}(x_1 D^{(\alpha)^{(1/2)}} x_2)$$

is an isomorphism of $\mathscr{H}_1^{(\alpha)} \times \mathscr{H}_{2s}^{(\alpha)}$ onto the subspace $\overline{BQ_\alpha \mathscr{H}}$. $U^{(\alpha)}$ defines a product representation of $\overline{BQ_\alpha \mathscr{H}} = \mathscr{T}_1^{(\alpha)} \times \mathscr{T}_2^{(\alpha)}$. $U^{(\alpha)}$ therefore defines a pair of isomorphisms $U_1^{(\alpha)}: \mathscr{H}_1^{(\alpha)} \to \mathscr{T}_1^{(\alpha)}$ and $U_2^{(\alpha)}: \mathscr{H}_{2s}^{(\alpha)} \to \mathscr{T}_2^{(\alpha)}$ by $U^{(\alpha)} = U_1^{(\alpha)} \times U_2^{(\alpha)}$. On the subspace

$$\mathscr{H}_s = \sum_{\alpha}{}' \oplus \mathscr{H}_1^{(\alpha)} \times \mathscr{H}_{2s}^{(\alpha)}$$

the relation

$$B = \sum_{\alpha}{}' Q_\alpha (U_1^{(\alpha)} \times U_2^{(\alpha)} D^{(2)^{(1/2)}}) Q_\alpha$$

is satisfied; B is the null operator on the subspace in $\mathscr{H}$ which is orthogonal to $\mathscr{H}_s$. An operator A in M has the following form

$$A = \sum_{\alpha}{}' \tilde{Q}_\alpha(\tilde{A}^{(\alpha)} \times 1)\tilde{Q}_\alpha$$

with respect to the decomposition of the invariant subspace $\overline{B\mathscr{H}}$ described by (14.11):

$$\overline{B\mathscr{H}} = \sum_{\alpha}{}' \oplus \mathscr{T}_1^{(\alpha)} \times \mathscr{T}_2^{(\alpha)},$$

where $\tilde{Q}_\alpha$ is the projection operator on $\mathscr{T}_1^{(\alpha)} \times \mathscr{T}_2^{(\alpha)}$.

In addition, for $A^{(\alpha)}$ in (14.8) we obtain

$$U_1^{(\alpha)}A^{(\alpha)} = \tilde{A}^{(\alpha)}U_1^{(\alpha)}.$$

Therefore the irreducible spaces $\mathscr{H}_1^{(\alpha)}$ and $\mathscr{T}_1^{(\alpha)}$ are isomorphic. An isomorphism of $(M)\mathscr{H}_s$ onto $(M)\overline{B\mathscr{H}}$ is defined by the $U^{(\alpha)} = U_1^{(\alpha)} \times U_2^{(\alpha)}$.

For finite-dimensional $\mathscr{H}$ it is necessary to verify that an operator system M which contains both A and A^+ is always completely reducible (in an invariant, reducible subspace there must be invariant subspaces of smaller dimension). In general this is not the case for infinite dimensional $\mathscr{H}$, so that, in applications, it is necessary to obtain a special proof of the complete reducibility of M.

15 The Spaces $\mathscr{B}(\mathscr{H}_1, \mathscr{H}_2, \ldots)$ and $\mathscr{B}'(\mathscr{H}_1, \mathscr{H}_2, \ldots)$

The results of the previous sections can easily be extended to sequences of Hilbert spaces. Here we shall only give a brief outline of the proofs.

Suppose that we are given a sequence $\mathscr{H}_1, \mathscr{H}_2, \ldots$ of Hilbert spaces. Let us define $\mathscr{H}$ as a disjoint union of the sets $\mathscr{H}_n$, that is,

$$\mathscr{H} = \bigcup_n \mathscr{H}_n. \tag{15.1}$$

$\mathscr{H}_n$ is therefore a subset of $\mathscr{H}$. We find that, in the sense of subsets of $\mathscr{H}$, $\mathscr{H}_n \cap \mathscr{H}_m = \varnothing$ for $n \neq m$. $\mathscr{H}$ is clearly not a linear vector space! An inner product between two elements of $\mathscr{H}$ is therefore only defined if the elements belong to the same subset $\mathscr{H}_n$. Otherwise, two elements which belong to different $\mathscr{H}_n$ are said to be orthogonal.

An orthonormal set in $\mathscr{H}$ is a set $\{x_\nu\}$ of pairwise orthogonal elements of $\mathscr{H}$ for which $\|x_\nu\| = 1$. Such a set is said to be complete if its intersection with $\mathscr{H}_n$ is, for each n, a complete orthonormal basis for $\mathscr{H}_n$. Each $x \in \mathscr{H}$ may be expressed in terms of a complete orthonormal basis as follows: x belongs to one of the $\mathscr{H}_n$ and may therefore be expressed in terms of the portion $\{x_\nu^{(n)}\}$ of the complete orthonormal basis belonging to $\mathscr{H}_n$ in the form described in §2.

A closed subspace of $\mathscr{H}$ is a subset $\mathscr{T}$ for which all the $\mathscr{T}_n = \mathscr{T} \cap \mathscr{H}_n$ are closed subspaces of $\mathscr{H}_n$. $\mathscr{T}$ is therefore the disjoint union of the $\mathscr{T}_n$, that is,

$\mathscr{T} = \bigcup_n \mathscr{T}_n$. Let $\mathscr{T}_n^\perp$ denote the orthogonal subspace of $\mathscr{T}_n$ in $\mathscr{H}_n$. We now define $\mathscr{T}^\perp = \bigcup_n \mathscr{T}_n^\perp$. Let $\mathscr{T}^{(1)} = \bigcup_n \mathscr{T}_n^{(1)}$ and $\mathscr{T}^{(2)} = \bigcup_n \mathscr{T}_n^{(2)}$. The greatest lower bound (infimum) $\mathscr{T}^{(1)} \wedge \mathscr{T}^{(2)}$ of $\mathscr{T}^{(1)}$ and $\mathscr{T}^{(2)}$ is given by $\mathscr{T}^{(1)} \wedge \mathscr{T}^{(2)} = \mathscr{T}^{(1)} \cap \mathscr{T}^{(2)} = \bigcup_n (\mathscr{T}_n^{(1)} \cap \mathscr{T}_n^{(2)})$; the least upper bound (supremum) $\mathscr{T}^{(1)} \vee \mathscr{T}^{(2)}$ of $\mathscr{T}^{(1)}$ and $\mathscr{T}^{(2)}$ is given by $\mathscr{T}^{(1)} \vee \mathscr{T}^{(2)} = \bigcup_n (\mathscr{T}_n^{(1)} \vee \mathscr{T}_n^{(2)})$. It is easy to verify that the set of closed subspaces of $\mathscr{H}$ is a complete orthocomplemented lattice.

An operator A in $\mathscr{H}$ is defined as a sequence $A = (A_1, A_2, \ldots)$ of operators A_n in $\mathscr{H}_n$ which satisfies the condition: For $x \in \mathscr{H}_n$ $Ax = A_n x$. A is said to be bounded if there exists a c for which $\|Ax\| < c\|x\|$. It follows that

$$\|A\| \stackrel{\text{def}}{=} \sup_{\|x\| \leq 1} \|Ax\| = \sup_n \|A_n\|.$$

We denote the set of bounded operators in $\mathscr{H}$ by $\mathscr{A}(\mathscr{H}_1, \mathscr{H}_2, \ldots)$. It is easy to show that $\mathscr{A}(\mathscr{H}_1, \mathscr{H}_2, \ldots)$ is a Banach algebra, provided that the product of A and B is defined by $AB = (A_1 B_1, A_2 B_2, \ldots)$; the adjoint A^+ of A is defined by $A^+ = (A_1^+, A_2^+, \ldots)$. A is said to be self-adjoint if $A^+ = A$.

The set $\mathscr{A}_r(\mathscr{H}_1, \mathscr{H}_2, \ldots)$ of all self-adjoint operators of $\mathscr{A}(\mathscr{H}_1, \mathscr{H}_2, \ldots)$ is an ordered Banach space where $A > 0$ if $A_n > 0$ for all n. The unit sphere of $\mathscr{A}_r(\mathscr{H}_1, \mathscr{H}_2, \ldots)$ is the order interval $[-\mathbf{1}, \mathbf{1}]$ where $\mathbf{1}$ is the unit operator $\mathbf{1}x = x$ for all x. $\mathscr{A}_r(\mathscr{H}_1, \mathscr{H}_2, \ldots)$ is therefore an order unit space. For $A > 0$ we define the positive square root:

$$A^{1/2} = (A_1^{1/2}, A_2^{1/2}, \ldots),$$

where the $A_n^{1/2}$ are positive roots of A_n.

If $\mathscr{T} = \bigcup_n \mathscr{T}_n$ is a closed subspace of $\mathscr{H}$ then to each $x \in \mathscr{H}$ there is a corresponding $q \in \mathscr{T}$ given by $q = P_n x$ for $x \in \mathscr{T}_n$. Thus $q = Px$ where $P = P_1, P_2, \ldots)$ where P_n are projection operators onto $\mathscr{T}_n$. It follows that $P^+ = P$ and $P^2 = P$. Conversely, if for $P \in \mathscr{A}_r(\mathscr{H}_1, \mathscr{H}_2, \ldots)$ $P^2 = P$, then a closed subspace of $\mathscr{H}$ is defined by $\mathscr{T} = P\mathscr{H}$ and $P = (P_1, P_2, \ldots)$, $P_n^2 = P_n$ and $\mathscr{T} = \bigcup_n \mathscr{T}_n$ where $\mathscr{T}_n = P_n \mathscr{H}_n$. For $x \in \mathscr{H}$, $\|x\| = 1$ the projection operator on the subspace generated by x, P_x, is given as follows (for $x \in \mathscr{H}_n$):

$$P_x = (0, 0, \ldots, P_x^{(n)}, 0, \ldots)$$

The relationships derived in §6 may be directly applied onto the set of projection operators in $\mathscr{A}_r(\mathscr{H}_1, \mathscr{H}_2, \ldots)$.

An operator $V \in \mathscr{A}(\mathscr{H}_1, \mathscr{H}_2, \ldots)$ is said to be isometric if for $V = (V_1, V_2, \ldots)$ all V_n are isometric. U is said to be unitary if U and U^+ are isometric. All the results from §7 are directly applicable to the isometric and unitary operators in $\mathscr{A}(\mathscr{H}_1, \mathscr{H}_2, \ldots)$.

From §8 it follows that if $A \in \mathscr{A}_r(\mathscr{H}_1, \mathscr{H}_2, \ldots)$ where $A = (A_1, A_2, \ldots)$ and if $E_n(\lambda)$ is the spectral family of A_n and if $E(\lambda) = (E_1(\lambda), E_2(\lambda), \ldots)$ then

$$A = \int \lambda \, dE(\lambda),$$

where the integral is norm-convergent. A similar result holds for unitary operators.

An operator $A = (A_1, A_2, \ldots)$ is said to be compact if all the A_n are compact. For a compact operator it follows that (9.2) holds.

Let x_ν be a complete orthonormal basis for $\mathscr{H}$; let $A \in \mathscr{A}_{r+}(\mathscr{H}_1, \mathscr{H}_2, \ldots)$. (11.1) is independent of the choice of the complete orthonormal basis; we denote it by $\operatorname{tr}(A)$. For $A = (A_1, A_2, \ldots)$

$$\operatorname{tr}(A) = \sum_n \operatorname{tr}(A_n). \tag{15.2}$$

Let $\mathscr{B}(\mathscr{H}_1, \mathscr{H}_2, \ldots)$ denote the set of operators from $\mathscr{A}_r(\mathscr{H}_1, \mathscr{H}_2, \ldots)$ for which

$$\|A\|_s = \operatorname{tr}(\sqrt{A^2}) = \operatorname{tr}(A_+) + \operatorname{tr}(A_-) = \sum_n \operatorname{tr}(\sqrt{A_n^2})$$

$$= \sum_n \operatorname{tr}(A_{n+}) + \sum_n \operatorname{tr}(A_{n-})$$

is finite. Thus we obtain

$$\|A\|_s = \sum_n \|A_n\|_s,$$

where $\|A_n\|_s$ is the norm in $\mathscr{B}(\mathscr{H}_n)$. $\mathscr{B}(\mathscr{H}_1, \mathscr{H}_2, \ldots)$ is norm-separable because all the $\mathscr{B}(\mathscr{H}_n)$ are norm-separable.

It is easy to show that $\mathscr{B}(\mathscr{H}_1, \mathscr{H}_2, \ldots)$ is an ordered Banach space and is, in addition, a base norm space where the basis K is the set of all $W = (W_1, W_2, \ldots) > 0$ for which

$$\operatorname{tr}(W) = \sum_n \operatorname{tr}(W_n) = 1.$$

For $B \in \mathscr{A}_r(\mathscr{H}_1, \mathscr{H}_2, \ldots)$ $l(B) = \operatorname{tr}(AB)$ is a bounded linear form over $\mathscr{B}(\mathscr{H}_1, \mathscr{H}_2, \ldots)$ for which $\operatorname{tr}(AB) \leq \|A\|_s\|B\|$, where the latter follows directly from $\operatorname{tr}(AB) = \sum_n \operatorname{tr}(A_n B_n)$.

Conversely, if l is a bounded linear form over $\mathscr{B}(\mathscr{H}_1, \mathscr{H}_2, \ldots)$ then, from

$$A = (A_1, A_2, \ldots) = (A_1, 0, \ldots) + (0, A_2, 0, \ldots) + \ldots,$$

where this sum converges in the norm, it follows that

$$l(A) = \sum_n l_n(A_n),$$

where $l_n(A_n)$ is a bounded linear form over $\mathscr{B}(\mathscr{H}_n)$. From $|l(A)| \leq c\|A\|_s$ we obtain $|l_n(A_n)| \leq c\|A_n\|_s$. According to §11 we obtain $l_n(A_n) = \operatorname{tr}(A_n B_n)$ where $B_n \in \mathscr{L}_r(\mathscr{H}_n)$ and $\|B_n\| \leq c$. Therefore we obtain

$$l(A) = \sum_n l_n(A_n) = \sum_n \operatorname{tr}(A_n B_n) = \operatorname{tr}(AB),$$

where $B = (B_1, B_2, \ldots)$ and $\|B\| \leq c$.

The dual Banach space $\mathscr{B}'(\mathscr{H}_1, \mathscr{H}_2, \ldots)$ which corresponds to $\mathscr{B}(\mathscr{H}_1, \mathscr{H}_2, \ldots)$ may therefore be identified with $\mathscr{A}_r(\mathscr{H}_1, \mathscr{H}_2, \ldots)$. By an extension of the results obtained in §11 we find that $\mathscr{B}'(\mathscr{H}_1, \mathscr{H}_2, \ldots)$ is separable in the $\sigma(\mathscr{B}'(\ldots), \mathscr{B}(\ldots))$-topology. In addition, the results from §11 about affine functionals over K remain valid for $\mathscr{B}(\mathscr{H}_1, \mathscr{H}_2, \ldots)$.

It is easy to verify that Gleason's theorem is applicable to the set $G(\mathcal{H}_1, \mathcal{H}_2, \ldots)$ of projection operators from $\mathscr{A}_r(\mathcal{H}_1, \mathcal{H}_2, \ldots)$.

Since almost all theorems for a single Hilbert space are applicable to $\mathcal{H} = \bigcup_n \mathcal{H}_n$, in the text we have referred readers to those sections in AIV where the theorems have been proven for the case of a single Hilbert space. We shall assume that the reader is able to extend these theorems to $\mathcal{H} = \bigcup_n \mathcal{H}_n$ using the contents of §15.

Here it is important to note that it is often useful to imbed $\mathcal{H} = \bigcup_n \mathcal{H}_n$ into a Hilbert space

$$\mathcal{H}_S = \sum_n \oplus \mathcal{H}_n,$$

that is, the elements of $\mathcal{H}_S$ are defined as sequences $x = (x_1, x_2, \ldots)$ where $x_n \in \mathcal{H}_n$ and $\sum_n \|x_n\|^2 < \infty$. The inner product in $\mathcal{H}_S$ is defined by

$$\langle x, y \rangle = \sum_n \langle x_n, y_n \rangle.$$

An element $x \in \mathcal{H} = \bigcup_n \mathcal{H}_n$ is an element of one of the $\mathcal{H}_n$ and is identified with $(0, 0, \ldots, x, \ldots)$ where x is in the nth position. We may identify $A = (A_1, A_2, \ldots) \in \mathscr{A}(\mathcal{H}_1, \mathcal{H}_2, \ldots)$ with the following operator in $\mathscr{L}(\mathcal{H}_S)$:

$$A(x_1, x_2, \ldots) = \sum_n A_n x_n.$$

In this way $\mathscr{A}(\mathcal{H}_1, \mathcal{H}_2, \ldots)$, $\mathscr{A}_r(\mathcal{H}_1, \mathcal{H}_2, \ldots)$ become subsets of $\mathscr{L}(\mathcal{H}_S)$ and $\mathscr{L}_r(\mathcal{H}_S)$ respectively, and, as it is easy to show, $\mathscr{B}(\mathcal{H}_1, \mathcal{H}_2, \ldots)$ becomes a subset of $\mathscr{B}(\mathcal{H}_S)$. If P_n is the projection operator onto the subspace $\mathcal{H}_n$ of $\mathcal{H}_S$ then $\mathscr{A}_r(\mathcal{H}_1, \mathcal{H}_2, \ldots)$ is precisely the set of those operators in $\mathscr{L}_r(\mathcal{H}_S)$ which commute with all P_n. In the same manner, $\mathscr{B}(\mathcal{H}_1, \mathcal{H}_2, \ldots)$ is the set of all operators in $\mathscr{B}(\mathcal{H}_S)$ which commute with all P_n. The basis $K(\mathcal{H}_1, \mathcal{H}_2, \ldots)$ of the base norm space $\mathscr{B}(\mathcal{H}_1, \mathcal{H}_2, \ldots)$ is the subset of all those operators from $K(\mathcal{H}_S)$ (the basis for the base norm space $\mathscr{B}(\mathcal{H}_S)$) which commute with all the P_n.

Of course, the imbedding described above has no physical meaning for the physically interpretable structure introduced in III, §3 and §5. This imbedding can, however, lead to important computational techniques.

References

[1] G. Ludwig. *Die Grundstrukturen Einer Physikalischen Theorie*. Berlin–Heidelberg–New York: Springer-Verlag, 1978.

[2] G. Ludwig. *Einführung in die Grundlagen der Theoretischen Physik*, Vols. I–IV. Braunschweig: Vieweg, 1974–1979.

[3] G. Ludwig. Quantum theory as a theory of interactions between macroscopic systems which can be described objectively. *Erkenntnis*, **16**, 359–387 (1981).

[4] N. Bourbaki. *Theory of Sets*. Paris: Hermann; Reading, Mass.: Addison-Wesley, 1968.

[5] G. Ludwig. *Makroskopische Systeme und Quantenmechanik*. Notes in Math. Phys., Vol. 5; Marburg, 1972.

[6] G. Ludwig. *Meß- und Präparierprozesse*. Notes in Math. Phys., Vol. 6; Marburg, 1972.

[7] G. Ludwig. Measuring and preparing processes. In *Foundation of Quantum Mechanics and Ordered Linear Spaces*. Springer Lecture Notes in Physics, Vol. 29, 1974.

[8] H. Neumann. A mathematical model for a set of microsystems. *Int. J. Theor. Phys.*, **17**, 3, 1978.

[9] M. Drieschner. *Voraussage-Wahrscheinlichkeit-Objekt*. Springer Lecture Notes in Physics, Vol. 99, 1979.

[10] V. S. Varadarajan. *Geometry of Quantum Theory*, Vol. II. New York: Van Nostrand Reinhold, 1970.

[11] O. M. Nikodým. Sure l'extistance d'une mesure parfaitement additive et non séparable. *Mém. Acad. Roy. Belgique*, **XVII**, 1939.

[12] P. Jordan. *Anschauliche Quantentheorie*. Berlin–Heidelberg–New York: Springer-Verlag, 1936.

[13] G. Ludwig. *An Axiomatic Basis for Quantum Mechanics*. New York–Heidelberg–Berlin: Springer-Verlag, 1983.

[14] M. Jammer. *The Philosophy of Quantum Mechanics*. New York: Wiley, 1977.
E. Scheibe. *The Logical Analysis of Quantum Mechanics*. New York: Pergamon Press, 1973.

[15] L. Kanthack. In preparation.

[16] P. Mittelstaedt. *Quantum Logic*. Dordrecht: Reidel, 1978.

[17] G. Ludwig. *Deutung des Begriffs "physikalische Theorie" und axiomatische Grundlegung der Hilbertraumstruktur der Quantenmechanik durch Hauptsätze des Messens*. Springer Lecture Notes in Physics, Vol. 4, 1970.

[18] H. Neumann. *Idealizations of Preparation and Registration Procedures*. In preparation.

[19] O. Melsheimer and R. Werner. In preparation.

[20] H. J. Schmidt. Coordinatization of certain convex sets in axiomatic quantum theory. Doctoral thesis; Marburg Univ., 1975.

[21] H. Neumann. Classical systems and observables in quantum mechanics. *Comm. Math. Phys.*, **23**, 100 (1971).
H. Neumann. *Classical Systems in Quantum Mechanics and Their Representations in Topological Spaces*. Notes in Math. Phys., Vol. 10; Marburg, 1972.
H. Neumann. *On the Representation of Classical Systems*. Springer Lecture Notes in Physics, Vol. 29, 1974, pp. 316–321.

[22] K. Kraus. Position observables of the photon. In *The Uncertainty Principle and Foundations of Quantum Mechanics*, W. C. Price and S. S. Chissick (Eds.). New York: Wiley, 1977.
H. Neumann. Transformation properties of observables. *Helv. Phys. Acta*, **45**, 811 (1972).

[23] W. Band and J. L. Park. Quantum state determination: Quorum for a particle in one dimension. *Amer. J. Phys.*, **47**, 188 (1979).

[24] L. S. Pontrjagin. *Topological Groups*. London: Gordon & Breach, 1966.

[25] H. Yamabe. A generalization of a theorem of Gleason. *Ann. Math.* (2), **58**, 351 (1953).

[26] A. Böhm. *The Rigged Hilbert Space and Quantum Mechanics*. Springer Lecture Notes in Physics, Vol. 78, 1978.

[27] G. Ludwig. The connection between the objective description of macrosystems and quantum mechanics of "many particles". In *Essays in Honor of Wolfgang Yourgrau*, Alwyn van der Merve (Ed.). New York: Plenum Press, 1982.

[28] G. W. Mackey. *Induced Representations and Quantum Mechanics*. Reading, Mass.: Benjamin, 1968.

[29] A. Hartkämper and H. Neumann. Private communication.

[30] S. Sakai. *C*-algebras and W*-algebras*. New York–Heidelberg–Berlin: Springer-Verlag, 1971.
F. W. Shultz. Pure states as a dual object for C^*-algebras. *Commun. Math. Phys.*, **82**, 497 (1982).

J. Dixmier. *C*-algebras*. Amsterdam: North-Holland, 1977.
W. Arveson. *An Introduction to C*-algebras*. New York–Heidelberg–Berlin: Springer-Verlag, 1976.

[31] D. Kappos. *Probability Algebras and Stochastic Spaces*. New York: Academic Press, 1969.

[32] N. Bourbaki. *General Topology*. Paris: Hermann; Reading, Mass: Addison-Wesley, 1966.

[33] H. H. Schaefer. *Topological Vector Spaces*. New York–Heidelberg–Berlin: Springer-Verlag, 1971.
G. Jameson. *Ordered Linear Spaces*. Springer Lecture Notes in Math., Vol. 141, 1970.
K. Ng. *Partially Ordered Topological Vector Spaces*. Oxford: Clarendon Press, 1973.
R. Cristescu. *Ordered Vector Spaces and Linear Operators*. Tunbridge Wells, Kent: Abacus Press, 1976.

[34] P. R. Halmos. *Introduction to Hilbert Space*. New York: Chelsea, 1957.
M. Reed and B. Simon. *Methods of Modern Mathematical Physics*, Vol. I, §II. New York: Academic Press, 1972.
J. Weidmann. *Linear Operators in Hilbert Space*. New York–Heidelberg–Berlin: Springer-Verlag, 1980.

[35] M. H. Stone. *Linear Transformations in Hilbert Space*. Providence, RI: American Mathematical Society, 1932.
M. Reed and B. Simon. *Methods of Modern Mathematical Physics*, Vol. I, §VIII. New York: Academic Press, 1972.

[36] W. O. Amrein. *Nonrelativistic Quantum Dynamics*. Dordrecht: Reidel, 1981.

[37] G. Ludwig. *Die Grundlagen der Quantenmechanik*. Berlin–Heidelberg–New York: Springer-Verlag, 1954.

[38] G. Birkhoff and J. von Neumann. The logic of quantum mechanics. *Ann. Math.*, **37**, 823 (1936).

[39] G. W. Mackey. *Mathematical Foundation of Quantum Mechanics*. Reading, Mass.: Benjamin, 1963.

[40] J. M. Jauch. *Foundation of Quantum Mechanics*. Reading, Mass.: Addison-Wesley, 1973.

[41] C. Piron. *Foundation of Quantum Physics*. New York: Benjamin, 1976.

[42] S. P. Gudder. *Stochastic Methods in Quantum Physics*. Amsterdam: North-Holland, 1979.

[43] D. J. Foulis and C. H. Randall. Operational statistics, I: Basis concepts. *J. Math. Phys.*, **13**, 1667–1675 (1972).
C. H. Randall and D. J. Foulis. Operational statistics, II: Manuals of operations and their logics. *J. Math. Phys.*, **14**, 1472–1480 (1973).
D. J. Foulis and C. H. Randall. Empirical logic and tensor products; and Operational statistics and tensor products. In: *Interpretations and Foundations of Quantum Theory*, H. Neumann (Ed.). Mannheim: B. I. Wissenschaftsverlag, 1981; and references cited therein.

[44] N. Zierler. Axioms for nonrelativistic quantum mechanics. *Pacific J. Math.*, **11**, 1151 (1961).

[45] G. Ludwig. The measuring process and an axiomatic foundation of quantur mechanics (Appendix of this article). In: *Rendiconti della Scuola Internazional die Fisica "Eurico Fermi", IL Corso*. New York: Academic Press, 1971.
G. Ludwig and H. Neumann. Connections between different approaches to th foundations of quantum mechanics. In *Interpretations and Foundations of Quantur Theory*, H. Neumann (Ed.). Mannheim: B. I. Wissenschaftsverlag, 1981.

[46] H. Gerstberger, H. Neumann and R. Werner. Makroskopische Kausalität un relativistische Quantenmechanik. In: *Grundlagenprobleme der modernen Physik* J. Nitsch, J. Pfarr and E. W. Stachow (Eds.). Mannheim: B. I. Wissenschaftsverlag 1981.
H. Neumann and R. Werner. Causality between preparation and registratio processes in relativistic quantum theory. To appear.

[47] G. Ludwig. Axiomatische Basis einer physikalischen Theorie und theoretisch Begriffe. *Z. Allg. Wiss.*, **XII/1**, 55 (1981).

[48] G. Ludwig. Der Meßprozeß. *Z. Phys.*, **135**, 483–511 (1953).
G. Ludwig. Zur Deutung der Beobachung in der Q. M. *Phys. Bl.*, **11**, 489–49 (1955).
G. Ludwig. Zum Ergodensatz und zum Begriff der makroskopischen Observabler *Z. Naturforsch.*, **12a**, 662–663 (1957).
G. Ludwig. Zum Ergodensatz und zum Begriff der makroskopischen Obser vablen, I. *Z. Phys.*, **150**, 346–375 (1958).
G. Ludwig. Zum Ergodensatz und zum Begriff der makroskopischen Obser vablen, II. *Z. Phys.*, **152**, 98–115 (1958).
G. Ludwig. Axiomatic quantum statistics of macroscopic systems (Ergodi theory). *Rendiconti della Scuola Internazionale de Fisica "Enrico Fermi", XI Corso*. New York: Academic Press, 1960, pp. 57–132.
G. Ludwig. Gelöste und ungelöste Probleme des Meßprozesses in der Quanten mechanik. In: *Werner Heisenberg und die Physik unserer Zeit*. Braunschweig 1961 pp. 150–181.
G. Ludwig. Zur Begründung der Thermodynamik auf Grund der Quanten mechanik. *Z. Phys.*, **171**, 476–486 (1963).
G. Ludwig. Zur Begründung der Thermodynamik auf Grund der Quanten mechanik II, Masterequation. *Z. Phys.*, **173**, 232–240 (1963).
G. Ludwig. Versuch einer axiomatischen Grundlegung der Quantenmechanik unc allgemeinerer physikalischer Theorie. *Z. Phys.*, **181**, 233–260 (1964).
G. Ludwig. An axiomatic foundation of quantum mechanics on a nonsubjective basis. In: *Quantum Theory and Reality*. New York–Heidelberg–Berlin: Springer Verlag, 1967, pp. 98–104.
G. Ludwig. Attempt of an axiomatic foundation of quantum mechanics and more general theories, II. *Commun. Math. Phys.*, **4**, 331–348 (1967).
G. Ludwig. Hauptsätze über das Messen als Grundlage der Hilbert–Raum-Struktur der Quantenmechanik. *Z. Naturforsch.*, **22a**, 1303–1323 (1967).
G. Ludwig. Ein weiterer Hauptsatz über das Messen als Grundlage der Hilbert-Raum-Struktur der Quantenmechanik. *Z. Naturforsch.*, **22a**, 1324–1327 (1967).
G. Ludwig. Attempt of an axiomatic foundation of quantum mechanics and more general theories, III. *Commun. Math. Phys.*, **9**, 1–12 (1968).
G. Dähn. Attempt of an axiomatic foundation of quantum mechanics and more general theories, IV. *Commun. Math. Phys.*, **9**, 192–211 (1968).
P. Stolz. Attempt of an axiomatic foundation of quantum mechanics and more general theories, V. *Commun. Math. Phys.*, **11**, 303–313 (1969).

G. Ludwig. [17].
P. Stolz. Attempt of an axiomatic foundation of quantum mechanics and more general theories, VI. *Commun. Math. Phys.*, **23**, 117–126 (1971).
G. Ludwig. The measuring process and an axiomatic foundation of quantum mechanics. In: *Foundations of Quantum Mechanics. Rendiconti della Scuola Internazionale di Fisica "Enrico Fermi", IL Corso*. New York: Academic Press, 1971, pp. 287–317.
G. Ludwig. *A Physical Interpretation of an Axiom within an Axiomatic Approach to Quantum Mechanics and a New Formulation of this Axiom as a General Covering Condition*. Notes in Math. Phys., Vol. 1; Marburg, 1971.
G. Ludwig. *Transformationen von Gesamtheiten und Effekten*. Notes in Math. Phys., Vol. 4; Marburg, 1971. 61 Seiten.
G. Ludwig. [5].
G. Ludwig. [6].
G. Ludwig. An improved formulation of some theorems and axioms in the axiomatic foundation of the Hilbert space structure of quantum mechanics. *Commun. Math. Phys.*, **26**, 78–86 (1972).
G. Ludwig. Why a new approach to find quantum theory? In: *The Physicist's Conception of Nature*. Dordrecht–Boston: Reidel, 1973. Seite 702–708.
G. Ludwig. [7].
G. Ludwig. *Measurement as a Process of Interaction Between Macroscopic Systems*. Notes in Math. Phys., Vol. 14; Marburg, 1974.
G. Ludwig. A theoretical description of single microsystems. In: *The Uncertainty Principle and Foundations of Quantum Mechanics*, W. C. Price and S. S. Chissick, (Eds.). New York: Wiley, 1977.
G. Ludwig. An axiomatic basis of quantum mechanics. In: *Interpretations and Foundations of Quantum Theory*, H. Neumann (Ed.). Mannheim: B. I. Wissenschaftsverlag, 1981.

List of Frequently Used Symbols

$\mathcal{Q}$	set of preparation procedures	22
$\lambda_{\mathcal{Q}}$	probability function for $\mathcal{Q}$	22
$\mathcal{R}_0$	set of registration methods	23
$\lambda_{\mathcal{R}_0}$	probability function for $\mathcal{R}_0$	23
$\mathcal{R}$	set of registration procedures	23
$\mathcal{S}$	set of selection procedures generated by $\mathcal{Q}$, $\mathcal{R}$	26
$\lambda_{\mathcal{S}}$	probability function for $\mathcal{S}$	26
$\mathcal{F}$	set of effect processes	31
$\mathcal{K}$, K	sets of ensembles	42, 56
$\mathcal{L}$, L	sets of effects	43, 57
φ	canonical map $\mathcal{Q}' \to \mathcal{K} \subset K$	43, 57
ψ	canonical map $\mathcal{F} \to \mathcal{L} \subset L$	43, 57
μ	probability function $\mathcal{K} \times \mathcal{L} \to [0, 1]$ and $K \times L \to [0, 1]$	48
Str(A)	dispersion of measurement value for observable A	195
co A	smallest convex set containing the set A	362
$\overline{\text{co}}$ A	smallest convex set containing the set A (closed in the norm topology)	56, 140, 362
$\overline{\text{co}}^{\sigma}$ A	smallest convex set containing the set A (closed in the $\sigma(X, X')$-topology)	362

List of Axioms

Index

Texts and Monographs in Physics

J. Kessler: **Polarized Electrons** (1976).
W. Rindler: **Essential Relativity: Special, General, and Cosmological**, Revised Second Edition (1977).
K. Chadan and P.C. Sabatier: **Inverse Problems in Quantum Scattering Theory** (1977).
C. Truesdell and S. Bharatha: **The Concepts and Logic of Classical Thermodynamics as a Theory of Heat Engines: Rigourously Constructed upon the Foundation Laid by S. Carnot and F. Reech** (1977).
R.D. Richtmyer: **Principles of Advanced Mathematical Physics.** Volume I (1978). Volume II (1981).
R.M. Santilli: **Foundations of Theoretical Mechanics.** Volume I: The Inverse Problem in Newtonian Mechanics (1978). Volume II: Birkhoffian Generalization of Hamiltonian Mechanics (1983).
A. Böhm: **Quantum Mechanics** (1979).
H. Pilkuhn: **Relativistic Particle Physics** (1979).
M.D. Scadron: **Advanced Quantum Theory and Its Applications Through Feynman Diagrams** (1979).
O. Bratteli and D.W. Robinson: **Operator Algebras and Quantum Statistical Mechanics.** Volume I: C*- and W*-Algebras. Symmetry Groups. Decomposition of States (1979). Volume II: Equilibrium States. Models in Quantum Statistical Mechanics (1981).
J.M. Jauch and F. Rohrlich: **The Theory of Photons and Electrons: The Relativistic Quantum Field Theory of Charged Particles with Spin One-half**, Second Expanded Edition (1980).
P. Ring and P. Schuck: **The Nuclear Many-Body Problem** (1980).
R. Bass: **Nuclear Reactions with Heavy Ions** (1980).
R.G. Newton: **Scattering Theory of Waves and Particles**, Second Edition (1982).
G. Ludwig: **Foundations of Quantum Mechanics I** (1983).
G. Gallavotti: **The Elements of Mechanics** (1983).
F.J. Yndurain: **Quantum Chromodynamics: An Introduction to the Theory of Quarks and Gluons** (1983).